Advances in Biocompatible and Biodegradable Polymers-Volume II

Advances in Biocompatible and Biodegradable Polymers-Volume II

Editors

José Miguel Ferri
Vicent Fombuena Borràs
Miguel Fernando Aldás Carrasco

MDPI • Basel • Beijing • Wuhan • Barcelona • Belgrade • Manchester • Tokyo • Cluj • Tianjin

Editors
José Miguel Ferri
Department of Nuclear and Chemical Engineering
Universitat Politècnica de València
Alcoy
Spain

Vicent Fombuena Borràs
Department of Nuclear and Chemical Engineering
Universitat Politècnica de València
Alcoy
Spain

Miguel Fernando Aldás Carrasco
Center of Polymers Applied Research (CIAP)
Escuela Politécnica Nacional
Quito
Ecuador

Editorial Office
MDPI
St. Alban-Anlage 66
4052 Basel, Switzerland

This is a reprint of articles from the Special Issue published online in the open access journal *Polymers* (ISSN 2073-4360) (available at: https://www.mdpi.com/journal/polymers/special_issues/biocompat_biodegrad_polym).

For citation purposes, cite each article independently as indicated on the article page online and as indicated below:

LastName, A.A.; LastName, B.B.; LastName, C.C. Article Title. *Journal Name* **Year**, *Volume Number*, Page Range.

Volume 2
ISBN 978-3-0365-7406-6 (Hbk)
ISBN 978-3-0365-7407-3 (PDF)

Volume 1-2
ISBN 978-3-0365-7402-8 (Hbk)
ISBN 978-3-0365-7403-5 (PDF)

Contents

About the Editors ix

Preface to "Advances in Biocompatible and Biodegradable Polymers-Volume II" xi

Hui Qian
Major Factors Influencing the Size Distribution Analysis of Cellulose Nanocrystals Imaged in Transmission Electron Microscopy
Reprinted from: *Polymers* **2021**, *13*, 3318, doi:10.3390/polym13193318 1

Mosab Kaseem, Zeeshan Ur Rehman, Shakhawat Hossain, Ashish Kumar Singh and Burak Dikici
A Review on Synthesis, Properties, and Applications of Polylactic Acid/Silica Composites
Reprinted from: *Polymers* **2021**, *13*, 3036, doi:10.3390/polym13183036 15

Yen-Liang Liu, Chun-Che Yen, Tzu-Shang Thomas Liu, Chih-Hung Chang, Tiffany Ting-Fang Shih and Jyh-Horng Wang, et al.
Safety and Efficacy of Kartigen® in Treating Cartilage Defects: A Randomized, Controlled, Phase I Trial
Reprinted from: *Polymers* **2021**, *13*, 3029, doi:10.3390/polym13183029 35

Aiah A. El-Rashidy, Sara El Moshy, Israa Ahmed Radwan, Dina Rady, Marwa M. S. Abbass and Christof E. Dörfer, et al.
Effect of Polymeric Matrix Stiffness on Osteogenic Differentiation of Mesenchymal Stem/Progenitor Cells: Concise Review
Reprinted from: *Polymers* **2021**, *13*, 2950, doi:10.3390/polym13172950 49

Kalpana Settu, Pin-Tzu Chiu and Yu-Ming Huang
Laser-Induced Graphene-Based Enzymatic Biosensor for Glucose Detection
Reprinted from: *Polymers* **2021**, *13*, 2795, doi:10.3390/polym13162795 81

Julia Dreier, Christian Brütting, Holger Ruckdäschel, Volker Altstädt and Christian Bonten
Investigation of the Thermal and Hydrolytic Degradation of Polylactide during Autoclave Foaming
Reprinted from: *Polymers* **2021**, *13*, 2624, doi:10.3390/polym13162624 91

Roxana Gheorghita, Liliana Anchidin-Norocel, Roxana Filip, Mihai Dimian and Mihai Covasa
Applications of Biopolymers for Drugs and Probiotics Delivery
Reprinted from: *Polymers* **2021**, *13*, 2729, doi:10.3390/polym13162729 99

Antje Vollrath, Christian Kretzer, Bärbel Beringer-Siemers, Blerina Shkodra, Justyna A. Czaplewska and Damiano Bandelli, et al.
Effect of Crystallinity on the Properties of Polycaprolactone Nanoparticles Containing the Dual FLAP/mPEGS-1 Inhibitor BRP-187
Reprinted from: *Polymers* **2021**, *13*, 2557, doi:10.3390/polym13152557 131

Chrysanthos Maraveas, Ilker S. Bayer and Thomas Bartzanas
Recent Advances in Antioxidant Polymers: From Sustainable and Natural Monomers to Synthesis and Applications
Reprinted from: *Polymers* **2021**, *13*, 2465, doi:10.3390/polym13152465 145

Flávia P. Morais, Ana M. M. S. Carta, Maria E. Amaral and Joana M. R. Curto
An Innovative Computational Strategy to Optimize Different Furnish Compositions of Tissue Materials Using Micro/Nanofibrillated Cellulose and Biopolymer as Additives
Reprinted from: *Polymers* **2021**, *13*, 2397, doi:10.3390/polym13152397 . . . . **177**

Aina Perez-Nakai, Alejandro Lerma-Canto, Ivan Domingez Candela, Daniel Garcia-Garcia, Jose Miguel Ferri and Vicent Fombuena
Comparative Study of the Properties of Plasticized Polylactic Acid with Maleinized Hemp Seed Oil and a Novel Maleinized Brazil Nut Seed Oil
Reprinted from: *Polymers* **2021**, *13*, 2376, doi:10.3390/polym13142376 . . . . **195**

Abril Fonseca-García, Carolina Caicedo, Enrique Javier Jiménez-Regalado, Graciela Morales and Rocio Yaneli Aguirre-Loredo
Effects of Poloxamer Content and Storage Time of Biodegradable Starch-Chitosan Films on Its Thermal, Structural, Mechanical, and Morphological Properties
Reprinted from: *Polymers* **2021**, *13*, 2341, doi:10.3390/polym13142341 . . . . **213**

Shen Su
Prediction of the Miscibility of PBAT/PLA Blends
Reprinted from: *Polymers* **2021**, *13*, 2339, doi:10.3390/polym13142339 . . . . **223**

Silvia Lajewski, Annika Mauch, Kalman Geiger and Christian Bonten
Rheological Characterization and Modeling of Thermally Unstable Poly(3-hydroxybutyrate-co-3-hydroxyvalerate) (PHBV)
Reprinted from: *Polymers* **2021**, *13*, 2294, doi:10.3390/polym13142294 . . . . **233**

Olivia A. Attallah, Marija Mojicevic, Eduardo Lanzagorta Garcia, Muhammad Azeem, Yuanyuan Chen and Shumayl Asmawi, et al.
Macro and Micro Routes to High Performance Bioplastics: Bioplastic Biodegradability and Mechanical and Barrier Properties
Reprinted from: *Polymers* **2021**, *13*, 2155, doi:10.3390/polym13132155 . . . . **243**

Mohammed Naffakh
Nanocomposite Materials Based on TMDCs WS_2 Modified Poly(L-Lactic Acid)/Poly(Vinylidene Fluoride) Polymer Blends
Reprinted from: *Polymers* **2021**, *13*, 2179, doi:10.3390/polym13132179 . . . . **269**

Jie-Mao Wang, Hao Wang, Erh-Chiang Chen, Yun-Ju Chen and Tzong-Ming Wu
Role of Organically-Modified Zn-Ti Layered Double Hydroxides in Poly(Butylene Succinate-Co-Adipate) Composites: Enhanced Material Properties and Photodegradation Protection
Reprinted from: *Polymers* **2021**, *13*, 2181, doi:10.3390/polym13132181 . . . . **287**

Miguel Aldas, José Miguel Ferri, Dana Luca Motoc, Laura Peponi, Marina Patricia Arrieta and Juan López-Martínez
Gum Rosin as a Size Control Agent of Poly(Butylene Adipate-Co-Terephthalate) (PBAT) Domains to Increase the Toughness of Packaging Formulations Based on Polylactic Acid (PLA)
Reprinted from: *Polymers* **2021**, *13*, 1913, doi:10.3390/polym13121913 . . . . **299**

Ivan Dominguez-Candela, Jose Miguel Ferri, Salvador Cayetano Cardona, Jaime Lora and Vicent Fombuena
Dual Plasticizer/Thermal Stabilizer Effect of Epoxidized Chia Seed Oil (*Salvia hispanica* L.) to Improve Ductility and Thermal Properties of Poly(Lactic Acid)
Reprinted from: *Polymers* **2021**, *13*, 1283, doi:10.3390/polym13081283 . . . . **319**

Ana Ibáñez-García, Asunción Martínez-García and Santiago Ferrándiz-Bou
Recyclability Analysis of Starch Thermoplastic/Almond Shell Biocomposite
Reprinted from: *Polymers* **2021**, *13*, 1159, doi:10.3390/polym13071159 . . . . **335**

Liliana Mititelu-Tartau, Maria Bogdan, Daniela Angelica Pricop, Beatrice Rozalina Buca, Loredana Hilitanu and Ana-Maria Pauna, et al.
Biocompatibility and Pharmacological Effects of Innovative Systems for Prolonged Drug Release Containing Dexketoprofen in Rats
Reprinted from: *Polymers* **2021**, *13*, 1010, doi:10.3390/polym13071010 . . . . **351**

Sarieh Momeni, Erfan Rezvani Ghomi, Mohamadreza Shakiba, Saied Shafiei-Navid, Majid Abdouss and Ashkan Bigham, et al.
The Effect of Poly (Ethylene glycol) Emulation on the Degradation of PLA/Starch Composites
Reprinted from: *Polymers* **2021**, *13*, 1019, doi:10.3390/polym13071019 . . . . **365**

Huihui Yuan, Chenli Xue, Jiaqian Zhu, Zhaogang Yang and Minbo Lan
Preparation and Antifouling Property of Polyurethane Film Modified by PHMG and HA Using Layer-by-Layer Assembly
Reprinted from: *Polymers* **2021**, *13*, 934, doi:10.3390/polym13060934 . . . . **385**

Di Sheng Lai, Azlin Fazlina Osman, Sinar Arzuria Adnan, Ismail Ibrahim, Awad A. Alrashdi and Midhat Nabil Ahmad Salimi, et al.
On the Use of OPEFB-Derived Microcrystalline Cellulose and Nano-Bentonite for Development of Thermoplastic Starch Hybrid Bio-Composites with Improved Performance
Reprinted from: *Polymers* **2021**, *13*, 897, doi:10.3390/polym13060897 . . . . **399**

Junsik Bang, Hyunju Lee, Yemi Yang, Jung-Kwon Oh and Hyo Won Kwak
Nano/Micro Hybrid Bamboo Fibrous Preforms for Robust Biodegradable Fiber Reinforced Plastics
Reprinted from: *Polymers* **2021**, *13*, 636, doi:10.3390/polym13040636 . . . . **421**

Alka Prasher, Roopali Shrivastava, Denali Dahl, Preetika Sharma-Huynh, Panita Maturavongsadit and Tiffany Pridgen, et al.
Steroid Eluting Esophageal-Targeted Drug Delivery Devices for Treatment of Eosinophilic Esophagitis
Reprinted from: *Polymers* **2021**, *13*, 557, doi:10.3390/polym13040557 . . . . **433**

Saw Yin Yap, Srimala Sreekantan, Mohd Hassan, Kumar Sudesh and Ming Thong Ong
Characterization and Biodegradability of Rice Husk-Filled Polymer Composites
Reprinted from: *Polymers* **2021**, *13*, 104, doi:10.3390/polym13010104 . . . . **453**

About the Editors

José Miguel Ferri

Dr. José Miguel Ferri (h-index 14—Scopus) works at the Department of Nuclear and Chemical Engineering of the Escuela Politécnica Superior de Alcoy (EPSA) of the Universitat Politècnica de València, (UPV) Alcoy, Spain. He holds degrees in Technical Engineering, Industrial Mechanical Engineering, and Materials Engineering. He joined the Technological Institute of Materials (ITM) in 2014 as a research technician and obtained an international PhD in engineering and industrial production in 2017. His field of research concerns the development of new metallic materials (Mg foams for the medical sector), polymeric materials (the improvement of the thermal, mechanical, and processing properties of biodegradable polymers), and composite materials (metallic matrices for the electronics and aeronautics sectors and polymeric matrices for the aeronautics and construction sectors).

Vicent Fombuena Borràs

Dr. Vicent Fombuena Borràs (h-index 18—Scopus) works at the Department of Nuclear and Chemical Engineering of the Escuela Politécnica Superior de Alcoy (EPSA) of the Universitat Politècnica de València, (UPV) Alcoy, Spain. He holds degrees in Technical Engineering, Industrial Chemical Engineering, and Materials Engineering. He joined the Technological Institute of Materials (ITM) in 2008 as a research technician and obtained an international PhD in engineering and industrial production in 2012. His research work concerns the implementation of a circular economic model in the polymer industry. He has focused his efforts on obtaining multiple active compounds from seeds and agroforestry residues for application in thermoplastic and thermosetting polymers. He has extensive experience in developing techniques for the chemical, thermal, and mechanical characterization of materials.

Miguel Fernando Aldás Carrasco

Miguel Fernando Aldás Carrasco (h-index 10—Scopus) has been a lecturer and researcher at the Escuela Politécnica Nacional (Quito, Ecuador) since 2008. He actively collaborates with the Centro de Investigaciones Aplicadas a Polímeros (Center of Polymers Applied Research) (CIAP) and the Faculty of Chemical Engineering and Agroindustry. He holds degrees in Chemical Engineering from Escuela Politécnica Nacional (Quito, Ecuador), a Master in Innovative Materials degree from Université Claude Bernard Lyon 1 (Lyon, France), and a Doctorate in engineering from Universitat Politècnica de València (Valencia, Spain). He has worked under the direction of Prof. Juan López Martínez and Dr. Marina P. Arrieta, constituting a portion of the team of researchers at the Materials Technology Institute (ITM-UPV) at the Alcoy campus within the Development of Materials for Sustainable Structures (DEMES) group. His field of research concerns the use of natural additives for biopolymers and biodegradable materials, the development of thermoplastic starch-based materials, the study of the degradation of synthetic and biodegradable plastic materials, and the technology behind processing techniques.

Preface to "Advances in Biocompatible and Biodegradable Polymers-Volume II"

Among the strategies for reducing the negative effects on the environment effected by the uncontrolled consumption and low potential for the recovery of conventional plastics, the synthesis of new biodegradable and recyclable plastics represents one of the most promising methods for minimizing the negative effects of conventional non-biodegradable plastics. The spectrum of existing biodegradable materials is still very narrow; thus, to achieve greater applicability, research is being carried out on biodegradable polymer mixtures, the synthesis of new polymers, and the incorporation of new stabilizers for thermal degradation, alongside the use of other additives such as antibacterials or new and more sustainable plasticizers. Some studies analyze direct applications, such as shape memory foams, new cartilage implants, drug release, etc.

The reader can find several studies on the degradation of biodegradable polymers under composting conditions; however, novel bacteria that degrade polymers considered non-biodegradable in other, unusual conditions (such as conditions of high salinity) are also presented.

José Miguel Ferri, Vicent Fombuena Borràs, and Miguel Fernando Aldás Carrasco
Editors

polymers

MDPI

Article

Major Factors Influencing the Size Distribution Analysis of Cellulose Nanocrystals Imaged in Transmission Electron Microscopy

Hui Qian

Nanotechnology Research Center, National Research Council of Canada, 11421 Saskatchewan Drive, Edmonton, AB T6G 2M9, Canada; Hui.Qian@nrc-cnrc.gc.ca

Abstract: Size distributions of cellulose nanocrystals (CNCs), extracted from softwood pulp via strong sulfuric acid hydrolysis, exhibit large variability when analyzed from transmission electron microscopy (TEM) images. In this article, the causes of this variability are studied and discussed. In order to obtain results comparable with those reported, a reference material of CNCs (CNCD-1) was used to evaluate size distribution. CNC TEM specimens were prepared as-stained and dried with a rapid-flushing staining method or hydrated and embedded in vitreous ice with the plunge-freezing method. Several sets of bright-field TEM (BF-TEM), annular dark-field scanning TEM (ADF-STEM) and cryogenic-TEM (cryo-TEM) images were acquired for size distribution analysis to study the contributing factors. The rapid-flushing staining method was found to be the most effective for contrast enhancement of CNCs, not only revealing the helical structure of single CNCs but also resolving the laterally jointed CNCs. During TEM specimen preparation, CNCs were fractionated on TEM grids driven by the coffee-ring effect, as observed from contrast variation of CNCs with a stain-depth gradient. From the edge to the center of the TEM grids, the width of CNCs increases, while the aspect ratio (length to width) decreases. This fractionated dispersion of CNCs suggests that images taken near the center of a droplet would give a larger mean width. In addition to particle fractionation driven by the coffee-ring effect, the arrangement and orientation of CNC particles on the substrate significantly affect the size measurement when CNC aggregation cannot be resolved in images. The coexistence of asymmetric cross-section CNC particles introduces a large variation in size measurement, as TEM images of CNCs are mixed projections of the width and height of particles. As a demonstration of how this contributes to inflated size measurement, twisted CNC particles, rectangular cross-section particles and end-to-end jointed CNCs were revealed in reconstructed three-dimensional (3D) micrographs by electron tomography (ET).

Keywords: nanoparticle size distribution; cellulose nanocrystals; negative staining; cryo-TEM ADF-STEM; electron tomography; coffee-ring effect; fractionation

Citation: Qian, H. Major Factors Influencing the Size Distribution Analysis of Cellulose Nanocrystals Imaged in Transmission Electron Microscopy. *Polymers* **2021**, *13*, 3318. https://doi.org/10.3390/polym13193318

Academic Editors: José Miguel Ferri, Vicent Fombuena Borràs and Miguel Fernando Aldás Carrasco

Received: 3 September 2021
Accepted: 26 September 2021
Published: 28 September 2021

1. Introduction

Cellulose nanocrystals (CNCs) are crystalline particles derived from naturally abundant plant or animal cellulose sources (wood, cotton, tunicate and bacteria, etc.) via strong acid hydrolysis. Depending on the origin of bulk cellulose and acid hydrolysis conditions, the crystallinity, shape and geometric dimensions of extracted CNCs exhibit great variety. In general, CNCs come in rod, ribbon or whisker-like shapes, with lengths ranging from tens of nanometers to several microns and widths ranging from 3 to 50 nm [1–4]. Compared with bulk cellulose with greater amorphous fractions, CNCs exhibit a higher aspect ratio (length-to-width) with the reactive surface of hydroxyl side groups, a greater axial elastic modulus and unique liquid crystalline properties. These properties, along with their natural abundance and biocompatibility, make CNCs attractive for many industrial applications, such as sustainable energy and electronics, biomedical engineering, water treatment, etc. [5–8].

CNCs are reminiscent of crystalline regions within elementary fibrils, and the cross-sections of terminating surfaces are either square, rectangle or hexagon [9]. However, CNCs derived following acid hydrolysis display considerable variability in crystallinity and morphology despite being from the same cellulosic source. The processing conditions during the exaction and post-drying of CNCs may result in the variability of CNC products [3,10]. On the other hand, the characterization of products, as a bridging procedure between production and utilization, may also provide inconsistent results due to varying characterization techniques. This variability makes further functionalization and application of CNCs inconsistent. Therefore, the development of consistent, reliable and accurate measurement protocols is critical to understand the various processes required for optimizing CNC production and utilization.

The structure and size distribution of CNCs, as significant physicochemical properties, have been characterized using light scattering and electron microscopy for several decades. Since the early 1950s, transmission electron microscopy (TEM) has been used to reveal the morphology of CNCs extracted from cotton, ramie and bacterial cellulose [1,11]. Samira et al. performed comprehensive characterization using TEM/cryogenic-TEM, atomic force microscopy (AFM) and small- and wide-angle X-ray scattering (SAXS and WAXS) to determine the shape and size distribution of CNCs from several cellulose sources by [4]. However, challenges in CNC size distribution analysis still remain, and a standard analysis protocol has yet to be established, although numerous efforts have been made in several aspects, including analysis methods and material treatment [12–14]. Johnston's team recently reported interlaboratory comparisons of CNC size distributions measured with both TEM and AFM. The results show great variability between the participating laboratories, and a skew-normal distribution method was proposed to accommodate the variability from different datasets [15,16]. Interestingly, the mean width (7.5 nm) measured in TEM is still approximately twice the mean height (3.4 nm) measured in AFM. This may indicate that not all CNC particles are composed of single crystallites, CNC particles are composed of single crystallites in different sizes or CNCs are laterally jointed bundles. More recently, both the width and height of CNCs were measured using AFM images with an internal calibration standard to evaluate the broadening effect of the AFM tip on width measurement. The results show 28% of CNCs with an approximately symmetric transverse cross-section (square) and the remainder with an asymmetric cross-section [17]. However, a validation procedure is required for width measurement in AFM, which may introduce another source of variability from gold nanoparticles used as the internal calibration standard.

Unlike AFM topography images, TEM images are the projection of three-dimensional (3D) objects along the electron-beam direction on a camera or detector as a two-dimensional (2D) image. Therefore, the arrangement and orientation of CNCs on substrate/TEM grids will affect the projected image, which is used for size measurement. As illustrated in the schematic diagram (Figure S1) in the Supplementary Materials, the width measured from a projected image is equal to the height when a rod-shaped particle has a symmetric cross-section. However, with an asymmetric cross-section, they may be different, depending on which side lies on the substrate. Therefore, the measured width distribution from TEM images is a mix of the width and height of 3D particles. In addition to the orientation on the substrate, CNC particle arrangement may also introduce variation in width and length measurements when individual particles cannot be resolved. Furthermore, sessile droplets on TEM grids may cause the CNC particles to be distributed fractionally during drying, leading to an inaccurate size distribution due to the missing representation of particles from certain regions of the TEM grids. In this article, several image sets with different imaging conditions were processed to evaluate the size distribution of CNCs. The major factors influencing the size distribution of CNCs will be discussed in depth based on observations and analysis.

2. Materials and Methods

2.1. TEM Specimen Preparation

The CNC material used is a National Research Council Canada-certified reference (NRCC), CNCD-1 [12,15–18]. The CNCs were extracted with sulfuric acid hydrolysis of softwood pulp, followed by neutralization and sodium exchange, purification and spray drying. The CNC aqueous solution was prepared by dispersing dry CNC powders in deionized water with a concentration of 0.02% (*w*/*v*). Vortex or sonication of the CNCs in aqueous solution was performed before preparing TEM specimens.

Continuous and perforated carbon-film-supported TEM grids were treated with glow discharge in air for 15 s at 15 mA current (PELCO easiGlow™, Ted Pella, Inc., Redding, CA, USA) before applying samples. CNC TEM specimens were prepared under four different conditions as described below. The first specimen (Sp1) was CNCs negatively stained with 2% uranyl acetate (UA) on TEM grids. Briefly, a 3 μL droplet of CNC aqueous solution was deposited on an ultra-thin continuous carbon-film TEM grid. Excess solution was blotted off with filter paper from the edge of the grid or the top of the droplet after 10 s incubation. The TEM grid was then tilted 30 to 45 degrees, rapidly washed with two running UA droplets and placed with a 2 μL droplet. After about 30 s of incubation, UA was blotted away with filter paper from the edge of the grids, and a thin layer of stain was left to air-dry on the TEM grids. The second specimen (Sp2) was pristine CNCs deposited on carbon TEM grids with the same preparation procedure as Sp1, except that the droplet incubation time was 60 s without UA staining. The third specimen (Sp3) was negatively stained CNCs on carbon TEM grids with gold nanoparticles as fiducial markers for electron tomography (ET). Colloid gold nanoparticles with a 5 nm diameter (0.1 mg/mL) were deposited on carbon-coated TEM grids and air-dried prior to CNC deposition. The same procedure for Sp1 was then followed to obtain negatively stained CNCs.

The fourth specimen (Sp4) was CNCs embedded in vitreous ice, and the cryo-TEM specimen was prepared using the plunge-freezing method [19]. Briefly, the plunge freezer Dewar/chamber (EMS-002 Rapid Immersion Freezer, Electron Microscopy Sciences, Hatfield, PA, USA) was precooled with liquid nitrogen for 30 min. Compressed ethane gas was liquefied in a Falcon tube surrounded with liquid nitrogen. The liquid ethane was then poured into a cryogen container immersed in a liquid nitrogen chamber with a temperature of −184 °C. The container was ready for use when the ethane was lightly frozen at the bottom and the walls of the container. One 4 μL droplet of CNC aqueous solution was placed on the perforated carbon side of a TEM grid facing up for 2 min. The tweezer securing the TEM grid was then mounted to the releasing anvil. Excess solution was blotted using filter paper from the back side of the TEM grid for 4 s, and the TEM grid was rapidly plunged into the liquid ethane. The frozen CNCs on the TEM grid were kept in liquid nitrogen and transferred for cryogenic-TEM (cryo-TEM) imaging.

2.2. Image Acquisition

All bright-field TEM (BF-TEM) and annular dark-field scanning TEM (ADF-STEM) images of CNCs were obtained using a JEOL 2200FS TEM/STEM at a 200 kV accelerating voltage with an in-column Omega energy filter. BF-TEM images with a pixel size of 2048 × 2048 were acquired on a Gatan slow-scan camera with pixel resolution at selected magnifications calibrated with a standard specimen (MAG*I*CAL, Electron Microscopy Science, USA). For imaging CNCs embedded in vitreous ice, the frozen specimen was transferred to a cryo-TEM holder (Gatan Inc., Pleasanton, CA, USA) while being kept below −170 °C. During TEM imaging, the specimen was kept at −180 °C, and the electron beam was blanked between image acquisitions to minimize beam damage on CNCs. The beam dose for each image was 2 e^-/A^2. Ten to fifteen TEM images of CNCs from different regions on the TEM grid were acquired at each magnification.

In STEM mode, ADF-STEM images of CNCs were acquired using a 0.5 nm probe with a pixel dwelling time of 10 μs and an image size of 512 × 512 pixels. The pixel resolution was 0.5 nm, which was the same as the scanning step to avoid oversampling between pixels. Approximately 15 ADF-STEM images of CNCs taken from different areas were used for size measurement.

The TEMography® software package (Frontier Inc., Tokyo, Japan) was used for the acquisition of tilt-serial BF TEM images, fine alignment of 2D images, reconstruction of 3D data, rendering and visualization of CNCs in 3D. The tilt-serial dataset was collected by tilting the holder along the x-axis from $-70°$ to $+70°$ with a $2°$ increment and a defocus value of 1 μm. Gold nanoparticles were used as fiducial markers and focusing objects during image acquisition and fine alignment.

2.3. Size Measurement and Distribution Analysis

ImageJ (Version 1.53c), an open-source software, was used for image processing [20]. Single CNC particles or particles with clear separation outlines were manually selected for size measurement. The size of CNC particles was measured directly from TEM/STEM images. As shown in Figure 1, the length (L) is defined as the distance in a straight line between the two furthest points along a single CNC particle (major axis), and the width (W) (minor axis) is measured at the midpoint of its length unless a particle is clearly asymmetric. An ImageJ macro was used to open sequential images automatically after manual measurement of the length and width of selected CNCs on an active image. Measurement results were exported in a data format suitable for further statistical analysis in OriginPro 2017 or Excel. Approximately 200 particles were measured to give adequate statistical confidence. Histograms of length and width distributions for each dataset were plotted, and Gaussian fitting was applied for descriptive statistics. Box plots labeled with median, mean and one standard deviation (SD) above and below the mean were also used for comparing size distributions in datasets of stained CNCs and CNCs in vitreous ice.

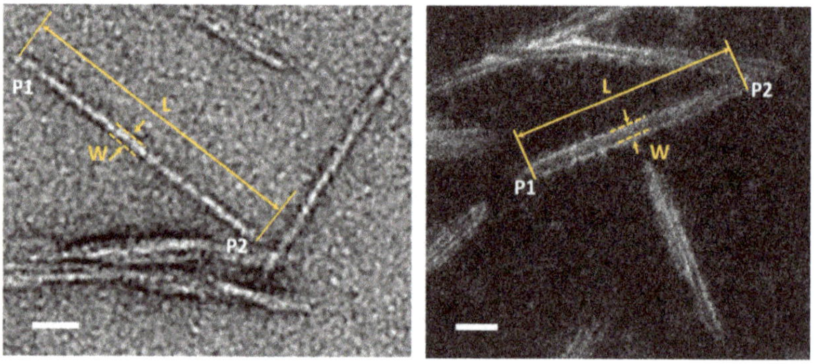

Figure 1. The length (L) is defined as the straight line between the furthest two points (P1 & P2) along the individual CNC particle; the width (W) is defined as the distance between two edges of CNC particle, as indicated in BF-TEM image (**left**) and ADF-STEM image of CNCs (**right**). Scale bars are 20 nm.

3. Results and Discussion

3.1. Contrast of CNC Particles in a TEM Image

CNCs are low-density (1.6 g/cm^3), high-aspect-ratio and electron-beam-sensitive biopolymers. The mass contrast of CNC nanoparticles in a normal BF-TEM image is very weak (Figure 2a). To enhance the contrast, conventional negative staining is widely used, and generally, 1–2% aqueous uranyl acetate is used as the staining reagent. However, due to the different staining methods/protocols, inconsistent quality and properties of the supporting films on TEM grids, as well as individual operator skills, staining results vary even for CNCs deposited on the same TEM grids [21]. As shown in Figure S2, a typical region of CNCs stained with 2% uranyl acetate has a gradient stain depth, resulting in CNCs with different contrasts. The very shallow stain cannot cover the overall contour of

CNCs, resulting in a weak contrast (Figure 2b). The CNCs in the deep stain appear bright with a fairly uniform dark background (Figure 2d), whereas those in the shallow stain are outlined by the stain (dark outlines), and the helical structure of the CNC particles is revealed clearly (Figure 2c). Despite this stain-depth variation, the rapid-flushing staining method described in the TEM specimen preparation was found to be the most suitable method among all those tested. To ensure that particles selected for image processing and size measurement were as homogenous as possible, regions of CNCs with similar stain depths were chosen as one dataset.

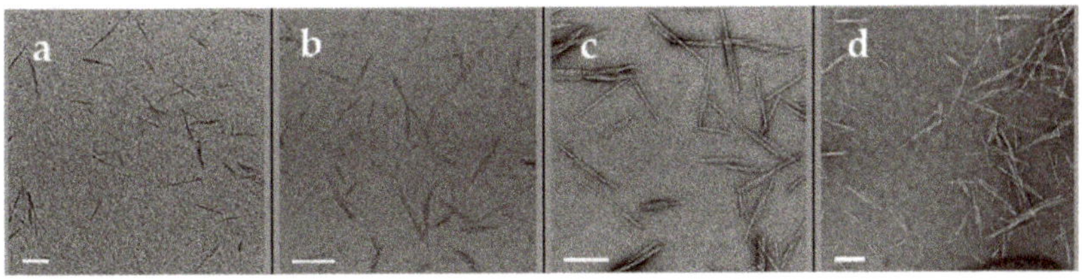

Figure 2. Variation in stain-depth of CNCs shown in BF-TEM images. (**a**) unstained; (**b**) very shallow; (**c**) shallow and (**d**) deep stained CNCs, respectively. Scale bars: 100 nm.

In addition to heavy-metal staining such as UA, the contrast of CNCs can also be enhanced with other approaches such as electron energy filtering, objective lens defocusing and phase-plate insertion in the BF-TEM mode. As shown in Figure 3a, a zero-loss BF-TEM image of unstained CNCs was taken with a 10 eV energy filter and a defocus of −3 μm, in which the helical-like structure of CNCs was revealed, which indicates that the helical structure in stained CNCs is not an artifact. In the last decade, phase-plate imaging has also been explored in contrast enhancement of soft materials. However, considering the various phase-plate types, phase-shift interpretation and access to instruments with a phase plate, it might be challenging to standardize the size distribution with this approach. In addition to the above approaches of contrast enhancement, STEM imaging is an important method of contrast tuning for polymers and biomolecules [22,23]. The contrast of an ADF-STEM image is related to mass thickness and atomic numbers in the specimen. The pristine CNCs in the ADF-STEM images give high contrast, the individual CNC particles are distinguishable and some CNC particles (pointed by arrows) with nonuniform width along their long axis can be revealed (Figure 3b). On the other hand, electron-beam-induced contaminations can easily build up during scanning and blur the image at high resolution. For CNCs stained with uranyl acetate, edges of CNCs exhibit sharp contrast from the stain with a shallow depth in the ADF-STEM image, as shown in Figure 3d. It is easier to identify the individual CNCs than those in the BF-TEM images (Figure 3c) taken from an adjacent area, even though the helical or twist structure of CNCs is not revealed in the ADF-STEM images, which may not be important for size measurement.

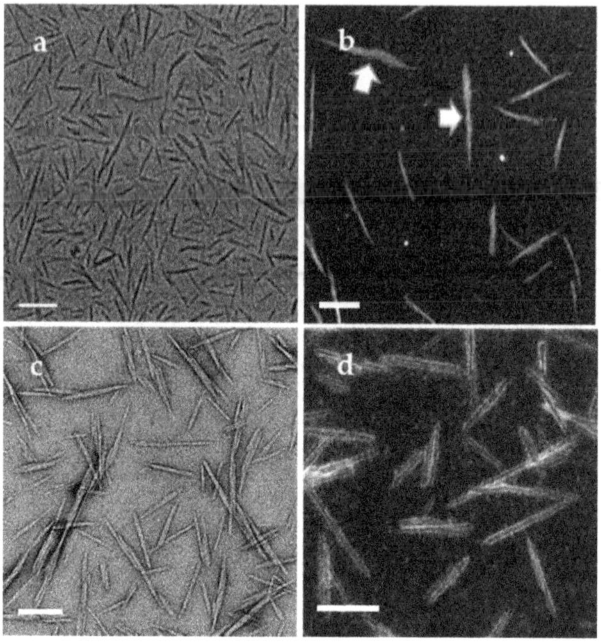

Figure 3. (**a**) under-focused (−3 µm) BF-TEM and (**b**) ADF-STEM images of unstained CNCs; (**c**) BF-TEM and (**d**) ADF-STEM image of stained CNCs. Scale bars: 100 nm.

With the contrast enhancement methods discussed above, sets of BF-TEM and ADF-STEM images of stained and unstained CNCs deposited on carbon TEM grids (Sp1 and Sp2 specimens) were acquired. The length and width of selected CNCs imaged at different conditions were measured, and the descriptive statistics of their distributions are summarized in Table S1. For curiosity, a set of BF-TEM images of CNCs was taken for each of the areas with a gradient stain depth, indicated as Zones A1, A2 and A3 in Figure S2, and the image resolution was 0.41 nm/pixel, 0.41 nm/pixel and 0.29 nm/pixel, respectively. The mean length of CNCs in Zones A1, A2 and A3 is 100.5 nm, 88.8 nm and 79.6 nm, and the mean width is 5.8 nm, 6.4 nm and 6.7 nm, respectively. Moving toward inward areas with a shallow stain (a similar stain depth as in Zone A3), an image set of ADF-STEM for Zone A4 was taken with an image resolution of 0.5 nm/pixel. The mean length and width of CNCs in Zone A4 are 76.3 nm and 7.0 nm, respectively. The histograms of length and width distributions are all included in Figure S3. Box plots of length, width and aspect ratio (length to width) for Zones A1 to A4 are shown in Figure 4. From Zones A1 to A4 with decreasing stain depths, the CNC width increases, while the CNC length and aspect ratio (AR) decrease, which suggests a correlation between size distribution and stain depth or imaging zone. Thus, the question to be answered is: does the stain depth affect the size of CNCs, or are CNC particles distributed differentially or fractionally on the substrate by size during droplet drying?

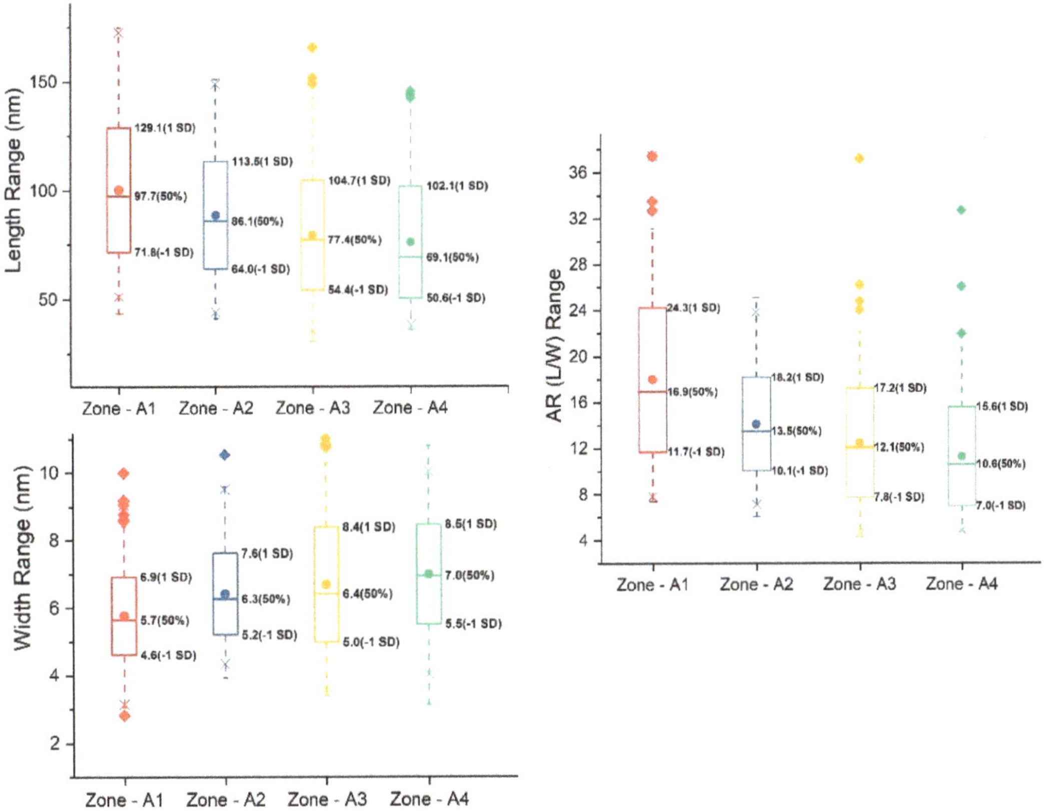

Figure 4. Box plots of length, width and aspect ratio (L/W) distribution of stained.CNCs at zones A1 to A4 with deceasing stain-depth. The line and dot in the box are the median and mean, respectively. The values of the median and one standard deviation (SD) above and below the mean are labelled to the left of boxes.

3.2. Dispersion of CNCs Adsorbed on Carbon-Film-Supported TEM Grids

As known, when a sessile droplet containing dispersed solid dries on a solid surface, a characteristic "coffee ring" is formed due to capillary flow toward the droplet edge [24]. In general, during the evaporation of a sessile droplet, the distribution of solids can be affected by capillary flow direction [24], Marangoni flows [25], particle−substrate interactions, particle–particle interaction and particle shape [26]. During TEM specimen preparation, the coffee-ring effect starts once the initial droplet is placed on TEM grids, and the degree of this effect depends on CNC concentration, incubation time, droplet volume and film surface properties. In the initial droplet, rod-shaped particles can be fractionated by particle size along the capillary flow direction when the contact lines of droplets recede on the substrate [27,28]. Particles with a high AR can be orderly arranged near the ring (droplet) edge with their major axis parallel to the contact lines [29]. After a certain incubation time, the initial capillary flow direction is interrupted by shear force from blotting, and the initial droplet is divided into many smaller droplets scattered on the carbon film surface, which might cause redistribution of the CNC particles on the substrate. In subsequent staining, the staining reagent may change the CNC distribution and orientation, but the lateral movement of CNCs should be confined within small areas. After drying, coffee rings might be formed, resulting in gradient thickness of stain. In Specimen Sp1 of stained CNCs, 3 μL of CNC solution and a 10 s incubation time for the initial droplet were used to minimize the coffee-ring effect so that CNCs of all sizes are included for quantitative analysis. However, it is difficult to void the coffee-ring effect completely, as seen in the gradient depth of stain in Figure S2. The results from Zone A1 to Zone A4 with decreasing stain depth, showing that the mean width increases while the mean length and the mean AR decrease, are due to the coffee-ring effect fractionating CNC dispersion on the substrate, instead of different stain thicknesses.

To avoid the influence of staining reagent on CNC distribution, unstained CNCs dried on TEM grids (Specimen Sp2) were prepared with the same volume of the initial droplet (3 μL) as Sp1 but with a longer incubation time (60 s). As shown in Figure 5a, well-dispersed CNCs are surrounded by more concentrated CNCs lying along their major axis in a coffee-ring pattern (only part of the ring is shown). The major axis of CNCs was redirected to follow the blotting direction rather than being parallel to the pinned line. The size distribution of CNCs in the center area is shown as Dataset A5 in Table S1. The mean length and width are 85.6 nm and 8.9 nm, respectively. The mean length is comparable to those in Datasets A1 to A4, while the mean width is much larger. Such a large mean width was also reported by Johnston's team with a consensus distribution of 7.7 ± 2.2 nm [15]. In their case, a 10 μL volume and a 4 min incubation time were used for the initial droplet of CNCs, and an extra washing step with water was added before the subsequent staining of CNCs. The larger droplet volume and longer incubation time made the fractionation and ordered arrangement of CNCs on the substrate driven by the coffee-ring effect more predominant. The washing step may have removed or redistributed some loose CNCs, but the subsequent staining should not affect the dispersion of CNCs already adsorbed on the carbon film surface. Due to the polydispersity of CNCD-1 materials, the fractionated dispersion of CNCs caused by the coffee-ring effect may introduce variability of size distribution when CNCs are imaged and selected from different regions on TEM grids, especially in the radial direction (TEM grid center to edge). The measured mean width of CNCs adsorbed on the center of the TEM grids is larger than that near the edge. Unfortunately, the location of images taken for each laboratory was not specified. In general, areas near the center of the TEM grids would be used for imaging as the default location, likely resulting in biased size distributions.

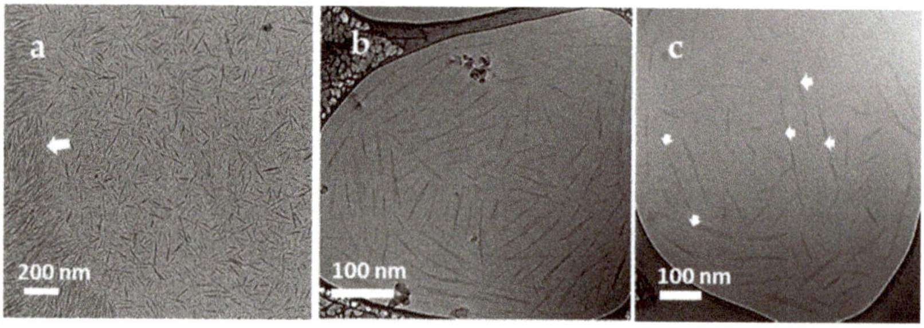

Figure 5. (**a**) coffee-ring formed while unstained CNCs dried on carbon film and more CNCs are concentrated on the edge (white arrow) with well-dispersed CNCs in the center; (**b**) cryo-TEM images of CNCs embedded in vitreous ice taken from the areas near the rim of TEM grids; CNCs lined up parallel to the hole edge and filled across the hole; and (**c**) less ordered CNCs embedded in the ice near the hole edge taken from near the center of TEM grids; possible end-to-end or slightly side-by-side arranged CNCs are pointed by arrows.

To further prove this coffee-ring effect or the dispersion of CNCs within a droplet, Specimen Sp4 was prepared using the plunge-freezing method to vitrify the CNC aqueous solution without a further drying process. Compared with Specimens Sp1 and Sp2, the incubation time of Sp4 was longer so that most CNCs could settle down within the droplet. Despite blotting from the backside of the droplet, the majority of CNCs near the edge of the droplet were arranged with their major axis parallel to each other (Figure 5b) and dispersed across the entire carbon film hole. CNCs near the center of the TEM grids were less ordered and dispersed than those near the edge of the carbon film hole (Figure 5c). Cryo-TEM images were acquired from regions near the center of the TEM grids as Datasets A6 and near the edge as Dataset A7, respectively. The size distribution in each dataset is summarized in Table S1. The histograms of length and width distributions are shown in Figure S4. The mean width and length with standard deviation for A6 and A7 are 7.0 ± 1.1 nm and 112.4 ± 40.6 nm, 6.2 ± 1.2 nm and 96.4 ± 30.0 nm, respectively. The population of CNCs near the center (A6) has a larger length and width than that near the edge (A7), as shown in the box plots in Figure 6. The ARs in both regions are similar, in contradiction to the dried CNC specimen, Sp1. The blotting direction and TEM grid surface used for the cryo-TEM specimen and the dried CNC specimen were different, which may interfere with the CNC distribution differently during blotting. As illustrated in Figure S5, shear force is induced once the filter paper touches the edge of the continuous carbon-film-coated TEM grid, causing CNCs near the air–water interface to drain along the shear force direction. However, when blotting from the backside of perforated film, the shear force induced flow is through the holes so that CNCs near the air–water interface are drawn within thinned aqueous film rather than being removed [30,31]. CNCs within frozen aqueous film can be dispersed at different heights along the beam direction, which is different from CNCs drying on continuous film after blotting. As shown in Figure 5c, the larger particles marked by arrows might be composed of multiple end-to-end or slightly side-by-side CNC particles or overlapped projections from two particles embedded in different vertical positions, contributing to size inflation for both length and width.

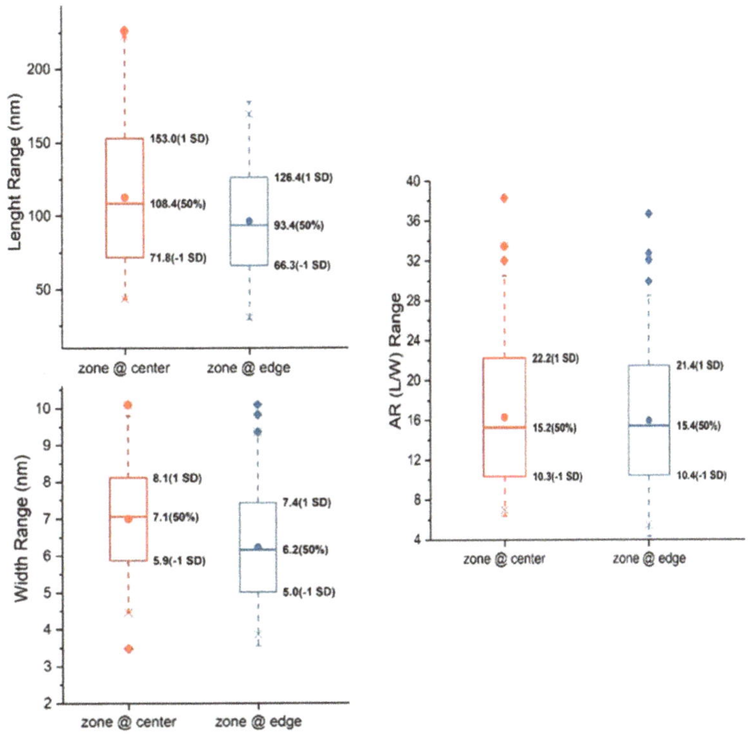

Figure 6. Box plots of length, width and aspect ratio (L/W) distribution of CNCs embedded in vitreous ice near the center and edge of TEM grids. The line and dot in the box is the median and mean, respectively. The values of the median and one standard deviation (SD) above and below the mean are labelled to the left of boxes.

3.3. Orientation of CNCs on Carbon-Film-Supported TEM Grids

As discussed above, CNCs on TEM grids could be fractionated during TEM specimen preparation, resulting in variability of size distribution when CNCs images are taken from different regions in the radial direction. In addition to the fractionated dispersion, the orientation of CNCs on the substrate also has a significant influence on size measurement from TEM images, as illustrated in Figure S1. In the CNCD-1 materials, about 28% of CNCs are particles with a square or a symmetric cross-section, while the remainder are asymmetric with one axis 2–3 times longer than the other, as analyzed in AFM [17]. When these asymmetric particles adsorb on the substrate surface with random orientations, the width distribution measured from TEM images may have large heterogeneity, as the TEM images are mixed projections of width and height from the minor-axial side of CNCs. Due to the strong hydrogen bond, CNCs tend to be laterally or twistedly jointed, and the widest side is normally the preferred orientation on the substrate. Figure 7 shows a typical example of aggregated CNCs in one region, which are either in a lateral (NP1, NP2), vertical and lateral (NP3) or twisted arrangement (NP4). NP3 and NP4 may be composed of more than two CNCs, and the width of NP4 along the long axis is not uniform. All these arrangements and orientations of CNCs on the substrate cause the variability of size measurement both in TEM and AFM. With proper staining of CNCs, the separation line between laterally jointed CNCs can be resolved in TEM. As shown in the line profile of NP1 in Figure 7, two particles with widths of 5.34 nm and 5.64 nm are well revealed and resolved. However, for unstained CNCs dried or embedded in vitreous ice or overstained CNCs, the separation between two CNC particles may not be revealed, resulting in inflated width measurement in TEM. In AFM, due to its limited resolving power in the lateral direction, it is difficult to resolve or separate laterally jointed CNCs. Therefore, the mean width measured with AFM may be overestimated when the orientation of CNCs is on the wider side. When CNC particles are arranged end-to-end or slightly side-by-side as illustrated in Figure S1, the length measurement will be affected if the separation cannot be resolved in either TEM or AFM.

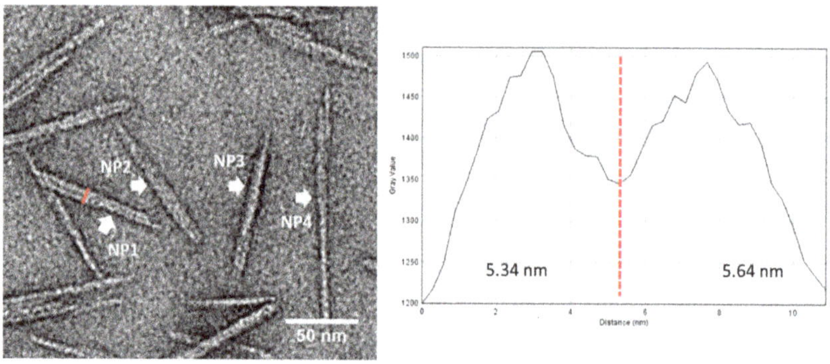

Figure 7. Right: Aggregated CNCs with different orientations marked by arrows (NP1, NP2, NP3 & NP4) on carbon film supported TEM grids. **Left**: The line profile of NP1 composed of two laterally jointed CNCs with width of 5.34 nm and 5.64 nm, respectively.

To reveal particle orientation and arrangement on the substrate, typical regions of CNC particles on Specimen Sp3 were used for ET. Figure 8a shows a reconstructed 3D micrograph of two geometrically twisted CNCs, which are identified as one individual CNC particle (marked with an orange arrow) in the BF-TEM image (Figure 8b). These two CNCs are screw-like or helical along the long axis, as revealed in the reconstructed volume views. The length and width of these two CNCs measured from the reconstructed 3D data are 57 nm and 3.5 nm and 60 nm and 3.8 nm, respectively. Moreover, in the 2D TEM image (Figure 8b), the width measured for this twisted particle is 5.7 nm, which is used for width distribution analysis. In the same field of view, there is another particle marked with a white arrow. Tilting the particle along the long axis by about 20 degrees showed the particle is actually composed of three end-to-end-connected single CNC particles (Figure 9). Thus, the length measurement for size distribution will give a larger value.

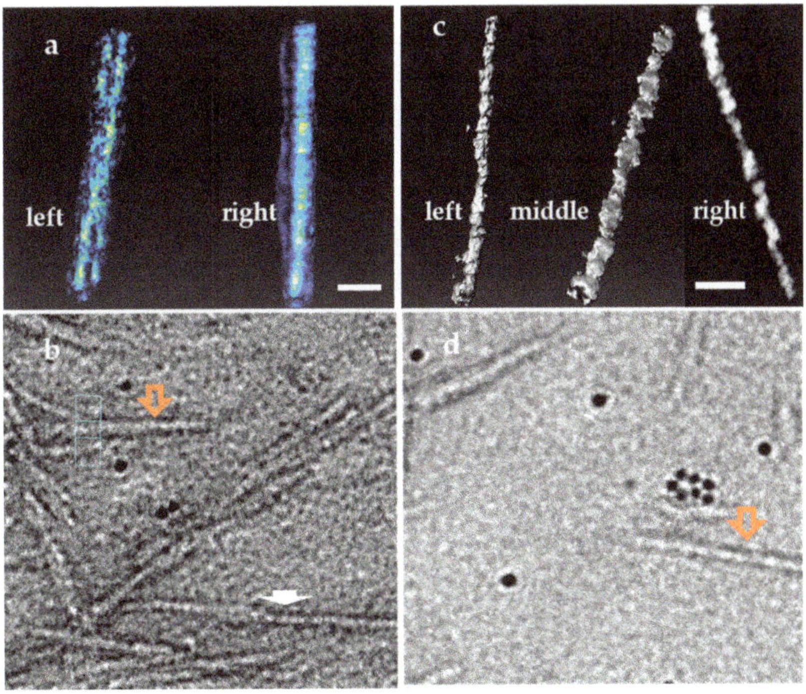

Figure 8. Geometrically twisted CNCs bundles and ribbon-liked CNCs coexists with rod-like CNCs (Sp3 specimen). Scale bars 10 nm. (**a**) Reconstructed 3D volume view of two helical CNCs twisted with an angle (left) and overlapped view (right); (**b**) 2D BF-TEM image of the twisted CNCs (marked as orange arrow) used for 3D reconstruction; (**c**) Reconstructed 3D isosurface of a ribbon-like CNC viewed at an angle with the narrowest width (left), widest width (middle) and the slice view (right); (**d**) BF-TEM image of the CNC used for 3D reconstruction (marked as orange arrow).

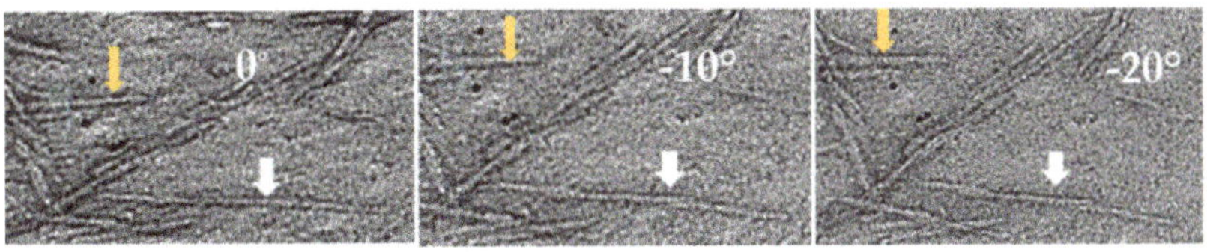

Figure 9. The end-to-end arranged CNC particles is shown as one particle in BF-TEM (**Left**); as two particles (**Middle**) tilted at 10° along long-axis; and as three particles tilted at 20° (**Right**).

In addition to the particle–particle interactions and arrangement on TEM specimens, the intrinsic shape of CNC particles also contributes to size measurement variability. Ribbon-like CNCs marked with an orange arrow in Figure 8d coexist in the same materials, as shown in Figure 8c of a typical reconstructed 3D isosurface view. The reconstructed 3D data reveal that the CNC particle is helical-like, and its width is not uniform along the long axis. The length, height (thickness) and width of this particle are 60 nm, 3.6 nm (left in Figure 8c) and 5.7 nm (right in Figure 8c), respectively. Thus, this asymmetric or rectangular cross-section suggests that CNC particles with a large width may be composed of more than one single crystallite.

Therefore, the shape, geometrical variation and orientation of CNCs on the substrate all contribute to the size measurement. The size measured from 2D images may not be the true parameter of CNC particles; rather, it is the average size of the projected images of all possible shapes and geometrical arrangements of CNCs deposited on the substrate, resulting in large variability of size distribution.

4. Conclusions

High-quality CNC images with good contrast and resolution can mitigate human bias in particle selection and size measurement. The traditional method of staining with heavy metal to enhance contrast for soft materials or biopolymers is still a very effective approach for CNCs. The rapid-flushing staining method has demonstrated that it can reveal the helical structure of single CNCs and separate laterally jointed CNC particles (Figure 2c,d, Figures 3c and 7). It can also reduce the preferential orientation of CNCs on the substrate by running droplets. AFD-STEM imaging is also a good approach for contrast enhancement, as shown in Figure 3, revealing the pristine structure of CNCs. To overcome beam-induced contamination and damage in future work, cooling the TEM holder at liquid nitrogen temperature can be performed when imaging unstained CNCs in ADF-STEM. ADF-STEM images of CNCs with a shallow stain depth give a sharp contrast of CNC particle edges, which may be helpful in particle identification and measurement.

Ideal statistical analysis of size distribution should include CNC particles of all sizes with equal representation. Due to the polydispersity of CNCs and their propensity to agglomerate, it is very challenging to obtain well-dispersed single CNC particles across TEM grids. During TEM specimen preparation, CNC particles could be fractionated across the substrate due to the coffee-ring effect, as seen in dried and hydrated cryo-TEM specimens. Droplet volume, drop-casting incubation time and substrate surface properties will affect the degree of fractionation. A smaller droplet volume and a shorter incubation time are recommended to minimize the coffee-ring effect while preparing CNCs dried on TEM grids. CNCs from different regions of TEM grids, especially in the radial direction, are recommended to be imaged and analyzed for size distribution. Specimen Sp1 is a typical example of fractionated CNCs on a substrate. The size distribution was obtained using the combined datasets from Zones A1 to A3, giving a mean width and length of 6.3 ± 1.3 nm and 89.9 ± 26.2 nm, respectively. Histograms of length and width and descriptive statistics are shown in Figure S6. Compared with the reference values in the certificate (mean width of 7.3 ± 1.8 nm and length of 87.0 ± 35 nm), the length distribution agrees well, but not for the width distribution. The higher mean width of the reference is likely due to lower CNC representation from the edge of droplets with a lower mean width in the measurement sampling, as the center of the droplet is typically the default starting measurement location when TEM grids are loaded inside the instrument. In addition to fractionation, the arrangement and orientation of CNC particles on the substrate significantly affect the size measurement when CNC aggregates cannot be resolved in images. The large mean width in Dataset A5 of unstained CNCs indicates that a large population of laterally jointed or twisted CNCs lying with the widest side on the substrate was included in the analysis. Furthermore, when a large CNC population with an asymmetric cross-section coexists with a symmetric population, the random CNC orientations on the substrate contribute to the variability of the width distribution because the measured width in the TEM images contains the width and height of 3D CNC particles.

In short, to obtain an accurate size measurement of CNCs from TEM images, high-quality CNC TEM specimens and images are essential. To avoid unintended bias of size distribution, imaging and analyzing CNCs from all regions across the TEM grids is recommended. For future work, correlated AFM and TEM imaging can be explored for size distribution analysis of CNCs deposited on the same substrate, which may provide more insight into cross-section shape distribution.

Supplementary Materials: The following are available online at https://www.mdpi.com/article/10.3390/polym13193318/s1, Figure S1: Schematics of the transverse cross-section, orientation on substrate and projected image on camera of rod-shaped nanoparticles and the possible particle–particle arrangements, Figure S2: Stain-depth variation of CNCs deposited on TEM grid, Table S1: Length and width distributions measured from BF-TEM, ADF-STEM and cryo-TEM images of CNCs, Figure S3: Histograms of length, width and aspect ratio distributions of CNCs measured from images taken in Zones A1, A2 and A3 of stained CNCs on TEM grids (Specimen Sp1), Figure S4: Histograms of length and width distributions for unstained CNCs from TEM specimen Sp2 and cryo-TEM specimen Sp4, Figure S5: Initial droplet of CNC aqueous solution on continuous carbon-film-coated TEM grids and perforated carbon-film-coated TEM grids, Figure S6: Size distribution of CNCs by combining Datasets A1, A2 and A3 measured from Specimen Sp1.

Funding: This research received no external funding.

Institutional Review Board Statement: Not applicable.

Informed Consent Statement: Not applicable.

Data Availability Statement: The data presented in this study are available on request from the corresponding author.

Acknowledgments: The author would like to thank Shan Zou with the Metrology Research Center, NRC, for providing the CNCD-1 material and many insightful discussions on AFM imaging.

Conflicts of Interest: The author declares no conflict of interest.

References

1. Ranby, B.G. Physico-chemical investigations on bacterial cellulose. *Ark. Kemi* **1952**, *4*, 249.
2. De Rodriguez, N.L.G.; Thielemans, W.; Dufresne, A. Sisal cellulose whiskers reinforced polyvinyl acetate nanocomposites. *Cellul.* **2006**, *13*, 261–270. [CrossRef]
3. Beck-Candanedo, S.; Roman, M.; Gray, D.G. Effect of Reaction Conditions on the Properties and Behavior of Wood Cellulose Nanocrystal Suspensions. *Biomacromolecules* **2005**, *6*, 1048–1054. [CrossRef] [PubMed]
4. Elazzouzi-Hafraoui, S.; Nishiyama, Y.; Putaux, J.-L.; Heux, L.; Dubreuil, F.; Rochas, C. The Shape and Size Distribution of Crystalline Nanoparticles Prepared by Acid Hydrolysis of Native Cellulose. *Biomacromolecules* **2008**, *9*, 57–65. [CrossRef] [PubMed]
5. Zhu, H.; Luo, W.; Ciesielski, P.N.; Fang, Z.; Zhu, J.Y.; Henriksson, G.; Himmel, M.E.; Hu, L. Wood-Derived Materials for Green Electronics, Biological Devices, and Energy Applications. *Chem. Rev.* **2016**, *116*, 9305–9374. [CrossRef]
6. Jorfi, M.; Foster, E.J. Recent advances in nanocellulose for biomedical applications. *J. Appl. Polym. Sci.* **2015**, *132*, 41719. [CrossRef]
7. Sunasee, R.; Hemraz, U.D.; Ckless, K. Cellulose nanocrystals: A versatile nanoplatform for emerging biomedical applications. *Expert Opin. Drug Deliv.* **2016**, *13*, 1243–1256. [CrossRef]
8. Karim, Z.; Claudpierre, S.; Grahn, M.; Oksman, K.; Mathew, A.P. Nanocellulose based functional membranes for water cleaning: Tailoring of mechanical properties, porosity and metal ion capture. *J. Membr. Sci.* **2016**, *514*, 418–428. [CrossRef]
9. Moon, R.J.; Martini, A.; Nairn, J.; Simonsen, J.; Youngblood, J. Cellulose nanomaterials review: Structure, properties and nanocomposites. *Chem. Soc. Rev.* **2011**, *40*, 3941–3994. [CrossRef]
10. Hamad, W.Y.; Hu, T.Q. Structure-process-yield interrelations in nanocrystalline cellulose extraction. *Can. J. Chem. Eng.* **2010**, *88*, 392–402. [CrossRef]
11. Mukherjee, S.; Woods, H. X-ray and electron microscope studies of the degradation of cellulose by sulphuric acid. *Biochim. Biophys. Acta (BBA)-Bioenerg.* **1953**, *10*, 499–511. [CrossRef]
12. Jakubek, Z.; Chen, M.; Couillard, M.; Leng, T.; Liu, L.; Zou, S.; Baxa, U.; Clogston, J.D.; Hamad, W.Y.; Johnston, L.J. Characterization challenges for a cellulose nanocrystal reference material: Dispersion and particle size distributions. *J. Nanoparticle Res.* **2018**, *20*, 98. [CrossRef]
13. Bai, W.; Holbery, J.; Li, K. A technique for production of nanocrystalline cellulose with a narrow size distribution. *Cellul.* **2009**, *16*, 455–465. [CrossRef]
14. Beck, S.; Bouchard, J.; Berry, R. Dispersibility in Water of Dried Nanocrystalline Cellulose. *Biomacromolecules* **2012**, *13*, 1486–1494. [CrossRef]
15. Meija, J.; Bushell, M.; Couillard, M.; Beck, S.; Bonevich, J.; Cui, K.; Foster, J.; Will, J.; Fox, D.; Cho, W.; et al. Particle Size Distributions for Cellulose Nanocrystals Measured by Transmission Electron Microscopy: An Interlaboratory Comparison. *Anal. Chem.* **2020**, *92*, 13434–13442. [CrossRef] [PubMed]
16. Bushell, M.; Meija, J.; Chen, M.; Batchelor, W.; Browne, C.; Cho, J.-Y.; Clifford, C.A.; Al-Rekabi, Z.; Vanderfleet, O.M.; Cranston, E.D.; et al. Particle size distributions for cellulose nanocrystals measured by atomic force microscopy: An interlaboratory comparison. *Cellulose* **2021**, *28*, 1387–1403. [CrossRef]

17. Chen, M.; Parot, J.; Hackley, V.A.; Zou, S.; Johnston, L.J. AFM characterization of cellulose nanocrystal height and width using internal calibration standards. *Cellulose* **2021**, *28*, 1933–1946. [CrossRef]
18. Available online: http://www.nrc.ca/crm (accessed on 18 June 2021).
19. Qian, H.; Jia, Y.; McCluskie, M.J. Application of cryogenic transmission electron microscopy for evaluation of vaccine delivery carriers. In *Vaccine Delivery Technology: Methods and Protocols*; Methods in Molecular Biology; Pfeifer, B., Hill, A., Eds.; Humana: New York, NY, USA, 2020; Volume 2183, ISBN 978-1-0716-0794-7.
20. Rasband, W.S. U.S. National Institute of Health, Bethesda, Maryland, USA. Available online: https://imagej.nih.gov/ij/ (accessed on 18 June 2021).
21. Scarff, C.A.; Fuller, M.J.G.; Thompson, R.F.; Iadaza, M.G. Variations on Negative Stain Electron Microscopy Methods: Tools for Tackling Challenging Systems. *J. Vis. Exp.* **2018**, *10*, e57199. [CrossRef] [PubMed]
22. Wall, J.S.; Hainfeld, J.F. Mass Mapping with the Scanning Transmission Electron Microscope. *Annu. Rev. Biophys. Biophys. Chem.* **1986**, *15*, 355–376. [CrossRef]
23. Williams, D.B.; Cater, C.B. *Transmission Electron Microscopy: A Text Book for Material Science*; Springer: New York, NY, USA, 2009; Chapter 22.
24. Deegan, R.D.; Bakajin, O.; Dupont, T.F.; Huber, G.; Nagel, S.R.; Witten, T.A. Capillary flow as the cause of ring stains from dried liquid drops. *Nature* **1997**, *389*, 827–829. [CrossRef]
25. Hu, A.H.; Larson, R.G. Marangoni Effect Reverses Coffee-Ring Depositions. *J. Phys. Chem. B* **2006**, *110*, 7090–7094. [CrossRef] [PubMed]
26. Yunker, P.J.; Still, T.; Lohr, M.A.; Yodh, A.G. Suppression of the coffee-ring effect by shape-dependent capillary interactions. *Nat. Cell Biol.* **2011**, *476*, 308–311. [CrossRef] [PubMed]
27. Fan, F.; Stebe, K.J. Size-Selective Deposition and Sorting of Lyophilic Colloidal Particles on Surfaces of Patterned Wettability. *Langmuir* **2005**, *21*, 1149–1152. [CrossRef] [PubMed]
28. Ryu, S.-A.; Kim, J.Y.; Kim, S.Y.; Weon, B.M. Drying-mediated patterns in colloid-polymer suspensions. *Sci. Rep.* **2017**, *7*, 1079. [CrossRef] [PubMed]
29. Dugyala, V.R.; Basavaraj, M.G. Evaporation of Sessile Drops Containing Colloidal Rods: Coffee-Ring and Order–Disorder Transition. *J. Phys. Chem. B* **2015**, *119*, 3860–3867. [CrossRef]
30. Zheng, Y.; Lin, Z.; Zakin, J.L.; Talmon, Y.; Davis, H.T.; Scriven, L.E. Cryo-TEM Imaging the Flow-Induced Transition from Vesicles to Threadlike Micelles. *J. Phys. Chem. B* **2000**, *104*, 5263–5271. [CrossRef]
31. Glaeser, R.M. Proteins, interfaces, and cryo-EM grids. *Curr. Opin. Colloid Interface Sci.* **2018**, *34*, 1–8. [CrossRef] [PubMed]

polymers

MDPI

Review

A Review on Synthesis, Properties, and Applications of Polylactic Acid/Silica Composites

Mosab Kaseem [1,*], Zeeshan Ur Rehman [2], Shakhawat Hossain [3], Ashish Kumar Singh [4] and Burak Dikici [5,*]

1 Department of Nanotechnology and Advanced Materials Engineering, Sejong University, Seoul 05006, Korea
2 School of Materials Science & Engineering, Changwon National University, Changwon 641-773, Korea; Zeeshan.physics@gmail.com
3 Department of Industrial and Production Engineering, Jashore University of Science and Technology, Jashore 7408, Bangladesh; shakhawat.ipe@just.edu.bd
4 Department of Applied Sciences, Bharati Vidyapeeth's College of Engineering, New Delhi 110063, India; ashish.singh.rs.apc@itbhu.ac.in
5 Department of Metallurgical and Materials Engineering, Ataturk University, Erzurum 25240, Turkey
* Correspondence: mosabkaseem@sejong.ac.kr (M.K.); burakdikici@atauni.edu.tr (B.D.)

Abstract: Polylactic acid (PLA)/silica composites as multifunctional high-performance materials have been extensively examined in the past few years by virtue of their outstanding properties relative to neat PLA. The fabrication methods, such as melt-mixing, sol–gel, and in situ polymerization, as well as the surface functionalization of silica, used to improve the dispersion of silica in the polymer matrix are outlined. The rheological, thermal, mechanical, and biodegradation properties of PLA/silica nanocomposites are highlighted. The potential applications arising from the addition of silica nanoparticles into the PLA matrix are also described. Finally, we believe that a better understanding of the role of silica additive with current improvement strategies in the dispersion of this additive in the polymer matrix is the key for successful utilization of PLA/silica nanocomposites and to maximize their fit with industrial applications needs.

Keywords: polylactic acid; silica; composites; thermal stability; toughness; biodegradability; 3D printing applications

Citation: Kaseem, M.; Ur Rehman, Z.; Hossain, S.; Singh, A.K.; Dikici, B. A Review on Synthesis, Properties, and Applications of Polylactic Acid/Silica Composites. *Polymers* **2021**, *13*, 3036. https://doi.org/10.3390/polym13183036

Academic Editors: José Miguel Ferri, Vicent Fombuena Borràs and Miguel Fernando Aldás Carrasco

Received: 19 August 2021
Accepted: 7 September 2021
Published: 8 September 2021

1. Introduction

Owing to its versatility and good performance, polylactic acid (PLA) as a biodegradable polymer is regarded as one of the most favorable polymers to substitute petroleum-based plastics [1–3]. The low toughness and poor thermal stability of PLA, however, restrict the usage of PLA in applications, such as food packaging and medical implants [3]. To improve the performance, therefore, several approaches, such as blending with other polymers, mixing with inorganic fillers, and copolymerization, have been suggested by several research groups [4–10]. In particular, the mixing with inorganic additives, such as TiO_2, carbon nanotube, ZnO, Al_2O_3, MgO, and SiO_2, was reported to be a useful method to boost the performance of neat PLA [5,7].

Silicon dioxide (SiO_2), commonly referred to as silica, which may exist in the amorphous and crystalline structure, was found to be a useful filler for improving the mechanical performance of polymeric materials [11,12]. The silica, as an additive, is used in coatings, food, and biomedical applications. For example, silica as a bio-safe additive can be incorporated along with silver nanoparticles to improve the antibacterial properties and corrosion properties of biomaterials [13–23]. The amorphous silica nanoparticles extracted from natural sources, such as rice husk and fly ash, can be used as a nucleating agent when added to thermoplastic polymers in very low amounts, owing to their large surface area [24]. The improving dispersibility of silica within the PLA matrix during fabrication of PLA/silica composites would lead to enhancing the mechanical, thermal, and rheological properties of neat PLA [21]. The presence of silanol groups, siloxane bridges, and hydroxyl groups on the silica would facilitate the surface functionalization of silica for biomedical, catalysis, and sensor applications [12].

Therefore, by taking into consideration the attractive properties of silica, the fabrication of PLA/silica composites would be a beneficial procedure to enhance the performance of PLA. To the best of our knowledge, this is the first systematic article deliberating the recent progress in PLA/silica. Thus, this work aims to discuss the recent developments in the field of PLA/silica composites in terms of fabrication, properties, and applications.

2. Synthesis of PLA/Silica Composites

The direct melt-mixing, solution mixing, sol–gel process, and in situ polymerization methods are considered as the main methods utilized for the synthesis of PLA/silica composites. The melt-mixing method involves the direct mixing of PLA with silica nanoparticles, while the sol–gel process can be performed either in the existence of PLA or simultaneously during the polymerization of lactic acid monomers [25]. The solution mixing method starts from the dissolution of polymers in a suitable solvent with nanoparticles together with the evaporation of the solvent, or precipitation [25]. As for the in situ polymerization method, the silica nanoparticles should be distributed in the monomers before polymerization [25]. In addition, it is worth mentioning that the freezing-drying process would also be a promising method to fabricate PLA composites. In the freezing-drying process, the colloidal dispersion of the aerogel precursors is frozen, with the liquid component freezing into different morphologies depending on a variety of factors, such as the precursor concentration, type of liquid, temperature of freezing, and freezing container [25]. However, a lack of works was found on the utilization of this method to fabricate PLA/silica composites.

The surface modification by physical or chemical methods is a common procedure to increase the compatibility between PLA and silica nanoparticles [26–30]. The functionalization of hydrophilic silica nanoparticles, which is usually conducted on the reactive silanol end-groups, would improve the hydrophobicity of silica nanoparticles, improving their dispersion in the PLA matrix. This process, i.e., the formation of hydrophobic-fumed silica, can be obtained by chemical treatment of hydrophilic silica with silanes or siloxanes. As a result of the improved dispersion of silica particles in the PLA matrix, the rheological properties of the composites are improved. In general, the large surface area and the smooth surface of silica nanoparticles could increase the interactions not only between silica nanoparticles but also between PLA and silica nanoparticles. Thus, good physical interactions between the silica and PLA matrix can be achieved, leading to significant enhancements in composites properties. Several functionalization (coupling) agents, such as tetraethyloxysilane (TEOS) and γ-glycidoxypropyltrimethoxysilanes (GOPTMS) were used to improve the dispersibility of silica nanoparticles in the PLA matrix. In the study of Hakim et al. [26], a reactive extrusion method was utilized for the melt-mixing of PLA with 2.5 wt.% of silica nanoparticles. The schematic illustrations of SiO_2 before and after surface modification by organic chains are displayed in Figure 1. As shown in Figure 1b, the surface -OH groups of both surface-modified silica types were mostly substituted by functionalized organic chains. Thus, many PLA chains could be grafted on one silica nanoparticle, which would have risen the local shear field applied to the agglomerated nanoparticles during mixing, thereby improving the dispersibility of silica nanoparticles in the polymer matrix. However, transmission electron microscopy (TEM) observations implied that the uniformity and dispersibility of the nanoparticles under the experimental processing conditions were found not to be affected by the enhanced interfacial interaction of both surface-treated silica types (Figure 1c–e).

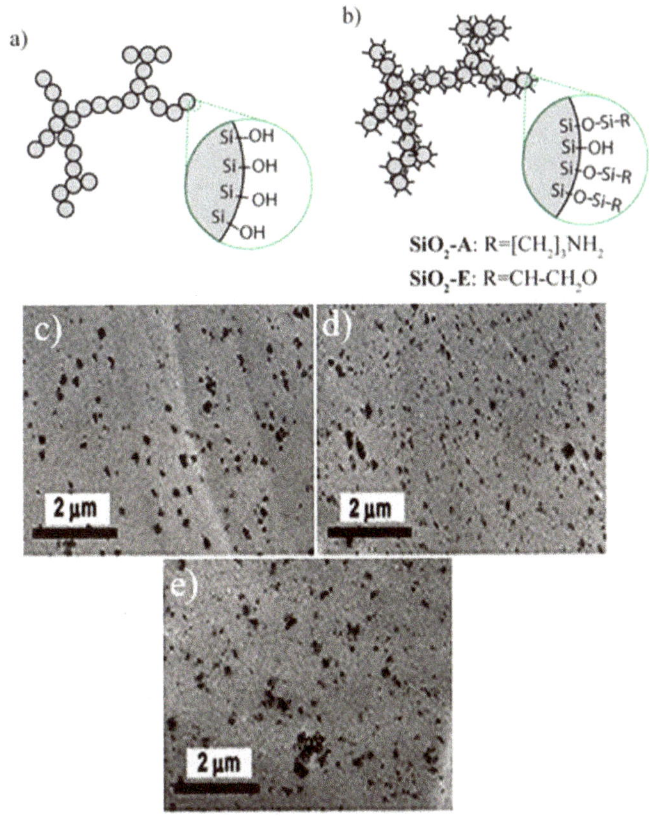

Figure 1. Schematic representation of the (**a**) surface-unmodified silica and (**b**) surface-modified silica aggregate. (**c**–**e**) TEM images of the PLA composites containing (**c**) un-modified SiO_2, (**d**) SiO_2-A, and (**e**) SiO_2-E [26].

The fabrication of PLA/silica composites via the sol–gel method was first reported by Yan et al. [31], who aimed to obtain plasticized composites. The fabrication process of the composites involved an in situ synthesis of silica nanoparticles via condensation reactions of TEOS and GOPTMS in the presence of PLA and polyethylene glycol as a plasticizer in tetrahydrofuran. Hydrochloric acid was utilized as a catalyst during the synthesis method. The infrared results proved the formation of a silica network structure within the PLA matrix, while the results of mechanical tests indicated that the incorporation of 4 wt.% of silica nanoparticles would increase the tensile strength of PLA from 15 to 18 MPa. In another work, Yan et al. [32] polymerized L-lactic acid (PLLA) in the presence of silica without the use of catalysts, but in solution. In this case, the silica surface was not organically treated. The polycondensation was carried out in toluene to remove the water formed by azeotropic dehydration. The result is silica grafted with PLLA oligomers. The authors were able to witness the grafting by infrared spectroscopy characterization (unfortunately, the molar masses are not specified). The grafted silica was then dispersed in PLLA and its good dispersion helped to improve mechanical properties, as compared to the PLLA composites with non-grafted silica.

Wu et al. [32] also polycondensed L-lactic acid in the presence of silica nanoparticles, but this time in bulk. The authors first mixed an aqueous solution of L-Lactic acid (LA) with an acidic silica sol containing silica particles of 12 nm. The mixture was dehydrated under vacuum with sonication treatment to well disperse the particles. After complete drying of the mixture, the polycondensation was performed under vacuum conditions to remove the water formed. The grafting occurred as above, with the polycondensation using the SiOH groups on the surface of the silica nanoparticles. The results showed that the molar mass of the grafted PLLA was about 31,100 g·mol^{-1}. Liu et al. [33] fabricated PLA/silica composites through ring-opening polymerization of lactide initiated by modified silica nanoparticles in the presence of stannous octoate as a catalyst. The experimental conditions, such as the weight ratio of silica to the lactide monomer, reaction temperature, and reaction time were optimized to be 1:20, 140 °C, and 72 h, respectively. The morphological observations revealed that silica nanoparticles tended to be distributed uniformly within the PLA matrix, improving compatibility between PLA and silica nanoparticles. Accordingly, the thermal and mechanical properties of the composites were improved significantly in comparison to that of pure PLA. In other works [24,34,35], the melt-mixing method to fabricate PLA/silica composites was carried out at 175 °C. In comparison to neat PLA, the PLA/silica composites containing various contents of silica (from 1 to 10 wt.%) exhibited significant improvements in the thermal stability and the barrier properties against nitrogen and oxygen gases, which were connected with the establishment of the silica network structure as approved by Fourier-transform infrared (FTIR) and rheological results. In the work of Liu and co-workers [36], amino-functionalized nano-SiO_2 (m@g-SiO_2) was first prepared through a coupling reaction on the surface of silica nanoparticles before its melt-mixing with PLA. Molecular dynamics simulation was carried out to discover the correlation between PLA and silica nanoparticles before and after the organic functionalization of silica. The crystallization behavior and the mechanical properties of PLA exhibited significant improvements after the melt-mixing with functionalized silica nanoparticles which were attributed to the fact the organic modification of silica nanoparticles would lead to enhancements in the interaction energy and mobility of PLA chains.

To improve the dispersion of silica in the PLA matrix, an in situ melt condensation of l-lactic acid was carried out in the presence of silica nanoparticles [37]. However, when the content of silica exceeded 10 wt.%, some agglomerated nanoparticles appeared in the TEM images of the prepared composites. The agglomeration of silica nanoparticles would be attributed to the strong van der Waals forces, which tended to reduce the physical properties of the obtained composites. Thus, it is believed that the physical mixtures of PLA and organo-modified SiO_2 resulted in the separation in discrete phases, leading to inferior mechanical properties [38,39]. Zou and coworkers tried to mix silica nanoparticles with a copolymer made of PLA and epoxidized soybean oil [40]. The FTIR and thermogravimetric analysis (TGA) results confirmed the reactions between silica and epoxidized soybean oil, which, in turn, led to improve crystallization behavior and mechanical properties of PLA, while Sepulveda et al. [41] demonstrated that the direct grafting of L-lactic acid oligomer onto the silica surface through its silanol groups was a good strategy to enhance the physical, thermal, and mechanical properties of PLA/silica composites. As such, Zhu et al. [42] found that the surface functionalization of fumed silica nanoparticles oleic acid by oleic acid would help to improve the rheological, thermal, and mechanical properties of PLA/silica composites prepared via the melt-mixing method, which was linked to the good interfacial adhesion in the composites containing functionalized-silica nanoparticles.

3. Rheological Properties

Understanding the rheological properties is very important due to their considerable effects on molecular weight, morphology, chain structure, and chain motions [43–49]. Basilissi et al. [50] demonstrated that the melt viscosity of PLA/silica composites fabricated by bulk ring-opening polymerization can be improved by the silane-based modification of silica nanoparticles. Li et al. [51] reported that both the storage modulus and biodegradation rate of PLA tended to be improved by the addition of silica nanoparticles into the PLA matrix. The formation of a silica network structure was responsible for storage modulus improvements, while the enhancement in the biodegradation rate was ascribed to the easy release of silica aggregates from the PLA matrix. Nerantzaki et al. [52] prepared a series of poly(DL-lactide) (PDLLA)/SiO_2 composites by a novel two-step technique (ring-opening polymerization (ROP)—polycondensation). The concentration of SiO_2 was varied from 2.5 to 5, 10, and 20 wt.%. The results of this work are presented in Figure 2. Based on the intrinsic viscosity results, it was demonstrated that the average molecular weight of PDLLA (Mn $\approx$ 38,097 g mol^{-1}) tended to reduce with an increase in the silica content. Moreover, the average Mn values of PLA were found to decrease when the content of silica increased from 2.5 to 20 wt.%. This finding was elucidated based on the fact that the addition of silica to the PDLLA matrix would hinder the increment in the molecular weight of the polymer. Besides, silica nanoparticles can interact with DL-lactide through their silanol groups. In another study [53], the silica was functionalized with TEOS and GOPTMS through graft-condensation reaction. Afterward, the functionalized silica was melt-mixed with PLA by reactive extrusion technique. In comparison to neat PLA, the addition of functionalized silica to the PLA matrix caused considerable improvements in values of the complex viscosity and the storage modulus of the PLA/silica composites. The inclusion of silica was expected to enhance the hydrolysis resistance by the formation of a stable silica network. Although the molecular weight of PLA was reduced by 12 wt.% during the processing, a slighter reduction of less than 10% was reported upon the addition of TEOS, which also resulted in increasing the value of zero-shear rate viscosity (η_0) obtained via the Cross model.

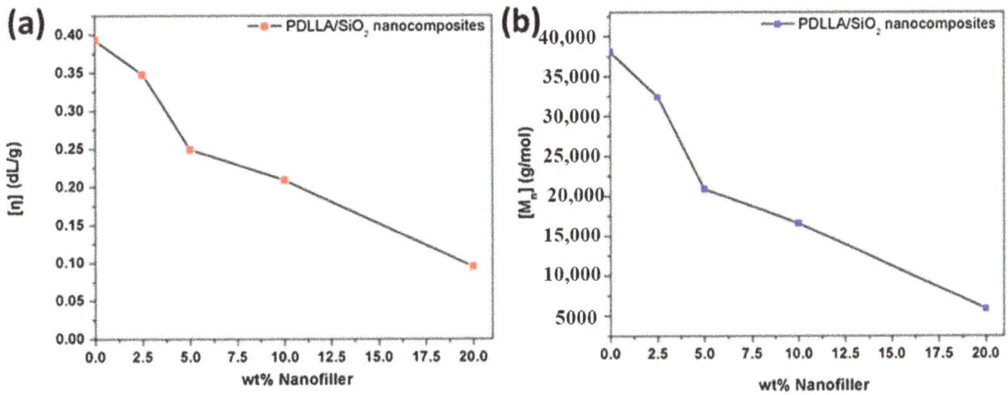

Figure 2. Effect of silica content on the (**a**) intrinsic viscosity and (**b**) average molecular weight of PDLLA [52].

As reported by Hao's group [54], the inclusion of 1.1, 2.8, 5.8, and 9.0 vol.% of silica particles into the PLA matrix led to the improvement in the viscoelastic properties of PLA. Here, silica with different sizes (same density), namely, silica 300 (7 nm), silica OX50 (40 nm), and silica 63 (9000 nm), were used. As shown in Figure 3, the low-frequency G′ increased with the addition of silica and reached an approximately frequency-independent plateau at the high content (above 2.8 vol.%). Moreover, it was observed that the effect induced by silica 63 on the storage modulus (G0) and complex viscosity (η^*) of PLA/silica 63 composites was extremely weak. With the increased silica loading, G0 varied negligibly in the high-frequency region and just a slight increase was found at the low-frequency range. Moreover, the times obtained from the plots of storage modulus (G′(t)/G′ onset versus time) at 180 °C in the case of neat PLA were found to be 5000 s, while longer times of 8000 s were reported for PLA/silica composites. The results obtained in this work clearly show that the rheological properties of PLA/silica composites are strongly affected not only by the particle size but also by the particle content.

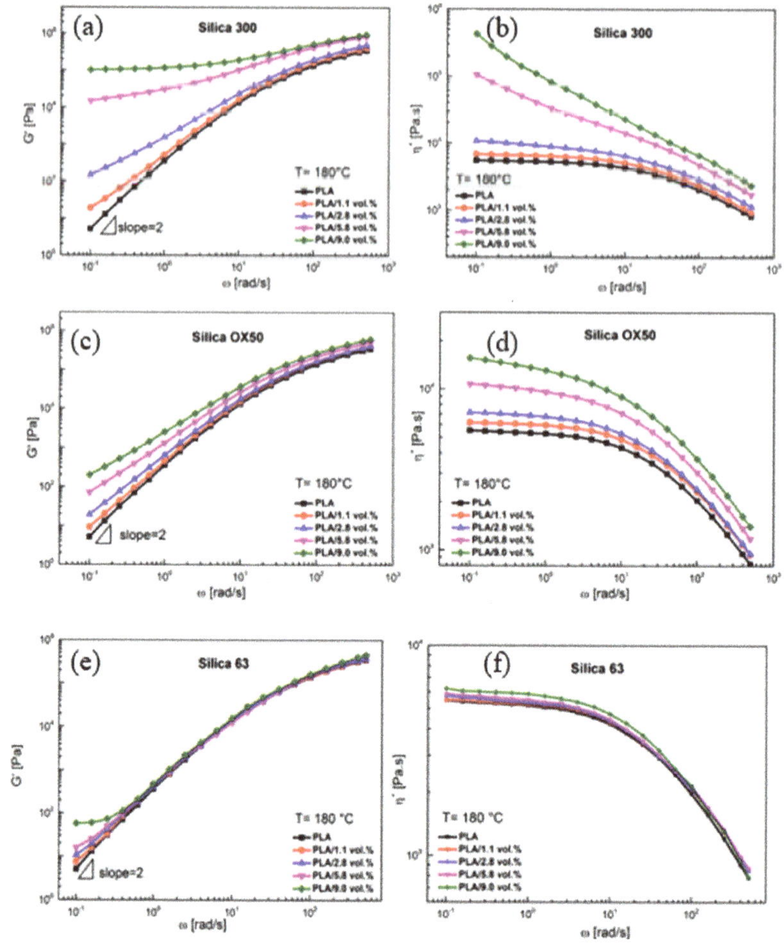

Figure 3. The influence of silica type on the rheological properties of PLA/silica composites: (**a**,**b**) PLA/silica 300; (**c**,**d**) PLA/silica OX50; (**e**,**f**) PLA/silica 63 [54].

4. Thermal Properties

The thermal stability of PLA can be improved by the incorporation of silica nanoparticles. In the works of Wen et al. [24,55], PLLA/silica composites were fabricated via a melt-mixing method. The thermal stability of PLA showed significant improvement with the addition of silica which was linked to the barrier effect of the silica network structure, which was also responsible for the improvements in the rheological properties of PLA discussed in the previous chapter. Zhang et al. [56] reported experimentally and theoretically, using a molecular dynamics simulation, that the thermal properties of PLA tended to be enhanced with the addition of silica into the polymer matrix during the processing in a twin-screw extruder. The results indicate that the glass transition temperature (T_g) of PLA was increased by 1.34 °C, while the thermal stability was increased by 12 °C when 2 wt.% of fumed silica was added into the PLA matrix. Klonos and Pissis [57] examined the thermal behavior of PLA/composites by taking the role of H-bonds, formed due to the reaction between the carbonyl groups in PLLA with hydroxyl groups on the silica surface, into account. Examinations of findings involved a combination of assessments on initially amorphous and on semicrystalline (annealed) samples. No change in the T_g by the silica was noted by differential scanning calorimetry (DSC), whereas the heat capacity step decreased in the PNCs. The segmental relaxation in the broadband dielectric relaxation spectroscopy (DRS) became, however, quicker and weaker in the PLA/silica composites. They reported that T_g and the temperature difference between onset (T_{onset}) and end (T_{end}) of the event ($DT_{GT} = T_{end} + T_{onset}$) tended to increase after annealing of crystallization, by 6–8 K and by a factor of ~2, respectively, as compared to the amorphous samples, due to constraints imposed by crystallites and heterogeneities. As for semicrystalline composites, T_g was increased by ~2 K, as compared to neat PLA. Simultaneously, DT_{GT} decreased slightly with increasing silica content. The authors attributed these results to the changes in semicrystalline morphology (Figure 4), polymer diffusion, and porosity/dispersion of silica particles.

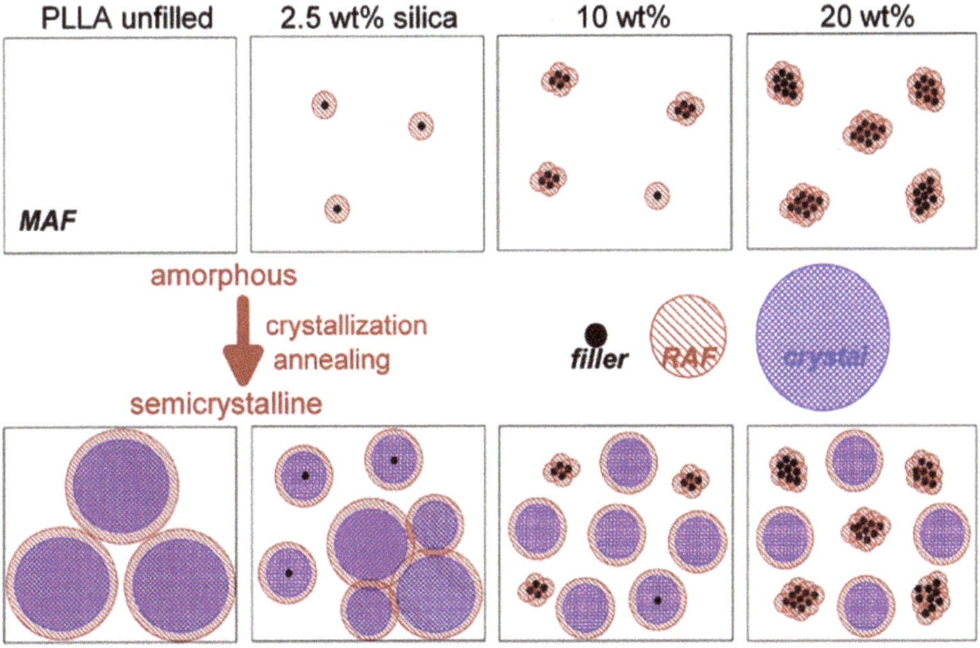

Figure 4. The proposed distribution of the silica in the PLA matrix before and after annealing of crystallization [57].

In the work of Bouamer et al. [58], a casting method was used to fabricate hybrid PLA composites with silica and AlO particles. In this regard, 10 wt.% of either silica or AlO particles was incorporated into the PLA matrix to fabricate PLA/silica and PLA/AlO composites, respectively, while 5 wt.% of SiO_2 and 5 wt.% of AlO particles were added to the PLA matrix to formulate PLA/silica/AlO composites. Based on the XRD patterns, it was reported that the crystallinity of the PLA acid film could be increased with the addition of silica/AlO particles into the PLA matrix. This result was attributed to the nucleating role of silica and AlO particles, which would facilitate the crystallization of PLA. However, Santos et al. [59] reported that the crystallinity of PLA was not affected by the inclusion of silica into the PLA matrix, while the dual addition of silica and cellulose into the PLA matrix resulted in significant improvements in the crystallization behavior of PLA. This finding was ascribed to the synergism between the two types of nanoparticles in which the agglomeration of cellulose nanoparticles could be prevented in the presence of silica. Thus, the nucleation was related not only to the chemical nature of the particles but also to the increased contact surface between the silica and cellulose nanoparticles. Zou et al. [60] demonstrated that the inclusion of a proper amount of silica nanoparticles (nucleating agent) would cause a reduction in the nucleation barrier shortening the nucleation period of PLA. The higher crystallinity was obtained in the composites containing 1.5 wt.% of silica nanoparticles. Such an amount of silica was expected to increase the interfacial compatibility and crystallinity of the composites thus enhancing the thermal stability. According to Prapruddivongs et al. [61], the crystallinity of the PLA film would have been affected by the type of silica added to the PLA matrix. Here, two types of silica, such as commercial silica ($CSiO_2$) and silica extracted from rice husks ($RSiO_2$), were added. The DSC thermograms of PLA composites (containing Triallyl isocyanurate and dicumyl peroxide) indicated that the silica nanoparticles affected the crystallinity and the melting behavior of PLA by impeding the chemical crosslinking reactions which were reflected by the change in the FTIR functional band of silica/CPLA composites at 1685 cm^{-1}.

Prapruddivongs and coworkers [62] studied the properties of PLA/silica and chemically crosslinked PLA (CrPLA)/silica composites prepared via the melt-mixing method in the presence of triallyl isocyanate and dicumyl peroxide as crosslinking agents. Here, two types of silica, such as $CSiO_2$ and $RSiO_2$, were used. Irrespective of the silica type, the thermal properties of the PLA/silica CrPLA/silica composites were improved. The addition of 1 wt.% of $CSiO_2$ and $RSiO_2$ led to an increase in the T50 value (the temperature associated with loss of 50 wt.%) of PLA from 321 °C to 339 °C and 342 °C, respectively. It was found also that the degradation temperature of PLA composites tended to increase in the presence of triallyl isocyanate and dicumyl peroxide. Moreover, the thermal stability of CrPLA/$RSiO_2$ composites was better than the thermal stability of CrPLA/$CSiO_2$, implying that $RSiO_2$ was an efficient additive for enhancing the thermal stability of PLA and CrPLA. Vidakis and coworkers [63] prepared PLA/silica composites via the melt-mixing of PLA with different contents of silica, such as 0.5, 1, 2, and 4 wt.%. Although the mass loss for the PLA/4 wt.% silica composite was lower than for PLA/1 wt.% silica, the overall thermal stability of the PLA seemed not to be influenced by the existence of the silica filler, which would be related to the strong H-bond interactions between Si–OH in silica nanoparticles with PLA chains. This is, in turn, restricted their release into the environment.

Lv et al. [64] examined the thermal properties of the PLA/silica composites prepared via the melt-mixing method. As displayed in Figure 5a, the increase in the silica content increased the degradation temperature of PLA/silica composites. For example, the inclusion of 10 wt.% of silica led to increasing the degradation temperature by 15 °C, indicating a significant increase in the thermal stability of PLA was achieved. From Figure 5b, it was found that the cold crystallization temperature (T_{cc}) of neat PLA acid tended to be shifted to lower temperatures upon the inclusion of silica nanoparticles, suggesting that plying the silica nanoparticle worked as nucleating agents or it could hinder crystallization from the melt. While the low-temperature melting peaks became weak, the high-temperature melting peaks showed a gradual increase with the increase in the silica content in the composites. This result would suggest that the inclusion of silica causes a reduction in the defective crystals, increasing the perfect crystals for the PLA phase. According to Ge et al. [65], the dispersibility of silica nanoparticles in the PLA matrix could be improved when the content of silica was less than 3 wt.%. Accordingly, the crystallinity of PLA was improved. The best crystallization behavior was obtained when 1 wt.% of silica was added to the PLA matrix. However, although some agglomerated nanoparticles were observed, the thermal stability of PLA was enhanced by the addition of silica nanoparticles.

The effect of silica content on the T_g of PLA/silica composites was explored by Pilić and coworkers [66]. It was reported that the addition of low amounts of silica, such as 0.2 and 0.5 wt.%, into the PLA matrix led to an increase in the T_g of neat PLA (47.6 °C) to 48.9 and 50.6 °C, respectively. This behavior was ascribed to the fact that the chain mobility throughout the PLA matrix volume tended to be decreased in the presence of silica nanoparticles, while the high loadings of silica, such as 1, 2, 3, and 5 wt.%, resulted in a decrease in the values of T_g in comparison to that in neat PLA, which was assigned to the agglomeration of silica nanoparticles within the PLA matrix which affected the chain mobility of the polymer. However, Wen et al. [56] reported that the addition of various loadings of silica, such as 1, 3, 5, 7, and 10 w.%, did not affect the value of T_g of PLA, suggesting that the impacts of silica on the chain mobility of PLA were insignificant.

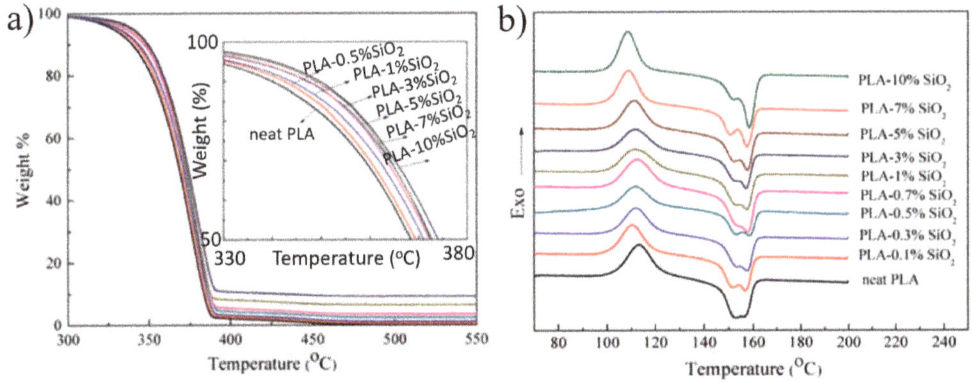

Figure 5. The effects of SiO_2 content on the thermal behavior of neat PLA and PLA/silica composites: (**a**) thermogravimetric analysis; (**b**) crystallization and melting behavior [64].

Techawinyutham et al. [67] examined the thermal stability of PLA and porous silica-containing capsicum oleoresin (SiCO)-modified PLA composites before and after the accelerating weathering test. The results imply that the accelerated weathering test induced the photolysis and partial hydrolysis, which resulted in improvements in the crystallization behavior of PLA. In addition, the accelerated weathering reduced the storage and loss moduli due to the increased chain mobility caused by the chain scission of the polymer chains.

Wu et al. [68] first grafted PLA on the surface of silica nanoparticles, then examined the thermal properties of PLA/PLA-grafted silica composites. Based on the DSC results, it was found that the addition of PLA-grafted silica could accelerate the crystallization rate of PLA. Besides, PLA/PLA-grafted silica composites exhibited typical homopolymer-like behavior in the final structure, regardless of the PLA-grafted silica content. The PLLA/silica composites containing 2.5 wt.% of silica exhibited a shielding effect to the evolution of gases that were released during the decomposition, improving mostly the initial stages of thermal degradation [69]. The thermal stability of PLA could be increased by 20 °C when low amounts of stearic acid-modified silica nanoparticles (0.1–1.5 wt.%) were added to the PLA matrix [40]. Similarly, Khankrua et al. [70] demonstrated that the presence of 5 wt.% of silica nanoparticles in the polymer matrix could cause significant improvements in the thermal stability of PLA. The effect of untreated silica nanoparticles on the thermal stability of PLA was examined by Basilissi et al. [51]. They found that the thermal stability of the composites tended to improve with the increase in the silica content. For instance, the temperature corresponding to 5 wt.% weight loss in the case of PLA composites containing 2 wt.% silica was higher by 70 °C than the counterpart corresponded to neat PLA, which was connected to the formation of silica network structure in the composites. The formation of a silica network structure hindered the diffusion of volatile decomposition products out of PLA and the diffusion of oxygen into the matrix. Similar results were reported by Wen et al. [24], who fabricated PLA/silica composites via the melt-mixing method and found that the thermal stability of PLA could be increased by the addition of $\leq$5 wt.% silica. However, the higher loading of silica ($\geq$5 wt.%) would cause an agglomeration of the nanoparticles in the polymer matrix due to the weak dispersion, which led to a reduction in thermal stability of PLA. In contrast, Lv et al. [64] found that the thermal stability of PLA/silica composites made by melt-mixing method tended to increase with the increase in the silica content—even when high loadings of silica, such as 7 and 10 wt.%, were added to the PLA matrix. Similarly, Mustapa et al. [71] postulated that the thermal stability of PLA could be increased with the addition of 2.5 and 7.5 wt.% of silica nanoparticles. In another works by Mustapa and coworkers [72–74], it was reported that the melting temperature and crystallization behavior were affected by the addition of silica nanoparticles, which acted as nucleation sites that induced the crystallization phenomenon of PLA. Lai and Li [75] functionalized the silica by the melt-mixing with polyurethane and found that the functionalized silica could greatly trigger the nucleation and crystallization behavior of PLA, when compared to the counterpart composites with non-functionalized silica. The influence of silica particles on the thermal properties of PLA composites is summarized in Table 1.

Table 1. The influence of silica incorporation on the thermal properties of PLA composites.

Silica (wt.%)	T_g (°C)	T_{cc} (°C)	T_{onset} (°C)	T_{max} (°C)	ΔH_m (J/g)	Ref.
10	-	-	342.3	370.9	-	[24]
5.0	59	135	-	-	8	[35]
2.0	64.5	0	-	-	31.32	[36]
0.5	52.3	102.8	113.5	-	54.3	[49]
3.0	61.22	107.87	162.60	168.70	34.57	[55]
2.0	52.23	-	-	-	-	[56]
2.5	60	-	-	-	37	[57]
1.5	-	-	-	300	-	[60]
4.0	60.8	108.3	146.6	153.6	24.9	[61]
4.0	-	-	273	374	-	[63]
10	-	-	-	370	-	[64]
5.0	-	120	-	-	70	[65]
0.5	50.6	-	-	-	-	[66]
3.08	57.66	112.33	147.5	153.41	22.73	[67]
0.5	61.02	124.26	-	-	-	[68]
0.2	-	-	345.54	362.56	-	[70]
7.5	59.0	-	355.5	412.3	-	[71]
7.5	59.0	124.94	-	-	6.22	[72]
7.5	60.0	96.3	355.5	412.3	9.9	[73]
7.5	57.4	117	-	-	-	[74]

5. Mechanical Properties

Several research teams have aimed to improve the mechanical properties of PLA with the addition of silica to the polymer matrix. For example, He et al. [76] reported that the inclusion of silica nanoparticles is a good strategy to improve the mechanical properties of PLA. Indeed, the tensile strength (TS), Young's modulus (YM), and impact strength of the PLA/silica composites made by the melt-mixing method were greatly increased with the addition of silica nanoparticles, which was linked to the good dispersibility of the particles in the polymer matrix, as well as PLA-silica interactions [77]. The low loading of silica (below 5 wt.%) was found to be distributed homogenously within the PLA matrix, while some agglomerates were observed at the high loading (higher than 5 wt.%) [36]. Ahmed et al. [78] studied the impact of silica additive into PLA matrix-3D printing. When the silica addition amount was 10 wt.%, its TS, flexibility, YM, and other properties were greatly enhanced. The application of this kind of composite material is feasible, but once the silica content exceeds 15%, the performance is reduced. The improved mechanical properties promote the recycling of PLA while retaining its biodegradability. As such, YM and ultimate tensile stress (UTS) of PDLLA were increased by 106% and 63.7%, respectively, upon the addition of 3 wt.% silica into the polymer matrix [79].

Yan et al. [32] prepared PLA/silica composites via two steps: grafting of L-lactic acid oligomer onto the silica surface, followed by melt-mixing with PLA. This procedure led to fabricating composites with improved mechanical properties. In another work by Yan and co-workers [31], the TS of the plasticized PLA/silica composite could be increased to large values even in the presence of low loadings of silica. Although the addition of the silica nanoparticles accelerated the biodegradation rate of PLA, the resultant composites had improved mechanical properties, as reported by Georgiopoulos et al. [80].

To obtain better dispersion in the PLA matrix, silica nanoparticles with 24 nm in average diameter were functionalized via stearic acid [38]. Different contents of silica, such as 0, 0.1, 0.3, 0.5, 0.8, 1, and 1.5 wt.%, were mixed with PLA via an extrusion technique. The mechanical tests revealed that the addition of 0.8 wt.% of silica nanoparticles into the PLA matrix led to increasing the Izod impact strength from 1.57 to 5.13 kJ/m^2. The impact strength values of the PLA/silica composites increased with increasing silica content, then decreased when the silica content exceeded 0.8 wt.%. Moreover, the presence of nano-silica in the PLA matrix caused improvements in the EB and TS of PLA (Figure 6a). As shown in Figure 6b, dimple morphologies were observed on the fractural surfaces of PLA/silica composites, suggesting that neat PLA was less ductile than PLA/silica composites. According to the lubrication sliding and the ductile behavior, it was confirmed the toughness of PLA could be largely improved by the incorporation of silica nanoparticles. The effects of silica content (1, 3, and 5 wt.%) on the mechanical performance of PLA film were studied Zirak and Tabari [81]. It was found that the values of the TS and YM of PLA tended to be increased upon the incorporation of silica nanoparticles into the PLA matrix, regardless of the silica content. For example, the addition of 5 wt.% of silica into PLA led to an increase in the values of TS and YM for PLA from 29 MPa and 2.3 GPa to 43 MPa and 3.1 GPa, respectively, which was attributed to the reinforcing effect of SiO_2 nanoparticles. In contrast, the value of EB of PLA (17.5%) was reduced with the inclusion of silica nanoparticles, where a value of 10.5% was found in the composite containing 5 wt.% SiO_2.

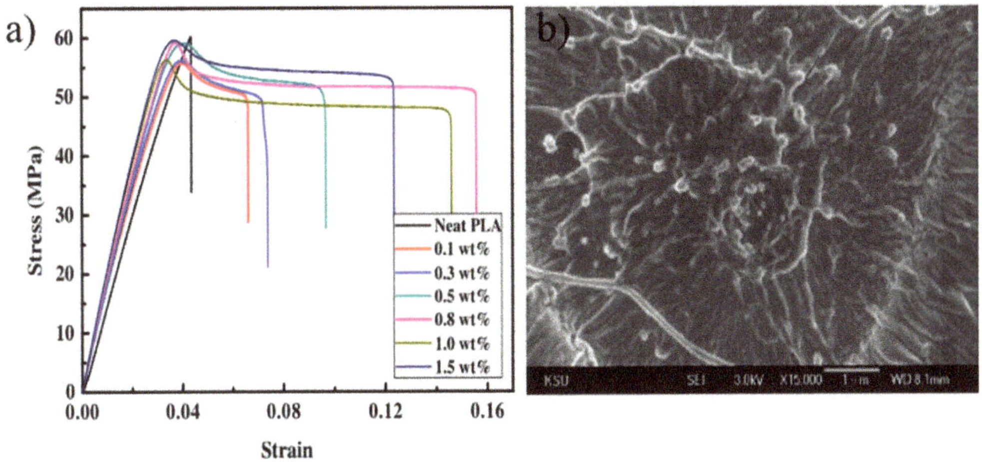

Figure 6. (**a**) The stress–strain curve of neat PLA and PLA composites. (**b**) The fracture morphology after the tensile test of PLA composite [38].

In Yan's research [32], it was approved that the grafting of L-lactic acid oligomer on the silica nanoparticles by condensation reaction without catalyst would result in the formation of a good filler, helping to increase the toughness of PLA. Chrissafis et al. [82] reported that the inclusion of 2.5 wt.% of fumed silica and montmorillonite into the PLA matrix would result in the formation of composites having higher values of TS and YM than neat

PLA. However, the addition of fumed silica and montmorillonite to PLA led to a reduction in the value of EB, implying that those two additives acted as reinforcing agents. Lai and Hsieh [83] examined the influence of the surface functionalization of silica on the mechanical properties of PLA. To form the modified silica, polyethylene glycol methyl ether was grafted on the silica in the presence of aminosilane. The interfacial interaction between silica and PLA was improved in the presence of modified silica; thereby, the value of TS of the composites with modified silica was higher than the counterpart composites containing unmodified silica.

Battegazzore et al. [84] used different contents, such as 5, 10, 20, and 30 wt.%, of silica powder, obtained by the conventional extraction method from rice husk, to fabricate PLA/silica nanocomposites via the melt-mixing method. Based on Archimedes law, the silica density was calculated to be 1.82 and 1.88 g·cm^{-3} for the extracted silica and commercial silica, respectively, while the density of PLA was 1.25 g·cm^{-3}. The density values were used to calculate the volumetric fractions. When the extracted silica nanoparticles were melted with PLA, the YM of neat PLA was increased, while the oxygen permeability of PLA was slightly reduced. As compared to the composites including 10 and 30 wt.% of commercial silica, the counterpart composites with extracted silica exhibited better mechanical properties. Considering the economic analysis of the whole process and by reusing the energy recovered from burning rice husk, the authors considered the composites containing 20 wt.% of extracted silica as economically sustainable materials.

It is worth mentioning that silica can improve the mechanical properties of other PLA composites. For example, the addition of silica can improve the jute/PLA interfacial adhesion, which results in the enhancement of the mechanical performances of the resultant composites [85].

On the other hand, the influence of other silica-based materials on the properties of PLA was explored by many research groups. For instance, the addition of glass fibers to the PLA matrix improved the mass flow rate and flexural modulus, while their effects on the impact and flexural strength, as well as thermal stability of PLA, were insignificant [86]. Besides, the TS and EB were reduced with the addition of glass fibers.

6. Biodegradbility and Other Properties

In general, significant research still needs to be performed to achieve the final target of ideal biodegradable PLA/silica composites that exhibit high performance and easy biodegradability when their roles are completed. The hydrophilic nature of silica nanoparticles is expected to affect the degradation of PLA. The hydroxyl groups in silica are bound together by hydrogen bonds and can assist the hydrogen bonding interaction with the functional groups in PLA or a covalent bonding with a macromolecular chain [87]. Thus, silica nanoparticles are expected to facilitate the hydrolysis or enzymatic attacks of ester groups of PLA, leading to a fast biodegradation rate. For instance, Li et al. [50] reported that the weight loss during the biodegradation process of PLA/silica composites, fabricated by melt-mixing method, was larger than that in neat PLA. Indeed, the incorporation of 9 wt.% of silica nanoparticles into the PLA matrix led to a biodegradation rate of 0.36 mg cm^{-2} h^{-1}, which was 6.5 times higher than that of neat PLA. Figure 7a shows the fast biodegradation of the PLA composites compared to neat PLA. This result was confirmed by the DSC curves shown in Figure 7b. As shown in Figure 7b, all samples were amorphous at a T_g of 60 °C. The improvements in the degradation rate were attributed to the easy release of silica particles from the PLA matrix. The hydrophilic silica facilitated the hydrolysis and enzymatic attack of ester groups of PLA chains. The biodegradation was more pronounced in the composite membrane after 2 months of in vitro tests [88]. On the other hand, it was reported that the flame retardant properties of PLA could be improved greatly by adding treated silica nanoparticles into the PLA matrix [64]. Indeed, the weight loss ratios of PLA/silica composites at different burning time intervals tended to decrease with the increase in the silica content, which was linked to the excellent dispersion of particles in the PLA matrix. So, the larger specific surface area in the composites provided a better effect of thermal insulation [64]. SiO_2-fluorinated PLA composites can be used as reversible and highly hydrophobic coatings to protect the exterior of buildings [89]. Similarly, it was reported that the flame retardancy of PLA could be improved by the addition of fumed silica and Ni_2O_3 into the polymer matrix [90].

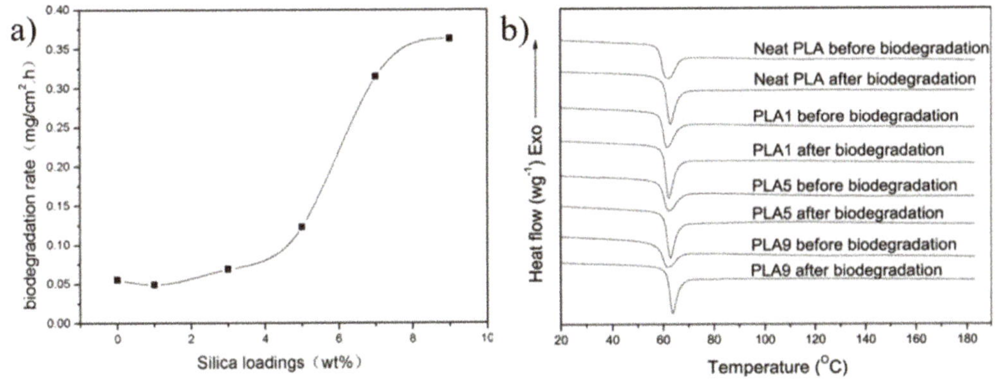

Figure 7. (**a**) The effect of silica content on the biodegradation rate of PLA/silica composites. (**b**) The DSC curves of the PLA/silica composites before and after degradation of 24 h [50].

A study has been reported by Zhao et al. [91], where *N*-halamine precursor with epoxy and hydantoin structures, 3-(4′-epoxyethyl-benzyl)-5, 5-dimethylhydantoin (EBDMH) was utilized for the *N*-halamine-modified silica nanoparticles. EBDMH was fabricated and immobilized onto aminofunctionalized silica nanoparticles to form EBDMH–SiO_2 nanoparticles (Figure 8a). PLA was mixed with EBDMH–SiO_2 nanoparticles via the melt-mixing method. The efficiency of the PLA/EBDMH–SiO_2 composite as antimicrobial material was evaluated against *S. aureus* and *E. coli*, respectively, and the obtained results are shown in Figure 8b. The PLA/EBDMH–SiO_2 composite exhibited excellent bactericidal efficiency. Indeed, the PLA/EBDMH–SiO_2 composite with a contact time of 10 min neutralized about 90.2% of *S. aureus* and 89.4% of *E. coli*. When the contact time was increased to 180 min, a kill efficiency of 99.97% and 99.91% against *S. aureus* and *E. coli*, respectively, was obtained. Due to improved biocompatibility between PLA and silica, as well as the excellent antibacterial efficiency, the PLA/EBDMH–SiO_2 composite could be utilized for hygienic product packaging and filters, as well as medical textiles.

The impact of silica nanoparticles on the interfacial tension between PLA and supercritical CO_2 at high temperature and high pressures was examined by Sarikhani and coworkers [92]. The addition of a low loading of silica (less than 2 wt.%) led to reducing the interfacial tension, while the interfacial tension tended to increase when the silica content was larger than 2 wt.%, which was linked to the fact that higher levels of silica originated attractive lateral capillary forces due to the perturbation of the PLA–CO_2 interface by particles. The interfacial interactions between PLA and silica nanoparticles were found to be decreased with an increase in the CO_2 content, which facilitated the adsorption behavior at higher pressures. Based on the experiments carried out by Seng and co-workers [93], it was reported that the addition of 1 wt.% of silanol treated-silica into the PLA matrix led to a decrease of 40% in the hygroscopicity of PLA, while the other loadings of silica caused improvements between 3 and 19% in the reduction in hygroscopicity. Chen et al. [94] reported that the hydrophilicity of PLA/silica composites tended to be improved with increasing the silica content, where the inclusion of 5 wt.% of silica into the PLA matrix led to a decrease in the water contact angle of PLA from 82° to 68°. In other words, the less silica in the composite, the larger the contact angle is, thereby, the higher the hydrophobicity of the composite surface is. Thus, the hydrolytic degradation ability of PLLA/silica composites was accelerated in the presence of higher loadings of silica in the polymer matrix.

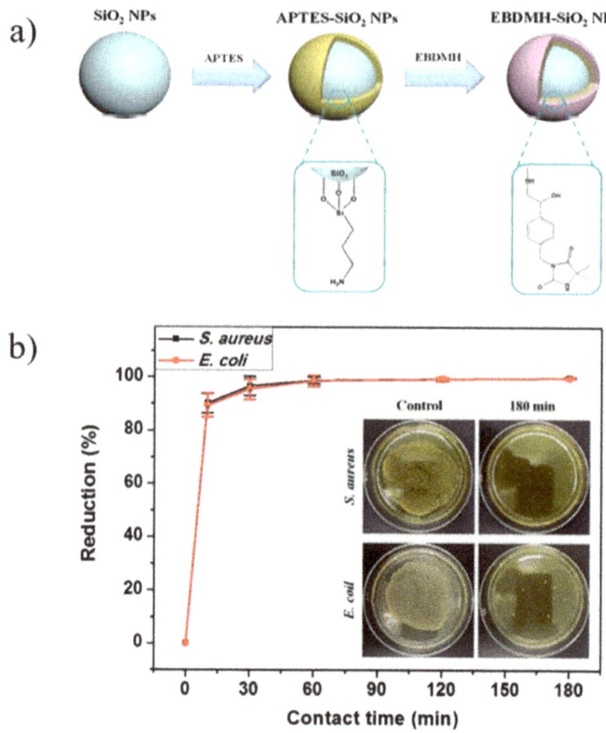

Figure 8. (**a**) Synthesis of *N*-halamine precursor-modified silica nanoparticles (EBDMH–SiO_2 NPs). (**b**) Antibacterial tests of PLA/EBDMH–SiO_2-9 plastic sheets after chlorination against *S. aureus* and *E. coli*. In the inset are photographs showing the bacterial culture plates of *S. aureus* and *E. coli* upon 180 min contact with the control and PLA/EBDMH–SiO_2-9 plastic sheets [91].

Kosowska and Szatkowski [95] examined the impacts of silica addition on the ultraviolet aging of PLA nonwovens obtained by the electrospinning technique. The inclusion of silica nanoparticles greatly increased the percentages of crystalline and amorphous phases in the fabricated films and altered the photodegradation mechanism. In the work of Jin-Bo [96], a novel porous membrane composed of PLA with the addition of SiO_2–CaO by the sol–gel method was designed. The surface potentials of the composite membranes became more negative with higher SiO_2–CaO contents. Apatite with an orderly ring structure nucleated and grew on the surface of the composite membranes after immersion in SBF for 7 days, implying that the incorporation of SiO_2–CaO significantly improves the bioactivity of PLA. Based on the dielectric spectroscopy study [97], it was reported that the lifetime and thermal stability of the electric state in PLA can be increased with the addition of SiO_2, where the optimal concentration of silica for negative corona electrets was found to be 4 wt.%.

7. Potential Applications of PLA/Silica Composites

Owing to its high resistance to heat and good thermal stability, silica has received considerable attention in 3D-printing applications. The enhanced melt viscosity of PLA/silica composites could facilitate the molding and processing of these materials which could lead to a variety of applications [54]. Since the usage of PLA used in the bone tissue applications would not be able to withstand the high-load resistance associated with such applications, the incorporation of silica would make PLA a suitable candidate in these applications, as the flexural modulus and the tensile strength of PLA tended to be improved significantly with the incorporation of a silica filler. For this reason, it was reported that the biorenewable PLA/SiO_2 composites can be used as promising materials in the bone repair processes in animal models [98]. As indicated by cell-culture measurements conducted by Abe et al. [99], PLA/silica composite could exhibit excellent cytocompatibility. Interestingly, the ROS-responsive LDLR peptides-conjugated PLA-coated mesoporous silica nanoparticles were reported to have an important opportunity for oxidative stress therapy in the central nervous system [100]. Moreover, the good thermal stability and mechanical properties of PLA/silica composites would make these materials suitable for structural applications where high thermal stability and high mechanical strength are needed [71]. Thanks to their physical, thermal, rheological, and mechanical properties, the PLA/silica composites can be used in 3D-printing applications, as indicated by Thongsang and coworkers [101]. Moreover, the PLA/silica composites have a great opportunity to be used as smart and active packaging. According to Jaikaew [102], the CO_2/O_2 permeability ratio of PLA/silica composite films can be tuned by varying the types of silica particles and their compositions. Light transmission reduction in both the UV and visible regions is achieved in PLA/modified-silica bio-composite films. As approved by Opaprakasit et al. [103], the films obtained from PLA/modified silica composites have strong potential to be used as biodegradable packaging materials with tunable gas permeability.

8. Future Outlooks of PLA/Silica Composites

The state of knowledge discussed above clearly shows that the fabrication, properties, and applications of PLA/silica composites will keep growing in the future. We thought that future trends of the research on PLA/silica composites should be focused on several relatively new directions, as follows:

- Although PLA/silica composites have been utilized in 3D-printing applications, more research would be needed in this area, as the main challenge in 3D-printing applications is enhancing the flexibility, as well as controlling the viscosity, of PLA.
- The dispersion of silica nanoparticles in the PLA matrix should be improved, as various undiscovered applications in various fields, such as the aerospace, energy, and chemical industries, are expected to be available shortly.
- The utilization of PLA/silica composites to produce hierarchically porous PLA materials for biomedical applications, such as tissue scaffolding, would be a promising research area shortly. The processing conditions, however, should be selected carefully to successfully fabricate such materials.
- Even PLA/silica composites exhibited some levels of biodegradability; they are still far from being considered a solution for plastic waste accumulation.

9. Conclusions

Based on the brief discussion reported in this work, the inclusion of silica nanoparticles into the PLA matrix is one of the promising methods to enhance the performance of PLA while retaining its biodegradability. The fabrication methods, namely, melt blending and in situ polymerization, as well as the organo-modification of silica nanoparticles to increase their affinity to hydrophobic PLA, would play a pivotal role in the final performance of PLA/silica composites. Indeed, the surface functionalization of silica nanoparticles had critical impacts on the rheological performance of the silica within the polymer matrix. The thermal stability, biodegradability, YM, and TS of PLA were greatly improved by the inclusion of silica nanoparticles into the polymer matrix. Such improved properties qualified the PLA/silica composites to be suitable candidates in food packaging, 3D printing, and biomedical applications.

Author Contributions: M.K.: Conceptualization, writing—original draft, writing-review and editing; funding acquisition; Z.U.R., S.H., A.K.S. and B.D.: writing—review and editing. All authors have read and agreed to the published version of the manuscript.

Funding: This work was supported by the National Research Foundation of Korea (NRF) grant funded by the Korea government (MSIT) (No. NRF-2019R1G1A1099335).

Institutional Review Board Statement: Not applicable.

Informed Consent Statement: Informed consent was obtained from all subjects involved in the study. The data presented in this study are available on request from the corresponding authors.

Data Availability Statement: The data presented in this study are available on request from the corresponding author.

Conflicts of Interest: The authors declare no conflict of interest.

References

1. Chandra, R.; Rustgi, R. Biodegradable polymers. *Prog. Polym. Sci.* **1998**, *23*, 1273–1335. [CrossRef]
2. Briassoulis, D. An overview on the mechanical behaviour of biodegradable agricultural films. *J. Polym. Environ.* **2004**, *12*, 65–81. [CrossRef]
3. Hamad, K.; Kaseem, M.; Ayyoob, M.; Joo, J.; Deri, F. Polylactic acid blends: The future of green, light and tough. *Prog. Polym. Sci.* **2018**, *85*, 83–127. [CrossRef]
4. Hamad, K.; Kaseem, M.; Yang, H.W.; Deri, F.; Ko, Y.G. Properties and medical applications of polylactic acid: A review. *Express Polym. Lett.* **2015**, *9*, 435–455. [CrossRef]
5. Kaseem, M.; Hamad, K.; Deri, F.; Ko, Y.G. A review on recent researches on polylactic acid/carbon nanotube composites. *Polym. Bull.* **2017**, *74*, 2921–2937. [CrossRef]
6. Kaseem, M.; Ko, Y.G. Melt flow behavior and processability of polylactic acid/polystyrene (PLA/PS) polymer blends. *J. Polym. Environ.* **2017**, *25*, 994–998. [CrossRef]
7. Kaseem, M.; Hamad, K.; Rehman, Z.U. Review of recent advances in polylactic acid/TiO_2 composites. *Materials* **2019**, *12*, 3659. [CrossRef] [PubMed]
8. Hamad, K.; Kaseem, M.; Deri, F. Preparation and characterization of binary and ternary blends with poly(lactic acid), polystyrene, and acrylonitrile-butadiene-styrene. *J. Biomater. Nanobiotechnol.* **2012**, *3*, 405–412. [CrossRef]
9. Kaseem, M. Poly(lactic acid) composites. *Materials* **2019**, *12*, 3586. [CrossRef]
10. Hamad, K.; Ko, Y.G.; Kaseem, M.; Deri, F. Effect of acrylonitrile–butadiene–styrene on flow behavior and mechanical properties of polylactic acid/low density polyethylene blend. *Asia-Pac. J. Chem. Eng.* **2014**, *9*, 349–353. [CrossRef]
11. Kadhum, S.A. The Effect of two types of nano-particles (ZnO and SiO_2) on different types of bacterial growth. *Biomed. Pharmacol. J.* **2017**, *10*, 1701–1708. [CrossRef]
12. Li, H.; Liebscher, M.; Curosu, I.; Choudhury, S.; Hempel, S.; Davoodabadi, M.; Dinh, T.T.; Yang, J.; Mechtcherine, V. Electrophoretic deposition of nano-silica onto carbon fiber surfaces for an improved bond strength with cementitious matrices. *Cem. Concr. Compos.* **2020**, *114*, 103777. [CrossRef]
13. Singh, L.; Karade, S.; Bhattacharyya, S.; Yousuf, M.; Ahalawat, S. Beneficial role of nanosilica in cement based materials A review. *Constr. Build. Mater.* **2013**, *47*, 1069–1077. [CrossRef]
14. Sikora, P.; Chougan, M.; Cuevas, K.; Liebscher, M.; Mechtcherine, V.; Ghaffar, S.H.; Stephan, D. The Effects of Nano-and Micro-Sized Additives on 3D Printable Cementitious and Alkali-Activated Composites: A Review. *Appl. Nanosci.* **2021**, 1–19. Available online: https://link.springer.com/article/10.1007/s13204-021-01738-2 (accessed on 16 August 2021).
15. Ren, Z.; Zhou, W.; Qing, Y.; Duan, S.; Pan, H.; Zhou, Y.; Li, N. Microwave absorption and mechanical properties of SiCf/SiOC composites with SiO2 fillers. *Ceram. Int.* **2021**, *47*, 8478–8485. [CrossRef]
16. Kaseem, M.; Choi, K.; Ko, Y.G. A highly compact coating responsible for enhancing corrosion properties of Al-Mg-Si alloy. *Mater. Lett.* **2017**, *196*, 316–319. [CrossRef]
17. Liu, W.; Wu, X.; Zhan, H.; Yan, F. Synthesis of bioactive poly(ethylene glycol)/SiO_2-CaO-P_2O_5 hybrids for bone regeneration. *Mater. Sci. Eng. C* **2012**, *32*, 707–711. [CrossRef]
18. Jia, H.; Hou, W.; Wei, L.; Xu, B.; Liu, X. The structures and antibacterial properties of nano-SiO_2 supported silver/zinc–silver materials. *Dent. Mater.* **2008**, *24*, 244–249. [CrossRef]
19. Petousis, M.; Tzounis, L.; Papageorgiou, D.; Vidakis, N. Decoration of SiO_2 and Fe_3O_4 nanoparticles onto the surface of MWCNT-grafted glass fibers: A simple approach for the creation of binary nanoparticle hierarchical and multifunctional composite interphases. *Nanomaterials* **2020**, *10*, 2500. [CrossRef]
20. Fatimah, S.; Kamil, M.P.; Kwon, J.H.; Kaseem, M.; Ko, Y.G. Dual incorporation of SiO_2 and ZrO_2 nanoparticles into the oxide layer on 6061 Al alloy via plasma electrolytic oxidation: Coating structure and corrosion properties. *J. Alloys Compd.* **2017**, *707*, 358–364. [CrossRef]
21. Tsai, P.-A.; Chiu, W.-M.; Lin, C.-E.; Wu, J.-H. Fabrication and characterization of PLA/SiO_2/Al_2O_3 composites prepared by Sol-Gel process. *Polym. Plast. Technol. Eng.* **2013**, *52*, 1488–1495. [CrossRef]
22. Kim, G.H.; Hwang, S.W.; Jung, B.N.; Kang, D.; Shim, J.K.; Seo, K.H.E. Effect of PMMA/Silica hybrid particles on interfacial adhesion and crystallization properties of Poly(Lactic Acid)/Block acrylic elastomer composites. *Polymers* **2020**, *12*, 2231. [CrossRef]

23. Zehra, T.; Kaseem, M.; Hossain, S.; Ko, Y.G. Fabrication of a protective hybrid coating composed of TiO_2, MoO_2, and SiO_2 by plasma electrolytic oxidation of titanium. *Metals* **2021**, *8*, 1182. [CrossRef]
24. Wen, X.; Zhang, K.; Wang, Y.; Han, L.; Han, C.; Zhang, H.; Chen, S.; Dong, L. Study of the thermal stabilization mechanism of biodegradable poly(L-lactide)/silica nanocomposites. *Polym. Int.* **2011**, *60*, 202–210. [CrossRef]
25. Basu, A.; Nazarkovsky, M.; Ghadi, R.; Khan, W.; Domb, A.J. Poly(lactic acid)-based nanocomposites. *Polym. Adv. Technol.* **2017**, *28*, 919–930. [CrossRef]
26. Hakim, R.H.; Cailloux, J.; Santana, O.O.; Bou, J.; S_anchez-Soto, M.; Odent, J.; Raquez, J.M.; Dubois, P.; Carrasco, F.; Maspoch, M.L. PLA/SiO_2 composites: Influence of the filler modifications on the morphology, crystallization behavior, and mechanical properties. *J. Appl. Polym. Sci.* **2017**, *134*, 45367. [CrossRef]
27. Yan, S.; Yin, J.; Yang, J.; Chen, X. Structural characteristics and thermal properties of plasticized poly(L-lactide)-silica nanocomposites synthesized by sol-gel method. *Mater. Lett.* **2007**, *61*, 2683–2686. [CrossRef]
28. Dash, S.; Mishra, S.; Patel, S.; Mishra, B.K. Organically modified silica: Synthesis and applications due to its surface interaction with organic molecules. *Adv. Colloid Interface Sci.* **2008**, *140*, 77–94. [CrossRef]
29. Parida, S.K.; Dash, S.; Patel, S.; Mishra, B.K. Adsorption of organic molecules on silica surface. *Adv. Colloid Interface Sci.* **2006**, *121*, 77–110. [CrossRef] [PubMed]
30. Bagwe, R.P.; Hilliard, L.R.; Tan, W. Surface modification of silica nanoparticles to reduce aggregation and nonspecific binding. *Langmuir* **2006**, *22*, 4357–4362. [CrossRef] [PubMed]
31. Mai, Y.-W.; Yu, Z.-Z. *Polymer Nanocomposites*; Woodhead Publishing Limited: Cambridge, UK, 2006.
32. Yan, S.; Yin, J.; Yang, Y.; Dai, Z.; Ma, J.; Chen, X. Surface-grafted silica linked with L-Lactic acid oligomer: A novel nanofiller to improve the performance of biodegredable poly(L-Lactide). *Polymer* **2007**, *48*, 1688–1694. [CrossRef]
33. Wu, L.; Cao, D.; Huang, Y.; Li, B.G. Poly(L-lactic acid)/SiO_2 nanocomposites via in situ melt polycondensation of L-lactic acid in the presence of acidic silica sol: Preparation and characterization. *Polymer* **2008**, *49*, 742–748. [CrossRef]
34. Liu, L.Z.; Ma, H.J.; Zhu, X.S. Preparation and properties of polylactide/nano-silica in situ composites. *Pigment. Resin. Technol.* **2010**, *39*, 27–31. [CrossRef]
35. Chang, C.C.; Tsai, Y.L.; Chiu, J.J.; Chen, H. Preparation of phase change materials microcapsules by using PMMA network-silica hybrid shell via sol-gel process. *J. Appl. Polym. Sci.* **2009**, *112*, 1688–1694. [CrossRef]
36. Fukushima, K.; Tabuani, D.; Abbate, C.; Arena, M.; Rizzarelli, P. Preparation, characterization and biodegradation of biopolymer nanocomposites based on fumed silica. *Euro. Polym. J.* **2011**, *47*, 139–152. [CrossRef]
37. Luo, D.; Zhen, W.; Dong, C.; Zhao, L. Performance and multi-scale investigation on the phase miscibility of poly(lactic acid)/amided silica nanocomposites. *Inter. J. Biol. Macromol.* **2021**, *177*, 271–283. [CrossRef] [PubMed]
38. Ma, Z.; Wu, L.; Peng, B. Poly(L-lactic acid)/silicone dioxide nanocomposites prepared via in situ melt polycondensation of L-lactic acid in the presence of acidic silica sol: Dispersion stability of nanoparticles during dehydration/oligomerization. *J. Appl. Polym. Sci.* **2012**, *124*, 3980–3987. [CrossRef]
39. Wu, J.H.; Yen, M.S.; Kuo, M.C.; Chen, B.H. Physical properties and crystallization behavior of silica particulates reinforced poly(lactic acid) composites. *Mater. Chem. Phys.* **2013**, *142*, 726–733. [CrossRef]
40. Ray, S.S.; Okamoto, M. Polymer/layered silicate nanocomposites: A review from preparation to processing. *Prog. Polym. Sci.* **2003**, *28*, 1539–1641.
41. Zou, J.; Ma, T.; Zhang, J.; He, W.; Huang, F. Preparation and characterization of PLLA–ESO/surface-grafted silica nanocomposites. *Polym. Bull.* **2011**, *67*, 1261–1271. [CrossRef]
42. Sepulveda, J.; Villegas, C.; Torres, A.; Vargas, E.; Rodriguez, F.; Baltazar, S.; Prada, A.; Rojas, A.; Romero, J.; Faba, S.; et al. Effect of functionalized silica nanoparticles on the mass transfer process in active PLA nanocomposite films obtained by supercritical impregnation for sustainable food packaging. *J. Supercrit. Fluids* **2020**, *161*, 104844. [CrossRef]
43. Zhu, A.; Diao, H.; Rong, Q.; Cai, A. Preparation and properties of polylactide–silica nanocomposites. *J. Appl. Polym. Sci.* **2010**, *116*, 2866–2873. [CrossRef]
44. Kaseem, M.; Hamad, K.; Deri, F. Slit die rheology of thermoplastic starch during extrusion process. *Inter. J. Polym. Mater.* **2013**, *17*, 51–60. [CrossRef]
45. Kaseem, M.; Hamad, K.; Deri, F. On-line rheological and mechanical measurement of ABS/starch composites. *Inter. J. Polym. Mater.* **2013**, *62*, 260–264. [CrossRef]
46. Kaseem, M.; Hamad, K.; Park, J.H.; Ko, Y.G. Rheological properties of ABS/wood composites. *Euro. J. Wood. Wood. Prod.* **2015**, *73*, 701–703. [CrossRef]
47. Kaseem, M.; Hamad, K.; Yang, H.W.; Deri, F.; Ko, Y.G. Melt rheology of poly(vinylidene fluoride) (PVDF)/low density polyethylene (LDPE) blends. *Poly. Sci. Part. A* **2015**, *57*, 233–238. [CrossRef]
48. Kaseem, M.; Hamad, K.; Deri, F.; Ko, Y.G. Effect of wood fibers on the rheological and mechanical properties of polystyrene/wood composites. *J. Wood. Chem. Technol.* **2017**, *37*, 251–260. [CrossRef]
49. Kaseem, M.; Hamad, K.; Deri, F. Preparation and studying properties of thermoplastic starch/acrylonitrile–butadiene–styrene blend. *Int. J. Plast. Technol.* **2012**, *16*, 39–49. [CrossRef]
50. Basilissi, L.; di Silvestro, G.; Farina, H.; Ortenzi, M.A. Synthesis and characterization of PLA nanocomposites containing nanosilica modified with different organosilanes II: Effect of the organosilanes on the properties of nanocomposites: Thermal characterization. *J. Appl. Polym. Sci.* **2013**, *128*, 3057–3063. [CrossRef]

51. Li, Y.; Han, C.; Bian, J.; Han, L.; Dong, L.; Gao, G. Rheology and biodegradation of polylactide/silica nanocomposites. *Polym. Compos.* **2012**, *33*, 1719–1727. [CrossRef]
52. Nerantzaki, M.; Prokopiou, L.; Bikiaris, D.N.; Patsiaoura, D.; Chrissafis, K.; Klonos, P.; Kyritsis, A.; Pissis, P. In situ prepared poly(DL-lactic acid)/silica nanocomposites: Study of molecular composition, thermal stability, glass transition and molecular dynamics. *Thermochim. Acta* **2018**, *669*, 16–29. [CrossRef]
53. Pandis, C.; Trujillo, S.; Matos, J.; Madeira, S.; Ródenas-Rochina, J.; Kripotou, S.; Kyritsis, A.; Mano, J.F.; Ribelles, J.L.G. Porous polylactic acid–silica hybrids: Preparation, characterization, and study of mesenchymal stem cell osteogenic differentiation. *Macromol. Biosci.* **2015**, *15*, 262. [CrossRef]
54. Hao, X.; Kaschta, J.; Schubert, D.W. Viscous and elastic properties of polylactide melts filled with silica particles: Effect of particle size and concentration. *Compos. Part. B* **2016**, *89*, 44–53. [CrossRef]
55. Wen, X.; Lin, Y.; Han, C.; Zhang, K.; Ran, X.; Li, Y.; Dong, L. Thermomechanical and optical properties of biodegradable poly(L-lactide)/silica nanocomposites by melt compounding. *J. Appl. Polym. Sci.* **2009**, *114*, 3379–3388. [CrossRef]
56. Zhang, J.; Lou, J.; Ilias, S.; Krishnmachari, P.; Yan, J. Thermal properties of poly(lactic acid) fumed silica nanocomposites: Experiments and molecular dynamics simulations. *Polymer* **2008**, *49*, 2381–2386. [CrossRef]
57. Klonos, P.; Pissis, P. Effects of interfacial interactions and of crystallization on rigid amorphous fraction and molecular dynamics in polylactide/silica nanocomposites: A methodological approach. *Polymer* **2017**, *112*, 228–243. [CrossRef]
58. Bouamer, A.; Benrekaa, N.; Younes, A. Characterization of polylactic acid ceramic composites synthesized by casting method. *Mater. Today Proc.* **2021**, *42*, 2959–2962. [CrossRef]
59. Santos, F.A.D.; Tavares, M.I.B. Development and characterization of hybrid materials based on biodegradable PLA matrix, microcrystalline cellulose and organophilic silica. *Polímeros* **2014**, *24*, 561–566. [CrossRef]
60. Zuo, Y.; Chen, K.; Li, P.; He, X.; Li, W.; Wu, Y. Effect of nano-SiO2on the compatibility interface and properties of polylactic acid-grafted-bamboofiber/polylactic acid composite. *Inter. J. Biol. Macromol.* **2020**, *157*, 177–186. [CrossRef]
61. Prapruddivongs, C.; Wongpreede, M.A.T. Effect of silica resources on the biodegradation behavior of poly (lactic acid)and chemical crosslinked poly (lactic acid) composites. *Polym. Test.* **2018**, *71*, 87–94. [CrossRef]
62. Prapruddivongs, C.; Rukrabiab, J.; Kulwongwit, N.; Wonpreedee, T. Effect of surface-modified silica on the thermal and mechanical behaviors of poly(lactic acid) and chemically crosslinked poly(lactic acid) composites. *J. Thermoplast. Compos. Mater.* **2020**, *33*, 1692–1706. [CrossRef]
63. Vidakis, N.; Petousis, M.; Velidakis, E.; Mountakis, N.; Tzounis, L.; Liebscher, M.; Grammatikos, S.A. Enhanced mechanical, thermal and antimicrobial properties of additively manufactured polylactic Acid with optimized nano silica content. *Nanomaterials* **2021**, *11*, 1012. [CrossRef] [PubMed]
64. Lv, H.; Song, S.; Sun, S.; Ren, L.; Zhang, H. Enhanced properties of poly(lactic acid) with silica nanoparticles. *Polym. Adv. Technol.* **2016**, *27*, 1156–1163. [CrossRef]
65. Ge, H.; Zhu, Z.; Yin, H.; Zhang, X.; Wang, R. Fabrication and properties of sc-PLA/SiO_2, Fabrication and Properties of sc-PLA/SiO_2 Composites. *IOP Conf. Ser. Mater. Sci. Eng.* **2014**, *62*, 012031. [CrossRef]
66. Pilic, B.M.; Radusin, T.I.; Ristic, I.S.; Silvestre, C.; Lazic, V.L.; Balos, S.S.; Duraccio, D. Hydrophobic silica nanoparticles as reinforcing filler for poly (lactic acid) polymer matrix. *Hem. Ind.* **2016**, *70*, 73–80. [CrossRef]
67. Techawinyutham, L.; Siengchin, S.; Dangtungee, R.; Parameswaranpillai, J. Influence of accelerated weathering on the thermo-mechanical, antibacterial, and rheological properties of polylactic acid incorporated with porous silica-containing varying amount of capsicum oleoresin. *Compos. Part. B* **2019**, *175*, 107108. [CrossRef]
68. Wu, F.; Lan, X.; Ji, D.; Liu, Z.; Yang, W.; Yang, W.; Yang, M. Grafting polymerization of polylactic acid on the surface of nano-SiO2 and properties of PLA/PLA-grafted-SiO2 nanocomposites. *J. Appl. Polym. Sci.* **2013**, *129*, 3019–3027. [CrossRef]
69. Chrissafis, K.; Bikiaris, D. Can nanoparticles really enhance thermal stability of polymers? Part I: An overview on thermal decomposition of addition polymers. *Thermochim. Acta* **2011**, *523*, 1–24. [CrossRef]
70. Khankrua, R.; Pivsa-Art, S.; Hiroyuki, H.; Suttiruengwong, S. Thermal and mechanical properties of biodegradable polyester/silica nanocomposites. *Energy Procedia* **2013**, *34*, 705–713. [CrossRef]
71. Mustapa, I.R.; Ansari, M.N.M.; Shanks, R.A. Poly(Lactic acid) Hemp composites combined with nano-silica. *TechConnect Briefs* **2011**, *3*, 7136–7138.
72. Mustapa, I.R.; Shanks, R.A.; Kong, I. Poly(lactic acid)-hemp-nanosilica hybrid composites: Thermomechanical, thermal behavior and morphological properties. *Inter. J. Adv. Sci. Eng. Technol.* **2013**, *3*, 192–199.
73. Mustapa, I.R.; Shanks, R.A.; Kong, I. Multiple melting behavior of poly(lactic acid)-hemp-silica composites using modulated-temperature differential scanning calorimetry. *J. Polym. Eng.* **2014**, *34*, 895–903. [CrossRef]
74. Mustapa, I.R.; Shanks, R.A.; Kong, I. Melting behaviour and dynamic mechanical properties of poly(lactic acid)-hemp-nanosilica composites. *Asian Trans. Basic Appl. Sci.* **2013**, *3*, 29–37.
75. Lai, S.M.; Li, P.W. Effect of thermoplastic polyurethane-modified silica on melt- blended poly(Lactic Acid) (PLA) nanocomposites. *Poly Poly Compos.* **2017**, *25*, 583–592. [CrossRef]
76. He, H.; Pang, Y.; Duan, Z.; Luo, N.; Wang, Z. The Strengthening and toughening of biodegradable poly (lactic acid) using the SiO2-PBA core—Shell nanoparticle. *Materials* **2019**, *12*, 2510. [CrossRef]
77. Lei, W.C.; Zhang, M.Q.; Rong, M.Z.; Friedrich, K. Silica nanoparticles filled polypropylene: Effects of particle surface treatment, matrix ductility and particle species on mechanical performance of the composites. *Comp. Sci. Tech.* **2005**, *65*, 635–645.

78. Ahmed, W.; Siraj, S.; Al-Marzouqi, A.H. 3D Printing PLA waste to produce ceramic based particulate reinforced composite using abundant silica-sand: Mechanical properties characterization. *Polymers* **2020**, *12*, 2579. [CrossRef]
79. Régibeau, N.; Tilkin, R.G.; Grandfils, C.; Heinrichs, B. Preparation of poly-d,l-lactide based nanocomposites with polymer-grafted silica by melt blending: Study of molecular, morphological, and mechanical properties. *Polym. Compos.* **2021**, *42*, 955–972. [CrossRef]
80. Georgiopoulos, P.; Kontou, E.; Meristoudi, A.; Pispas, S.; Chatzinikolaidou, M. The effect of silica nanoparticles on the thermomechanical properties and degradation behavior of polylactic acid. *J. Biomater. Appl.* **2004**, *29*, 662–674. [CrossRef]
81. Zirak, M.F.; Tabari, M. PLA-SiO2 nanocomposite films: Morphological and mechanical properties and specific end-use characteristics. *Nanomed Res. J.* **2018**, *3*, 140–145.
82. Chrissafis, K.; Pavlidou, E.; Paraskevopoulos, K.; Beslikas, T.; Nianias, N.; Bikiaris, D. Enhancing mechanical and thermal properties of PLLA ligaments with fumed silica nanoparticles and montmorillonite. *J. Therm. Anal. Calorim.* **2011**, *105*, 313–323. [CrossRef]
83. Lai, S.M.; Hsieh, T.T. Preparation and properties of polylactic acid (PLA)/silica nanocomposites. *J. Macromol. Sci. Part. B* **2016**, *3*, 211–228. [CrossRef]
84. Battegazzore, D.; Bocchini, S.; Alongi, J.; Frache, A. Rice husk as bio-source of silica: Preparation and characterization of PLA–silica bio-composites. *RSC Adv.* **2014**, *4*, 54703–54712. [CrossRef]
85. Song, X.; Fang, C.; Li, Y.; Zhang, Y.; Wang, P. The characterization of mechanical properties of Jute/PLA composites with modified nano Sio2 by coupling agent. *Res. Sq.* **2021**. [CrossRef]
86. Malinowski, R.; Janczak, K.; Rytlewski, P.; Kaczor, A.R.; Moraczewski, K.; Żuk, T. Influence of glass microspheres on selected properties of polylactide composites. *Compos. Part B* **2015**, *76*, 13–19. [CrossRef]
87. Bikiaris, D.N.; Vassiliou, A.A. *Nanocomposite Coatings and Nanocomposites Materials*; Trans Tech Publications Ltd.: Freienbach, Switzerland, 2009.
88. Stodolak-Zych, E.; Szumera, M.; Blazewicz, M. Osteoconductive nanocomposite materials for bone regeneration. *Mater. Sci. Forum* **2013**, *730*, 38–43. [CrossRef]
89. Pedna, A.; Pinho, L.; Frediani, P.; Mosquera, M.J. Obtaining SiO2–fluorinated PLA bionanocomposites with application as reversible and highly-hydrophobic coatings of buildings. *Prog. Org. Coat.* **2016**, *90*, 91–100. [CrossRef]
90. Gong, J.; Tian, N.; Wen, X.; Chen, X.; Liu, J.; Jiang, Z.; Mijowska, E.; Tang, T. Synergistic effect of fumed silica with Ni2O3 on improving flame retardancy of poly(lactic acid). *Polym. Deg. Stab.* **2014**, *104*, 18–27. [CrossRef]
91. Zhao, Y.; Wei, B.; Wu, M.; Zhang, H.; Yao, J.; Chen, X.; Shao, Z. Preparation and characterization of antibacterial poly(lactic acid) nanocomposites with N-halamine modified silica. *Inter. J. Biol. Macromol.* **2020**, *155*, 1468–1477. [CrossRef]
92. Sarikhani, K.; Jeddi, K.; Thompson, R.B.; Park, C.B.; Chen, P. Adsorption of surface-modified silica nanoparticles to the interface of melt poly (lactic acid) and supercritical carbon dioxide. *Langmuir* **2015**, *31*, 5571–5579. [CrossRef]
93. Seng, C.T.; Noum, S.Y.A.E.; Sivanesan, S.K.A.; Yu, L.J. Reduction of hygroscopicity of PLA filament for 3D printing by introducing nano silica as filler. *AIP Conf. Proceed.* **2020**, *2233*, 020024. [CrossRef]
94. Chen, H.M.; Wang, Y.P.; Chen, J.; Yang, J.H.; Zhang, N.; Huang, T.; Wang, Y. Hydrolytic degradation behavior of poly(L-lactide)/SiO2 composites. *Polym. Degrad. Stab.* **2013**, *98*, 2672–2679. [CrossRef]
95. Kosowska, K.; Szatkowsk, P. Infuence of ZnO, SiO2 and TiO2 on the aging process of PLA fbers produced by electrospinning method. *J. Therm. Anal. Calorim.* **2020**, *140*, 1769–1778. [CrossRef] [PubMed]
96. Bo, L.J.; Yong, L.X.; Feng, L.W.; Hao, Z.J. Preparation and characterization of bioactive poly (Lactic Acid)/SiO_2-CaO composite membranes. *J. Inorg. Mater.* **2011**, *26*, 998–1002.
97. Galikhanov, E.; Lounev, I.; Guzhova, A.; Gusev, Y.; Galikhanov, M.; Vasilyeva, M. Study of electret state in polylactic acid with nanosized filler by dielectric spectroscopy. *AIP Conf. Proc.* **2016**, *1748*, 020007. [CrossRef]
98. Majola, A.; Vainionpää, S.; Vihtonen, K.; Vasenius, J.; Törmälä, P.; Rokkanen, P. Intramedullary Fixation of Cortical Bone Osteotomies with Self-Reinforced Polylactic Rods in Rabbits. *Int. Orthop.* **1992**, *16*, 101–108. [CrossRef]
99. Abe, S.; Sasaki, A.; Iwadera, N.; Akasaka, T.; Uo, M.; Watari, F. Biodistribution and biocompatibility of poly(lactic acid)-coated SiO_2 particle. *Nano Biomed.* **2011**, *3*, 300–305.
100. Shen, Y.; Cao, B.; Snyder, N.R.; Woeppel, K.M.; Eles, J.R.; Cui, X.T. ROS responsive resveratrol delivery from LDLR peptide conjugated PLA-coated mesoporous silica nanoparticles across the blood-brain barrier. *J. Nanobiotechnol.* **2018**, *16*, 13. [CrossRef]
101. Thongsang, S.; Sontikaew, S.; Kachapol, K.; Kaewkate, T.; Aunchanlung, W. Comparision of filler types in polylactic acide composites for 3D printing applications. *Matter Int. J. Sci. Technol.* **2020**, *5*, 98–109. [CrossRef]
102. Jaikaew, N.; Petchsuk, A.; Opaprakasit, P. Preparation and properties of polylactide bio-composites with surface-modified silica particles. *Chiang Mai J. Sci.* **2018**, *45*, 2059–2068.
103. Opaprakasit, P.; Boonpa, S.; Jaikaew, N.; Petchsuk, A.; Tangboriboonrat, P. Preparation of surface-modified silica particles from rice husk ash and its composites with degradable polylactic acid. *Macromol. Symp.* **2015**, *354*, 48–54. [CrossRef]

polymers

MDPI

Article

Safety and Efficacy of Kartigen® in Treating Cartilage Defects: A Randomized, Controlled, Phase I Trial

Yen-Liang Liu [1,2], Chun-Che Yen [3], Tzu-Shang Thomas Liu [4], Chih-Hung Chang [5,6], Tiffany Ting-Fang Shih [7], Jyh-Horng Wang [8], Ming-Chia Yang [9], Feng-Huei Lin [10] and Hwa-Chang Liu [8,11,*]

1 Master Program for Biomedical Engineering, College of Biomedical Engineering, China Medical University, Taichung 406040, Taiwan; allen.liu@cmu.edu.tw
2 Graduate Institute of Biomedical Sciences, College of Medicine, China Medical University, Taichung 406040, Taiwan
3 Kartigen Biomedical Inc., Taipei 100047, Taiwan; cyenslk@gmail.com
4 Southern California Bone and Joint Clinic, Apple Valley, CA 92307, USA; redcometsports@gmail.com
5 Department of Orthopaedic Surgery, Far Eastern Memorial Hospital, New Taipei 220216, Taiwan; orthocch@mail.femh.org.tw
6 Graduate School of Biotechnology and Bioengineering, Yuan Ze University, Taoyuan 320315, Taiwan
7 Department of Medical Imaging and Radiology, National Taiwan University Hospital, Taipei 100225, Taiwan; ttfshih@ntu.edu.tw
8 Department of Orthopaedic Surgery, National Taiwan University Hospital, Taipei 100225, Taiwan; jhwang@ntuh.gov.tw
9 Biomedical Technology and Device Research Laboratories, Industrial Technology Research Institute, Hsinchu 310401, Taiwan; s1979329@gmail.com
10 Department of Biomedical Engineering, College of Engineering, National Taiwan University, Taipei 106319, Taiwan; double@ntu.edu.tw
11 Department of Orthopaedic Surgery, Taiwan Adventist Hospital, Taipei 105404, Taiwan
* Correspondence: hcliu@ntuh.gov.tw

Citation: Liu, Y.-L.; Yen, C.-C.; Liu, T.-S.T.; Chang, C.-H.; Shih, T.T.-F.; Wang, J.-H.; Yang, M.-C.; Lin, F.-H.; Liu, H.-C. Safety and Efficacy of Kartigen® in Treating Cartilage Defects: A Randomized, Controlled, Phase I Trial. *Polymers* **2021**, *13*, 3029. https://doi.org/10.3390/polym13183029

Academic Editors: José Miguel Ferri, Vicent Fombuena Borràs and Miguel Fernando Aldás Carrasco

Received: 30 July 2021
Accepted: 31 August 2021
Published: 7 September 2021

Abstract: Here, we aimed to investigate the safety and preliminary efficacy of Kartigen®, a matrix with autologous bone marrow mesenchymal stem cell-derived chondrocyte precursors embedded in atelocollagen. As a surgical graft, Kartigen® was implanted onto the cartilage defects at the weight-bearing site of the medial femoral condyle of the knee. Fifteen patients were enrolled and stratified into two groups, undergoing either Kartigen® implantation (n = 10) or microfracture (control group, n = 5). The primary endpoint was to evaluate the safety of Kartigen® by monitoring the occurrence of adverse events through physician queries, physical examinations, laboratory tests, and radiological analyses for 2 years. There were no infections, inflammations, adhesions, loose body, or tumor formations in the Kartigen®-implanted knees. The preliminary efficacy was assessed using the International Knee Documentation Committee (IKDC) score, visual analog scale, and second-look arthroscopy. The postoperative IKDC scores of the Kartigen® group significantly improved in the 16th week (IKDC = 62.1 ± 12.8, p = 0.025), kept increasing in the first year (IKDC = 78.2 ± 15.4, $p < 0.005$), and remained satisfactory in the second year (IKDC = 73.6 ± 13.8, $p < 0.005$), compared to the preoperative condition (IKDC = 47.1 ± 17.0), while the postoperative IKDC scores of the control group also achieved significant improvement in the 28th week (IKDC = 68.5 ± 6.1, p = 0.032) versus preoperative state (IKDC = 54.0 ± 9.1). However, the IKDC scores decreased in the first year (IKDC = 63.5 ± 11.6) as well as in the second year (IKDC = 52.6 ± 16.4). Thirteen patients underwent second-look arthroscopy and biopsy one year after the operation. The Kartigen® group exhibited integration between Kartigen® and host tissue with a smooth appearance at the recipient site, whereas the microfracture group showed fibrillated surfaces. The histological and immunohistochemical analyses of biopsy specimens demonstrated the columnar structure of articular cartilage and existence of collagen type II and glycosaminoglycan mimic hyaline cartilage. This study indicates that Kartigen® is safe and effective in treating cartilage defects.

Keywords: cartilage defect; knee; Kartigen®; chondrocyte precursors; stem cell therapy

1. Introduction

Cartilage defects are highly prevalent joint disorders and leading causes of disability and chronic pain in the world [1]. Articular cartilage defects are mainly attributed to trauma, osteoarthritis, osteonecrosis, and osteochondritis dissecans. Chondral lesions were found in around 60% of the patients undergoing knee arthroscopy [2–4].

Back in 1743, Hunter described the challenge of cartilage repair, stating that "once the cartilage is destroyed, it never recovers [5,6]." His observation still holds today. The avascular characteristics of cartilage constrain its self-regeneration from injury. If left untreated, the damaged cartilage gradually progresses into severe osteoarthritis. Through decades of effort, multiple surgical treatments have been developed to promote cartilage healing, such as abrasion arthroplasty [7], microfracture [8,9], and mosaicplasty [10,11]. However, these surgical approaches are usually associated with fibrocartilage formation [12,13], limited tissue sources [14], and donor-site morbidity [15], and its long-term efficacy remains controversial [16].

Cell therapy for cartilage repair was proposed in the 1980s by Robert Langer and Charles Vacanti using the approaches of tissue engineering. They clearly defined that a fine reconstruction of cartilage defect must include selected cells for transplantation, excipients, which are seeded with selected cells, and functional restoration of defect areas as before [17]. The cellular therapeutic innovation was realized in 1994. Autologous chondrocyte implantation (ACI) was introduced to treat cartilage defects in the knee [18]. In the past two decades, ACI has demonstrated its efficacy for knee osteoarthritis. Two systematic reviews concluded that, relative to microfracture and mosaicplasty, ACI may be the best option for large defects in active young patients who have had the symptoms for a short period and have not undergone a chondral surgery before [13,19]. In addition, a scaffold-based ACI, matrix-induced autologous chondrocyte implantation (MACI), outperformed microfracture in a 2-year randomized study [20]. Still, only a few products are available on the market: Carticel (FDA-approved in 1997), Chondron (South Korea MFDS-approved in 2001), and MACI (FDA-approved in 2016). The scarcity implies that some limitations remain, such as the limited source of chondrocytes, donor site morbidity, uncertain hyaline cartilage formation, low recovery of the recipient site, and questionable longevity of these implants or their derivative tissues [21].

The discovery of adult stem cells aroused a paradigm shift in regenerative medicine. The features of self-renewal and multipotency of stem cells make them ideal cell sources for cellular therapy. A variety of stem cell-based therapeutic innovations have been developed using mesenchymal stem cells (MSCs) derived from bone marrow [22], adipose tissue [23], synovium [24], peripheral blood [25], or periosteum [26]. Several clinical trials have demonstrated the safety and therapeutic efficacy of autologous MSC implantation for cartilage repair [27–30]. However, the implantation of undifferentiated MSCs cannot guarantee certain chondrogenesis in vivo, which might lead to heterogeneity of regenerated tissues.

The chondrogenesis of MSCs can be guided using growth factors [31] or biophysical/biomechanical stimuli [32] to improve the functional properties of the derived neocartilage tissues, including mature matrix formation [29,33]. Our previous study identified a unique population of chondrocyte precursors (CPs) derived from bone marrow mesenchymal stem cells (BMSCs) during chondrogenic induction [34]. These atelocollagen-embedded CPs (Kartigen®) can secrete glycosaminoglycan (GAG) and collagen type II but without lacunae formation.

A variety of biomaterials have been used for cartilage tissue engineering, including collagen [35], alginate [36], poly-lactic-glycolic acid [37], and tri-copolymer [38]. In this study, we selected atelocollagen because it is a low-immunogenic derivative of collagen obtained by the removal of N– and C–terminal telopeptide components [39]. Atelocollagen has been broadly applied in the regeneration of cartilage [34,40,41], intervertebral disc [42], cornea [43], periodontal tissues [44], and skin [45] to serve as a carrier for cell delivery and to provide an appropriate microenvironment for tissue regeneration.

A 9-year follow-up trial demonstrated that Kartigen® integrated with the host tissue and resulted in the formation of hyaline-like cartilage, thereby improving the impaired knee functions [34]. Due to the lack of a randomized control group in the previous study, we initiated this controlled and randomized trial to evaluate the safety and efficacy of Kartigen for repairing cartilage defects in the weight-bearing site of medial femoral condyles through the comparison with the microfracture treatment.

2. Materials and Methods

2.1. Ethical Approval

This study was an open-label, controlled, randomized, single-center, phase I clinical trial to evaluate the clinical safety of Kartigen® and its clinical improvements versus microfracture. This study was approved by the Taiwan Food and Drug Administration (TFDA, study number 1076026300) and by the Institutional Review Board of Taiwan Adventist Hospital (IRB number: 105-B-09). According to the guidelines of the Declaration of Helsinki, informed consent was obtained from each subject.

2.2. Study Population

This study enrolled 15 patients with cartilage defects at the weight-bearing site of the medial femoral condyle. The size of cartilage defects ranged from 0.6 to 4 cm^2. Patients were enrolled in this study between September 2018 and June 2020. Patients were selected according to inclusion and exclusion criteria in Table S1. They were randomly allocated into groups of Kartigen® implantation ($n = 10$) or microfracture treatment ($n = 5$). Based on our previous study [34], the seeding density of CPs was 1.6–3.3 $\times$ 10^6 cells/cm^2, and the total cell number was less than 1.32 $\times$ 10^7 cells. The schedule of study procedures and assessments is shown in Table S2.

2.3. Manufacture of Kartigen®

The manufacture of Kartigen® was described previously [34]. In brief, cell culture was performed under standard operative procedures at the cell processing unit (CPU) of Kartigen Biomedical Inc. (Taipei, Taiwan), following the Good Tissue Practice regulations. Fifteen mL of heparinized blood was aspirated from the iliac crest, collected in sterile 50-mL tubes. The MSCs were isolated using the density-gradient medium Ficoll (Cat. No. 17-5446-52, GE Healthcare, Little Chalfont, UK). The nucleated cells were collected from the interface, washed twice in PBS, and then suspended in Dulbecco's Modified Eagle's Medium (DMEM-LG, Cat. No. 31600, Gibco, Carlsbad, CA, USA) supplemented with the patient's serum. BMSC were expanded to a sufficient number, enough to fill the cartilage defects of patients. The required cell number was estimated according to the size of the defects. Before the BMSCs were induced into CPs, a small portion of BMSCs were used to assess the number, viability, and the immunophenotype of BMSCs, and to undergo sterility testing. The cell number and viability were quantified using trypan blue staining and an automated cell counter. BMSC's immunophenotype must be fulfilled by the minimal ISCT criteria: More than 95% of BMSCs are CD90-positive, and less than 2% of BMSCs are CD34-negative. Sterility testing must be negative for aerobic and anaerobic bacteria, mycoplasma, and <0.5 EU/mL endotoxin. The expanded cells were then seeded in excipient (Kartigen®) and cultured in CPs' induction medium.

2.4. Surgical Operation

The knee lesions were assessed arthroscopically. The defect size was measured through the arthroscope. For the Kartigen® implantation group, the required cell number of CPs was estimated before the collection of bone marrows. We aimed to implant 2 $\times$ 10^6 cells/cm^2 in the defect. The cartilage defect at the medial femoral condyle was approached either by mini-medial arthrotomy of 2 to 3 cm in length or by the arthroscope only. After debridement, the cartilage defect was filled with Kartigen® and then sealed with fibrin glue (TISSEEL, Baxter AG). For microfracture, we followed a standard arthroscopic procedure as reported by Steadman et al. [46]. The multiple holes (microfractures) were about 3–4 mm apart, to preserve the structure and function of the subchondral plate.

2.5. Rehabilitation

Both groups of patients underwent the same rehabilitation regimen. The operated knee was kept in 20- to 30-degree flexion for 72 h, and then the knee was allowed to move freely. Partial weight-bearing was started 24 h after the operation, and full weight-bearing was allowed 4 weeks after the operation.

2.6. Safety Evaluation

A series of evaluations were assessed to assure the safety of the study for 2 years after treatments. These evaluations included physicians' queries, physical examinations, laboratory tests, and radiological studies. The Common Terminology Criteria for Adverse Events was used to elicit and report toxic effects [47]. We specifically monitored the possible severe adverse events, including the severity of pain, infection of the operation site, joint adhesion, the abnormal gross appearance of the knee, active range of motion, loose body formation, and tumorigenesis.

2.7. Efficacy Evaluation

The preliminary efficacy was assessed using the International Knee Documentation Committee (IKDC) scoring system [48], visual analog scale (VAS), and arthroscopy. Arthroscopy was performed 1 year after the operation to observe the changes of defects. At the same time, a biopsy specimen (2 mm in diameter) was taken for histological analysis.

2.8. Arthroscopic and Histological Analysis

Second-look arthroscopy was done with the patients' written consent. Alcian blue staining and immunohistology were applied to evaluate the existence of GAG and collagen type II. The International Cartilage Repair Society (ICRS) arthroscopic assessment scale was used to evaluate the degree of cartilage repair [49,50]. Alcian blue staining (Cat. No. B8438, Sigma-Aldrich, Saint Louis, MO, USA) was used to assess the amount of GAG accumulation in the implanted tissues. Nuclear Fast Red solution (Cat. No. H-3403, Vector Laboratories, Burlingame, CA, USA) was used for counterstaining. The immunohistochemistry was applied to quantify the protein expression level of collagen type II. An anti-collagen type II antibody (Cat. No. Ab34712, Abcam, Cambridge, UK) and UltraVision Quanto Detection System (Cat. No. TL-060-QHL, Thermo Scientific, Fremont, CA, USA) were used for the immunohistochemistry, according to the manufacturer's instructions. All samples were processed and stained using the same procedure.

2.9. Statistical Methods

The statistical analyses of IKDC scores and VAS scores were performed using t-test and Wilcoxon signed-rank test, respectively. A two-tailed paired t-test was used to test the difference of IKDC scores at different time points (before and after the operation) of the same treatment group. To compare the efficacy of treatments, we used a one-tailed t-test to test the null hypothesis: IKDC scores of microfracture treatment were higher than that of Kartigen® implantation. The statistical significance of the t-tests was set as $p < 0.05$. All statistical analyses were performed using Microsoft Excel.

3. Results

3.1. Demography

Fifteen patients were enrolled and randomized into two groups. The control group underwent microfracture, and the study group underwent Kartigen® implantation. No patients were withdrawn from this trial. The flow chart of the study is shown in Figure S1. The demographic information of the patients is shown in Table 1. There was no significant difference in age between the two groups, but the defect sizes were significantly different.

Table 1. Demography of the patients.

Group	Kartigen® Group (n = 10)	Microfracture (n = 5)
Gender (F:M)	5:5	2:3
Age (year)	22–77 (54.8 ± 18.0)	55–77 (67.8 ± 8.5)
Defect size (cm^2)	1.3–4.0 (2.9 ± 0.8) *	0.6–1.5 (1.0 ± 0.4)
Defect treatment	Kartigen®	Microfracture

* Note: The defect size of the Kartigen® group was significantly larger than that of microfracture group ($p < 0.005$).

3.2. Safety Assessment

In the Kartigen® group, three patients exhibited three treatment-emergent adverse events (TEAEs): renal stone with hematuria, cervical spondylosis, and upper respiratory tract infection, respectively. In the control group, two patients demonstrated four TEAEs, including urinary tract infection, cataract, prostate hypertrophy, and coronary artery stenosis. All seven TEAEs were not treatment-related. These TEAEs are summarized in Table S3. By X-ray examination and MRI study, there was no inflammation, joint adhesion, loose body, or tumorigenesis in the Kartigen®-implanted knee and microfracture-treated knee. Physical examination and laboratory tests revealed no infection among the 15 patients. Moreover, neither mortality nor complications were noted after operations in this study. Some laboratory values were abnormal, such as blood sugar, GOT, GPT, and urinary red blood cell in both groups. However, they were felt to be reflective of each patient's underlying medical condition and not as results of participation in the study.

3.3. Efficacy Assessment

We applied IKDC subjective knee evaluation criteria to evaluate the knee function before and after the treatments. Figure 1 showed a tendency that the IKDC scores continuously improved in the Kartigen® group, but the control group achieved the highest IKDC score in the 28th week and gradually decreased. In the study group, the IKDC score before Kartigen® implantation was 47.1 ± 17.0 (mean ± standard deviation) and then significantly improved to 62.1 ± 12.8 (p-value = 0.025) 16 weeks after implantation. The IKDC scores continuously increased to 78.2 ± 15.4 with a statistical significance (p-value < 0.005) one year after the operation and still maintained at 73.6 ± 13.8 2 years after the operation. While IKDC scores slightly decreased in the second year, the scores were still significantly higher than that of preoperation (p-value < 0.005). Meanwhile, although the control group demonstrated significantly improved IKDC scores from 54.0 ± 9.1 (preoperation) to 68.5 ± 6.1 (28 weeks after the operation) with a p-value of 0.032, the score decreased to 63.5 ± 11.6 1 year after the operation and to 52.6 ± 16.4 2 years after the operation. The comparison of IKDC scores between the Kartigen® group and the microfracture group in the second year exhibited statistical significance with the p-value of 0.029. Still, a longer follow-up will be needed to evaluate the long-term efficacy and longevity of the regenerated cartilage.

The VAS was used to measure the pain before and after treatments. Both the Kartigen® implantation and microfracture treatment effectively achieved pain relief at the 10th week after the operation (Figure 2). However, the Kartigen® implantation exhibited consistent pain relief through the 1-year follow-up, which was validated statistically using Wilcoxon signed-rank test with p-values < 0.005.

3.4. Arthroscopy and Histological Analysis

The regeneration of cartilage defect was grossly observed by knee arthroscopy, and a specimen of 2 mm in diameter was taken 1 year after surgery. Nine out of 10 patients in the Kartigen® group and 4 out of 5 patients in the control group consented to second-look arthroscopy. The gross appearance of regenerated cartilage tissues was examined under arthroscope and compared with their corresponding arthroscopic images taken before treatment (Figure 3). Eight out of 9 of the Kartigen® group exhibited smooth and elastic surfaces at the recipient sites (Figure 3A), but none of the control group showed normal or near-normal appearance. Only fibrillated surface was noted (Figure 3B). The ICRS scale is shown in Table 2 with the classification of grade I (normal tissue), grade II (nearly normal tissue), grade III (abnormal), grade IV (severely abnormal).

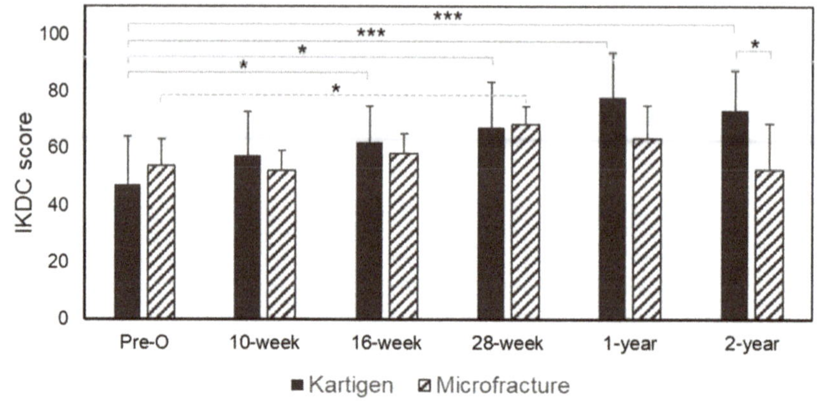

Figure 1. Assessment of knee functions of daily activities by IKDC scores. The knee functions of the patients were significantly improved 16 weeks after the Kartigen® implantation and 28 weeks after microfracture, respectively. The paired t-test was conducted to determine the difference between the means of the IKDC scores before and after the treatments at different time points among the same groups. Two years after operation, the IKDC score of the Kartigen® group was higher than the control group. The error bars stand for standard deviations. Asterisks represent the statistical significance: * $p < 0.05$, *** $p < 0.005$.

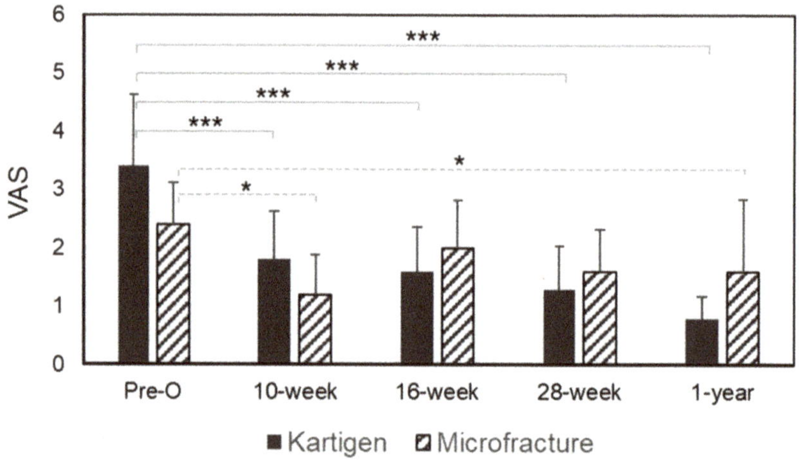

Figure 2. VAS assessment. The pain relief was effective and significant in the Kartigen®-treated group, as VAS scores decreased since the 10th week after operation, but the VAS scores flatulated in the control group. The error bars stand for standard deviations. Asterisks represent the statistical significance: * $p < 0.05$, *** $p < 0.005$.

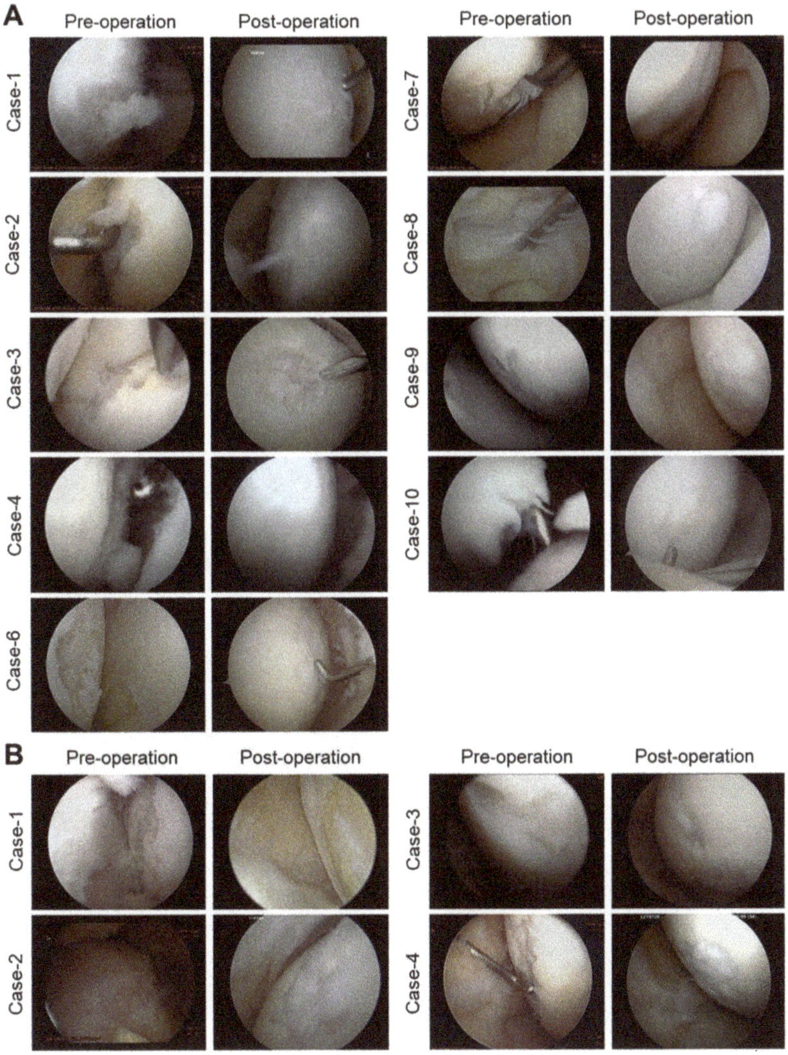

Figure 3. Arthroscopic images before and 1 year after operation. Images of Kartigen® group (**A**) and microfracture group (**B**).

Table 2. ICRS arthroscopic assessment.

ICRS Cartilage Repair Assessment			
Kartigen® Group (*n* = 10)		**Microfracture Group (*n* = 5)**	
Grade I	3	Grade I	0
Grade II	5	Grade II	0
Grade III	1	Grade III	3
Grade IV	0	Grade IV	1
Total	9	Total	4

The Kartigen® group demonstrated better results in both arthroscopic observations and histological analysis (Figure 4). In the control group, the regenerated cartilage did not exhibit columnar chondrocyte distribution but only fibrocartilage (Figure 4A). GAG and collagen type II were observed in the biopsy specimens, and a total of 12 specimens (Kartigen®, $n = 8$; control, $n = 4$) are presented in Figure 4B,C. New chondrocytes at defect had the same appearance as the original ones (Figure 4B). However, the chondrocytes in the repaired defect in the Kartigen® group were smaller and denser than those in the original cartilage.

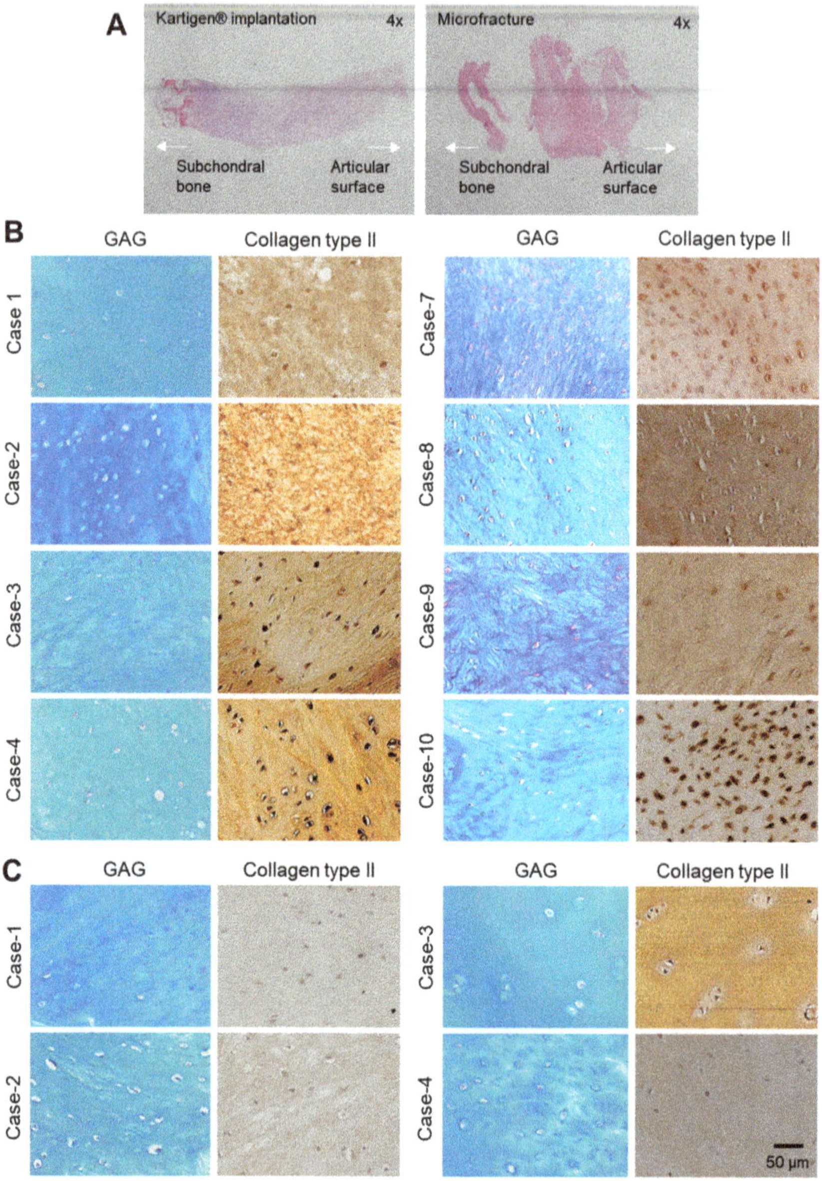

Figure 4. Immunohistochemical analyses of biopsy specimens. (**A**) Whole-slide histological images of regenerated cartilage under 4× objective lens. Alcian blue staining and immunochemical study show expression levels of GAG/Collagen type II, respectively, in Kartigen group (**B**) and microfracture group (**C**).

4. Discussion

The development of Kartigen® was started from our previous study using atelocollagen-embedded chondrogenic BMSCs to repair the full thickness of cartilage defect in miniature pigs [51]. The promising results encouraged us to initiate a case series study to test the safety and efficacy of Kartigen® for treating cartilage defects. In the 9 years of follow-up study, improvement of knee functions was satisfactory and sustainable [34]. Most importantly, infection, inflammation, joint adhesion, loose body, and tumor formation were not found in the Kartigen®-implanted knees. However, in the case series study, the contralateral knees of the same patients were used as a control group. To further validate the therapeutic efficacy of Kartigen® in comparison with standard surgical treatments, such as microfracture, a new controlled and randomized phase I clinical trial was conducted. As expected, no severe adverse events were found, and all the TEAEs were not treatment-related. Although clinical outcome was similar for Kartigen implantation and microfracture in 1-year follow-up, a sustained improvement of knee functions was found in the Kartigen® group in the 2 years follow-up.

Microfracture was introduced for focal articular cartilage repair in the 1980s and soon became a standard treatment for cartilage defects [9,46,52]. However, its results are usually associated with fibrocartilage production [12,13], and its long-term efficacy remains poor [16,53]. Mosaicplasty provides a better improvement than microfracture, but there are still several disadvantages, such as the donor-site morbidity for autografts [54], limited sources of grafts, a technical challenge in leveling the graft during operation [55], and poor integration of the implant into host tissue [56,57]. On the contrary, the procedure of Kartigen® implantation is technically simple and can be carried out with a small arthrotomy or using arthroscope. Furthermore, there is no donor site morbidity, and sources of cells are not limited.

The ability of engrafted cells to integrate into the recipient site and participate in the repair process is crucial for successful clinical outcomes. It has been reported that intra-articular injection of autologous MSCs reduced the degeneration of cartilage defects and provided pain relief [58], but integration of injected MSCs into damaged cartilage defects is unclear and doubtful. On the contrary, using atelocollagen as a cell carrier, Kartigen® prevents cell loss, is easy for implantation for cartilage defects of any shape, and achieves uniform cell distribution at the recipient site. These advantages contribute to the efficacy and durability of Kartigen® implantation. In addition, atelocollagen has been proven to enable a gradual proliferation and matrix synthesis of chondrocytes, which allow chondrocytes to maintain their phenotype for up to 4 weeks in vitro [40]. Atelocollagen gel can also support cell proliferation, matrix synthesis, and chondrogenic differentiation of MSCs [42]. Recently, the Adachi group reported the efficacy of repairing osteochondral defects with minced cartilage embedded in atelocollagen gel [41]. Our previous clinical trial with 9 years of follow-up also demonstrated the safety and efficacy of atelocollagen in repairing cartilage defects together with CPs [34].

Unlike previous studies using mature chondrocytes for cartilage repair in ACI [59], our study showed CPs in Kartigen® exhibiting sufficient integration capacity. Under arthroscopic examination, the integration between the graft and the recipient site was complete (Figure 3). The histological analyses of the biopsy specimens also demonstrated the integration of the implanted tissue into the surrounding articular cartilage (Figure 4). The accumulation of GAG and collagen type II was confirmed in the biopsy specimens (Figure 4). However, there are several limitations to this study. Because this is a phase I study, the sample size number is small. The 2 years of follow-up are not long enough to reach a final conclusion for cartilage defect repair.

Recent advances in 3D bioprinting [60–63] have enabled reconstructions of functional living cartilage to recapitulate the complexity and architecture [64,65] of an articular surface. We are looking into partnerships to integrate that technology and potentially further improve the outcome with Kartigen® implantation. This may result in more favorable biomechanical properties at the recipient sites and allow earlier weight-bearing and range of motion without concern of graft detachment. Since MSCs are multipotent, additional biochemical and biomechanical stimulations can delicately manipulate the chondrogenic differentiation and maturation of seeded cells [66,67] to improve the functional properties of the derived neo-cartilage tissues [29]. The integration of 3D bioprinting and biomimetic in vitro chondrogenesis will drive advanced therapeutic innovations for cartilage repairs.

5. Conclusions

In this study, Kartigen® containing CPs was proven safe and free of adverse events, such as infection, inflammation, joint adhesion, loose body, or tumor formation.

Supplementary Materials: The following are available online at https://www.mdpi.com/article/10.3390/polym13183029/s1. Table S1: Inclusion and exclusion criteria, Table S2: Scheduled visits of the study design, Table S3: Adverse events, Figure S1: Flowchart of this clinical trial.

Author Contributions: H.-C.L. and F.-H.L. conceived the project. H.-C.L., Y.-L.L., C.-C.Y. and T.-S.T.L. wrote the article. C.-C.Y. and M.-C.Y. supervised the preparation of Kartigen® and conducted immunohistochemical staining. H.-C.L. performed the surgery and collected clinical information. C.-H.C. and J.-H.W. interpreted the arthroscopic and histology images. T.T.-F.S. assessed the radiological study (MRI). H.-C.L. performed arthroscopy. H.-C.L. supervised the project. All authors have read and agreed to the published version of the manuscript.

Funding: The study was funded by Kartigen Biomedical Inc.

Institutional Review Board Statement: The study was conducted according to the guidelines of the Declaration of Helsinki, and approved by the the Taiwan Food and Drug Administration (TFDA, study number 1076026300) and by the Institutional Review Board of Taiwan Adventist Hospital (IRB number: 105-B-09).

Informed Consent Statement: Informed consent was obtained from all patients involved in the study. Written informed consent was obtained from the patients.

Acknowledgments: We would like to convey our sincere appreciation to the data and safety monitoring board consisting of Yi-Sheng Chan at Chang Gung Memorial Hospital, Chun-Hsiung Huang at Mackay Memorial Hospital, and Chih-Hao Chang at National Taiwan University Hospital. We thank Ying-Tai Jin at Taiwan Adventist Hospital for assessing the immunohistochemical analyses.

Conflicts of Interest: C.C. Yen is a manager of Kartigen Biomedical Inc.

References

1. James, S.L.; Abate, D.; Abate, K.H.; Abay, S.M.; Abbafati, C.; Abbasi, N.; Abbastabar, H.; Abd-Allah, F.; Abdela, J.; Abdelalim, A.; et al. Global, regional, and national incidence, prevalence, and years lived with disability for 354 diseases and injuries for 195 countries and territories, 1990–2017: A systematic analysis for the global burden of disease study 2017. *Lancet* **2018**, *392*, 1789–1858. [CrossRef]
2. Curl, W.W.; Krome, J.; Gordon, E.S.; Rushing, J.; Smith, B.P.; Poehling, G.G. Cartilage injuries: A review of 31,516 knee arthroscopies. *Arthrosc. J. Arthrosc. Relat. Surg.* **1997**, *13*, 456–460. [CrossRef]
3. Hjelle, K.; Solheim, E.; Strand, T.; Muri, R.; Brittberg, M. Articular cartilage defects in 1000 knee arthroscopies. *Arthrosc. J. Arthrosc. Relat. Surg.* **2002**, *18*, 730–734. [CrossRef] [PubMed]
4. Årøen, A.; Løken, S.; Heir, S.; Alvik, E.; Ekeland, A.; Granlund, O.G.; Engebretsen, L. Articular cartilage lesions in 993 consecutive knee arthroscopies. *Am. J. Sports Med.* **2004**, *32*, 211–215. [CrossRef] [PubMed]
5. Hunter, W. Of the structure and disease of articulating cartilages. *Clin. Orthop. Relat. Res.* **1995**, *317*, 3–6.
6. Hunter, W. VI. Of the structure and diseases of articulating cartilages. *Philos. Trans.* **1743**, *42*, 514–521.
7. Johnson, L.L. Arthroscopic abrasion arthroplasty: A review. *Clin. Orthop. Relat. Res.* **2001**, *391*, S306–S317. [CrossRef]
8. Mithoefer, K.; Williams, R.J., III; Warren, R.F.; Potter, H.G.; Spock, C.R.; Jones, E.C.; Wickiewicz, T.L.; Marx, R.G. The microfracture technique for the treatment of articular cartilage lesions in the knee: A prospective cohort study. *J. Bone Jt. Surg.* **2005**, *87*, 1911–1920. [CrossRef]
9. Steadman, J.R.; Rodkey, W.G.; Singleton, S.B.; Briggs, K.K. Microfracture technique forfull-thickness chondral defects: Technique and clinical results. *Oper. Tech. Orthop.* **1997**, *7*, 300–304. [CrossRef]
10. Hangody, L.; Kish, G.; Kárpáti, Z.; Udvarhelyi, I.; Szigeti, I.; Bély, M. Mosaicplasty for the treatment of articular cartilage defects: Application in clinical practice. *Orthopedics* **1998**, *21*, 751–756. [CrossRef] [PubMed]
11. Bugbee, W.D.; Convery, F.R. Osteochondral allograft transplantation. *Clin. Sports Med.* **1999**, *18*, 67–75. [CrossRef]

12. Hunziker, E.B. Articular cartilage repair: Are the intrinsic biological constraints undermining this process insuperable? *Osteoarthr. Cartil.* **1999**, *7*, 15–28. [CrossRef] [PubMed]
13. Oussedik, S.; Tsitskaris, K.; Parker, D. Treatment of articular cartilage lesions of the knee by microfracture or autologous chondrocyte implantation: A systematic review. *Arthrosc. J. Arthrosc. Relat. Surg.* **2015**, *31*, 732–744. [CrossRef] [PubMed]
14. Bartha, L.; Vajda, A.; Duska, Z.; Rahmeh, H.; Hangody, L. Autologous osteochondral mosaicplasty grafting. *J. Orthop. Sports Phys. Ther.* **2006**, *36*, 739–750. [CrossRef]
15. Gomoll, A.; Farr, J.; Gillogly, S.; Kercher, J.; Minas, T. Surgical management of articular cartilage defects of the knee. *J. Bone Jt. Surg.* **2010**, *92*, 2470–2490.
16. Knutsen, G.; Engebretsen, L.; Ludvigsen, T.C.; Drogset, J.O.; Grøntvedt, T.; Solheim, E.; Strand, T.; Roberts, S.; Isaksen, V.; Johansen, O. Autologous chondrocyte implantation compared with microfracture in the knee: A randomized trial. *J. Bone Jt. Surg.* **2004**, *86*, 455–464. [CrossRef] [PubMed]
17. Vacanti, J.P. Beyond transplantation: Third annual samuel jason mixter lecture. *Arch. Surg.* **1988**, *123*, 545–549. [CrossRef]
18. Brittberg, M.; Lindahl, A.; Nilsson, A.; Ohlsson, C.; Isaksson, O.; Peterson, L. Treatment of deep cartilage defects in the knee with autologous chondrocyte transplantation. *N. Engl. J. Med.* **1994**, *331*, 889–895. [CrossRef]
19. Harris, J.D.; Siston, R.A.; Pan, X.; Flanigan, D.C. Autologous chondrocyte implantation: A systematic review. *J. Bone Jt. Surg.* **2010**, *92*, 2220–2233. [CrossRef] [PubMed]
20. Basad, E.; Ishaque, B.; Bachmann, G.; Stürz, H.; Steinmeyer, J. Matrix-induced autologous chondrocyte implantation versus microfracture in the treatment of cartilage defects of the knee: A 2-year randomised study. *Knee Surg. Sports Traumatol. Arthrosc.* **2010**, *18*, 519–527. [CrossRef]
21. Iwasa, J.; Engebretsen, L.; Shima, Y.; Ochi, M. Clinical application of scaffolds for cartilage tissue engineering. *Knee Surg. Sports Traumatol. Arthrosc.* **2009**, *17*, 561–577. [CrossRef]
22. Wakitani, S.; Imoto, K.; Yamamoto, T.; Saito, M.; Murata, N.; Yoneda, M. Human autologous culture expanded bone marrow mesenchymal cell transplantation for repair of cartilage defects in osteoarthritic knees. *Osteoarthr. Cartil.* **2002**, *10*, 199–206. [CrossRef] [PubMed]
23. Koh, Y.-G.; Choi, Y.-J.; Kwon, S.-K.; Kim, Y.-S.; Yeo, J.-E. Clinical results and second-look arthroscopic findings after treatment with adipose-derived stem cells for knee osteoarthritis. *Knee Surg. Sports Traumatol. Arthrosc.* **2015**, *23*, 1308–1316. [CrossRef]
24. Sekiya, I.; Muneta, T.; Horie, M.; Koga, H. Arthroscopic transplantation of synovial stem cells improves clinical outcomes in knees with cartilage defects. *Clin. Orthop. Relat. Res.* **2015**, *473*, 2316–2326. [CrossRef] [PubMed]
25. Saw, K.-Y.; Anz, A.; Jee, C.S.Y.; Merican, S.; Ng, R.C.S.; Roohi, S.A.; Ragavanaidu, K. Articular cartilage regeneration with autologous peripheral blood stem cells versus hyaluronic acid: A randomized controlled trial. *Arthrosc. J. Arthrosc. Relat. Surg.* **2013**, *29*, 684–694. [CrossRef] [PubMed]
26. Roberts, S.J.; van Gastel, N.; Carmeliet, G.; Luyten, F.P. Uncovering the periosteum for skeletal regeneration: The stem cell that lies beneath. *Bone* **2015**, *70*, 10–18. [CrossRef]
27. Nejadnik, H.; Hui, J.H.; Feng Choong, E.P.; Tai, B.-C.; Lee, E.H. Autologous bone marrow–derived mesenchymal stem cells versus autologous chondrocyte implantation: An observational cohort study. *Am. J. Sports Med.* **2010**, *38*, 1110–1116. [CrossRef]
28. Wakitani, S.; Okabe, T.; Horibe, S.; Mitsuoka, T.; Saito, M.; Koyama, T.; Nawata, M.; Tensho, K.; Kato, H.; Uematsu, K. Safety of autologous bone marrow-derived mesenchymal stem cell transplantation for cartilage repair in 41 patients with 45 joints followed for up to 11 years and 5 months. *J. Tissue Eng. Regen. Med.* **2011**, *5*, 146–150. [CrossRef]
29. Makris, E.A.; Gomoll, A.H.; Malizos, K.N.; Hu, J.C.; Athanasiou, K.A. Repair and tissue engineering techniques for articular cartilage. *Nat. Rev. Rheumatol.* **2015**, *11*, 21. [CrossRef]
30. Teo, A.Q.A.; Wong, K.L.; Shen, L.; Lim, J.Y.; Toh, W.S.; Lee, E.H.; Hui, J.H.P. Equivalent 10-year outcomes after implantation of autologous bone marrow–derived mesenchymal stem cells versus autologous chondrocyte implantation for chondral defects of the knee. *Am. J. Sports Med.* **2019**, *47*, 2881–2887. [CrossRef] [PubMed]
31. Pittenger, M.F.; Mackay, A.M.; Beck, S.C.; Jaiswal, R.K.; Douglas, R.; Mosca, J.D.; Moorman, M.A.; Simonetti, D.W.; Craig, S.; Marshak, D.R. Multilineage potential of adult human mesenchymal stem cells. *Science* **1999**, *284*, 143–147. [CrossRef] [PubMed]
32. Engler, A.J.; Sen, S.; Sweeney, H.L.; Discher, D.E. Matrix elasticity directs stem cell lineage specification. *Cell* **2006**, *126*, 677–689. [CrossRef]
33. Liu, H.-C. Tissue engineering of cartilage: The road a group of researchers have traveled. *J. Orthop. Sci.* **2008**, *13*, 396. [CrossRef]
34. Liu, H.-C.; Liu, T.-S.T.; Liu, Y.-L.; Wang, J.-H.; Chang, C.-H.; Shih, T.T.-F.; Lin, F.-H. Atelocollagen-embedded chondrocyte precursors as a treatment for grade-4 cartilage defects of the femoral condyle: A case series with up to 9-year follow-up. *Biomolecules* **2021**, *11*, 942. [CrossRef] [PubMed]
35. Zheng, M.-H.; Willers, C.; Kirilak, L.; Yates, P.; Xu, J.; Wood, D.; Shimmin, A. Matrix-induced autologous chondrocyte implantation (maci®): Biological and histological assessment. *Tissue Eng.* **2007**, *13*, 737–746. [CrossRef]
36. Wang, C.-C.; Yang, K.-C.; Lin, K.-H.; Liu, Y.-L.; Liu, H.-C.; Lin, F.-H. Cartilage regeneration in scid mice using a highly organized three-dimensional alginate scaffold. *Biomaterials* **2012**, *33*, 120–127. [CrossRef]
37. Uematsu, K.; Hattori, K.; Ishimoto, Y.; Yamauchi, J.; Habata, T.; Takakura, Y.; Ohgushi, H.; Fukuchi, T.; Sato, M. Cartilage regeneration using mesenchymal stem cells and a three-dimensional poly-lactic-glycolic acid (plga) scaffold. *Biomaterials* **2005**, *26*, 4273–4279. [CrossRef]

38. Chang, C.-H.; Liu, H.-C.; Lin, C.-C.; Chou, C.-H.; Lin, F.-H. Gelatin–chondroitin–hyaluronan tri-copolymer scaffold for cartilage tissue engineering. *Biomaterials* **2003**, *24*, 4853–4858. [CrossRef]
39. Ochiya, T.; Takahama, Y.; Nagahara, S.; Sumita, Y.; Hisada, A.; Itoh, H.; Nagai, Y.; Terada, M. New delivery system for plasmid DNA in vivo using atelocollagen as a carrier material: The minipellet. *Nat. Med.* **1999**, *5*, 707–710. [CrossRef] [PubMed]
40. Uchio, Y.; Ochi, M.; Matsusaki, M.; Kurioka, H.; Katsube, K. Human chondrocyte proliferation and matrix synthesis cultured in atelocollagen® gel. *J. Biomed. Mater. Res.* **2000**, *50*, 138–143. [CrossRef]
41. Matsushita, R.; Nakasa, T.; Ishikawa, M.; Tsuyuguchi, Y.; Matsubara, N.; Miyaki, S.; Adachi, N. Repair of an osteochondral defect with minced cartilage embedded in atelocollagen gel: A rabbit model. *Am. J. Sports Med.* **2019**, *47*, 2216–2224. [CrossRef]
42. Sakai, D.; Mochida, J.; Yamamoto, Y.; Nomura, T.; Okuma, M.; Nishimura, K.; Nakai, T.; Ando, K.; Hotta, T. Transplantation of mesenchymal stem cells embedded in atelocollagen® gel to the intervertebral disc: A potential therapeutic model for disc degeneration. *Biomaterials* **2003**, *24*, 3531–3541. [CrossRef]
43. Li, F.; Carlsson, D.; Lohmann, C.; Suuronen, E.; Vascotto, S.; Kobuch, K.; Sheardown, H.; Munger, R.; Nakamura, M.; Griffith, M. Cellular and nerve regeneration within a biosynthetic extracellular matrix for corneal transplantation. *Proc. Natl. Acad. Sci. USA* **2003**, *100*, 15346–15351. [CrossRef] [PubMed]
44. Kawaguchi, H.; Hirachi, A.; Hasegawa, N.; Iwata, T.; Hamaguchi, H.; Shiba, H.; Takata, T.; Kato, Y.; Kurihara, H. Enhancement of periodontal tissue regeneration by transplantation of bone marrow mesenchymal stem cells. *J. Periodontol.* **2004**, *75*, 1281–1287. [CrossRef]
45. Widjaja, W.; Maitz, P. The use of dermal regeneration template (pelnac®) in acute full-thickness wound closure: A case series. *Eur. J. Plast. Surg.* **2016**, *39*, 125–132. [CrossRef]
46. Steadman, J.R.; Rodkey, W.G.; Rodrigo, J.J. Microfracture: Surgical technique and rehabilitation to treat chondral defects. *Clin. Orthop. Relat. Res.* **2001**, *391*, S362–S369. [CrossRef] [PubMed]
47. U.S. Department of Health and Human Services. *Common Terminology Criteria for Adverse Events (Ctcae), Version 5.0.*; U.S. Department of Health and Human Services: Washington, DC, USA, 2017.
48. Irrgang, J.J.; Anderson, A.F. Development and validation of health-related quality of life measures for the knee. *Clin. Orthop. Relat. Res. (1976–2007)* **2002**, *402*, 95–109. [CrossRef]
49. Brittberg, M.; Peterson, L. Introduction of an articular cartilage classification. *ICRS Newsl.* **1998**, *1*, 5–8.
50. Peterson, L.; Minas, T.; Brittberg, M.; Nilsson, A.; Sjögren-Jansson, E.; Lindahl, A. Two-to 9-year outcome after autologous chondrocyte transplantation of the knee. *Clin. Orthop. Relat. Res. (1976–2007)* **2000**, *374*, 212–234. [CrossRef]
51. Chang, C.H.; Kuo, T.F.; Lin, F.H.; Wang, J.H.; Hsu, Y.M.; Huang, H.T.; Loo, S.T.; Fang, H.W.; Liu, H.C.; Wang, W.C. Tissue engineering-based cartilage repair with mesenchymal stem cells in a porcine model. *J. Orthop. Res.* **2011**, *29*, 1874–1880. [CrossRef]
52. Steadman, J.R.; Rodkey, W.G.; Briggs, K.K. Microfracture to treat full-thickness chondral defects: Surgical technique, rehabilitation, and outcomes. *J. Knee Surg.* **2002**, *15*, 170–176. [PubMed]
53. Knutsen, G.; Drogset, J.O.; Engebretsen, L.; Grøntvedt, T.; Ludvigsen, T.C.; Løken, S.; Solheim, E.; Strand, T.; Johansen, O. A randomized multicenter trial comparing autologous chondrocyte implantation with microfracture: Long-term follow-up at 14 to 15 years. *J. Bone Jt. Surg.* **2016**, *98*, 1332–1339. [CrossRef] [PubMed]
54. Cole, B.J.; Pascual-Garrido, C.; Grumet, R.C. Surgical management of articular cartilage defects in the knee. *J. Bone Jt. Surg.* **2009**, *91*, 1778–1790.
55. Nakagawa, Y.; Mukai, S.; Yabumoto, H.; Tarumi, E.; Nakamura, T. Serial changes of the cartilage in recipient sites and their mirror sites on second-look imaging after mosaicplasty. *Am. J. Sports Med.* **2016**, *44*, 1243–1248. [CrossRef]
56. Alparslan, L.; Winalski, C.S.; Boutin, R.D.; Minas, T. Postoperative Magnetic Resonance Imaging of Articular Cartilage Repair. *Semin. Musculoskelet. Radiol.* **2001**, *5*, 345–364. [CrossRef] [PubMed]
57. Choi, Y.S.; Potter, H.G.; Chun, T.J. Mr imaging of cartilage repair in the knee and ankle. *Radiographics* **2008**, *28*, 1043–1059. [CrossRef]
58. Lee, W.S.; Kim, H.J.; Kim, K.I.; Kim, G.B.; Jin, W. Intra-articular injection of autologous adipose tissue-derived mesenchymal stem cells for the treatment of knee osteoarthritis: A phase iib, randomized, placebo-controlled clinical trial. *Stem Cells Transl. Med.* **2019**, *8*, 504–511. [CrossRef]
59. Miot, S.; Brehm, W.; Dickinson, S.; Sims, T.; Wixmerten, A.; Longinotti, C.; Hollander, A.; Mainil-Varlet, P.; Martin, I. Influence of in vitro maturation of engineered cartilage on the outcome of osteochondral repair in a goat model. *Eur. Cell Mater.* **2012**, *23*, 222–236. [CrossRef]
60. Murphy, S.V.; Atala, A. 3d bioprinting of tissues and organs. *Nat. Biotechnol.* **2014**, *32*, 773–785. [CrossRef] [PubMed]
61. Moroni, L.; Burdick, J.A.; Highley, C.; Lee, S.J.; Morimoto, Y.; Takeuchi, S.; Yoo, J.J. Biofabrication strategies for 3d in vitro models and regenerative medicine. *Nat. Rev. Mater.* **2018**, *3*, 21–37. [CrossRef] [PubMed]
62. Foresti, R.; Rossi, S.; Selleri, S. Bio composite materials: Nano functionalization of 4d bio engineered scaffold. In Proceedings of the 2019 IEEE International Conference on BioPhotonics (BioPhotonics), Taipei, Taiwan, 15–18 September 2019; IEEE: Piscataway, NJ, USA, 2019; pp. 1–2.
63. Foresti, R.; Rossi, S.; Pinelli, S.; Alinovi, R.; Sciancalepore, C.; Delmonte, N.; Selleri, S.; Caffarra, C.; Raposio, E.; Macaluso, G. In-vivo vascular application via ultra-fast bioprinting for future 5d personalised nanomedicine. *Sci. Rep.* **2020**, *10*, 1–13.
64. Cui, X.; Breitenkamp, K.; Finn, M.; Lotz, M.; D'Lima, D.D. Direct human cartilage repair using three-dimensional bioprinting technology. *Tissue Eng. Part A* **2012**, *18*, 1304–1312. [CrossRef] [PubMed]

65. De Mori, A.; Peña Fernández, M.; Blunn, G.; Tozzi, G.; Roldo, M. 3d printing and electrospinning of composite hydrogels for cartilage and bone tissue engineering. *Polymers* **2018**, *10*, 285. [CrossRef] [PubMed]
66. Gonçalves, A.M.; Moreira, A.; Weber, A.; Williams, G.R.; Costa, P.F. Osteochondral tissue engineering: The potential of electrospinning and additive manufacturing. *Pharmaceutics* **2021**, *13*, 983. [CrossRef] [PubMed]
67. O'Conor, C.J.; Case, N.; Guilak, F. Mechanical regulation of chondrogenesis. *Stem Cell Res. Ther.* **2013**, *4*, 1–13. [CrossRef] [PubMed]

polymers

MDPI

Review

Effect of Polymeric Matrix Stiffness on Osteogenic Differentiation of Mesenchymal Stem/Progenitor Cells: Concise Review

Aiah A. El-Rashidy [1,2], Sara El Moshy [2,3], Israa Ahmed Radwan [2,3], Dina Rady [2,3], Marwa M. S. Abbass [2,3], Christof E. Dörfer [4] and Karim M. Fawzy El-Sayed [2,4,5,*]

1 Biomaterials Department, Faculty of Dentistry, Cairo University, Cairo 11562, Egypt; aiah.abdelwahab@dentistry.cu.edu.eg
2 Stem Cells and Tissue Engineering Research Group, Faculty of Dentistry, Cairo University, Cairo 11562, Egypt; sarah.mahmoud@dentistry.cu.edu.eg (S.E.M.); esraa.ahmed@dentistry.cu.edu.eg (I.A.R.); dina.radi@dentistry.cu.edu.eg (D.R.); marwa.magdy@dentistry.cu.edu.eg (M.M.S.A.)
3 Oral Biology Department, Faculty of Dentistry, Cairo University, Cairo 11562, Egypt
4 Clinic for Conservative Dentistry and Periodontology, School of Dental Medicine, Christian Albrechts University, 24105 Kiel, Germany; doerfer@konspar.uni-kiel.de
5 Oral Medicine and Periodontology Department, Faculty of Dentistry, Cairo University, Cairo 11562, Egypt
* Correspondence: karim.fawzy@gmail.com; Tel.: +20-2-236-349-65; Fax: +20-2-236-463-75

Abstract: Mesenchymal stem/progenitor cells (MSCs) have a multi-differentiation potential into specialized cell types, with remarkable regenerative and therapeutic results. Several factors could trigger the differentiation of MSCs into specific lineages, among them the biophysical and chemical characteristics of the extracellular matrix (ECM), including its stiffness, composition, topography, and mechanical properties. MSCs can sense and assess the stiffness of extracellular substrates through the process of mechanotransduction. Through this process, the extracellular matrix can govern and direct MSCs' lineage commitment through complex intracellular pathways. Hence, various biomimetic natural and synthetic polymeric matrices of tunable stiffness were developed and further investigated to mimic the MSCs' native tissues. Customizing scaffold materials to mimic cells' natural environment is of utmost importance during the process of tissue engineering. This review aims to highlight the regulatory role of matrix stiffness in directing the osteogenic differentiation of MSCs, addressing how MSCs sense and respond to their ECM, in addition to listing different polymeric biomaterials and methods used to alter their stiffness to dictate MSCs' differentiation towards the osteogenic lineage.

Keywords: mesenchymal stem cells; polymers; matrix; stiffness; osteoblasts; differentiation

Citation: El-Rashidy, A.A.; El Moshy, S.; Radwan, I.A.; Rady, D.; Abbass, M.M.S.; Dörfer, C.E.; Fawzy El-Sayed, K.M. Effect of Polymeric Matrix Stiffness on Osteogenic Differentiation of Mesenchymal Stem/Progenitor Cells: Concise Review. *Polymers* **2021**, *13*, 2950. https://doi.org/10.3390/polym13172950

Academic Editor: José Miguel Ferri

Received: 16 July 2021
Accepted: 5 August 2021
Published: 31 August 2021

Publisher's Note: MDPI stays neutral with regard to jurisdictional claims in published maps and institutional affiliations.

1. Introduction

Stem/progenitor cells are characterized by their outstanding differentiation potential into multiple types of specialized cell lineages, relying on their pluri- or multipotency, while maintaining their self-replicating ability [1–10]. Among the different stem/progenitor cell types, mesenchymal stem/progenitor cells (MSCs) have been widely used in tissue engineering, cell transplantation, and immunotherapy [11–14]. MSCs were initially isolated from the bone marrow, but can be currently isolated from almost every tissue in the body [15]. MSCs niches are located in different sites, including umbilical cord blood [16], menses blood [17], dental tissues [18], synovial fluid [19], adipose tissues [14], and dental tissues [4]. MSCs reside adjacent to vessel walls, near perivascular regions, on the endosteal surfaces of trabecular bone, and within the interfibrillar spaces [11].

Proliferation and differentiation of MSCs can be triggered by certain growth factors and chemicals, inducing specific genetic events, affecting the release of transcriptional factors, which regulate the differentiation of MSCs into specific lineages [14,17]. Additionally, biomaterial scaffolds can create a microenvironment that provides MSCs with appropriate conditions for directed differentiation [14]. MSCs further can secrete various immunomodulatory molecules, including cytokines, chemokines, and growth factors, which provide the self-regulated regenerative microenvironment for different injured tissues or organs [13,17].

Regeneration and healing of bone injuries, particularly in large bony defects, is a complicated process [11]. Based on the multipotency of MSCs, they can give rise to either osteoblasts, chondrocytes, myoblasts, or adipocytes in response to key transcriptional regulators that control the primary commitment and most of the follow-up differentiation [20]. MSCs further interact with the components of their local microenvironment (niche) of the extracellular matrix (ECM) [21].

ECM was earlier believed to be an inert matrix that only provides physical support to cells; later, the important role of ECM in various cellular processes was introduced [22]. The MSCs niche provides extrinsic signals including growth factors, ECM, and those released due to contact with other cells. Through these signals, the MSCs' niche could regulate the stem/progenitor cells' fate [23,24]. In this context, interactions of MSCs with their niche are reciprocal; thus, MSCs are capable of remodeling the niche in response to signals received from it [24].

Several transcription factors are involved in the osteogenic differentiation pathway, including runt-related transcription factor 2 (Runx2), osterix (Osx, or SP7), Smad, and β-catenin [25–27]. Runx2 expressing cells are defined as pre-osteoblasts, a heterogeneous population of cells that includes all cells transitioning from progenitors to mature osteoblasts. A three-stage differentiation of the pre-osteoblasts then follows. The first stage involves cells' proliferation and expression of transforming growth factor-beta receptor 1 (TGF-βR1), fibronectin, collagen, and osteopontin. The second stage involves the initiation of cellular differentiation and maturation of the ECM with alkaline phosphatase (ALP) and collagen expression. In the final stage, the ECM is enriched with osteocalcin, which promotes matrix mineralization [20]. Runx2 guides MSCs differentiation into osteoblasts besides inhibition of other differentiation pathways, particularly adipogenic and chondrogenic ones [28,29]. Various signaling pathways, such as bone morphogenic proteins (BMPs), Notch, and Wnt signaling pathways, could regulate Runx2 expression.

BMPs are famous for their capability to induce bone formation. They activate intracellular Smad, which translocates to the nucleus and acts as a transcription factor besides promoting Runx2 expression [30]. BMP9 stimulates the activation of Smad1/5/8 in MSCs cells. Moreover, Smad4 knockdown decreases the nuclear translocation of Smad1/5/8 and inhibits osteogenic differentiation [31]. Hence, Smad is of great importance, and its interaction with Runx2 is essential for osteogenic differentiation. Mutation of the C-terminal domain of Runx2 disrupts Runx2–Smad transcriptional activities, which leads to the suppression of osteogenic differentiation [32].

Osx is an essential transcription factor for osteogenic differentiation and subsequent bone formation. In Osx null mice, no bone formation occurred; additionally, in the Runx2 null mice, no expression of Osx was noted, indicating that Osx acts as a downstream of Runx2 and emphasizing its role in MSCs osteogenic differentiation and bone formation [33]. Moreover, Wnt signaling pathway activation in MSCs induces Osx expression and suppresses peroxisome proliferator-activated receptor γ (PPAR-γ) [34]. Moreover, β-catenin has a competitive inhibitory relationship with PPAR-γ, where activation of one of them leads to the degradation of the other [35]. Therefore, Wnt/β-catenin signaling activation shifts MSCs' commitment towards osteogenesis at the expense of adipogenic differentiation [34].

β-catenin further plays a critical role in MSCs' osteogenic differentiation. Its absence blocks the osteogenic differentiation and allows for the chondrogenic differentiation of MSCs [36]. Wnt signaling is essential for the β-catenin function. Wnt signaling accumulates β-catenin in the cytoplasm and translocates it into the nucleus, activating the transcription of downstream genes. The absence of Wnt signaling leads to the degradation of β-catenin and interferes with MSCs' osteogenic differentiation [37]. The sensitivity of β-catenin to matrix stiffness during the differentiation of adipose-derived stromal cells (ASCs) has been demonstrated [38]. β-catenin increased nuclear translocation with increased matrix stiffness and enhanced the expression of Runx2, thus stimulating osteogenesis.

Stem/progenitor cells' behavior is largely affected by extracellular signals from the microenvironment, including chemical and mechanical cues from the ECM [39]. Unlike chemical cues, the mechanical properties of the microenvironment act as signals that are consistent along with time and space, thus providing long-range stimulation to cells over long periods and over relatively long distances. Recent literature has focused on the paramount role of the ECM mechanical properties in controlling stem/progenitor cells' behavior, including maintaining their potency, self-renewal and differentiation, migration, proliferation, and interaction with other cells [39,40]. Matrix-related mechanical stimuli, including strain, shear stress, matrix rigidity, and topography, could impact stem/progenitor cell phenotypes through controlling gene transcription and signaling pathways [40,41].

The extracellular-signal-regulated kinase (ERK) and p38 are members of the mitogen-activated protein kinase (MAPK) enzymes family that is concerned with mechanotransduction pathways [42]. ERK is a potent regulator of MSCs' differentiation, as mechanical stimulation activates ERK through integrin focal adhesion complexes and the initiation of MAPK–ERK signaling cascade [43]. Besides ERK, p38 is involved in MSCs' differentiation. The p38-MAPK signaling activity has been identified as an essential factor for osteoblastic differentiation [44–46]. Ras-mediated signaling has been further presented as a master key that affects multiple intracellular pathways, including ERK, PI3K/AKT, and Smad [47,48]. Inhibition of Ras (RasN17) significantly downregulates AKT, ERK, and Smad1/5/8 activation, as well as osteogenic markers' expression. Conversely, active Ras (RasV12) has little effect on osteogenic markers' expression [49]. Consequently, inducing transcription factors to control and guide MSCs' differentiation has become an essential strategy for guided tissue regeneration [26]. Interference between signaling pathways through interaction between different transcription factors can drive MSCs towards specific cell linage; for example, osteogenic signaling can inhibit the adipogenic signaling pathway, and vice versa [41].

Matrix stiffness has a profound impact on MSCs' behavior. The adhesion, proliferation, and spreading capacity of umbilical cord MSCs varied when cultured on polyacrylamide gels coated with fibronectin with different stiffness (Young's modulus: 13–16, 35–38, 48–53, and 62–68 kPa) [50]. Maximum spreading of MSCs was observed with increased matrix stiffness. The soft matrix promoted adipogenic differentiation with high expression of PPARγ and C/EBPα, while MSCs cultured on the 48–53 kPa matrix differentiate into muscle cells with increased expression of MOYG. On the other hand, MSCs cultured on stiff matrix differentiate into osteoblast with increased expression of ALP, collagen type I, Runx2, and osteocalcin [50]. Additionally, bone-marrow MSCs cultured on fibronectin-coated polyacrylamide hydrogels with different stiffnesses, ranging from 13 to 68 kPa, demonstrated enhanced adhesion, spreading and proliferation upon increasing matrix stiffness [51]. On 62–68 kPa, MSCs exhibited a polygonal morphology with a more extensive spreading area and high expression of Runx2, ALP, and osteopontin. These data highlight the critical role of matrix stiffness in regulating MSCs behavior which aids in the development of new biomaterials for tissue regeneration.

Insights into how stem/progenitor cells sense signals from the ECM and how they respond to these signals at the molecular level have become an area of increasing research [21,22,52]. Lately, stem/progenitor cells were shown to be capable of sensing and responding to the structural and functional cues of the matrix [22,52], such as the topography of the ECM components, adhesive properties of the ECM, and ECM stiffness [24,53]. The cells adhere to the ECM via several specific cell-surface receptors, known as integrins [21,22]. Integrins transmit signals from ECM to the cells, thus affecting the proliferation and differentiation of stem/progenitor cells through mechanotransduction of signals [21,22]. It is suggested that the cells use actomyosin filaments (stress fibers) contractility for reciprocal interactions with their matrix [23]. When cells are grown in vitro, extensive efforts to mimic the in vivo microenvironments have been made to control and direct stem/progenitor cell commitment into specific cell lineages required for regenerative medicine.

Natural and synthetic polymeric materials could offer versatile matrices, which are biocompatible and biodegradable, with tunable characteristics, precise control of their topography, and ease of processing [54,55]. Biomaterial stiffness, which determines the material's resistance to deformation in response to an applied force, is a vital property in tissue engineering. ECM stiffness is calculated by dividing the load by the elastic deformation of the matrix [56], is denoted by the elastic modulus or Young's modulus (E), and represents the resistance that a cell feels when it deforms the ECM [57].

ECM stiffness guides stem/progenitor cells' differentiation down corresponding tissue lineages [58]. Osteogenic differentiation of MSCs was shown to be favored on more rigid substrate, while adipogenic differentiation is favored on softer substrates [21]. Such control of MSCs fate by matrix stiffness was shown to be complementary to, and even synergistic with, the regulatory effects of specialized cell culture media commonly used to direct mesenchymal stem/progenitor cell differentiation into specific lineages [23]. Various biomaterials coupled with different methods of controlling stiffness are employed to develop specific stiffness ranges for regulating MSCs differentiation in vitro. Controlling of substrates' stiffness could be tuned through adjusting the biomaterial composition, the amount/concentration/ratio of material components, the degree of crosslinking, and the reaction conditions [56]. Taking into consideration that the bulk stiffness of most native tissues is much lower than that of plastic or glassware used for in vitro tissue culture [24], the development of biomimetic polymeric matrices of tunable stiffness, mimicking native tissues, allowed new data to reveal more details on the impact of mechanical cues of the microenvironment, especially ECM stiffness, on cellular properties [58].

In this review, we highlight the regulatory role of matrix stiffness in directing the osteogenic differentiation of MSCs, addressing how MSCs sense and respond to their ECM, in addition to listing different polymeric biomaterials commonly used in vitro and methods used to alter their stiffness to dictate MSCs differentiation towards the osteogenic lineage. Moreover, through the current review, we aim to elucidate the effect of ECM stiffness on the MSCs' osteogenic potential and the underlying mechanism, which is of particular importance during the process of designing new materials for bone-tissue regeneration.

2. MSCs and Mechanotransduction

MSCs can sense and assess the stiffness of extracellular substrates [59,60]. The ability of stem/progenitor cells to sense changes in the surrounding environment is known as mechanosensation. Stem/progenitor cells can also transduce mechanical stimuli in the surrounding environment into biochemical signals to induce cellular responses through the process of mechanotransduction [61]. Mechanotransduction is the mechanism underlying the increased osteogenic differentiation of MSCs on stiffer matrix [62]. The process of mechanotransduction in stem/progenitor cells is mediated through focal adhesion, associated integrins, and cellular cytoskeleton, in addition to mechanosensitive ion channels.

2.1. Focal Adhesion and Integrins

Focal adhesions are complexes of highly specialized proteins and macromolecules that can attach cells to the ECM, allowing them to sense and respond to mechanical stimuli [63]. Focal adhesion is composed of the transmembrane protein integrin, which has intracellular and extracellular domains (Figure 1). Integrin's intracellular domain is linked to the actin cytoskeleton via cytoplasmic adapter proteins, which include the actin-binding proteins α-actinin, vinculin, and talin [64]. The integrin extracellular domain binds to ECM components, such as collagen, laminin, and fibronectin, via its extracellular domain, thereby establishing a mechanical connection between ECM and intracellular cytoskeleton components [64–66]. Fifty different proteins have been associated with focal adhesion [67], including intracellular proteins as focal adhesion kinase (FAK) and p130Cas [68,69].

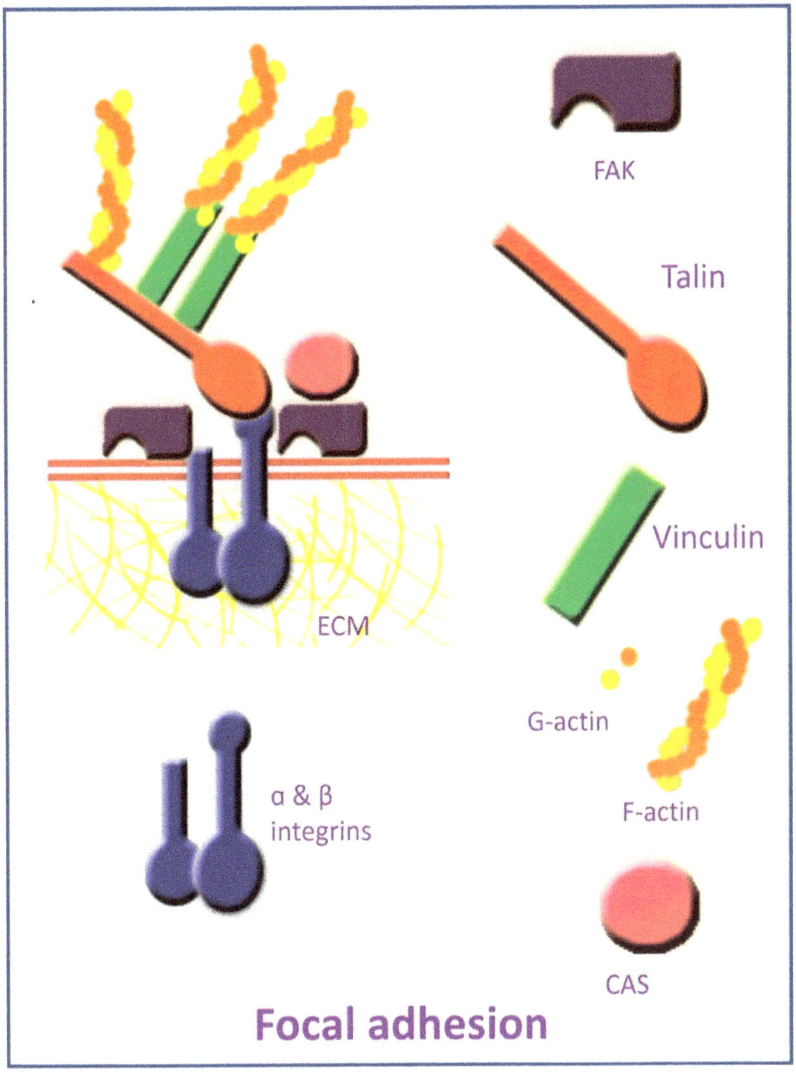

Figure 1. Elements of focal adhesion.

Integrins are alpha and beta subunits heterodimers existing in different combinations [64]. There are 18α and 8β subunits, which account for 24 different integrin heterodimers in mammals specific to an exact set of ECM ligands [70,71]. Through their intracellular and extracellular domains, integrins are capable of joining intracellular cytoskeleton with the external environment, thereby creating mechanical integration between ECM and intracellular cytoskeleton [72]. They can transmit cellular signals to the ECM, and reciprocally can convey signals from the ECM intracellularly [73], triggering an intracellular signaling pathway, resulting in alteration of cellular migration, proliferation, and differentiation [74].

It is noteworthy that MSCs' surface integrins' expression can influence MSCs' lineage commitment [75]. Further, matrix stiffness can influence integrin expression on MSCs, which can dictate and direct stem/progenitor cell fate [76]. Undifferentiated MSCs were found to mostly express α1, α3, αV, β1, and β2 integrins, while α2, α4, α5, α6, β3, β4, and β5 were expressed to a lesser extent [77]. MSCs' osteogenic differentiation was reported to be associated with upregulation of integrin α5 expression on MSCs' surface in response to ECM stimuli [77–80]. Integrin α5 upregulation promotes osteogenesis through activation of FAK via the ERK1/2 MAPKs and PI3K signaling [79]. MSCs' expression of integrin subunits α2 [62,76], α1, αV, and β3 was also upregulated with increased ECM stiffness, favoring osteogenic differentiation [62], while α5 and β1 expression was upregulated in the matrix with lower stiffness [76]. Additionally, activation of MSCs expression of α5β1 and αVβ3 integrin complexes in response to ECM morphology was associated with enhanced osteogenic differentiation [81]. On the other hand, osteogenic differentiation was associated with reduced expression of α1, α3, α4, β3, and β4 integrin subunits [77], while MSCs' adipogenic differentiation was associated with the upregulation of α6 and reduction of α2, α4, α3, β3, and β4 integrin subunits expression [77]. Increased integrin α5 expression can also inhibit both adipogenic and chondrogenic differentiation, while promoting MSCs osteogenic differentiation [82].

The binding of integrin to ECM components triggers the intracytoplasmic assembly of focal adhesion proteins, including talin, FAK, p130Cas, and vinculin, first forming focal complexes, which then grow, giving rise to focal adhesions, linking actin fibers to ECM components [83]. Formation of focal adhesion, with the associated triggering of intracellular signaling pathways, is essential for MSCs migration, proliferation and differentiation [84–86].

MSCs lineage commitment and osteogenic differentiation in response to ECM mechanical cues, including matrix stiffness, involve upregulation of focal adhesion formation. Increasing matrix stiffness can promote number [84] and area of focal adhesions [87]. In turn, upregulation of focal adhesion number and size [60,88,89] has been linked to increased osteogenic differentiation of MSCs. Additionally, tightly packed focal adhesion can stimulate osteogenic differentiation [90].

2.2. *Cytoskeleton Elements*

Cytoskeletal-related proteins are responsible for the ability of stem/progenitor cells to respond to mechanical cues, including stiffness of the ECM [85]. In addition to their role in providing a cellular structural framework, repolarization of cytoskeleton elements in response to mechanical stimuli, transmits the signals from the ECM to the nucleus, resulting eventually in altered gene expression [64,91,92]. Structural elements of the cellular cytoskeleton include microfilaments, intermediate filaments, and microtubules [64].

The actin cytoskeleton is responsible for the maintenance of cell shape, motility, and contractility. They also act as mechanical sensors for the extracellular environment [93]. It is formed of F-actin, which is a helical polymer of G-actin coupled with actin-binding and actin-bundling proteins, such as α-actinin, vinculin, and talin [94] (Figure 2). Actin cytoskeleton forms a web in association with cellular junctions and forms a core of microvilli, filopodia, and lamellipodia [64]. Actin perinuclear cap is a dome-like structure formed of contractile actin filament and phosphorylated myosin, covering the top of the nucleus and connected to the nucleus through linkers of nucleoskeleton and cytoskeleton (LINC) protein complexes [95,96]. This actin perinuclear cap provides a mechanism through which mechanical signals, transduced through focal adhesion, can reach the nucleus to induce cellular responses [95,97]. Ultimately, the actin perinuclear cap is responsible for conveying signals regarding matrix stiffness to the nucleus [97].

The binding of MSCs to a stiff substrate induces actin polymerization, as evident by an increased ratio of F-actin to G-actin, forming actin stress fibers, which trigger intracellular-signaling pathways [67]. Stress fibers are actomyosin complex composed of F-actin and myosin-2 stabilized by crosslinking proteins [98] (Figure 2). The process of actin polymerization is regulated by the FAK signaling pathway [99]. Actin polymerization and stress fibers formation are essential for establishing cell to ECM interaction [100]. Polymerization of actin dictates lineage commitment of MSCs, as actin depolymerization was noticed during adipogenic differentiation [101]. On the contrary, actin polymerization combined with an increased ratio of F-actin to G-actin upregulated osteogenic differentiation [99,102–105]. On the other hand, disruption of actin polymerization can reduce osteogenic differentiation [103]. Increased osteogenic differentiation on stiffer substrates was also associated with increased expression of F-actin [78], in addition to actin-binding protein (vinculin) [106].

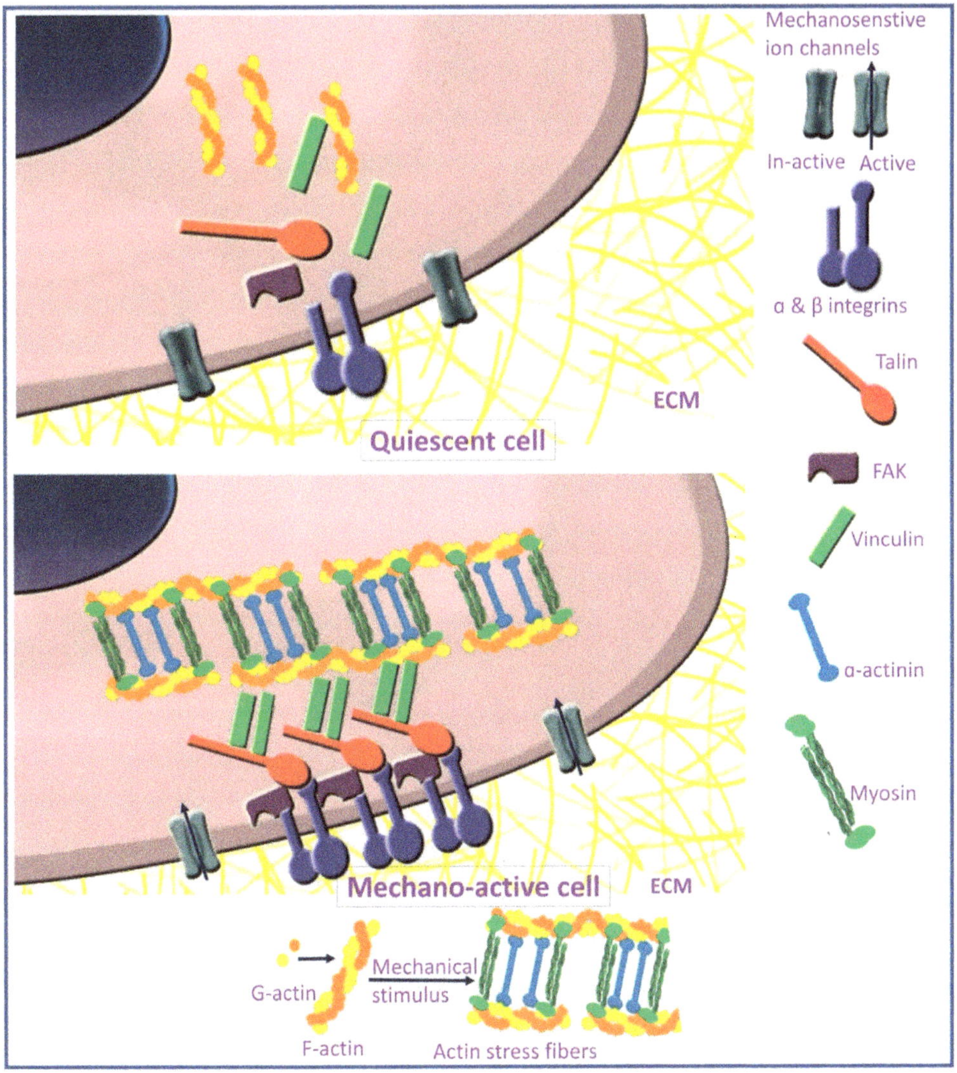

Figure 2. Focal adhesion formation and stress fibers assembly in mechano-active cells.

Actin filaments can also interact with other components of the cellular cytoskeleton as intermediate filaments [64]. Intermediate filaments have a diameter of about 10 nm and have a role in maintaining cell shape and cellular junctions [64]. F-actin promotes intermediate filaments and vinculin assembly and disassembly, which are required for the process of osteogenesis through the transient receptor potential melastatin 7–osterix axis [107].

2.3. Mechanosensitive Ion Channels

Mechanosensitive ion channels are a further mechanism implicated in MSCs' mechanotransduction on stiff matrices. These ion channels are sensitive to substrate stiffness. Upon mechanical stimulation, they allow the intracellular influx of ions and can form complexes with stress fibers, eliciting intracellular signaling pathways [108]. Mechanical stimulation affects cell differentiation through a change in calcium influx through activated channels [109]. The change in the calcium influx results in the activation of the MAPK signaling pathway [110].

2.4. MSCs' Aging and Mechanosensitivity

Different age-dependent changes in MSCs were reported, such as decreased proliferation ability [111] and osteogenic differentiation potential [112–114]. Moreover, age-associated bone loss was linked to the reduced osteogenic potential of MSCs [115]. Aged multipotent progenitor cells lose their sensitivity to alterations in polyacrylamide substrates, while younger multipotent progenitor cells showed a lineage-dependent response to stiffness [116]. The effect of MSCs aging on their mechanosensitivity was investigated by comparing the response of child (11 to 12 years old) and adult MSCs (20–30 years old) to variations in stiffness (10 and 300 kPa) of type I collagen-coated polyacrylamide substrates [117]. Child MSCs revealed more mechanosensitive (increased nuclear-translocation of YAP), improved angiogenesis (enhanced endothelial tubule formation), and osteogenesis (increased alkaline phosphatase activity and mineralization) on stiff substrates as compared to adult MSCs. Based on a customized PCR array, an age-dependent, stiffness-induced upregulation of NOX1, VEGFR1, VEGFR2, WIF1, and JNK3 in child MSCs compared to adults MSCs [117]. Understanding the mechanism behind the age-altered mechanosensitivity of MSCs may open up new avenues to identify potential therapeutic targets to reproduce the enhanced osteogenic and angiogenic potential of adults with bone degeneration and disease.

3. The Role of Matrix Stiffness in Triggering MSCs' Osteogenic Differentiation

Matrix stiffness regulates the MSCs' differentiation into mature specific cells by activating transcription factors that upregulate genes responsible for the initiation and progression of particular cell-linage differentiation. The singling pathways involved in MSCs' osteogenic differentiation are illustrated in (Figure 3). Rigid matrices led to increased MSCs spreading and improved actomyosin contractility, promoting osteogenic differentiation. This enhanced potential was accompanied by increased Runx2, β-catenin, and Smad, implying the significant impact of mechanosensing the matrix stiffness and its role in determining the cell fate [41,50]. The relation between Runx2 expression, owing to mechanosensation with actomyosin contractility, was confirmed by inhibiting myosin, which caused a decrease in Runx2 expression [118]. However, the effect of matrix stiffness on MSCs' differentiation disappeared at the monolayer state [49].

The hippo pathway is one of the signaling pathways involved in MSCs' differentiation and is regulated by intra- and extracellular signals [119]. The downstream effectors of the hippo signaling pathway are yes-associated protein (YAP) and transcriptional co-activator with PDZ-binding motif (TAZ) [120]. YAP and TAZ transduce signals necessary for determining MSCs' fate. The control of the Hippo pathway is through phosphorylation and nuclear translocation of YAP/TAZ [121]. Additionally, matrix stiffness can control the localization and activity of YAP/TAZ [122,123], which is identified through the structural and functional regulation of the cell cytoskeleton to adjust cellular tension [124]. The stresses sensed by MSCs are transmitted to the nucleus and lead to an increase in the nuclear membrane tension, causing expansion of the nuclear pores, which promote sudden nuclear inflow of YAP [125]. In MSCs cultured on a rigid matrix (40 kPa) undergoing osteogenic differentiation, YAP/TAZ has been localized in the nucleus. In comparison, MSCs cultured on a soft matrix (0.7 kPa), YAP/TAZ persisted in the cytoplasm, directing MSCs to undergo adipogenic differentiation [122]. Moreover, YAP knocking-down resulted in inhibition of osteogenesis and enhancement of adipogenesis [126]. During MSCs' osteogenic differentiation, TAZ functions as a co-activator of Runx2 to stimulate osteogenesis and inhibits PPAR-γ, which reduces adipogenic differentiation [127]. These findings highlight the significant role of YAP/TAZ as a potent regulator of stiffness-induced osteogenic differentiation.

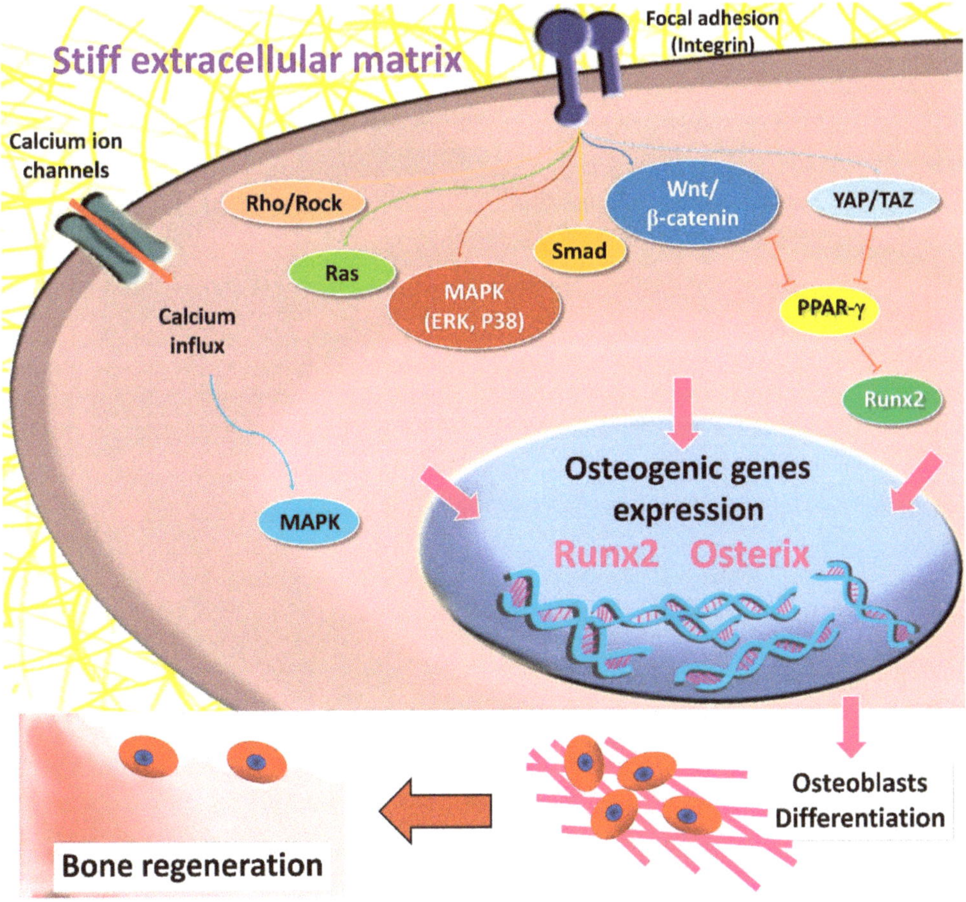

Figure 3. Signaling pathways involved in stiffness induced MSCs' osteogenic differentiation.

MSCs fate is also directed through actomyosin contractility and activated Rho/Rho kinase (ROCK) signaling [88], along with mechanotransduction mediated by focal adhesion and integrin [128]. In response to increased stiffness, activated Rho stimulates actomyosin stress fiber assembly [129], which causes an increase in cell contractility and activation of ERK, promoting osteogenic differentiation [130]. Furthermore, Rho combined with the actin cytoskeleton is essential to maintain nuclear YAP/TAZ in MSCs [122]. Activation of FAK via ROCK signaling led to upregulation of osteogenic marker Runx2, ALP, and matrix mineralization denoting osteogenesis of human adipose stem/progenitor cells [131]. In addition, the inhibition of FAK and ROCK signaling caused an upregulation of adipogenic markers. Furthermore, matrix stiffness modulates MSCs' osteogenic differentiation through the Ras pathway, which is accompanied by an increase in the phosphorylation levels of Smad1/5/8, AKT and ERK [49]. Ras (RasN17) inhibition resulted in a significant decrease of Smad1/5/8, AKT, and ERK activity, as well as osteogenic markers' expression [49].

Cells on stiff matrices develop high cytoskeletal tension, which is evidenced by enhanced actin stress fibers and large spread area. Below a compressive modulus of 25 kPa, regardless of the adhesive ligand presented, there is not enough cytoskeletal tension to promote osteogenic lineage differentiation [88]. Based on these results, it has been postulated that, unless a cell develops cytoskeletal tension exceeding a certain threshold stiffness (substrates with moduli of $\geq$ 25 kPa), osteogenic differentiation will not occur and the cell would need the presence of an osteogenic ligand for Runx2 expression for further differentiation to take place. On the other hand, MyoD1 (a marker for myoblasts) expression demonstrated less ECM dependence compared with Runx2, as it was markedly expressed in cells cultivated on substrates with stiffnesses higher than 9 kPa, regardless of the protein coating [132]. Additionally, on soft poly(acrylamide-co-acrylic acid) substrates (E = 15.4 kPa) that mimic muscle elasticity when grafted with arginine–glycine–aspartate (RGD) peptide sequence, MSCs were directed to a spindle-shaped morphology similar to C2C12 myoblasts, while stiffer matrices (E = 47.5 kPa) that mimic osteoid tissue's crosslinked collagen yield the cells in polygonal morphology, similar to MC3T3-E1 pre-osteoblasts [118].

Inflammation can further counteract the inductive effect of matrix stiffness on osteogenic differentiation. Periodontal ligament stem cells (PDLSCs) cultured with the inflammatory cytokine interleukin (IL)-1β on gelatin/methacrylate hydrogels with different matrix stiffness showed a marked reduction in matrix stiffness-dependent osteogenic differentiation and expression of osteocalcin, as well as Runx2. This was through the activation of p38 signaling pathways, which were activated by IL-1β [133]. Further, macrophages encapsulated in gelatin/methacrylate hydrogels with high stiffness showed a high tendency to polarize toward the pro-inflammatory M1 phenotype, which was associated with a negative impact on the osteogenic differentiation of bone-marrow mesenchymal stem cells (BMMSCs) [134].

4. Matrix-Dependent MSCs' Osteogenic Differentiation

Several natural and synthetic polymeric biomaterials are currently used in tissue engineering and regenerative medicine, serving as biomimetic matrices for in vitro culturing [135]. These biopolymers can be generally divided into two classes: natural and synthetic polymers. Natural polymers include alginate, collagen, gelatin, hyaluronic acid, elastin, actin, keratin, albumin, chitosan, and others. They are characterized by their inherent bioactivity and ability to mimic natural tissues, yet they suffer from possible immunogenicity, structural complexity, and poor mechanical properties. Chitosan can be used for increasing energy storage of α-cobalt molybdate (CoMoO) nano-flakes in the presence of a crosslinking agent such as citric acid [136]. Compared to natural polymers, synthetic polymers have higher mechanical properties, are readily available, with tunable physicochemical properties and degradation rate, but lack natural tissue resemblance [55,137]. Major synthetic polymers used include polyethylene glycol (PEG), polydimethylsiloxane (PDMS), polyesters, polyacrylamide, vinyl polymers, and self-assembling peptides, in addition to poly (lactic acid) (PLA), poly (glycolic acid) (PGA), poly (lactide-co-glycolic acid) (PLGA), and others [55,135].

4.1. *Natural Polymers*

4.1.1. Alginate

Alginates extracted from seaweeds and algae are composed of β-1,4-linked blocks of β-D-mannuronic acid (M) and its C-5 epimer α-L-guluronic acid [138]. Alginates are widely used polysaccharides for hydrogelation in tissue engineering as they can be gelated easily through the addition of divalent cations [139]. Since alginate could act as a template for binding of manganese ions, the presence of a high concentration of alginate in the electrolytic manganese dioxide altered the morphology from spindle-shaped to cactus-shaped [140].

Spatially modulating the mechanical properties in an alginate bioink, 3D printed constructs were postulated to regulate MSCs' fate. Micro-CT-based imaging with 3D bioprinting and bioreactor system were utilized to fabricate 3D human MSCs-laden porous bone-like scaffolds with varying compressive moduli based on two unmodified polymers (alginate and gelatin). Softer scaffolds with low alginate concentration (0.8% alginate, 0.66 ± 0.08 kPa) revealed accelerated and enhanced osteogenic differentiation with upregulated ALP activity than stiffer scaffolds (1.8% alginate, 5.4 ± 1.2 kPa). In the presence of the osteogenic differentiation medium, cells on soft scaffolds exhibited osteoblastic and early osteocyte-related gene expression and showed a 3D cellular network within the mineralized matrix [141].

Increasing alginate molecular weight, as well as increasing the crosslinking ratio, produces a significantly stiffer bioink. Upon bioprinting cylindrical MSCs laden constructs with spatially variable mechanical stiffness from the core to the periphery, more MSCs underwent osteogenic differentiation within the stiffer regions of the printed constructs as evident by increased ALP staining [142]. In contrast to most studies, an investigation demonstrated that, under basal conditions and in the absence of RGD ligands, alginate hydrogel with bimodal molecular weight distribution (50% LMW and 50% HMW) and 1 wt.% polymer concentration of low-stiffness 3D matrices (tan $\partial \approx 0.4$–0.6) provided a permissive environment for human MSCs osteogenic differentiation and expressed high levels of ALP and osteocalcin as compared to the stiffer 2 wt.% alginate hydrogel with the presence of RGD ligands [143].

4.1.2. Collagen

Collagen is considered to be the most abundant protein in mammals [144]. Being the main ECM protein, collagen, together with soluble factors, may act as a niche for MSCs osteogenesis and bone mineralization [145]. Mechanical properties of collagen fibers vary depending upon their location in different tissues. Thus, the cells can sense local fibrillar microenvironments with different physical cues. Collagen gels were engineered to attain varying fiber stiffness (from 1.1 to 9.3 kPa), while maintaining bulk stiffness below 200 Pa, by changing the polymerization temperature to 4 °C (Col-4), 21 °C (Col-21), and 37 °C (Col-37) without changing the density of the collagen. A polymerization temperature of 4 °C led to shorter, thicker, and stiffer collagen fibers (Col-4), with limited fiber recruitment and force transmission and fewer focal adhesions. Cells grown on Col-4 showed much slower spreading as compared to Col-37 with similar bulk stiffness but with more flexible and longer fibers that can be easily remodeled. Human MSCs cultured on Col-4 revealed a much lower ratio of osteogenic differentiation (21.1%) compared to that on Col-37 with (34.1%) ALP positive reactivity [146].

Matrix stiffness possesses a high impact on cellular bioactivity regardless of the presence or absence of growth factors. This was proved by culturing porcine adipose-derived stem cells (ADSCs) on sequentially integrated benzophenone photo-immobilization and carbodiimide, crosslinking collagen–glycosaminoglycan, with stiffness ranging from 2.85 to 5 MPa, in the presence or absence of covalently immobilized platelet-derived growth factor (PDGF-BB) and BMP-2 [147].

In the presence of osteogenic culture media, human MSCs coated with three bilayers of collagen/alginate nanofilm with relatively high stiffness 24 and 53 MPa revealed an augmented osteogenic differentiation efficiency with a significant increase in ALP by activating transcriptional co-activators with the PDZ binding motif through extracellular signal-related kinase and p38-MAPK signaling [148].

Nanoparticulate mineralized collagen–glycosaminoglycan scaffolds chemically crosslinked with 1-ethyl-3-(3-dimethylaminopropyl) carbodiimide and N-hydroxysuccinimide had a higher range of elastic moduli 3.90 ± 0.36 kPa as compared to non-crosslinked materials. Cultured human MSCs on crosslinked substrates showed higher expression of osteogenic genes and proteins compared to non-crosslinked versions. This was maintained via the mechanotransduction mediators YAP/TAZ and the Wnt signaling pathway [149].

MSCs seeded within a 3D collagen gel with an elastic modulus of ~108 Pa stimulated by vibrations of nanoscale amplitude in a vibrational bioreactor showed increased expression of Runx2, collagen I, ALP, osteopontin, osteocalcin, and BMP-2. This indicated that the viscoelastic properties of the collagen gel allowed the transfer of high-frequency vibrations to the cells seeded in 3D [150].

In the absence of any differentiation supplementation, MSCs grew on stiffer (1.5 kPa) dehydrothermal and 1-ethyl-3-3-dimethyl aminopropyl carbodiimide (EDC) crosslinked collagen–glycosaminoglycan scaffolds showed the greatest level of Runx2 expression, while substrates with lower stiffness (0.5 kPa) resulted in significant elevation of SOX9 expression, indicating that MSCs are directed towards a chondrogenic lineage [151].

Additionally, 3D scaffolds with the highest proportion of collagen in collagen and hydroxyapatite mixture coated on the decellularized cancellous bone with various stiffness (13.00 ± 5.55 kPa, 13.87 ± 1.51 kPa, and 37.7 ± 19.6 kPa) exhibited the highest stiffness that, in turn, promoted higher expressions of osteopontin and osteocalcin [152].

4.1.3. Gelatin

Gelatin is a natural, biocompatible, non-immunogenic, hydrophilic, and biodegradable collagen derivative [153–155]. It is acquired via acid or alkaline hydrolysis of collagen into single molecules [156]. Being derived from a natural source, gelatin is characterized by having RGD cell-binding motifs that can enhance cellular attachment [157]; it can also promote cell proliferation and differentiation [158,159]. One of the major disadvantages of gelatin is its low mechanical properties [153,160], in addition to its thermal instability [161].

Gelatin crosslinking through the addition of chemical groups can help reduce these shortcomings [161]. Stiffness of gelatin hydrogel scaffold can be controlled through changing the ratio of crosslinking agents, such as methacryloyl, giving rise to gelatin–methacrylate (GelMA) [133,162,163], transglutaminase [134], or EDC [164], and through the incorporation of variable additives as starch [162] or polyethylene glycol diacrylate (PEGDA) [163]. Increased gelatin hydrogel scaffold degree of crosslinking and matrix stiffness were positively associated with increased osteogenic differentiation of MSCs [133,134,162–164]. Crosslinked gelatin hydrogel can also be modified to enhance osteogenic potential through incorporation of the bisphosphonate alendronate [163]. GelMA hydrogel with tunable stiffness was constructed by using different GelMA concentrations 10, 12, and 14 wt to yield hydrogel with stiffness 25.75 ± 1.21 kPa, 59.71 ± 8.87 kPa, and 117.82 ± 9.83 kPa, respectively. Osteogenic differentiation of PDLSCs, as well as expression of osteocalcin and Runx2, showed a significant increase with increasing matrix stiffness through activation of ERK1/2 signaling pathway [133].

Gelatin/starch-based hydrogel was fabricated with tunable stiffness from crosslinked gelatin with variable degrees of methacrylation (GelMA; 31%, 72%, and 95%) covalently bound to variable ratios of pentenoates modified starch (10 v% starch and 20 v% starch). Increasing the degree of methacrylation and combining crosslinked gelatin with starch, with subsequent increase in matrix stiffness, effectively promoted osteogenic differentiation of adipose stem cells (ASCs), as was evident by an increased ALP expression. GelMA 95% combined with starch showed the highest degree of osteogenic differentiation, while the highest degree of adipogenic differentiation was observed on the least crosslinked and most flexible gelatin hydrogel (GelMA 31%) [162].

Three-dimensional porous gelatin scaffolds crosslinked using EDC further demonstrated an increase in the elastic modulus from ~0.6 to ~2.5 kPa, without any change in the scaffold internal structure. BMMSCs cultured on EDC-crosslinked gelatin scaffolds with increased stiffness showed an increased osteogenic differentiation as evidenced by increased Runx2 and osteocalcin expression in vitro. Subcutaneous implantation of EDC-crosslinked gelatin scaffold loaded with BMMSCs transfected with adenovirus encoding BMP-2 in mice demonstrated an increased bone formation in vivo, as compared to the control, non-crosslinked scaffold with low stiffness [164].

Transglutaminase-crosslinked gelatin scaffold with variable stiffness was constructed, using gelatin concentrations of 3%, 6%, and 9%. The 9% gelatin gave rise to the highest stiffness (60.54 ± 10.45 kPa), while 3% gelatin resulted in the lowest stiffness (1.58 ± 0.42 kPa). BMMSCs encapsulated in the hydrogel with the highest stiffness demonstrated the highest osteogenic differentiation as revealed by ALP activity, calcified nodule formation, expression of SP7 transcription factor-2, Osx, Runx2, and osteocalcin [134].

Augmentation of GelMA with alendronate and PEGDA showed a positive effect on osteogenic differentiation of BMMSCs. GelMA and alendronate were added at different concentrations and grafted on gelatin molecules, followed by further crosslinking, using 20 wt.% PEGDA to improve hydrogel scaffold stiffness from 4 to 40 kPa. Osteogenic differentiation of grafted BMMSCs was promoted on stiffer hydrogel with higher alendronate concentration, as evident by upregulation of ALP activity, collagen type I, and osteocalcin expression, as well as calcium deposition [163].

4.1.4. Decellularized Matrix and Demineralized Bone

Decellularized cell-derived matrices (dCDMs) could further provide a way to mimic natural tissues. It has been reported that aligned dCDMs could contribute to producing a more homogeneous environment, which resulted in a uniform response of cells to the biophysical environment, displaying a highly homogeneous phenotype and can undergo differentiation if properly stimulated [165]. Substrates displaying linear topographic patterns were obtained by replica molding, using PDMS, while flat PDMS substrates were produced by using a polystyrene dish as a control.

Cell-derived matrices (CDMs) were attained by cultivating MC3T3-E1 pre-osteoblasts in the presence of ascorbic acid for two weeks on linear or flat surfaces. MC3T3-E1 cultivated on nanopatterned substrates produced an aligned fibrillar matrix, whose microarchitectural features remained intact after the decellularization process. Atomic force microscope measurements performed on bare dCDMs revealed very low Young's moduli in the range of (0.01–0.1 kPa) that was increased by genipin crosslinking to reach (i.e., 0.1–1.5 kPa). These matrices were further seeded with murine MSCs and cultivated in the presence of either osteogenic or adipogenic media for two weeks. Both the aligned and random dCDMs promoted murine MSC adhesion and proliferation in their pristine state, while maintaining high levels of stemness markers, with a more homogeneous population of undifferentiated cells, were seen on aligned dCDMs. On the pristine dCDMs, MSCs promptly underwent adipogenic differentiation when stimulated with induction media, while they were minimized in the presence of osteogenic medium, due to very low stiffness. On the contrary, MSCs responded consistently on stiff dCDMs, displaying a significant adipogenic and osteogenic differentiation potential [165].

In another study, 3D demineralized bone matrices with the same 3D microstructure (porosity and pore size), but with various compressive moduli (high: 66.06 ± 27.83 MPa, medium: 26.90 ± 13.16 MPa, and low: 0.67 ± 0.14 MPa), were fabricated by controlling the decalcification duration (1 h, 12 h, and 5 d, respectively). Low-stiffness scaffolds promoted BMMSCs' osteogenic differentiation. Subcutaneous implantation in a rat model further revealed efficient improvement of cells' infiltration and deposition of collagen fibers, in addition to upregulated positive osteocalcin and osteopontin expression, as well as angiogenesis upon utilizing the low-stiffness scaffolds. Further implantation in a femoral condylar defect rabbit model supported the previous findings and revealed that stromal-cell-derived factor-1α/CXC chemokine receptor (SDF-1α/CXCR4) signaling pathway was essential for the stiffness-mediated stem/progenitor recruitment and osteogenic differentiation during bone repair [166].

4.1.5. Hyaluronic Acid

Hyaluronic acid (HA) is a linear non-sulfated polysaccharide made up of repeated disaccharide molecules in alternating patterns (D-glucuronic acid and N-acetyl-D-glucosamine). This pattern is linked through interchanging β-1,4 and β-1,3 glycosidic bonds. HA is a fundamental component of ECM that regulates various cellular biological processes, such as migration, adhesion, proliferation, and differentiation, through binding with a specific receptor on the target cell [167,168]. Owing to its exceptional biocompatible, biodegradable and non-immunogenic properties, HA is clinically used for drug delivery and tissue regeneration [9,169,170]. As being a natural extracellular component, HA mimics the typical ECM and could initiate signaling pathways responsible for osteogenesis [171,172]. Moreover, the physicochemical and biological properties of HA could be altered by chemical modification [173].

Through adjusting the crosslinker (PEGTA) density, a series of hydrogels with different biochemical and biomechanical properties were developed by utilizing a thiol-functionalized HA and a thiol-functionalized recombinant human gelatin. Human BMMSCs were cultured on the hydrogels with different stiffness (storage modulus (G') and corresponding PEGTA concentrations, namely 0.15 kPa (0.25%), 1.5 kPa (1.75%), and 4 kPa (2.5%), in adipogenic and osteogenic conditions. Adipogenic differentiation was confirmed by gene expression of lipoprotein lipase (LPL), as well as PPARγ2, with similar LPL expression levels demonstrated on the hydrogels with varying stiffness, whereas PPARγ2 expression was markedly enhanced upon increasing hydrogel stiffness. Cells exhibited spindle-shape morphology on the 0.15 kPa hydrogel, while displaying elongated and cuboidal appearance, similar to osteoblasts on greater stiffness hydrogels. Human MSCs cultured on the 1.5 kPa hydrogel significantly expressed osteopontin, while those cultured on the 4 kPa hydrogel revealed a significant upregulation in the expression of the late osteogenic gene (bone sialoprotein) [174].

In a novel HA hydrogel platform, ligation of the HAVDI adhesive peptide sequence from the extracellular N-cadherin domain 1 and the RGD adhesive motif from fibronectin led to Rac1-GTP-dependent reductions in the attachment of myosin IIA to the focal adhesions. This lack of myosin IIA incorporation into focal adhesions hindered the maturation of these adhesions with increasing substrate stiffness (E = 5, 10, and 15 kPa) and thereby decreased traction force generation on the underlying substrate. These alterations in the mechanical state of the MSCs further reduced mechanosensitive YAP/TAZ translocation to the nucleus, herewith attenuating the signaling pathways involved in mesenchymal development, including cell proliferation and osteogenic differentiation [175].

An in vitro culture system for osteochondral tissue engineering was developed, using HA gels with various stiffness (G' ranging from 10 to 45 Pa) attained by mixing Glycosil®, a thiol-modified hyaluronan gel with the crosslinking agent PEG at ratios from (1:1 to 7:1). The co-differentiation media (a ratio of 50% chondrogenic:50% osteogenic) proved to be suitable for appropriate chondrogenic and osteogenic differentiation of human MSCs. On the stiffest matrix (HA:PEG construct at a 2:1 ratio), the three chondrogenic markers (aggrecan, collagen II, and sox 9) were expressed by the differentiated human MSCs cultured for 21 days [176].

Moreover, human BMMSCs were initially entrapped in a HA carrying sulfhydryl groups and a hydrophilic polymer bearing both acrylate and tetrazine groups with the shear elastic modulus (G') $=180 \pm 42$ Pa. The stiffness of the matrix was increased ($G' = 520 \pm 80$ Pa) by adding HA conjugated with multiple copies of trans-cyclooctene (TCO) to the human MSCs-laden gel culture media. The 3D matrix tagged with a TCO-cell-adhesive motif promoted the cells to undergo remarkable actin polymerization, changing from a rounded phenotype to a spindle morphology with long processes. After an additional seven days of culture in the modified media, quantitative analysis showed that RGD tagging enhanced cellular expression of matrix metalloproteinase 1, whereas it decreased the expression of tenascin C and collagen I/III. RGD tagging, however, was not sufficient alone to induce chondrogenic, adipogenic, fibroblastic/myofibroblastic, or osteogenic differentiation [177].

Photo-crosslinked methacrylated HA hydrogels incorporating fragmented polycaprolactone (PCL) nanofibers with compression modulus 3122.5 ± 43.7 Pa promoted osteogenic differentiation of adipose-derived stem/progenitor cells incorporated into the composite hydrogel. The biomarkers collage type 1, ALP, and Runx2 were significantly expressed in the hydrogels containing nanofibers. In addition, the results of alizarin red staining confirmed osteogenic differentiation [178].

4.1.6. Fibrin

Fibrin is one of the natural biopolymers that offers many advantages based on its excellent biocompatibility and cell-adhesion properties [179]. However, fibrin has low mechanical properties that can be modified by adjusting the concentration and ionic strength of fibrinogen to obtain a polymeric substrate mimicking native ECM [55]. A high concentration of fibrinogen and thrombin resulted in a stiffer fibrin matrix, as compared to fibrin with lower fibrinogen and thrombin concentration, as altering these two components allowed the tuning of fibrin elasticity. Microfluidic biochips coated with stiff fibrin substrates modified with gold-nanowires-enhanced osteogenic differentiation of human amniotic mesenchymal stem cells (AMSCs) and led to significant elevation in collagen type I levels and matrix mineralization (calcium deposition), while softer fibrin matrices with lower fibrinogen and thrombin concentration enhanced human AMSCs chondrogenic differentiation [180].

4.2. Synthetic Polymers

4.2.1. Polyethylene Glycol

Polyethylene glycol (PEG) is one of the most widely used synthetic polymers in the tissue engineering field. It is characterized by being chemically and biologically inert and by the high hydrophilicity of the polymer backbone. There is a wide range of polymer architectures and lengths that are commercially or synthetically accessible [181].

MSCs cultured on 3D thixotropic PEG-silica nanocomposite gel with high stiffness ($\geq$75 Pa) expressed the highest level of Runx2 and osteocalcin. Additionally, RGD cell-adhesion peptide sequence immobilization in the gel of 75 Pa stiffness promoted ~13% higher expression of the osteogenic transcription factor [182]. Rat BMMSCs cultured on relatively soft (130 kPa) and stiff (3170 kPa) PEG hydrogels with RGD nano-spacings of 49 and 135 nm, incubated in the mixed osteogenic and adipogenic medium, exhibited a higher density of adherent MSCs, and osteogenesis was promoted on stiffer hydrogels. When the hydrogel stiffness was controlled, the large RGD nano-spacing was beneficial for osteogenesis, while the small RGD nano-spacing generated more adipogenesis [183].

Human MSCs were encapsulated in a multilayer PEG-based hydrogel composed of a soft cartilage-like layer of chondroitin sulfate (48 kPa) and low RGD concentrations, a stiff bone-like layer 345 kPa with high RGD concentrations, and an intermediate interfacial layer with 100 kPa. The recorded stiffness of the multilayer hydrogel was 90 kPa. Opposite to static conditions, dynamic mechanical stimulation generated a high expression of collagens with collagen II in the cartilage-like layer, collagen X in the interfacial layer, and collagen I in the bone-like layer with the presence of mineral deposits in the bone layer [184].

PEG/silk fibroin/HA (PEG/SF/HA) scaffold was prepared with varying HA concentrations, which influenced scaffold stiffness (80.98 to 190.51 kPa). PEG/SF/HA containing 50 mg HA cultured with rat BMMSCs enhanced cell adhesion, viability, the expression of all the osteogenesis-related markers in vitro and promoted superior calvarial defect repair in vivo, through modulating gene and protein expression levels [185].

Additionally, human MSCs seeded on regularly and randomly patterned photodegradable PEG hydrogel surfaces with different stiff-to-soft ratios from ~2–3 kPa to ~10–12 kPa displayed higher cell spread, elongated morphologies, and superior YAP activation and osteogenic differentiation on the regularly patterned regions, as compared to those cultured on random patterns [186]. High PEG substrate stiffness (~25 kPa) and $\alpha5\beta1$ integrin signaling stimulated by c(RRETAWA) induced osteogenic differentiation of human MSCs [187].

4.2.2. Polydimethylsiloxane

Polydimethylsiloxane (PDMS) is characterized by its biocompatibility, flexibility, optical clarity, and elastic tunability [188]. Dental follicle stem cells (DFCs) [189], and human exfoliated deciduous teeth (SHED) [190] were cultured on elastic PDMS substrates. Different stiffnesses, ranging from 11 to 93 kPa, were attained by changing the Sylgard®'s crosslinker to base ratios (1:55, 1:45, and 1:35 by weight) [189,190]. Coating PDMS with fibronectin caused a slight increase in ALP-activity of DFCs and continuous expression of cementoblast marker CP23 on standard cell culture dishes [189]. Osteogenic differentiation of SHED and DFCs was not supported by similar grades of ECM stiffness. In a study that involved adding osteogenic differentiation medium to DFCs on PDMS, DFCs revealed a significantly higher ALP activity and accumulation of calcium on the softest substrate (PDMS 1:55) [189], while SHED demonstrated high osteogenic differentiation on PDMS (1:35) stiffer substrate [190].

ASCs were cultured on soft and stiff PDMS substrates with moduli of elasticity ranging from (0.046 $\pm$ 0.02 MPa) and (1.014 $\pm$ 0.15 MPa), respectively. Stiff substrate enhanced the directed differentiation of ASCs into osteogenic lineages as evidenced by positive ALP stain. This enhancement was supplemented with the upregulated expression of Runx2 and Osx transcriptional factors [38].

Osteogenic differentiation of rat MSCs incubated in osteogenic medium grown on PDMS, with stiffness gradients that ranged from 0.19 to 3.10 MPa, utilizing a temperature gradient during curing, was proven to be strongly influenced by substrate stiffness and the ECM macromolecules pre-adsorbed onto the substrates. Calcein Blue (CB)-positive bone-nodule-like colonies were only observed on the stiff end of PDMS coated with fibronectin and gelatin, while oxygen-plasma-treated surfaces were entirely devoid of CB-positive colonies after 1 week of osteoinductive culture [191].

4.2.3. Vinyl Polymers

A variety of functionalized vinyl monomers are commercially available or can be synthetically customized, rendering vinyl polymer-based hydrogels useful as structurally diverse scaffolds [181]. The osteogenic capability of 3D porous scaffolds composed of polytetrafluoroethylene (PTFE) and polyvinyl alcohol (PVA) with and without graphene oxide (GO) nanoparticles was investigated. These two scaffolds were fabricated through chemical crosslinking with small amounts of boric acids and a controlled freeze-drying method. The scaffolds exhibited randomly oriented nanofibers of 2 and 650 nm and compressive moduli of 620 and 130 kPa, respectively. Human ADSCs seeded on stiffer PTFE/PVA/GO scaffolds revealed a significant elevation in ALP activity, calcium deposition, and osteogenic related genes expression as compared to the softer scaffold without graphene oxide [192].

Cylindrical PV alcohol (PVA)/HA hydrogel prepared with a liquid nitrogen–contacting gradual freezing–thawing method to produce hydrogel with a wide range stiffness gradient (between ~20 kPa and ~200 kPa). Human BMMSCs cultured on PVA/HA hydrogel favored certain stiffness ranges to get differentiated into specific cell lineages: ~20 kPa for nerve cell, ~40 kPa for muscle cell, ~80 kPa for chondrocyte, and ~190 kPa for osteoblast [193]. Moreover, a minimal hydrogel matrix stiffness of 4.47 kPa was recognized to activate transcriptional co-activator TAZ and induce MSCs' osteogenic differentiation [194].

4.2.4. Polyesters

Polyesters are popular polymers that contain ester groups in the polymer backbones, enabling them to produce biomedical hydrogels that can undergo biodegradation [181]. Poly(ether-ester-urethane) (PEEU) containing poly (ρ-dioxanone) (PPDO) and PCL segments can be electrospun into fiber meshes. PEEU fiber meshes were tailored by varying the PPDO:PCL weight ratio, thus affecting their stiffness. Human ADSCs cultured on the stiffer fiber meshes (e.g., PEEU70) significantly demonstrated enhanced osteogenic differentiation with higher levels of osteocalcin expression and ALP activity. Moreover, higher levels of HA were detected on the stiffer fiber meshes [195].

Hydrophilic degradable porous 3D nanocomposite scaffolds composed of PCL, Poly (2-hydroxyethylmethacrylate) (PHEMA), and Apacite (apatite–calcite) nanostructures (15 and 25 wt.%) with mechanical values (E ~ 7.109 MPa and σ ~ 0.414 MPa) provided a balanced microenvironment that resulted in osteogenic induction of human BMMSCs. Von Kossa staining, calcium content, and ALP results confirmed the highest bone cells' differentiation on PCLPHEMA/25% Apacite nanocomposites [196].

4.2.5. Polyacrylamide

Polyacrylamide formed from only acrylamide subunits is nonionic. Copolymerizing it with other monomers such as 2-acrylamido-2-methylpropane sulfonate or acrylate forms anionic polyacrylamide, while cationic polyacrylamide could be synthesized upon copolymerization with dimethyl diallyl ammonium. Polyacrylamide substrate is bio-inert; thus, its surface must be conjugating with adhesive ECM proteins to allow for cell attachment [197,198]. Polyacrylamide is widely utilized in literature as a model for investigating the mechanoregulatory role of substrate stiffness combined or uncombined with other parameters in osteogenic differentiation. The stiffness of polyacrylamide hydrogels is commonly modified by altering the concentration of acrylamide monomer or bis-acrylamide crosslinker [62].

Upon seeding human MSCs on 250-Pa polyacrylamide gels coated with a mixture of collagen type 1 and fibronectin, the progression of the cells throughout the cell cycle was prohibited despite the presence of serum. Conversely, the quiescent cells reentered the cell cycle when presented on a stiff polyacrylamide substrate (7.5 kPa). Moreover, the non-proliferative cells revealed an adipogenic differentiation potential upon culturing on 250-Pa gels in adipogenic media or an osteogenic potential into osteoblasts if transferred to a stiff substrate in the presence of osteogenic media [199]. Micropatterned polyacrylamide gels were fabricated with varying stiffness (10 to 40 kPa), using PDMS stamps coated with fibronectin. MSCs cultured on protein-coated gels revealed a stiffness-dependent osteogenic markers' expression (Runx2 and osteopontin) with a maximum expression at 30 kPa [200]. Osteogenic differentiation as revealed by Runx2 expression was upregulated significantly only on collagen I-coated gels with high stiffness (80 kPa), while myogenic differentiation, as ascertained by MyoD1 expression, occurred on all gel–protein coated matrices that had a stiffness of 9 kPa. Peak MyoD1 expression was demonstrated on gels with a modulus of 25 kPa coated with fibronectin. Polyacrylamide hydrogels prepared with variable stiffnesses, ranging from 13 to 68 kPa, through varying the concentrations of bis-acrylamide (0.1%, 0.5%, and 0.7%), showed a difference in the gel morphology. Under scanning electron microscopy, gels with low stiffness (13–16 kPa) appeared flat and non-porous. On the other hand, higher stiffness matrices (48–53 kPa and 62–68 kPa) showed multiple small porosities. Such inherent porosities of polyacrylamide hydrogels could enhance the flow of culture media and better mimic the natural cellular environment, as compared to plastic and glass substrates. Moreover, BMMSCs cultured on 62–68 kPa fibronectin-coated polyacrylamide hydrogels demonstrated a polygonal morphology and revealed an osteogenic phenotype with significantly high levels of ALP, Runx2, and osteopontin [51].

The modulatory effect of extracellular matrix type and density on the mechanotransduction of stem/progenitor cells and the correlated integrin involved in signals translocation were assessed through conjugating each of the four major cell adhesion ECM proteins (fibronectin, collagen I, collagen IV, and laminin) on polyacrylamide hydrogels with tunable stiffness (soft, 3 kPa; and stiff, 38 kPa). The results revealed that increasing ECM ligand density alone can induce YAP nuclear translocation without changing substrate stiffness with a different optimized ligand density. Using antibody-blocking techniques for $\alpha v\beta 3$-, $\alpha 5$-, and $\alpha 2\beta 1$-integrins revealed the involvement of $\alpha v\beta 3$-, $\alpha 5$-, and $\alpha 2\beta 1$-integrins with fibronectin, while $\alpha 5$-integrin was further associated with collagen type I and IV. On the contrary, laminin was associated with $\alpha 5$- and $\alpha 2\beta 1$-integrins. Moreover, altering ECM type resulted in modulation of human MSC osteogenesis confirmed by quantitative real-time (qRT)-PCR for Runx2 and ALP without changing substrate stiffness [201].

The mechanotransduction role of FAK, α5/β1 integrin and Wnt-signaling pathways mediated by stiff matrices, in regulating osteogenic differentiation of human MSCs cultured on 62–68 kPa fibronectin-coated polyacrylamide hydrogels were further investigated. Throughout osteogenesis, gene and protein expressions of integrin α5/β1 were enhanced, together with the expression of signaling molecules FAK, p-ERK, p-Akt, GSK-3β, p-GSK-3β, and β-catenin. Antibody blocking of integrin α5 significantly downregulated the stiffness-induced expression of osteogenic markers (Runx2, alpha-1 type I collagen, and BGLAP) with associated downregulated expression of ERK, p-ERK, FAK, and β-catenin protein. Reciprocally, GSK-3β, p-GSK-3β, Akt, and p-Akt expressions were upregulated. The presence of the Akt inhibitor Triciribine reduced the expression of p-Akt and p-GSK-3β, whereas Akt, GSK-3β, and β-catenin were unchanged. These results emphasized the role of p-Akt in regulating the expression of p-GSK-3β on 62–68 kPa ECM during osteogenesis [78].

MSCs cultivated on polyacrylamide hydrogels with elasticity (7.0 ± 1.2 and 42.1 ± 3.2 kPa) and coated with type I collagen in osteogenic medium revealed enhanced osteogenic differentiation potential on stiff substrates, with an upregulated expression of Runx2, type I collagen, and osteocalcin genes. On stiff matrices, Western blot analysis revealed an increase in mechanotransducers involved in osteogenic differentiation ROCK, FAK, and ERK1/2, whereas their inhibition resulted in decreased osteogenic markers' expression. Furthermore, α2-integrin was upregulated on stiff matrices during osteogenesis, and its knockdown by siRNA hindered the osteogenic phenotype through FAK, ROCK, and ERK1/2. Therefore, it could be concluded that α2-integrin is involved in osteogenesis mediated by matrix stiffness [62].

Additionally, upon culturing human MSCs on poly acrylamide-co-acrylic acid hydrogels grafted with RGDs, myogenic differentiation occurred at 13–17 kPa, while osteogenic differentiation was revealed at 45–49 kPa stiffness confirmed further with positive protein immunostaining of MyoD, as well as Osx, osteocalcin, and Runx2. Stiffer matrices grafted with BMP-2 mimetic peptide (E = 47.5 kPa) also induced osteoblast lineage commitment, having a similar effect as the ones grafted with RGDs. On the contrary, the osteogenic effect of BMP-2 mimetic peptides on MSCs was inhibited on very soft microenvironments (0.76–3.21 kPa) due to F-actin cytoskeleton reorganization that inhibited BMP-induced smad1/5/8 phosphorylation and subsequent differentiation of the cells into osteoblast-like cells [118].

Umbilical cord (UC) MSCs attained similar behavior upon being cultivated on different stiffness (13–16, 35–38, 48–53, and 62–68 kPa) polyacrylamide gels coated with fibronectin. Quantitative RT-PCR results showed that soft matrices promoted adipogenic differentiation, as evident by upregulated expressions of adipocytic markers (PPARγ and C/EBPα). On the contrary, stiff matrices (48–53 kPa) enhanced the tendency of the cells to differentiate into muscles, as demonstrated by enhanced expression of desmin and MOYG. High stiffness substrates (62–68 kPa) significantly promoted the expression of osteogenic markers, such as Runx2, collagen type I, ALP, and osteocalcin [50].

Moreover, the effect of mechanical loading and biomaterial stiffness on MSCs differentiation was investigated upon cultivating MSCs in osteogenic and adipogenic media on soft (23 ± 0.3 kPa) and stiff (111 ± 2 kPa) polyacrylamide as compared to PDMS (1.5 ± 0.07 MPa) either strained with 8% cyclic strain at 1 Hz or unstrained. Without strain, the expression of ALP was markedly higher on PDMS than on both polyacrylamide types. With 8% cyclic strain, ALP expression was upregulated in all groups, with the highest expression in soft polyacrylamide. Moreover, adipogenesis was the highest on the unstrained soft polyacrylamide, while it was significantly decreased on soft and stiff polyacrylamide when strained [202].

4.2.6. Self-Assembling Peptides

Human MSCs were encapsulated within a 3D culture and grown on top of 2D culture biomimetic self-assembling peptide (SAP) hydrogel containing 1 mg/mL RGDs-functionalized peptide (KFE–RGD) at the shear moduli of 0.25, 1.25, 5, and 10 kPa. Changes in adipogenic and osteogenic gene expression were relatively modest with no visual signs of differentiation as mineral deposition. The cells maintained a fibroblast-like phenotype throughout the culture period. However, on introducing 1:1 mixed adipogenic/osteogenic induction medium, the stiff matrices of 10 kPa induced the most efficient osteogenesis, with alizarin red-stained calcium deposits [203].

4.2.7. Other Polymers

Indirectly 3D-printed "stiffness memory" poly(urea-urethane) (PUU)/POSS elastomeric nano-hybrid scaffolds with thermo-responsive mechanical properties that soften at body temperature by inverse self-assembling have been developed. The initial stiffness and subsequent stiffness relaxation (>10 kPa) of the scaffolds directed the proliferation and differentiation of human BMMSCs towards the osteogenic lineages on stiffer scaffolds over 4 weeks, as measured by immunohistochemistry, histology, ELISA, and qPCR, while soft substrates (<1 kPa) promoted MSCs' chondrogenic differentiation [204]. Table 1 lists the key studies investigating the effect of polymeric matrix stiffness on osteogenic differentiation of mesenchymal stem/progenitor cells.

Table 1. Key studies on the effect of polymeric matrix stiffness on osteogenic differentiation of mesenchymal stem/progenitor cells.

Study	Cell Source	Polymer	Modification	Modulus of Elasticity	Results
			Alginate		
Zhang et al., 2020 [141]	hMSCs	Alginate–gelatin scaffold	3D bioprinted porous scaffolds different alginate concentration (0.8%alg and 1.8%alg) and different initial cell seeding density (1.67, 5, and 15 M cells/mL)	Soft scaffold 0.66 ± 0.08 kPa Stiff scaffold 5.4 ± 1.2 kPa	UpregulatedALP-activity-related, 3D-bone-like-tissue-related, osteoblast-related, and early osteocyte-related gene expression
Freeman and Kelly, 2017 [142]	MSCs	Alginate hydrogel	3D bioprinting matrix with varying alginate molecular weight and cross linker ratio		Osteogenic differentiation with increased ALP staining
Maia et al., 2014 [143]	hMSCs	Alginate hydrogel	3D matrix with bimodal molecular weight distribution at different polymer concentrations (1 and 2 wt.%) and RGD densities (0, 100 or 200 µM)	2 wt.% hydrogels (tan ∂ = 0.2), 1 wt.% hydrogels (tan ∂ = 0.4–0.6).	1 wt.% alginate hydrogel matrices upregulated hMSCs osteogenic differentiation and expressed high levels of ALP and OCN
			Collagen		
Xie et al., 2017 [146]	hMSCs	Collagen gel	Varying polymerization temperature 4, 21, and 37 °C.	Fiber stiffness: 1.1 to 9.3 kPa Bulk stiffness: 16.4 to 151.5 Pa	Collagen gel polymerized at 37 °C resulted in 34.1% ALP positive staining
Banks et al., 2014 [147]	ADSCs	Collagen–glycosaminoglycan (CG)	Chemical Crosslinking with EDAC and NHS Covalent immobilization of PDGF-BB and BMP-2 by benzophenone photolithography	2.85 to 5 MPa	Upregulated expression of collagen 1, ALP, and OCN with increased stiffness
Hwang et al., 2019 [148]	hMSCs	Three bilayers of collagen/alginate nano film		24 and 53 MPa	Increase in alkaline phosphatase activity

Table 1. *Cont.*

Study	Cell Source	Polymer	Modification	Modulus of Elasticity	Results
Zhou et al., 2021 [149]	hMSCs	Nano-particulate mineralized collagen glycosaminoglycan	Chemical crosslinking with EDAC and NHS	3.90 −/+ 0.36 kPa	Increase in expression of ALP, collagen 1, and Runx2
Tsimbouri et al., 2017 [150]	MSCs	Collagen gel	3D collagen gel culture on the vibrational bioreactor	~108 Pa	Increased expression of Runx2, collagen I, ALP, OPN, OCN, and BMP2.
Murphy et al., 2012 [151]	MSCs	Collagen/ gly-cosaminoglycan	DHT and EDAC crosslinking	0.5, 1, and 1.5 kPa	Osteogenic differentiation with Runx2 expression
Chen et al., 2015 [152]	Rat MSCs	3D scaffold collagen and hydroxyapatite	Coated on decellularized cancellous bone	13.00 ± 5.55 kPa, 13.87 ± 1.51 kPa, and 37.7 ± 19.6 kPa	Highest scaffold stiffness promoted higher expressions of OPN and OC
Chen et al., 2017 [205]	Rat MSCs	Collagen and hydroxyapatite, coated on decellularized cancellous bone	3D oscillatory perfusion bioreactor system	6.74 ± 1.16 kPa-8.82 ± 2.12 kPa-23.61 ± 8.06 kPa	Osteogenic differentiation of MSCs
Gelatin					
Wan et al., 2019 [133]	PDLSCs	Gelatin	Crosslinked with variable concentrations of methacryloyl	GelMA concentrations of 10, 12, and 14 wt% stiffness 25.75 ± 1.21, 59.71 ± 8.87, and 117.82 ± 9.83 kPa, respectively	Increasing matrix stiffness increased osteogenic differentiation of PDLSCs, with upregulated expression of OCN and Runx2
He et al., 2018 [134]	BMMSCs	Gelatin 3%, 6%, and 9%.	Crosslinked with transglutaminase	9% gelatin gave rise to the highest stiffness (60.54 ± 10.45 kPa), while 3% gelatin resulted in the lowest stiffness (1.58 ± 0.42 kPa)	BMMSCs encapsulated in hydrogel with highest stiffness demonstrated the highest osteogenic differentiation
Van Nieuwenhove et al., 2017 [162]	ADSCs	Gelatin with variable degrees of methacrylation (GelMA 31%, GelMA 72%, and GelMA 95%)	Covalently bound to variable ratios of pentenoates modified starch (10 v% starch and 20 v% starch)		Increase in matrix stiffness promoted osteogenic differentiation of ADSCs
Jiang et al., 2015 [163]	BMMSCs	GelMA encapsulating alendronate	Crosslinked by PEG diacrylate	stiffness increased from 4 to 40 kPa	Increased osteogenic differentiation of BMMSCs on stiffer hydrogel with higher alendronate concentration with upregulated ALP, collagen I, OCN, and calcium deposition

Table 1. *Cont.*

Study	Cell Source	Polymer	Modification	Modulus of Elasticity	Results
Sun et al., 2014 [164]	BMMSCs	Three-dimensional porous gelatin scaffolds	Crosslinked using EDC	Crosslinked scaffold demonstrated an increase in the elastic modulus from w 0.6 to ≈ 2.5 kP without any change in the scaffold internal structure	Increased stiffness increased osteogenic differentiation evidenced by increased Runx2 and OCN in vitro and increased bone formation in vivo
Decellularized matrix and Demineralized Bone					
Ventre et al., 2019 [165]	Murine MSCs	Decellularized MC3T3-E1-cell-derived matrix on replica from PDMS	Genipin crosslinking	Young's modulus increased from (0.01–0.1 kPa) to (0.1–1.5 kPa).	MSCs on stiff dCDMs, revealed significant adipogenic and osteogenic differentiation potentials
Hu et al., 2018 [166]	BMMSC	Demineralized bone matrices	Controlling the decalcification duration (1 h, 12 h, and 5 d, respectively)	High: 66.06 ± 27.83 MPa, Medium: 26.90 ± 13.16 MPa Low: 0.67 ± 0.14 MPa	Low stiffness scaffolds promoted osteogenesis in vitro. Subcutaneous implantation in a rat model and in a femoral condylar defect rabbit model revealed positive OCN and OPN expression
Hyaluronic acid (HA)					
Zhao et al., 2014 [174]	hBMMSCs	Thiol functionalized hyaluronic acid (HA) and thiol functionalized recombinant human gelatin	Crosslinked by poly (ethylene glycol) tetra-acrylate	0.15, 1.5, and 4 kPa	Change in cell morphologies with different stiffness. Cells cultured on the 4 kPa hydrogel revealed an enhanced expression of late osteogenic genes
Cosgrove et al., 2016 [175]	Juvenile bovine MSCs	Methacrylated HA hydrogel	Ligation of the HAVDI adhesive peptide sequence from N-cadherin domain 1 and RGD from fibronectin	5, 10, and 15 kPa	Lack of myosin IIA incorporated into focal adhesions hindered their maturation with increasing substrate stiffness and decreased osteogenesis
Dorcemus et al., 2017 [176]	hMSCs-bone-marrow-derived	Thiol-modified hyaluronan gel	Crosslinked by PEG at ratios ranging from 1:1 to 7:1	Storage moduli from 10 to 45 Pa	Differences between the top (cartilage-forming) and bottom (bone-forming) regions of the scaffold proved its capability for osteochondral engineering

Table 1. *Cont.*

Study	Cell Source	Polymer	Modification	Modulus of Elasticity	Results
Hao et al., 2018 [177]	hMSCs-bone-marrow-derived	HA carrying sulfhydryl groups and a hydrophilic polymer bearing both acrylate and tetrazine groups	Matrix metalloprotease -degradable peptidic crosslinker and adding HA conjugated with multiple copies of trans-cyclooctene (TCO)	(G′) = 180 ± 42 Pa increased to G′ = 520 ± 80 Pa	The 3D matrix tagged with a TCO- motif promoted the cells to undergo change from a rounded to spindle phenotype
			Fibrin		
Hashemzadeh et al., 2019 [180]	hADSCs	Fibrin hydrogels embedding gold nanowires	Altering fibrinogen and thrombin concentration and incorporation of gold nanowires		With high fibrinogen and thrombin concentration, gold nanowires, promoted osteogenic differentiation
			Polyethylene glycol (PEG)		
Pek et al., 2010 [182]	MSCs	Thixotropic polyethylene glycol–silica (PEG–silica) nano composite gel	3D cell culture Cell-adhesion peptide RGD (Arg–Gly–Asp) sequence immobilization	≥75 Pa	Higher expression of the osteogenic transcription factor
Ye et al., 2015 [183]	Rat BMMSCs	PEG	PEG hydrogels with RGD nano-spacings of 49 and 135 nm and incubated in mixed osteogenic and adipogenic medium	Soft hydrogels (130 kPa) and stiff hydrogels (3170 kPa)	Stiff hydrogels promoted osteogenesis. Large RGD nano-spacing promoted osteogenesis
Steinmetz et al., 2015 [184]	hMSCs	Multilayer PEG-based hydrogel	Simple sequential photopolymerization- high RGD concentrations- dynamic mechanical stimulation	345 kPa	Collagen I generation with mineral deposits were evident
Yang et al., 2020 [185]	Rat BMMSCs	PEG/silk fibroin/HA scaffold	Varying HA concentration	80.98 to 190.51 kPa	Expression of all the osteogenesis-related markers in vitro and superior calvarial defect repair in vivo
Yang et al., 2016 [186]	hMSCs	PEG hydrogel	Regularly and randomly patterned photodegradable hydrogel	~10–12 kPa	Osteogenic differentiation of MSCs cultured on random patterns
Gandavarapu et al., 2014 [187]	hMSCs	PEG hydrogels	functionalized with c(RRETAWA) hydrogels through α5 integrins	~25 kPa	Osteogenic differentiation of hMSCs
			Polydimethylsiloxane (PDMS)		
Xie et al., 2018 [38]	ASCs	PDMS		1.014 ± 0.15 MPa	Osteogenic differentiation by ALP stain and upregulation of Runx2 and Osx transcriptional factors
Viale-Bouroncle et al., 2014 [189]	DFCs	PDMS	Coating PDMS with fibronectin and cultured in osteogenic differentiation medium	11 kPa	High ALP activity and accumulation of calcium on the soft substrate

Table 1. *Cont.*

Study	Cell Source	Polymer	Modification	Modulus of Elasticity	Results
Viale-Bouroncle et al., 2012 [190]	SHED	PDMS	Adding osteogenic differentiation medium	93 kPa	High osteogenic differentiation
Wang et al., 2012 [191]	Rat MSCs	PDMS	Osteogenic medium with temperature gradient curing	0.19 to 3.10 MPa	Calcein Blue–positive bone-nodule-like colonies
Vinyl polymers					
Khoramgah et al., 2020 [192]	hADSCs	Poly tetra fluoro ethylene (PTFE) and PVA with and without graphene oxide nanoparticles	3D porous scaffolds-chemical crosslinking with small amounts of boric acids–controlled freeze-drying method	620 and 130 kPa	Elevation in ALP activity, calcium deposition, and osteogenic-related genes expression
Oh et al., 2016 [193]	hBMMSCs	Cylindrical PVA/HA hydrogel	Liquid nitrogen—contacting gradual freezing–thawing method	~20 kPa and ~200 kPa	Stiffness of ~190 kPa led to osteoblast differentiation
Polyesters					
Sun et al., 2019 [195]	hADSCs	Poly(ether-ester-urethane) (PEEU) containing PPDO and PCL segments	Electrospun into fiber meshes with varying PPDO to PCL weight ratios	2.6 ± 0.8 MPa (PEEU40), 3.2 ± 0.9 MPa (PEEU50), 4.0 ± 0.9 MPa (PEEU60) 4.5 ± 0.8 MPa (PEEU70)	Enhanced osteogenic differentiation of hADSCs with higher levels of OCN, ALP, and hydroxyapatite detected on the stiffer fiber meshes
Self-assembling peptides					
Hogrebe and Gooch, 2016 [203]	hMSCs	Biomimetic self-assembling peptide hydrogel containing 1 mg/mL RGD-functionalized peptide (KFE–RGD)	hMSCs were encapsulated within 3D culture and grown on top of 2D culture Adding 1:1 mixed adipogenic/osteogenic induction medium	(G′) 10 kPa	Osteogenesis induction and alizarin red-stained calcium deposits
Other Polymers					
Olivares-Navarrete et al., 2017 [76]	MSCs	Methyl acrylate/methyl methacrylate polymer	Altering monomer concentration.	0.1 MPa to 310 MPa	Chondrogenic and osteogenic differentiation when grown on substrates with less than 10 MPa stiffness
Wu et al., 2018 [204]	hBMMSCs	Poly(urea-urethane) (PUU)/POSS elastomeric nano-hybrid scaffolds	Thermoresponsive scaffolds indirectly 3D printed by inverse self-assembling	>10 kPa	Osteogenic differentiation

5. Conclusions

The world of biomaterials, specifically polymers, lingers to impact the biomedicine field. Various materials are currently under investigation to produce ECM with varying stiffness for tissue engineering. Matrix stiffness regulates the MSCs' differentiation into mature specific cells by activating transcription factors that upregulate the genes responsible for the initiation and progression of particular cell linage differentiation. Rigid matrices lead to increased MSCs spreading and improved actomyosin contractility, promoting osteogenic differentiation. This enhanced potential is accompanied by increased Runx2, β-catenin, and Smad, implying the significant impact of mechanosensing of the matrix stiffness and its role in determining cell fate. The comprehensive signaling mechanisms by which micro-environmental stiffness controls the lineage specification of MSCs are still unknown. Additional research is needed to understand the variety of potential signaling forces involved in MSCs' osteogenic differentiation that can lead to the development of new therapeutic modalities addressing bone disorders. Moreover, we believe that further research on the implicated mechanical and physical factors, such as topographic changes and external mechanical forces affecting cell properties, including cell shape and colony size, can offer a broader understanding of cell-fate determination.

Author Contributions: All authors contributed equally. All authors have read and agreed to the published version of the manuscript.

Funding: This publication was funded by the Christian Albrechts University of Kiel, Germany.

Institutional Review Board Statement: Not applicable.

Informed Consent Statement: Not applicable.

Data Availability Statement: Not applicable.

Conflicts of Interest: The authors declare no conflict of interest.

Abbreviations

3D	Three-dimensional
ADSCs	Adipose derived stromal cells
ALP	Alkaline phosphatase
ASCs	adipose derived stromal cells
BMMSCs	Bone marrow mesenchymal stem cells
BMP-2	Bone morphogenetic protein 2
dCDMs	Decellularized cell derived matrix
DFCs	Dental follicle stem cells
ECM	Extracellular matrix
EDAC	1-ethyl-3-(3-dimethylami-nopropyl) carbodiimide
GC-TRS	Glycol chitin-based thermo-responsive hydrogel scaffold
GelMA	Gelatin with variable degrees of methacrylation
HA	Hyaluronic acid
HA	Hydroxyapatite
hADSCs	Human adipose derived mesenchymal stem cells
hBMMSCs	Human Bone marrow mesenchymal stem cells
hMSCs	Human mesenchymal stem cells
MSCs	Mesenchymal stem cells
NHS	N-hydroxysuccinimide
OCN	Osteocalcin
ON	Osteonectin
OPN	osteopontin
PCL	poly (ε-caprolactone)
PDLSCs	periodontal ligament stem cells
PDMS	Polydimethylsiloxane
PEEU	Poly (ether-ester-urethane)
PEG	Polyethylene glycol
PEGDA	Polyethylene glycol diacrylate
PEGMC	Polyethylene glycol-maleate-citrate
PPDOP	poly (ρ-dioxanone)

PV	polyvinyl
PVA	polyvinyl alcohol
qRT-PCR	Quantitative Reverse transcription real-time polymerase chain reaction
RGD	Arginine-glycine-aspartic acid
Runx2	Runt-related transcription factor-2
SCAP	Stem cells of the apical papilla
SDF-1α	Stromal derived factor-1alpha
SHED	Stem cells isolated from human exfoliated deciduous teeth

References

1. Singh, V.K.; Saini, A.; Kalsan, M.; Kumar, N.; Chandra, R. Describing the Stem Cell Potency: The Various Methods of Functional Assessment and In silico Diagnostics. *Front. Cell Dev. Biol.* **2016**, *4*, 134. [CrossRef]
2. El-Sayed, K.M.; Paris, S.; Graetz, C.; Kassem, N.; Mekhemar, M.; Ungefroren, H.; Fandrich, F.; Dorfer, C. Isolation and characterisation of human gingival margin-derived STRO-1/MACS(+) and MACS(-) cell populations. *Int. J. Oral Sci.* **2015**, *7*, 80–88. [CrossRef]
3. Fawzy El-Sayed, K.M.; Ahmed, G.M.; Abouauf, E.A.; Schwendicke, F. Stem/progenitor cell-mediated pulpal tissue regeneration: A systematic review and meta-analysis. *Int. Endod. J.* **2019**, *52*, 1573–1585. [CrossRef] [PubMed]
4. Fawzy El-Sayed, K.M.; Dorfer, C.; Fandrich, F.; Gieseler, F.; Moustafa, M.H.; Ungefroren, H. Adult mesenchymal stem cells explored in the dental field. *Adv. Biochem. Eng. Biotechnol.* **2013**, *130*, 89–103. [CrossRef] [PubMed]
5. Fawzy El-Sayed, K.M.; Dorfer, C.E. Gingival Mesenchymal Stem/Progenitor Cells: A Unique Tissue Engineering Gem. *Stem. Cells Int.* **2016**, *2016*, 7154327. [CrossRef]
6. Fawzy El-Sayed, K.M.; Elahmady, M.; Adawi, Z.; Aboushadi, N.; Elnaggar, A.; Eid, M.; Hamdy, N.; Sanaa, D.; Dorfer, C.E. The periodontal stem/progenitor cell inflammatory-regenerative cross talk: A new perspective. *J. Periodontal Res.* **2019**, *54*, 81–94. [CrossRef] [PubMed]
7. Fawzy El-Sayed, K.M.; Elsalawy, R.; Ibrahim, N.; Gadalla, M.; Albargasy, H.; Zahra, N.; Mokhtar, S.; El Nahhas, N.; El Kaliouby, Y.; Dorfer, C.E. The Dental Pulp Stem/Progenitor Cells-Mediated Inflammatory-Regenerative Axis. *Tissue Eng. Part B Rev.* **2019**, *25*, 445–460. [CrossRef]
8. Fawzy El-Sayed, K.M.; Jakusz, K.; Jochens, A.; Dorfer, C.; Schwendicke, F. Stem Cell Transplantation for Pulpal Regeneration: A Systematic Review. *Tissue Eng. Part B Rev.* **2015**, *21*, 451–460. [CrossRef]
9. Fawzy El-Sayed, K.M.; Mekhemar, M.K.; Beck-Broichsitter, B.E.; Bahr, T.; Hegab, M.; Receveur, J.; Heneweer, C.; Becker, S.T.; Wiltfang, J.; Dorfer, C.E. Periodontal regeneration employing gingival margin-derived stem/progenitor cells in conjunction with IL-1ra-hydrogel synthetic extracellular matrix. *J. Clin. Periodontol.* **2015**, *42*, 448–457. [CrossRef] [PubMed]
10. Fawzy El-Sayed, K.M.; Paris, S.; Becker, S.T.; Neuschl, M.; De Buhr, W.; Sälzer, S.; Wulff, A.; Elrefai, M.; Darhous, M.S.; El-Masry, M.; et al. Periodontal regeneration employing gingival margin-derived stem/progenitor cells: An animal study. *J. Clin. Periodontol.* **2012**, *39*, 861–870. [CrossRef]
11. Wang, X.; Wang, Y.; Gou, W.; Lu, Q. Role of mesenchymal stem cells in bone regeneration and fracture repair: A review. *Int. Orthop.* **2013**, *37*, 2491–2498. [CrossRef]
12. Zhou, L.L.; Liu, W.; Wu, Y.M.; Sun, W.L.; Dorfer, C.E.; Fawzy El-Sayed, K.M. Oral Mesenchymal Stem/Progenitor Cells: The Immunomodulatory Masters. *Stem. Cells Int.* **2020**, *2020*, 1327405. [CrossRef]
13. Fu, X.; Liu, G.; Halim, A.; Ju, Y.; Luo, Q.; Song, G. Mesenchymal Stem Cell Migration and Tissue Repair. *Cells* **2019**, *8*, 784. [CrossRef] [PubMed]
14. Patel, D.M.; Shah, J.; Srivastava, A.S. Therapeutic potential of mesenchymal stem cells in regenerative medicine. *Stem. Cells Int.* **2013**, *2013*, 496218. [CrossRef]
15. Via, A.G.; Frizziero, A.; Oliva, F. Biological properties of mesenchymal Stem Cells from different sources. *Muscles Ligaments Tendons J.* **2012**, *2*, 154–162.
16. Lee, O.K.; Kuo, T.K.; Chen, W.-M.; Lee, K.-D.; Hsieh, S.-L.; Chen, T.-H. Isolation of multipotent mesenchymal stem cells from umbilical cord blood. *Blood* **2004**, *103*, 1669–1675. [CrossRef]
17. Ding, D.C.; Shyu, W.C.; Lin, S.Z. Mesenchymal stem cells. *Cell Transplant.* **2011**, *20*, 5–14. [CrossRef] [PubMed]
18. Rodríguez-Lozano, F.J.; Bueno, C.; Insausti, C.L.; Meseguer, L.; Ramírez, M.C.; Blanquer, M.; Marín, N.; Martínez, S.; Moraleda, J.M. Mesenchymal stem cells derived from dental tissues. *Int. Endod. J.* **2011**, *44*, 800–806. [CrossRef]
19. Jorgenson, K.D.; Hart, D.A.; Krawetz, R.; Sen, A. Production of Adult Human Synovial Fluid-Derived Mesenchymal Stem Cells in Stirred-Suspension Culture. *Stem. Cells Int.* **2018**, *2018*, 8431053. [CrossRef]
20. Rutkovskiy, A.; Stensløkken, K.-O.; Vaage, I.J. Osteoblast Differentiation at a Glance. *Med. Sci. Monit. Basic Res.* **2016**, *22*, 95–106. [CrossRef] [PubMed]
21. Assis-Ribas, T.; Forni, M.F.; Winnischofer, S.M.B.; Sogayar, M.C.; Trombetta-Lima, M. Extracellular matrix dynamics during mesenchymal stem cells differentiation. *Dev. Biol.* **2018**, *437*, 63–74. [CrossRef] [PubMed]
22. Jhala, D.; Vasita, R. A Review on Extracellular Matrix Mimicking Strategies for an Artificial Stem Cell Niche. *Polym. Rev.* **2015**, *55*, 561–595. [CrossRef]
23. Even-Ram, S.; Artym, V.; Yamada, K.M. Matrix control of stem cell fate. *Cell* **2006**, *126*, 645–647. [CrossRef]

24. Watt, F.M.; Huck, W.T.S. Role of the extracellular matrix in regulating stem cell fate. *Nat. Rev. Mol. Cell Biol.* **2013**, *14*, 467–473. [CrossRef]
25. Gaur, T.; Lengner, C.J.; Hovhannisyan, H.; Bhat, R.A.; Bodine, P.V.; Komm, B.S.; Javed, A.; Van Wijnen, A.J.; Stein, J.L.; Stein, G.S. Canonical WNT signaling promotes osteogenesis by directly stimulating Runx2 gene expression. *J. Biol. Chem.* **2005**, *280*, 33132–33140. [CrossRef]
26. Almalki, S.G.; Agrawal, D.K. Key transcription factors in the differentiation of mesenchymal stem cells. *Differentiation* **2016**, *92*, 41–51. [CrossRef]
27. Long, F. Building strong bones: Molecular regulation of the osteoblast lineage. *Nat. Rev. Mol. Cell Biol.* **2012**, *13*, 27–38. [CrossRef]
28. Komori, T. Regulation of osteoblast differentiation by transcription factors. *J. Cell Biochem.* **2006**, *99*, 1233–1239. [CrossRef] [PubMed]
29. Zhao, Z.; Zhao, M.; Xiao, G.; Franceschi, R.T. Gene transfer of the Runx2 transcription factor enhances osteogenic activity of bone marrow stromal cells in vitro and in vivo. *Mol. Ther.* **2005**, *12*, 247–253. [CrossRef]
30. Fujii, M.; Takeda, K.; Imamura, T.; Aoki, H.; Sampath, T.K.; Enomoto, S.; Kawabata, M.; Kato, M.; Ichijo, H.; Miyazono, K. Roles of bone morphogenetic protein type I receptors and Smad proteins in osteoblast and chondroblast differentiation. *Mol. Biol. Cell* **1999**, *10*, 3801–3813. [CrossRef]
31. Xu, D.-J.; Zhao, Y.-Z.; Wang, J.; He, J.-W.; Weng, Y.-G.; Luo, J.-Y. Smads, p38 and ERK1/2 are involved in BMP9-induced osteogenic differentiation of C3H10T1/2 mesenchymal stem cells. *BMB Rep.* **2012**, *45*, 247–252. [CrossRef]
32. Javed, A.; Bae, J.-S.; Afzal, F.; Gutierrez, S.; Pratap, J.; Zaidi, S.K.; Lou, Y.; Van Wijnen, A.J.; Stein, J.L.; Stein, G.S. Structural coupling of Smad and Runx2 for execution of the BMP2 osteogenic signal. *J. Biol. Chem.* **2008**, *283*, 8412–8422. [CrossRef]
33. Nakashima, K.; Zhou, X.; Kunkel, G.; Zhang, Z.; Deng, J.M.; Behringer, R.R.; De Crombrugghe, B. The novel zinc finger-containing transcription factor osterix is required for osteoblast differentiation and bone formation. *Cell* **2002**, *108*, 17–29. [CrossRef]
34. Kang, S.; Bennett, C.N.; Gerin, I.; Rapp, L.A.; Hankenson, K.D.; MacDougald, O.A. Wnt signaling stimulates osteoblastogenesis of mesenchymal precursors by suppressing CCAAT/enhancer-binding protein α and peroxisome proliferator-activated receptor γ. *J. Biol. Chem.* **2007**, *282*, 14515–14524. [CrossRef]
35. Lecarpentier, Y.; Claes, V.; Duthoit, G.; Hébert, J.-L. Circadian rhythms, Wnt/beta-catenin pathway and PPAR alpha/gamma profiles in diseases with primary or secondary cardiac dysfunction. *Front. Physiol.* **2014**, *5*, 429. [CrossRef]
36. Day, T.F.; Guo, X.; Garrett-Beal, L.; Yang, Y. Wnt/β-catenin signaling in mesenchymal progenitors controls osteoblast and chondrocyte differentiation during vertebrate skeletogenesis. *Dev. Cell* **2005**, *8*, 739–750. [CrossRef] [PubMed]
37. Huelsken, J.; Behrens, J. The Wnt signalling pathway. *J. Cell Sci.* **2002**, *115*, 3977–3978. [CrossRef]
38. Xie, J.; Zhang, D.; Zhou, C.; Yuan, Q.; Ye, L.; Zhou, X. Substrate elasticity regulates adipose-derived stromal cell differentiation towards osteogenesis and adipogenesis through β-catenin transduction. *Acta Biomater.* **2018**, *79*, 83–95. [CrossRef] [PubMed]
39. Kshitiz; Afzal, J.; Chang, H.; Goyal, R.; Levchenko, A. Mechanics of Microenvironment as Instructive Cues Guiding Stem Cell Behavior. *Curr. Stem. Cell Rep.* **2016**, *2*, 62–72. [CrossRef]
40. Kshitiz; Park, J.; Kim, P.; Helen, W.; Engler, A.J.; Levchenko, A.; Kim, D.-H. Control of stem cell fate and function by engineering physical microenvironments. *Integr. Biol.* **2012**, *4*, 1008–1018. [CrossRef] [PubMed]
41. Han, S.B.; Kim, J.K.; Lee, G.; Kim, D.H. Mechanical Properties of Materials for Stem Cell Differentiation. *Adv. Biosyst.* **2020**, *4*, 2000247. [CrossRef]
42. Cargnello, M.; Roux, P.P. Activation and Function of the MAPKs and Their Substrates, the MAPK-Activated Protein Kinases. *Microbiol. Mol. Biol. Rev.* **2011**, *75*, 50–83. [CrossRef]
43. MacQueen, L.; Sun, Y.; Simmons, C.A. Mesenchymal stem cell mechanobiology and emerging experimental platforms. *J. R. Soc. Interface* **2013**, *10*, 20130179. [CrossRef] [PubMed]
44. Greenblatt, M.B.; Shim, J.-H.; Zou, W.; Sitara, D.; Schweitzer, M.; Hu, D.; Lotinun, S.; Sano, Y.; Baron, R.; Park, J.M.; et al. The p38 MAPK pathway is essential for skeletogenesis and bone homeostasis in mice. *J. Clin. Investig.* **2010**, *120*, 2457–2473. [CrossRef]
45. Thouverey, C.; Caverzasio, J. The p38α MAPK positively regulates osteoblast function and postnatal bone acquisition. *Cell Mol. Life Sci.* **2012**, *69*, 3115–3125. [CrossRef]
46. Rodríguez-Carballo, E.; Gámez, B.; Sedó-Cabezón, L.; Sánchez-Feutrie, M.; Zorzano, A.; Manzanares-Céspedes, C.; Rosa, J.L.; Ventura, F. The p38α MAPK function in osteoprecursors is required for bone formation and bone homeostasis in adult mice. *PLoS ONE* **2014**, *9*, e102032. [CrossRef]
47. Chakladar, A.; Dubeykovskiy, A.; Wojtukiewicz, L.J.; Pratap, J.; Lei, S.; Wang, T.C. Synergistic activation of the murine gastrin promoter by oncogenic Ras and beta-catenin involves SMAD recruitment. *Biochem. Biophys. Res. Commun.* **2005**, *336*, 190–196. [CrossRef]
48. Yamashita, M.; Otsuka, F.; Mukai, T.; Otani, H.; Inagaki, K.; Miyoshi, T.; Goto, J.; Yamamura, M.; Makino, H. Simvastatin antagonizes tumor necrosis factor-alpha inhibition of bone morphogenetic proteins-2-induced osteoblast differentiation by regulating Smad signaling and Ras/Rho-mitogen-activated protein kinase pathway. *J. Endocrinol.* **2008**, *196*, 601–613. [CrossRef]
49. Xue, R.; Li, J.Y.; Yeh, Y.; Yang, L.; Chien, S. Effects of matrix elasticity and cell density on human mesenchymal stem cells differentiation. *J. Orthop. Res.* **2013**, *31*, 1360–1365. [CrossRef]
50. Xu, J.; Sun, M.; Tan, Y.; Wang, H.; Wang, H.; Li, P.; Xu, Z.; Xia, Y.; Li, L.; Li, Y. Effect of matrix stiffness on the proliferation and differentiation of umbilical cord mesenchymal stem cells. *Differentiation* **2017**, *96*, 30–39. [CrossRef]

51. Sun, M.; Chi, G.; Li, P.; Lv, S.; Xu, J.; Xu, Z.; Xia, Y.; Tan, Y.; Xu, J.; Li, L.; et al. Effects of Matrix Stiffness on the Morphology, Adhesion, Proliferation and Osteogenic Differentiation of Mesenchymal Stem Cells. *Int. J. Med. Sci.* **2018**, *15*, 257–268. [CrossRef]
52. Ahmed, M.; Ffrench-Constant, C. Extracellular Matrix Regulation of Stem Cell Behavior. *Curr. Stem Cell Rep.* **2016**, *2*, 197–206. [CrossRef]
53. Guilak, F.; Cohen, D.M.; Estes, B.T.; Gimble, J.M.; Liedtke, W.; Chen, C.S. Control of stem cell fate by physical interactions with the extracellular matrix. *Cell Stem Cell* **2009**, *5*, 17–26. [CrossRef]
54. Zhang, N.; Kohn, D.H. Using polymeric materials to control stem cell behavior for tissue regeneration. *Birth Defects Res. Part. C Embryo Today Rev.* **2012**, *96*, 63–81. [CrossRef] [PubMed]
55. Abbass, M.M.S.; El-Rashidy, A.A.; Sadek, K.M.; Moshy, S.E.; Radwan, I.A.; Rady, D.; Dörfer, C.E.; Fawzy El-Sayed, K.M. Hydrogels and Dentin–Pulp Complex Regeneration: From the Benchtop to Clinical Translation. *Polymers* **2020**, *12*, 2935. [CrossRef] [PubMed]
56. Lv, H.; Wang, H.; Zhang, Z.; Yang, W.; Liu, W.; Li, Y.; Li, L. Biomaterial stiffness determines stem cell fate. *Life Sci.* **2017**, *178*, 42–48. [CrossRef] [PubMed]
57. Engler, A.J.; Sen, S.; Sweeney, H.L.; Discher, D.E. Matrix elasticity directs stem cell lineage specification. *Cell* **2006**, *126*, 677–689. [CrossRef]
58. Lv, H.; Li, L.; Sun, M.; Zhang, Y.; Chen, L.; Rong, Y.; Li, Y. Mechanism of regulation of stem cell differentiation by matrix stiffness. *Stem Cell Res. Ther.* **2015**, *6*, 103. [CrossRef]
59. Guo, R.; Lu, S.; Merkel, A.R.; Sterling, J.A.; Guelcher, S.A. Substrate Modulus Regulates Osteogenic Differentiation of Rat Mesenchymal Stem Cells through Integrin β1 and BMP Receptor Type IA. *J. Mater. Chem. B* **2016**, *4*, 3584–3593. [CrossRef]
60. Trappmann, B.; Gautrot, J.E.; Connelly, J.T.; Strange, D.G.; Li, Y.; Oyen, M.L.; Cohen Stuart, M.A.; Boehm, H.; Li, B.; Vogel, V.; et al. Extracellular-matrix tethering regulates stem-cell fate. *Nat. Mater.* **2012**, *11*, 642–649. [CrossRef]
61. Naqvi, S.M.; McNamara, L.M. Stem Cell Mechanobiology and the Role of Biomaterials in Governing Mechanotransduction and Matrix Production for Tissue Regeneration. *Front. Bioeng. Biotechnol.* **2020**, *8*, 597661. [CrossRef]
62. Shih, Y.R.; Tseng, K.F.; Lai, H.Y.; Lin, C.H.; Lee, O.K. Matrix stiffness regulation of integrin-mediated mechanotransduction during osteogenic differentiation of human mesenchymal stem cells. *J. Bone Miner. Res.* **2011**, *26*, 730–738. [CrossRef]
63. Miller, A.E.; Hu, P.; Barker, T.H. Feeling Things Out: Bidirectional Signaling of the Cell–ECM Interface, Implications in the Mechanobiology of Cell Spreading, Migration, Proliferation, and Differentiation. *Adv. Healthc. Mater.* **2020**, *9*, 1901445. [CrossRef] [PubMed]
64. Nanci, A. *Ten Cate's Oral Histology: Development, Structure and Function*, 8th ed.; El Sevier Mosby: St Louis, Hong Kong, 2013; pp. 81–83.
65. Ciobanasu, C.; Faivre, B.; Le Clainche, C. Actomyosin-dependent formation of the mechanosensitive talin-vinculin complex reinforces actin anchoring. *Nat. Commun.* **2014**, *5*, 3095. [CrossRef] [PubMed]
66. Martino, F.; Perestrelo, A.R.; Vinarský, V.; Pagliari, S.; Forte, G. Cellular Mechanotransduction: From Tension to Function. *Front. Physiol.* **2018**, *9*, 824. [CrossRef]
67. Ohashi, K.; Fujiwara, S.; Mizuno, K. Roles of the cytoskeleton, cell adhesion and rho signalling in mechanosensing and mechanotransduction. *J. Biochem.* **2017**, *161*, 245–254. [CrossRef]
68. Chen, H.; Cheng, C.Y. *Focal Adhesion Kinase*; Springer International Publishing: New York, NY, USA, 2018; pp. 1800–1812.
69. Bradbury, P.M.; Turner, K.; Mitchell, C.; Griffin, K.R.; Middlemiss, S.; Lau, L.; Dagg, R.; Taran, E.; Cooper-White, J.; Fabry, B.; et al. The focal adhesion targeting domain of p130Cas confers a mechanosensing function. *J. Cell Sci.* **2017**, *130*, 1263–1273. [CrossRef]
70. Takada, Y.; Ye, X.; Simon, S. The integrins. *Genome Biol.* **2007**, *8*, 215. [CrossRef] [PubMed]
71. Hynes, R.O. Integrins: Bidirectional, allosteric signaling machines. *Cell* **2002**, *110*, 673–687. [CrossRef]
72. Sun, Z.; Guo, S.S.; Fässler, R. Integrin-mediated mechanotransduction. *J. Cell Biol.* **2016**, *215*, 445–456. [CrossRef] [PubMed]
73. Kim, M.; Carman, C.V.; Springer, T.A. Bidirectional transmembrane signaling by cytoplasmic domain separation in integrins. *Science* **2003**, *301*, 1720–1725. [CrossRef]
74. Ellert-Miklaszewska, A.; Poleszak, K.; Pasierbinska, M.; Kaminska, B. Integrin Signaling in Glioma Pathogenesis: From Biology to Therapy. *Int. J. Mol. Sci.* **2020**, *21*, 888. [CrossRef] [PubMed]
75. Petrie, T.A.; Raynor, J.E.; Dumbauld, D.W.; Lee, T.T.; Jagtap, S.; Templeman, K.L.; Collard, D.M.; García, A.J. Multivalent integrin-specific ligands enhance tissue healing and biomaterial integration. *Sci. Transl. Med.* **2010**, *2*, 45ra60. [CrossRef]
76. Olivares-Navarrete, R.; Lee, E.M.; Smith, K.; Hyzy, S.L.; Doroudi, M.; Williams, J.K.; Gall, K.; Boyan, B.D.; Schwartz, Z. Substrate Stiffness Controls Osteoblastic and Chondrocytic Differentiation of Mesenchymal Stem Cells without Exogenous Stimuli. *PLoS ONE* **2017**, *12*, e0170312. [CrossRef]
77. Frith, J.E.; Mills, R.J.; Hudson, J.E.; Cooper-White, J.J. Tailored integrin-extracellular matrix interactions to direct human mesenchymal stem cell differentiation. *Stem. Cells Dev.* **2012**, *21*, 2442–2456. [CrossRef]
78. Sun, M.; Chi, G.; Xu, J.; Tan, Y.; Xu, J.; Lv, S.; Xu, Z.; Xia, Y.; Li, L.; Li, Y. Extracellular matrix stiffness controls osteogenic differentiation of mesenchymal stem cells mediated by integrin α5. *Stem. Cell Res. Ther.* **2018**, *9*, 52. [CrossRef] [PubMed]
79. Hamidouche, Z.; Fromigué, O.; Ringe, J.; Häupl, T.; Vaudin, P.; Pagès, J.C.; Srouji, S.; Livne, E.; Marie, P.J. Priming integrin alpha5 promotes human mesenchymal stromal cell osteoblast differentiation and osteogenesis. *Proc. Natl. Acad. Sci. USA* **2009**, *106*, 18587–18591. [CrossRef] [PubMed]

80. Zhang, D.; Ni, N.; Wang, Y.; Tang, Z.; Gao, H.; Ju, Y.; Sun, N.; He, X.; Gu, P.; Fan, X. CircRNA-vgll3 promotes osteogenic differentiation of adipose-derived mesenchymal stem cells via modulating miRNA-dependent integrin α5 expression. *Cell Death Differ.* **2021**, *28*, 283–302. [CrossRef] [PubMed]
81. Lee, J.; Abdeen, A.A.; Tang, X.; Saif, T.A.; Kilian, K.A. Geometric guidance of integrin mediated traction stress during stem cell differentiation. *Biomaterials* **2015**, *69*, 174–183. [CrossRef]
82. Liu, J.; DeYoung, S.M.; Zhang, M.; Zhang, M.; Cheng, A.; Saltiel, A.R. Changes in integrin expression during adipocyte differentiation. *Cell Metab.* **2005**, *2*, 165–177. [CrossRef] [PubMed]
83. Lock, J.G.; Wehrle-Haller, B.; Strömblad, S. Cell-matrix adhesion complexes: Master control machinery of cell migration. *Semin. Cancer Biol.* **2008**, *18*, 65–76. [CrossRef]
84. Gavazzo, P.; Viti, F.; Donnelly, H.; Oliva, M.A.G.; Salmeron-Sanchez, M.; Dalby, M.J.; Vassalli, M. Biophysical phenotyping of mesenchymal stem cells along the osteogenic differentiation pathway. *Cell Biol. Toxicol.* **2021**, *2021*, 1–19. [CrossRef]
85. Saidova, A.A.; Vorobjev, I.A. Lineage Commitment, Signaling Pathways, and the Cytoskeleton Systems in Mesenchymal Stem Cells. *Tissue Eng. Part B Rev.* **2020**, *26*, 13–25. [CrossRef] [PubMed]
86. Byron, A.; Frame, M.C. Adhesion protein networks reveal functions proximal and distal to cell-matrix contacts. *Curr. Opin. Cell Biol.* **2016**, *39*, 93–100. [CrossRef] [PubMed]
87. Han, S.J.; Bielawski, K.S.; Ting, L.H.; Rodriguez, M.L.; Sniadecki, N.J. Decoupling substrate stiffness, spread area, and micropost density: A close spatial relationship between traction forces and focal adhesions. *Biophys. J.* **2012**, *103*, 640–648. [CrossRef]
88. McBeath, R.; Pirone, D.M.; Nelson, C.M.; Bhadriraju, K.; Chen, C.S. Cell shape, cytoskeletal tension, and RhoA regulate stem cell lineage commitment. *Dev. Cell* **2004**, *6*, 483–495. [CrossRef]
89. Lavenus, S.; Berreur, M.; Trichet, V.; Pilet, P.; Louarn, G.; Layrolle, P. Adhesion and osteogenic differentiation of human mesenchymal stem cells on titanium nanopores. *Eur. Cell Mater.* **2011**, *22*, 84–96. [CrossRef]
90. Frith, J.E.; Mills, R.J.; Cooper-White, J.J. Lateral spacing of adhesion peptides influences human mesenchymal stem cell behaviour. *J. Cell Sci.* **2012**, *125*, 317–327. [CrossRef]
91. Rosenberg, N. The role of the cytoskeleton in mechanotransduction in human osteoblast-like cells. *Hum. Exp. Toxicol.* **2003**, *22*, 271–274. [CrossRef]
92. Wang, N.; Naruse, K.; Stamenović, D.; Fredberg, J.J.; Mijailovich, S.M.; Tolić-Nørrelykke, I.M.; Polte, T.; Mannix, R.; Ingber, D.E. Mechanical behavior in living cells consistent with the tensegrity model. *Proc. Natl. Acad. Sci. USA* **2001**, *98*, 7765–7770. [CrossRef] [PubMed]
93. Harris, A.R.; Jreij, P.; Fletcher, D.A. Mechanotransduction by the actin cytoskeleton: Converting mechanical stimuli into biochemical signals. *Annu. Rev. Biophys.* **2018**, *47*, 617–631. [CrossRef]
94. Pollard, T.D.; Borisy, G.G. Cellular motility driven by assembly and disassembly of actin filaments. *Cell* **2003**, *112*, 453–465. [CrossRef]
95. Kim, D.H.; Wirtz, D. Cytoskeletal tension induces the polarized architecture of the nucleus. *Biomaterials* **2015**, *48*, 161–172. [CrossRef] [PubMed]
96. Khatau, S.B.; Hale, C.M.; Stewart-Hutchinson, P.J.; Patel, M.S.; Stewart, C.L.; Searson, P.C.; Hodzic, D.; Wirtz, D. A perinuclear actin cap regulates nuclear shape. *Proc. Natl. Acad. Sci. USA* **2009**, *106*, 19017–19022. [CrossRef]
97. Kim, D.H.; Khatau, S.B.; Feng, Y.; Walcott, S.; Sun, S.X.; Longmore, G.D.; Wirtz, D. Actin cap associated focal adhesions and their distinct role in cellular mechanosensing. *Sci. Rep.* **2012**, *2*, 555. [CrossRef]
98. Tojkander, S.; Gateva, G.; Lappalainen, P. Actin stress fibers—Assembly, dynamics and biological roles. *J. Cell Sci.* **2012**, *125*, 1855–1864. [CrossRef] [PubMed]
99. Huang, C.H.; Chen, M.H.; Young, T.H.; Jeng, J.H.; Chen, Y.J. Interactive effects of mechanical stretching and extracellular matrix proteins on initiating osteogenic differentiation of human mesenchymal stem cells. *J. Cell Biochem.* **2009**, *108*, 1263–1273. [CrossRef]
100. Svitkina, T. The Actin Cytoskeleton and Actin-Based Motility. *Cold Spring Harb. Perspect. Biol.* **2018**, *10*, a018267. [CrossRef] [PubMed]
101. Chen, L.; Hu, H.; Qiu, W.; Shi, K.; Kassem, M. Actin depolymerization enhances adipogenic differentiation in human stromal stem cells. *Stem Cell Res.* **2018**, *29*, 76–83. [CrossRef]
102. Rodríguez, J.P.; González, M.; Ríos, S.; Cambiazo, V. Cytoskeletal organization of human mesenchymal stem cells (MSC) changes during their osteogenic differentiation. *J. Cell Biochem.* **2004**, *93*, 721–731. [CrossRef]
103. Sonowal, H.; Kumar, A.; Bhattacharyya, J.; Gogoi, P.K.; Jaganathan, B.G. Inhibition of actin polymerization decreases osteogeneic differentiation of mesenchymal stem cells through p38 MAPK pathway. *J. Biomed. Sci.* **2013**, *20*, 71. [CrossRef]
104. Sen, B.; Xie, Z.; Case, N.; Thompson, W.R.; Uzer, G.; Styner, M.; Rubin, J. mTORC2 regulates mechanically induced cytoskeletal reorganization and lineage selection in marrow-derived mesenchymal stem cells. *J. Bone Miner. Res.* **2014**, *29*, 78–89. [CrossRef]
105. Chen, L.; Shi, K.; Frary, C.E.; Ditzel, N.; Hu, H.; Qiu, W.; Kassem, M. Inhibiting actin depolymerization enhances osteoblast differentiation and bone formation in human stromal stem cells. *Stem Cell Res.* **2015**, *15*, 281–289. [CrossRef]
106. Zhang, T.; Lin, S.; Shao, X.; Zhang, Q.; Xue, C.; Zhang, S.; Lin, Y.; Zhu, B.; Cai, X. Effect of matrix stiffness on osteoblast functionalization. *Cell Prolif.* **2017**, *50*. [CrossRef] [PubMed]
107. Khan, A.U.; Qu, R.; Fan, T.; Ouyang, J.; Dai, J. A glance on the role of actin in osteogenic and adipogenic differentiation of mesenchymal stem cells. *Stem Cell Res. Ther.* **2020**, *11*, 283. [CrossRef]

108. Kobayashi, T.; Sokabe, M. Sensing substrate rigidity by mechanosensitive ion channels with stress fibers and focal adhesions. *Curr. Opin. Cell Biol.* **2010**, *22*, 669–676. [CrossRef] [PubMed]
109. Wu, Z.; Wong, K.; Glogauer, M.; Ellen, R.P.; McCulloch, C.A.G. Regulation of Stretch-Activated Intracellular Calcium Transients by Actin Filaments. *Biochem. Biophys. Res. Commun.* **1999**, *261*, 419–425. [CrossRef] [PubMed]
110. Iqbal, J.; Zaidi, M. Molecular regulation of mechanotransduction. *Biochem. Biophys. Res. Commun.* **2005**, *328*, 751–755. [CrossRef]
111. Kretlow, J.D.; Jin, Y.-Q.; Liu, W.; Zhang, W.J.; Hong, T.-H.; Zhou, G.; Baggett, L.S.; Mikos, A.G.; Cao, Y. Donor age and cell passage affects differentiation potential of murine bone marrow-derived stem cells. *BMC Cell Biol.* **2008**, *9*, 1–13. [CrossRef]
112. Stenderup, K.; Justesen, J.; Clausen, C.; Kassem, M. Aging is associated with decreased maximal life span and accelerated senescence of bone marrow stromal cells. *Bone Miner.* **2003**, *33*, 919–926. [CrossRef]
113. Asumda, F.Z.; Chase, P.B. Age-related changes in rat bone-marrow mesenchymal stem cell plasticity. *BMC Cell Biol.* **2011**, *12*, 1–11. [CrossRef]
114. Shen, J.; Tsai, Y.-T.; DiMarco, N.M.; Long, M.A.; Sun, X.; Tang, L. Transplantation of mesenchymal stem cells from young donors delays aging in mice. *Sci. Rep. Cetacean Res.* **2011**, *1*, 1–7. [CrossRef] [PubMed]
115. Lin, T.h.; Gibon, E.; Loi, F.; Pajarinen, J.; Córdova, L.A.; Nabeshima, A.; Lu, L.; Yao, Z.; Goodman, S.B. Decreased osteogenesis in mesenchymal stem cells derived from the aged mouse is associated with enhanced NF-κB activity. *J. Orth Res.* **2017**, *35*, 281–288. [CrossRef]
116. Pelissier, F.A.; Garbe, J.C.; Ananthanarayanan, B.; Miyano, M.; Lin, C.; Jokela, T.; Kumar, S.; Stampfer, M.R.; Lorens, J.B.; LaBarge, M.A. Age-Related Dysfunction in Mechanotransduction Impairs Differentiation of Human Mammary Epithelial Progenitors. *Cell Rep.* **2014**, *7*, 1926–1939. [CrossRef]
117. Barreto, S.; Gonzalez-Vazquez, A.; Cameron, A.R.; Cavanagh, B.; Murray, D.J.; O'Brien, F.J. Identification of the mechanisms by which age alters the mechanosensitivity of mesenchymal stromal cells on substrates of differing stiffness: Implications for osteogenesis and angiogenesis. *Acta Biomater.* **2017**, *53*, 59–69. [CrossRef] [PubMed]
118. Zouani, O.F.; Kalisky, J.; Ibarboure, E.; Durrieu, M.-C. Effect of BMP-2 from matrices of different stiffnesses for the modulation of stem cell fate. *Biomaterials* **2013**, *34*, 2157–2166. [CrossRef] [PubMed]
119. Ma, S.; Meng, Z.; Chen, R.; Guan, K.-L. The Hippo pathway: Biology and pathophysiology. *Annu Rev. Biochem.* **2019**, *88*, 577–604. [CrossRef] [PubMed]
120. Yamaguchi, H.; Taouk, G.M. A Potential role of YAP/TAZ in the interplay between metastasis and metabolic alterations. *Front. Oncol.* **2020**, *10*, 928. [CrossRef] [PubMed]
121. Dobrokhotov, O.; Samsonov, M.; Sokabe, M.; Hirata, H. Mechanoregulation and pathology of YAP/TAZ via Hippo and non-Hippo mechanisms. *J. Transl. Med.* **2018**, *7*, 1–14. [CrossRef]
122. Dupont, S.; Morsut, L.; Aragona, M.; Enzo, E.; Giulitti, S.; Cordenonsi, M.; Zanconato, F.; Le Digabel, J.; Forcato, M.; Bicciato, S. Role of YAP/TAZ in mechanotransduction. *Nature* **2011**, *474*, 179–183. [CrossRef]
123. Zhao, B.; Li, L.; Wang, L.; Wang, C.-Y.; Yu, J.; Guan, K.-L. Cell detachment activates the Hippo pathway via cytoskeleton reorganization to induce anoikis. *Genes Dev.* **2012**, *26*, 54–68. [CrossRef] [PubMed]
124. Aragona, M.; Panciera, T.; Manfrin, A.; Giulitti, S.; Michielin, F.; Elvassore, N.; Dupont, S.; Piccolo, S. A mechanical checkpoint controls multicellular growth through YAP/TAZ regulation by actin-processing factors. *Cell* **2013**, *154*, 1047–1059. [CrossRef] [PubMed]
125. Elosegui-Artola, A.; Andreu, I.; Beedle, A.E.; Lezamiz, A.; Uroz, M.; Kosmalska, A.J.; Oria, R.; Kechagia, J.Z.; Rico-Lastres, P.; Le Roux, A.-L. Force triggers YAP nuclear entry by regulating transport across nuclear pores. *Cell* **2017**, *171*, 1397–1410.e1314. [CrossRef] [PubMed]
126. Lorthongpanich, C.; Thumanu, K.; Tangkiettrakul, K.; Jiamvoraphong, N.; Laowtammathron, C.; Damkham, N.; U-pratya, Y.; Issaragrisil, S. YAP as a key regulator of adipo-osteogenic differentiation in human MSCs. *Stem Cell Res. Ther.* **2019**, *10*, 402. [CrossRef] [PubMed]
127. Hong, J.H.; Hwang, E.S.; McManus, M.T.; Amsterdam, A.; Tian, Y.; Kalmukova, R.; Mueller, E.; Benjamin, T.; Spiegelman, B.M.; Sharp, P.A.; et al. TAZ, a transcriptional modulator of mesenchymal stem cell differentiation. *Science* **2005**, *309*, 1074–1078. [CrossRef]
128. Du, J.; Chen, X.; Liang, X.; Zhang, G.; Xu, J.; He, L.; Zhan, Q.; Feng, X.Q.; Chien, S.; Yang, C. Integrin activation and internalization on soft ECM as a mechanism of induction of stem cell differentiation by ECM elasticity. *Proc. Natl. Acad. Sci. USA* **2011**, *108*, 9466–9471. [CrossRef]
129. Chrzanowska-Wodnicka, M.; Burridge, K. Rho-stimulated contractility drives the formation of stress fibers and focal adhesions. *J. Cell Biol.* **1996**, *133*, 1403–1415. [CrossRef]
130. Arnsdorf, E.J.; Tummala, P.; Kwon, R.Y.; Jacobs, C.R. Mechanically induced osteogenic differentiation–the role of RhoA, ROCKII and cytoskeletal dynamics. *J. Cell Sci.* **2009**, *122*, 546–553. [CrossRef]
131. Hyväri, L.; Ojansivu, M.; Juntunen, M.; Kartasalo, K.; Miettinen, S.; Vanhatupa, S. Focal Adhesion Kinase and ROCK Signaling Are Switch-Like Regulators of Human Adipose Stem Cell Differentiation towards Osteogenic and Adipogenic Lineages. *Stem Cells Int.* **2018**, *2018*, 2190657. [CrossRef]
132. Rowlands, A.S.; George, P.A.; Cooper-White, J.J. Directing osteogenic and myogenic differentiation of MSCs: Interplay of stiffness and adhesive ligand presentation. *Am. J. Physiol. Cell Physiol.* **2008**, *295*, C1037–C1044. [CrossRef]

133. Wan, W.; Cheng, B.; Zhang, C.; Ma, Y.; Li, A.; Xu, F.; Lin, M. Synergistic Effect of Matrix Stiffness and Inflammatory Factors on Osteogenic Differentiation of MSC. *Biophys. J.* **2019**, *117*, 129–142. [CrossRef]
134. He, X.T.; Wu, R.X.; Xu, X.Y.; Wang, J.; Yin, Y.; Chen, F.M. Macrophage involvement affects matrix stiffness-related influences on cell osteogenesis under three-dimensional culture conditions. *Acta Biomater.* **2018**, *71*, 132–147. [CrossRef]
135. Tang, X.; Thankappan, S.K.; Lee, P.; Fard, S.E.; Harmon, M.D.; Tran, K.; Yu, X. Chapter 21—Polymeric Biomaterials in Tissue Engineering and Regenerative Medicine. In *Natural and Synthetic Biomedical Polymers*; Kumbar, S.G., Laurencin, C.T., Deng, M., Eds.; Elsevier: Oxford, UK, 2014; pp. 351–371.
136. Ramkumar, R.; Sundaram, M.M. A biopolymer gel-decorated cobalt molybdate nanowafer: Effective graft polymer cross-linked with an organic acid for better energy storage. *New J. Chem.* **2016**, *40*, 2863–2877. [CrossRef]
137. Yang, J.M.; Olanrele, O.S.; Zhang, X.; Hsu, C.C. Fabrication of Hydrogel Materials for Biomedical Applications. In *Novel Biomaterials for Regenerative Medicine*; Chun, H.J., Park, K., Kim, C.-H., Khang, G., Eds.; Springer Singapore: Singapore, 2018; pp. 197–224.
138. Lee, K.Y.; Mooney, D.J. Alginate: Properties and biomedical applications. *Prog. Polym. Sci.* **2012**, *37*, 106–126. [CrossRef]
139. Lee, H.-p.; Gu, L.; Mooney, D.J.; Levenston, M.E.; Chaudhuri, O. Mechanical confinement regulates cartilage matrix formation by chondrocytes. *Nat. Mater.* **2017**, *16*, 1243–1251. [CrossRef] [PubMed]
140. Wickramaarachchi, K.; Sundaram, M.M.; Henry, D.J.; Gao, X. Alginate Biopolymer Effect on the Electrodeposition of Manganese Dioxide on Electrodes for Supercapacitors. *ACS Appl. Energy Mater.* **2021**, *4*, 7040–7051. [CrossRef]
141. Zhang, J.; Wehrle, E.; Adamek, P.; Paul, G.R.; Qin, X.-H.; Rubert, M.; Müller, R. Optimization of mechanical stiffness and cell density of 3D bioprinted cell-laden scaffolds improves extracellular matrix mineralization and cellular organization for bone tissue engineering. *Acta Biomater.* **2020**, *114*, 307–322. [CrossRef] [PubMed]
142. Freeman, F.E.; Kelly, D.J. Tuning alginate bioink stiffness and composition for controlled growth factor delivery and to spatially direct MSC fate within bioprinted tissues. *Sci. Rep.* **2017**, *7*, 1–12. [CrossRef]
143. Maia, F.R.; Fonseca, K.B.; Rodrigues, G.; Granja, P.L.; Barrias, C.C. Matrix-driven formation of mesenchymal stem cell–Extracellular matrix microtissues on soft alginate hydrogels. *Acta Biomater.* **2014**, *10*, 3197–3208. [CrossRef]
144. Yener, B.; Acar, E.; Aguis, P.; Bennett, K.; Vandenberg, S.L.; Plopper, G.E. Multiway modeling and analysis in stem cell systems biology. *BMC Syst. Biol.* **2008**, *2*, 1–17. [CrossRef]
145. Stein, G.S.; Lian, J.B.; Owen, T.A. Relationship of cell growth to the regulation of tissue-specific gene expression during osteoblast differentiation. *FASEB J.* **1990**, *4*, 3111–3123. [CrossRef]
146. Xie, J.; Bao, M.; Bruekers, S.p.M.; Huck, W.T. Collagen gels with different fibrillar microarchitectures elicit different cellular responses. *ACS Appl. Mater. Interfaces* **2017**, *9*, 19630–19637. [CrossRef] [PubMed]
147. Banks, J.M.; Mozdzen, L.C.; Harley, B.A.; Bailey, R.C. The combined effects of matrix stiffness and growth factor immobilization on the bioactivity and differentiation capabilities of adipose-derived stem cells. *Biomaterials* **2014**, *35*, 8951–8959. [CrossRef] [PubMed]
148. Hwang, J.-H.; Han, U.; Yang, M.; Choi, Y.; Choi, J.; Lee, J.-M.; Jung, H.-S.; Hong, J.; Hong, J.-H. Artificial cellular nano-environment composed of collagen-based nanofilm promotes osteogenic differentiation of mesenchymal stem cells. *Acta Biomater.* **2019**, *86*, 247–256. [CrossRef]
149. Zhou, Q.; Lyu, S.; Bertrand, A.A.; Hu, A.C.; Chan, C.H.; Ren, X.; Dewey, M.J.; Tiffany, A.S.; Harley, B.A.; Lee, J.C. Stiffness of Nanoparticulate Mineralized Collagen Scaffolds Triggers Osteogenesis via Mechanotransduction and Canonical Wnt Signaling. *Macromol. Biosci.* **2021**, *21*, 2000370. [CrossRef]
150. Tsimbouri, P.M.; Childs, P.G.; Pemberton, G.D.; Yang, J.; Jayawarna, V.; Orapiriyakul, W.; Burgess, K.; Gonzalez-Garcia, C.; Blackburn, G.; Thomas, D. Stimulation of 3D osteogenesis by mesenchymal stem cells using a nanovibrational bioreactor. *Nat. Biomed. Eng.* **2017**, *1*, 758–770. [CrossRef]
151. Murphy, C.M.; Matsiko, A.; Haugh, M.G.; Gleeson, J.P.; O'Brien, F.J. Mesenchymal stem cell fate is regulated by the composition and mechanical properties of collagen–glycosaminoglycan scaffolds. *J. Mech. Behav. Biomed. Mater.* **2012**, *11*, 53–62. [CrossRef]
152. Chen, G.; Dong, C.; Yang, L.; Lv, Y. Interfaces 3D scaffolds with different stiffness but the same microstructure for bone tissue engineering. *ACS Appl. Mater.* **2015**, *7*, 15790–15802. [CrossRef] [PubMed]
153. Radhakrishnan, J.; Krishnan, U.M.; Sethuraman, S. Hydrogel based injectable scaffolds for cardiac tissue regeneration. *Biotechnol. Adv.* **2014**, *32*, 449–461. [CrossRef] [PubMed]
154. Echave, M.C.; Saenz del Burgo, L.; Pedraz, J.L.; Orive, G. Gelatin as Biomaterial for Tissue Engineering. *Curr. Pharm. Des.* **2017**, *23*, 3567–3584. [CrossRef] [PubMed]
155. Su, K.; Wang, C. Recent advances in the use of gelatin in biomedical research. *Biotechnol. Lett.* **2015**, *37*, 2139–2145. [CrossRef]
156. Gasperini, L.; Mano, J.F.; Reis, R.L. Natural polymers for the microencapsulation of cells. *J. R. Soc. Interface* **2014**, *11*, 20140817. [CrossRef] [PubMed]
157. Weng, Y.J.; Ren, J.R.; Huang, N.; Wang, J.; Chen, J.Y.; Leng, Y.X.; Liu, H.Q. Surface engineering of Ti–O films by photochemical immobilization of gelatin. *Mater. Sci Eng. C* **2008**, *28*, 1495–1500. [CrossRef]
158. Zhu, Y.; Gao, C.; He, T.; Shen, J. Endothelium regeneration on luminal surface of polyurethane vascular scaffold modified with diamine and covalently grafted with gelatin. *Biomaterials* **2004**, *25*, 423–430. [CrossRef]

159. Salamon, A.; van Vlierberghe, S.; van Nieuwenhove, I.; Baudisch, F.; Graulus, G.-J.; Benecke, V.; Alberti, K.; Neumann, H.-G.; Rychly, J.; Martins, J.C.; et al. Gelatin-Based Hydrogels Promote Chondrogenic Differentiation of Human Adipose Tissue-Derived Mesenchymal Stem Cells In Vitro. *Materials* **2014**, *7*, 1342–1359. [CrossRef]
160. Zhou, D.; Ito, Y. Inorganic material surfaces made bioactive by immobilizing growth factors for hard tissue engineering. *RSC Adv.* **2013**, *3*, 11095–11106. [CrossRef]
161. Sun, M.; Sun, X.; Wang, Z.; Guo, S.; Yu, G.; Yang, H. Synthesis and Properties of Gelatin Methacryloyl (GelMA) Hydrogels and Their Recent Applications in Load-Bearing Tissue. *Polymers* **2018**, *10*, 1290. [CrossRef]
162. Van Nieuwenhove, I.; Salamon, A.; Adam, S.; Dubruel, P.; Van Vlierberghe, S.; Peters, K. Gelatin- and starch-based hydrogels. Part B: In vitro mesenchymal stem cell behavior on the hydrogels. *Carbohydr. Polym.* **2017**, *161*, 295–305. [CrossRef]
163. Jiang, P.; Mao, Z.; Gao, C. Combinational effect of matrix elasticity and alendronate density on differentiation of rat mesenchymal stem cells. *Acta Biomater.* **2015**, *19*, 76–84. [CrossRef] [PubMed]
164. Sun, H.; Zhu, F.; Hu, Q.; Krebsbach, P.H. Controlling stem cell-mediated bone regeneration through tailored mechanical properties of collagen scaffolds. *Biomaterials* **2014**, *35*, 1176–1184. [CrossRef] [PubMed]
165. Ventre, M.; Coppola, V.; Natale, C.F.; Netti, P.A. Aligned fibrous decellularized cell derived matrices for mesenchymal stem cell amplification. *J. Biomed. Mater. Res. A* **2019**, *107*, 2536–2546. [CrossRef]
166. Hu, Q.; Liu, M.; Chen, G.; Xu, Z.; Lv, Y. Demineralized Bone Scaffolds with Tunable Matrix Stiffness for Efficient Bone Integration. *ACS Appl. Mater. Interfaces* **2018**, *10*, 27669–27680. [CrossRef]
167. Lee, J.Y.; Spicer, A.P. Hyaluronan: A multifunctional, megaDalton, stealth molecule. *Curr. Opin. Cell Biol.* **2000**, *12*, 581–586. [CrossRef]
168. Garg, H.G.; Hales, C.A. *Chemistry and Biology of Hyaluronan*; Elsevier: Amsterdam, The Netherlands, 2004.
169. Fawzy El-Sayed, K.M.; Dahaba, M.A.; Aboul-Ela, S.; Darhous, M.S. Local application of hyaluronan gel in conjunction with periodontal surgery: A randomized controlled trial. *Clin. Oral Investig.* **2012**, *16*, 1229–1236. [CrossRef]
170. Xing, F.; Li, L.; Zhou, C.; Long, C.; Wu, L.; Lei, H.; Kong, Q.; Fan, Y.; Xiang, Z.; Zhang, X. Regulation and directing stem cell fate by tissue engineering functional microenvironments: Scaffold physical and chemical cues. *Stem Cells Int.* **2019**, *2019*. [CrossRef]
171. Zhai, P.; Peng, X.; Li, B.; Liu, Y.; Sun, H.; Li, X. The application of hyaluronic acid in bone regeneration. *Int. J. Biol. Macromol.* **2020**, *151*, 1224–1239. [CrossRef]
172. Itano, N. Simple primary structure, complex turnover regulation and multiple roles of hyaluronan. *J. Biochem.* **2008**, *144*, 131–137. [CrossRef] [PubMed]
173. Anseth, K.S.; Burdick, J.A. New directions in photopolymerizable biomaterials. *MRS Bull.* **2002**, *27*, 130–136. [CrossRef]
174. Zhao, W.; Li, X.; Liu, X.; Zhang, N.; Wen, X. Effects of substrate stiffness on adipogenic and osteogenic differentiation of human mesenchymal stem cells. *Mater. Sci. Eng. C* **2014**, *40*, 316–323. [CrossRef] [PubMed]
175. Cosgrove, B.D.; Mui, K.L.; Driscoll, T.P.; Caliari, S.R.; Mehta, K.D.; Assoian, R.K.; Burdick, J.A.; Mauck, R.L. N-cadherin adhesive interactions modulate matrix mechanosensing and fate commitment of mesenchymal stem cells. *Nat. Mater.* **2016**, *15*, 1297–1306. [CrossRef] [PubMed]
176. Dorcemus, D.L.; George, E.O.; Dealy, C.N.; Nukavarapu, S.P. Harnessing External Cues: Development and Evaluation of an In Vitro Culture System for Osteochondral Tissue Engineering. *Tissue Eng. Part A* **2017**, *23*, 719–737. [CrossRef] [PubMed]
177. Hao, Y.; Song, J.; Ravikrishnan, A.; Dicker, K.T.; Fowler, E.W.; Zerdoum, A.B.; Li, Y.; Zhang, H.; Rajasekaran, A.K.; Fox, J.M.; et al. Rapid Bioorthogonal Chemistry Enables in Situ Modulation of the Stem Cell Behavior in 3D without External Triggers. *ACS Appl Mater. Interfaces* **2018**, *10*, 26016–26027. [CrossRef] [PubMed]
178. Patel, M.; Koh, W.-G.J.P. Composite Hydrogel of Methacrylated Hyaluronic Acid and Fragmented Polycaprolactone Nanofiber for Osteogenic Differentiation of Adipose-Derived. *Stem Cells* **2020**, *12*, 902. [CrossRef]
179. Ducret, M.; Montembault, A.; Josse, J.; Pasdeloup, M.; Celle, A.; Benchrih, R.; Mallein-Gerin, F.; Alliot-Licht, B.; David, L.; Farges, J.C. Design and characterization of a chitosan-enriched fibrin hydrogel for human dental pulp regeneration. *Dent. Mater.* **2019**, *35*, 523–533. [CrossRef] [PubMed]
180. Hashemzadeh, H.; Allahverdi, A.; Ghorbani, M.; Soleymani, H.; Kocsis, Á.; Fischer, M.B.; Ertl, P.; Naderi-Manesh, H. Gold Nanowires/Fibrin Nanostructure as Microfluidics Platforms for Enhancing Stem Cell Differentiation: Bio-AFM Study. *Micromachines* **2020**, *11*, 50. [CrossRef]
181. Spicer, C.D. Hydrogel scaffolds for tissue engineering: The importance of polymer choice. *Polym. Chem.* **2020**, *11*, 184–219. [CrossRef]
182. Pek, Y.S.; Wan, A.C.; Ying, J.Y. The effect of matrix stiffness on mesenchymal stem cell differentiation in a 3D thixotropic gel. *Biomaterials* **2010**, *31*, 385–391. [CrossRef] [PubMed]
183. Ye, K.; Wang, X.; Cao, L.; Li, S.; Li, Z.; Yu, L.; Ding, J. Matrix stiffness and nanoscale spatial organization of cell-adhesive ligands direct stem cell fate. *Nano Lett.* **2015**, *15*, 4720–4729. [CrossRef]
184. Steinmetz, N.J.; Aisenbrey, E.A.; Westbrook, K.K.; Qi, H.J.; Bryant, S.J. Mechanical loading regulates human MSC differentiation in a multi-layer hydrogel for osteochondral tissue engineering. *Acta Biomater.* **2015**, *21*, 142–153. [CrossRef] [PubMed]
185. Yang, Y.; Feng, Y.; Qu, R.; Li, Q.; Rong, D.; Fan, T.; Yang, Y.; Sun, B.; Bi, Z.; Khan, A.U. Synthesis of aligned porous polyethylene glycol/silk fibroin/hydroxyapatite scaffolds for osteoinduction in bone tissue engineering. *Stem. Cell Res. Ther.* **2020**, *11*, 1–17. [CrossRef]

186. Yang, C.; DelRio, F.W.; Ma, H.; Killaars, A.R.; Basta, L.P.; Kyburz, K.A.; Anseth, K.S. Spatially patterned matrix elasticity directs stem cell fate. *Proc. Natl. Acad. Sci. USA* **2016**, *113*, E4439–E4445. [CrossRef]
187. Gandavarapu, N.R.; Alge, D.L.; Anseth, K.S. Osteogenic differentiation of human mesenchymal stem cells on α5 integrin binding peptide hydrogels is dependent on substrate elasticity. *Biomater. Sci.* **2014**, *2*, 352–361. [CrossRef]
188. Chou, S.-Y.; Cheng, C.-M.; LeDuc, P.R. Composite polymer systems with control of local substrate elasticity and their effect on cytoskeletal and morphological characteristics of adherent cells. *Biomaterials* **2009**, *30*, 3136–3142. [CrossRef]
189. Viale-Bouroncle, S.; Völlner, F.; Möhl, C.; Küpper, K.; Brockhoff, G.; Reichert, T.E.; Schmalz, G.; Morsczeck, C. Soft matrix supports osteogenic differentiation of human dental follicle cells. *Biochem. Biophys. Res. Commun.* **2011**, *410*, 587–592. [CrossRef]
190. Viale-Bouroncle, S.; Gosau, M.; Küpper, K.; Möhl, C.; Brockhoff, G.; Reichert, T.E.; Schmalz, G.; Ettl, T.; Morsczeck, C. Rigid matrix supports osteogenic differentiation of stem cells from human exfoliated deciduous teeth (SHED). *Differentiation* **2012**, *84*, 366–370. [CrossRef]
191. Wang, P.-Y.; Tsai, W.-B.; Voelcker, N.H. Screening of rat mesenchymal stem cell behaviour on polydimethylsiloxane stiffness gradients. *Acta Biomater.* **2012**, *8*, 519–530. [CrossRef]
192. Khoramgah, M.S.; Ranjbari, J.; Abbaszadeh, H.-A.; Mirakabad, F.S.T.; Hatami, S.; Hosseinzadeh, S.; Ghanbarian, H. Freeze-dried multiscale porous nanofibrous three dimensional scaffolds for bone regenerations. *BioImpacts BI* **2020**, *10*, 73. [CrossRef]
193. Oh, S.H.; An, D.B.; Kim, T.H.; Lee, J.H. Wide-range stiffness gradient PVA/HA hydrogel to investigate stem cell differentiation behavior. *Acta Biomater.* **2016**, *35*, 23–31. [CrossRef]
194. Hwang, J.-H.; Byun, M.R.; Kim, A.R.; Kim, K.M.; Cho, H.J.; Lee, Y.H.; Kim, J.; Jeong, M.G.; Hwang, E.S.; Hong, J.H. Extracellular matrix stiffness regulates osteogenic differentiation through MAPK activation. *PLoS ONE* **2015**, *10*, e0135519. [CrossRef]
195. Sun, X.; Tung, W.; Wang, W.; Xu, X.; Zou, J.; Gould, O.E.; Kratz, K.; Ma, N.; Lendlein, A. The effect of stiffness variation of electrospun fiber meshes of multiblock copolymers on the osteogenic differentiation of human mesenchymal stem cells. *Clin. Hemorheol. Microcirc.* **2019**, *73*, 219–228. [CrossRef]
196. Shams, M.; Karimi, M.; Heydari, M.; Salimi, A. Nanocomposite scaffolds composed of Apacite (apatite-calcite) nanostructures, poly (ε-caprolactone) and poly (2-hydroxyethylmethacrylate): The effect of nanostructures on physico-mechanical properties and osteogenic differentiation of human bone marrow mesenchymal stem cells in vitro. *Mater. Sci. Eng. C Mater. Biol. Appl.* **2020**, *117*, 111271. [CrossRef]
197. Tang, X.; Ali, M.Y.; Saif, M.T.A. A Novel Technique for Micro-patterning Proteins and Cells on Polyacrylamide Gels. *Soft Matter* **2012**, *8*, 7197–7206. [CrossRef] [PubMed]
198. Daliri, K.; Pfannkuche, K.; Garipcan, B. Effects of physicochemical properties of polyacrylamide (PAA) and (polydimethylsiloxane) PDMS on cardiac cell behavior. *Soft Matter* **2021**, *17*, 1156–1172. [CrossRef]
199. Winer, J.P.; Janmey, P.A.; McCormick, M.E.; Funaki, M. Bone marrow-derived human mesenchymal stem cells become quiescent on soft substrates but remain responsive to chemical or mechanical stimuli. *Tissue Eng. Part A* **2009**, *15*, 147–154. [CrossRef]
200. Lee, J.; Abdeen, A.A.; Huang, T.H.; Kilian, K.A. Controlling cell geometry on substrates of variable stiffness can tune the degree of osteogenesis in human mesenchymal stem cells. *J. Mech. Behav. Biomed. Mater.* **2014**, *38*, 209–218. [CrossRef] [PubMed]
201. Stanton, A.E.; Tong, X.; Yang, F. Extracellular matrix type modulates mechanotransduction of stem cells. *Acta Biomater.* **2019**, *96*, 310–320. [CrossRef]
202. Gungordu, H.I.; Bao, M.; van Helvert, S.; Jansen, J.A.; Leeuwenburgh, S.C.G.; Walboomers, X.F. Effect of mechanical loading and substrate elasticity on the osteogenic and adipogenic differentiation of mesenchymal stem cells. *J. Tissue Eng. Regen. Med.* **2019**, *13*, 2279–2290. [CrossRef]
203. Hogrebe, N.J.; Gooch, K.J. Direct influence of culture dimensionality on human mesenchymal stem cell differentiation at various matrix stiffnesses using a fibrous self-assembling peptide hydrogel. *J. Biomed. Mater. Res. A* **2016**, *104*, 2356–2368. [CrossRef]
204. Wu, L.; Magaz, A.; Wang, T.; Liu, C.; Darbyshire, A.; Loizidou, M.; Emberton, M.; Birchall, M.; Song, W. Stiffness memory of indirectly 3D-printed elastomer nanohybrid regulates chondrogenesis and osteogenesis of human mesenchymal stem cells. *Biomaterials* **2018**, *186*, 64–79. [CrossRef]
205. Chen, G.; Xu, R.; Zhang, C.; Lv, Y. Responses of MSCs to 3D scaffold matrix mechanical properties under oscillatory perfusion culture. *ACS Appl. Mater.* **2017**, *9*, 1207–1218. [CrossRef]

polymers

MDPI

Article

Laser-Induced Graphene-Based Enzymatic Biosensor for Glucose Detection

Kalpana Settu *, Pin-Tzu Chiu and Yu-Ming Huang

Department of Electrical Engineering, National Taipei University, New Taipei City 23741, Taiwan; s410787003@gm.ntpu.edu.tw (P.-T.C.); idris.herondale@gmail.com (Y.-M.H.)
* Correspondence: kalpana@mail.ntpu.edu.tw; Tel.: +886-2867-4111-1 (ext. 68806)

Abstract: Laser-induced graphene (LIG) has recently been receiving increasing attention due to its simple fabrication and low cost. This study reports a flexible laser-induced graphene-based electrochemical biosensor fabricated on a polymer substrate by the laser direct engraving process. For this purpose, a 450 nm UV laser was employed to produce a laser-induced graphene electrode (LIGE) on a polyimide substrate. After the laser engraving of LIGE, the chitosan–glucose oxidase (GOx) composite was immobilized on the LIGE surface to develop the biosensor for glucose detection. It was observed that the developed LIGE biosensor exhibited good amperometric responses toward glucose detection over a wide linear range up to 8 mM. The GOx/chitosan-modified LIGE biosensor showed high sensitivity of 43.15 μA mM^{-1} cm^{-2} with a detection limit of 0.431 mM. The interference studies performed with some possible interfering compounds such as ascorbic acid, uric acid, and urea exhibited no interference as there was no difference observed in the amperometric glucose detection. It was suggested that the LIGE-based biosensor proposed herein was easy to prepare and could be used for low-cost, rapid, and sensitive/selective glucose detection.

Keywords: biosensor; laser-induced graphene; polyimide; glucose; enzyme

Citation: Settu, K.; Chiu, P.-T.; Huang, Y.-M. Laser-Induced Graphene-Based Enzymatic Biosensor for Glucose Detection. *Polymers* **2021**, *13*, 2795. https://doi.org/10.3390/polym13162795

Academic Editors: José Miguel Ferri and Claudio Gerbaldi

Received: 5 July 2021
Accepted: 17 August 2021
Published: 20 August 2021

1. Introduction

Numerous novel and cutting-edge technologies and materials are necessary to satisfy the new trends and requisites of analytical systems as the needs for environmental, biomedical, food and beverage analysis are progressing very quickly. The development of biosensors has evolved as one of the most promising research directions to overcome these challenges. Therefore, biosensor-based techniques have recently started being applied for the determination of different clinically, environmentally and biologically active materials [1–3]. In this regard, the design of biosensors in nanoscience/nanotechnology, environmental, medicine and food monitoring has been significantly increased during the past decade for their extensive applications. These advanced technologies have assisted the construction of highly sensitive, selective, customizable, and portable sensors for the determination of various clinically significant materials such as glucose, etc. [4]. The progress of such glucose biosensors has an inordinate significance in diagnosing and controlling diabetes mellitus, which is considered a worldwide public health problem. Diabetes mellitus would increase the risk of heart disease, kidney failure, blindness, postoperative and wound infections [5,6].

Diabetes mellitus has increased worldwide over the past five decades. Diabetes is a medical condition in which patients experience glucose concentration diverging from the normal range of 80–120 mg/dL (4.4–6.6 mM) [7]. In 2019, the International Diabetes Federation (IDF) assessments indicated that approximately 463 million adults have diabetes, and it might rise to 700 million by 2045 [8]. Diabetic patients are required to perform glucose testing several times a day to maintain normal glucose levels. Hence, the rapid quantification of glucose concentration in bodily fluids is vital for diagnosing and treating diabetic patients. For this purpose, the design of an easy, rapid and low-cost technology for the determination of glucose is essential in clinical diagnosis [9,10].

Glucose biosensors have significantly contributed to the detection of glucose levels in diabetic patients [11,12]. Studies have indicated that among various biosensors, glucose oxidase (GOx) enzyme-based electrochemical biosensors were considered to offer good selectivity and sensitivity for glucose detection [13,14]. Amperometry is the widely used electrochemical technique for glucose detection. Amperometric sensors could provide several advantages, such as ease of use, short analysis time, high sensitivity, and higher signal-to-noise ratio compared to other sensors [14–16]. The common idea applied for the development of amperometric biosensors is the efficiency of charge transfer, which can be better enhanced. Additionally, the biocompatibility issues of the sensors could be resolved by modifying electrodes with polymers such as chitosan or hydrogels [17]. In addition, various features of the electrodes could easily be altered by selecting the optimal chemical and electrochemical parameters during the effective electrode modifications [18,19].

The amperometric glucose biosensor generally uses an enzyme glucose oxidase (GOx), which catalyzes glucose oxidation at the electrode and provides high selectivity in glucose detection. Most enzymatic amperometric biosensors are based on disposable screen-printed enzyme electrode strips [20–22]. However, the wastage of materials might occur during the screen-printing process, limiting the applications of screen-printed electrodes.

Graphene, a carbon-based nanomaterial, has gained substantial attention in many areas. In terms of electrochemical properties, graphene could provide high conductivity with a remarkable heterogeneous electron transfer rate [23,24]. In 2014, it was found that polymers such as polyimide (PI) could be directly converted into porous graphene using a CO_2 laser machine with a 10.6 μm wavelength [25]. In addition to infrared CO_2 (10.6 μm) laser, visible laser [26–31] and ultraviolet laser [32] have also been successfully used to synthesize laser-induced graphene (LIG). The laser-irradiation of the PI film caused the photo-thermal generation of the graphene due to the local heating of the film. Upon heating the film, the carbon atoms bonded with oxygen (C–O, C=O) and nitrogen (C–N) atoms via sp^3 and sp^2 hybridization breakdown and rearranged to form several layers of sp^2 hybridized carbon atoms of graphene [25,33]. The laser induction of graphene has been performed in ambient conditions without any material wastage. In addition, the shape/pattern of LIG could also be easily customized by computer design, which holds great promise toward developing glucose biosensors.

Recently, Pereira et al. demonstrated the electrochemical response of GOx adsorbed on a CO_2 laser-scribed LIG [34]. The GOx enzyme adsorbed on LIG remained catalytically active even after running the cyclic voltammetry up to +1.0 V for glucose detection. The LIG electrodes facilitated the direct electron transfer between the GOx and the electrode surface without mediators.

In this study, we fabricated a laser-induced graphene electrode (LIGE) by simple direct laser engraving with the UV laser on polyimide tape. The LIGE surface was immobilized with GOx/Chitosan composite for selective detection on glucose. Amperometric measurement was used to quantify the glucose concentration with the developed LIGE enzymatic biosensor. The novelty of the present work lies in the detection of glucose with enhanced sensitivity using a simple, low-cost LIGE-based biosensor.

2. Materials and Methods

2.1. Chemicals and Instruments

Glucose, uric acid, ascorbic acid, chitosan, and glucose oxidase (GOx, from Aspergillus niger, Type X-S, lyophilized powder, 118,000 units/g solid) were purchased from Sigma-Aldrich Corp. (St. Louis, MO, USA). A single-sided Kapton® polyimide tape with a film thickness of ~30.4 μm and a width of 50 mm was obtained from STAREK Scientific Co., Ltd. (Taipei, Taiwan). Photo/printing paper (HYA300, A4—120 gm^{-2}, 0.15 mm) was purchased from a local book store. All the electrochemical measurements were conducted using a portable potentiostat (PalmSens 4, PalmSens, Houten, The Netherlands). Raman spectroscopic study was conducted using a micro-Raman spectrometer (JASCO NRS-4100; Laser 532 nm) with a spectral resolution of 2 cm^{-1}. Data processing/plotting was performed using Origin 9.1 software (OriginLab Inc., Northampton, MA, USA).

2.2. Fabrication of LIGE Sensor

A 3-electrode system was designed using AutoCAD software with a 3 mm diameter of working electrode and laser-inscribed to graphene-based electrodes. Kapton® polyimide tape was pasted onto a paper substrate and cleaned with isopropanol and deionized water. Then, the designed pattern made in graphic software was inscribed on the surface of the Kapton tape using a laser engraving machine (HANLIN 7WLS, 7 W, 450 nm) to form highly conductive graphene electrodes, as shown in Figure 1. The resistance of the graphene-based electrode was optimized by adjusting the laser power intensity (22% of the machine's maximum power), engraving depth (5%), the distance between the laser head and the polyimide substrate (~13 cm). The duration for fabricating a complete LIGE sensor was 2.8 min.

Figure 1. LIG 3-electrode system on polyimide tape fabricated by laser inscribing.

2.3. Immobilization of GOx/Chitosan Composite on the LIGE

The glucose biosensor was prepared by immobilizing the glucose oxidase and chitosan hydrogel homogeneous biocomposite on the LIGE surface. The resulting biocomposite could retain the enzyme bioactivity at considerably extreme conditions [35]. Five milligrams of GOx and three milligrams of chitosan were dissolved in 0.5 mL of deionized water and stirred for 5 min [36]. Subsequently, 5 µL of the mixture was cast onto the surface of the LIGE working electrode. Then, the LIGE sensor was kept in a refrigerator at 4 °C for 24 h.

2.4. Electrochemical Measurements

All the electrochemical measurements were carried out using PalmSens 4 potentiostat (PalmSens, Houten, The Netherlands) at room temperature. The electrochemical redox characteristics of the LIGE were measured by Cyclic voltammetry (CV) with different concentrations of potassium ferri (III)cyanide ($K_3[Fe(CN)_6]$) in 50 mM of phosphate-buffered solution (PBS). CV measurements were performed at a scan rate of 50 mV/s with a potential range from −0.8 to +0.8 V. Chronoamperometry (CA) experiments for glucose detection with LIGE were performed in 50 mM PBS at the fixed applied voltage of 0.8 V for 60 s. The detection principle of glucose is based on the electron transfer mechanism. GOx reacts with glucose in the presence of O_2 and produces gluconolactone and H_2O_2. A change in electrical current occurs at the electrode surface during these reactions due to the electron transfer. Additionally, the resulting current response is proportional to the number of glucose molecules present in the sample.

2.5. Optimization of Applied Potential and pH

CA measurements were used to determine the optimal applied potential and pH for glucose detection. The CA potential was optimized by varying the potential from 0.3 V to 1.3 V (5 mM Glucose, pH 7), and the resulting CA current was sampled at 60 s. CA measurements were performed with a LIGE biosensor at 5 mM glucose solution with the applied potential of 0.8 V by varying the pH of the phosphate-buffered solution from a pH of 5 to 9, and the optimal pH was found.

2.6. Interference Study

The response of the LIGE biosensor for glucose detection was evaluated in the presence of potential interferences such as 0.1 mM ascorbic acid, 0.1 mM uric acid, and 3 mM urea (pH 7; 5 mM glucose; 0.8 V).

3. Results and Discussion

3.1. Characterization of LIGE

3.1.1. Raman Spectra

In this study, a graphene three-electrode system for electrochemical sensing applications was developed by direct laser inscribing on polymer substrate (Polyimide). The prepared LIGE was characterized with Raman spectra, as shown in Figure 2. The Raman spectrum consists of G band at ca. 1592 cm^{-1} related to the E_{2g} phonon of the sp^2 carbon atoms, and D band at ca. 1340 cm^{-1} corresponds to the disordered grain boundaries [37,38]. Two other bands were observed at 2697 and 2900 cm^{-1}. The band at ca. 2700 cm^{-1} is known as the 2D band, an indicator of the number of graphene layers. A sharp peak will appear at ca. 2700 cm^{-1} for monolayer graphene. Here, the broadened band was observed, which would be attributed to the prepared graphene containing many layers with some defects. The band that appeared at 2900 cm^{-1} is called an S3 band, which is a second-order peak derived from the D–G peak combination. The band intensity ratio of S3–2D is proportional to the reduction in defects [38]. This Raman spectra result indicated that the obtained black material on polyimide substrate was carbon-based graphene.

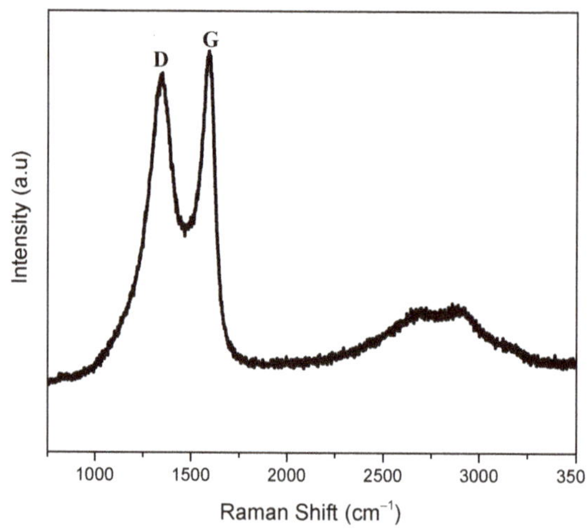

Figure 2. Raman spectra of LIGE.

3.1.2. Electrochemical Characterization

Before developing the glucose biosensor with LIGE, validating the LIGE sensor towards electrochemical sensing was necessary. The ferri/ferrocyanide ($Fe(CN)_6^{3-/4-}$) redox couple is one of the most widely used electron mediators for electrochemical reactions [39]. The performance of an electrochemical sensor towards an electron mediator was considered most relevant to general biochemical sensing applications. Thus, the electrochemical efficacy of the LIGE sensor was evaluated using cyclic voltammetry responses in different concentrations of ferricyanide redox mediator ($K_3[Fe(CN)_6]$), as shown in Figure 3a. As seen from Figure 3a, the oxidation peaks' current increased from 35.495 to 65.043 μA when the ferricyanide concentration ranged from 0.5 to 2.5 mM. The oxidation peak current showed an excellent linear relationship with different ferricyanide concentrations, as shown in Figure 3b. The linear regression equation was $y = 14.54x + 28.69$ $(R^2 = 0.998)$, where y and x are the height of oxidation peak current (μA) and ($K_3[Fe(CN)_6]$) concentration (mM), respectively. The fabricated LIGE provided a favorable response for varying ferricyanide concentrations, indicating excellent electrocatalytic properties. Moreover, the reproducibility of all CV responses was within 5% RSD (relative standard deviation) (n = 4). These results demonstrated the remarkable electrocatalytic response of the fabricated LIGE sensor.

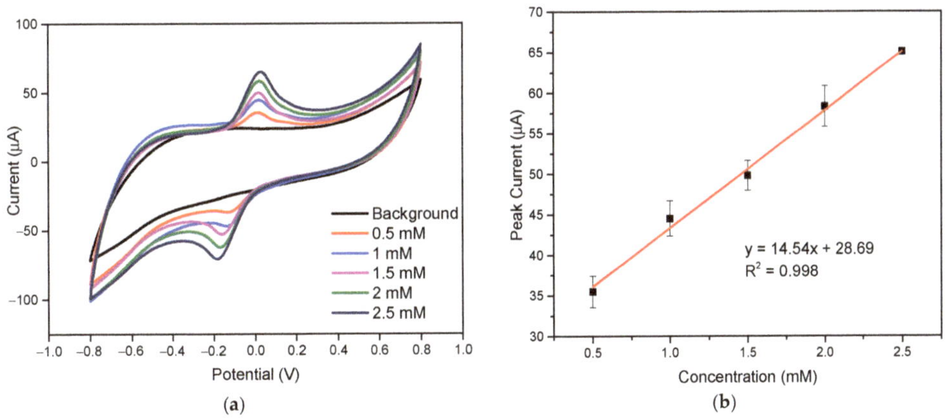

Figure 3. (**a**) CV responses of ferricyanide solutions with varying concentrations; and (**b**) oxidation current peaks vs. concentration. Scan rate was 50 mV s^{-1}.

3.2. Characterization of GOx/Chitosan Immobilized LIGE

Cyclic voltammetry measurement was performed to confirm the LIGE immobilization with GOx/Chitosan. Figure 4 shows the cyclic voltammograms of potassium ferricyanide at bare LIGE and GOx/chitosan composite-modified LIGE. It can be seen that after the immobilization of GOx/chitosan composite onto the LIGE surface, the peak current decreased to 24.325 from 58.336 μA of the bare LIGE. The electron transfer kinetics of $[Fe(CN)6]^{4-}/[Fe(CN)6]^{3-}$ is significantly hindered after the LIGE surface was modified with GOx/chitosan. This result confirmed that the GOx/chitosan was successfully immobilized on the LIGE surface.

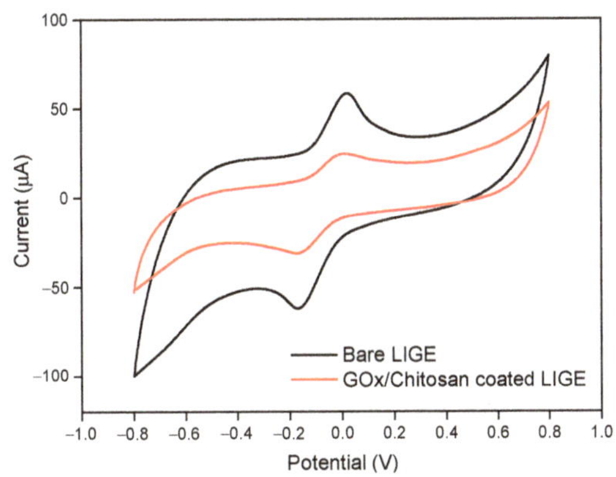

Figure 4. Cyclic voltammograms on bare LIGE and GOx/chitosan-modified LIGE in the presence of 2 mM potassium ferricyanide.

3.3. Amperometric Detection of Glucose by the Proposed LIGE

The chronoamperometry technique was employed to detect glucose using GOx/Chitosan coated LIGE sensor at a constant oxidation potential of +0.8 V. Figure 5a depicts the chronoamperometric responses of the LIGE biosensor with glucose concentrations ranging from 0 to 10 mM. The current response increased with increasing glucose concentrations. The steady-state current response at 60 s was chosen for the detection of glucose concentration. The amperometric current response of the LIGE biosensor exhibited a linear relationship with the glucose concentrations ranging from 0 to 8 mM, and the current began to level off at a glucose concentration higher than 8 mM as shown in Figure 5b. The linear regression equation was $y = 3.05x + 8.54$, with a coefficient of determination R^2 = 0.97 and a sensitivity of 43.15 μA mM^{-1} cm^{-2}. The limit of detection was calculated according to the $3s_a/b$ criterion, where b was the slope of the calibration curve, and s_a was the estimated standard deviation of the y-intercepts of the regression line [3]. The detection limit calculated was 0.431 mM. As seen from Figure 5b, the linear part of the calibration curve includes the normal glucose levels (4.4 to 6.6 mM) in the human blood. Thus, this study could offer a simple approach for the clinical glucose measurement with a disposable LIGE-based biosensor. The performance of the proposed biosensor was compared with other reported glucose biosensors, as shown in Table 1. The developed LIGE-based enzymatic glucose biosensor exhibited good analytical characteristics towards glucose detection such as good linearity and high sensitivity. Moreover, the fabrication and detection procedures of the proposed LIGE-based biosensor were also simple, rapid, and cost-effective.

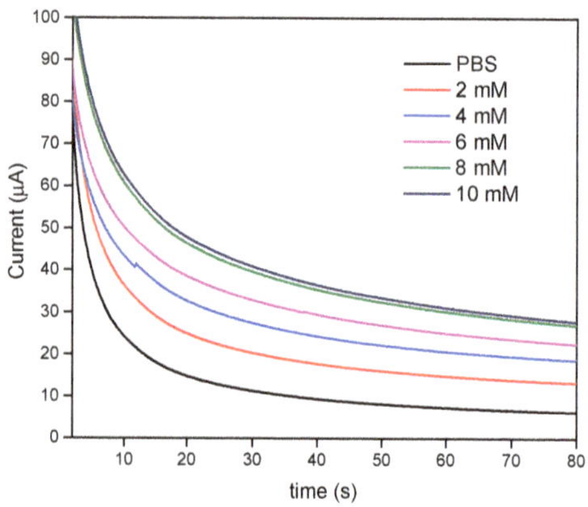

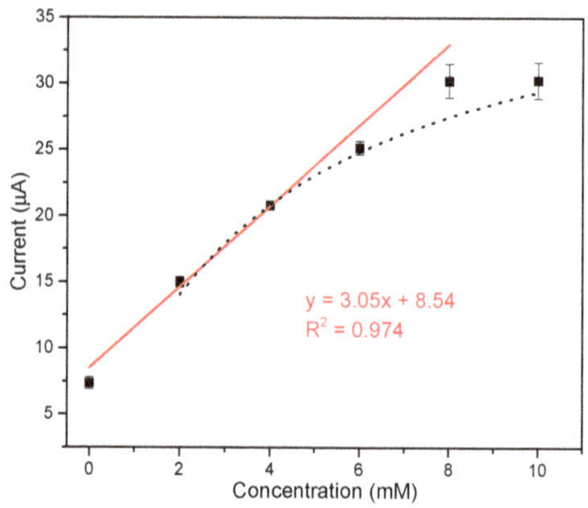

Figure 5. (**a**) Chronoamperometry response with different concentrations of glucose; (**b**) the relationship between the glucose concentration and the chronoamperometric current response at 60 s.

Table 1. Comparison of the analytical performance of glucose biosensors.

Glucose Biosensor [a]	Sensitivity (μA mM^{-1} cm^{-2})	Linear Range (mM)	LOD (μM)	Reference
GOx/Chitosan-modified LIGE	43.15	0–8	431	This work
GC/MWCNT/Fe_3O_4/PDA–GOx	5.04	2–20	2.25	[40]
LSG/PBSE/PtNPs/GOx	12.64	0.005–3.2	2.57	[41]
MoS_2/Chitosan/GOx-Gelatin/PGE	0.8 (μA mM^{-1})	0.01–0.8	3.18	[42]
CPE/GOx-SiO_2/Lig	0.78	0.5–9	145	[19]
Au–Cys–GA–Gox	2.65	1.5–7	940	[43]
PPy/GOD/SPCE	0.21	0–5	-	[44]

[a] GC—glassy carbon electrode; MWCNT—multi-walled carbon nanotubes; Fe_3O_4/PDA—magnetite/polydopamine; LSG—laser-scribed graphene; PBSE—pyrenebutanoic acid-succinimide ester; PtNPs—platinum nanoparticles; MoS_2—molybdenum disulfide; PGE—pencil graphite electrode; CPE—carbon paste electrode; SiO_2/Lig—silica/lignin; Cys—cysteine; GA—glutaraldehyde; PPy—polypyrrole; SPCE—screen-printed carbon electrodes; GOD/GOx—Glucose oxidase.

3.4. Michaelis–Menten Kinetics

The maximum response current (I_{max}) and the apparent Michaelis–Menten constant (K_m^{app}) were used to analyze the relationship between chronoamperometric signals and enzymatic reaction. As shown in Figure 5b, when glucose concentration exceeds 6 mM, a response plateau was observed with the characteristic of Michaelis–Menten kinetic mechanism. From the calibration plot (Figure 5b dotted line), the current response showed hyperbolic dependence on glucose concentration and was in good agreement with Michaelis–Menten kinetics [45]. The kinetic parameters, the maximum current generated during the enzymatic reaction (I_{max}) and the apparent Michaelis constant (K_m^{app}) are the corresponding a and b parameters of hyperbolic function $y = ax/(b + x)$ [46]. The apparent Michaelis–Menten constant (K_m^{app}) is an indication of enzymatic mimics–substrate kinetics. From the hyperbolic calibration plot (Figure 5b dotted line), the I_{max} and Michaelis constant K_m^{app} were 40.34 μA and 3.75 mM, respectively. The value of K_m^{app} is consistent with the reported value (K_m^{app} = 3.84 mM) for other GOx immobilized on the chitosan complex over triangular silver nanoprisms/platinum biosensor [47].

3.5. Optimization of Applied Potential and Buffer pH

Chronoamperometry measurements were used to determine the optimal applied potential and pH for glucose detection with the developed enzymatic LIGE biosensor. Figure 6a shows the chronoamperometric response of the LIGE biosensor at 60 s with different applied potential values ranging from 0.3 to 1.3 V. The results showed that the current increased with increasing applied potential from 0.3 to 0.8 V and currents tended to level off when the potential increased beyond 0.8 V. Thus, 0.8 V was selected as the optimized potential for amperometric glucose detection. Figure 6b illustrates the chronoamperometry current response of the biosensor as a function of the pH of PBS containing 2 mM glucose. The current responses at pH 5, pH 6, and pH 7 were almost similar. Considering the pH of a physiological buffer, pH 7 was chosen for the glucose detection experiments.

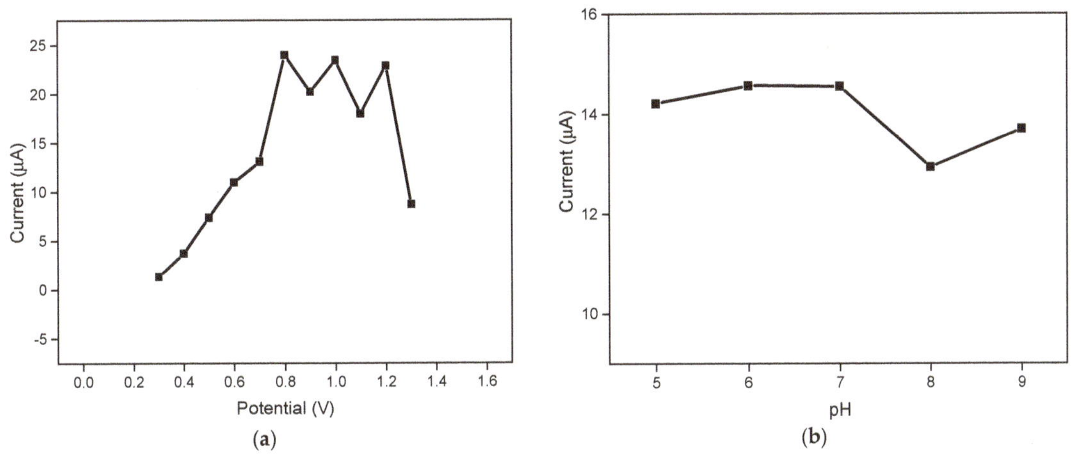

Figure 6. (**a**) Chronoamperometry response at 60 s in different applied potential with 5 mM glucose; and (**b**) Chronoamperometry response at 60 s in different buffer pH with 2 mM glucose.

3.6. Interference Study

The developed LIGE-based enzymatic biosensor was evaluated with possible interferences by comparing the chronoamperometric responses before and after adding some interferents such as ascorbic acid (0.1 mM), uric acid (0.1 mM), and urea (3 mM) in 5 mM glucose. As shown in Figure 7, the chronoamperometric current responses for glucose without and with interferents showed practically no interference. The LIGE was modified with GOx, which is the standard enzyme for biosensors and it has relatively higher selectivity for glucose [48]. Hence, the LIGE biosensor was suggested to possess good selectivity due to the specificity of the GOx enzyme.

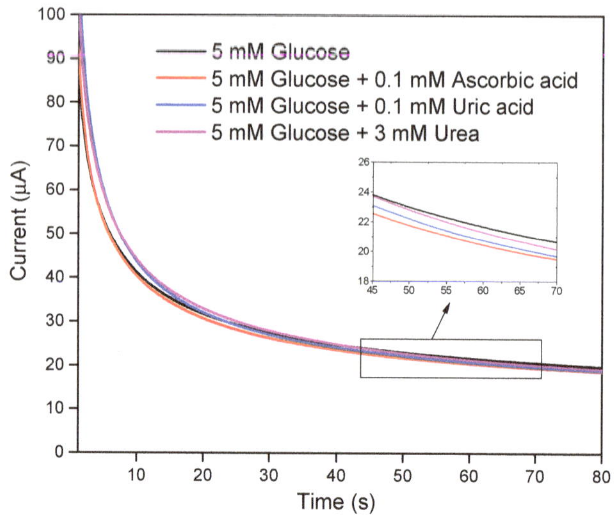

Figure 7. Chronoamperograms of 5 mM glucose with/without interferences. The inset shows the zoomed part of the result from 45 to 70 s.

3.7. Stability and Reproducibility of Biosensor

The stability of the developed GOx/chitosan-modified LIGE biosensor was evaluated by measuring the amperometric current response in the presence of 5 mM glucose over 25 days stored at 4 °C in a refrigerator. The biosensor exhibited ~90% stability for 10 days, and the response remained approximately 72–85% after 10 days. The reproducibility of the developed biosensor was assessed from the current response of different biosensors prepared independently. In this work, all the measurements were taken from at least three independent sensors ($n \geq 3$), and the reproducible signals were obtained with the RSD less than 6%.

4. Conclusions

We developed a simple laser-induced graphene-based enzymatic biosensor for glucose detection. The proposed detection strategy could offer an easy and low-cost route to mass-produce sensitive biosensing electrodes. The chronoamperometric measurements successfully detected the glucose over a linear range from 0 to 8 mM with a detection limit of 0.431 mM. The biosensor response was not affected by interfering compounds (ascorbic acid, uric acid and urea) and demonstrated the high specificity and selectivity of this LIGE biosensor in glucose detection. The proposed LIGE biosensor holds excellent promise in point-of-care diagnosis. Our future study aims to validate the biosensor response in human blood samples for real-life applications.

Author Contributions: K.S.: conceptualization, supervision, validation, writing—review and editing; P.-T.C.: methodology, validation, data curation, writing—original draft; Y.-M.H.: validation, data curation. All authors have read and agreed to the published version of the manuscript.

Funding: This research was supported under Grants MOST 110-2222-E-305-001 and MOST 107-2221-E-305-012-MY3 by the Ministry of Science and Technology, Taiwan, and Grant 109-NTPU_ORDA-F-008 by National Taipei University, Taiwan.

Institutional Review Board Statement: Not applicable.

Informed Consent Statement: Not applicable.

Conflicts of Interest: The authors declare no conflict of interest.

References

1. Ramanavicius, S.; Ramanavicius, A. Charge Transfer and Biocompatibility Aspects in Conducting Polymer-Based Enzymatic Biosensors and Biofuel Cells. *Nanomaterials* **2021**, *11*, 371. [CrossRef] [PubMed]
2. Ramanavicius, S.; Ramanavicius, A. Conducting Polymers in the Design of Biosensors and Biofuel Cells. *Polymers* **2021**, *13*, 49. [CrossRef]
3. Ridhuan, N.S.; Razak, K.A.; Lockman, Z. Fabrication and Characterization of Glucose Biosensors by Using Hydrothermally Grown ZnO Nanorods. *Sci. Rep.* **2018**, *8*, 13722. [CrossRef] [PubMed]
4. German, N.; Ramanaviciene, A.; Ramanavicius, A. Dispersed Conducting Polymer Nanocomposites with Glucose Oxidase and Gold Nanoparticles for the Design of Enzymatic Glucose Biosensors. *Polymers* **2021**, *13*, 2173. [CrossRef]
5. Batool, R.; Rhouati, A.; Nawaz, M.H.; Hayat, A.; Marty, J.L. A review of the construction of nano-hybrids for electro-chemical biosensing of glucose. *Biosensors* **2019**, *9*, 46. [CrossRef]
6. Labib, M.; Sargent, E.H.; Kelley, S.O. Electrochemical Methods for the Analysis of Clinically Relevant Biomolecules. *Chem. Rev.* **2016**, *116*, 9001–9090. [CrossRef] [PubMed]
7. Blunt, B.A.; Barrett-Connor, E.; Wingard, D.L. Evaluation of Fasting Plasma Glucose as Screening Test for NIDDM in Older Adults: Rancho Bernardo Study. *Diabetes Care* **1991**, *14*, 989–993. [CrossRef]
8. *IDF DIABETES ATLAS*, 9th ed.; International Diabetes Federation, 2019; pp. 1–176. Available online: https://diabetesatlas.org/en/resources/ (accessed on 1 July 2021).
9. German, N.; Ramanaviciene, A.; Ramanavicius, A. Formation and Electrochemical Evaluation of Polyaniline and Polypyrrole Nanocomposites Based on Glucose Oxidase and Gold Nanostructures. *Polymers* **2020**, *12*, 3026. [CrossRef]
10. German, N.; Ramanaviciene, A.; Ramanavicius, A. Formation of polyaniline and polypyrrole nanocomposites with em-bedded glucose oxidase and gold nanoparticles. *Polymers* **2019**, *11*, 377. [CrossRef]
11. Gopalan, A.; Muthuchamy, N.; Lee, K. A novel bismuth oxychloride-graphene hybrid nanosheets based non-enzymatic photo-electrochemical glucose sensing platform for high performances. *Biosens. Bioelectron.* **2017**, *89*, 352–360. [CrossRef]
12. Sridara, T.; Upan, J.; Saianand, G.; Tuantranont, A.; Karuwan, C.; Jakmunee, J. Non-Enzymatic Amperometric Glucose Sensor Based on Carbon Nanodots and Copper Oxide Nanocomposites Electrode. *Sensors* **2020**, *20*, 808. [CrossRef]
13. Liu, S.; Ju, H. Reagentless glucose biosensor based on direct electron transfer of glucose oxidase immobilized on colloidal gold modified carbon paste electrode. *Biosens. Bioelectron.* **2003**, *19*, 177–183. [CrossRef]
14. Malitesta, C.; Palmisano, F.; Torsi, L.; Zambonin, P.G. Glucose fast-response amperometric sensor based on glucose oxidase immobilized in an electropolymerized poly(o-phenylenediamine) film. *Anal. Chem.* **1990**, *62*, 2735–2740. [CrossRef]
15. Gorton, L.; Bremle, G.; Csöregi, E.; Jönsson-Pettersson, G.; Persson, B. Amperometric glucose sensors based on immobilized glucose-oxidizing enzymes and chemically modified electrodes. *Anal. Chim. Acta* **1991**, *249*, 43–54. [CrossRef]
16. Koudelka, M.; Gernet, S.; De Rooij, N. Planar amperometric enzyme-based glucose microelectrode. *Sens. Actuators* **1989**, *18*, 157–165. [CrossRef]
17. Tajik, S.; Beitollahi, H.; Nejad, F.G.; Dourandish, Z.; Khalilzadeh, M.A.; Jang, H.W.; Venditti, R.A.; Varma, R.S.; Shokouhimehr, M. Recent Developments in Polymer Nanocomposite-Based Electrochemical Sensors for Detecting Environmental Pollutants. *Ind. Eng. Chem. Res.* **2021**, *60*, 1112–1136. [CrossRef]
18. Andriukonis, E.; Celiesiute-Germaniene, R.; Ramanavicius, S.; Viter, R.; Ramanavicius, A. From microorganism-based amperometric biosensors towards microbial fuel cells. *Sensors* **2021**, *21*, 2442. [CrossRef]
19. Jędrzak, A.; Rębiś, T.; Klapiszewski, Ł.; Zdarta, J.; Milczarek, G.; Jesionowski, T. Carbon paste electrode based on functional GOx/silica-lignin system to prepare an amperometric glucose biosensor. *Sens. Actuators B Chem.* **2018**, *256*, 176–185. [CrossRef]
20. Crouch, E.; Cowell, D.C.; Hoskins, S.; Pittson, R.W.; Hart, J.P. Amperometric, screen-printed, glucose biosensor for analysis of human plasma samples using a biocomposite water-based carbon ink incorporating glucose oxidase. *Anal. Biochem.* **2005**, *347*, 17–23. [CrossRef]
21. Gao, Q.; Guo, Y.; Zhang, W.; Qi, H.; Zhang, C. An amperometric glucose biosensor based on layer-by-layer GOx-SWCNT conjugate/redox polymer multilayer on a screen-printed carbon electrode. *Sens. Actuators B Chem.* **2011**, *153*, 219–225. [CrossRef]
22. Guan, W.-J.; Li, Y.; Chen, Y.-Q.; Zhang, X.-B.; Hu, G.-Q. Glucose biosensor based on multi-wall carbon nanotubes and screen printed carbon electrodes. *Biosens. Bioelectron.* **2005**, *21*, 508–512. [CrossRef]
23. Altuntas, D.B.; Tepeli, Y.; Anik, U. Graphene-metallic nanocomposites as modifiers in electrochemical glucose biosensor transducers. *2D Mater.* **2016**, *3*, 034001. [CrossRef]
24. Pumera, M. Graphene in biosensing. *Mater. Today* **2011**, *14*, 308–315. [CrossRef]
25. Lin, J.; Peng, Z.; Liu, Y.; Ruiz-Zepeda, F.; Ye, R.; Samuel, E.L.G.; Yacaman, M.J.; Yakobson, B.I.; Tour, J.M. Laser-induced porous graphene films from commercial polymers. *Nat. Commun.* **2014**, *5*, 5714. [CrossRef]
26. Zhang, Z.; Song, M.; Hao, J.; Wu, K.; Li, C.; Hu, C. Visible light laser-induced graphene from phenolic resin: A new approach for directly writing graphene-based electrochemical devices on various substrates. *Carbon* **2018**, *127*, 287–296. [CrossRef]

27. Romero, F.J.; Salinas-Castillo, A.; Rivadeneyra, A.; Albrecht, A.; Godoy, A.; Morales, D.P.; Rodriguez, N. In-Depth Study of Laser Diode Ablation of Kapton Polyimide for Flexible Conductive Substrates. *Nanomaterials* **2018**, *8*, 517. [CrossRef]
28. Stanford, M.G.; Zhang, C.; Fowlkes, J.D.; Hoffman, A.; Ivanov, I.N.; Rack, P.D.; Tour, J.M. High-Resolution Laser-Induced Graphene. Flexible Electronics beyond the Visible Limit. *ACS Appl. Mater. Interfaces* **2020**, *12*, 10902–10907. [CrossRef]
29. Tao, L.-Q.; Tian, H.; Liu, Y.; Ju, Z.-Y.; Pang, Y.; Chen, Y.-Q.; Wang, D.-Y.; Tian, X.-G.; Yan, J.-C.; Deng, N.-Q.; et al. An intelligent artificial throat with sound-sensing ability based on laser induced graphene. *Nat. Commun.* **2017**, *8*, 14579. [CrossRef]
30. Bobinger, M.R.; Romero, F.J.; Salinas-Castillo, A.; Becherer, M.; Lugli, P.; Morales, D.P.; Rodríguez, N.; Rivadeneyra, A. Flexible and robust laser-induced graphene heaters photothermally scribed on bare polyimide substrates. *Carbon* **2019**, *144*, 116–126. [CrossRef]
31. Cai, J.; Lv, C.; Watanabe, A. Cost-effective fabrication of high-performance flexible all-solid-state carbon mi-cro-supercapacitors by blue-violet laser direct writing and further surface treatment. *J. Mater. Chem. A* **2016**, *4*, 1671–1679. [CrossRef]
32. Carvalho, A.F.; Fernandes, A.J.S.; Leitao, C.; Deuermeier, J.; Marques, A.; Martins, R.; Fortunato, E.; Costa, F.M. Laser-Induced Graphene Strain Sensors Produced by Ultraviolet Irradiation of Polyimide. *Adv. Funct. Mater.* **2018**, *28*, 1805271. [CrossRef]
33. Tehrani, F.; Bavarian, B. Facile and scalable disposable sensor based on laser engraved graphene for electrochemical detection of glucose. *Sci. Rep.* **2016**, *6*, 1–10. [CrossRef] [PubMed]
34. Pereira, S.; Santos, N.; Carvalho, A.; Fernandes, A.; Costa, F. Electrochemical Response of Glucose Oxidase Adsorbed on Laser-Induced Graphene. *Nanomaterials* **2021**, *11*, 1893. [CrossRef]
35. Luo, X.-L.; Xu, J.-J.; Du, Y.; Chen, H.-Y. A glucose biosensor based on chitosan–glucose oxidase–gold nanoparticles bio-composite formed by one-step electrodeposition. *Anal. Biochem.* **2004**, *334*, 284–289. [CrossRef] [PubMed]
36. Yoon, H.; Nah, J.; Kim, H.; Ko, S.; Sharifuzzaman, M.; Barman, S.C.; Xuan, X.; Kim, J.; Park, J.Y. A chemically modified laser-induced porous graphene based flexible and ultrasensitive electrochemical biosensor for sweat glucose detection. *Sens. Actuators B Chem.* **2020**, *311*, 127866. [CrossRef]
37. Si, Y.; Samulski, E.T. Synthesis of water soluble graphene. *Nano Lett.* **2008**, *8*, 1679–1682. [CrossRef]
38. Johra, F.T.; Lee, J.-W.; Jung, W.-G. Facile and safe graphene preparation on solution based platform. *J. Ind. Eng. Chem.* **2014**, *20*, 2883–2887. [CrossRef]
39. Sriprachuabwong, C.; Karuwan, C.; Wisitsorrat, A.; Phokharatkul, D.; Lomas, T.; Sritongkham, P.; Tuantranont, A. Inkjet-printed graphene-PEDOT:PSS modified screen printed carbon electrode for biochemical sensing. *J. Mater. Chem.* **2012**, *22*, 5478–5485. [CrossRef]
40. Kuznowicz, M.; Jędrzak, A.; Rębiś, T.; Jesionowski, T. Biomimetic magnetite/polydopamine/β-cyclodextrins nanocomposite for long-term glucose measurements. *Biochem. Eng. J.* **2021**, *174*, 108127. [CrossRef]
41. Hossain, F.; Slaughter, G. Flexible electrochemical uric acid and glucose biosensor. *Bioelectrochemistry* **2021**, *141*, 107870. [CrossRef]
42. Altuntaş, D.B.; Kuralay, F. MoS2/Chitosan/GOx-Gelatin modified graphite surface: Preparation, characterization and its use for glucose determination. *Mater. Sci. Eng. B* **2021**, *270*, 115215. [CrossRef]
43. Lović, J.; Stevanović, S.; Nikolić, N.D.; Petrović, S.; Vuković, D.; Prlainović, N.; Mijin, D.; Ivić, M.A. Glucose Sensing Using Glucose Oxidase-Glutaraldehyde-Cysteine Modified Gold Electrode. *Int. J. Electrochem. Sci.* **2017**, *12*, 5806–5817. [CrossRef]
44. Weng, B.; Morrin, A.; Shepherd, R.; Crowley, K.; Killard, A.J.; Innis, P.C.; Wallace, G.G. Wholly printed polypyrrole na-noparticle-based biosensors on flexible substrate. *J. Mater. Chem. B* **2014**, *2*, 793–799. [CrossRef] [PubMed]
45. Gao, Z.; Lin, Y.; He, Y.; Tang, D. Enzyme-free amperometric glucose sensor using a glassy carbon electrode modified with poly(vinyl butyral) incorporating a hybrid nanostructure composed of molybdenum disulfide and copper sulfide. *Microchim. Acta* **2017**, *184*, 807–814. [CrossRef]
46. Anusha, J.; Raj, C.J.; Cho, B.-B.; Fleming, A.T.; Yu, K.-H.; Kim, B.C. Amperometric glucose biosensor based on glucose oxidase immobilized over chitosan nanoparticles from gladius of *Uroteuthis duvauceli*. *Sens. Actuators B Chem.* **2015**, *215*, 536–543. [CrossRef]
47. Shi, W.; Ma, Z. Amperometric glucose biosensor based on a triangular silver nanoprisms/chitosan composite film as immobilization matrix. *Biosens. Bioelectron.* **2010**, *26*, 1098–1103. [CrossRef]
48. Bankar, S.; Bule, M.V.; Singhal, R.; Ananthanarayan, L. Glucose oxidase—An overview. *Biotechnol. Adv.* **2009**, *27*, 489–501. [CrossRef]

polymers MDPI

Communication

Investigation of the Thermal and Hydrolytic Degradation of Polylactide during Autoclave Foaming

Julia Dreier [1], Christian Brütting [2], Holger Ruckdäschel [2], Volker Altstädt [2] and Christian Bonten [1,*]

1 Institut für Kunststofftechnik, University of Stuttgart, Pfaffenwaldring 32, 70569 Stuttgart, Germany; julia.dreier@ikt.uni-stuttgart.de
2 Department of Polymer Engineering, University of Bayreuth, Universitätsstraße 30, 95477 Bayreuth, Germany; christian.bruetting@uni-bayreuth.de (C.B.); ruckdaeschel@uni-bayreuth.de (H.R.); volker.altstaedt@uni-bayreuth.de (V.A.)
* Correspondence: christian.bonten@ikt.uni-stuttgart.de

Abstract: Polylactide (PLA) is one of the most important bioplastics worldwide and thus represents a good potential substitute for bead foams made of the fossil-based Polystyrene (PS). However, foaming of PLA comes with a few challenges. One disadvantage of commercially available PLA is its low melt strength and elongation properties, which play an important role in foaming. As a polyester, PLA is also very sensitive to thermal and hydrolytic degradation. Possibilities to overcome these disadvantages can be found in literature, but improving the properties for foaming of PLA as well as the degradation behavior during foaming have not been investigated yet. In this study, reactive extrusion on a twin-screw extruder is used to modify PLA in order to increase the melt strength and to protect it against thermal degradation and hydrolysis. PLA foams are produced in an already known process from the literature and the influence of the modifiers on the properties is estimated. The results show that it is possible to enhance the foaming properties of PLA and to protect it against hydrolysis at the same time.

Keywords: polylactide; biofoam; hydrolysis; degradation

Citation: Dreier, J.; Brütting, C.; Ruckdäschel, H.; Altstädt, V.; Bonten, C. Investigation of the Thermal and Hydrolytic Degradation of Polylactide during Autoclave Foaming. *Polymers* **2021**, *13*, 2624. https://doi.org/10.3390/polym13162624

Academic Editor: José Miguel Ferri

Received: 29 June 2021
Accepted: 30 July 2021
Published: 6 August 2021

1. Introduction

Degradation describes any type of mechanism in which there is a reduction in molecular weight and consequently a shortening of the polymer chains. These include hydrolysis, enzymatic oxidation, photooxidation and auto-oxidation. Mechanical and thermal as well as UV radiation-induced degradation can also occur. All these processes take place without the presence of microorganisms, which is why they are called abiotic degradation processes. This is their difference from biodegradation, which involves microorganisms. The abiotic degradation processes can lead to fragmentation and the formation of small microparticles, which in turn can be metabolized by microorganisms [1,2].

PLA, as with many other bioplastics, is very sensitive to thermal and hydrolytic degradation, which is characteristic of all polyesters. In this process, cleavage takes place at the hydrolyzable groups, such as esters, by water molecules. The structural formula of PLA is shown below in Figure 1, with the functional groups color-coded.

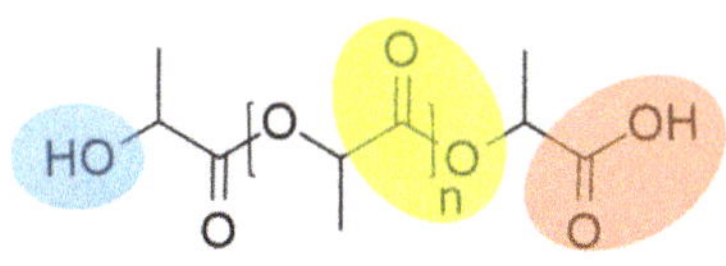

Figure 1. Chemical structure of PLA with its functional groups.

In hydrolysis, a distinction must be made between two different types, acidic and basic ester hydrolysis. Both reactions can occur in PLA and it must be noted whether the reaction takes place within the chain or at the end groups and under which conditions. Hydrolysis is influenced by several parameters, including on the one hand the prevailing environmental conditions such as water activity, temperature, pH and time [3,4]. On the other hand, the degree of crystallization, molar mass, size and geometry of the samples, stereo complex formation, number of acid end groups and hydrophobicity also play a crucial role [5–8]. A special feature in PLA is also the occurrence of the so-called autocatalyzed hydrolysis. The mechanism is similar to that of acid ester hydrolysis. The proton of the carboxyl group catalyzes the hydrolysis reaction. The proton activates the carbonyl group and makes it more susceptible to attack by water molecules. Under neutral conditions, a slower degradation takes place and in alkaline media, a faster degradation takes place than in acidic environments [1,4,9]. Hydrolysis often occurs during processing (for example extrusion or injection molding) under the influence of high temperatures. One way to counteract this is to pre-dry the PLA pellets very well before processing or to add stabilizers during processing [6,7,10]. However, hydrolysis can occur not only at high temperatures, but also at relatively moderate temperatures around 60 °C and under the influence of increased humidity, such as under industrial compost conditions [11].

Hydrolytic degradation does not necessarily end in complete decomposition of the material. However, it must be considered in order to identify and minimize downgrading of the polymer properties. The literature contains numerous papers and patents that counteract undesirable degradation with chemical modifiers [3,6–8,10,12–17]. The substance classes epoxides, carbodiimides and phosphorous acid esters have shown the most promise to date [10,13]. Most of these modifiers react with the end groups of PLA to inhibit the hydrolysis. The determination of the acid value is a method to verify possible reactions between modifiers and PLA and degradation through processing. Although there are some ways to prevent the hydrolysis of PLA, to the best of our knowledge, this was not investigated for foaming. Especially in bead foaming, PLA has to undergo different processing steps, such as compounding, foaming and welding.

Often, biopolymers such as PLA are said to be potential alternatives in packaging applications. Here, expanded PS (EPS) and expanded Polypropylene (EPP) are the market leaders due to their possible complex geometries compared with their low densities. In order to compete with these materials, PLA bead foams have to be made and fused together. Standau et al. [18] showed different ways of producing bead foams. As an example, expanded PLA (EPLA) can be made by a stirring autoclave process described by Nofar [19]. This process is characterized by a water-polymer mixture which is processed at temperatures far above 100°C, which leads to a tremendous degradation and therefore a loss in mechanical stability. To be competitive with polyolefine bead foams, the used polylactides need to be modified in order to prevent degradation during processing.

As a conclusion, PLA suffers from hydrolytic degradation during processing. Therefore, several modifiers have been used to increase the molecular structure and prevent the PLA from degradation during processing. Modified samples have been processed in a stirring autoclave process according to the literature to evaluate the stabilizing effect of the modifiers.

2. Experimental

2.1. Materials

In this study, PLA Ingeo 2003 D from NatureWorks Ltd. (Minnetonka, USA) was modified with two different chemical modifiers. Dicumyl peroxide (DCUP) from Sigma Aldrich was used as a melt strength enhancer and poly carbodiimide from Lanxess, Cologne, Germany as a hydrolysis stabilizer. Both modifiers were incorporated by reactive extrusion with a ZSK 26 twin screw extruder (Coperion GmbH, Stuttgart, Germany). The concentrations were selected based on pre-trials with 0.1 and 0.2 wt% for the peroxide and 0.5 wt% for the poly carbodiimide.

2.2. Methods

2.2.1. Rheological Investigation

Using a Discovery HR-2 plate-to-plate rheometer from TA Instruments, USA (plate diameter 25 mm; plate spacing 1 mm), the linear viscoelastic deformation behavior of PLA and the modified grades was determined. Prior to measurement, the materials were dried overnight at 40 °C in a vacuum oven. The viscosity measurements were then carried out under a nitrogen atmosphere. To determine the thermal stability and to investigate if the chemical reactions were completed, time-sweep measurements were performed at a constant angular velocity of 1 rad/s, a shear stress amplitude of 5 % and a measurement temperature of 180 °C. The measurement time was 1800 s. A frequency sweep was performed to determine the viscoelastic flow behavior. Here, at a shear stress amplitude of 5% and a temperature of 180 °C, the angular velocity is varied from 500 to 0.01 rad/s.

2.2.2. Size Exlcusion Chromatography (SEC)

An Agilent 1260 (Agilent, Waldbronn, Germany), equipped with a PSS-SDV precolumn (10 µm) 50 mm × 8 mm and three linear PSS-SDV columns (10 µm) 300 mm × 8 mm (with pore sizes of 103 Å, 105 Å and 106 Å), from PSS (Mainz, Germany) was used to determine the molar mass. Measurements were made in chloroform at a temperature of 30 °C and a flow rate of 1.0 mL/min. Using a PSS SECurtiy 1260 differential detector (PSS, Mainz, Germany), the molar mass could be calculated. In order to investigate the influence of the modification and the chemical constitution, GPC measurements were performed using a PSS DVD 1260 viscosity detector (PSS, Mainz, Germany).

2.2.3. Acid Value Determination

The acid number was determined according to EN ISO 2114:2000 to verify whether the acid end group had been modified by reactive extrusion. For the subsequent determination of the acid number, 3 or 6 g PLA were dissolved in 100 mL chloroform (Carl Roth, Karlsruhe/Germany) and stirred for at least three hours. The glass electrode (type 6.0229.100 from Methrom (Filderstadt, Germany)) was then placed in the analysis vessel. While stirring, the methanolic KOH solution was added stepwise (0.5 mL steps or 0.2 mL steps around the equivalence point) and the voltage value of the potentiometer ("AL15" from Tintometer GmbH—Lovibond Water Testing, Dortmund, Germany) was noted. The measurement was stopped as soon as the voltage value of the potentiometer changed only slightly after the addition of the methanolic KOH solution. The potentiometer "AL15" was used (Tintometer GmbH—Lovibond Water Testing, Dortmund, Germany).

2.2.4. Foaming

A self-made stirring autoclave was used for processing the polymer granules. Here, the materials were processed as it is known for EPLA bead foams in the literature [19]. Deionized water and the pellets were put in the autoclave. Afterwards, it was sealed and a pressure of 50 bar of CO_2 was applied. The setup was heated up to a temperature of 120 °C, saturated for 30 min and released out of the autoclave. The gained materials after this processing were used to evaluate the degradation behavior of the samples.

3. Results

3.1. SEC Results

The used PLA was modified on a twin-screw extruder with DCUP (0.1 wt%/0.2 wt%) and PDCI (0.5 wt%). After the reactive modification, the effectiveness of the modifiers was verified. One way to investigate this is by the method of SEC, which can be used to determine the molar mass and a molar mass distribution. The principle is based on the separation of molecules according to their molecular size. Figure 2 shows the molar mass and polydispersity index (PDI) of all produced compounds.

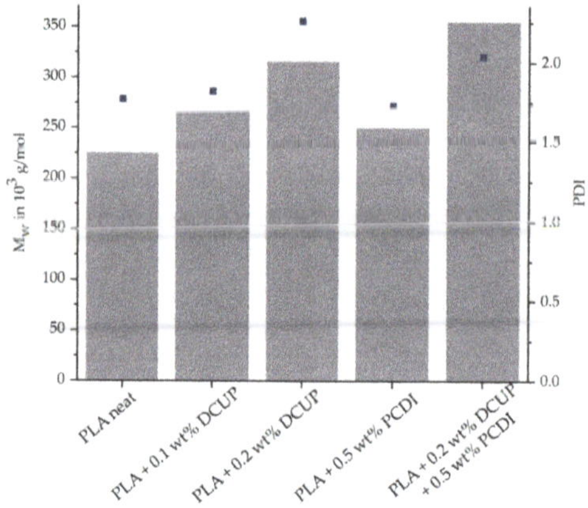

Figure 2. SEC-Results of all PLA compounds.

The organic peroxide led, as well as the PDCI, to an increased molar mass. As the amount of DCUP increases, this effect is enhanced. The addition of PCDI leads also to an increase in molar mass but is less pronounced than DCUP. The reason why the molar mass does not increase that much is that PCDI only reacts with the end groups of PLA. The combination of DCUP and PCDI showed the highest increase in molar mass. The results show on the one hand that a successful modification took place during the reactive extrusion on the twin-screw extruder and thus a chemical reaction between the two components has occurred. Furthermore, the increase in molar mass indicates a change in the linear structure of the PLA, which will be considered in more detail in the following rheological investigations.

3.2. Rheological Investigation

Figure 3 shows the measured frequency sweeps of the unmodified PLA compared to the modified ones.

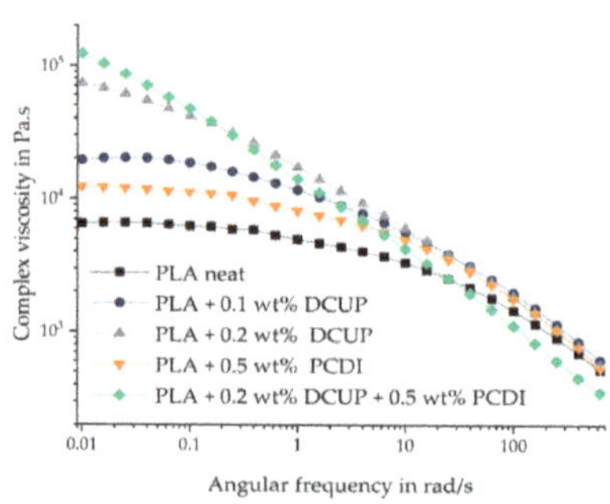

Figure 3. Rheological measurements of the PLA compounds.

All materials show a typical shear thinning behavior. Pure, unmodified PLA shows a chain progression typical for linear polymers. Modification with DCUP leads to an increase in zero viscosity, but the curves still show typical shear thinning behavior and form a Newtonian plateau in the low-frequency range, except PLA with 0.2. wt% DCUP and 0.5 wt% PCDI. With higher amount of DCUP, a higher zero shear viscosity is observed. The increase in viscosity can be explained by the change in the molecular chain structure of the modified PLA. DCUP decomposes during reactive extrusion to form radicals that can attack the PLA chain leading to a chain extension or/and branching. It has already been described in the literature that DCUP can lead to branching or cross-linking of polymer chains [12]. The longer chains and branching in the modified PLA result in entanglements that act like a physical network, leading to longer relaxation times. This leads to the fact that the Newtonian range is only indicated at low frequencies. The addition of 0.2 wt% DCUP is more effective than that of 0.1 wt% because a higher number of possible reactive groups are available which leads to an amount of extensions. PCDI was added as a hydrolysis stabilizer. As seen with the SEC results before, the addition of 0.5 wt% PCDI leads to an increase in molar mass und therefore also to an increase in the complex viscosity compared to neat PLA. Consequently, the combination of DCUP with PCDI results in the highest complex viscosity of all materials and to a change in the curve flow due to a complementation of the chain extension and the prevention to degradation.

Timesweep measurements were used to observe the thermal stability. The thermal stability is crucial for the processing, especially for the foaming process of polymers. In addition, information is obtained about the conversion of the chemical reaction during the reactive extrusion. Degradation reactions as well as post-reactions are generally undesirable and should be avoided or at least taken into account. The normalized complex viscosity as a function of time is shown in Figure 4.

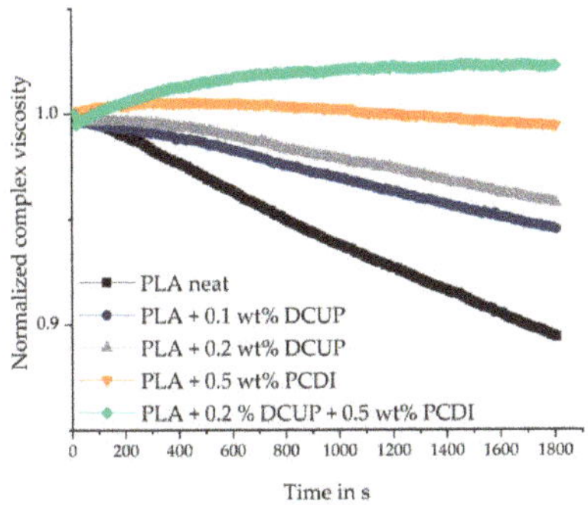

Figure 4. Normalized time-dependent viscosity of the compounds.

The time sweep was done for 30 min at isothermal temperature of 180 °C for all investigated compounds. It can be seen that the neat PLA has the highest degree of degradation, followed by the different peroxide-containing compounds. The PCDI stabilized material and the modified stabilized modifications show the highest relative viscosity indicating no or even a low degradation during the timesweep. After the first 400 s, for all materials except those with PCDI, the normalized complex viscosity decreases due to degradation processes. The use of PCDI led to the desired stabilization process. However, only a slight decrease in complex viscosity for the PLA with 0.5 wt% PCDI can be observed after 30 min. Additionally, the reaction regarding the chain extension proceeded completely and no post-reactions could be observed. The combination of DCUP and PCDI has the highest complex viscosity and is stable over the time of 1800 s.

3.3. *Acid Value Determination*

The determination of the acid value is a method to verify if chemical reactions at the end groups of PLA occurred. Table 1 shows the results of the acid value of the different PLA compounds before and after the foaming.

Table 1. Determined acid values in mmol/kg.

Compound	Acid Value before Foaming	Acid Value after Foaming
PLA neat	18.78 ± 0.04	30.60 ± 0.13
PLA + 0.1 wt% DCUP	20.80 ± 0.25	31.87 ± 1.20
PLA + 0.2 wt% DCUP	26.63 ± 0.19	34.83 ± 0.31
PLA + 0.5 wt% PCDI	0.72 ± 0.03	2.03 ± 0.98
PLA + 0.2 wt% DCUP + 0.5 wt% PCDI	1.15 ± 0.20	5.67 ± 0.31

The higher the acid value, the higher the amount of carboxyl end groups of PLA and therefore more possibilities for a hydrolytic reaction are given. PCDI can react with these end groups and lead to a small acid value. This means that PCDI is capturing the carboxyl end groups, which hinders the hydrolytic reaction. The results before indicated that a reaction between PLA and PCDI happened. The acid values before foaming were determined after the compounding step. The addition of 0.5 wt% PCDI leads to the smallest acid value. It confirms the other results and the expected reaction of PLA and PCDI. The combination of both modifiers has a higher acid value of 1.15 but it is still in a range where hardly any end groups exist in the whole polymer chains. As a comparison the neat PLA showed an acid value of about 18.8 mmol/Kg. The conditions during the foaming led for all materials to an increase of the acid value indicating a chain scission during the processing. This is not surprising, looking at the processing conditions (PLA pellets in water at 120 °C for 30 min). As it is known from literature, hydrolysis of PLA takes place around 60 °C and under humidity. Therefore, it can be concluded that the processing conditions for the foaming contain the ideal surroundings for hydrolysis.

4. Conclusions

PLA was successfully modified via reactive extrusion. By using 0.2 wt% of an organic peroxide, the molar mass and complex viscosity could be increased. To prevent PLA from undergoing hydrolysis, a stabilizer was used. To receive good foaming properties as well as a thermally and hydrolytically stable PLA, the two modifiers were combined. The combination of 0.2 wt% DCUP and 0.5 wt% PCDI leads to the highest molar mass and the best thermal stabilization. By using the acid value determination, it could be shown that the stabilizer reacts with the end groups of PLA and that several processing steps such as extrusion and foaming can lead to a degradation of PLA. As a result, it is indispensable to prevent PLA from hydrolysis to ensure that there is no downgrade of the properties of PLA. In future works, different modifiers and stabilizers could be investigated and also the following processing steps of bead foams, such as welding.

Author Contributions: Conceptualization, J.D.; methodology, J.D. and C.B. (Christian Brütting); software, J.D., validation, J.D. and C.B. (Christian Brütting); formal analysis, J.D.; investigation, J.D. and C.B. (Christian Brütting); resources, J.D.; data curation, J.D.; writing—original draft preparation, J.D. and C.B. (Christian Brütting); writing—review and editing, J.D. and C.B. (Christian Brütting); visualization, J.D.; supervision, V.A., H.R. and C.B. (Christian Bonten); funding acquisition, V.A., H.R. and C.B. (Christian Bonten). All authors have read and agreed to the published version of the manuscript.

Funding: This research was funded by German Research Foundation (DFG) (392851979).

Institutional Review Board Statement: Not applicable.

Informed Consent Statement: Not applicable.

Data Availability Statement: Not applicable.

Acknowledgments: The authors kindly like to thank the scientific staff members who contributed with inspiring discussions and all technicians and students for their support during the experiments.

Conflicts of Interest: The authors declare no conflict of interest.

References

1. Lucas, N.; Bienaime, C.; Belloy, C.; Queneudec, M.; Silvestre, F.; Nava-Saucedo, J.-E. Polymer biodegradation: Mechanisms and estimation techniques. *Chemosphere* **2008**, *73*, 429–442. [CrossRef] [PubMed]
2. Farah, S.; Anderson, D.G.; Langer, R. Physical and mechanical properties of PLA, and their functions in widespread applications—A comprehensive review. *Adv. Drug Deliv. Rev.* **2016**, *107*, 367–392. [CrossRef] [PubMed]
3. Elsawy, M.A.; Kim, K.-H.; Park, J.-W.; Deep, A. Hydrolytic degradation of polylactic acid (PLA) and its composites. *Renew. Sustain. Energy Rev.* **2017**, *79*, 1346–1352. [CrossRef]
4. Haider, T.P.; Völker, C.; Kramm, J.; Landfester, K.; Wurm, F.R. Kunststoffe der Zukunft? Der Einfluss von bioabbaubaren Polymeren auf Umwelt und Gesellschaft. *Angew. Chem.* **2019**, *131*, 50–63. [CrossRef]
5. Lim, L.-T.; Auras, R.; Rubino, M. Processing technologies for poly(lactic acid). *Prog. Polym. Sci.* **2008**, *33*, 820–852. [CrossRef]
6. Al-Itry, R.; Lamnawar, K.; Maazouz, A. Improvement of thermal stability, rheological and mechanical properties of PLA, PBAT and their blends by reactive extrusion with functionalized epoxy. *Polym. Degrad. Stab.* **2012**, *97*, 1898–1914. [CrossRef]
7. Yang, L.; Chen, X.; Jing, X. Stabilization of poly(lactic acid) by polycarbodiimide. *Polym. Degrad. Stab.* **2008**, *93*, 1923–1929. [CrossRef]
8. Janorkar, A.V.; Metters, A.T.; Hirt, D.E. Modification of Poly(lactic acid) Films: Enhanced Wettability from Surface-Confined Photografting and Increased Degradation Rate Due to an Artifact of the Photografting Process. *Macromolecules* **2004**, *37*, 9151–9159. [CrossRef]
9. Gorrasi, G.; Pantani, R. Effect of PLA grades and morphologies on hydrolytic degradation at composting temperature: Assessment of structural modification and kinetic parameters. *Polym. Degrad. Stab.* **2013**, *98*, 1006–1014. [CrossRef]
10. Ehrenstein, G.W.; Pongratz, S. *Beständigkeit von Kunststoffen*; Edition Kunststoffe; Hanser: München, Germany, 2007; ISBN 978-3-446-21851-2.
11. Mohanty, A.K.; Drzal, L.T.; Misra, M. *Natural Fibers, Biopolymers, and Biocomposites*; Taylor & Francis: Boca Raton, FL, USA, 2005; ISBN 9781280400148.
12. Badia, J.D.; Ribes-Greus, A. Mechanical recycling of polylactide, upgrading trends and combination of valorization techniques. *Eur. Polym. J.* **2016**, *84*, 22–39. [CrossRef]
13. Cicero, J.A.; Dorgan, J.R.; Dec, S.F.; Knauss, D.M. Phosphite stabilization effects on two-step melt-spun fibers of polylactide. *Polym. Degrad. Stab.* **2002**, *78*, 95–105. [CrossRef]
14. Göttermann, S.; Standau, T.; Weinmann, S.; Altstädt, V.; Bonten, C. Effect of chemical modification on the thermal and rheological properties of polylactide. *Polym. Eng. Sci.* **2017**, *57*, 1242–1251. [CrossRef]
15. Karst, D.; Yang, Y. Molecular modeling study of the resistance of PLA to hydrolysis based on the blending of PLLA and PDLA. *Polymer* **2006**, *47*, 4845–4850. [CrossRef]
16. Oliveira, M.; Santos, E.; Araújo, A.; Fechine, G.J.; Machado, A.V.; Botelho, G. The role of shear and stabilizer on PLA degradation. *Polym. Test.* **2016**, *51*, 109–116. [CrossRef]
17. Rasal, R.M.; Janorkar, A.V.; Hirt, D.E. Poly(lactic acid) modifications. *Prog. Polym. Sci.* **2010**, *35*, 338–356. [CrossRef]
18. Standau, T.; Zhao, C.; Castellón, S.M.; Bonten, C.; Altstädt, V. Chemical Modification and Foam Processing of Polylactide (PLA). *Polymers* **2019**, *11*, 306. [CrossRef] [PubMed]
19. Nofar, M.; Ameli, A.; Park, C.B. A novel technology to manufacture biodegradable polylactide bead foam products. *Mater. Des.* **2015**, *83*, 413–421. [CrossRef]

polymers

MDPI

Review

Applications of Biopolymers for Drugs and Probiotics Delivery

Roxana Gheorghita [1,2], Liliana Anchidin-Norocel [1], Roxana Filip [3], Mihai Dimian [2,4] and Mihai Covasa [1,5,*]

1 Department of Health and Human Development, Stefan cel Mare University of Suceava, 720229 Suceava, Romania; roxana.puscaselu@usm.ro (R.G.); liliana.norocel@usm.ro (L.A.-N.)
2 Integrated Center for Research, Development and Innovation in Advanced Materials, Nanotechnologies, and Distributed Systems for Fabrication and Control, Stefan cel Mare University of Suceava, 720229 Suceava, Romania; dimian@usm.ro
3 Hipocrat Clinical Laboratory, 720003 Suceava, Romania; roxana.filip@usm.ro
4 Department of Computers, Electronics and Automation, Stefan cel Mare University of Suceava, 720229 Suceava, Romania
5 Department of Basic Medical Sciences, College of Osteopathic Medicine, Western University of Health Sciences, Pomona, CA 91766, USA
* Correspondence: mcovasa@westernu.edu

Abstract: Research regarding the use of biopolymers has been of great interest to scientists, the medical community, and the industry especially in recent years. Initially used for food applications, the special properties extended their use to the pharmaceutical and medical industries. The practical applications of natural drug encapsulation materials have emerged as a result of the benefits of the use of biopolymers as edible coatings and films in the food industry. This review highlights the use of polysaccharides in the pharmaceutical industries and as encapsulation materials for controlled drug delivery systems including probiotics, focusing on their development, various applications, and benefits. The paper provides evidence in support of research studying the use of biopolymers in the development of new drug delivery systems, explores the challenges and limitations in integrating polymer-derived materials with product delivery optimization, and examines the host biological/metabolic parameters that can be used in the development of new applications.

Keywords: drug; controlled release; polysaccharide; probiotics

Citation: Gheorghita, R.; Anchidin-Norocel, L.; Filip, R.; Dimian, M.; Covasa, M. Applications of Biopolymers for Drugs and Probiotics Delivery. *Polymers* **2021**, *13*, 2729. https://doi.org/10.3390/polym13162729

Academic Editors: José Miguel Ferri, Vicent Fombuena Borràs and Miguel Fernando Aldás Carrasco

Received: 30 June 2021
Accepted: 11 August 2021
Published: 15 August 2021

1. Introduction

Biopolymers are generated by living organisms [1–5] and are defined as biologically degradable polymers [6]. They represent possible materials for the replacement of synthetic plastics due to an increased interest in developing environmental sustainability [7]. Biopolymers have a structural backbone with carbon, oxygen, and nitrogen atoms which makes them easily biodegradable. Biodegradation breaks them down into carbon dioxide, water, humic matter (organic macromolecular material), biomass, and other natural substances; thus, these materials are naturally recycled through biological processes [3].

A classification system based on their origin, synthesis and processing of different biodegradable polymers [8–12] is presented in Figure 1 in the form of a social network analysis. It divides the biopolymers in four major categories: extracted from biomass products (agrobiopolymers), from microorganisms, and from biotechnological and petrochemical products. Biopolymers from biomass products have diverse compounds such as polysaccharides (starches, celluloses, alginates, pectins, gums, and chitosan) [13]; proteins of animal origin (whey, collagen, and gelatin); proteins of vegetal origin (zein, soya, and wheat gluten) [14,15]; and lipids (bees wax, carnauba wax, and free fatty acids) [16,17]. Most biopolymers can be extracted from natural sources such as plants, animals, and microorganisms including algae and agro-wastes [18]. Agro-sources of biopolymers include bananas, maize, potatoes, tapioca, yams, rice, corn, wheat, cotton, sorghum, and barley [19,20], while animal sources are derived from cattle, pigs, and other products. Agro-waste-based sources include apple pomace [21], tomato pomace, pineapple [22], orange and lemon peels, wheat straw, rice husks [23,24], paper wastes, crops, wood, and green wastes, while the marine sources are sponges [25], corals, lobsters, fishes, and shrimps [26]. Biomaterials manufactured from these products are described as stretchy, soft, and gel-like, with many characteristics of both solids and fluids. It is known that biopolymers can be smart and flexible materials even in living organisms [6] because they have a structure that is constantly manipulated either in response to environmental changes or by enzymes throughout different stages of the organism's lifecycle [27].

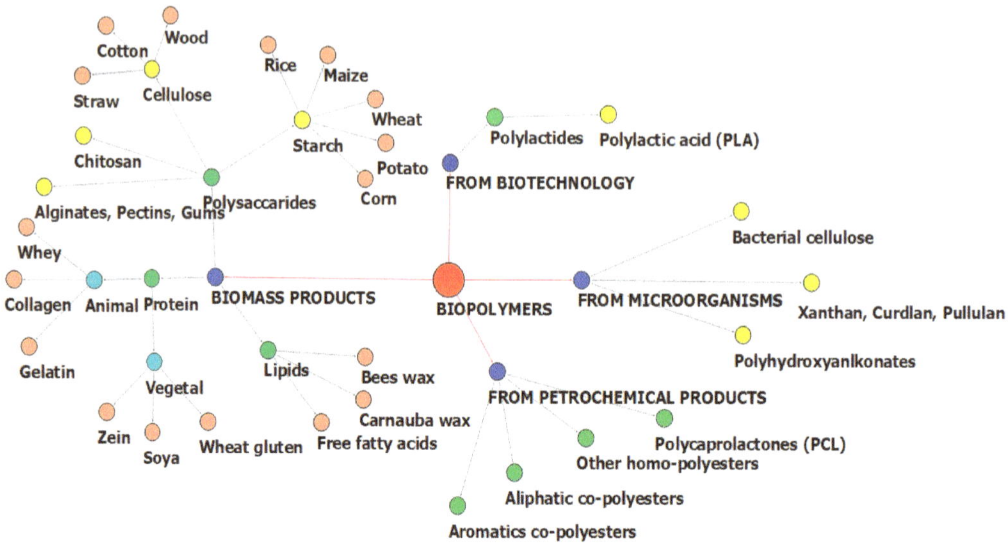

Figure 1. A social network graphical illustration of biopolymers classification.

The biopolymer composites can be prepared by several methods such as extrusion, electrospinning, grafting, different types of molding [28], solvent casting, melt blending, intercalation, filament winding, phase separation [29], laser printing, and film stacking [5].

Currently, the manufactured design and optimization of biopolymers through mathematical models are very advantageous because they improve their physical, chemical, electrical, and mechanical properties in order to increase resistance in humid, warm, or cold storage conditions and for applications that require specific features [30].

Although significant work has been done on the use of a wide range of biopolymers, most of them have been based on polysaccharides due to their improved properties compared to other categories such as proteins or lipids. Thus, this review focuses on the ability of biopolymers to be used successfully in the pharmaceutical industry as encapsulating agents, particularly for delivery of drugs and probiotics. Specifically, the paper describes the use of alginate, chitosan, agar, starch, and cellulose by focusing on the properties and characteristics that make them suitable candidates for product delivery, offering advantages over the chemically derived polymers. The following will also be presented: the development of encapsulated substances based on biopolymers; challenges and limitations such as the encapsulation process, shelf life, controlled release of embedded drugs, protection, and viability of live strains; and the rate of release at different pH mediums of the gastrointestinal fluids.

2. Biopolymers vs. Conventional Synthetic Materials

Several studies have been conducted concerning the utilization of biopolymers with the aim of developing sustainable packaging materials. Although significant improvements have been made, there is still considerable debate over economic considerations, environmental concerns, and product packaging performance [31].

Living organisms produce a variety of polymers as a significant part of their morphological, cellular, and dry matter. These biopolymers play vital roles in the life cycle of organisms including the preservation and expression of genetic information, catalysis of reaction [32], energy or other nutrients, sensing of abiotic and biotic factors, protecting against the attack of other cells, storage of carbon, and negotiation of the adhesion to the surface of other organisms [4].

Biopolymers present important features such as biodegradability [33,34], biocompatibility [35], sustainability [36], bioresorbability [37], flexibility [38], antibacterial activity [6], renewability [39], and stability [2]. They are also less toxic [40], non-immunogenic [41], non-carcinogenic, non-thrombogenic, carbon neutral, and have the advantage of easy extraction [42]. These properties are directly influenced by parameters such as the type of material used as the structural matrix (charge distribution, molecular mass, and conformation), film developing conditions (concentration, pH, solvent, temperature, etc.) and category and concentration of additives (antimicrobials, crosslinking agents, plasticizers, antioxidants, etc.) [43].

Until recently, conventional synthetic materials have become part of most materials in our lives, including those present in beverages and food, clothes, daily used instruments, and baby-toys, and even in biomedical applications such as surgical equipment, drug delivery systems, and cosmetic personal care materials. Some studies have associated these materials with potential adverse health problems, particularly in pregnant women and newborn infants. To this end, hormonally active agents are a group of polymeric chemicals that have been associated with critical health issues such as cancerous tumors, congenital disabilities, and other disorders [40]. Furthermore, people have become more aware of the effects of chemically derived compounds and are more cautious in their use of conventional synthetic materials due to their effects on health and the environment. Today's consumers are more informed and sophisticated in their preferences and choices, increasingly looking for natural and vegan alternate products with high biocompatibility and low environmental implications. Furthermore, increasing efforts and research on the management of plastic waste on Earth are aimed towards finding eco-friendly alternatives to plastics [5]. Such eco-friendly alternatives can be represented by biopolymers, which are disposed in the environment and easily degradable through the enzymatic actions of microorganisms [44].

Compared with conventional synthetic materials that have a simpler and more random structure, these biopolymers are complex molecular assemblies that adopt defined and precise 3D configuration and structures [45]. Based on the composition and chemical structure of biopolymers, they are almost identical to the macromolecules of the native extracellular environment [46]. Many characteristics differentiate between the two types of materials, which are summarized in Table 1. Biopolymers have multiple advantages over conventional plastics due to their low/no toxicity, biodegradability, sustainability, biocompatibility, and extreme hydrophilicity. Furthermore, their morphology and chemical modifications can have a significant impact on their rate of biodegradation [47], an important feature in the development of new applications for food, biomedical, and pharmaceutical industries. Conversely, synthetic materials have a low cost and high thermal and mechanical properties that make them more usable than biopolymers.

Some applications of biopolymers have used mixtures with synthetic materials (such as polyethylene and polyvinyl alcohol), plasticizers (sorbitol and glycerin), nitrogenous bases, and others, thus obtaining a partially biodegradable material [7].

Table 1. Characteristics of biopolymers vs. synthetic polymers.

Characteristic of Materials	Biopolymers	Synthetic Polymers	References
Main source	Agro-resources	Petroleum and gas	[48]
Biodegradability/environmentally friendly	YES	NO/slow	[49,50]
Structure	Well defined	Stochastic	[48]
Chemical backbone structure	Carbon, oxygen, and nitrogen	Mostly carbon	[48]
Dispersity	Unity	>1	[51]
Physicochemical resistance	Low	High	[52]
Toxicity	Low	High	[41]
Thermal stability	Low	High	[52]
Mechanical properties	Low	High	[53]
Sustainability	High	Low	[52]
Availability	High	Decreasing	[52]
Cost	High (depends on the type)	Low	[54]

Although biopolymers have many advantages, there are a number of limitations in their processing, starting from the extraction and all the way up to the final product. First, being a completely natural product, biopolymers' final properties depend largely on the raw material. This can vary greatly due to the origin, climatic conditions, location, harvesting, and processing. Therefore, the world production of biopolymers cannot always maintain the same sustainability. To date, no universal acceptable procedures have been developed for the collection and manufacturing of biopolymer powders from vegetable materials. This is important both for the safety as well as the quality and performance of the final product. Second, because the production of biopolymers is still in its infancy, the production costs are quite high [55]. However, the elimination of recycling and waste taxes through world legislation mitigates some of the high costs. Third, the production of biopolymers necessitates special equipment other than those currently used. The development of such equipment and protocols requires time, additional costs, and trained staff. However, given that biopolymer processing technology is relatively easy and accessible, some existing equipment has been adapted for this purpose, thus reducing the costs [56]. Finally, the lower performance of biopolymers compared to conventional materials may limit their use, although continuous research advances in the technology and material combinations show great improvements in their characteristics that are comparable to conventional materials currently used [57].

3. Applications of Biopolymers

Recent research demonstrates the potential applications of biopolymers as materials for manufacturing medical devices [58]. The most suitable characteristics for suggesting these biomaterials are molecular weight, lubricity, material chemistry, water absorption degradation, shape and structure, solubility [59], hydrophilicity/hydrophobicity [60], erosion mechanism [61], and surface energy [62]. Besides these, other applications of biopolymers, such as those presented in Figure 2, are found in industries such as pharmaceutical preparations with encapsulation; food (edible film packaging and emulsifier) [63]; agriculture, which includes sustainable activities, methods for water recovery, and materials used as soil conditioner; the cosmetics industry (especially hydrogels) [64]; and water treatment substances, biosensors, and even data storage elements [1]. In these industries, polysaccharide-based materials have been developed under different forms such as films, membranes, fibers, hydrogels, food casing [65], sponges, and air gels [66].

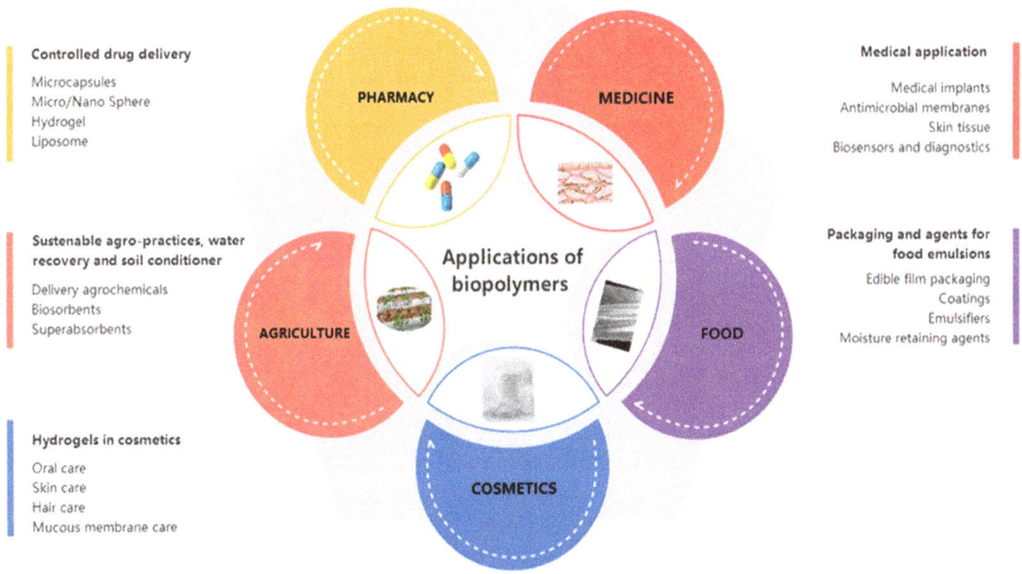

Figure 2. Applications of biopolymers.

Packaging in the bio-medical industry is a method that allows for the closure of a pharmaceutical product from its fabrication to its end use. In pharmaceutical packaging, biopolymers are used to protect pills, nutraceuticals, drugs, surgical devices, powders, and liquids [63]. Pharmaceutical packaging has an impact on the isolation and ensures the safety, identity, and convenience of using the products. Packaging should be compatible with the patient's condition, be free of adverse effects on his/her health, and be environmentally safe [64].

As mentioned above, biopolymers can be also used in the preparation of edible packaging films for food products [65]. These films made from biomaterials can be ingested with the food because they are prepared from polysaccharides and proteins. Edible films have received special attention in the last years because of their alternative potential to replace synthetic materials, which could minimize packaging waste and reduce environmental pollution [66]. As a food packaging material, it can also improve the antimicrobial effect of packaging [67], shelf-life heat resistance, flexibility, mechanical strength, and barrier properties [68]. Edible films/coatings are currently used in a variety of other applications including collagen casings for sausages, chocolate coatings for fruits, and coatings for chocolates and other items [69]. Furthermore, biopolymers are used as emulsifiers and as both thickening and moisture-retaining agents in the food industry with the goal of improving the stability and physicochemical properties of food emulsions [70,71]. Finally, biopolymers have been extensively used in the delivery of bioactive compounds such as probiotics that are susceptible to degradation during preparation, storage, or under the adverse environmental conditions of the human gut. Similarly, they have been used for applications in the pharmaceutical industries as a delivery agent to improve drug stability and bioavailability. In the following sections, we will discuss the applications of biopolymers in drugs and probiotics delivery.

3.1. Biopolymers in the Pharmaceutical Industry

Due to their special properties, biopolymers have slowly begun to replace conventional materials. Whereas in the beginning they were mainly used in the food industry, their application in other related industries took place relatively quickly. In the pharmaceutical industry, they were initially used for the same purpose as in the food industry, which is as thickening and emulsifying agents, host molecules, bulking agents, or fibers. In addition, their use in cosmetics has increased substantially. According to existing data, it is estimated that the world biopolymer market will reach approximately USD 10 billion by the end of 2021, an increase by approximately 17% between 2017 and 2021. The largest market segment is owned by Western Europe with approximately 41.5% of the global market [72]. In biomedicine, polymers have been used successfully both experimentally and in in vivo applications, wound dressing, tissue engineering, drug delivery, or in medical devices such as electronics, sensors, and batteries. Furthermore, due to their physical, thermal, mechanical, and optical properties, biopolymers are ideal materials widely used for food and pharmaceutical applications [5] (Figure 3).

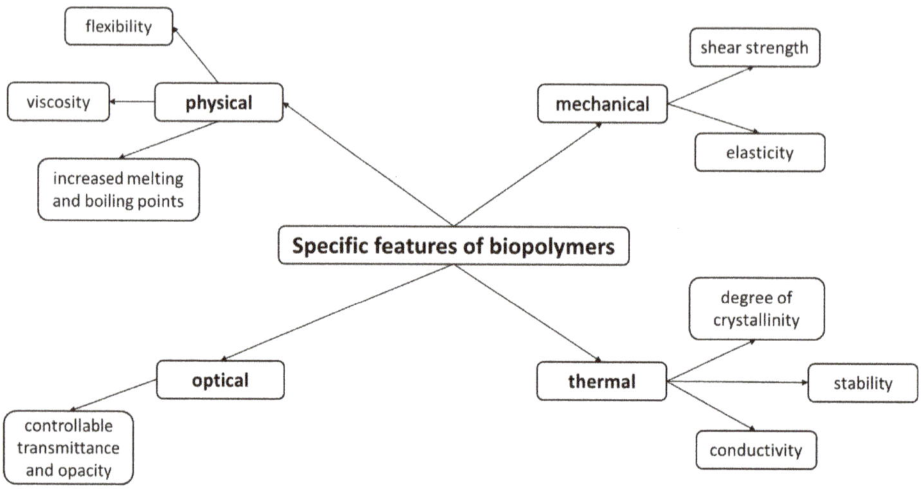

Figure 3. Specific features of biopolymers.

The composition and matrix of biopolymers can be manipulated in order to obtain the appropriate functional properties such as microstructure, permeability, and chargeability that are dependent on the internal structure of the polymer. Electrical characteristics influence the bonding of particles in the biopolymer matrix and their capacity to aggregate. The biopolymer fractions that prevent aggregation are the ones with a high electrical charge [58]. Based on these properties, biopolymers are used successfully to obtain nanoparticles, nanoemulsions, nanogels, or hydrogels with applications in the biomedical industry as carrier systems. Among them, polysaccharides are the most widely used category of biopolymers, either individually or in mixtures with other biopolymers to replace the synthetic materials or exist in addition to them.

3.1.1. Biopolymers for Controlled Drug Release

Encapsulation involves the protection of living cells from destruction by entrapment in biopolymer membranes and it is applied in micro and macrocapsules [73]. It is the procedure by which one or more materials, representing the active part or core material, is embedded or coated with another material or system, which is actually a mantle, shell, carrier, or encapsulant [74] (Figure 4).

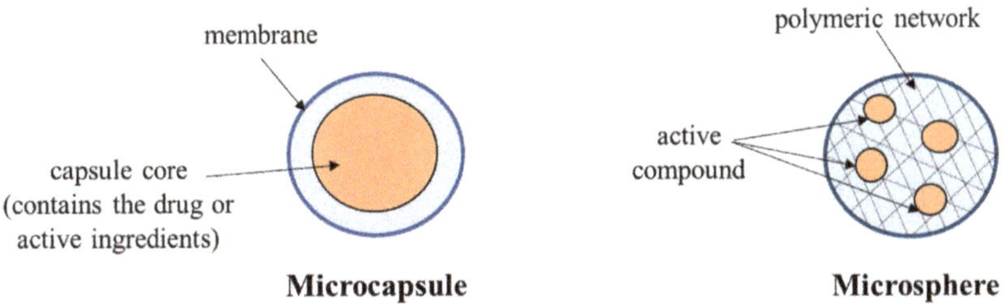

Figure 4. Graphical representation of drug encapsulation (adapted from Madene et al. [74] with permission from the publisher).

A specific feature of macrocapsules is the relatively large difference between the surface area and the volume. Thus, it is necessary to use a large number of nutrients to obtain an appropriate diffusion gradient for the entry of nutrients. This aspect overlaps with the necessary nutrition for the cells. In macrocapsules, living cells are entrapped in large diffusion chambers formed as flat sheets, hollow fibers, and disks with semi-permeable properties [75]. Macrocapsules can be used as intra or extra-vascular devices [76]. In intravascular devices, cells are connected to the bloodstream as a shunt, oriented outside the artificial capillaries. They are found in the vicinity of blood circulation, assisting with the rapid transfer of therapeutic and nutritional substances such as oxygen. The biggest disadvantage is the potential for developing thrombosis. Therefore, research is moving towards their use as extravascular devices with cells entrapped within semi-permeable diffusion chambers and placed transdermally or in the peritoneal cavity without the need for direct circulatory access. This involves a minor surgery and permits a quick and easy substitution in case of graft failure or when the transplant has to be replaced for other reasons [73]. Microcapsules allow for a fast transfer of beneficial substances and accurately mimics the release of substances such as glucose or insulin. Due to their benefits, most studies focused on developing microcapsules with low or non-inflammatory responses. This feature is used successfully in the treatment of endocrine diseases [77].

Many biocompatible polymers have been used as encapsulation materials. For this, a biopolymer must meet certain criteria: (i) stable and not interacting with the drug it contains; (ii) not interfering with the function and cellular viability; non-toxic, inexpensive, and biodegradable; (iii) both the biopolymer and its degradation products must be non-antagonistic to the host; (iv) molecular weight, solubility characteristics, glass transition temperature, microstructure, and chemical functionality should allow for proper drug diffusion and release; (v) biosafe and biocompatible; and (vi) when biocompatibility needs to be improved, the biopolymer should be combined with other compounds for a synergistic effect.

Depending on the mechanism that controls the release of the active agent from the delivery system, the controlled-release modalities may be different. Thus, biopolymer erosion, diffusion, and swelling, followed by diffusion or degradation, may occur [78]. The erosion mechanisms involve: (i) hydrolysis of hydrogels, an important feature for the controlled release of macromolecules; (ii) solubilization of water-insoluble biopolymers by reactions with groups pendant from the polymer covalently bonded atoms; and (iii) cleavage of hydrolytically labile bonds within the biopolymer covalently bonded atoms. The diffusion process occurs when an encapsulated drug or other active agent crosses the outer membrane of the capsule through the biopolymer used for the controlled-release device. In the case of diffusion-controlled systems, the drug delivery system must be stable in the biological environment and must maintain its size and shape through the swelling or degradation [79]. For example, when biopolymers are combined with other bioactive agents, the drug must be able to diffuse through their molecular structure or through pores when it reaches the biological environment. At this stage, it is very important that there are no changes to the biopolymer itself. Swelling-controlled release devices are those systems that, although dry in the initial phase, will swell when they reach the body and come into contact with fluids or water. The swelling ability of the biopolymers can be triggered by changing the environmental conditions of the delivery system. This is one of the most important and useful features of the biopolymers because, by changing the pH or temperature, the release of drugs or incorporated active substances can be controlled [80]. Finally, the biodegradation of a biopolymer in the body is a natural process through which the active ingredient is completely eliminated.

Synthetic polymers have long been of interest for use as encapsulating agents of various therapeutic substances. Although they show improved pharmacokinetics compared to small molecule drugs, their accumulation in the body has raised toxicity issues [81]. With the reorientation of the medical industry towards the use of biopolymers, the major issue is the selection of the right compounds based on the need and desired effects. Not all biopolymers are suitable as encapsulating agents for drugs. It is important that they release the active substance to the target area at the right time in a safe manner and without side effects, especially considering that the predominant routes are oral or intravenous administration [82]. The most used biopolymers and which are the focus of this review are those based on polysaccharides, such as sodium alginate, chitosan, agar, starch, and cellulose. They react synergistically with other biopolymers and polymers, have low toxicity and non-immunogenic behavior, and are compatible with tissues and cells. These polysaccharides are stable in vitro and in vivo, and are used in the development of microcapsules, microspheres, or nanocapsules. When tested in vivo, they showed high biocompatibility and biodegradability, facilitating treatment, minimizing side effects, and improving the health condition. Their high solubility is a plus for their use as disintegrants in water-soluble tablets. For example, when used in tablets, the coating of chitosan and starch improved their visual appearance, protected the drug from degradation, and masked the unpleasant taste of the incorporated substance [83]. When used as capsule material, gelatin was replaced with alginate, a vegan version, or with cellulose, for hard capsules. The main biopolymers that are widely used and presented in this review are alginate, chitosan, agar, starch, and cellulose.

Alginate and Its Use for Drug Delivery

Alginate is probably the most widely used biopolymer in the pharmaceutical and medical industries. This is due to its special encapsulation properties and role in wound healing. It was first isolated in the 1980s and since then it became a multifunctional compound in many applications. Thus, it is obtained at low cost and is a renewable and readily available, biodegradable, non-toxic, biocompatible, and, importantly, mucoadhesive and non-immunogenic compound [84]. It is recognized in the pharmaceutical industry as an excipient and is used to treat reflux esophagitis [85].

Structurally, alginate is a hidrosoluble polysaccharide formed from alternative blocks of 1–4 linked α-L-guluronic acid and β-D-mannuronic acid residues. It contains varying lengths of G-blocks, M-blocks, and/or MG/GM-blocks. High G content alginates have the ability to form stiffer, brittle, and more porous gels, but with increased strength, while high M content alginates tend to obtain more elastic and weaker gels [86].

Alginate is obtained from brown algae and is found as alginic acid sodium, calcium, and magnesium salts. The algae species used for the extraction of trading alginates are *Macrocystis pyrifera*, *Laminaria hyperborea*, *Saccharina japonica*, and *Ascophyllum nodosum*. It can be synthesized by various species of bacteria such as *Azotobacter vinelandii* and various *Pseudomonas* species, although they are not commercially available [72]. Alginate extraction is achieved by a relatively simple process. First, the raw material from the algae is ground and washed with acid followed by the extraction with hot alkali. The alginic acid is obtained after the extract has been filtered, precipitated with calcium, and acidified. The required salt form of alginate is obtained by treating insoluble alginic acid with metallic carbonates, oxides, or hydroxides [87]. Alginate biocompatibility has been extensively studied, with data showing that oral administration of alginate does not trigger many immune responses, in addition to the finding that it is non-toxic and biodegradable [88].

By contrast, intravenous administration of most commercial alginates can lead to adverse body reactions and fibrosis [89].

As an encapsulating agent, alginate was first used in the treatment of diabetes in the encapsulation of pancreatic islet cells [90]. Since then, it has been used for both macro and microencapsulation, and for other endocrine and recombinant cells for the release of therapeutic gene products such as growth hormones or human clotting factor IX (Table 2). It is also used in bioartificial organs such as they kidneys or for the protection of hepatocytes or parathyroids. Alginate gels cannot provide immunoprotection because they are too porous. Therefore, in most applications, alginate gels must be coated with cationic polymers of synthetic origin. For alginate-based coatings, the most used cationic polymers are poly-L-lysine and poly-L-ornithine, but lately polyethylene glycol (PEG), glutaraldehyde, chitosan, and agarose have also been applied. Occasionally, other substances are used to reduce permeability, to ensure mechanical stability, and to increase the durability of the capsules; however, PEG remains the most used coating material. Another way to stabilize alginate gels is through the application of covalent crosslinking molecules, although this method of encapsulation interferes with the functional viability of the cells and can lead to cell toxicity [91].

Table 2. Alginate use for drug delivery.

Biopolymers	Entrapped Substances	Applications	Benefits	References
Alginate	Rifampicin	Drug delivery carriers	Nanoparticles are pH sensitive with the highest release of the active substance occurring at a pH of 7.4. Toxicity and safety tests were excellent with no systemic toxicity after oral administration of nanoparticles.	[92]
Alginate	Ibuprofen	Drug delivery system	Controlled drug release was maintained for 4 h (67.53% of the drug formulation).	[93]
Alginate and methylcellulose	Indomethacin	Drug delivery carrier	Controlled drug release. There was no interaction between the loaded drug and the polymers.	[94]
Alginate	Metformin hydrochloride	Drug delivery system	Good release time; microspheres may be used in the treatment of diabetes.	[95]
Alginate	Diclofenac sodium	Controlled-release microparticles	The drug: polymer (ratio 1:3) was obtained by emulsification and the drug release followed zero order kinetics, optimum for controlled drug release delivery.	[96]
Alginate and sodium carboxymethylcellulose	Ceftriaxone sodium	Multiarticulate beads	The use of the biopolymer matrix decreased drug release in gastric conditions but sustained it at intestinal pH. The beads swelled at pH 1.2 but particle diffusion and erosion occurred at pH 6.8.	[97]
Alginate	Furosemide	Controlled drug delivery beads	Drug release was controlled due to the thicker membrane and reduced beads swelling. Release of Furosemide depends on the conditions of the coating treatment.	[98]
Alginate	Isoniazid	Oral drug delivery	Microspheres were present in the intestinal lumen 4 h after administration and were detectable in the intestine after 24 h of oral administration. Approximately 26% of the drug was released in the gastrointestinal fluid (pH 1.2) in 6 h and 71.25% in the simulated intestinal fluid (pH 7.4) in 30 h.	[99]

Table 2. *Cont.*

Biopolymers	Entrapped Substances	Applications	Benefits	References
Alginate	Nicotinic acid	Aerogels	The release of the drug was prolonged when the core was surrounded by several alginate-based membranes. Due to coating, 50% of the drug was released within 4 h.	[100]
Alginate, carboxymethylcellulose, and chitosan	Amoxicillin	Coated beads	In gastric pH conditions, the drug release was prolonged from 61 min to up to 8 h.	

Thus, based on its characteristics, alginate seems to be the most suitable biopolymer used for drug encapsulation (Table 2). This is due to its specific properties, especially as a matrix for controlled drug delivery devices. In addition, alginate is cheap and readily available, is accepted for consumption in quantum statis doses, is nontoxic, and ensures the protection of the mucous membranes of the upper gastrointestinal tract [101].

Chitosan and Its Use for Drug Delivery

Chitosan is a polysaccharide found in shellfish, fungi, annelids, mollusks, and insects. It is the second most outspread natural polysaccharide on Earth, after cellulose. Commercially, it is produced from chitin, being a poly β (1 → 4) -2-amino-2-deoxy-β-D-glucan deacetylated chitin. It has a strong affinity for polyanions, contains reactive NH_3^+ and OH^- groups, and is soluble in acidic aqueous solutions. It is nontoxic, odorless, bio-compatible, and biodegradable. Due to its antibacterial properties, chitosan is used for microencapsulation, in particular for cells that require a cationic environment. Numerous applications in drug delivery include drug targeting systems for oral, nasal, ocular, and transdermal routes [102]. For this purpose, chitosan has been used in the development of gels, films, oral tablets, beads, and microspheres [103].

The capacity of chitosan as an encapsulating agent is greatly influenced by its molecular weight, degree of deacetylation and crystallinity, and extent of ionization/the free amino group. Thus, when the amino group at the 2-position of glucosamine units of chitosan is the main site for the immobilization of thiol groups, it results in thiolated chitosan. The thiolated chitosan derivatives are chitosan-cysteine, chitosan-thiolactic acid, chitosan-thioglycolicacid, chitosan-homocystenine, chitosan-N-acetylcysteine, and chitosan-glutathione. The thiolated chitosan has been used for anticancer drugs because it offers efficient mucoadhesivity, membrane permeation, and an enhancing capability and improved inhibition for P-glycoprotein [104]. Phosphorylated chitosan and its derivatives have different features such as high hidrosolubility and a metal chelating tendency, used in tissue regeneration, drug delivery intermediates, fuel cells, and in the food industry [105].

Structurally, chitosan is composed of free amine groups in media with a pH over 7.5 and protonated amines are formed in media with a lower pH. These pH-sensitive characteristics make chitosan-based compounds suitable in controlled-release technologies. Under well-established conditions, chitosan microcapsules containing the drug as an active ingredient permits its slow release at the target site [106]. For example, when encapsulated in chitosan, lipophilic drugs were effectively released into the intestinal tract [107]. When used as a vehicle to encapsulate vaccines, it allowed for their controlled release and delivery to targeted sites [108].

There are a number of advantages in using chitosan in the pharmaceutical industry for drug delivery, such as: (i) the controlled release of encapsulated substances; (ii) the elimination of toxic agents in the development process (due to dissolution in aqueous solution); (iii) crosslinking readily available free amino groups; and (iv) improved membrane absorption by mixing cationic chitosan with an anionic material [109]. When used as a coating agent for nanoparticles for the treatment of brain disease, chitosan protects against enzyme degradation, controls release, and improves bioavailability. In addition, it enhances drug permeability across the blood–brain barrier by affecting tight junctions [110].

Equally important are its hemostatic, bacteriostatic, anticholesterol, anticarcinogenic, and fungal characteristics.

In addition, chitosan has good bioadhesive properties and slows down the drug release in the nasal cavity, thus increasing bioavailability and the transfer of drugs from the nasal cavity to the brain [111]. When chitosan was used in membrane development, it increased permeability to acidic drugs. It is insoluble at a pH greater than 6.5 and prevents the burst effect of the release in the first segments of the gastrointestinal tract [112]. It has also been used successfully for antiviral and antibiotic encapsulation, as seen from Table 3.

Table 3. Chitosan use for drug delivery.

Biopolymers	Entrapped Substances	Applications	Benefits	References
Chitosan	Acyclovir	Drug delivery system	The grafting efficiency was 94% and the slow release of the drug was prolonged to 12 h.	[113]
Chitosan	Chlorhexidine diacetate	Buccal tablets	The tablets dissolved more quickly in vitro than chlorhexidine powder and both maintained and even improved the antimicrobial activity of the drug, particularly against *Candida albicans*, due to antimicrobial activity of the polymer itself.	[114]
Chitosan aspartate, glutamate, and hydrochloride	Vancomycin	Peptidic model drug	The sustained release from the microspheres minimized its solubilization in the upper gastrointestinal tract.	[115]
Chitosan	Tetracycline	Controlled drug system	The concentration of drug released was above the minimum limit required for the inhibition of *Staphylococcus aureus*.	[116]
Chitosan and oleic acid	Camptothecin	Controlled drug system	The encapsulation efficiency was about 78%. When its effectiveness in the simulated gastrointestinal fluids was tested, the drug was released slowly into the gastric environment. In intestinal fluids, the drug release was controlled. The drug embedded in chitosan was 75% protected from hydrolysis.	[117]
Chitosan	Satranidazole	Subgingival films for periodontitis	The drug was released for 96 h; the population of gram-positive bacteria was reduced.	[118]
Chitosan and alginate	Amygdalin	Drug delivery system	The controlled amygdalin release was performed for 10 h; the maximum amygdalin released was 70.46% at pH 3.1, 81.86% at pH 5.0, and 86.03% at pH 7.4.	[119]
Chitosan and graphene	Isosfamide	Sustained drug microspheres	The drug diffusion was the most controlled for when isosfamide was entrapped in microspheres.	[120]
Chitosan and xanthan gum	Ciprofloxacin	Controlled-release hydrogel	The entrapment efficiency of the prepared hydrogel increased with the drug increasing concentration and the maximum was reached at 93.8%.	[121]
Chitosan	Interferon-α	Nanoparticles for oral delivery	In mice, the nanoparticles were found in plasma at 1 h after administration, unlike the commercial interferon which could not be detected.	[122]

Chitosan has been extensively used as a matrix for extended drug release, especially due to the simple obtaining procedure, low cost, and biocompatibility. The biocompatibility of chitosan also derives from the fact that it is already part of the human food chain due to its presence in numerous fungi [123]. Chitosan increases the solubility of insoluble drugs when used in mixtures with inorganic nanoparticles, forming a stable complex with safe delivery to the specific site. It was effective when encapsulated hemoglobin, astaxanthin, quercetin, vaccines, or vitamins. Besides its applicability in drug delivery, chitosan is also used in wound dressing, tissue engineering, bioimaging, biosensors, and packaging, among other uses [124].

Agar and Its Use for Drug Delivery

Agar is a long-chain biopolymer obtained from species of algae from the Rhodophyceae class, most commonly found in *Gelidium sp.* and *Gracilaria sp.* It represents the supporting structure of algae and is composed of a mixture of agarose and agaropectin, the gelling and the non-gelling fraction, respectively [125]. Agaropectin is usually removed during processing in order to obtain an agar powder with higher gel strength. Agarose is composed of repetitive units of D-galactose and 3-6, anhydro-L-galactose, linked by alternating α- (1 → 3) and β- (1 → 4) glycosidic bonds.

The ratio of agarose to agaropectin depends largely on seaweed growth, the environmental condition of seaweed growth, extraction methods, and rheological and gelling properties. These changes affect the final mechanical properties of the gels [126]. Agar quality can be significantly improved by modification, which is the most widely used chemical method. It involves hydroxypropylation, acetylation, etherification, and oxidation, the last one being the most commonly used [127]. Due to its gelling capacity, gel reversibility, and high hysteresis, agar is intensely used in various applications, mainly in the food industry, due to its ability to form gel and have an odorless taste. The most important agar evaluation index is gel strength, an important feature for pricing and developing new applications. The easiest way to improve agar characteristics is to remove the sulfate groups with hydrogen peroxide. Thus, after modification, the viscosity, ash content, and sulfate content decreased. Conversely, the gel strength, whiteness, and transparency increased after modification, in contrast to gelling, melting, and dissolving temperatures that decreased after modification [128]. Unlike other biopolymers, agar has been widely used as an encapsulating agent for probiotics since 1988 [129]. The method followed a simple way of encapsulation, which involves the use of drug microparticles and their dispersing at high temperatures in a hydrophilic liquid vehicle. After cooling, due to the transition to ambient temperature, the beads solidify. The same encapsulation method is currently used both for agar and other biopolymers used for this purpose [130]. The method depends on dropping a hot hydrophilic polymeric solution on the top of a cooled organic liquid, such as ethyl acetate which is a non-toxic compound, during which the polymer and the incorporated drug are insoluble. Usually, when only agar is used as an encapsulating agent, the release of the drug occurs in two phases. The first and faster phase leads to the release of 10–20% of the drug, based on the agar content of the beads. The second is a slower and more prolonged phase and becomes even slower as dissolution proceeds. In the first phase, the drug presents in a molecular state on the surface and is released in the outer layer of the bead so that in the second phase, its release is due to dissolution from the solid core. When used as an encapsulating agent, the larger the mass of agar in the beads, the denser the matrix formed and the lower the transfer of drug molecules through the beads. Similarly, the beads that contain a lower percentage of agar in the composition have a higher water content, which explains the rapid rate of drug release [131]. Therefore, agar can be used for the development of sustained-release dosage systems because it is a natural, inert, non-toxic, renewable, biocompatible, and inexpensive material.

Starch and Its Use for Drug Delivery

Starch is one of the most abundant renewable biopolymers on Earth and is non-allergenic, GRAS (generally recognized as safe), and cheap [132]. It is found in peas, corn, rice, wheat, potato, and beans [133]. Starch granules vary in size, shape, particle size distribution, and in the amylose–amylopectin ratio depending on the botanical origin and maturity [134]. The high encapsulation efficiency was reached when the amylose:amylopectin ratio was 25:75 [135]. Starch granules are composed of amylose and amylopectin, free fatty acids and lysophopholipids, proteins, phosphate esters, and water [136]. Amylose is the linear fraction and is composed of glucopyranose units linked by α-(1,4)-glycosidic linkages, while amylopectin is a highly branched polymer with short α-(1,4)-glycosidic chains linked by α-(1,6)-glycosidic branching points [137]. Although amylopectin has a high viscosity and is a good thickening agent, it produces very weak gels with poor mechanical properties [138]. Starch is a biopolymer available in the form of powders, hydrogels, films, and sponges [139]. Due to its low cost, physicochemical features, biodegradability, and biocompatibility, native and modified starch has been widely used in the food, chemical, pharmaceutical, and environmental industries [140].

In the pharmaceutical and medical industries, starch has been used as a pharmaceutical excipient, a tablet super disintegrant (immediate release tablet formulations), and a controlled/sustained-release polymer or as plasma volume expander, useful for patients suffering from trauma, heavy blood loss, or in cancer treatment [141]. Research has focused on the ability of native starch to be dissolved by pancreatic enzymes after oral ingestion, followed by absorption from the small intestine into the systemic circulation. There is also a resistant part of starch that is not digested in the small intestine and is fermented by colonic bacteria. When used as an encapsulant for drugs, it is combined with other biopolymers precisely to limit or attenuate enzymatic degradation in the stomach, thus facilitating the absorption of an adequate amount of the therapeutic agent [142]. In pharmacotherapy, the main objective of such a system is to provide controlled drug release and prevent fluctuations of active substances in the blood in order to maintain drug plasma concentration within the optimal range, in accordance with therapeutic recommendations.

Starch with high crystallinity levels has been explored as an encapsulation matrix. In order to be used successfully in drug delivery and other industries, starch can be modified so that the physicomechanical properties are adjusted to maximize its use. Starch can be modified by chemical, physical, enzymatic, and genetic processes. Of these, chemical process is used most frequently due to its non-disintegrating nature and potential increase in the functionality of the modified starch. The applications of starch as an encapsulating agent of active substances are presented in Table 4.

Table 4. Applications of starch for drug delivery.

Biopolymers	Entrapped Substances	Applications	Benefits	References
Corn starch	Chlorhexidine gluconate	Long-term drug delivery system	In vitro drug release was observed for 21 days and it inhibited *Staphylococcus aureus* growth.	[134]
Corn starch	Clonidine	Transdermal patches	Transdermal patches with a single dose of 30 µg hydrogel had an effect 15 min after application in treated mice.	[143]
Corn starch	Salicylic acid	Hydrogel membrane	The diffusion of the drug through the membrane was 4.11×10^{-6} cm^2/s. The hydrogel was tested as an artificial skin for transferring nutrients or medicines, or for healing substances to the target area.	[144]
Corn starch/ethylene vinyl alcohol blend	Non-steroid anti-inflammatory agent	Drug delivery carriers	In vitro tests showed an immediate burst effect, followed by a slower, controlled release of the drug that lasted up to 10 days.	[145]
Potato starch	Ibuprofen, benzocaine, and sulphapyridine	Starch-based stable carriers	Encapsulation altered starch digestion; resistant starch was available in the colon for fermentation.	[146]
Glutinous rice starch, sodium alginate, and calcium chloride	Metformin hydrochloride	Hydrogel beads for controlled drug delivery	The initial drug entrapment efficiency was very low for the metformin hydrochloride because of its high solubility. Encapsulation improved it when combined with pre-gelatinized starch gel.	[147]
Ensete ventricosum starch	Epichlorohydrin	Drug-release sustaining pharmaceutical excipient	The in vitro drug release profile showed a minimum burst release, followed by a sustained release for 12 h.	[148]
Starch-clay composites	Tramadol	Tablet formulations	The controlled drug release of tramadol from starch-clay biocomposites was achieved in approximately 350 min.	[149]
Starch-chitosan	Hydroxyurea	Cancer therapy	The drug release was sensitive to pH and increased in the acid environment. The drug/starch/chitosan had a toxicity effect and, at certain concentrations, killed cancer cells.	[150]
PVA-corn starch hydrogel	Erythromycin	Wound dressing	The release of erythromycin from the PVA/corn starch network was higher than the drug containing PVA hydrogel (after 1800 min, released 76.7 mg of the total drug).	[151]

Table 4. *Cont.*

Biopolymers	Entrapped Substances	Applications	Benefits	References
Corn starch-sponge matrix	Uranine, indomethacin, and nifedipine	Sustained-release capsule	After intraduodenal administration, 2.5% of the capsule exhibited a sustained release of the drug in the plasma.	[152]
Starch-poly-ε-caprolactone	Dexamethasone	Drug delivery and tissue engineering applications	The drug from the outermost layer of the microparticles was quickly released. In vitro tests showed a sustained-release pattern for 30 days.	[153]
High-amylose starch-microcrystalline cellulose	Ranitidine hydrochloride	Gastric-floating drug delivery systems	In vitro tests indicated that the system with 3:7 (wt./wt.) starch/cellulose ratio maintained the buoyancy for more than a day; the drug release was 45.87% in the first hour, followed by a sustained release for up to 10 h.	[154]
Maize starch	Probiotics, e.g., *Lactobacillus plantarum*	Microencapsulated probiotic	In low acid environments, *L. plantarum* encapsulated in the starch matrix was more stable. After simulated digestion and heating treatments, the cells maintained their high viability, unlike formulations with native starch	[155]

Most starch-based drug delivery systems have been developed with starches extracted from potato, maize, corn, cassava, and wheat [118]. As shown in Table 4, starch is a viable source of biopolymer, used as an encapsulating agent for controlled drug delivery systems. In its unmodified form, starch is not as effective as a drug delivery system due to poor mechanical properties, such as low shear stress resistance or high retrogradation and syneresis, thermal decomposition, reduced processability, and solubility in common organic solvents [156]. However, after modification, starch can be used successfully for this purpose. For example, modifying starch in order to obtain resistant starch has led to its use for improving the gut microbiota population with a role in modulating signaling pathways associated with anti-inflammation, anti-diabetes, and anti-obesity [157]. Resistant starch, due to its high amylose content and low amylopectin, has been recognized as a healthy food for humans and animals. It can be considered prebiotic and may reach the colon due to its resistance to digestion by pancreatic enzymes in the small intestine [158]. Therefore, encapsulation has been suggested as the best approach to improve prebiotic–probiotic symbiosis.

Cellulose and Its Use for Drug Delivery

Cellulose, a natural polymer, is the most renewable and abundant polysaccharide. Cellulose has been used as an immunoprotective macrocapsule because it does not form a hydrogel and it is mostly applied in inert diffusion chambers. As an encapsulating agent, it is beneficial for cytotoxic epithelial cells in the treatment of pancreatic cancer, insulin-producing cell lines (HIT-T15), embryonic kidney cells, and hybridoma cells. It is recognized as a new nanovehicle for oral colorectal cancer treatment with high drug release at a neutral pH compared to acid pH, being proposed as a safe oral delivery system for controlled colon cancer treatment [159].

Cellulose is the structural part of the cell wall of green plants, algae, or oomycetes. It is part of the polysaccharide group and is composed of a linear chain of β (1 → 4)-linked D-glucose units. Considering it has an amphiphilic character, it can be used as a surfactant and/or stabilizer at the water–oil interface in pickering emulsions [160]. Cellulose is insoluble in water and most organic solvents. The cellulose derivative, carboxy-methyl-cellulose (CMC), contains carboxymethyl groups bound to the OH-groups of glucopyranose monomers on the cellulose backbone. CMC is mostly applied as a matrix molecule and, in order to ensure mechanical stability and immunoprotection, requires surface coating. CMC has been used as an encapsulating agent for probiotics, but due to the hydrophilicity of the cellulose derivatives, physical degradation occurs when passing through the digestive system. Combined with alginate, it provides a better medium system for probiotics with enhanced tolerance at low pH and a more durable delivery of probiotic cells. Long-term storage stability depends on low water activity and low temperature. The most used dehydration methods to reduce water activity are freeze-drying, spray-drying, vacuum-drying, convective air-drying, and fluidized bed-drying. Among all, freeze-drying is the best method for preserving cells' viability because it reduces the damage to biological structures by eliminating water through sublimation [155].

Cellulose crystals have been used in combination with chitosan to encapsulate vitamin C. Stability of vitamin C is highly dependent on light, pH, and the dissolved oxygen in the environment, but is maintained due to the encapsulation with cellulose and chitosan crystals, and this may be a way to preserve highly unstable compounds during long-term storage in functional systems [161]. Similarly, nanofibrillated cellulose, combined with soybean oil-in-water emulsion and whey protein isolate, was used to encapsulate vitamin D3. Vitamin D3 encapsulation efficiency has improved with increasing emulsifier concentrations. Increasing the concentration of nanofibrillated cellulose has improved the stability and efficiency of encapsulation against environmental stresses (pH changes, salt addition, and thermal processing). The procedure may be the basis for more suitable encapsulation technologies for liposoluble vitamins in emulsion-based food products [162].

In addition to encapsulating vitamins, cellulose and cellulose derivatives have been used as agents to encapsulate drugs and probiotics with active substances. For example, ethyl cellulose nanoparticles have been shown to be effective in encapsulating clarithromycin (3:1 weight ratio of ethyl cellulose:clarithromycin). Once encapsulated, clarithromycin was more effective against *Helicobacter pylori* gastric infections. Tests performed in vivo on laboratory mice have clearly indicated better elimination of bacteria from the stomach by encapsulated clarithromycin compared to the nonencapsulated drug [163]. Ethyl cellulose and microcrystalline cellulose were also used for the encapsulation of antihypertensive drugs. This is important considering that, in the standard method of manufacturing microspheres involving emulsification and solvent evaporation, the solvents used are usually dichloromethane or chloroform, which are hazardous for the environment. Therefore, less toxic substances such as ethyl acetate are used to prepare the microspheres. Furthermore, the drug release from the microspheres is faster than the tableted ones, suggesting that tableting of the microparticulate systems may be optimal [163].

3.2. Biopolymers in Probiotic Encapsulation and Delivery

Due to their special properties, biopolymers have been used to encapsulate probiotics. Probiotics are living organisms with benefits on the hosts' health if ingested in adequate amounts [164]. According to the International Scientific Association for Probiotics and Prebiotics (ISAPP), a sufficient amount of probiotics with a beneficial effect on the hosts' health involves ingesting 1×10^9 CFU per serving [165]. Unlike probiotics, prebiotics are nutrients, usually high-fiber foods, providing the substrate that is selectively utilized by the hosts' microorganisms, conferring a health benefit [165]. Most probiotics in the human body form the commensal intestinal microbiota with a role in increasing resistance to infections and boosting host immune system, glucose and lipid metabolism, degradation of complex carbohydrates, and synthesis of vitamins and bile acid [166]. Although the effects of probiotics on various diseases is still debatable, several studies showed beneficial effects in the treatment and prevention of infectious diseases. For example, strains of *Lactobacillus plantarum*, *Lactobacillus casei*, or *Lactobacillus paracasei* had antifungal, antibacterial, and antioxidant effects. Other strains have been shown to have anti-inflammatory effects, to lessen the risk of osteoporosis, maintain cholesterol levels, and prevent the proliferation of cancer cells [167]. Therefore, due to their beneficial effects and health claims, there has been a worldwide explosion of probiotic-based health products in the form of dietary supplements [168]. As such, the global probiotic market is soon reaching USD 50 billion and, with that, the range of probiotics-containing products and associated health claims continue to expand rapidly. Currently, in Europe, the probiotic market is subject to regulatory requirements and compliance with rules and regulations in order to meet certain standards for product registration and use [169].

The use of encapsulation technologies of probiotics has been intensely studied in order to increase probiotics' viability throughout manipulation, storage, commercialization, and incorporation in food and pharmaceutical products so that these cells are viable during their transit and residence in the gastrointestinal tract. Therefore, improving probiotic survival and resistance to adverse conditions through encapsulation is paramount to their effectiveness in health and disease conditions. To prove their effectiveness, encapsulated probiotic strains have been incorporated into a wide range of food products such as yoghurt [170], cheeses [171], frozen dairy desserts [172], beverages [173], and meat products [174], increasing their therapeutic effects [168]. Once encapsulated, probiotics embedded in food matrices maintained their viability even two months under refrigeration [175]. They can also be mixed in a single microcapsule or in dual core capsules with separation microcompartments [176] and a combination of at least two strains can improve their effect [174]. In order to scale up production, several industrial partnerships between food producers and probiotic companies have been formed. For example, Christian Hansen and Dos Pinos developed probiotic ice cream, Balchem Encapsulates and Rosell Institute developed probiotic raisins and bars, and Dannon uses probiotics encapsulated in their products [177]. Encapsulated probiotics have been effective in irritable bowel syndrome [178], colitis, abdominal pain [179], and other gut or metabolic conditions characterized by microbiota dysbiosis [180].

Several encapsulation methods have been developed and used.

(1) Microencapsulation represents a physicochemical or mechanical process used to trap a substance (active agent) into a coating material (defined as wall material). In this way, spherical particles have a thin and strong but semipermeable membrane with a diameter from nanometers to a few millimeters [174]. The purpose of the procedure is to protect compounds or viable cells against environmental agents that can destroy the core [181].

(2) Spray-drying technique is suitable for industrial applications on a large scale, involving atomization of a liquid mixture and the solvent is evaporated at contact with hot air or gas.

(3) Lyophilization involves freezing the cells with the material used for encapsulation (usually at freezing temperatures), followed by vacuum elimination of water at a pressure between 0.05 to 0.1 mBar and temperature between −50 °C to −30 °C. To preserve and stabilize the activity of lyophilized probiotics, cryoprotectants are added, such as lactose, trehalose, sorbitol, sucrose, milk protein, or skim milk.

(4) Extrusion is the most common technique to use biopolymers as encapsulation materials. The method involves obtaining a hydrocolloid solution, followed by the addition of microorganisms, formation of droplets using a syringe needle (pilot scale) or an extruder (industrial scale), and their release into a hardening solution (typically calcium chloride) [182].

(5) Emulsion is when a small volume of a hydrocolloid suspension containing microorganisms (discontinuous phase) is added to a larger volume of vegetable oil (continuous phase). Using an emulsifier, the mixture is homogenized. After emulsion formation, it can be insolubilized to manufacture gel capsules. The big disadvantage of this method is that the particles obtained vary greatly in shape and size, although bead sizes can be reduced by mechanical homogenization [183].

(6) Spray–freeze-drying is a combined procedure that involves steps used in lyophilization (freeze-drying) and spray-drying. The advantage is that it provides capsules with a controlled size and higher specific surface area, unlike those obtained by spray-drying. The disadvantages of the method refers to high costs (approximately 50 times higher than the classic spray-drying version), long processing times, and the high-energy requirement.

(7) Layer-by-layer is technology based on alternating coating layers of cationic (e.g., chitosan) with anionic (e.g., alginate) biopolymers on cells via electrostatic interaction [184]. It has the advantage of enhanced bacterial viability throughout the gastrointestinal tract, along with the survival of probiotic cells against acidic and bile salt insults, mucoadhesion and growth on intestinal tissues, and in vivo survival [179].

In addition to ensuring cell viability along the gastrointestinal tract, the stability of probiotics during storage is also very important. In this regard, encapsulation has proved to be an effective method. For this, the material used to encapsulate microorganisms is the first and most important factor in maintaining their viability [185]. It improves the survival of probiotics during manufacturing processes, especially heat processing [186] and storage [187]. An important aspect of this process is cytotoxicity. According to ISO 10993-5 [188], a material used for encapsulation is potentially cytotoxic when cell viability decreases below 70% after exposure [189]. In this regard, polysaccharides such as alginate, starch, chitosan, and cellulose, as well as other biopolymers or chemicals, have low or no toxicity and do not affect cell viability. On the contrary, they maintained cellular stability for a long time, particularly when kept in refrigerated or frozen conditions.

Although the encapsulation method has many advantages, there are still several aspects that must be consider. These are: (i) biosafety concerns preventing clinical translation of the cell microencapsulation; (ii) concerns regarding the manipulation and extraction procedures that must be refined in order to be as minimally invasive as possible; (iii) concerns regarding the optimization of cost effectiveness; and (iv) concerns regarding the consideration of internationally accepted regulations for the use of probiotics. Therefore, applications of biopolymers for the coating of encapsulated strains for the purpose of protection in the intestinal gastrointestinal tract or as carriers for direct encapsulation of microorganisms should involve procedures that facilitate high bacterial viability.

3.2.1. Alginate's Use for Probiotic Delivery

Among biomaterials used for encapsulation, alginate is the most widely used due to its strong gelling properties and ability to coat within a short time. Additionally, as a dietary fiber, alginate strengthens the functionality of probiotics used in several diseases such as diabetes or obesity [190]. As seen from Table 5, alginate has proven to be a good microencapsulation agent by extending cell viability in refrigeration and freezing conditions, as well as in adverse gastric and intestinal environments.

Table 5. Utilization of alginate as a probiotic encapsulating material.

Biopolymers	Encapsulated Strain	Encapsulation Method	Benefits	References
Alginate and gelatin	*Lactobacillus rhamnosus*	Extrusion	The cells of *L. rhamnosus* survived in beads with 10^5 CFU/g after four months (initially 10^9 CFU/g).	[191]
Alginate	*Lactococcus lactis* spp. *cremoris*	Extrusion	No release of bacteria in the stomach simulated condition (first 120 min) or the survival in the intestinal fluid until 240 min.	[192]
Alginate	*Bifidobacterium pseudocatenulatum*	Extrusion	None of the uncoated probiotic cells survived after immersion in the simulated small intestine fluid. By contrast, 5.6 log10 CFU/g of viable probiotic cells remained in the tested microgels.	[193]
Alginate	*Staphylococcus succinus* and *Enterococcus fecium*	Extrusion	The encapsulated cells showed 98.75–88.75% of viability in simulated gastric fluids. Survival was constant throughout the storage time and decreased from 8.1 log CFU/mL to 7.9 log CFU/mL after 30 days of storage at 4 °C.	[194]
Alginate and milk	*Lactobacillus bulgaricus*	Extrusion	The viability of the encapsulated probiotic was the same after 120 min of incubation in an acid medium (simulated gastric fluid with pH 2.5). The viability of encapsulated *L. bulgaricus* was kept at 8 log CFU/g after 120 min of incubation at pH 2.0. Stability of the encapsulated probiotic can be preserved for one month after storage at 4 °C.	[195]

Table 5. *Cont.*

Biopolymers	Encapsulated Strain	Encapsulation Method	Benefits	References
Alginate and starch	*Lactobacillus fermentum*	Lyophilization	The survival rate of the probiotic was significantly higher for microparticles blended with starch than those with no starch.	[196]
Alginate, chitosan, and locust beam	*Lactobacillus rhamnosus*	Freeze-drying	In contrast to the alginate-based capsules, the alginate locust beam capsules improved stress tolerance (6× for freeze drying, 100× for thermotolerance, and 10× for acid).	[197]
Alginate and chitosan	*Saccharomyces cerevisiae* Y235	Emulsification	The viable microencapsulated cells were kept at 7.00 log CFU/g after six months at −20 °C and remained 6.29 log CFU/g after incubation in SGF for 2 h and in SIF for 12 h, reaching the standard value (10^6–10^7 CFU/g).	[198]
Alginate and chitosan	*Bifidobacterium pseudocatenulatum*		The highest stability of *B. pseudocatenulatum* was at the highest concentrations of alginate (4.41 g/100 mL) and chitosan (0.56 g/100 mL). Resistance of alginate–chitosan capsule in SGF was better than in SIF.	[199]
Alginate and chitosan	*Bifidobacterium breve*	Layer-by-layer	Three-layer coated matrix was the best method to increase viability from <3 log CFU/mL, seen in encapsulated cells, up to a maximum of 8.84 ± 0.17 log CFU/mL upon exposure to in vitro gastric conditions. Multilayer-coated alginate released their loads to the intestine with a gradual delivery over 240 min.	[200]
Alginate, starch, and chitosan	*Lactobacillus acidophilus*	Extrusion	Biopolymers ensured better stability of probiotics after exposure to SGF and SIF with 6.35 log CFU/g, while lower counts were noticed for freeze-dried microcapsules. During storage, cell viability of the probiotics stored in the freeze-dried form was up to six logs for 30 days and 135 days in the moist form when kept at room temperature.	[201]
Alginate, chitosan, and xanthan gum	*Lactobacillus plantarum*	Extrusion	Sequential incubation of biopolymers in SGF and SIF facilitated high survival of *L. plantarum* (95%) at pH < 2. Encapsulation improved storage stability of *L. plantarum* at 4 °C.	[202]

Abbreviations: CFU/g, colony-forming unit per gram; CFU/mL, colony-forming unit per milliliter; SGF, simulated gastric fluids; and SIF, simulated intestinal fluids.

Alginate is hemocompatible, does not accrue in organs, is water soluble, biodegradable, and can form gels under mild conditions. It develops gel at ambiental temperatures and prevents the destruction of the activity of thermolabile drugs. By cross-linking with other agents, it forms insoluble gel that delays the drug release. However, alginates have low mechanical properties, therefore they must be reinforced by combining with other biopolymers or with various conjugates in order to obtain both ionically and covalently cross-linkable capsules.

3.2.2. Chitosan Use for Probiotic Delivery

Like other polymers, chitosan has been used to encapsulate probiotics. The best-performing formulas were identified in combinations of chitosan with other biopolymers such as alginate, agar, or gelatin. The most used combination is with alginate, in which chitosan is used as a final layer of microcapsules (Table 6). This is because at pH 7, chitosan that is positively charged develops strong bonds with gelatin and agar, which are negatively charged [203]. What sets chitosan apart is its antibacterial properties as it is a cationic polysaccharide. The disadvantage of chitosan, however, is the need for solubilization in an acidic environment. Usually, acetic acid is used to solubilize the powder and obtain the coating-forming solution without turbidity, which occurs when a compound has not been fully solubilized [204]. Chitosan cannot be used individually as an encapsulating agent with a role in maintaining cell viability. This is because it increases cell membrane permeability, leading, in the end, to cell loss [205]. Due to this, chitosan is mixed with other natural substances when used as an encapsulating agent.

Table 6. Chitosan use for probiotic encapsulation.

Biopolymers	Encapsulated Strain	Encapsulation Method	Benefits	References
Chitosan, agar, and gelatin	*Lactobacillus plantarum*	Emulsification	Particles with a diameter of approximately 6 mm did not solubilize in SGF 20 min after exposure. Cell viability in the biopolymer-free formula decreased completely after 2 h, unlike coated particles whose viability was 9.2 CFU/g after 2 h.	[206]
Chitosan and alginate	vaccine with *Lactobacillus plantarum*	Extrusion	The oral vaccine containing *L. plantarum*, used against spring viremia of carp virus, was effective even after 56 days due to the encapsulation.	[207]
Chitosan and xanthan gum	*Pediococcus acidilactici*	Extrusion	The encapsulated cells maintained their cell viability for 8 h in the gastrointestinal fluid with maximum release occurring after 24 h. The encapsulated cells maintained their viability for three days when tested in deionized water.	[208]
Chitosan and alginate	*Bifidobacterium breve*	Extrusion	In an acidic medium (pH 2), cell viability was maintained for 1 h. As pH increased (4 and 5), cell viability increased to 120 min. After 2 h, the swelling ratio decreased, a sign that the microcapsules began to disintegrate. Chitosan maintained cellular stability at pH 4 and 5, and alginate at pH 2.	[209]
Chitosan and alginate	*Lactobacillus reuteri DSM 17938*	Vibration technology	Unencapsulated cells were more labile to gastrointestinal stress conditions (reduction by 2.09 log cycles after 3 h). The encapsulated ones resisted better with a reduction of 0.82 log cycles.	[210]
Chitosan and alginate	*Saccharomyces boulardii*	Extrusion	Encapsulation of strains with chitosan and alginate facilitated maintenance of cell viability up to 6 h after administration in mice.	[211]
Chitosan and alginate	*L. acidophilus and L. casei*	Extrusion	Galactooligosaccharides potentiated the effect of microencapsulation. Cell viability was reduced by 3.1 logs for *L. acidophilus* and 2.9 logs for *L. casei* when tested at a very low pH (1.55) of SIF.	[212]
Chitosan and alginate	*Bacteria strain 4.1.Z (B. amyloliquefaciens, B. subtilis,* and *B. methylotrophicus)*	Vibration and extrusion	After lyophilization, the microcapsules maintained their viability (10^6–10^7 CFU/g) for about two months under refrigeration. Chitosan maintained the integrity of capsules for 24 h.	[213]
Chitosan and alginate	*Lactobacillus reuteri KUB-AC5*	Emulsification	The viability of non-encapsulated cells decreased in 40 min from 8 logs CFU/mL to <4 log CFU/mL, being completely eliminated after 1 h. The encapsulated cells were much more stable with a reduction of 1 log CFU/mL after 180 min at pH 1.8.	[214]
Chitosan and hydrochloride-alginate	*Bacillus licheniformis*	Orifice-polymerization method	The chitosan coating protected the microcapsules; cell release (6.19 CFU/mL) in 1 h in SGF (pH 2) and 4 h in the simulated intestinal fluid (pH 6).	[215]

Abbreviations: CFU/g, colony-forming unit per gram; CFU/mL, colony-forming unit per milliliter; SGF, simulated gastric fluids; and SIF, simulated intestinal fluids.

Although the effectiveness of chitosan encapsulation has been demonstrated, it does not appear to be the best biopolymer for probiotic encapsulation. Besides the fact that it cannot be used individually and produces turbidity, the obtained microparticles are usually larger in size, more porous, wrinkled [213], sticky, and the aggregation is worsened [211]. However, utilization of chitosan should not be limited as chitosan improves potential bioadhesion and facilitates the controlled release of bacteria [213].

3.2.3. Agar Use for Probiotic Delivery

Agar is one of the polysaccharides intensively used to obtain tablets or other formulas of drugs released in the gastrointestinal tract, but is less used as an encapsulating agent for probiotics. Although there have been attempts to microencapsulate with agar, when used in combination with other biopolymers (Table 6), research on its use is limited. This is due to the higher ability to obtain films and lesser ability to facilitate the development of coatings. These applications are mainly due to the ability of agar to form viscous solutions by solubilizing the powder in water at very high temperatures (over 90 °C) in order to obtain a termoreversible gel. Agar cannot produce gel at lower temperatures and high temperatures affect the viability of microorganisms. To date, no working method that could involve the solubilization of agar powder in liquids with lower temperatures has been developed. When used as an encapsulating material for essential oils, the temperature of the film-forming solution is lowered to 40 °C, after which essential oils are incorporated [216,217]. Agar-based films have low mechanical properties such as low tensile strength and poor elasticity. Therefore, in the development of films, it is preferred to mix it with other polysaccharides, proteins, or lipids.

3.2.4. Starch Use for Probiotic Delivery

Starch can be used as an encapsulating agent but it has weaker characteristics than alginate, chitosan, or cellulose. In the pharmaceutical field, starch is used mainly for encapsulating drugs or active substances when molded into tablets or oral formulations. This is because starch is strongly hydrophilic and is easily dissolved in liquids at ambient temperatures. However, it has the ability to form very small microspheres constituted in resistant aggregates, allowing for a better protection of the core [218]. For example, when used as an encapsulating agent for *L. plantarum*, rice starch maintained cell viability both at 4 °C (refrigeration conditions) and at temperatures above 50 °C. However, as an encapsulating agent, starch proved to be more effective when mixed with other compounds. For example, the combination of starch and alginate resulted in microcapsules with increased probiotics resistance to simulated gastric conditions (Table 7).

Table 7. Starch as an encapsulation material for probiotics.

Biopolymers	Encapsulated Strain	Encapsulation Method	Benefits	References
Rice starch	*Lactobacillus casei*, *Lactobacillus brevis*, and *Lactobacillus plantarum*	Extrusion	The viability of encapsulated cells (8.27/8.46/7.65 log CFU/g) was kept constant for two months at refrigeration. In contrast, non-encapsulated cells lost their viability by approximately 3 log CFU/g during storage.	[219]
Starch and pectin	*Lactobacillus plantarum*	Extrusion	Cell viability was reduced from 10 log CFU/g to 1 log CFU/g for free cells maintained for 2 h in gastric conditions (pH 1.5–3). Cells encapsulated in pectin had higher viability (4.6 log CFU/g) but the best protection was observed with the addition of starch to which the viability increased to 6.94 log CFU/g.	[220]
Starch from corn and rice	*Lactobacillus plantarum*	Freeze-drying	Encapsulated cells showed thermal stability and maintained their integrity for 35 min at 55 °C. Unencapsulated cells subjected to the same treatment lost their viability by 63% after only 10 min of exposure to 55 °C.	[221]
Starch, alginate, chitosan, and inulin	*Lactobacillus casei* and *Bifidobacterium bifidum*	Emulsification	Encapsulated *L. casei* and *B. bifidum* lost their viability when subjected to simulated gastric conditions for 120 min. Cell viability decreased from 25.10×10^{10} CFU/mL to 6.30×10^{6} CFU/mL for *L. casei*. Encapsulated *B. bifidum* lost 4.65 log/mL of the bacterial culture, while the unencapsulated form had undetectable cell viability after 90 min.	[222]
Starch and alginate	*Lactobacillus fermentum*	Emulsification	Encapsulated in the matrix, cells maintained viability when stored at 4 °C for 45 days. In environmental conditions, however, cells showed a decrease of 1.7 log after 24 h, with complete loss after 2 weeks.	[196]

Table 7. *Cont.*

Biopolymers	Encapsulated Strain	Encapsulation Method	Benefits	References
Starch	*Lactobacillus paracasei*	Electrospinning	Tested at different storage temperatures (4, 25, and 37 °C), *L. paracasei* cells maintained their initial viability of 13.6 × 10 CFU/mL when stored for three weeks at 4 °C and 25 °C but not at 37 °C. Unencapsulated cells lost about 90% of their viability regardless of the storage temperature.	[223]
Maize starch, maltodextrin, and gum arabic	*Lactobacillus acidophilus*	Spray-drying	After 30days of storage at room temperature, only strains encapsulated with maltodextrin, namely gum arabic, maintained their cell viability of 10^6 CFU/g. After 60 days, no encapsulating material prevented the loss of cell viability. Of the tested coatings, starch least protected the bacterial strains.	[224]
Taro and rice starch	*Lactobacillus paracasei*	Spray-drying	When stored, the taro–starch encapsulated strains were more stable; cells maintained their viability for a month, both at temperatures of 4 °C and 25 °C.	[218]
Cassava starch and alginate	*Lactobacillus brevis*	Emulsification	Encapsulation efficiency was higher than 89%. In gastrointestinal conditions, cell viability was better for microcapsules than free cells (96.07% compared to 76.51%). After 5 h of maintenance in the same conditions, viability of *L. brevis* encapsulated cells was 8.69 log CFU/mL, unlike the non-encapsulated ones with 6.87 log CFU/mL.	[225]
Starch and alginate	*Lactobacillus casei*	Extrusion	The addition of 2% starch to the alginate-based film-forming solution increased cell viability from 4×10^8 to 3.1×10^{11}. Increasing starch did not change the results. Tested under simulated gastrointestinal conditions, cell viability was maintained for up to 6 h.	[226]

3.2.5. Cellulose Use for Probiotic Delivery

Carboxymethyl cellulose (CMC) is one of the most widely used forms of cellulose due to the fact that it is the most affordable in terms of spread and cost. Although it is a compound that prevents lipid oxidation and reduces oxygen permeability due to the small number of hydroxyl groups in the structure, CMC is a highly water-soluble compound. This limits its use as an encapsulation material for probiotics as it is degraded in the digestive system. Therefore, when used in combination with other biopolymers such as polysaccharides (carrageenan, alginate, chitosan, and starch), proteins (gelatin), or other natural compounds (inulin), it increased the viability of encapsulated anaerobic probiotics by 36% [227]. The microencapsulation characteristics of CMC have been improved due to the addition of other substances such as gelatin or carrageenan. In general, the encapsulated probiotics maintained their viability for 120 min in simulated gastric conditions, regardless of the type of probiotic encapsulated. When lyophilization was used as the encapsulation method, cellulose and alginate maintained the viability of *L. plantarum* for approximately 160 days in refrigeration conditions (Table 8).

Table 8. Cellulose as a probiotic encapsulating material.

Biopolymers	Encapsulated Strain	Encapsulation Method	Benefits	References
CMC and gelatin	*Lactobacillus rhamnosus*	Emulsification	After 120 min of exposure to SGF and SIF, cell viability was maintained at approximately 77.5% (4 log CFU/mL) in capsules and 60% (5 log CFU/mL) in free cells.	[228]

Table 8. *Cont.*

Biopolymers	Encapsulated Strain	Encapsulation Method	Benefits	References
CMC and κ-carrageenan	*Lactobacillus plantarum*	Extrusion	Cellular stability was greatly improved for encapsulated samples: in an acidic medium (pH 2), it decreased from 10 log CFU/g to 0 after 90 min (non-encapsulated cells) and to about 8 log CFU/g after 120 min (encapsulated cells). During storage for 30 days at 4 °C, cell stability changed from 10 log CFU/g to 2 log CFU/g (free cells) and from 10 log CFU/g to 7 log CFU/g (encapsulated cells).	[229]
Cellulose and pectin	lactic acid bacteria	High-pressure microfluidization	Viability of non-encapsulated cells decreased from 9.56 to 5.29 log CFU/mL in an acid medium, while encapsulation protected cells (decrease of 1.88 log CFU/mL after 2 h in the same conditions).	[230]
CMC and inulin	*Lactobacillus plantarum*	Casting	Cell viability decreased during storage whether or not probiotics were encapsulated.	[227]
CMC and rice bran	*Lactobacillus reuteri*	Emulsification	After heat exposure (85 °C, 25 s), cell viability decreased by more than 57%, although *L. reuteri* is a thermotolerant bacterium. However, the survival rate of encapsulated cells was approximately 6 log CFU/g.	[231]
CMC and chitosan	*Lactobacillus rhamnosus*	Extrusion	Microencapsulated strains were stable at pH 2–4; at the highest pH value tested (12.5), all microcapsules disintegrated.	[232]
Cellulose, alginate, starch, and lecithin	*Lactobacillus rhamnosus*	Extrusion	Under gastric conditions, viability of encapsulated cells was 37% higher than that of free ones. Encapsulation had a positive effect on storage, in which viability decreased by 1.23 log (25 degrees) and 1.08 log (4 degrees), unlike free cells in which stability decreased by 3.17 and 1.93.	[233]
Cellulose and alginate	*Lactobacillus plantarum*	Extrusion and lyophilization	Lyophilized encapsulated cells showed the best stability in the simulated gastrointestinal conditions: gradual release of 2.6×10^6 CFU/mL for 210 min. When refrigerated, encapsulated cells maintained viability for up to 160 days.	[155]
Cellulose and alginate	*Lactobacillus plantarum*	Extrusion-dripping	After 120 min in SGF, viability of non-encapsulated cells decreased by 66.6%, while encapsulated strains had a 58.4% better viability than that of free cells. The addition of cellulose protected the capsules from the action of pH.	[234]

Abbreviations: CMC, carboxymethyl cellulose; SGF, simulated gastric fluids; SIF, simulated intestinal fluids; CFU/g, colony-forming unit per gram; and CFU/mL, colony-forming unit per milliliter.

4. Challenges and Limitations

The rapid adoption of the use of biopolymers is still hindered by several factors. First, research thus far has been conducted primarily in vitro; therefore, more in vivo and clinical trials are needed to demonstrate the health benefits of biopolymers and the biocompatibility in various biomedical applications, in particular when used as encapsulation materials for drug delivery. When used for the treatment of various diseases, more studies are needed to assess the appropriate compound: either alone or in combination to achieve the desired payload in a highly regulated and site-specific manner at therapeutically relevant concentrations [227].

A second challenge is to obtain materials with properties similar or better to synthetic products by improving end-use mechanical properties, kinetics and release, thermal resistance, and barrier properties. For example, some products exhibit low mechanical properties, rapid degradation, and high hydrophilic capacity especially in humid or adverse environments, rendering their application unviable.

Although numerous studies have examined the use of encapsulated probiotics, there is a need for in depth interdisciplinary research that includes microbiologists, medical doctors, and biomaterial, food, agro and chemical engineers. This will lead to better and more efficient prototypes of probiotic encapsulating formulations, to the identification of the most specific/effective probiotic strains, and to the most suitable polymeric carriers applied for product manufacturing. In addition, this will lead to the optimization of the entire process based on the natural characteristics and sensitivity of the selected strain, and will identify ways on how to develop the best formula based on in vitro, in vivo, and pre-clinical techniques considering the release, manufacturing, packaging, transportation, and storage of capsules. Third, the challenges related to costs, economic aspects, and the gap between policy and implementation of the new technologies on a global level need to be addressed in this rapidly emerging field.

5. Conclusions

Although research in this area has revolutionized the biomedical and pharmaceutical industries, significant work still lies ahead if we are to effect changes not only at the individual level but also to bring sustained and affordable environmental changes. Biopolymers have proven effective as encapsulation materials for controlled drug release systems. Significant progress has been made on the biocompatibility, biodegradability, and mechanical and thermal properties of the materials involved. However, challenges still remain in developing target-specific carriers that are biocompatible with various delivery routes for providing sustainable release at the target site. Strategies have been proposed to improve stability of polymers, product kinetics, and release time as well as clinical efficiency. For biomedical applications, it is important to develop uniform guidelines for polymer applications in order to improve versatility and safety and avoid contamination. From the evidence presented thus far, it is obvious that among biopolymers, polysaccharides-based applications are the most used in the field due to their protective, physicochemical, and low immunogenicity characteristics. Although they present some limitations, the ability to react synergistically with other biopolymers or other natural or synthetic substances make their applications widely used. More studies evaluating the technical parameter optimization, efficiency of encapsulation with different formulations, and product-loading capacity concerning viability and metabolic activity should be undertaken. For example, studies examining the functional interactions between the polymer networks and the coating materials, in order to improve capsule stability, product metabolic activity, release time, and viability, should be high on the list. Notwithstanding current limitations from the host perspective, the use of polymers, particularly polysaccharides-based, will continue to expand with an eye towards improving polysaccharide-drug interactions, the optimization of pharmacokinetics and pharmacodynamics, and the compatibility of the polysaccharide with the target tissue. Nevertheless, while more research is needed, polymer-based applications are of great benefit for delivering small molecules that are highly effective, biopotent, and safe.

Author Contributions: All authors have contributed equally to this work. All authors have read and agreed to the published version of the manuscript.

Funding: The work was supported by the project titled "The analysis of interrelationship between gut microbiota and the host with applications in the prevention and control of type 2 diabetes", co-financed by the European Regional Development Fund through Competitiveness Operational Program under the contract number 120/16.09.2016.

Institutional Review Board Statement: Not applicable.

Informed Consent Statement: Not applicable.

Data Availability Statement: The data presented in this study are available upon request from the corresponding author.

Conflicts of Interest: The authors declare no conflict of interest.

References

1. Rebelo, R.; Fernandes, M.; Fangueiro, R. Biopolymers in medical implants: A brief review. *Procedia Eng.* **2017**, *200*, 236–243. [CrossRef]
2. Bala, I.A.; Abdullahi, M.R.; Bashir, S.S. A review on formulation of enzymatic solution for biopolymer hydrolysis. *J. Chem.* **2017**, *6*, 9–13.
3. Yadav, P.; Yadav, H.; Shah, V.G.; Shah, G.; Dhaka, G. Biomedical biopolymers, their origin and evolution in biomedical sciences: A systematic review. *J. Clin. Diagnostic Res.* **2015**, *9*, 21–25. [CrossRef]
4. Aggarwal, J.; Sharma, S.; Kamyab, H.; Kumar, A. The Realm of Biopolymers and Their Usage: An Overview. Available online: http://www.jett.dormaj.com (accessed on 24 June 2021).
5. Udayakumar, G.P.; Muthusamy, S.; Selvaganesh, B.; Sivarajasekar, N.; Rambabu, K.; Banat, F.; Sivamani, S.; Sivakumar, N.; Hosseini-Bandegharaei, A.; Show, P.L. Biopolymers and composites: Properties, characterization and their applications in food, medical and pharmaceutical industries. *J. Environ. Chem. Eng.* **2021**, *9*, 105322. [CrossRef]
6. Pattanashetti, N.A.; Heggannavar, G.B.; Kariduraganavar, M.Y. Smart biopolymers and their biomedical applications. *Procedia Manuf.* **2017**, *12*, 263–279. [CrossRef]
7. Rendón-Villalobos, R.; Ortíz-Sánchez, A.; Tovar-Sánchez, E.; Flores-Huicochea, E. The role of biopolymers in obtaining environmentally friendly materials. In *Composites from Renewable and Sustainable Materials*; InTech: Houston, TX, USA, 2016.
8. Nath, K.; Bhattacharyya, S.K.; Das, N.C. Biodegradable polymeric materials for EMI shielding. In *Materials for Potential EMI Shielding Applications*; Elsevier: Amsterdam, The Netherlands, 2020; pp. 165–178.
9. Kumar, S.; Thakur, K. Bioplastics—Classification, production and their potential food applications. *J. Hill Agric.* **2017**, *8*, 118. [CrossRef]
10. Prajapati, S.K.; Jain, A.; Jain, A.; Jain, S. Biodegradable polymers and constructs: A novel approach in drug delivery. *Eur. Polym. J.* **2019**, *120*, 109191. [CrossRef]
11. Arora, S. Biopolymers as Packaging Material in Food and Allied Industry Value Addition of Makhana and Its by-Products View Project. 2018. Available online: https://www.researchgate.net/publication/342765641 (accessed on 1 February 2021).
12. Díez-Pascual, A.M. Synthesis and applications of biopolymer composites. *Int. J. Mol. Sci.* **2019**, *20*, 2321. [CrossRef] [PubMed]
13. Qasim, U.; Osman, A.I.; Al-Muhtaseb, A.H.; Farrell, C.; Al-Abri, M.; Ali, M.; Vo, D.V.N.; Jamil, F.; Rooney, D.W. Renewable cellulosic nanocomposites for food packaging to avoid fossil fuel plastic pollution: A review. *Environ. Chem. Lett.* **2021**, *19*, 613–641. [CrossRef]
14. Hamouda, T. Sustainable packaging from coir fibers. In *Biopolymers and Biocomposites from Agro-Waste for Packaging Applications*; Elsevier: Amsterdam, The Netherlands, 2021; pp. 113–126.
15. Shivam, P. Recent developments on biodegradable polymers and their future trends. *Int. Res. J. Sci. Eng.* **2016**, *4*, 17–26.
16. Chen, H.; Wang, J.; Cheng, Y.; Wang, C.; Liu, H.; Bian, H.; Pan, Y.; Sun, J.; Han, W. Application of protein-based films and coatings for food packaging: A review. *Polymers* **2019**, *11*, 2039. [CrossRef]
17. Swain, S.K.; Pattanayak, A.J.; Sahoo, A.P. *Functional Biopolymer Composites*; Springer: Berlin, Germany, 2018; pp. 159–182.
18. Varma, K.; Gopi, S. Biopolymers and their role in medicinal and pharmaceutical applications. In *Biopolymers and Their Industrial Applications*; Elsevier: Amsterdam, The Netherlands, 2021; pp. 175–191.
19. Carvalho, A.J.F. Starch: Major sources, properties and applications as thermoplastic materials. In *Monomers, Polymers and Composites from Renewable Resources*; Elsevier: Amsterdam, The Netherlands, 2008; pp. 321–342. ISBN 9780080453163.
20. Temesgen, S.; Rennert, M.; Tesfaye, T.; Nase, M. Review on spinning of biopolymer fibers from starch. *Polymers* **2021**, *13*, 1121. [CrossRef] [PubMed]
21. Gustafsson, J.; Landberg, M.; Bátori, V.; Åkesson, D.; Taherzadeh, M.J.; Zamani, A. Development of bio-based films and 3D objects from apple pomace. *Polymers* **2019**, *11*, 289. [CrossRef]
22. Liu, W.; Misra, M.; Askeland, P.; Drzal, L.T.; Mohanty, A.K. "Green" composites from soy based plastic and pineapple leaf fiber: Fabrication and properties evaluation. *Polymer* **2005**, *46*, 2710–2721. [CrossRef]
23. Maraveas, C. Production of sustainable and biodegradable polymers from agricultural waste. *Polymers* **2020**, *12*, 1127. [CrossRef] [PubMed]
24. Heredia-Guerrero, J.A.; Heredia, A.; Domínguez, E.; Cingolani, R.; Bayer, I.S.; Athanassiou, A.; Benítez, J.J. Cutin from agro-waste as a raw material for the production of bioplastics. *J. Exp. Bot.* **2017**, *68*, 5401–5410. [CrossRef] [PubMed]
25. Khrunyk, Y.; Lach, S.; Petrenko, I.; Ehrlich, H. Progress in modern marine biomaterials research. *Mar. Drugs* **2020**, *18*, 589. [CrossRef]
26. Kaur, S.; Dhillon, G.S. The versatile biopolymer chitosan: Potential sources, evaluation of extraction methods and applications. *Crit. Rev. Microbiol.* **2014**, *40*, 155–175. [CrossRef]
27. Torres, F.G.; Troncoso, O.P.; Pisani, A.; Gatto, F.; Bardi, G. Natural polysaccharide nanomaterials: An overview of their immunological properties. *Int. J. Mol. Sci.* **2019**, *20*, 5092. [CrossRef]
28. Velu, R.; Calais, T.; Jayakumar, A.; Raspall, F. A comprehensive review on bio-nanomaterials for medical implants and feasibility studies on fabrication of such implants by additive manufacturing technique. *Materials* **2020**, *13*, 92. [CrossRef]
29. Chaitanya, S.; Singh, I. Processing of PLA/sisal fiber biocomposites using direct- and extrusion-injection molding. *Mater. Manuf. Processes* **2017**, *32*, 468–474. [CrossRef]

30. Khanna, S.; Srivastava, A.K. A simple structured mathematical model for biopolymer (PHB) production. *Biotechnol. Prog.* **2005**, *21*, 830–838. [CrossRef] [PubMed]
31. Ncube, L.K.; Ude, A.U.; Ogunmuyiwa, E.N.; Zulkifli, R.; Beas, I.N. Environmental impact of food packaging materials: A review of contemporary development from conventional plastics to polylactic acid based materials. *Materials* **2020**, *13*, 4994. [CrossRef]
32. Wankhade, V. Animal-derived biopolymers in food and biomedical technology. In *Biopolymer-Based Formulations: Biomedical and Food Applications*; Elsevier: Amsterdam, The Netherlands, 2020; pp. 139–152. ISBN 9780128168981.
33. Parker, G. Measuring the environmental performance of food packaging: Life cycle assessment. In *Environmentally Compatible Food Packaging*; Elsevier: Amsterdam, The Netherlands, 2008; pp. 211–237. ISBN 9781845691943.
34. Shankar, S.; Rhim, J.-W. Bionanocomposite films for food packaging applications. In *Reference Module in Food Science*; Elsevier: Amsterdam, The Netherlands, 2018.
35. Hassan, M.E.; Bai, J.; Dou, D.Q. Biopolymers; Definition, classification and applications. *Egypt. J. Chem.* **2019**, *62*, 1725–1737. [CrossRef]
36. Soldo, A.; Miletić, M.; Auad, M.L. Biopolymers as a sustainable solution for the enhancement of soil mechanical properties. *Sci. Rep.* **2020**, *10*, 1–13. [CrossRef] [PubMed]
37. Song, J.; Winkeljann, B.; Lieleg, O. Biopolymer-based coatings: Promising strategies to improve the biocompatibility and functionality of materials used in biomedical engineering. *Adv. Mater. Interfaces* **2020**, *7*, 2000850. [CrossRef]
38. Gabor, D.; Tita, O. Biopolymers used in food packaging: A review. *Acta Univ. Cibiniensis Ser. E Food Technol.* **2012**, *16*, 3–19.
39. Sadasivuni, K.K.; Saha, P.; Adhikari, J.; Deshmukh, K.; Ahamed, M.B.; Cabibihan, J.J. Recent advances in mechanical properties of biopolymer composites: A review. *Polym. Compos.* **2020**, *41*, 32–59. [CrossRef]
40. Jummaat, F.; Bashir Yahya, E.; Khalil HPS, A.; Adnan, A.S.; Mohammed Alqadhi, A.; Abdullah, C.K.; Sofea, A.A.; Olaiya, N.G.; Abdat, M.; Hps, K. The role of biopolymer-based materials in obstetrics and gynecology applications: A review. *Polymers* **2021**, *13*, 633. [CrossRef]
41. Reddy, M.S.B.; Ponnamma, D.; Choudhary, R.; Sadasivuni, K.K. A comparative review of natural and synthetic biopolymer composite scaffolds. *Polymers* **2021**, *13*, 1105. [CrossRef]
42. Olivia, M.; Jingga, H.; Toni, N.; Wibisono, G. Biopolymers to improve physical properties and leaching characteristics of mortar and concrete: A review. In *IOP Conference Series: Materials Science and Engineering*; Institute of Physics Publishing: Bristol, UK, 2018; Volume 345, p. 012028.
43. Gurgel Adeodato Vieira, M.; Altenhofen da Silva, M.; Oliveira dos Santos, L.; Masumi Beppu, M. Natural-based plasticizers and biopolymer films: A review. *Eur. Polym. J.* **2011**, *47*, 254–263. [CrossRef]
44. Polman, E.M.N.; Gruter, G.J.M.; Parsons, J.R.; Tietema, A. Comparison of the aerobic biodegradation of biopolymers and the corresponding bioplastics: A review. *Sci. Total Environ.* **2021**, *753*, 131953. [CrossRef]
45. Mohan, S.; Oluwafemi, O.S.; Kalarikkal, N.; Thomas, S.; Songca, S.P. Biopolymers—Application in nanoscience and nanotechnology. In *Recent Advances in Biopolymers*; InTech: Houston, TX, USA, 2016.
46. Imre, B.; Pukánszky, B. Compatibilization in bio-based and biodegradable polymer blends. *Eur. Polym. J.* **2013**, *49*, 1215–1233. [CrossRef]
47. Jayanth, D.; Kumar, P.S.; Nayak, G.C.; Kumar, J.S.; Pal, S.K.; Rajasekar, R. A review on biodegradable polymeric materials striving towards the attainment of green environment. *J. Polym. Environ.* **2018**, *26*, 838–865. [CrossRef]
48. Cziple, F.A.; Velez Marques, A.J. Biopolymers Versus Synthetic Polymers. Available online: https://creativepegworks.wordpress.com/2015/12/01/biopolymers-vs-synthetic-polymers/ (accessed on 26 June 2021).
49. Simionescu, B.C.; Ivanov, D. Natural and synthetic polymers for designing composite materials. In *Handbook of Bioceramics and Biocomposites*; Springer: Berlin, Germany, 2016; pp. 233–286. ISBN 9783319124605.
50. Gowthaman, N.S.K.; Lim, H.N.; Sreeraj, T.R.; Amalraj, A.; Gopi, S. Advantages of biopolymers over synthetic polymers. In *Biopolymers and Their Industrial Applications*; Elsevier: Amsterdam, The Netherlands, 2021; pp. 351–372.
51. Luh, T.-Y.; Yang, H.-C.; Lin, N.-T.; Lin, S.-Y.; Lee, S.-L.; Chen, C.-H. OMCOS for functional polymers-double-stranded DNA-like polymers. *Pure Appl. Chem.* **2008**, *80*, 819–829. [CrossRef]
52. Colmenares, J.C.; Kuna, E. Photoactive hybrid catalysts based on natural and synthetic polymers: A comparative overview. *Molecules* **2017**, *22*, 790. [CrossRef] [PubMed]
53. Ryder, K.; Ali, M.A.; Carne, A.; Billakanti, J. The potential use of dairy by-products for the production of nonfood biomaterials. *Crit. Rev. Environ. Sci. Technol.* **2017**, *47*, 621–642. [CrossRef]
54. Maraveas, C. Environmental sustainability of greenhouse covering materials. *Sustainability* **2019**, *11*, 6129. [CrossRef]
55. Mtibe, A.; Motloung, M.P.; Bandyopadhyay, J.; Ray, S.S. Synthetic biopolymers and their composites: Advantages and limitations—An overview. *Macromol. Rapid Commun.* **2021**, *42*, 2100130. [CrossRef] [PubMed]
56. Ortelli, S.; Costa, A.L.; Torri, C.; Samorì, C.; Galletti, P.; Vineis, C.; Varesano, A.; Bonura, L.; Bianchi, G. Innovative and sustainable production of biopolymers. In *Factories of the Future: The Italian Flagship Initiative*; Springer: Berlin, Germany, 2019; pp. 131–148. [CrossRef]
57. Adeyeye, O.A.; Sadiku, E.R.; Reddy, A.B.; Ndamase, A.S.; Makgatho, G.; Sellamuthu, P.S.; Perumal, A.B.; Nambiar, R.B.; Fasiku, V.O.; Inrahim, I.D.; et al. The use of biopolymers in food packaging. In *Green Polymers and Their Nanocomposites*; Springer: Singapore, 2019; pp. 137–158. [CrossRef]

58. Wróblewska-Krepsztul, J.; Rydzkowski, T.; Michalska-Pożoga, I.; Thakur, V.K. Biopolymers for biomedical and pharmaceutical applications: Recent advances and overview of alginate electrospinning. *Nanomaterials* **2019**, *9*, 404. [CrossRef]
59. Ferraris, S.; Spriano, S.; Scalia, A.C.; Cochis, A.; Rimondini, L.; Cruz-Maya, I.; Guarino, V.; Varesano, A.; Vineis, C. Topographical and biomechanical guidance of electrospun fibers for biomedical applications. *Polymers* **2020**, *12*, 2896. [CrossRef]
60. Dhandayuthapani, B.; Yoshida, Y.; Maekawa, T.; Kumar, D.S. Polymeric scaffolds in tissue engineering application: A review. *Int. J. Polym. Sci.* **2011**, *2011*, 290602. [CrossRef]
61. Ogueri, K.S.; Jafari, T.; Escobar, I.J.L.; Laurencin, C.T. Polymeric biomaterials for scaffold-based bone regenerative engineering. *Regen. Eng. Transl. Med.* **2019**, *5*, 128–154. [CrossRef]
62. Bhatia, S. Natural polymers vs. synthetic polymer. In *Natural Polymer Drug Delivery Systems*; Springer: Berlin, Germany, 2016; pp. 95–118.
63. Innovative Packaging of Medicines-IJAHR. Available online: https://www.ipinnovative.com/journals/IJAHR/article-full-text/10648 (accessed on 27 June 2021).
64. Zadbuke, N.; Shahi, S.; Gulecha, B.; Padalkar, A.; Thube, M. Recent trends and future of pharmaceutical packaging technology. *J. Pharm. Bioallied Sci.* **2013**, *5*, 98–110. [CrossRef]
65. Gheorghita Puscaselu, R.; Amariei, S.; Norocel, L.; Gutt, G. New edible packaging material with function in shelf life extension: Applications for the meat and cheese industries. *Foods* **2020**, *9*, 562. [CrossRef]
66. Janes, M.E.; Dai, Y. Edible films for meat, poultry and seafood. In *Advances in Meat, Poultry and Seafood Packaging*; Elsevier: Amsterdam, The Netherlands, 2012; pp. 504–521. ISBN 9781845697518.
67. Avila-Sosa, R.; Palou, E.; López-Malo, A. Essential oils added to edible films. In *Essential Oils in Food Preservation, Flavor and Safety*; Elsevier: Amsterdam, The Netherlands, 2016; pp. 149–154. ISBN 9780124166448.
68. Zhao, Y.; Mcdaniel, M. Sensory quality of foods associated with edible film and coating systems and shelf-life extension. In *Innovations in Food Packaging*; Elsevier: Amsterdam, The Netherlands, 2005; pp. 434–453. ISBN 9780123116321.
69. Min, S.; Krochta, J.M. Antimicrobial films and coatings for fresh fruit and Vegetables. In *Improving the Safety of Fresh Fruit and Vegetables*; Elsevier: Amsterdam, The Netherlands, 2005; pp. 454–492. ISBN 9781855739567.
70. Mcclements, D.J. Biopolymers in food emulsions. In *Modern Biopolymer Science*; Elsevier: Amsterdam, The Netherlands, 2009; pp. 129–166. ISBN 9780123741950.
71. Ahmed, Z.; Ahmad, A. Biopolymer produced by the lactic acid bacteria: Production and practical application. In *Microbial Production of Food Ingredients and Additives*; Elsevier: Amsterdam, The Netherlands, 2017; pp. 217–257.
72. Martau, G.A.; Mihai, M.; Vodnar, D.C. The use of chitosan, alginate, and pectin in the biomedical and food sector-biocompatibility, bioadhesiveness, and biodegradability. *Polymers* **2019**, *11*, 1837. [CrossRef]
73. De Vos, P.; Lazarjani, H.A.; Poncelet, D.; Faas, M.M. Polymers in cell encapsulation from an enveloped cell perspective. *Adv. Drug Deliv. Rev.* **2014**, *67–68*, 15–34. [CrossRef]
74. Madene, A.; Jacquot, M.; Scher, J.; Desobry, S. Flavour encapsulation and controlled release—A review. *Int. J. Food Sci. Technol.* **2006**, *41*, 1–21. [CrossRef]
75. Therapeutic Applications of Cell Microencapsulation—Google Cărți. Available online: https://books.google.ro/books?id=7u9d2fkYHvMC&pg=PA1&hl=ro&source=gbs_toc_r&cad=3#v=onepage&q&f=false (accessed on 9 August 2021).
76. Biopolymer Nanoparticles: A Review of Prospects for Application as Carrier for Therapeutics and Diagnostics—Pharmaceutical Research and Allied Sciences. Available online: https://ijpras.com/article/biopolymer-nanoparticles-a-review-of-prospects-for-application-as-carrier-for-therapeutics-and-diagnostics (accessed on 9 August 2021).
77. Lee, M.K.; Bae, Y.H. Cell transplantation for endocrine disorders. *Adv. Drug Deliv. Rev.* **2000**, *42*, 103–120. [CrossRef]
78. Sharma, K.; Singh, V.; Arora, A. Natural biodegradable polymers as matrices intransdermal drug delivery. *Int. J. Drug Dev. Res.* **2011**, *3*, 85–103.
79. DJ, M. Designing biopolymer microgels to encapsulate, protect and deliver bioactive components: Physicochemical aspects. *Adv. Colloid Interface Sci.* **2017**, *240*, 31–59. [CrossRef]
80. Altomare, L.; Bonetti, L.; Campiglio, C.E.; de Nardo, L.; Draghi, L.; Tana, F.; Farè, S. Biopolymer-based strategies in the design of smart medical devices and artificial organs. *Int. J. Artif. Organ* **2018**, *41*, 337–359. [CrossRef]
81. Schmaljohann, D. Thermo- and pH-responsive polymers in drug delivery. *Adv. Drug Deliv. Rev.* **2006**, *58*, 1655–1670. [CrossRef]
82. George, A.; Shah, P.A.; Shrivastav, P.S. Natural biodegradable polymers based nano-formulations for drug delivery: A review. *Int. J. Pharm.* **2019**, *561*, 244–264. [CrossRef] [PubMed]
83. Polymers in Pharmaceutical Drug Delivery System: A Review. Available online: https://www.researchgate.net/publication/285986312_Polymers_in_pharmaceutical_drug_delivery_system_A_review (accessed on 26 July 2021).
84. Paques, J.P.; Van Der Linden, E.; Van Rijn, C.J.M.; Sagis, L.M.C. Preparation methods of alginate nanoparticles. *Adv. Colloid Interface Sci.* **2014**, *209*, 163–171. [CrossRef]
85. Severino, P.; da Silva, C.F.; Andrade, L.N.; de Lima Oliveira, D.; Campos, J.; Souto, E.B. Alginate nanoparticles for drug delivery and targeting. *Curr. Pharm. Des.* **2019**, *25*, 1312–1334. [CrossRef]
86. De, S.; Robinson, D. Polymer relationships during preparation of chitosan-alginate and poly-l-lysine-alginate nanospheres. *J. Control. Release* **2003**, *89*, 101–112. [CrossRef]
87. Smith, A.M.; Miri, T. Alginates in Foods. In *Practical Food Rheology: An Interpretive Approach*; John Wiley&Sons: Hoboken, NJ, USA, 2010; pp. 113–132. ISBN 9781405199780.

88. Espevik, T.; Otterlei, M.; Skjåk-Braek, G.; Ryan, L.; Wright, S.D.; Sundan, A. The involvement of CD14 in stimulation of cytokine production by uronic acid polymers. *Eur. J. Immunol.* **1993**, *23*, 255–261. [CrossRef]
89. De Vos, P.; De Haan, B.; Pater, J.; Van Schilfgaarde, R. Association between capsule diameter, adequacy of encapsulation, and survival of microencapsulated rat islet allografts. *Transplantation* **1996**, *62*, 893–899. [CrossRef] [PubMed]
90. Lim, F.; Sun, A.M. Microencapsulated islets as bioartificial endocrine pancreas. *Science* **1980**, *210*, 908–910. [CrossRef]
91. Gattás-Asfura, K.M.; Stabler, C.L. Chemoselective cross-linking and functionalization of alginate via Staudinger ligation. *Biomacromolecules* **2009**, *10*, 3122–3129. [CrossRef] [PubMed]
92. Thomas, D.; KurienThomas, K.; Latha, M.S. Preparation and evaluation of alginate nanoparticles prepared by green method for drug delivery applications. *Int. J. Biol. Macromol.* **2020**, *154*, 888–895. [CrossRef] [PubMed]
93. Design and Characterization of Ibuprofen Loaded Alginate Microspheres Prepared by Ionic Gelation Method. Available online: https://www.researchgate.net/publication/327743699_Design_and_Characterization_of_Ibuprofen_Loaded_Alginate_Microspheres_Prepared_by_Ionic_Gelation_Method (accessed on 24 July 2021).
94. Bose, P.S.; Reddy, P.S.; Nagaraju, R. Preparation and evaluation of indomethacin loaded alginate microspheres. *Ceska Slov. Farm.* **2016**, *65*, 104–110.
95. Hariyadi, D.M.; Hendradi, E.; Erawati, T.; Jannah, E.N.; Febrina, W. Influence of drug-polymer ratio on physical characteristics and release of metformin hydrochloride from metformin alginate microspheres. *Trop. J. Pharm. Res.* **2018**, *17*, 1229–1233. [CrossRef]
96. Ahmed, M.M.; El-Rasoul, S.A.; Auda, S.H.; Ibrahim, M.A. Emulsification/internal gelation as a method for preparation of diclofenac sodium–sodium alginate microparticles. *Saudi Pharm. J.* **2013**, *21*, 61–69. [CrossRef] [PubMed]
97. Patel, N.; Lalwani, D.; Gollmer, S.; Injeti, E.; Sari, Y.; Nesamony, J. Development and evaluation of a calcium alginate based oral ceftriaxone sodium formulation. *Prog. Biomater.* **2016**, *5*, 117–133. [CrossRef]
98. Setty, C.M.; Sahoo, S.S.; Sa, D.B. Alginate-coated alginate-polyethyleneimine beads for prolonged release of furosemide in simulated intestinal fluid. *Drug Dev. Ind. Pharm.* **2008**, *31*, 435–446. [CrossRef] [PubMed]
99. Rastogi, R.; Sultana, Y.; Aqil, M.; Ali, A.; Kumar, S. Alginate microspheres of isoniazid for oral sustained drug delivery. *Int. J. Pharm.* **2007**, *334*, 71–77. [CrossRef]
100. Veronovski, A.; Knez, Ž.; Novak, Z. Preparation of multi-membrane alginate aerogels used for drug delivery. *J. Supercrit. Fluids* **2013**, *79*, 209–215. [CrossRef]
101. Lee, S.; Kim, Y.C.; Park, J.H. Zein-alginate based oral drug delivery systems: Protection and release of therapeutic proteins. *Int. J. Pharm.* **2016**, *515*, 300–306. [CrossRef] [PubMed]
102. Varshosaz, J. The promise of chitosan microspheres in drug delivery systems. *Expert Opin. Drug Deliv.* **2007**, *4*, 263–273. [CrossRef]
103. Bakshi, P.S.; Selvakumar, D.; Kadirvelu, K.; Kumar, N.S. Chitosan as an environment friendly biomaterial—A review on recent modifications and applications. *Int. J. Biol. Macromol.* **2020**, *150*, 1072–1083. [CrossRef]
104. Negm, N.A.; Hefni, H.H.H.; Abd-Elaal, A.A.A.; Badr, E.A.; Abou Kana, M.T.H. Advancement on modification of chitosan biopolymer and its potential applications. *Int. J. Biol. Macromol.* **2020**, *152*, 681–702. [CrossRef]
105. Jayakumar, R.; Reis, R.L.; Mano, J.F. Chemistry and applications of phosphorylated chitin and chitosan. *E Polym.* **2006**, *6*, 1–16. [CrossRef]
106. Raza, Z.A.; Khalil, S.; Ayub, A.; Banat, I.M. Recent developments in chitosan encapsulation of various active ingredients for multifunctional applications. *Carbohydr. Res.* **2020**, *492*, 108004. [CrossRef] [PubMed]
107. Hussain, S.A.; Abdelkader, H.; Abdullah, N.; Kmaruddin, S. Review on micro-encapsulation with Chitosan for pharmaceuticals applications. *MOJ Curr. Res. Rev.* **2018**, *1*, 77–84. [CrossRef]
108. Koppolu, B.P.; Smith, S.G.; Ravindranathan, S.; Jayanthi, S.; Suresh Kumar, T.K.; Zaharoff, D.A. Controlling chitosan-based encapsulation for protein and vaccine delivery. *Biomaterials* **2014**, *35*, 4382–4389. [CrossRef]
109. Prabaharan, M. Review paper: Chitosan derivatives as promising materials for controlled drug delivery. *J. Biomater. Appl.* **2008**, *23*, 5–36. [CrossRef] [PubMed]
110. Yu, S.; Xu, X.; Feng, J.; Liu, M.; Hu, K. Chitosan and chitosan coating nanoparticles for the treatment of brain disease. *Int. J. Pharm.* **2019**, *560*, 282–293. [CrossRef] [PubMed]
111. Wang, X.; Chi, N.; Tang, X. Preparation of estradiol chitosan nanoparticles for improving nasal absorption and brain targeting. *Eur. J. Pharm. Biopharm.* **2008**, *70*, 735–740. [CrossRef]
112. Panos, I.; Acosta, N.; Heras, A. New drug delivery systems based on chitosan. *Curr. Drug Discov. Technol.* **2008**, *5*, 333–341. [CrossRef]
113. Rokhade, A.P.; Patil, S.A.; Aminabhavi, T.M. Synthesis and characterization of semi-interpenetrating polymer network microspheres of acrylamide grafted dextran and chitosan for controlled release of acyclovir. *Carbohydr. Polym.* **2007**, *67*, 605–613. [CrossRef]
114. Giunchedi, P.; Juliano, C.; Gavini, E.; Cossu, M.; Sorrenti, M. Formulation and in vivo evaluation of chlorhexidine buccal tablets prepared using drug-loaded chitosan microspheres. *Eur. J. Pharm. Biopharm.* **2002**, *53*, 233–239. [CrossRef]
115. Cerchiara, T.; Luppi, B.; Bigucci, F.; Petrachi, M.; Orienti, I.; Zecchi, V. Controlled release of vancomycin from freeze-dried chitosan salts coated with different fatty acids by spray-drying. *J. Microencapsul.* **2003**, *20*, 473–478. [CrossRef]
116. Govender, S.; Pillay, V.; Chetty, D.J.; Essack, S.Y.; Dangor, C.M.; Govender, T. Optimisation and characterisation of bioadhesive controlled release tetracycline microspheres. *Int. J. Pharm.* **2005**, *306*, 24–40. [CrossRef]

117. Almeida, A.; Araújo, M.; Novoa-Carballal, R.; Andrade, F.; Gonçalves, H.; Reis, R.L.; Lúcio, M.; Schwartz, S.; Sarmento, B. Novel amphiphilic chitosan micelles as carriers for hydrophobic anticancer drugs. *Mater. Sci. Eng. C* **2020**, *112*, 110920. [CrossRef]
118. Nair, S.; Anoop, K. Design and in vitro evaluation of controlled release Satranidazole subgingival films for periodontitis therapy. *Recent Res. Sci. Technol.* **2014**, *2*, 6–11.
119. Sohail, R.; Abbas, S.R. Evaluation of amygdalin-loaded alginate-chitosan nanoparticles as biocompatible drug delivery carriers for anticancerous efficacy. *Int. J. Biol. Macromol.* **2020**, *153*, 36–45. [CrossRef] [PubMed]
120. Shariatinia, Z.; Mazloom-Jalali, A. Chitosan nanocomposite drug delivery systems designed for the ifosfamide anticancer drug using molecular dynamics simulations. *J. Mol. Liq.* **2019**, *273*, 346–367. [CrossRef]
121. Hanna, D.H.; Saad, G.R. Encapsulation of ciprofloxacin within modified xanthan gum-chitosan based hydrogel for drug delivery. *Bioorg. Chem.* **2019**, *84*, 115–124. [CrossRef] [PubMed]
122. Cánepa, C.; Imperiale, J.C.; Berini, C.A.; Lewicki, M.; Sosnik, A.; Biglione, M.M. Development of a drug delivery system based on chitosan nanoparticles for oral administration of interferon-α. *Biomacromolecules* **2017**, *18*, 3302–3309. [CrossRef]
123. Sathiyaseelan, A.; Saravanakumar, K.; Anand Mariadoss, A.V.; Wang, M.H. Biocompatible fungal chitosan encapsulated phytogenic silver nanoparticles enhanced antidiabetic, antioxidant and antibacterial activity. *Int. J. Biol. Macromol.* **2020**, *153*, 63–71. [CrossRef] [PubMed]
124. Ali, A.; Ahmed, S. A review on chitosan and its nanocomposites in drug delivery. *Int. J. Biol. Macromol.* **2018**, *109*, 273–286. [CrossRef]
125. Mostafavi, F.S.; Zaeim, D. Agar-based edible films for food packaging applications—A review. *Int. J. Biol. Macromol.* **2020**, *159*, 1165–1176. [CrossRef]
126. Bertasa, M.; Dodero, A.; Alloisio, M.; Vicini, S.; Riedo, C.; Sansonetti, A.; Scalarone, D.; Castellano, M. Agar gel strength: A correlation study between chemical composition and rheological properties. *Eur. Polym. J.* **2020**, *123*, 109442. [CrossRef]
127. Zhang, C.; An, D.; Xiao, Q.; Weng, H.; Zhang, Y.; Yang, Q.; Xiao, A. Preparation, characterization, and modification mechanism of agar treated with hydrogen peroxide at different temperatures. *Food Hydrocoll.* **2020**, *101*, 105527. [CrossRef]
128. Xiao, Q.; An, D.; Zhang, C.; Weng, H.; Zhang, Y.; Chen, F.; Xiao, A. Agar quality promotion prepared by desulfation with hydrogen peroxide. *Int. J. Biol. Macromol.* **2020**, *145*, 492–499. [CrossRef] [PubMed]
129. El-Helw, A.E.R.; El-Said, Y. Preparation and characterization of agar beads containing phenobarbitone sodium. *J. Microencapsul.* **1988**, *5*, 159–163. [CrossRef]
130. Evandro, M.; Denis, R.; Zenia, A.; Emre, K.; Denis, P. Oil encapsulation in core-shell alginate capsules by inverse gelation. I: Dripping methodology. *J. Microencapsul.* **2017**, *34*, 82–90. [CrossRef]
131. Rivadeneira, J.; Audisio, M.; Gorustovich, A. Films based on soy protein-agar blends for wound dressing: Effect of different biopolymer proportions on the drug release rate and the physical and antibacterial properties of the films. *J. Biomater. Appl.* **2018**, *32*, 1231–1238. [CrossRef]
132. Zhu, F. Encapsulation and delivery of food ingredients using starch based systems. *Food Chem.* **2017**, *229*, 542–552. [CrossRef] [PubMed]
133. Queiroz, V.M.; Kling, I.C.S.; Eltom, A.E.; Archanjo, B.S.; Prado, M.; Simão, R.A. Corn starch films as a long-term drug delivery system for chlorhexidine gluconate. *Mater. Sci. Eng. C* **2020**, *112*, 110852. [CrossRef]
134. Qi, X.; Tester, R.F. Starch granules as active guest molecules or microorganism delivery systems. *Food Chem.* **2019**, *271*, 182–186. [CrossRef] [PubMed]
135. Hoyos-Leyva, J.D.; Bello-Pérez, L.A.; Alvarez-Ramirez, J.; Garcia, H.S. Microencapsulation using starch as wall material: A review. *Food Rev. Int.* **2018**, *34*, 148–161. [CrossRef]
136. Santana, Á.L.; Angela, M.; Meireles, A. New starches are the trend for industry applications: A review. *Food Public Health* **2014**, *2014*, 229–241. [CrossRef]
137. Chen, J.; Wang, Y.; Liu, J.; Xu, X. Preparation, characterization, physicochemical property and potential application of porous starch: A review. *Int. J. Biol. Macromol.* **2020**, *148*, 1169–1181. [CrossRef]
138. Šárka, E.; Dvořáček, V. New processing and applications of waxy starch (a review). *J. Food Eng.* **2017**, *206*, 77–87. [CrossRef]
139. Hemamalini, T.; Giri Dev, V.R. Comprehensive review on electrospinning of starch polymer for biomedical applications. *Int. J. Biol. Macromol.* **2018**, *106*, 712–718. [CrossRef]
140. Wang, B.; Sui, J.; Yu, B.; Yuan, C.; Guo, L.; Abd El-Aty, A.M.; Cui, B. Physicochemical properties and antibacterial activity of corn starch-based films incorporated with *Zanthoxylum bungeanum* essential oil. *Carbohydr. Polym.* **2021**, *254*, 117314. [CrossRef]
141. Singh, A.; Nath, L.; Singh, A. Pharmaceutical, food and non-food applications of modified starches: A critical review. *Electron. J. Environ. Agric. Food Chem.* **2010**, *9*, 1214–1221.
142. Rodrigues, A.; Emeje, M. Recent applications of starch derivatives in nanodrug delivery. *Carbohydr. Polym.* **2012**, *87*, 987–994. [CrossRef]
143. Saboktakin, M. In-vitro and in-vivo characterizations of novel modified starch transdermal drug delivery system. *MOJ Proteom. Bioinform.* **2017**, *5*, 132–135. [CrossRef]
144. Pal, K. Starch based hydrogel with potential biomedical application as artificial skin. *Afr. J. Biomed. Res.* **2006**, *9*, 23–29. [CrossRef]
145. Malafaya, P.B.; Elvira, C.; Gallardo, A.; San Román, J.; Reis, R.L. Porous starch-based drug delivery systems processed by a microwave route. *J. Biomater. Sci. Polym. Ed.* **2001**, *12*, 1227–1241. [CrossRef]

146. Janaswamy, S. Encapsulation altered starch digestion: Toward developing starch-based delivery systems. *Carbohydr. Polym.* **2014**, *101*, 600–605. [CrossRef] [PubMed]
147. Sachan, N.; Bhattacharya, A. Modeling and characterization of drug release from glutinous rice starch based hydrogel beads for controlled drug delivery. *Int. J. Health Res.* **2010**, *2*, 55398. [CrossRef]
148. Tesfay, D.; Abrha, S.; Yilma, Z.; Woldu, G.; Molla, F. Preparation, optimization, and evaluation of epichlorohydrin cross-linked enset (ensete ventricosum (welw.) cheeseman) starch as drug release sustaining excipient in microsphere formulation. *BioMed Res. Int.* **2020**, *2020*, 2147971. [CrossRef]
149. Alabi, C.O.; Singh, I.; Odeku, O.A. Evaluation of starch-clay composites as a pharmaceutical excipient in tramadol tablet formulations. *Polim. Med.* **2020**, *50*, 33–40. [CrossRef]
150. Starch-Chitosan Modified Blend as Long-Term Controlled Drug Release For Cancer Therapy. Available online: http://staff.uokufa.edu.iq/publication_details.php?khalidahk.abbas&recordID=6799 (accessed on 26 June 2021).
151. Mehrizi, A.A. Synthesis and Characterization of Erythromycin Loaded PVA-Corn Starch Hydrogel Wound Dressing. Available online: https://www.academia.edu/15627564/Synthesis_and_characterization_of_Erythromycin_loaded_PVA_corn_starch_hydrogel_wound_dressing (accessed on 26 June 2021).
152. Shibata, N.; Nishumura, A.; Naruhashi, K.; Nakao, Y.; Miura, R. Preparation and pharmaceutical evaluation of new sustained-release capsule including starch-sponge matrix (SSM). *Biomed. Pharmacother.* **2010**, *64*, 352–358. [CrossRef]
153. Balmayor, E.R.; Tuzlakoglu, K.; Azevedo, H.S.; Reis, R.L. Preparation and characterization of starch-poly-ε-caprolactone microparticles incorporating bioactive agents for drug delivery and tissue engineering applications. *Acta Biomater.* **2009**, *5*, 1035–1045. [CrossRef]
154. Xu, J.; Tan, X.; Chen, L.; Li, X.; Xie, F. Starch/microcrystalline cellulose hybrid gels as gastric-floating drug delivery systems. *Carbohydr. Polym.* **2019**, *215*, 151–159. [CrossRef] [PubMed]
155. Li, W.; Liu, L.; Tian, H.; Luo, X.; Liu, S. Encapsulation of *Lactobacillus plantarum* in cellulose based microgel with controlled release behavior and increased long-term storage stability. *Carbohydr. Polym.* **2019**, *223*, 115065. [CrossRef]
156. Ab'lah, N.; Wong, T.W. Starch as oral colon-specific nano- and microparticulate drug carriers. In *Polymer Science and Innovative Applications*; Elsevier: Amsterdam, The Netherlands, 2020; pp. 287–330.
157. Yang, X.; Darko, O.K.; Huang, Y.; He, C.; Yang, H.; He, S.; Li, J.; Li, J.; Hocker, B.; Yin, Y. Resistant starch regulates gut microbiota: Structure, biochemistry and cell signalling. *Cell Physiol. Biochem.* **2017**, *42*, 306–318. [CrossRef] [PubMed]
158. Fuentes-Zaragoza, E.; Sánchez-Zapata, E.; Sendra, E.; Sayas, E.; Navarro, C.; Fernández-Lõpez, J.; Pérez-Alvarez, J.A. Resistant starch as prebiotic: A review. *Starch Staerke* **2011**, *63*, 406–415. [CrossRef]
159. Pooresmaeil, M.; Behzadi Nia, S.; Namazi, H. Green encapsulation of LDH(Zn/Al)-5-Fu with carboxymethyl cellulose biopolymer; new nanovehicle for oral colorectal cancer treatment. *Int. J. Biol. Macromol.* **2019**, *139*, 994–1001. [CrossRef]
160. Li, Y.; Yu, S.; Chen, P.; Rojas, R.; Hajian, A.; Berglund, L. Cellulose nanofibers enable paraffin encapsulation and the formation of stable thermal regulation nanocomposites. *Nano Energy* **2017**, *34*, 541–548. [CrossRef]
161. Baek, J.; Ramasamy, M.; Willis, N.C.; Kim, D.S.; Anderson, W.A.; Tam, K.C. Encapsulation and controlled release of vitamin C in modified cellulose nanocrystal/chitosan nanocapsules. *Curr. Res. Food Sci.* **2021**, *4*, 215–223. [CrossRef]
162. Mitbumrung, W.; Suphantharika, M.; Mcclements, D.J.; Winuprasith, T. Encapsulation of vitamin D3 in pickering emulsion stabilized by nanofibrillated mangosteen cellulose: Effect of environmental stresses. *J. Food Sci.* **2019**, *84*, 3213–3221. [CrossRef]
163. Pan-In, P.; Banlunara, W.; Chaichanawongsaroj, N.; Wanichwecharungruang, S. Ethyl cellulose nanoparticles: Clarithomycin encapsulation and eradication of H. pylori. *Carbohydr. Polym.* **2014**, *109*, 22–27. [CrossRef]
164. Serna-Cock, L.; Vallejo-Castillo, V. Probiotic encapsulation. *Afr. J. Microbiol. Res.* **2013**, *7*, 4743–4753. [CrossRef]
165. Salminen, S.; Collado, M.C.; Endo, A.; Hill, C.; Lebeer, S.; Quigley, E.M.M.; Sanders, M.E.; Shamir, R.; Swann, J.R.; Szajewska, H.; et al. The international scientific association of probiotics and prebiotics (ISAPP) consensus statement on the definition and scope of postbiotics. *Nat. Rev. Gastroenterol. Hepatol.* **2021**, *2021*, 1–19. [CrossRef]
166. Silva, D.R.; Sardi, J.C.O.; Pitangui, N.S.; Roque, S.M.; da Silva, A.C.B.; Rosalen, P.L. Probiotics as an alternative antimicrobial therapy: Current reality and future directions. *J. Funct. Foods* **2020**, *73*, 104080. [CrossRef]
167. Misra, S.; Pandey, P.; Mishra, H.N. Novel approaches for co-encapsulation of probiotic bacteria with bioactive compounds, their health benefits and functional food product development: A review. *Trends Food Sci. Technol.* **2021**, *109*, 340–351. [CrossRef]
168. Kailasapathy, K. Microencapsulation of Probiotic Bacteria 39 Microencapsulation of Probiotic Bacteria: Technology and Potential Applications. Available online: https://www.caister.com/bacteria-plant (accessed on 27 July 2021).
169. Burgain, J.; Gaiani, C.; Linder, M.; Scher, J. Encapsulation of probiotic living cells: From laboratory scale to industrial applications. *J. Food Eng.* **2011**, *104*, 467–483. [CrossRef]
170. Kailasapathy, K. Encapsulation technologies for functional foods and nutraceutical product development. *CAB Rev.* **2009**, *4*, 33. [CrossRef]
171. Viability and Survival of Free and Encapsulated Probiotic Bacteria in Cheddar Cheese | Western Sydney University ResearchDirect. Available online: https://researchdirect.westernsydney.edu.au/islandora/object/uws%3A5276 (accessed on 30 July 2021).
172. Homayouni, A.; Azizi, A.; Ehsani, M.R.; Yarmand, M.S.; Razavi, S.H. Effect of microencapsulation and resistant starch on the probiotic survival and sensory properties of synbiotic ice cream. *Food Chem.* **2008**, *111*, 50–55. [CrossRef]
173. Mcmaster, L.D.; Kokott, S.A. Micro-encapsulation of Bifidobacterium lactis for incorporation into soft foods. *World J. Microbiol. Biotechnol.* **2005**, *21*, 723–728. [CrossRef]

174. Muthukumarasamy, P.; Holley, R.A. Survival of Escherichia coli O157:H7 in dry fermented sausages containing microencapsulated probiotic lactic acid bacteria. *Food Microbiol.* **2007**, *24*, 82–88. [CrossRef] [PubMed]
175. Frakolaki, G.; Giannou, V.; Kekos, D.; Tzia, C. A review of the microencapsulation techniques for the incorporation of probiotic bacteria in functional foods. *Crit. Rev. Food Sci. Nutr.* **2021**, *61*, 1515–1536. [CrossRef]
176. Zhao, C.; Zhu, Y.; Kong, B.; Huang, Y. Dual-core prebiotic microcapsule encapsulating probiotics for metabolic syndrome. *ACS Appl. Mater. Interfaces* **2020**, *12*, 42586–42594. [CrossRef] [PubMed]
177. Zhang, Y.; Lin, J.; Zhong, Q. The increased viability of probiotic *Lactobacillus salivarius* NRRL B-30514 encapsulated in emulsions with multiple lipid-protein-pectin layers. *Food Res. Int.* **2015**, *71*, 9–15. [CrossRef]
178. Whorwell, P.J.; Morel, J.; Duane, C. Efficacy of an encapsulated probiotic Bifidobacterium infantis 35624 in women with irritable bowel syndrome. *Am. J. Gastroenterol.* **2006**, *101*, 1581–1590. [CrossRef]
179. Anselmo, A.C.; Mchugh, K.J.; Webster, J.; Langer, R.; Jaklenec, A. Layer-by-layer encapsulation of probiotics for delivery to the microbiome. *Adv. Mater.* **2016**, *28*, 9486–9490. [CrossRef] [PubMed]
180. Zheng, D.-W.; Li, R.-Q.; An, J.-X.; Xie, T.-Q.; Han, Z.-Y.; Xu, R.; Fang, Y.; Zhang, X.-Z. Prebiotics-encapsulated probiotic spores regulate gut microbiota and suppress colon cancer. *Adv. Mater.* **2020**, *32*, 2004529. [CrossRef] [PubMed]
181. Solanki, H.K.; Pawar, D.D.; Shah, D.A.; Prajapati, V.D.; Jani, G.K.; Mulla, A.M.; Thakar, P.M. Development of microencapsulation delivery system for long-term preservation of probiotics as biotherapeutics agent. *BioMed Res. Int.* **2013**, *2013*, 620719. [CrossRef]
182. Martín, M.J.; Lara-Villoslada, F.; Ruiz, M.A.; Morales, M.E. Microencapsulation of bacteria: A review of different technologies and their impact on the probiotic effects. *Innov. Food Sci. Emerg. Technol.* **2015**, *27*, 15–25. [CrossRef]
183. Capela, P.; Hay, T.K.C.; Shah, N.P. Effect of homogenisation on bead size and survival of encapsulated probiotic bacteria. *Food Res. Int.* **2007**, *40*, 1261–1269. [CrossRef]
184. Razavi, S.; Janfaza, S.; Tasnim, N.; Gibson, D.L.; Hoorfar, M. Nanomaterial-based encapsulation for controlled gastrointestinal delivery of viable probiotic bacteria. *Nanoscale Adv.* **2021**, *3*, 2699–2709. [CrossRef]
185. Darjani, P.; Hosseini Nezhad, M.; Kadkhodaee, R.; Milani, E. Influence of prebiotic and coating materials on morphology and survival of a probiotic strain of *Lactobacillus casei* exposed to simulated gastrointestinal conditions. *LWT Food Sci. Technol.* **2016**, *73*, 162–167. [CrossRef]
186. Anal, A.K.; Singh, H. Recent advances in microencapsulation of probiotics for industrial applications and targeted delivery. *Trends Food Sci. Technol.* **2007**, *18*, 240–251. [CrossRef]
187. Tripathi, M.K.; Giri, S.K. Probiotic functional foods: Survival of probiotics during processing and storage. *J. Funct. Foods* **2014**, *9*, 225–241. [CrossRef]
188. ISO 10993-5:2009(en), Biological Evaluation of Medical Devices—Part 5: Tests for In Vitro Cytotoxicity. Available online: https://www.iso.org/obp/ui/#iso:std:iso:10993:-5:ed-3:v1:en (accessed on 27 June 2021).
189. Singh, P.; Medronho, B.; dos Santos, T.; Nunes-Correia, I.; Granja, P.; Miguel, M.G.; Lindman, B. On the viability, cytotoxicity and stability of probiotic bacteria entrapped in cellulose-based particles. *Food Hydrocoll.* **2018**, *82*, 457–465. [CrossRef]
190. Puscaselu, R.G.; Lobiuc, A.; Dimian, M.; Covasa, M. Alginate: From food industry to biomedical applications and management of metabolic disorders. *Polymers* **2020**, *12*, 2417. [CrossRef] [PubMed]
191. Lopes, S.; Bueno, L.; De Aguiar Júnior, F.; Finkler, C.L.L. Preparation and characterization of alginate and gelatin microcapsules containing *Lactobacillus rhamnosus*. *An. Acad. Bras. Cienc.* **2017**, *89*, 1601–1613. [CrossRef] [PubMed]
192. Ramos, P.E.; Silva, P.; Alario, M.M.; Pastrana, L.M.; Teixeira, J.A.; Cerqueira, M.A.; Vicente, A.A. Effect of alginate molecular weight and *M/G* ratio in beads properties foreseeing the protection of probiotics. *Food Hydrocoll.* **2018**, *77*, 8–16. [CrossRef]
193. Zhang, Z.; Gu, M.; You, X.; Sela, D.A.; Xiao, H.; Mcclements, D.J. Encapsulation of bifidobacterium in alginate microgels improves viability and targeted gut release. *Food Hydrocoll.* **2021**, *116*, 106634. [CrossRef]
194. Sathyabama, S.; Kumar, M.R.; Devi, P.B.; Vijayabharathi, R.; Priyadharisini, V.B. Co-encapsulation of probiotics with prebiotics on alginate matrix and its effect on viability in simulated gastric environment. *LWT Food Sci. Technol.* **2014**, *57*, 419–425. [CrossRef]
195. Shi, L.E.; Li, Z.H.; Li, D.T.; Xu, M.; Chen, H.Y.; Zhang, Z.L.; Tang, Z.X. Encapsulation of probiotic lactobacillus bulgaricus in alginate'milk microspheres and evaluation of the survival in simulated gastrointestinal conditions. *J. Food Eng.* **2013**, *117*, 99–104. [CrossRef]
196. Martin, M.J.; Lara-Villoslada, F.; Ruiz, M.A.; Morales, M.E. Effect of unmodified starch on viability of alginate-encapsulated *Lactobacillus fermentum* CECT5716. *LWT Food Sci. Technol.* **2013**, *53*, 480–486. [CrossRef]
197. Cheow, W.S.; Kiew, T.Y.; Hadinoto, K. Controlled release of *Lactobacillus rhamnosus* biofilm probiotics from alginate-locust bean gum microcapsules. *Carbohydr. Polym.* **2014**, *103*, 587–595. [CrossRef]
198. Song, H.; Yu, W.; Liu, X.; Ma, X. Improved probiotic viability in stress environments with post-culture of alginate-chitosan microencapsulated low density cells. *Carbohydr. Polym.* **2014**, *108*, 10–16. [CrossRef]
199. Kamalian, N.; Mirhosseini, H.; Mustafa, S.; Manap, M.Y.A. Effect of alginate and chitosan on viability and release behavior of Bifidobacterium pseudocatenulatum G4 in simulated gastrointestinal fluid. *Carbohydr. Polym.* **2014**, *111*, 700–706. [CrossRef] [PubMed]
200. Cook, M.T.; Tzortzis, G.; Khutoryanskiy, V.V.; Charalampopoulos, D. Layer-by-layer coating of alginate matrices with chitosan-alginate for the improved survival and targeted delivery of probiotic bacteria after oral administration. *J. Mater. Chem. B* **2013**, *1*, 52–60. [CrossRef] [PubMed]

201. de Araújo Etchepare, M.; Raddatz, G.C.; de Moraes Flores, É.M.; Zepka, L.Q.; Jacob-Lopes, E.; Barin, J.S.; Ferreira Grosso, C.R.; de Menezes, C.R. Effect of resistant starch and chitosan on survival of *Lactobacillus acidophilus* microencapsulated with sodium alginate. *LWT Food Sci. Technol.* **2016**, *65*, 511–517. [CrossRef]
202. Fareez, I.M.; Lim, S.M.; Mishra, R.K.; Ramasamy, K. Chitosan coated alginate-xanthan gum bead enhanced pH and thermotolerance of *Lactobacillus plantarum* LAB12. *Int. J. Biol. Macromol.* **2015**, *72*, 1419–1428. [CrossRef]
203. Prata, A.S.; Grosso, C.R.F. Production of microparticles with gelatin and chitosan. *Carbohydr. Polym.* **2015**, *116*, 292–299. [CrossRef]
204. Morrison, I.D.; Ross, S. *Colloidal Dispersions: Suspensions, Emulsions, and Foams*; John Wiley&Sons: Hoboken, NJ, USA, 2002; p. 616.
205. Kwiecień, I.; Kwiecień, M. Application of polysaccharide-based hydrogels as probiotic delivery systems. *Gels* **2018**, *4*, 47. [CrossRef] [PubMed]
206. Albadran, H.A.; Monteagudo-Mera, A.; Khutoryanskiy, V.V.; Charalampopoulos, D. Development of chitosan-coated agar-gelatin particles for probiotic delivery and targeted release in the gastrointestinal tract. *Appl. Microbiol. Biotechnol.* **2020**, *104*, 5749–5757. [CrossRef]
207. Jia, S.; Zhou, K.; Pan, R.; Wei, J.; Liu, Z.; Xu, Y. Oral immunization of carps with chitosan–alginate microcapsule containing probiotic expressing spring viremia of carp virus (SVCV) G protein provides effective protection against SVCV infection. *Fish Shellfish Immunol.* **2020**, *105*, 327–329. [CrossRef]
208. Argin, S.; Kofinas, P.; Lo, Y.M. The cell release kinetics and the swelling behavior of physically crosslinked xanthan–chitosan hydrogels in simulated gastrointestinal conditions. *Food Hydrocoll.* **2014**, *40*, 138–144. [CrossRef]
209. Cook, M.T.; Tzortis, G.; Charalampopoulos, D.; Khutoryanskiy, V.V. Production and evaluation of dry alginate-chitosan microcapsules as an enteric delivery vehicle for probiotic bacteria. *Biomacromolecules* **2011**, *12*, 2834–2840. [CrossRef] [PubMed]
210. De Prisco, A.; Maresca, D.; Ongeng, D.; Mauriello, G. Microencapsulation by vibrating technology of the probiotic strain *Lactobacillus reuteri* DSM 17938 to enhance its survival in foods and in gastrointestinal environment. *LWT Food Sci. Technol.* **2015**, *61*, 452–462. [CrossRef]
211. Graff, S.; Hussain, S.; Chaumeil, J.-C.; Charrueau, C. Increased intestinal delivery of viable saccharomyces boulardii by encapsulation in microspheres. *Pharm. Res.* **2008**, *25*, 1290–1296. [CrossRef] [PubMed]
212. Krasaekoopt, W.; Watcharapoka, S. Effect of addition of inulin and galactooligosaccharide on the survival of microencapsulated probiotics in alginate beads coated with chitosan in simulated digestive system, yogurt and fruit juice. *LWT Food Sci. Technol.* **2014**, *57*, 761–766. [CrossRef]
213. Mirtič, J.; Rijavec, T.; Zupančič, Š.; Zvonar Pobirk, A.; Lapanje, A.; Kristl, J. Development of probiotic-loaded microcapsules for local delivery: Physical properties, cell release and growth. *Eur. J. Pharm. Sci.* **2018**, *121*, 178–187. [CrossRef]
214. Rodklongtan, A.; La-ongkham, O.; Nitisinprasert, S.; Chitprasert, P. Enhancement of *Lactobacillus reuteri* KUB-AC5 survival in broiler gastrointestinal tract by microencapsulation with alginate–chitosan semi-interpenetrating polymer networks. *J. Appl. Microbiol.* **2014**, *117*, 227–238. [CrossRef]
215. Wu, Q.X.; Xu, X.; Xie, Q.; Tong, W.Y.; Chen, Y. Evaluation of chitosan hydrochloride-alginate as enteric micro-probiotic-carrier with dual protective barriers. *Int. J. Biol. Macromol.* **2016**, *93*, 665–671. [CrossRef]
216. Çakmak, H.; Özselek, Y.; Turan, O.Y.; Fıratlıgil, E.; Karbancioğlu-Güler, F. Whey protein isolate edible films incorporated with essential oils: Antimicrobial activity and barrier properties. *Polym. Degrad. Stab.* **2020**, *179*, 109285. [CrossRef]
217. Zhang, B.; Liu, Y.; Wang, H.; Liu, W.; Cheong, K.; Teng, B. Characterization of seaweed polysaccharide-based bilayer films containing essential oils with antibacterial activity. *LWT Food Sci. Technol.* **2021**, *150*, 111961. [CrossRef]
218. Alfaro-Galarza, O.; López-Villegas, E.O.; Rivero-Perez, N.; Tapia- Maruri, D.; Jiménez-Aparicio, A.R.; Palma-Rodríguez, H.M.; Vargas-Torres, A. Protective effects of the use of taro and rice starch as wall material on the viability of encapsulated *Lactobacillus paracasei* subsp. Paracasei. *LWT Food Sci. Technol.* **2020**, *117*, 108686. [CrossRef]
219. Ashwar, B.A.; Gani, A.; Gani, A.; Shah, A.; Masoodi, F.A. Production of RS4 from rice starch and its utilization as an encapsulating agent for targeted delivery of probiotics. *Food Chem.* **2018**, *239*, 287–294. [CrossRef]
220. Dafe, A.; Etemadi, H.; Dilmaghani, A.; Mahdavinia, G.R. Investigation of pectin/starch hydrogel as a carrier for oral delivery of probiotic bacteria. *Int. J. Biol. Macromol.* **2017**, *97*, 536–543. [CrossRef]
221. Benavent-Gil, Y.; Rodrigo, D.; Rosell, C.M. Thermal stabilization of probiotics by adsorption onto porous starches. *Carbohydr. Polym.* **2018**, *197*, 558–564. [CrossRef]
222. Zanjani, M.A.K.; Ehsani, M.R.; Ghiassi Tarzi, B.; Sharifan, A. Promoting *Lactobacillus casei* and Bifidobacterium adolescentis survival by microencapsulation with different starches and chitosan and poly L-lysine coatings in ice cream. *J. Food Process. Preserv.* **2018**, *42*, e13318. [CrossRef]
223. Lancuški, A.; Aiman, A.A.; Avrahami, R.; Vilensky, R.; Vasilyev, G.; Zussman, E. Design of starch-formate compound fibers as encapsulation platform for biotherapeutics. *Carbohydr. Polym.* **2017**, *158*, 68–76. [CrossRef]
224. Reyes, V.; Chotiko, A.; Chouljenko, A.; Sathivel, S. Viability of *Lactobacillus acidophilus* NRRL B-4495 encapsulated with high maize starch, maltodextrin, and gum arabic. *LWT Food Sci. Technol.* **2018**, *96*, 642–647. [CrossRef]
225. Thangrongthong, S.; Puttarat, N.; Ladda, B.; Itthisoponkul, T.; Pinket, W.; Kasemwong, K.; Taweechotipatr, M. Microencapsulation of probiotic *Lactobacillus brevis* ST-69 producing GABA using alginate supplemented with nanocrystalline starch. *Food Sci. Biotechnol.* **2020**, *29*, 1475–1482. [CrossRef] [PubMed]

226. Sultana, K.; Godward, G.; Reynolds, N.; Arumugaswamy, R.; Peiris, P.; Kailasapathy, K. Encapsulation of probiotic bacteria with alginate–starch and evaluation of survival in simulated gastrointestinal conditions and in yoghurt. *Int. J. Food Microbiol.* **2000**, *62*, 47–55. [CrossRef]
227. Zabihollahi, N.; Alizadeh, A.; Almasi, H.; Hanifian, S.; Hamishekar, H. Development and characterization of carboxymethyl cellulose based probiotic nanocomposite film containing cellulose nanofiber and inulin for chicken fillet shelf life extension. *Int. J. Biol. Macromol.* **2020**, *160*, 409–417. [CrossRef] [PubMed]
228. Singh, P.; Medronho, B.; Miguel, M.G.; Esquena, J. On the encapsulation and viability of probiotic bacteria in edible carboxymethyl cellulose-gelatin water-in-water emulsions. *Food Hydrocoll.* **2018**, *75*, 41–50. [CrossRef]
229. Dafe, A.; Etemadi, H.; Zarredar, H.; Mahdavinia, G.R. Development of novel carboxymethyl cellulose/k-carrageenan blends as an enteric delivery vehicle for probiotic bacteria. *Int. J. Biol. Macromol.* **2017**, *97*, 299–307. [CrossRef] [PubMed]
230. Chen, B.; Lin, X.; Lin, X.; Li, W.; Zheng, B.; He, Z. Pectin-microfibrillated cellulose microgel: Effects on survival of lactic acid bacteria in a simulated gastrointestinal tract. *Int. J. Biol. Macromol.* **2020**, *158*, 826–836. [CrossRef]
231. Chitprasert, P.; Sudsai, P.; Rodklongtan, A. Aluminum carboxymethyl cellulose–rice bran microcapsules: Enhancing survival of *Lactobacillus reuteri* KUB-AC5. *Carbohydr. Polym.* **2012**, *90*, 78–86. [CrossRef] [PubMed]
232. Singh, P.; Medronho, B.; Alves, L.; da Silva, G.J.; Miguel, M.G.; Lindman, B. Development of carboxymethyl cellulose-chitosan hybrid micro- and macroparticles for encapsulation of probiotic bacteria. *Carbohydr. Polym.* **2017**, *175*, 87–95. [CrossRef]
233. Huq, T.; Fraschini, C.; Khan, A.; Riedl, B.; Bouchard, J.; Lacroix, M. Alginate based nanocomposite for microencapsulation of probiotic: Effect of cellulose nanocrystal (CNC) and lecithin. *Carbohydr. Polym.* **2017**, *168*, 61–69. [CrossRef] [PubMed]
234. Fareez, I.M.; Lim, S.M.; Zulkefli, N.A.A.; Mishra, R.K.; Ramasamy, K. Cellulose Derivatives Enhanced Stability of Alginate-Based Beads Loaded with *Lactobacillus plantarum* LAB12 against Low pH, High Temperature and Prolonged Storage. *Probiotics Antimicrob. Proteins* **2017**, *10*, 543–557. [CrossRef]

polymers

MDPI

Article

Effect of Crystallinity on the Properties of Polycaprolactone Nanoparticles Containing the Dual FLAP/mPEGS-1 Inhibitor BRP-187

Antje Vollrath [1,2], Christian Kretzer [3], Baerbel Beringer-Siemers [1], Blerina Shkodra [1,2], Justyna A. Czaplewska [1,2], Damiano Bandelli [1,2], Steffi Stumpf [1,2], Stephanie Hoeppener [1,2], Christine Weber [1,2], Oliver Werz [2,3] and Ulrich S. Schubert [1,2,*]

1 Laboratory of Organic Chemistry and Macromolecular Chemistry (IOMC), Friedrich Schiller University, Humboldtstraße 10, 07743 Jena, Germany; antje.vollrath@uni-jena.de (A.V.); baerbel.beringer-siemers@uni-jena.de (B.B.-S.); Blerina.Shkodra-pula@uni-jena.de (B.S.); justyna.czaplewska@uni-jena.de (J.A.C.); damianobandelli@gmail.com (D.B.); steffi-stumpf@uni-jena.de (S.S.); s.hoeppener@uni-jena.de (S.H.); christine.weber@uni-jena.de (C.W.)

2 Jena Center for Soft Matter (JCSM), Friedrich Schiller University, Philosophenweg 7, 07743 Jena, Germany; oliver.werz@uni-jena.de

3 Department of Pharmaceutical/Medicinal Chemistry, Institute of Pharmacy, Friedrich Schiller University, Philosophenweg 14, 07743 Jena, Germany; christian.kretzer@uni-jena.de

* Correspondence: ulrich.schubert@uni-jena.de

Citation: Vollrath, A.; Kretzer, C.; Beringer-Siemers, B.; Shkodra, B.; Czaplewska, J.A.; Bandelli, D.; Stumpf, S.; Hoeppener, S.; Weber, C.; Werz, O.; et al. Effect of Crystallinity on the Properties of Polycaprolactone Nanoparticles Containing the Dual FLAP/mPEGS-1 Inhibitor BRP-187. *Polymers* **2021**, *13*, 2557. https://doi.org/10.3390/polym13152557

Academic Editors: José Miguel Ferri, Vicent Fombuena Borràs and Miguel Fernando Aldás Carrasco

Received: 29 June 2021
Accepted: 29 July 2021
Published: 31 July 2021

Abstract: Seven polycaprolactones (PCL) with constant hydrophobicity but a varying degree of crystallinity prepared from the constitutional isomers ε-caprolactone (εCL) and δ-caprolactone (δCL) were utilized to formulate nanoparticles (NPs). The aim was to investigate the effect of the crystallinity of the bulk polymers on the enzymatic degradation of the particles. Furthermore, their efficiency to encapsulate the hydrophobic anti-inflammatory drug BRP-187 and the final in vitro performance of the resulting NPs were evaluated. Initially, high-throughput nanoprecipitation was employed for the εCL and δCL homopolymers to screen and establish important formulation parameters (organic solvent, polymer and surfactant concentration). Next, BRP-187-loaded PCL nanoparticles were prepared by batch nanoprecipitation and characterized using dynamic light scattering, scanning electron microscopy and UV-Vis spectroscopy to determine and to compare particle size, polydispersity, zeta potential, drug loading as well as the apparent enzymatic degradation as a function of the copolymer composition. Ultimately, NPs were examined for their potency in vitro in human polymorphonuclear leukocytes to inhibit the BRP-187 target 5-lipoxygenase-activating protein (FLAP). It was evident by Tukey's multi-comparison test that the degree of crystallinity of copolymers directly influenced their apparent enzymatic degradation and consequently their efficiency to inhibit the drug target.

Keywords: polycaprolactone (PCL); polyesters; hydrophobic-hydrophilic balance (HHB); nanoparticle formulation; nanoparticle crystallinity; FLAP antagonist; BRP-187

1. Introduction

With the first clinical approval of a polymer-based nano-drug in 1995 [1], interest in developing polymers as nanocarriers of (bio)pharmaceutical drugs has been steadily growing [2,3]. Due to their favorable characteristics, e.g., adjustable physical and mechanical properties, it is not surprising that polymer-based nanomaterials are now established in many areas of bionanotechnology. Polymers are widely used in delivery systems for therapeutics, in matrices for tissue engineering and, among others, in polymer-based composites for biomedical purposes [4–6].

The main criteria in designing polymers for therapeutic use have been based mainly on the biocompatibility and the biodegradability of the polymer backbone as well as the suitability of the polymer to be processed into a stable pharmaceutical formulation [7].

However, there are other parameters of equal importance to be considered to optimize a polymer for its application as a delivery vehicle. In fact, parameters such as molar mass, functional end-groups, hydrophobic-hydrophilic balance (HHB), melting temperature (T_m) and crystallinity strongly influence the drug loading and the drug release kinetics from the polymer matrix [8].

Independent investigations of the influence of the polymer crystallinity on the resulting particle characteristics (e.g., particle formation and degradation) while keeping the key properties of the system constant are rare or only provide partial conclusions typically due to influences of a third variable [9]. In particular, alterations of HHB are frequently accompanied by changes in crystallinity [10]. It is hence currently not fully understood if degradation or general performance of hydrophobic pharmapolymers in aqueous media are, in fact, strongly influenced by polymer crystallinity or if the hydrophobicity is the dominating factor.

Today, the most commonly used polymers for biomedical applications are polyesters, such as polylactide (PLA), poly(lactide-*co*-glycolide) (PLGA) and polycaprolactone (PCL) [11]. They are easy to access and offer a range of interesting advantages; i.e., (i) a complete hydrolytic and/or enzymatic biodegradation, (ii) a facile and controlled synthesis to obtain defined molar masses, (iii) various modification possibilities of the polymer structure, and (iv) commercial availability [12]. The advantages of PCL compared to other aliphatic polyesters include interesting thermal properties, higher durability and manufacturability, and a good compatibility with other polymers [13,14]. Thus, PCL represents a promising candidate to design materials with tailor-made properties [13,15,16]. Bandelli et al. recently demonstrated that copolymerization of the constitutional isomers ε-caprolactone (εCL) and δ-caprolactone (δCL) with a varying ratio of εCL and δCL can generate a library of five copolyesters featuring a constant HHB and similar molar masses in the range of 7 to 10 kDa, but the copolymers showed a varying crystallinity [17]. They are hence suitable materials to study the sole influence of crystallinity on the particle properties and performance. In this study, we utilized this library of poly(εCL-*ran*-δCL) to formulate drug-loaded nanoparticles (NPs). The aim was to investigate, firstly, whether such polymers provide suitable properties (particle size and polydispersity) to form an NP-based drug delivery system, and secondly, to study the effect of the crystallinity of the bulk polymers on the enzymatic degradation and the in vitro performance of the resulting NPs.

The anti-inflammatory drug BRP-187 (4-(4-chlorophenyl)-5-[4-(quinoline-2-ylmethoxy)-phenyl] isoxazol-3-carboxylic acid) is a dual inhibitor of the 5-lipoxygenase-activating protein (FLAP) and microsomal prostaglandin E2 synthase-1 (mPGES-1), which are crucial proteins within arachidonic acid (AA) metabolism. Inhibition of mPGES-1 and FLAP prevents the biosynthesis of pro-inflammatory prostaglandin (PG)E_2 and leukotrienes (LTs), respectively [18]. Several in vitro and in vivo studies with inhibitors of FLAP and/or mPGES-1 have demonstrated their efficient anti-inflammatory activity while exhibiting fewer adverse effects compared to the conventional non-steroidal anti-inflammatory drugs (NSAIDs) [19,20]. These observations suggest that dual inhibition of FLAP and/or mPGES-1, rather than blocking cyclooxygenase-1 or -2 pathways, might be a better strategy for intervention with inflammation. However, BRP-187 is a fatty acid-like molecule with poor water solubility and a strong tendency to bind plasma proteins [18]. Molecules exhibiting such properties typically cause challenges in reaching a sufficient bioavailability in vivo and require technological solutions to improve their pharmacokinetic drawbacks. We have previously demonstrated that encapsulating BRP-187 into PLGA NPs and acetalated dextran NPs enhanced its enzyme inhibition efficacy in vitro [21].

In the present study, we initially performed a high-throughput (HT) nanoprecipitation approach for the homopolymers PεCL and PδCL to screen a range of polymer and surfactant concentrations for the preparation of empty (unloaded) PCL particles. Once optimal formulation conditions were established, and drug-loaded NPs were prepared by batch nanoprecipitation of the PεCL and PδCL homopolymers as well as of the poly(εCL-*ran*-δCL) copolymers with BRP-187. PCL NPs with and without BRP-187 were characterized for their critical quality attributes, namely their particle size, polydispersity index (PDI), surface charge and drug loading. Other particle properties i.e., nanodispersion stability and NP degradation behavior were also investigated. Ultimately, BRP-187-NPs were studied in vitro in human polymorphonuclear leukocytes (PMNL) for their efficiency to inhibit the drug target FLAP, in comparison to the free BRP-187.

2. Methods

2.1. Materials

The PCL homopolymers and P(εCL-*ran*-δCL) copolymers were synthesized as previously reported [17]. Key characterization data are listed in Table S1 in the Supplementary Information (SI). For further details on their synthesis, the reader is referred to literature reports [17]. Polyvinylalcohol (PVA) (Mowiol 4-88), tetrahydrofuran (THF), dimethylsulfoxide (DMSO) and lipase from the yeast *Candida rugosa* were purchased from Sigma-Aldrich (Germany). BRP-187 was synthesized according to a published protocol [18]. Further materials are described in the specific experimental sections.

2.2. Automated High-Throughput Nanoprecipitation

Automated high-throughput nanoprecipitation was performed in a 96-well plate (Greiner Bio-One GmbH, Frickenhausen, Germany) utilizing a FasTrans liquid handling robot (Analytik Jena GmbH, Jena, Germany). Starting with a polymer stock solution of 10 mg mL^{-1} in THF, a dilution series with varying concentrations (0.25, 0.5, 1, 2, 3, 4, 5, 6, 7, 8, 9 and 10 mg mL^{-1}) was prepared. The polymer solutions (40 µL) were then automatically pipetted into 200 µL of either purified water (GenPure ultrapure water system, Thermo Scientific, Waltham, MA, USA) or PVA surfactant-containing aqueous solutions with a concentration of 0.25%, 0.5% or 1.0% (*w/v*). The resulting NP dispersions were mixed by pipetting up and down three times and then left for two hours for solvent evaporation. Each formulation was prepared twice. After solvent evaporation, the samples were diluted with pure water (1:2 ratio for a polymer concentration up to 4 mg ml^{-1} and 1:10 ratio for all NPs prepared with a polymer concentration above 4 mg mL^{-1}) and investigated via dynamic light scattering (as described in Section 2.4) [22].

2.3. Batch Nanoprecipitation

Polymer solutions with 5 mg mL^{-1} or 2.5 mg mL^{-1} were prepared in THF via batch nanoprecipitation. For the drug-loaded particles, 10 mg mL^{-1} of BRP-187 dissolved in DMSO were mixed with the polymer solution prior to formulation, which corresponded to 3% (*w/w*) of the drug to polymer mass. The drug stock solution was sonicated in an ultrasound water bath for 15 min at room temperature to ensure good dissolution. Particle formulation was carried out by injecting the polymer/drug solution into an aqueous phase containing 0.3% (*w/v*) PVA using a syringe pump (Aladdin AL1000-220, World Precision Instruments, Berlin, Germany) with a flow rate of 2 mL min^{-1} while stirring at 800 rpm. The solvent/non-solvent ratio was set to 1:8. The resulting particle suspensions were stirred for 24 h at room temperature for solvent evaporation and then centrifuged at 12.851 *x g* for 60 min at 20 °C using a Rotina 380 R centrifuge (Hettich Lab Technology, Tuttlingen, Germany). The supernatant was removed, and the NPs were redispersed in 2.5 mL pure water, vortexed and sonicated in an ultrasonic water bath for 30 min. The NPs were stored overnight at 4 °C and lyophilized in aliquots of 200 µL. After lyophilization, the mass of the NPs was determined using a precise analytical balance (MYA 11.4Y, Radwag Waagen, Hilden, Germany). The yield was calculated as follows: (mass of NPs recovered − mass of found PVA)/(mass of polymer + mass of drug) in the formulation × 100. To check reproducibility, five individual batches of the drug-loaded PCL particles were prepared and analyzed individually. The data provided represent the average values and the standard deviation of these five batches.

2.4. Dynamic Light Scattering (DLS) and Electrophoretic Light Scattering (ELS)

DLS measurements were performed utilizing a Nano ZS (Malvern Panalytical, Malvern, United Kingdom) with a laser wavelength of λ = 633 nm with non-invasive back-scatter (NIBS) technology [22]. The particle size is reported as the hydrodynamic diameter (d_H). The particle size distribution (PDI) was measured using pure water as a dispersant with a refractive index RI of 1330 and a viscosity of 0.8872 cP at 25 °C. Samples obtained from the automated HT-nanoprecipitation were measured at 25 °C in a micro cuvette (Brand GmbH, Wertheim, Germany) without any filtering step with the following settings: measurements of each sample were repeated three times for 10 sec at 25 °C. The samples obtained from batch nanoprecipitation were measured at a dilution of 1:10 up to 1:100 utilizing the following settings: five repeated measurements, each with five runs of 30 s. The zeta-potential of the lyophilized NPs was investigated by ELS using the same instrument at 25 °C with three repeated measurements.

The apparent degradation behavior of the NPs was analyzed by DLS by monitoring changes in the mean count rate at fixed measurement settings: measuring position at 4.65, attenuator factor 7 at 37 °C [21]. Before investigating, NPs were mixed with the enzyme solution (a lipase from Candida rugosa) in a 1:4 mass ratio of polymer to enzyme and incubated at 37 °C for pre-determined timepoints.

2.5. UV-Vis Spectroscopy Measurements

UV-Vis spectroscopy measurements were performed with the Infinite M200 Pro plate reader (Tecan Group, Männedorf, Switzerland). For determination of the encapsulation efficiency (EE) and the loading capacity (LC) of the BRP-187 in the PCL particles, lyophilized NPs were dissolved in DMSO, and the solutions were investigated in a flat-transparent 96-well quartz plate (Hellma, Jena, Germany) at λ = 316 nm with 3 × 3 multiple reads per well and a 2000 μm well border. A calibration curve of BRP-187 was obtained for each batch in the concentration range of 1.2 to 312.5 μg mL^{-1} with R^2 = 0.9997. The LC was calculated as follows: LC = (mass of drug recovered)/(mass of particle recovered) x 100. The EE was calculated as follows: EE = LC found/(mass of drug used) × 100. The determination of PVA in the NPs (%, *w/w*) was performed according to the published protocol [23].

2.6. Scanning Electron Microscopy (SEM)

A Sigma VP Field Emission Scanning Electron Microscope (Carl-Zeiss, Jena, Germany) equipped with an InLens detector with an accelerating voltage of 6 kV was used for electron microscopy imaging. Before the measurement, the samples were coated with a thin layer of platinum (4 nm) via sputter coating (CCU-010 HV, Safematic, Zizers, Switzerland).

2.7. Cell Isolation

The leukocytes isolation was performed according to a published protocol [21]. Leukocyte concentrates were prepared from peripheral blood obtained from healthy human adult donors that received no anti-inflammatory treatment for the last ten days (Institute of Transfusion Medicine, University Hospital Jena). The approval for the protocol was given by the ethical committee of the University Hospital Jena, and all methods were performed in accordance with the relevant guidelines and regulations. To isolate PMNL, the leukocyte concentrates were mixed with dextran (*Leuconostoc* spp. MW ~40,000, Sigma Aldrich, Taufkirchen, Germany) for sedimentation of erythrocytes and the supernatant was centrifuged on lymphocyte separation medium (Histopaque®-1077, Sigma Aldrich, Taufkirchen, Germany). Contaminating erythrocytes in the pelleted neutrophils were removed by hypotonic lysis (water). PMNL were then washed twice in ice-cold phosphate-buffered saline (PBS) and finally resuspended in PBS plus 0.1% of glucose and 1 mM $CaCl_2$.

2.8. Determination of FLAP-Dependent 5-LO Product Formation in PMNL

The evaluation of the effects on FLAP was performed according to our established protocol [21]. We assessed FLAP-dependent 5-LO product formation in human PMNL, cells (5×10^6 mL^{-1}) were pre-incubated with BRP-187 or NPs for indicated timepoints at 37 °C. The cells were stimulated with 2.5 μM Ca^{2+}-ionophore A23187 (Cayman, Ann Arbor, USA) for 10 min, and the incubation was stopped with 1 mL ice-cold methanol containing 200 ng mL^{-1} PGB_1 as an internal standard. Samples were subjected to solid phase extraction, and the formed lipid mediators (leukotriene B_4 (LTB_4), trans-isomers of LTB_4, 5-hydroxyeicosatetraenoic acid (5-HETE)) were separated and analyzed by reverse-phase high-performance liquid chromatography (RP-HPLC) as previously described [24]. Statistical analysis was performed with log-transformed values to obtain Gaussian-distributed data sets. Experiments were analyzed via one-way ANOVA and Tukey's multicomparison test with GraphPad Prism 9.1.2 (GraphPad, La Jolla, CA, USA).

2.9. Cell Viability

Freshly isolated PMNL were incubated with a control sample with 0.1% DMSO, BRP-187 (10 μM) or NPs containing the respective amount of BRP-187 (10 μM) at 37 °C in PBS containing 0.1% of glucose. After 5 h the cell suspension was subjected to a Vi-CELL XR cell counter (Beckman Coulter, Lahntal, Germany), for determination of cell viability by trypan blue staining.

3. Results and Discussion

In our previous study, five poly(εCL-*ran*-δCL) copolymers, herein named ε87-δ13, ε81-δ19, ε75-δ25, ε61-δ39 and ε45-δ55, and the two respective homopolymers PεCL and PδCL, herein referred to as ε100-δ0 and ε0-δ100, were synthesized exhibiting a constant HHB [17]. It was demonstrated that the HHB of the bulk polymers correlated with the HHB of the corresponding NPs when particles were prepared in THF using a polymer concentration of 1 mg mL^{-1} [17]. In the present study, the particle formation of the ε100-δ0 and ε0-δ100 was investigated over a wider range of polymer concentrations in THF ranging from 0.25 to 10 mg mL^{-1} using an automated pipetting robot that was adapted for the HT-nanoprecipitation [25]. Particles were formulated without surfactant as well as with PVA of different concentrations (0.25 to 1% (*w/v*)). Previous studies revealed that PVA of less than 0.5% (*w/v*) generated stable drug-loaded PLGA NPs, and it could be demonstrated that even concentrations of up to 5% (*w/v*) were generally non-toxic in vitro [26]. ε100-δ0 and ε0-δ100 homopolymers both formed NPs up to the highest tested polymer concentration of 10 mg mL^{-1} when PVA was used as a surfactant (SI, Figure S1). Even the lowest tested PVA concentration of 0.25% (*w/v*) was sufficient to obtain stable particle dispersions and ε100-δ0 NPs with a size of 150 to 300 nm and ε0-δ100 NPs with a particle size of 120 to 280 nm with PDI < 0.3. However, ε0-δ100 NPs prepared without surfactant failed to produce stable NP dispersions above concentrations of 0.5 mg mL^{-1} as indicated by a strong aggregation of the particles. This is not surprising since ε0-δ100 is above its glass transition temperature at room temperature, which could disturb the particle formation in the absence of a stabilizer. It is well-known that several factors influence the final NP properties, including the polymer concentration, the solvent used to dissolve the polymer and the type and the concentration of the surfactant [26–28]. THF was demonstrated to be a suitable solvent in the HT-screening, resulting in stable particle formation within a broad polymer concentration range when PVA was used as a surfactant. Hence, it was selected as solvent for the subsequently performed BRP-187 encapsulation experiments. All other formulation parameters for the preparation of PCL[BRP-187] NP were adapted from our previous study that described the encapsulation of BRP-187 into PLGA NPs [21]. The first batch nanoprecipitation with the drug and a polymer concentration of 5 mg mL^{-1} in THF yielded large particles with a diameter (d_H) of 400 to 600 nm with high LC values (SI, Table S2). However, the particles revealed significant aggregation after centrifugation and lyophilization, as indicated by the higher PDI values of 0.3 to 0.6. Hence, the initial polymer concentration was reduced to 2.5 mg mL^{-1} to optimize the dispersion stability and to decrease the particle size [28]. Particles within a size range of 200 to 260 nm and PDI values below 0.3 were obtained for all PCLs using a polymer concentration of 2.5 mg mL^{-1}

(Table 1, SI Table S4). It was further observed that empty NPs were approximately 30 to 50 nm smaller compared to the BRP-187-loaded NPs (SI, Table S3). The particle size of the empty NPs increased by approximately 40 to 80 nm when NPs were lyophilized and subsequently reconstituted in water (SI, Table S3). Similar tendencies were also observed for the PCL[BRP-187] NPs, although here the difference in size was on average only about 30 to 50 nm (Table 1), presumably caused by the strong affinity of the hydrophobic drug with the polymer matrix [29]. The particles were also investigated via SEM (Figure 1), which revealed individual or clustered particle populations within the particle size range as indicated by DLS measurements.

Table 1. Overview of PCL[BRP-187] NP properties prepared in THF using a polymer concentration of 2.5 mg mL^{-1}.

εCL/δCL (mol %)	T_m (°C)	X_c [a] (%)	d_H [b] (nm)	PDI [b]	ZP [c] (mV)	d_H [c] (nm)	PDI [c]	PVA % (*w*/*w*)	Yield [d] (%)	LC [e] (%)
ε100-δ0	69	73	229 ± 13	0.08 ± 0.02	−50 ± 1	268 ± 21	0.27 ± 0.09	4.5	87	1.5 ± 0.1
ε87-δ13	54	44	211 ± 5	0.08 ± 0.02	−38 ± 2	251 ± 13	0.30 ± 0.14	4.5	76	1.4 ± 0.5
ε81-δ19	52	38	218 ± 13	0.08 ± 0.02	−41 ± 1	267 ± 24	0.37 ± 0.27	4.7	81	1.4 ± 0.2
ε75-δ25	42	28	225 ± 13	0.16 ± 0.11	−34 ± 1	260 ± 23	0.42 ± 0.20	5.0	67	1.9 ± 0.6
ε61-δ39	24	4	209 ± 13	0.06 ± 0.12	−40 ± 1	223 ± 16	0.16 ± 0.10	8.2	61	1.7 ± 0.1
ε45-δ55	/ *	0	200 ± 13	0.10 ± 0.12	−32 ± 1	237 ± 61	0.18 ± 0.08	6.7	52	1.4 ± 0.2
ε0-δ100	/ *	8	259 ± 32	0.28 ± 0.14	−45 ± 2	262 ± 20	0.26 ± 0.26	5.5	54	3.2 ± 1.2

d_H represents the intensity-weighted distribution (n ≥ 4 batches) and zeta-potential (ZP) (n = 3 ELS measurements) * Amorphous or near amorphous polymers with glass transition temperature T_g below 37 °C [17]. [a] Bulk degree of crystallinity as determined by wide-angle X-ray scattering (WAXS) at room temperature. [b] NPs measured after purification. [c] NPs measured after lyophilization and subsequent resuspension in water. [d] Yield = (mass of NPs recovered − mass of found PVA)/(mass of polymer + mass of drug) in the formulation × 100. [e] Determined by UV-VIS spectroscopy at λ = 316 nm (n = 4) and calculated using LC = (mass of drug recovered)/(mass of particle recovered) × 100.

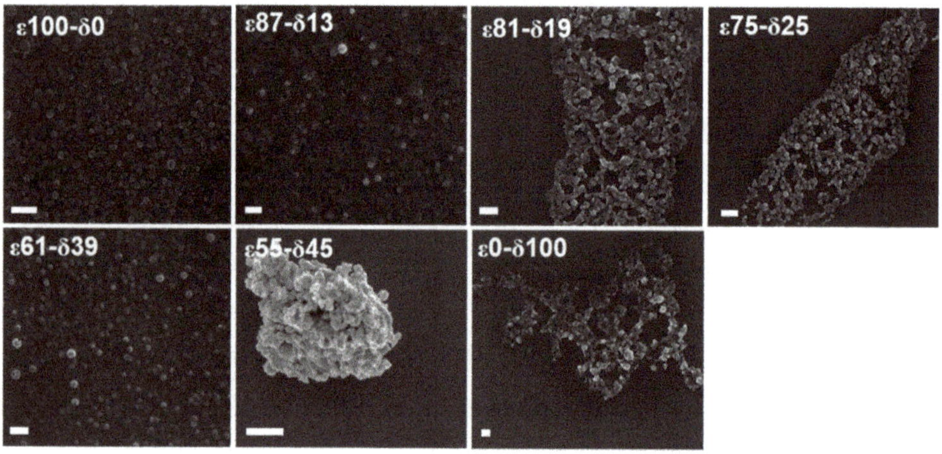

Figure 1. SEM micrographs of PCL[BRP-187] particles consisting of the homo- or copolymers with a varying composition. Scale bar = 1 μm.

The average LC of the PCL[BRP-187] NPs was between 1.4 and 1.9% for ε100-δ0 and the poly(εCL-*ran*-δCL) copolymers (Table 1) and similar to the LC values of PLGA NPs encapsulating the same drug [21]. The only exception was the ε0-δ100 homopolymer with an LC of 3.2%, probably due to its almost liquified state at room temperature. This resulted in a viscous dispersion with emulsion-like properties in which the drug was apparently entrapped during the purification process.

In general, the yield of both empty and drug-loaded PCL NPs decreased with increasing molar fraction of δCL (Figure 2A). In other words, NP yield increased with the degree of crystallinity of the polyester materials. Amorphous materials are frequently utilized as excipients in pharmaceutical formulations since they are known to increase the dissolution rate of insoluble drugs and to enhance their bioavailability [30]. However, their major disadvantage is seen in the fact that they exhibit high energy states at a molecular level and thus are prone to physical instabilities. In particular, such tendencies were observed with the NPs of the amorphous PδCL homopolymer, which displayed a higher polydispersity and the lowest yield. In technical terms, the low yield of the copolymers with a higher fraction of δCL could have resulted from their near-molten state at room temperature causing them to sediment at a lower rate due to their lower density. Thus, after 60 min of centrifugation, a lower amount of the NPs was recovered.

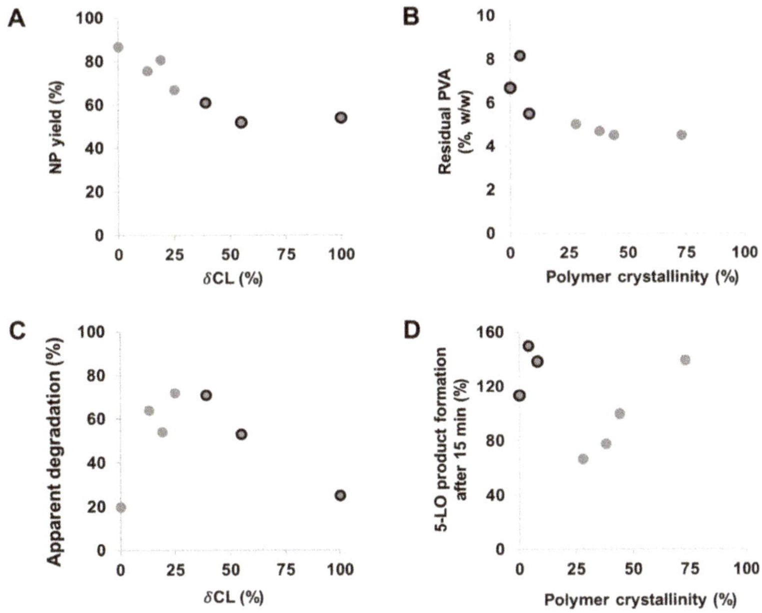

Figure 2. Influence of the δCL fraction on the yield of drug-loaded PCL NPs (**A**), influence of the polymer crystallinity on the residual PVA content of drug-loaded PCL NPs (**B**), apparent degradation represented by the normalized relative count rate (%) after 20 h plotted against the δCL fraction of the copolymers (**C**) and influence of polymer crystallinity on the efficiency of drug-loaded PCL NPs to inhibit 5-LO product formation (**D**). Black-circled data points represent PCL polymers with a degree of crystallinity below 10% and a T_g < 37 °C.

Furthermore, it was observed that the residual amount of PVA in the drug-loaded NPs was higher compared to the empty NPs for all PCL copolymers (Table 1 and SI, Table S3). As mentioned before, such differences are typically a result of strong drug–polymer interactions [31], and in this case, the interactions of the BRP-187 with the chains of PVA polymer. Moreover, the residual PVA content was noticeably higher for less crystalline copolymers with a higher δCL fraction and highest for the particles consisting of the PδCL homopolymer (Figure 2B). Apparently, the surfactant molecules tended to stick to the surface or were even incorporated into the particles formed from amorphous polyesters that are above their glass transition temperature during formulation. As soon as the materials were semicrystalline and below T_m, the degree of crystallinity did not influence the amount of residual PVA anymore. Besides providing dispersion stability, surfactants also influence the degradation rate of NPs since they adsorb at the surface of the particles forming a layer that protects from enzymatic hydrolysis to some degree [32]. Additional characterization experiments of the PCL[BRP-187] NPs were performed to investigate the degradation kinetics as well as the biological evaluation of the NP efficiency to inhibit the drug targets in vitro.

3.1. Degradation Studies

Among the aliphatic polyesters that are most commonly investigated for drug delivery applications, PCL has a superior thermal stability, with a decomposition temperature of 100 °C higher above that of the typical PLA- and PGA-based polymers [15]. Due to its high durability, PεCL has found a wide range of applications mainly for implantable medical devices [33,34], in which degradation occurs over two to four years [13]. However, to tailor their application for drug delivery purposes, faster degradation kinetics of the PεCL are desirable and can be achieved by copolymerization of εCL with its isomer δCL [9]. Introducing δCL repeating units to the PεCL polymer decreases its degree of crystallinity [17], and as such, it increases its rate of degradation as confirmed by investigations of films [35]. Figure 3 shows the enzymatic degradation of the PCL[BRP-187] particles incubated for 24 h at 37 °C as monitored by DLS. The apparent NPs degradation was inferred by monitoring changes in the sample concentration over time, as indicated by the count rate on the DLS under constant measurement settings [21]. In agreement with literature reports regarding film degradation, Figure 3 reveals that the degradation of the most crystalline ε100-δ0 was the slowest in the nanoparticulate state. ε100-δ0 is predominantly a semicrystalline material with a melting point considerably higher than the experimental temperature of 37 °C. It was noticed that except for the ε0-δ100 homopolymer, which degraded only about 25% after 24 h, the NP degradation rate generally increased with the amount of the δCL (Figure 3A). This was expected since the long-range order and the compact structure of crystalline materials requires higher levels of energy for degradation compared to the less organized molecular arrangement of amorphous materials [29,36].

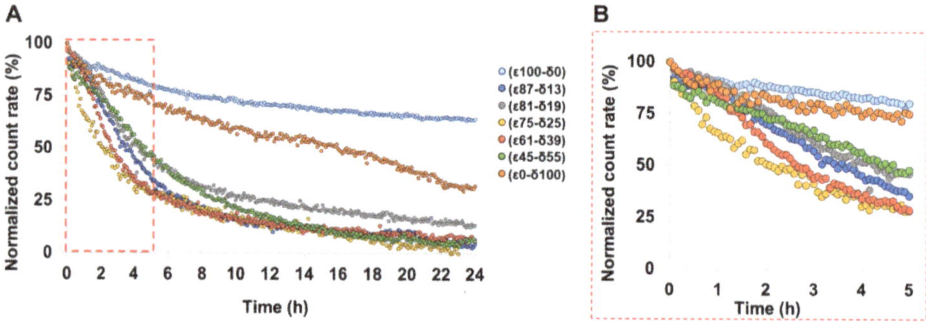

Figure 3. Normalized count rate of BRP-187-loaded PCL NPs incubated with *Candida rugosa* as measured by DLS for 24 h (**A**). (**B**) depicts a zoomed-in area into the data until 5 h.

This observation is further confirmed by other studies that have also demonstrated that the amorphous regions within bulk and/or films of PεCL polymer degraded faster compared to the crystalline regions [35,37,38]. Another study with similar observations argued that polyesters with higher crystallinity exhibit a slower degradation because in a densely packed crystal, it is more difficult for the enzymes to reach the cleavable bonds [39]. In general, our results revealed that all copolyester NPs featured an apparent degradability above 50% within 5 h (Figure 3B). A faster initial degradation was particularly observed for the ε75-δ25 and ε61-δ39 copolymers since they exhibit melting points (42 °C and 24 °C, respectively [17]) that are closer to the experimental temperature of 37 °C, which was chosen to simulate the conditions of the human body (Figures 2C and 3B) [8].

3.2. *In vitro* Performance of NPs

Although clear influences of the polymer crystallinity and physical state on NP formulation and enzymatic degradation were found, other effects might come into play in the more complex environment of a cell. The PCL[BRP-187] particles were hence tested in human PMNL for bioactivity. PMNL are the most abundant leukocytes in the blood and are a major source for FLAP-dependent 5-LO product biosynthesis, thus they are suitable cells for evaluation of various anti-LT agents. Note that FLAP as helper-protein of 5-LO has no enzyme activity that can be experimentally assessed, but instead assists 5-LO in LT formation by facilitating the access towards the substrate for the 5-LO enzyme. At first, the PCL[BRP-187] NPs were compared to the free drug for their influence on the cell viability of PMNL (cytotoxicity). No cytotoxic effects of the particles were found within a 5 h incubation as shown in the SI (Figure S2). These results were in agreement with previous studies that demonstrated PCL NPs to be biocompatible [40,41]. Considering their good biocompatibility, all PCL[BRP-187] particles as well as the free drug were studied for their efficiency to inhibit the drug target FLAP in PMNL and, thus, to prevent 5-LO product formation [42]. Therefore, a drug concentration of 0.3 μM was chosen for free and encapsulated BRP-187, which were investigated at different preincubation times (15 min, 1 h, 2 h and 5 h, respectively). As shown in Figure 4A, 5-LO product formation was clearly suppressed after 15 min of incubation with the PCL[BRP-187] particles to variant degrees, but essentially the particles performed as efficiently as the free drug. Apart from this, there was no significant difference between the different PCL[BRP-187] polymers at longer incubation time points (i.e., 1–5 h; Figure 4B and SI, Figure S3). More specifically, the NPs prepared with ε75-δ25 prevented the 5-LO product formation most after 15 min of incubation (Figure 4A). This observation also correlated with the fastest apparent degradation of the ε75-δ25 copolymer (Figure 3B), which might be promoted by its melting point of 42 °C, which is around the temperature of cell incubation (i.e., 37 °C). Karavelidis et al. reported that other polyesters with melting points around 37 °C exhibited a faster drug release [8]. It can be inferred that the rapid degradation of ε75-δ25 led to an accelerated release of the BRP-187, thereby considerably preventing the 5-LO product formation at early time points (Figure 2C,D). NPs formed from PCL with higher εCL fraction and hence higher T_m, as well as a higher degree of crystallinity, were less effective. As shown in Figure 2D, the 5-LO product formation was almost linearly dependent on the polymer crystallinity if only the semicrystalline materials are considered. The better performance of the NPs with lower crystallinity could be explained by two effects based on two different release mechanisms. Firstly, less crystalline materials with a larger fraction of amorphous domains enable a faster diffusion of the drug through the polymer matrix without barriers formed by crystalline domains [11,43]. Secondly, if the drug release is promoted through polymer degradation, these amorphous domains would most likely be more accessible for enzymes catalyzing the polyester hydrolysis [11].

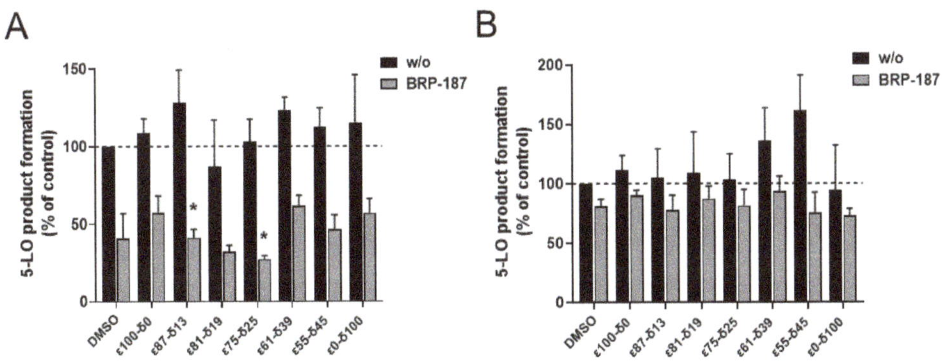

Figure 4. Measurement of 5-LO product formation as an indicator for the inhibition of the drug target FLAP by BRP-187 [36]. PMNL preincubated with DMSO, BRP-187 (0.3 μM), empty PCL particles (labeled as w/o) or PCL particles with BRP-187 (labeled with BRP-187; 0.3 μM respective BRP-187) for 15 min (**A**) or 5 h (**B**) at 37 °C. Values are given as 5-LO products as a percentage of control (DMSO) (n = 3). Statistical analysis was performed via one-way ANOVA and Tukeys multi comparison test with logarithmic trans-formed data (* $p < 0.05$).

Based on the apparent enzymatic degradation of the PCL NPs, a burst release of the drug is conceivable considering the immediate decrease in the count rate of at least 5 to 10% of all PCL NPs (Figure 3B). The slightly higher efficiency of the NPs to inhibit 5-LO product formation in PMNL within 15 min supports this idea (Figure 4A). For the polymers with a degree of crystallinity below 10% and T_g < 37 °C, namely ε61 δ39, ε55 δ45 and ε00-δ100, inhibition of 5-LO product formation is less apparent after 15 min of incubation (Figure 2D, black-circled data points). This is presumably because these polymers are molten and more viscous at 37 °C. As a consequence, they could delay the release of the drug and therefore hamper the drug action in the cells.

Furthermore, the coating effect of PVA might reduce the influence of the crystallinity of NPs or their intracellular drug release. It is reported that increasing amounts of residual surfactant decrease the cellular uptake of the NPs [44,45]. This could explain why the PCL copolymers with a higher fraction of δCL containing more residual PVA (Figure 2B) were less efficient to inhibit 5-LO product formation after 15 min of incubation (Figure 2C, Figure 4A) when compared to the PCL copolymers with a higher εCL fraction containing less residual PVA. However, no correlation was observed between suppression of 5-LO product formation and the PVA content in the formulation, showing that the trend cannot be generalized (SI, Figure S4).

4. Conclusions

A library of poly(εCL-*ran*-δCL) copolymers with a constant HHB but different degrees of crystallinity were used to encapsulate BRP-187 into polymer NPs. PCL[BRP-187] particles with a diameter of 200 to 300 nm were successfully produced, whereby a comparable drug-loading was observed with LC between 1.4% and 1.9%, with the exception of the PδCL homopolymer, which revealed a higher LC. It was evident that the degree of crystallinity directly influenced the enzymatic degradation rate of the PCL copolymer, whereby the degradation increased with an increasing fraction of δCL repeating units. In addition, increasing the amount of δCL in the polymer increased the amount of residual surfactant in the NP formulation but decreased the final NP yield. The release of bioactive BRP-187 from the PCL NPs was demonstrated in vitro in PMNL by inhibiting FLAP-dependent 5-LO product formation, whereby the inhibition efficiency was dependent on the degree of crystallinity of the copolymers used for the particle formulation. The NPs of ε75-δ25 revealed the fastest degradation and inhibited the 5-LO product formation more than the other copolymers after 15 min of incubation in PMNL; longer preincubation times (1 to 5 h) reduced the potency. In conclusion, although all PCL copolymers were suited to produce NPs, the ε75-δ25 copolymer can be considered as a more promising candidate to be further investigated for both its physicochemical properties and its performance in more complex biological models. When designing superior materials for NP-mediated drug delivery, it hence seems promising to rely on polymers that are in a solid state of matter at 37 °C but feature a low degree of crystallinity. However, it is not yet clear if these observations can be applied to other systems. Thus, our future research will concentrate on the encapsulation of other anti-inflammatory drugs in the polymer library with constant HHB to determine if the effect of polymer crystallinity of the present PCL systems can be transferred to other actives. In addition, we are currently establishing similar libraries mimicking the HHB of PLA to understand if our findings can be generalized in the field of polyester-based drug carrier materials.

Supplementary Materials: The following are available online at https://www.mdpi.com/article/10.3390/polym13152557/s1, Figure S1: Hydrodynamic diameter (intensity-weighted distribution, circles) and PDI (bars) of the homopolymer ε100-δ0 and ε0-δ100 NPs over a range of PVA concentration used in the formulation, Figure S2: Cell viability measured with a Beckman ViCell XR cell counter by trypan blue staining. A total of 1 × 107 PMNL were diluted in PBS plus 0.1% of glucose and incubated with DMSO, BRP-187 (10 μM), empty PCL particles (labeled as w/o) or PCL particles with BRP-187 (labeled with BRP-187; respective amount to 10 μM BRP-187) for 5 h at 37 °C. Values are given as 5-LO products as a percentage of control (DMSO) (n = 3), Figure S3: Measurement of 5-LO product formation as indicator for the inhibition of the drug target 5-lipoxygenase-activating protein (FLAP) by BRP-187.[37] A total of 5 × 106 polymorphonuclear leukocytes (PMNL) diluted in PBS containing 0.1% glucose and 1mM $CaCl_2$ were preincubated with DMSO, BRP-187 (0.3 µM), empty PCL particles (labeled as w/o) or PCL particles with BRP-187 (labeled as BRP-187; 0.3 µM respective BRP-187) for 1 h (A) and 2 h (B) at 37 °C and further stimulated with 2.5 µM A23187 for 10 min. The reaction was stopped with 1 mL ice-cold methanol containing 200 ng mL-1 PGB1 as internal standard. Lipid mediators were extracted via solid-phase extraction (SPE) and analyzed with HPLC. Values are given as 5-LO products (LTB4, its trans-isomers 4 and 5-HETE) as a percentage of control (DMSO) (n = 3), Figure S4: Influence of the residual PVA on the efficiency of drug-loaded PCL NPs on 5-LO inhibition. Black-circled data points represent PCL polymers with bulk degree of crystallinity below 10% and glass transition temperature Tg < 37 °C. Table S1: Molar mass and composition of the (co)polyesters. Details are described in a previous publication, Table S2: Properties of PCL[BRP-187] NPs formulated from THF utilizing polymer concentration of 5 mg mL^{-1} (n = 1 batch), Table S3: Particle properties of empty PCL NPs prepared in THF with c = 2.5 mg mL 1 (n = 2 batches) obtained by DLS and ELS measurements after purification and after lyophilization and subsequent resuspension (n = 2 for purified NPs, n = 1 for lyophilized NPs), Table S4: DLS intensity-weighted size distribution of PCL[BRP-187] NPs of one formulation round after purification, as well as after lyophilization and resuspension in water.

Author Contributions: Performance of experiments, A.V., C.K., B.B.-S., S.S.; writing —review and editing, A.V., C.K., B.S., C.W.; polymer synthesis, D.B.; synthesis of BRP-187, J.A.C.; supervision, S.H., C.W., O.W., U.S.S.; project administration, A.V. All authors have read and agreed to the published version of the manuscript.

Funding: This work was supported by the Deutsche Forschungsgemeinschaft (DFG), Collaborative Research Center SFB 1278 "PolyTarget", (project number 316213987, projects A04, A06 and Z01). The SEM facilities of the Jena Center for Soft Matter (JCSM) were established with a grant from the DFG (INST 275/307-1 FUGG).

Institutional Review Board Statement: Not applicable.

Informed Consent Statement: Not applicable.

Data Availability Statement: The data presented in this study are available on request from the corresponding author.

Conflicts of Interest: The authors declare no conflict of interest.

References

1. Barenholz, Y. Doxil[(R)]—the first fda-approved nano-drug: Lessons learned. *J. Control. Release* **2012**, *160*, 117–134. [CrossRef]
2. Mitragotri, S.; Burke, P.A.; Langer, R. Overcoming the challenges in administering biopharmaceuticals: Formulation and delivery strategies. *Nat. Rev. Drug Discov.* **2014**, *13*, 655–672. [CrossRef]
3. Murthy, S.K. Nanoparticles in modern medicine: State of the art and future challenges. *Int. J. Nanomed.* **2007**, *2*, 129–141.
4. Guo, Z.C.; Poot, A.A.; Grijpma, D.W. Advanced polymer-based composites and structures for biomedical applications. *Eur. Polym. J.* **2021**, *149*, 110388. [CrossRef]
5. Keshvardoostchokami, M.; Majidi, S.S.; Huo, P.P.; Ramachandran, R.; Chen, M.L.; Liu, B. Electrospun nanofibers of natural and synthetic polymers as artificial extracellular matrix for tissue engineering. *Nanomaterials* **2021**, *11*, 21. [CrossRef]
6. Shkodra, B.; Vollrath, A.; Schubert, U.S.; Schubert, S. Polymer-based nanoparticles for biomedical applications. In *Colloids for Nanobiotechnology—Synthesis, Characterization and Potential Applications*, 1st ed.; Parak, W., Feliu, N., Eds.; Elsevier: Amsterdam, The Netherlands, 2020; Volume 16, pp. 233–252.
7. Englert, C.; Brendel, J.C.; Majdanski, T.C.; Yildirim, T.; Schubert, S.; Gottschaldt, M.; Windhab, N.; Schubert, U.S. Pharmapolymers in the 21st century: Synthetic polymers in drug delivery applications. *Prog. Polym. Sci.* **2018**, *87*, 107–164. [CrossRef]
8. Karavelidis, V.; Karavas, E.; Giliopoulos, D.; Papadimitriou, S.; Bikiaris, D. Evaluating the effects of crystallinity in new biocompatible polyester nanocarriers on drug release behavior. *Int. J. Nanomed.* **2011**, *6*, 3021–3032.
9. Song, Q.L.; Xia, Y.N.; Hu, S.Y.; Zhao, J.P.; Zhang, G.Z. Tuning the crystallinity and degradability of PCL by organocatalytic copolymerization with delta-hexalactone. *Polymer* **2016**, *102*, 248–255. [CrossRef]
10. Schneiderman, D.K.; Hillmyer, M.A. Aliphatic polyester block polymer design. *Macromolecules* **2016**, *49*, 2419–2428. [CrossRef]
11. Kamaly, N.; Yameen, B.; Wu, J.; Farokhzad, O.C. Degradable controlled-release polymers and polymeric nanoparticles: Mechanisms of controlling drug release. *Chem. Rev.* **2016**, *116*, 2602–2663. [CrossRef]
12. Danhier, F.; Ansorena, E.; Silva, J.M.; Coco, R.; Le Breton, A.; Preat, V. PLGA-based nanoparticles: An overview of biomedical applications. *J. Control. Release* **2012**, *161*, 505–522. [CrossRef]

13. Woodruff, M.A.; Hutmacher, D.W. The return of a forgotten polymer-polycaprolactone in the 21st century. *Prog. Polym. Sci.* **2010**, *35*, 1217–1256. [CrossRef]
14. Hakkarainen, M.; Hoglund, A.; Odelius, K.; Albertsson, A.C. Tuning the release rate of acidic degradation products through macromolecular design of caprolactone-based copolymers. *J. Am. Chem. Soc.* **2007**, *129*, 6308–6312. [CrossRef] [PubMed]
15. Engelberg, I.; Kohn, J. Physicomechanical properties of degradable polymers used in medical applications—A comparative study. *Biomaterials* **1991**, *12*, 292–304. [CrossRef]
16. Lee, I.H.; Palombo, M.S.; Zhang, X.P.; Szekely, Z.; Sinko, P.J. Design and evaluation of a cxcr4 targeting peptide 4dv3 as an hiv entry inhibitor and a ligand for targeted drug delivery. *Eur. J. Pharm. Biopharm.* **2019**, *138*, 11–22.
17. Bandelli, D.; Muljajew, I.; Scheuer, K.; Max, J.B.; Weber, C.; Schacher, F.H.; Jandt, K.D.; Schubert, U.S. Copolymerization of caprolactone isomers to obtain nanoparticles with constant hydrophobicity and tunable crystallinity. *Macromolecules* **2020**, *53*, 5208–5217. [CrossRef]
18. Garscha, U.; Voelker, S.; Pace, S.; Gerstmeier, J.; Emini, B.; Liening, S.; Rossi, A.; Weinigel, C.; Rummler, S.; Schubert, U.S.; et al. BRP-187: A potent inhibitor of leukotriene biosynthesis that acts through impeding the dynamic 5-lipoxygenase/5-lipoxygenase-activating protein (FLAP) complex assembly. *Biochem. Pharmacol.* **2016**, *119*, 17–26. [CrossRef]
19. Koeberle, A.; Werz, O. Natural products as inhibitors of prostaglandin e-2 and pro-inflammatory 5-lipoxygenase-derived lipid mediator biosynthesis. *Biotechnol. Adv.* **2018**, *36*, 1709–1723. [CrossRef]
20. Koeberle, A.; Zettl, H.; Greiner, C.; Wurglics, M.; Schubert-Zsilavecz, M.; Werz, O. Pirinixic acid derivatives as novel dual inhibitors of microsomal prostaglandin e-2 synthase-1 and 5-lipoxygenase. *J. Med. Chem.* **2008**, *51*, 8068–8076. [CrossRef]
21. Shkodra-Pula, B.; Kretzer, C.; Jordan, P.M.; Klemm, P.; Koeberle, A.; Pretzel, D.; Banoglu, E.; Lorkowski, S.; Wallert, M.; Hoppener, S.; et al. Encapsulation of the dual FLAP/mPEGs-1 inhibitor brp-187 into acetalated dextran and plga nanoparticles improves its cellular bioactivity. *J. Nanobiotechnol.* **2020**, *18*, 73. [CrossRef] [PubMed]
22. Malvern Panalytical; Zetasizer Nano User Manual man0485. Available online: https://www.Malvernpanalytical.Com/de/learn/knowledge-center/user-manuals/man0485en (accessed on 8 July 2021).
23. Spek, S.; Haeuser, M.; Schaefer, M.M.; Langer, K. Characterisation of pegylated PLGA nanoparticles comparing the nanoparticle bulk to the particle surface using UV/Vis spectroscopy, SEC, H-1 nmr spectroscopy, and x-ray photoelectron spectroscopy. *Appl. Surf. Sci.* **2015**, *347*, 378–385. [CrossRef]
24. Werz, O.; Burkert, E.; Samuelsson, B.; Radmark, O.; Steinhilber, D. Activation of 5-lipoxygenase by cell stress is calcium independent in human polymorphonuclear leukocytes. *Blood* **2002**, *99*, 1044–1052. [CrossRef] [PubMed]
25. Perevyazko, I.Y.; Vollrath, A.; Pietsch, C.; Schubert, S.; Pavlov, G.M.; Schubert, U.S. Nanoprecipitation of poly(methyl methacrylate)-based nanoparticles: Effect of the molar mass and polymer behavior. *J. Polym. Sci. Pol. Chem.* **2012**, *50*, 2906–2913. [CrossRef]
26. Shkodra-Pula, B.; Grune, C.; Traege, A.; Vollrath, A.; Schuber, S.; Fischer, D.; Schubert, U.S. Effect of surfactant on the size and stability of plga nanoparticles encapsulating a protein kinase c inhibitor. *Int. J. Pharm.* **2019**, *566*, 756–764. [CrossRef] [PubMed]
27. Beck-Broichsitter, M.; Nicolas, J.; Couvreur, P. Solvent selection causes remarkable shifts of the "ouzo region" for poly(lactide-co-glycolide) nanoparticles prepared by nanoprecipitation. *Nanoscale* **2015**, *7*, 9215–9221. [CrossRef]
28. Homs, M.; Caldero, G.; Monge, M.; Morales, D.; Solans, C. Influence of polymer concentration on the properties of nano-emulsions and nanoparticles obtained by a low-energy method. *Colloids Surf. A Physicochem. Eng. Asp.* **2018**, *536*, 204–212. [CrossRef]
29. Sanchez, A.; Mejia, S.P.; Orozco, J. Recent advances in polymeric nanoparticle-encapsulated drugs against intracellular infections. *Molecules* **2020**, *25*, 3760. [CrossRef]
30. Newman, A.; Hastedt, J.E.; Yazdanian, M. New directions in pharmaceutical amorphous materials and amorphous solid dispersions. *AAPS Open* **2017**, *3*, 7. [CrossRef]
31. Du, S.; Li, W.S.; Wu, Y.R.; Fu, Y.; Yang, C.Q.; Wang, J. Comparison of the physical and thermodynamic stability of amorphous azelnidipine and its coamorphous phase with piperazine. *RSC Adv.* **2018**, *8*, 32756–32764. [CrossRef]
32. Lee, S.C.; Oh, J.T.; Jang, M.H.; Chung, S.I. Quantitative analysis of polyvinyl alcohol on the surface of poly(d,l-lactide-co-glycolide) microparticles prepared by solvent evaporation method: Effect of particle size and PVA concentration. *J. Control. Release* **1999**, *59*, 123–132. [CrossRef]
33. Hollander, J.; Genina, N.; Jukarainen, H.; Khajeheian, M.; Rosling, A.; Makila, E.; Sandler, N. Three-dimensional printed PCL-based implantable prototypes of medical devices for controlled drug delivery. *J. Pharm. Sci.* **2016**, *105*, 2665–2676. [CrossRef] [PubMed]
34. Manoukian, O.S.; Arul, M.R.; Sardashti, N.; Stedman, T.; James, R.; Rudraiah, S.; Kumbar, S.G. Biodegradable polymeric injectable implants for long-term delivery of contraceptive drugs. *J. Appl. Polym. Sci.* **2018**, *135*, 46068. [CrossRef]
35. Eldsater, C.; Erlandsson, B.; Renstad, R.; Albertsson, A.C.; Karlsson, S. The biodegradation of amorphous and crystalline regions in film-blown poly(epsilon-caprolactone). *Polymer* **2000**, *41*, 1297–1304. [CrossRef]
36. Pitt, G.G.; Gratzl, M.M.; Kimmel, G.L.; Surles, J.; Sohindler, A. Aliphatic polyesters ii. The degradation of poly (dl-lactide), poly (ε-caprolactone), and their copolymers in vivo. *Biomaterials* **1981**, *2*, 215–220. [CrossRef]
37. Cook, W.J.; Cameron, J.A.; Bell, J.P.; Huang, S.J. Scanning electron-microscopic visualization of biodegradation of polycaprolactones by fungi. *J. Polym. Sci. Pol. Lett.* **1981**, *19*, 159–165. [CrossRef]
38. Tilstra, L.; Johnsonbaugh, D. The biodegradation of blends of polycaprolactone and polyethylene exposed to a defined consortium of fungi. *J. Environ. Polym. Degrad.* **1993**, *1*, 10. [CrossRef]

39. Bandelli, D.; Helbing, C.; Weber, C.; Seifer, M.; Muljajew, I.; Jandt, K.D.; Schubert, U.S. Maintaining the hydrophilic hydrophobic balance of polyesters with adjustable crystallinity for tailor-made nanoparticles. *Macromolecules* **2018**, *51*, 5567–5576. [CrossRef]
40. Lukasiewicz, S.; Mikolajczyk, A.; Blasiak, E.; Fic, E.; Dziedzicka-Wasylewska, M. Polycaprolactone nanoparticles as promising candidates for nanocarriers in novel nanomedicines. *Pharmaceutics* **2021**, *13*, 191. [CrossRef]
41. Ortiz, R.; Prados, J.; Melguizo, C.; Arias, J.L.; Ruiz, M.A.; Alvarez, P.J.; Caba, O.; Luque, R.; Segura, A.; Aranega, A. 5-fluorouracil-loaded poly(epsilon-caprolactone) nanoparticles combined with phage e gene therapy as a new strategy against colon cancer. *Int. J. Nanomed.* **2012**, *7*, 95–107.
42. Radmark, O.; Werz, O.; Steinhilber, D.; Samuelsson, B. 5-lipoxygenase, a key enzyme for leukotriene biosynthesis in health and disease. *Biochim. Biophys. Acta* **2015**, *1851*, 9. [CrossRef] [PubMed]
43. Tallury, P.; Alimohammadi, N.; Kalachandra, S. Poly(ethylene-co-vinyl acetate) copolymer matrix for delivery of chlorhexidine and acyclovir drugs for use in the oral environment: Effect of drug combination, copolymer composition and coating on the drug release rate. *Dent. Mater.* **2007**, *23*, 404–409. [CrossRef] [PubMed]
44. Prabha, S.; Labhasetwar, V. Effect of residual polyvinyl alcohol on nanoparticle-mediated gene transfection in breast cancer cells. *Mol. Ther.* **2003**, *7*, S67.
45. Sahoo, S.K.; Panyam, J.; Prabha, S.; Labhasetwar, V. Residual polyvinyl alcohol associated with poly (d,l-lactide-co-glycolide) nanoparticles affects their physical properties and cellular uptake. *J. Control. Release* **2002**, *82*, 105–114. [CrossRef]

polymers

MDPI

Review

Recent Advances in Antioxidant Polymers: From Sustainable and Natural Monomers to Synthesis and Applications

Chrysanthos Maraveas [1,*], Ilker S. Bayer [2,*] and Thomas Bartzanas [1]

1 Department of Natural Resources and Agricultural Engineering, Agricultural University of Athens, 11855 Athens, Greece; t.bartzanas@aua.gr
2 Smart Materials, Istituto Italiano di Tecnologia, 16163 Genova, Italy
* Correspondence: c.maraveas@maraveas.gr (C.M.); Ilker.Bayer@iit.it (I.S.B.)

Citation: Maraveas, C.; Bayer, I.S.; Bartzanas, T. Recent Advances in Antioxidant Polymers: From Sustainable and Natural Monomers to Synthesis and Applications. *Polymers* **2021**, *13*, 2465. https://doi.org/10.3390/polym13152465

Academic Editor: José Miguel Ferri

Received: 5 July 2021
Accepted: 23 July 2021
Published: 27 July 2021

Abstract: Advances in technology have led to the production of sustainable antioxidants and natural monomers for food packaging and targeted drug delivery applications. Of particular importance is the synthesis of lignin polymers, and graft polymers, dopamine, and polydopamine, inulin, quercetin, limonene, and vitamins, due to their free radical scavenging ability, chemical potency, ideal functional groups for polymerization, abundance in the natural environment, ease of production, and activation of biological mechanisms such as the inhibition of the cellular activation of various signaling pathways, including NF-κB and MAPK. The radical oxygen species are responsible for oxidative damage and increased susceptibility to cancer, cardiovascular, degenerative musculoskeletal, and neurodegenerative conditions and diabetes; such biological mechanisms are inhibited by both synthetic and naturally occurring antioxidants. The orientation of macromolecules in the presence of the plasticizing agent increases the suitability of quercetin in food packaging, while the commercial viability of terpenes in the replacement of existing non-renewable polymers is reinforced by the recyclability of the precursors (thyme, cannabis, and lemon, orange, mandarin) and marginal ecological effect and antioxidant properties. Emerging antioxidant nanoparticle polymers have a broad range of applications in tumor-targeted drug delivery, food fortification, biodegradation of synthetic polymers, and antimicrobial treatment and corrosion inhibition. The aim of the review is to present state-of-the-art polymers with intrinsic antioxidant properties, including synthesis scavenging activity, potential applications, and future directions. This review is distinct from other works given that it integrates different advances in antioxidant polymer synthesis and applications such as inulin, quercetin polymers, their conjugates, antioxidant-graft-polysaccharides, and polymerization vitamins and essential oils. One of the most comprehensive reviews of antioxidant polymers was published by Cirillo and Iemma in 2012. Since then, significant progress has been made in improving the synthesis, techniques, properties, and applications. The review builds upon existing research by presenting new findings that were excluded from previous reviews.

Keywords: antioxidant; antioxidant polymers; lignin polymers; graft polymers; dopamine; polydopamine; inulin; quercetin; limonene; vitamins

1. Introduction

1.1. Definition of Antioxidants. Their Importance in Food Preservation

Antioxidants are a class of naturally occurring or synthetic compounds. The naturally occurring antioxidants include Vitamin C and Vitamin E (tocotrienols and tocopherols in general) [1–3]. Other classes are phenolic compounds and carotenoids [4]. Synthetic antioxidant molecules include α-lipoic acid, N-acetyl cysteine, melatonin, gallic acid, captopril, taurine, catechin, and quercetin [5]; these compounds are indispensable in the scavenging of free radical species. For example, gallic acid and poly Trolox ester polymers scavenge free radicals in the cytosolic cellular compartment [6]. In contrast, β-carotene, and vitamin E (tocopherol), are most effective against lipid peroxidation.

Wattamwar et al. [6] attribute the formation of reactive oxygen species to cellular processes, including inflammation—a process that results in the activation of endothelial cell macrophages, which in turn trigger the development of Nicotinamide adenine dinucleotide phosphate (NADPH) oxidase complex [6]. The NADPH complex is responsible for the conversion of molecular oxygen into O_2, a superoxide reactive anion, which is converted into H_2O_2 (hydrogen peroxide) following contact with superoxide dismutase [6]. The hydrogen peroxide molecule is a potent reactant, which binds with copper and iron metal ions, releasing hydroxyl radicals. Alternatively, $ONOO^-$ (peroxynitrite) is formed if the superoxide reactive anion is formed in a NO-rich environment. The reactive oxygen and nitrogen species formed through a sequence of reactions triggered by NADPH are responsible for cellular oxidative and lipid damage—phenomena that result in the release of oxidative stress markers (4-hydroxy-2-trans-nonenal, 3-nitrotyrosine (3NT), protein carbonyl). The primary function of melatonin, gallic acid, captopril, taurine, catechin, and quercetin, vitamin C, and E analogs is to bind to the markers to prevent structural damage to cellular proteins.

The biological importance of different antioxidants in food preservation and augmentation of body defense mechanisms depends on chemical properties. Pryor et al. [3] evaluated the performance of both natural and synthetic antioxidants in sodium dodecyl sulfate micelle solutions; it was noted that hydrogen bonding on the para and ether O_2 atoms predicted the reactivity of the antioxidants towards peroxyl radicals. Beyond H-bonding affinity towards reactive oxygen species was predicted by the presence of bulky tert-butyl groups [3]. Even though synthetic and natural antioxidants exhibit similar potency towards ROS, the selection of the chemicals in industrial applications is predicted by regulatory standards. European Parliament and Council Directive No.1333/2008 restricts the use of synthetic antioxidants in food preservation [4], except for 2, 4-dichlorophenoxyacetic acid (2,4-DA), 2-naphthol (2NL), 4-phenyl phenol (OPP), tert-butyl hydroquinone (TBHQ), butylated hydroxytoluene (BHT), and butylated hydroxyanisole (BHA) [4]. The listed synthetic antioxidants have found broad application in the food industry, particularly in preserving fruits and vegetables, due to their low cost, stability, performance, and widespread availability [3–5,7].

Vitamin C has been proven to exhibit superior activity in the quenching of reactive oxygen species and free radicals, resulting in the formation of ascorbyl radicals. The latter is a less potent radical compared to ROS based on the potential for oxidative damage [2]. Vitamin E exhibits a similar mechanism of action as Vitamin C. However, in the former case, the main mechanism of action involves protecting biological liquid compartments and the cleavage of the lipid peroxidation chain reactions. Alternatively, tocotrienols and tocopherols, in general, have been proven to contribute to the inactivation of ROS before regeneration by ascorbate [2]. Experimental evidence has also demonstrated the practical benefits of chain-breaking tocopherols in mitigating the auto-oxidation of polyunsaturated fatty acids. This process is responsible for cancer cell growth and atherosclerosis, as well as other life-threatening conditions [3]. Other unique biological functions associated with naturally occurring antioxidants include the modulation of the activity of specific enzymes, including mitogen-activated protein kinase, protein tyrosine kinase (PTK), protein tyrosine phosphatase (PTP), protein phosphatase 2A (PP2A), and protein kinase [2].

The commercial suitability of naturally occurring antioxidants is informed by various considerations beyond regulatory standards (European Parliament and Council Directive No.1333/2008) on food preservation, encompassing antimicrobial and antifungal activities [4], lipid oxidation reactions, and chemical reactions related to anti-oxidative effects [3–5]. The synergistic impact of different considerations helps explain why phenolic antioxidants are often preferred in food preservation (see Table 1).

The extraction techniques predict the selection of natural antioxidants for food preservation. Complex extraction techniques such as microwave-assisted extraction (MAE), ultrasound-assisted extraction, pressurized liquid extraction (PLE), high hydrostatic pressure (HHP), and supercritical fluid extraction (SFE) and methanol, ethanol, acetone, and water predict the cost of the extraction process and industrial application [1] (see Table 2).

Table 1. Active compounds, natural sources of antioxidants, and their use in different food matrices [4].

Natural Source	Main Active Compound	Food Matrix
Fennel and chamomile aqueous extracts	Phenolic compounds	Biscuits Cottage cheese Yogurt
Olive leaf and cakes extracts byproducts	Phenolic compounds	Antioxidant film
Litchi fruit pericarp extract	Phenolic compounds	Cooked nuggets
Green tea extract	Polyphenols	Sunflower oil
Cloves and cinnamon	Phenolic compounds	Meat samples
Tomato pomace extract	Carotenoids	Lamb steaks packaged
Ginkgo biloba leaves extract	Polyphenols	Pork meat
Cloudberry, beetroot, and willow herb	Flavonoid	Cooked pork patties
Canola olive oils, rice bran, and walnut	Polyphenols, vitamins E and B	Pork frankfurters
Coffee	Chlorophylls and carotenoids	Not defined
Wine	Phenolic compounds	Meat, fish, cereals

Table 2. Extraction techniques for different classes of naturally occurring antioxidants [4].

Extraction Process	Source	Antioxidant Extracted
Organic Solvents:		
Ethanol, dichloromethane, hexane	Coffee leaves	Chlorophylls and carotenoids
Ethanol, acetone, and water	*Baccharides* species	Phenolic content
	Sweet potato	Polyphenols and anthocyanins
Methanol, ethanol, and acetone	Spent grain	Phenolic content
	Spice herb	Phenolic content
	Peel of eggplant	Total phenolics, flavonoids, tannins, and anthocyanins
Supercritical fluid extraction (SFE)	Mango peel	Carotenoids
	Apple pomace	Phenolic compounds
	Myrtle leaves and berries	Phenolic acids, flavonoids, and anthocyanins
	Green algae	Carotenoids and phenolic compounds
	Cape gooseberry	Phenolic compounds and β-carotene
High Hydrostatic Pressure (HHP)	Red macroalgae	Proteins, polyphenols, and polysaccharides
	Tomato pulp	Flavonoids and lycopene
	Watercress	Phenolic acids and flavonoids from watercress
	Papaya seeds	Phenolic content
Pressurized liquid extraction (PLE)	Peppermint	Phenolic compounds and essential oils
	Carrot byproducts	Carotenoids
Ultrasound-assisted extraction	Green propolis	Phenolic compounds
Microwave-assisted extraction (MAE)	Pomegranate peels	Phenolic compounds
	Phalera macrocarpa fruit peel	Phenolic compounds

Natural and synthetic antioxidants are integral to reducing oxidative stress and imbalances in the cell redox reactions associated with reactive oxygen species. The formation of ROS is linked to the suppression of the innate antioxidant defense systems or the overproduction of oxygen species. Both mechanisms are common, considering that oxygen is ubiquitous in the environment, especially in biological processes involving aerobic organisms [2]. The visual illustration suggests that free radical scavenging ability is predicted by chemical structure and active components.

Supercritical Carbon Dioxide-Based Techniques

New evidence shows that the development of active packaging films is augmented by supercritical carbon dioxide (SC$-CO_2$) impregnation [8]. The observations made by Franco et al. [8] were collaborated by Lukic et al. [9] and Trucilo et al. [10], who documented the development of active food packaging products made of PLA/PCL combined with thymol and/or carvacrol and liposomes. The antioxidant and packaging properties of the active packaging materials were augmented by supercritical carbon dioxide (SC-CO_2) impregnation [9]. The choice of the supercritical CO_2 is grounded in its ability to yield good material properties based on its near-zero surface tension, low viscosity, and high diffusion coefficient [10]. The suitable material properties partly explain why SC-CO_2 has been extensively applied in membrane formation, desorption, micronization, and extraction. However, in contrast to Franco et al. [8], Lukic et al. [9], Trucilo et al. [10], Ozkan et al. [11] and Cejudo et al. [12] have argued that the utility of SC-CO_2 in active packaging films was dependent on the role of the material (co-solvent, anti-solvent, or swelling agent). On the one hand, SC-CO_2 can function as a supercritical anti-solvent (this is critical in micronization or co-precipitation processes). On the other hand, it can function as a solvent or co-solvent. Following the appraisal of the two techniques, it was clear that the SC-CO_2 supercritical anti-solvent process (SAS) is highly appropriate in packaging given it yields products with customized properties such as spherical nanoparticles, and nanostructured filaments with controlled mean size and particle size distribution.

1.2. Natural Phenolic Polymers (Lignin and Proanthocyanidins (PAs))

Lignin is a naturally occurring polymer/heterogeneous biomacromolecule that supports the connective tissues in plants [13–15]. The rigidity, tensile strength, resistance to chemical degradation, and pressure-resistance of lignin are linked to the phenyl propane created using coniferyl alcohol, p-coumaryl alcohol, and sinapyl alcohol units [16]—precursor-specific structural modifications of lignin translate to unique, different structural modifications. The microscopic-level changes improve the utility of the material in the production of plastics, paints, films, PU-based forms, nutritional supplements, resins and coatings, beverage additives, food additives, adhesive binders, carbon fiber composites, and enhancement of the structural properties of existing materials [17]. The application of lignin-based compounds in dopamine polymerization, synthesis of polydopamine and copolymers, polymerization of inulin, enzyme-catalyzed polyphenol, polymerization, antioxidant quercetin polymers, polyquercetin and quercetin copolymers, and antioxidant properties, and antioxidant terpene polymers is explored under Sections 2 and 3.

Proanthocyanidins are a class of natural phenolic compounds derived from natural phenols (flavan-3-ols) and are fundamental polyphenolic components of the human diet, including red wine, cocoa, dark chocolate, orange juices, grapefruit, parsley, black and green tea, onions, olives, broccoli, and apple skin [18,19]. The PAs are critical to the investigation of sustainable and natural monomers, synthesis, and applications of antioxidant polymers considering that they are the second most naturally occurring and abundant plant polyphenols after lignin [18]. In addition, PA compounds possess ideal chemical properties for industrial and biological applications. For example, cinnamon-derived polyphenols have been proven to have superior anti-inflammatory characteristics–a factor that is instrumental in the treatment of inflammation-related conditions, including inflammatory bowel disease, rheumatoid arthritis, asthma, cancer, diabetes, cardiovascular diseases, degenerative musculoskeletal diseases, and neurodegenerative conditions. The need for plant-based alternatives is supported by the limitations of traditional non-steroidal anti-inflammatory drugs (NSAIDs) [20], the manifestation of appropriate antioxidant properties [21,22], and a variety of naturally occurring precursors, including agro-wastes.

1.3. Recovery of Antioxidants from Agro-Wastes

The recovery of antioxidants from agro-wastes is a facile and scalable method in the manufacturing of a broad array of lignin-containing compounds for industrial and biological applications. Martinez-Avila et al. [23] noted that the fermentation of fruit peel waste generated high-quality phenolic antioxidants for industrial application. However, the recovery of antioxidants from agro-wastes is impacted by production-related constraints—traditional processes are associated with significant hazards to human health, low yields, and negative ecological effects; this explains why solid-state fermentation and solid-state shear pulverization have become the techniques of choice in the recovery of polyphenols from agro-wastes [23,24]. From a manufacturing perspective, the technical constraints are offset by the cost advantage. The cost of agro-waste is about USD (United stated dollars) 0.05–0.10, which is significantly low compared to synthetic antioxidants, which cost > USD 6 per kg (Table 3) [24]. Future advances in technology might resolve the low yields, hazards to human health, and negative ecological effects associated with toxic products.

Table 3. Cost of recovering antioxidants from selected agro-wastes and the phenol content [24].

Agro-Waste	Annual	Cost	Calculated Total Cost	Antioxidant	Total Phenol
	production (thousand tons)	($/kg)			content (mg GA/g)
Red Grape	100	0.050	5	αtocopherol, Gallic acid	~0.02
Turmeric				Curcumin	~3.1
Coffee grounds	0.005	0.050	0.00025	Melanoidin, Chalcogenic acid	17
Orange peel	20	0.050	1	Ascorbic acid and Flavonoids	~1.3
Irganox I10	Not known	6.36		Four substituted 2,6 di-tertiary butyl phenols	

In theory, it is feasible to extract antioxidants from nearly all commonly available agricultural wastes. Nonetheless, there is a strong preference for antioxidant-rich precursors such as orange peel, coffee grounds, turmeric shavings, and waste and grape pomace wastes, which exhibit superior performance in the reinforcement of plastic properties.

The antioxidants are incorporated into polyolefins to enhance their material properties, primarily recyclability, weathering resistance, and high-temperature processability [23,24], which vary in line with the chemical properties of the precursors/agro-waste materials. The presence of hydrogen-donating hindered phenols in the agro-waste materials is integral to the capture of alkyl peroxide radicals, a process that impedes the propagation of free radical reactions [24]. The precursor-specific chemical behavior underscores the need to select appropriate natural antioxidant molecules for antimicrobial treatment, biodegradation of polymers, tumor-targeted drug delivery, and food fortification.

1.4. Natural Antioxidant Macromolecules

Antioxidant polymers are a broad class of polymers, including polylactones [25], polymer nanoparticles integrated with living polymerization techniques [26], poly(b-malic acid) derivatives [27], lignin graft polymers, and polyphenols, which offer unparalleled capabilities in tumor-targeted drug delivery, food fortification [28], biodegradation of polymers, and antimicrobial treatment [29]. Emerging research evidence indicates that hydrophobic antioxidant polymers are effective corrosion inhibitors in steel structures.

The development of green, biocompatible, and biodegradable antioxidants is critical in polymer processing industries, packaging, health, cosmetics, textiles, and agricultural sectors. The transition to green materials is supported by growing environmental awareness. The need to explore antioxidant molecules is supported by widespread availability in nature, such as lignin and polyphenol adduct polysaccharides. Additionally, certain natural essential oils such as limonene also feature antioxidant properties and can be polymerized into renewable polylimonene while preserving their antioxidant properties. The antioxidant polymers (including polyfurfuryl alcohol and its derivatives) can be directly sourced from agricultural wastes. Polyphenols drawn from a variety of sources, including fruits and vegetables, dopamine, pyrogallol, tannic acid, and green tea catechins, undergo controlled oxidation and bio-inspired polymerization, which is vital in biomedical applications. For example, quercetin (a potent antioxidant) is polymerized to encapsulate cancer treatment medicines. Recently, edible flavors such as vanillin (phenolic aldehyde) have also been polymerized into an antioxidant copolymer known as polyoxalate co-vanillyl alcohol (PVAX) for biomedical applications. Vitamin C and other poly vitamins with potent antioxidant properties can facilitate the synthesis of ADMET, acyclic diene metathesis polymerization. The procedure involves a step-growth polymerization where α, ω-diene monomer is transformed in the presence of a ruthenium catalyst.

In contrast to synthetic antioxidants, natural antioxidants are naturally occurring compounds generated by plants, fungi, bacteria, and mammals, and are environmentally benign and largely biocompatible. Additionally, natural polymers are preferred due to the diminished risk of harmful byproduct formation in the course of their use relative to synthetic antioxidants [30]. Common antioxidant polymers include essential vitamins such as retinoids, ascorbic acid, tocopherols, and polyphenolic compounds epigallocatechin gallate (EGCG), quercetin, curcumin, and resveratrol [31]. The natural antioxidants exhibit unique material properties, including the ability to scavenge reactive oxygen species, targeted delivery, and long-term functionality/reactivity [32]. Such properties are vital agricultural, textiles, cosmetics, health, and packaging industries. Future applications require the development of materials with customized properties for specialized applications.

1.5. A Review of Antioxidant Molecule-Incorporating Polymers

The encapsulation of antioxidant molecules in conventional polymers is integral to the regulation of the potential toxicity, lability, solubility, and diffusion, and other antioxidant properties; this is particularly true for natural antioxidant polymers, which achieve high antioxidizing effectiveness at relatively low concentrations [33]. In other cases, the incorporation of antioxidant molecules into polymers has been proven useful in autosynergism and heterosynergism [33]. Lith and Ameer [32] and Sawant et al. [1] reported the successful encapsulation of vitamin C antioxidant polymer in PEO (or poly(ethylene glycol), PEG) micelles. The incorporation of vitamin C into PEG/PEO micelles was beneficial in the treatment of cancer. Preliminary in vivo experiments using mouse models demonstrated selective necrosis of breast cancer cells; the outcomes demonstrate the unique capabilities of antioxidant compounds embedded into polymers [1,32]. However, the potency of the antioxidant polymers was dependent on concentration—higher concentrations were established to be highly effective in overcoming cancer antioxidant defense mechanisms. Wang et al. [26] reported successful incorporation of paclitaxel into polyethylene glycol-b-poly(D, L-lactide) (PEG-PDLLA) micelle for tumor-targeted drug delivery. Preliminary clinical data show that the technique offers promising prospects in the treatment of tumors.

Beyond targeted drug delivery, the encapsulation of antioxidant molecules in conventional polymers was proven beneficial in agricultural and food packaging applications. The incorporation of primary phenolic antioxidants into polyolefins and polyethylene has been proven useful in the protection of the thermo-oxidative stability of polymers used in packing [30]. The stabilization of synthetic polymers exposed to natural elements, including UV radiation, shear forces, and higher temperatures, is integral to their durability post-production. However, there are ecological implications associated with the incorporation of natural phenolic antioxidants into polymers [30]. Some studies have suggested that the incorporation of antioxidant polymers into polymers poses significant environmental and health hazards. For example, the study found traces of plastics (polyethylene) in drinking water [34]. The problem was attributed to the byproducts of phenolic antioxidant additives used in pipeline production [34]. The compounds migrated to waterways via diffusion. The presence of polymers in drinking water elevates the risk of health complications associated with reactive oxygen species [32]. Even though the latter findings point to the negative effects of antioxidant molecules in conventional polymers, there are tangible benefits associated with the embedding of antioxidant species. The positive benefits in the agricultural, textiles, cosmetics, health and packaging industries help to offset the adverse effects.

2. Lignin Polymers and Antioxidant Properties

The review of lignin polymers as potential antioxidants for biodegradation of polymers and antimicrobial treatment [29], tumor-targeted drug delivery, food fortification [28], and diabetes treatment [15] is supported by the chemical composition of lignin structures, abundance in the natural environment, and ease of production; conservative estimates suggest that at least 70 million tons of lignin are generated from pulp [35]—of which 2% is converted to value-added products including lignin for biological, pharmaceutical and chemical industry applications [35]. The widespread availability of lignin in plant species predicts cost. The availability of lignin is predicted by the rate of lignin biosynthesis, which is responsible for biotic and abiotic stress management, organ/tissue development, and growth [36]; these processes are mediated by a broad class of enzymes and genes. Higher expression of CCR, CCoAOMT B, C3H, 4CL, and F5H was associated with pronounced tissue growth [37]. Since the genes predict the presence of G, S, and H units [16], the genetic composition of the different plant species strongly predicts the utility of the plant lignin in tumor-targeted drug delivery, food fortification [28], biodegradation of polymers, and antimicrobial treatment [29]. The observations made by Liu et al. [16] are valid, considering that the G, S, and H units which comprise sinapyl alcohol, coniferyl alcohol, and p-coumaryl alcohol, respectively, predict the rate of lignin monomer copolymerization. The function of different genes on plant species is heightened in Table 4.

Table 4. Function of selected genes [16].

Genes	Plant Species	Functions
4CL	*Oryza sativa* (*OsAAE3*)	Decreased lignin accumulation and increased sensitivity to rice blast
CCR	*Paspalum dilatatum* (*PdCCR1*)	Increased lignin content and altered lignin deposition
CAD	*Populus trichocarpa* (*PtrCAD1*)	Modified lignin content and structure
MOMT	*Arabidopsis thaliana* (*MOMT4*)	Depressed lignin biosynthesis and increased saccharification yields
COMT	*Sorghum bicolor* (*SbCOMT*)	Methylate the tricin precursors and participated in S lignin biosynthesis

The focus on selected plant species as precursors for lignin-based polyurethanes, dopamine polymerization, polydopamine and copolymers, polymerization of inulin, enzyme-catalyzed polyphenol polymerization, antioxidant quercetin polymers, poly-quercetin and quercetin copolymers, and antioxidant properties, antioxidant terpene polymers, polylimonine synthesis and properties, antioxidant random copolymers, and gallic acid grafted polymers is beyond the scope of this research inquiry. This means that the desired chemical properties of lignin-based compounds are considered in general without a specific emphasis on a particular plant species.

Lignin molecules have vast antioxidant properties due to the presence of certain functional groups such as p-hydroxy acetophenone extracted from oil palm fronds [38]. The antioxidant properties of lignin-derived polymers are predicted by the precursors. Different precursors have different antioxidant properties, which are predicted by the presence of carbohydrates, G + S phenols, Pi-conjugated carbons, ArC (Py-products with primary, secondary and tertiary carbons on the side chains), carboxylic groups, OH groups, and phenols [14]. The number of functional groups that predict chemical behavior considering the chemical structure of lignin is poorly understood—the aliphatic OH and phenol groups increase the probability for functionalization of the compound [10] and biological and pharmacological function. The antioxidant properties and higher binding affinity of the chemical functional groups in lignin enable the material to bind to bile acids in the intestines—a process that facilitates serum control.

The antioxidant properties are also integral to tumor suppression—animal models suggest that lignin reduces the adverse effects associated with different carcinogens, including 3,2-dimethyl-4-ami-biphenyl [15]. The phenolic compounds isolated from the different lignin structures have been proven to inhibit microbial growth through the inhibition of oxygen-mediated reactions, ATP depletion, and interference with the intracellular pH [35].

In other cases, organic functional groups within the lignin structure (carvacrol, thymol, and cinnamaldehyde) trigger bacterial lysis and damage the cell membrane [35]. The lignin compounds extracted from corn Stover have exhibited appropriate antioxidant properties in eliminating free radical initiators in the red blood cells such as AAPH (2,2 azobis (2-amidinopropane)) [35]. The trends and applications of lignin as a natural antioxidant supersede the antimicrobial effects; this means that natural lignin best functions as a free radical scavenger/borrower [35]. The utility of the antioxidant properties transcends pharmaceutical applications—the chemical antioxidant properties offer protective benefits to the skin and eyes; this means that topical formulations can be prepared from lignin extracts for cosmetic applications. Other potential commercial applications can be explored due to the heterogeneity of lignin structures drawn from diverse sources. Moreover, research has demonstrated that it is possible to customize the behavior of lignin in different applications to stimulate the desired biological function [35].

Even though diverse plants are a suitable source of lignin, only selected sources are used for commercial applications due to wide variations in the lignin composition by weight 10–40 wt%, the need to balance resource use and promote green economy [39]. In other cases, the herbaceous weight composition of biomass is lower 15–25% (*w*/*w*) [15]; this explains why most lignin is derived from byproducts during pulp production [15,39]. The production-related requirements have practical implications on the choice of lignin precursors for biological, polymer processing industries, agricultural, cosmetics, textiles, and agricultural applications. This observation is supported by the link between the precursor and chemical properties and production methods [35]. Certain sources are associated with a higher cost of production, low yield extraction, higher energy use, poor solubility, and presence of chemical impurities; this means that the choice of fractionation methods of technical lignin predicts commercial utility [35].

Even though precursor-specific chemical functional groups might be ideal in the development of lignin-based antioxidants for pharmaceutical, medical and packaging applications, other factors have to be taken into account, including the availability of the plant species and lignin content. The *Quercus* plant species have an extremely low lignin content (3.8%); this is in contrast to 54% in *Acacia auricuriformis* [36]. The high concentration of lignin in the latter justifies its use as a preferred source of antioxidants for a broad array of applications. However, there is significant information asymmetry about the performance of different plants with variable concentrations of lignin—most studies focus on selected plant species [14], with proven benefits. The bias towards selected plant species could have practical implications in commercial applications.

2.1. Lignin Graft Polymers and Copolymers

The development of lignin graft polymers and copolymers is integral to the sustainability of modern civilizations and human life. However, there are critical constraints in the synthesis of lignin-based high-performance materials [40]. Most lignin-based materials have been used in the development of low-value products such as pesticides, animal feeds, surfactants, binders, dyestuff dispersants, concrete additives, and materials for dust control, which account for about 2% of the total fraction of lignin produced globally [40]; this means that there is a clear mismatch between the rates of lignin production and the development of value-added products. A feasible approach for the synthesis of value-added lignin products is the derivatives that entail the development of lignin graft polymers and copolymers.

Modern plastics are primarily sourced from petroleum byproducts, but there are valid and emerging concerns about the environmental implications, greenhouse emissions, and destruction of marine and terrestrial environments [41–43]. The growing environmental concerns have validated the need to develop sustainable methods for reducing the carbon footprint associated with plastics; this has, in turn, triggered the development of bio-based plastics for agricultural and packaging applications [44–46]. A fundamental constraint includes the shortage of suitable precursors for bio-based plastics. Chitosan and corn are less abundant, pose a direct threat to commercial agriculture, and commercial viability is challenging [45,47], making it unfeasible to upscale operations. The highlighted challenges demonstrate the need for alternative and complementary sources of sustainable and biodegradable plastics; this informs the development of lignin graft polymers and copolymers for sustainable plastic materials. The focus on lignin in place of other forms of biomass is grounded on the natural abundance in plant and tree species [36,37]. The relative composition varies between 10 and 40 wt% [15,39], depending on the tree species, which is significant considering that about 70 million tons of pulp are generated [35]. The pulping byproducts are rich in lignin; this means that the precursor can be extracted using a scalable and low-cost method. Affordability is integral to the gradual phasing out of synthetic plastics with biodegradable materials in emerging and developed economies.

Lignin graft polymers have been synthesized using a variety of techniques, including blending natural lignin extracted from plants with commercial-off-the-shelf polymers [13] to augment the mechanical and thermal properties of the polymer blends. Copolymer blends have other desirable properties, including superior gene delivery properties, surfactant properties, UV absorber properties, super-plasticizer capabilities, and enhanced elasticity [16]. The realization of these properties is dependent on controlled polymerization and the selection of the backbone structure/polymers and the branch polymers [13,16,48]. The two are connected via covalent bonding.

The properties of the blends are predicted by a broad range of factors, including the polymerization techniques, click reactions, reversible-addition-fragmentation chain-transfer (RAFT) polymerization, and radical-mediated lignin-graft copolymerization [13,16,48]. Other feasible synthetic techniques include single-electron transfer living radical polymerization (SET-LRP), atom transfer radical polymerization (ATRP), macromolecular design via the interchange of xanthates (MADIX), and single-electron transfer-degenerative transfer living radical polymerization (SET-DTLRP) [49]. Even though there are multiple feasible copolymerization techniques, ATRP and RAFT are widely used.

The attainment of the desired properties during polymerization is contingent on graft density, the length of the grafts, and the functional groups on the graft polymers. The most commonly used grafting techniques (Figure 1) include grafting-through, grafting into, and grafting from [16]. The graft-from technique is characterized by the grafting of the polymers from active sites at the backbone polymer (lignin).

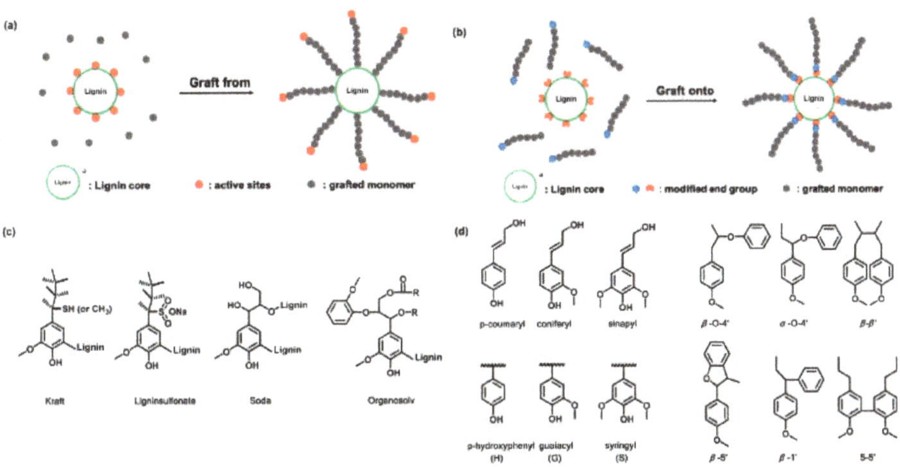

Figure 1. (**a**) An illustration of the graft-from synthesis of lignin copolymers; (**b**) a demonstration of graft-onto methods; (**c**) industrial lignin products; (**d**) sinapyl alcohol, coniferyl alcohol, and p-coumaryl alcohol units [16].

The grafted polymers emerge from the backbone as a result of radical polymerization, ring-open polymerization, RAFT, and atom transfer radical polymerization (ATRP) [16,48] (see Figure 2). Each polymerization technique has its benefits and constraints. For example, ATRP relies on an alkyl halide initiator, which is then radicalized, resulting in the polymerization of monomers [13]; this process is augmented by a transition metal catalyst. In contrast, RAFT polymerization is augmented by radicalization from the addition of a radical-generating species [13,16], while ROP is characterized by the opening of a cyclic molecule in the polymerization process; this results in the formation of a reactive species that undergo a chemical reaction with the next cyclic molecule [13]. The choice of either copolymerization process is predicted by a broad range of factors, including cost, the quality of the desired product. For example, Liu et al. [16] argue that RAFT is an ideal technique for newly made monomers, and ATRP is appropriate for commercially available acrylate monomers. In the former case, the lignin backbone is modified through the incorporation of an ester linkage to create a RAFT agent moiety; this procedure is subsequently followed by the polymerization of the vinyl monomers.

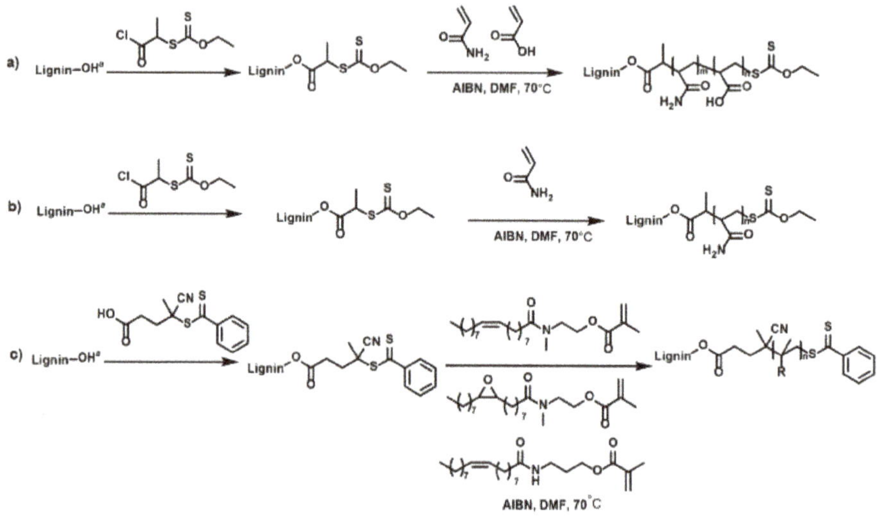

Figure 2. RAFT polymerization demonstrating: (**a**) acrylic acid and acrylamide graft-from polymerization; (**b**) RAFT polymerization of acrylamide; (**c**) soybean oil polymerization using the graft-from technique. The lignin compound is comprised of both phenolic hydroxyl and aliphatic hydroxyl groups [16].

The choice between RAFT and ATRP entails a tradeoff of material properties and the synthesis process. Lai et al. [50] suggest that RAFT is highly suitable for commercial applications relative to ATRP and other synthetic techniques [50] due to the incorporation of acrylic derivatives and organic substances. The bio-applications of RAFT polymerization documented by Boyer et al. [49] corroborate the assessment of Lai et al. [50] of the key benefits associated with RAFT. For instance, RAFT features most of the desirable aspects of traditional free radical polymerization, namely facile reaction conditions, tolerance of most functionalities, site-specific functionality, control of molecular weight distribution and molecular weight, and compatibility with a wide array of monomers. In contrast to traditional living radical polymerization (LRP), RAFT features site-specific functionality [50]. The commercial significance is offset by synthetic requirements.

The RAFT polymerization technique depends on the ability of RAFT agents to transport an ideal leaving group bonded to an S-atom to the reaction site; it is often challenging to attain such reaction specificity. In view of these constraints, the RAFT polymerization has low yields. There are also concerns about the synthesis of toxic compounds, namely

Grignard reagents [50]. The inadequacy of RAFT and ATRP techniques demonstrates that new methods are required to meet the need for the synthesis of next-generation lignin graft polymers and copolymers.

2.2. Lignin-Based Polyurethanes

The synthesis of lignin-based thermoplastic polyurethanes has been demonstrated through copolymerization [40]. Lignin precursor is incorporated into the backbone of the polyurethane structure, but most of the products are rarely recyclable or processable due to extensive crosslinking and networking within the polymer structure [7,40,51]; this means that the polydisperse nature of lignin molecules is a critical impediment to the synthesis of thermoplastic polyurethanes. Saito et al. [40] suggested that the challenge could be offset through the incorporation of tie molecules to link the hard segments or crystals. The synthesis of recyclable thermoplastic lignin-based polyurethanes could also be augmented by bridging the soft segment (characterized by a low glass transition temperature T_g) with the hard segments containing lignin to ensure that the final product features the mechanical rigidity of lignin and the soft rubbery properties of the soft segments. The weight ratio of lignin in polyurethane has a direct impact on the material properties—a higher lignin content translated to better stress–strain behavior (see Figure 3) [40]. The higher stress and strain tolerance and other desirable mechanical properties make lignin-based polyurethane materials appropriate for high-strength applications.

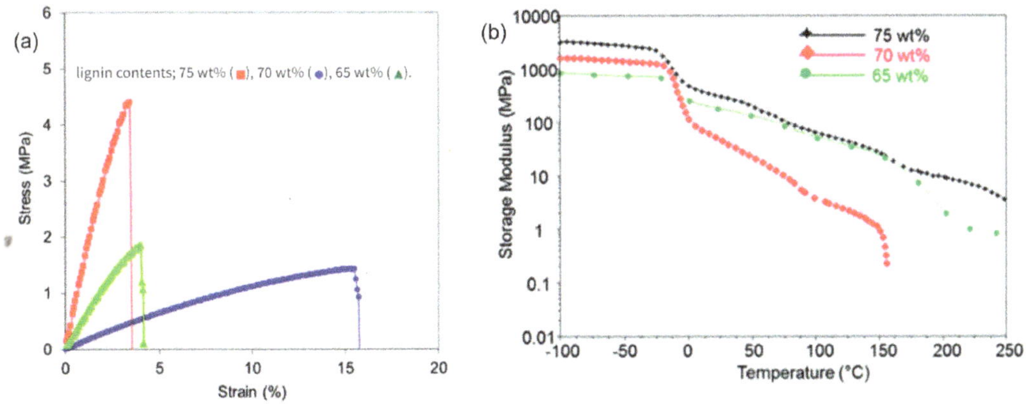

Figure 3. (**a**,**b**) Stress–strain behavior and storage modulus of lignin-based polyurethane thermoplastics [40].

Saito et al. [40] reported the successful synthesis of lignin-based polyurethanes made of telechelic polybutadiene soft segments. Additionally, Da Silva et al. [51] used a kraft lignin precursor to synthesize vanillin and lignin-based polyurethanes, while Cateto et al. [7] synthesized lignin-based polyurethanes using 4,40-methylene-diphenylene diisocyanate (MDI), polycaprolactone (PCL). The successful synthesis of lignin-based polyurethanes from a wide variety of precursors demonstrates that existing synthetic routes can be modified to suit different commercial applications, with or without a catalyst. However, specific adjustments must be made in the synthesis process, including the chemical modification via etherification and esterification reactions [7]. Additionally, oxypropylated lignin materials have been proven useful relative to non-oxypropylated materials in the formation of rigid polyurethane foams. However, there are critical constraints associated with chemical modification. For example, the costs are higher after esterification and etherification; this means that the industrial production of high-value products from lignin precursors is often offset by cost constraints [7], which can be partly offset through the development of new synthetic routes [7,41,51]. The cost constraints reported by Cateto et al. [7] were also acknowledged by Da Silva et al. [51], who argued that cost was not the only constraint.

Other issues relating to the food-versus-fuel dilemma and technical know-how must be resolved before the "novel systems become more competitive against the current industrial alternatives based on petrochemical resources" (p. 1273). The observations made by Cateto et al. [7], Da Silva et al. [51], and Saito et al. [40] illustrate that for lignocellulose feedstock (LCF), bio-refineries are preferred from a sustainability perspective. However, the future development of bio-based products would be catalyzed by conversion products, which offer competitive benefits relative to petrochemical products. This observation is supported by the chemical properties of lignin materials depicted in Table 5.

Table 5. Second-order kinetics of lignin-based materials [7].

Sample	*p* (30 min)	CV (%)	Linear Fit	The Domain of Validity (pNCO)	[NCO]0	Kapp 10^3 (L/mol/s)
(PCL400/1/0)	0.93	0.41	1 + 0.525t	0.93	4.72	1.87
A(PCL400/1/10)	0.91	0.19	1 + 0.491t	0.87	3.52	2.32
A(PCL400/1/15)	0.87	0.67	1 + 0.461t	0.82	3.54	2.16
A(PCL400/1/20)	0.84	0.43	1 + 0.469t	0.71	3.57	2.20
A(PCL400/1/25)	0.77	0.94	1 + 0.377t	0.66	3.59	1.76
A(PCL750/1/15)	0.84	1.30	1 + 0.350t	0.73	2.47	2.36
A(PCL750/1/20)	0.79	1.33	1 + 0.323t	0.66	2.56	2.08
A(PCL750/1/25)	0.74	1.18	1 + 0.327t	0.56	2.66	2.07
(PCL1000/1/0)	0.79	1.12	1 + 0.123t	0.79	1.73	1.16
A(PCL1000/1/10)	0.82	0.40	1 + 0.212t	0.75	1.98	1.77
A(PCL1000/1/15)	0.78	0.42	1 + 0.247t	0.63	2.11	1.98
A(PCL1000/1/20)	0.75	1.04	1 + 0.260t	0.57	2.23	1.94
A(PCL1000/1/25)	0.70	1.55	1 + 0.257t	0.48	2.35	1.84

The physicochemical properties and kinetic properties affirm the suitability of lignin antioxidants in a broad range of engineering and non-engineering applications. However, the formation of the products is dependent on a broad range of factors, including the yield from dopamine polymerization and the polydopamine and copolymers. The antioxidant properties of lignin-based polyurethanes are linked to the presence of an aromatic ring with hydroxyl and methoxy functional groups [50,51]. The functional groups are integral to the oxidation propagation reaction, which is inhibited by hydrogen donation [51]. According to the literature, the performance of lignin polymers was comparable to 2,2-diphenyl-1-picrylhydrazyl (DPPH), and butylated hydroxyl-toluene, among other commercial antioxidant polymers [50].

3. Dopamine Polymerization

The self-assembly of polymer molecules has gained significant research attention in research and development because of the versatility of the process, its practicability, and its ability to form products with a broad range of morphologies, including hierarchical assemblies, cylinders, vesicles, and micelles [52]. Polydopamine is a versatile organic molecule that can be deposited onto any substrate at the nanoscale; this means that the material might act as a suitable coat or primer [52–54]. The versatile chemistry of polydopamine is linked to the molecule's ability to form strong covalent bonds [54], physical interactions associated with π-interactions, hydrogen bonding, presence of different surface functional groups including amines, and thiols [54]. However, surface functionalization in isolation does not explain the broad commercial application of dopamine and polydopamine; the material exhibits appropriate antibacterial activities through gelation [54].

The extensive conjugation influences the application of polydopamine coatings in polymers, ceramics, and metals [53]. The extended application of polydopamine in the coating is augmented by extended bio-conjugation, which is integral in the bio-adhesion of protein molecules, adsorption-resistant surfaces, drug delivery systems, contrast agents, and bioadhesives. The adhesion properties of polydopamine are also vital in cardiovascular, diagnostic, and neurotechnology applications [53]. In other cases, the polydopamine coatings have been proven useful in the surface functionalization at the nanoscale to enhance the chemical and electronic properties [55].

The oxidative polymerization of dopamine primarily occurs in the presence of aminoethyl and catechol groups, which are catalysts for oxidative polymerization [56]—a process that is integral to the formation of PDA nano-coatings and polydopamine coatings [57] through the formation of the aromatic rings of dopamine. Recently, there has been a growing demand for nano-coatings developed using dopamine polymerization techniques; this could be attributed to the capability for secondary modification, generalizability, and simplicity of the synthetic process relative to other techniques [57].

Other unique applications of dopamine polymerization include the electrochemical analysis of dopamine, uric acid, and ascorbic acid using hollow nitrogen-doped carbon microspheres (HNCMS)-based glassy electrodes [58]. The application of dopamine polymerization in biosensors and bioelectronics documented by Xiao et al. [58] was corroborated by Kalimuthu and John [59], who reported successful electrochemical determination of xanthine, ascorbic acid, dopamine, and uric acid using 2-amino-1, 3, 4-thiadiazole (p-ATD)-modified glassy carbon electrodes.

The application of dopamine polymerization in biosensors and bioelectronics and the development of the desired products in manufacturing applications depends on a broad array of factors such as size control of polydopamine nodules and chemical composition of the precursors (carboxylic acid-containing compounds) [60]. The carboxylic-acid-containing compounds introduce an acidic environment, which yields products with unique chemical characteristics compared to dopamine polymerization under basic conditions.

Even though the findings reported by Chen et al. [61] seem to favor acidic dopamine polymerization, basic polymerization techniques have yielded stable products through the customization of the synthesis process. Du et al. [57] observed that the challenges associated with dopamine polymerization under basic environments could be offset by light-triggered regulation of light initiation and termination of dopamine polymerization; this was achieved through the incorporation of small quantities of antioxidant Vitamin C (sodium ascorbate). Vitamin C contributed to the inhibition of the polymerization process under basic conditions—a process that has remained a challenge in traditional synthesis. Du et al. [57] attributed the superior performance to Vitamin C's ability to delay/inhibit dopamine polymerization and reduce the reactive dopamine quinone. The inhibition process can be halted through UV irradiation. Once vitamin C is exposed to UV radiation, instantaneous dopamine polymerization is achieved. The customization of the process using a natural antioxidant and UV radiation helps to explain why it was feasible to attain optimal dopamine polymerization under basic conditions through a facile, scalable, and environmentally benign process.

Fichman and Schneider [54] reported a facile synthetic route for dopamine polymerization under basic conditions in the absence of Vitamin C. Optimal polymerization was achieved in this case through gelation of 1 wt% MAX1 peptide; this experiment was conducted in the presence of 10 mM dopamine at neutral pH [54], and molecular oxygen. The process resulted in the spontaneous polymerization of dopamine. The temperature was adjusted to room temperature to trigger a hydrophobic effect, which was instrumental in promoting MAX1 assembly. Considering that the final product exhibited a 66-fold improvement in mechanical rigidity relative to other materials synthesized using alternative synthetic routes [54], sodium ascorbate is not a prerequisite.

The reports of the positive synthesis of polydopamine and dopamine under basic and acidic conditions by Chen et al. [61], Du et al. [57], Fichman and Schneider [54], Kwon and Bettinger [53], and Qui et al. [52] do not address other emergent challenges such as the relationship between film thickness and solution pH, dopamine concentration and self-polymerization time optimization. The critical requirements for the process underscore the need to select a suitable pH range for the controlled synthesis of materials with the desired film thickness, mechanical rigidity, and industrial/biological performance.

3.1. Polydopamine and Copolymers

The chemistry of polydopamine (PDA) copolymers predicts the synthesis of derivative materials and promising applications in biomedicine, environmental, and energy applications. The application of PDA is predicted by the presence of the different functional groups (azido, alkyl-thiol, amino, carboxyl derivative groups, and carboxyl and alkyl groups) and the synthesis process [62]. There are three feasible mechanisms for polydopamine synthesis from derivatives, namely oxidative copolymerization of the mixtures (comprising of copolymers and monomers), and oxidative polymerization of DA analogs with additional functional groups (see Figure 4).

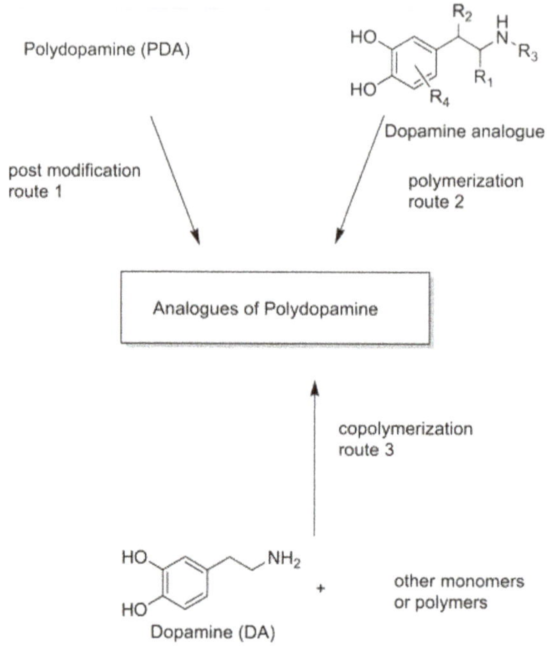

Figure 4. Polydopamine polymerization from different analogs [62].

Melanoidins

Melanoidin compounds are byproducts of the Maillard reaction and the reduction of amino acids, sugars and proteins during food preservation and processing [61]. In most cases, the compounds are drawn from coffee processing—the roasting of coffee bean components results in the generation of brown-colored nitrogenous compounds with high molecular weight [62], referred as melanoidins.

4. Polymerization of Inulin

Inulin is categorized as a fructan carbohydrate, which functions as a primary ingredient in industrial food processing. Inulin hydrogels have been employed as carriers for colonic drug targeting; such processes are fundamental to prebiotic fermentation, chocolate, and cheese production [63–65], as noted in the preceding sections. From a nutritional perspective, the functional properties of inulin and health beneficial effects include prebiotic effects, high dietary fiber content, and lower calorie value [65]. The physiological benefits linked to the consumption of inulin-rich foods are presented in Table 2. In industrial food production, inulin has been employed in fat replacement as a bulking agent [65], stabilizer, and fat replacer in cheese production. The observations made by Karimi et al. [65] are in agreement with Aidoo et al. [63], who reported the incorporation of inulin in sugar-free chocolate production (inulin functions as an alternative sweetener), and Guimarães et al. [64] research on inulin's role in prebiotic fermentation (especially in beverage stabilization). In other cases, inulin has been proven useful in the retro-gradation and gelatinization of wheat starch [66].

The potential applications of inulin are not confined to food production, given that inulin has been proven to be an ideal in vitro drug carrier of lignin derivatives. Vervoot et al. [67] noted that inulin was an ideal carrier for colonic drug targeting [67]; this is achieved through the surface functionalization of inulin with N, N, N′, N′-tetramethyl-ethylenediamine, and ammonium persulphate [67]. Colonic drug delivery is critical to the treatment of colon cancer and metabolic/bowel complications. Similar to Vervoot et al. [67], Kumar et al. [68] noted that inulin was a suitable bioactive polymer for pathogen-mimicking vaccine delivery systems. The biomedical applications of inulin in targeted drug delivery offer promising prospects in the treatment and management of life-threatening conditions, including cancer, Ebola, HIV/AIDS, malaria, and tuberculosis. The biomedical application of inulin is mediated by free-radical polymerization—a process that contributes to the incorporation of vinyl functional groups on the surface of inulin [67].

Beyond industrial applications, the polymerization of inulin has broader implications on the health of mammals. Wada et al. observed that the fermentation of indigestible carbohydrates in the diet by bowel microflora predicts the formation of short-chain fatty acids [69]. Acetate, propionate, and butyrate are the primary short-chain fatty acids (SCFA) derived from the digestion and fermentation of undigested carbohydrates and inulin in the human colon [70].

The formation of both short-chain and long-chain fatty acids has long-term effects on bowel physiology. The observations reflect long-standing beliefs and evidence concerning the relationship between improved health outcomes and high-fiber diets, causal epidemiological evidence, and metagenomics studies linking metabolic diseases to variations in the gut microbe and molecular signaling [71].

Even though there is growing use of inulin in biomedical and food-related industries, the extent of application depends on a broad range of factors, including the polymerization of inulin; this underscores the need for customized synthetic processes, including the reaction between N,N-dimethylformamide with glycidyl methacrylate in the presence of a catalyst (4-dimethylaminopyridine) [67]. The polymerization of inulin is integral to industrial application, primarily in the food and biomedical sectors, but there are critical technical constraints that impede successful polymerization.

5. Enzyme-Catalyzed Polyphenol Polymerization

Similar to the antioxidants for food production and biopharmaceutical applications, enzyme-catalyzed polyphenol polymerization reactions have a broad range of practical applications, including antioxidant, antibacterial, anticancer, cardioprotective, and neuroprotective properties [72]. However, the progression of most reactions is impeded under natural conditions; this justifies the need for enzyme-mediated reactions to attain the desired effects. In a recent study, Oliver et al. [72] noted that the impressive therapeutic activity of polyphenols in therapeutic agents and other clinical applications could not be realized without proper adjustment of the production processes and chemical properties. Under natural conditions, polyphenols exhibit poor membrane permeability, rapid metabolism, and poor bioavailability and UV degradation. The challenges have been partly resolved through amidification, esterification, free-radical grafting [73], step-growth, free radical, and enzyme-catalyzed reactions (direct polymerization of polyphenol monomers), and enzyme grafting-mediated conjugation with macromolecules.

Even though multiple techniques have been proposed to improve the industrial application of polyphenols, enzyme grafting and catalyzation of the reaction were given preference given polyphenol polymerization under natural conditions is constrained by the reaction mechanisms and chemical composition of the reactants, which impede synthesis using conventional methods [74,75]. The forward polymerization process is augmented by the inclusion of enzymes. The utility of laccase (an enzyme) in catechol, resorcinol, and hydroquinone selective polymerization was confirmed by Sun et al. [75] (see Figure 5). The process was phased; the initial phases entailed the formation of quinone intermediates through laccase catalysis, followed by oxidation and formation of covalent bonds. In the subsequent steps, carbon-carbon (C-C) and ether bonds were formed, linking catechol units, resorcinol, and hydroquinone units.

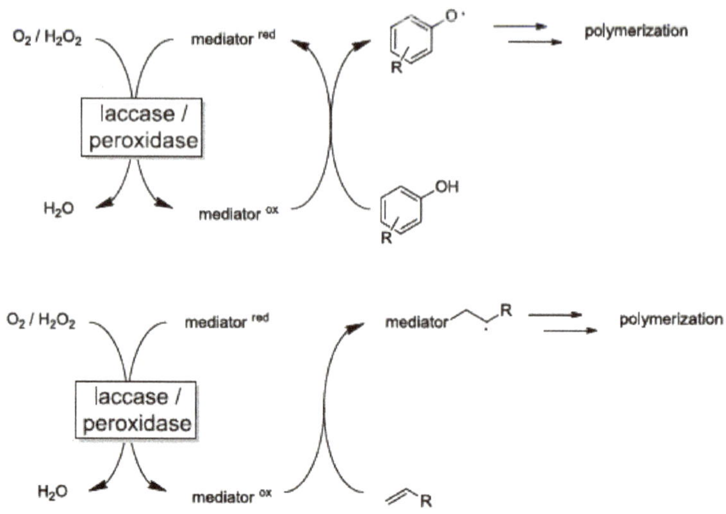

Figure 5. Illustration of laccase and peroxidase mediator systems (PMSs) in polymerization reactions [76].

A key benefit associated with the reaction process is the formation of reaction products under mild conditions and the elimination of toxic byproducts such as formaldehyde [74]. The merits of the laccase-catalyzed polymerization process are offset by the complexity of the reaction mechanism [76]. Hollman and Arends [76] argue that even though catalysts are critical to the progression of reactions, which are unfeasible using conventional methods, there are practical constraints such as unfavorable steric interactions between the polyphenols and the active sites within the enzyme. Alternatively, the suitability of the enzymes in reaction processes could be impaired by unfavorable redox reactions. Since enzymes in isolation do not address the challenges and limitations associated with the synthesis of new products, alternative processes must be developed to improve the yields in enzyme-catalyzed polyphenol polymerization reactions.

Other experiments have reported the practical benefits of using alternative enzymes, primarily hydrolases, to catalyze bond cleavage reactions and the reverse action of hydrolysis [74]. Similar to laccase hydrolyses, they catalyze polyphenol polymerization via bond-forming reaction. However, hydrolyses are suitable in oxidative polymerization reactions that are specific to peroxidase and laccase catalysts in the presence of phenol derivatives. Other factors that inform the choice of enzymes for polyphenol polymerization include inhibition of laccase-initiated polymerization and peroxidase-initiated polymerization and the use of molecular oxygen in place of hydrogen peroxide [76]. The focus on molecular oxygen was validated by the following considerations.

Even though hydrolyses, laccase, and peroxidase polymers have proven to be useful in polyphenol polymerization, the concerns associated with polyphenol polymerization should be addressed. Various mechanisms were proposed by Hollman and Arends [76] to address the catalyst-related problems; these include the ingenious use of a mediator as a radical transfer catalyst. However, the radical is not incorporated into the end product. The feasibility of this reaction has been demonstrated in the use of phenothiazines to address constraints associated with the sterically hindered/congested phenols and consequently catalyze the polymerization process. In other cases, the production constraints can be offset by the incorporation of ABTS or transition metal ions, particularly $Mn^{2+/3+}$, to facilitate the pyrrole polymerization process. If transition metal ion species and ABTS prove less useful, polyoxometallates can be employed. The complexities of polyphenol catalysis underscore the challenges that impact the synthesis and application of sustainable and natural monomers and antioxidants in food packaging, biopharmaceuticals, and agricultural applications.

The synthesis-related constraints raise fundamental questions about the commercial exploitation of naturally occurring antioxidants such as tocotrienols and tocopherols [1–3], phenolic compounds, carotenoids [4]. Even though polymerization of natural polyphenols is impacted by the generation of formaldehyde byproducts, unfavorable steric interactions between the polyphenols and the active sites within the enzyme, inhibition of laccase-initiated polymerization, the process offers practical benefits relative to the use of artificial antioxidant molecules such as α-lipoic acid, N-acetyl cysteine, melatonin, gallic acid, captopril, taurine, catechin, and quercetin [5], or the replacement of catalysts with alternative processes such as amidification, esterification, free-radical grafting [73], step-growth, and free radical-catalyzed reactions to improve the percentage yield (see Table 6).

Table 6. Yields, MW, and PD from acrylamide and styrene polymerization as initiators [76].

	Initiator: Styrene Polymerization	
Yield %	MW [$\times 10^{-3}$ g·mol^{-1}]	PD
17	27	2.1
59	68	2.0
14	80	2.0
14	97	2.2
14	57	1.6
	Initiator: Acrylamide Polymerization	
Yield %	MW [$\times 10^{-3}$ g·mol^{-1}]	PD
93	124	2.5
84	56	2.9
76	5	4.4
72	27	3.3
78	85	2.7
38	10.5	3.9

The outcomes documented by Hollman and Arends [76] validate the polymerization of polyphenols, quercetin, and other organic molecules with suitable antioxidant properties. A critical challenge concerns the shortcomings and tradeoffs of existing polymerization processes.

6. Antioxidant Quercetin Polymers

The polymerization of quercetin is part of a broader effort in recent years to enhance the performance of natural and synthetic antioxidant polymers via covalent insertion of antioxidant species (such as vitamins, curcumin, and quercetin) into polymeric chains [77]. The polymerization products have demonstrated appropriate properties in food packaging (as preserving agents) [78], biomedical and pharmaceutical applications, especially in the production of antioxidant-laden nanoparticles for cancer treatment [77,79,80], or contact lenses and hemodialysis membranes. In both applications, the potency of the antioxidants is contingent on successful polymerization. A distinct advantage of bioconjugates is the exploitation of unique benefits of different bio-conjugates, the slow rates of macromolecular system degradation, and better chemical and cellular stability.

Quercetin is a plant-derived aglycone flavonoid with antioxidant properties that are suitable in the food packaging industry and the production of targeted therapies for cancer, heart, liver, and lung complications [78,80,81]. The primary sources include tea, red wine, and common fruits and vegetables such as onions, berries, apples, red grapes, broccoli, and cherries [80]. The antioxidant properties influence medical and nutritional applications as nutraceutical/nutritional supplement to boost the immune system and elevate protection against cardiovascular and lung conditions, osteoporosis, and tumor growth [80,81]; this is achieved through the neutralization of reactive oxygen species and the mediation of the transduction pathways, enzymatic activity, and function of glutathione. The specific mechanisms for the prevention of liver damage include the inhibition of the cellular activation of various signaling pathways, including NF-κB and MAPK [80]. Moreover, the antioxidant inhibits the release and expression of apoptosis-related proteins triggered by LPS/d-GalN, suppresses oxidation marker-mediated production of LPS/d-GalN. Following the inhibition of the mechanisms listed above, quercetin catalyzes the antioxidant signal transduction pathways (particularly Nrf2/GCL/GSH) and other processes that increase the concentration of glutathione in the cells [79].

The chemical structure and properties of quercetin contribute to its bioavailability and solubility in cellular fluids and moderation of biochemical pathways. However, the commercial exploitation of the antioxidant properties of quercetin depends on the success of the polymerization process, which can be tailored to generate bioactive compounds such as template quercetin (QCT) nanoparticles for free-radical scavenging. The cellular function of the QCT nanoparticles was demonstrated through material characterization using transmission electron microscopy (TEM), dynamic light scattering, H-NMR spectroscopy, Fourier-transform infrared spectroscopy, and UV-Visible spectroscopy [80]. The demonstrated biopharmaceutical application of oxidation-triggered self-polymerization quercetin in the production of targeted therapies by Xu et al. [80] is in line with the findings of Sunoqrot et al. [79]. The latter study documented the successful bio-inspired polymerization of quercetin—a process that resulted in the synthesis of cancer therapies. The nanotechnology-based quercetin–curcumin therapeutics exhibited superior cytotoxic behavior against cancer tumors. However, the superior biological function was dependent on precision in the synthetic process.

In the current case, the quercetin–curcumin-loaded nano-medicine was synthesized in the presence of curcumin and thiol-terminated poly (ethylene glycol) (PEG)-mediated surface functionalization; this was integral in facilitating steric stabilization in a single reaction step. The reactants were exposed to dimethyl sulfoxide [79]—a universal solvent to obtain a homogenous solution. The next polymerization steps entailed the gradual addition of water. The single-step synthesis process preferred by Sunoqrot et al. [79] was confirmed to be efficient by Pouci et al. [77], who noted that the one-step synthesis route was integral to the synthesis of polymer antioxidant conjugates (that are less susceptible to degradation) without emitting toxic byproducts.

Even though Sunoqrot et al. [79] documented successful synthesis and clinical application of quercetin–curcumin-loaded nanoparticles for targeted cancer drug synthesis, the approach does not align with Zhang et al.'s [81] research on the antioxidant properties of quercetin. In contrast to Sunoqrot et al. [79], Zhang et al. [81] advocated the use of quercetin based on its greater reduction potential relative to curcumin. The reduction potential helps to predict the total antioxidant capacity (TAC), which is fourfold higher for quercetin compared to curcumin. The unique chemical and biological properties help explain the cellular action against LPS-induced reactive oxygen species. Considering that experimental data predict clinical applications, further research is necessary to assess the effectiveness of quercetin only and quercetin–curcumin-loaded nanoparticles for targeted cancer tumor treatment. The need to compare antioxidants' potential benefits in combined or individual treatments is consistent with the research of Zhang et al. [81].

The biomedical applications of quercetin documented by Xu et al. [80] are but a microcosm of the potential industrial applications considering that quercetin is indispensable in the food production industry, where it is incorporated as an additive in bio-polyether (PEO), bio-polyester (PLA), and commercial starch-based polymer (Mater-Bio) [78]; this depends on the extent of photo-stabilization and artificial photo-stabilization. The oxidation-triggered self-polymerization is depicted in Figure 6. The illustration demonstrates the role of incubation, oxidizing agent, and universal organic solvents in the synthesis of nanoparticles.

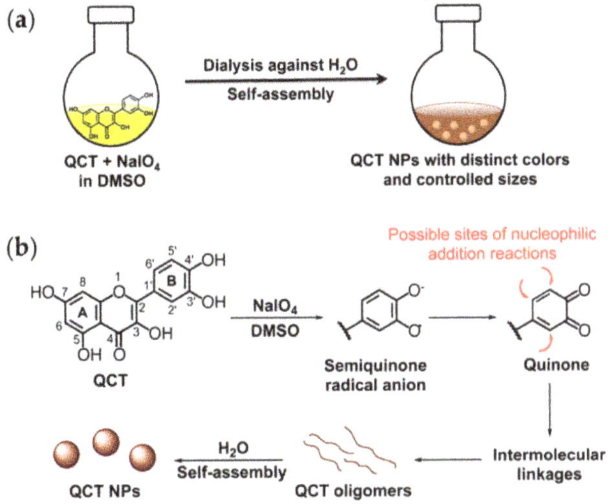

Figure 6. Synthesis process for template quercetin (QCT) nanoparticles [80] (**a**) Schematic of synthesis and (**b**) Potential chemical pathways of synthesis.

The impact of quercetin additives on the tensile strength, Young's Modulus, and elongation at break of bio-polyether, bio-polyester, and commercial starch-based polymer (Mater-Bio) is depicted in Table 7. The data demonstrate the influence of quercetin on stabilized and un-stabilized systems. The mechanical testing outcomes showed a significant improvement in the elastic modulus, tensile strength, and elongation at break after stabilization with quercetin additives (Q): Cyasorb® and synthetic Light Stabilizers (LS). In particular, a 5 wt% increase in elongation at break and the tensile strength resulted in a 10–20% improvement in the elastic modulus, tensile strength, and elongation at break. However, the microscale changes in the mechanical properties were attributed to different mechanisms.

Table 7. Changes in the mechanical properties (elastic modulus, tensile strength, and elongation at break) before and after stabilization with quercetin additives (Q): Cyasorb and synthetic Light Stabilizers (LS) [78].

Samples	E, MPa	TS, MPa	EB,%
PLA	1556 ±52	47.1 ± 2.5	9.3 ± 1.5
PLA/Q	1684 ±56	44.8 ± 2.2	14.8 ±2.0
PLA/LS	1843 ±62	51.2 ± 2.5	19.5 ±2.5
MB	105 ±5	24.2 ± 1.2	435 ±22
MB/Q	119 ±6	27.0 ± 1.3	470 ±24
MB/LS	120 ±6	24.6 ± 1.2	440 ±22
PEO	60.5 ± 2.3	3.5 ± 0.3	105 ± 5.0
PEO/Q	55.2 ± 2.5	3.1 ± 0.2	110 ± 6.0
PEO/LS	54.2 ± 2.3	3.2 ± 0.3	115 ± 7.5

On the one hand, the changes in the elongation at break of the stabilized bio-polyether, bio-polyester, and commercial starch-based polymer have been linked to the plasticizing effects triggered by the low MW stabilizing molecules such as quercetin [78]. On the other hand, the improvements in the tensile strength and elongation at break have been linked to higher rigidity of the stabilized films in relation to the machine direction [78]. The contrasting changes at the micro scale show that quercetin has unique effects on biofilms for food packaging—a phenomenon that introduces new challenges and benefits in the polymerization of quercetin for industrial applications. On a positive note, the higher orientation of the macromolecules in the presence of the plasticizing agent increases the suitability of quercetin in food packaging. It is deduced that other challenges could be addressed through innovations in polymerization.

The improvements in elastic modulus, tensile strength, and elongation at the break following stabilization with Q and LS by Morici et al. [78] are significant considering that the most common forms of quercetin polymerization are inspired by nature. The polymerization of polyquercetin offers further practical opportunities for developing customized therapeutics that influence glutathione (GSH) and enzymatic activity, moderate the signal transduction pathways, and diminish the availability of reactive oxygen species (ROS).

Polyquercetin and Quercetin Copolymers and Antioxidant Properties

Similar to quercetin, polyquercetin has superior antioxidant properties, depending on the synthetic route and ingredients. A facile and scalable micro-emulsion polymerization/crosslinking technique was developed by Sahiner [82], involved reacting quercetin molecules with glycerol diglycidyl ether (GDE) in the presence of l-α lecithin and cyclohexane, which act as the surfactant and organic phase, respectively. The process was proven to be useful in the synthesis of polyquercetin and quercetin copolymers with customizable antioxidant properties. In contrast to the methods proposed by Sahiner [82], Pivec et al. [83] explored an alternative technique comprising 0.5 g of flavonoid rutin (RU) hydrate in ultra-pure water. The mixture was subjected to enzymatic polymerization through the incorporation of Trametes Versicolor-based laccase (500 mg). The mixing was performed at room temperature, away from UV light for 24 h, after which ammonium sulfate was incorporated to enhance the precipitation process; this was followed by centrifugation, removal of the supernatant solution, and dialysis. A comparison of the two techniques shows that the microemulsion polymerization/crosslinking technique developed by Sahiner [82] was more facile and scalable compared to the lengthy procedures undertaken by Pivec et al. [83]. The focus on the synthetic route is grounded on the chemical structure–antioxidant activity relationship discussed in the preceding sections [77–79,82–85].

In contrast to quercetin molecules, which have found broad applications in targeted drug delivery and food packaging, polyquercetin has mainly been used as an electrode component for glass carbon electrodes modified by multi-walled carbon nanotubes; this was the case in studies conducted by Ziyatdinova et al. [84,85]. In both studies, the oxidation potentials of polyquercetin influenced the chronoamperometric determination of the antioxidant capacity and the quantification of gallic acid, catechin, and epigallocatechin gallate (EGCG) [84,85]. The variable industrial applications of quercetin and polyquercetin demonstrate the influence of polymerization and the choice of the crosslinking method.

7. Antioxidant Terpene Polymers

Similar to polyphenols, tocotrienols, tocopherols, quercetin, α-lipoic acid, N-acetyl cysteine, melatonin, polyquercetin, gallic acid, captopril, taurine, vitamins, and catechin terpenes are antioxidant polymers synthesized via natural and artificial polymerization processes [86–90]. However, the synthesis and commercial application of terpenes as antioxidants in food packaging and biomedical industries has attracted less research interest compared to polyphenols, vitamin C and D, and quercetin polymers—a phenomenon that is linked to the unique function and features of different terpenes compounds (D-limonene, β-ionone, geraniol, eugenol, myrcene) [91]. For example, each of the listed terpenes functions as an ideal insecticide (in citral mixtures) due to the antibacterial and antifungal properties (see Table 8); there are no notable side effects on human health given the diminished neurotoxicity risks.

Table 8. Classifications of terpenes and their potential application as pesticides [91].

Terpene Type	Function	Features
Limonene	Limonene enhances the properties of other terpenes.	Redistilled limonene has less odor, more stable than d-limonene
Beta-ionone	Antibacterial and antifungal properties	Beta-ionone has prophylactic value.
Geraniol	Like beta-ionone, antibacterial and antifungal properties.	Geraniol gives a pleasant fragrance.
Eugenol	Clove oil; anesthetic (anti-itch), antibacterial antifungal properties.	Contain a distinct fragrance which is like geraniol
Myrcene	Possesses antifungal, antibacterial properties	Famous for its fragrance properties

Despite the potential for application in various manufacturing applications, there has been minimal industry interest in terpene and terpene derivatives. Winnacker et al. [88] attributed this phenomenon to inadequate understanding of how terpenes could be combined with other natural and synthetic polymers to augment material characteristics (such as tensile strength, photo-degradation, elongation at break, antimicrobial properties, and biocompatibility) [88]. Moreover, even though the feedstock is cheap and widely available, there are gaps in knowledge concerning effective chemical processes to reduce material costs, and determining potential new applications for terpene derivatives and monomers remain underexplored [88].

The most widely applied procedures include radical initiator-free addition and solvation of thiols to terpenes such as R +/− and S−/− limonene and -β-pinene; chain and step-growth polymerization techniques involving acrylic monoterpenes myrcene (7-methyl-3-methylene-1,6-octadiene) and ocimenes and Bayer–Villiger oxidation of menthone into methide [89]. The application potential of terpenes includes the synthesis of terpene/fatty acid-based polyesters for industrial applications [89]. The reaction parameters, including conversion efficiency for polyester synthesis via copolymerization of propylene oxide (PO), are depicted in Table 9 [88].

Table 9. Mediated copolymerization of PO, time conversion, MW and Tg [88].

Entry Monomer	Cat.	[PO]/[2][b]	[ah-1 c]/[cat.-2][b]	Time [h]	Conversion[c] [%]	Mn[d] [kDa]	Mw/Mn[d]	cis[e] [%]	Tg[f] [8C]
1	cat.-2 a	500	100	0.75	86	11.4	1.25	>99	105
2	cat.-2 a	500	100	2.0	>99	18.8	1.66	52	91
3[g]	cat.-2 a	100	150	3.0	>99[h]	12.2	1.22	>99	102
4	cat.-2 b	500	100	2.0	76	9.0	1.18	>99	101
5	cat.-2 b	500	100	3.0	>99	14.6	1.72	10	78
6	cat.-2 c	500	100	1.0	63	7.4	1.11	>99	97
7	cat.-2 c	500	100	2.5	>99	10.2	1.16	97	103
8	cat.-2 c	500	100	5.0	>99	9.8	1.16	95	102
9	cat.-2 c	1000	200	4.0	>99	22.5	1.28	88	104
10	cat.-2 c	2000	400	9.0	>99	36.7	1.30	94	107
11	cat.-2 c	4000	800	20.5	>99	55.4	1.29	98	109

In other cases, terpenoid-based polymeric resins have been employed in stereolithographic 3D printing [87] and as starting materials in the production of sustainable polymers and advanced materials [88]. The renewed industry attention on terpenes and their derivatives is validated by the progress made in controlled polymerization, interesting structures, and the broad scope of industrial application, renewability, and abundant feedstock.

Polylimonine Synthesis and Properties

The synthesis of polylimonine has practical implications on the production of terpenes, given that the former classes of chemical compounds are optically active terpenes, which exhibit antioxidant behavior [92]. Commercial applications of limonene and polylimonene include food preservation, enhancing the flavor and aroma of foods [93]. A fundamental constraint in the development of polymeric products globally is the recyclability of the precursor and the ecological effects; this remains a priority concern in polylimonene synthesis, considering that most precursors are sourced from fossil fuels [94]. In light of this challenge, the synthesis of non-toxic and renewable plastics is a priority for manufacturers. Such plastics would be ideal substitutes for existing non-renewable polymers. Traditionally, ideal precursors for renewable polymers are starch, essential oils, and wood. However, there have been growing research interests in the production of renewable and non-toxic plastics from terpenes.

Terpenes are naturally found in citrus plants and essential oils. Other sources of terpenes include tea, thyme, cannabis, Spanish sage, and citrus fruits (lemon, orange, mandarin) [91]. Beyond the ubiquity of the precursors, terpenes (D-limonene in particular) are ideal based on their chemical structure and the presence of isoprene units; these units are key enablers for addition-based polymerization processes. The choice of D-limonene as a precursor in the current case is supported by the worldwide production capacity [94] and compatibility with microbial mediated synthesis processes, including PMD, Codon optimized MVK, and PMK (see Table 10).

Table 10. Microbial-based synthetic routes for polylimonene through microbial mediated synthesis processes [93].

Host	Engineering Design	Limonene Synthase Origin, (+/−)-Limonene, Accession Number
E. coli BLR (DE3) codon +	• Abies grandis tGPPS	*Mentha spicata*, (−), L13459
E. coli DH1 ΔacrAB	• HMGS and tHMGR from Staphylococcus aureus • Codon optimized MVK, PMK, and PMD from *S. cerevisiae* • AACT and IDI from *E. coli* • tGPPS from Abies grandis • One plasmid containing the mevalonate pathway genes and one plasmid with tGPPS and tLS • A. Borkumensis efflux pump	*M. spicata*, (−), L13459, codon-optimized
E. coli DH1	• HMGS and tHMGR from Staphylococcus aureus • Codon optimized MVK, PMK, and PMD from *S. cerevisiae* • AACT and IDI from *E. coli* • tGPPS from Abies grandis • One plasmid	*M. Spicata*, (−), accession number not clear, codon-optimized
E. coli BL21(DE3)	• HMGS and tHMGR, MVK, PMK, and PMD from *S. cerevisiae* • AACT and IDI from *E. coli* • tGPPS from Abies grandis/GPPS from Streptomyces sp. strain KO-3988	*M. Spicata*, (−), L13459, codon-optimized
Synechocystis sp. PCC 6803	• DXS, IDI, and CrtE from Synechocystis	*Schizonepeta tenuifolia*, enantio-selectivity not clear, AF282875
Synechococcus sp. PCC 7002	• Wild-type and ΔglgC background were compared	*M. Spicata*, (−), Q40322, codon-optimized
S. cerevisiae AE9	• Yeast FPPS (ERG20 K197G) mutated to partly produce GPP	*Perilla frutescens*, (−), KM015220 and Citrus limon, (+), AF514287
S. cerevisiae EPY210C	• tHMGR from *S. cerevisiae* • upc2–1 transcription factor	*M. spicata*, (−), L13459 and C. Limon, (+), AF514287, codon-optimized

The biotechnology-based production of limonene from microorganisms does not address production-related constraints, including the molar mass loss/lack of sufficient molar mass, selection of an acceptable monomer for conversion. Most of the precursors for terpenes have practical drawbacks and limitations—a factor influencing the choice of orange peel-based precursors—a precursor proven useful in the synthesis of polylimonene through recycling polymerization [94]. Other sources of terpenes include tea, thyme, cannabis, Spanish sage, and citrus fruits (lemon, orange, mandarin) [91]. Beyond the ubiquity of the precursors, terpenes (D-limonene in particular) are ideal.

Under standard conditions, Coelho and Vieira synthesized polylimonene by placing D-limonene in a solution containing Ziegler-Natta catalysts and Lewis acids in the first stage. The molar mass of the mixed species was below 1000 g/mol [94]. The second stage entailed the initiation of free-radical polymerization using benzoyl peroxide. The concentration of the free radical initiator is adjusted accordingly to attain the desired conversion efficiency. In the current case, the desired conversion efficiency was 12% [94]. The reduced copolymerization rates demonstrate there is an inverse link between the D-limonene ratio, molar masses, and concentration of styrene, methyl methacrylate, and n-butyl acrylate).

8. Antioxidant Random Copolymers

The synthesis of antioxidant copolymers such as dual-function amphiphilic random copolymers is critical to the suppression of medical-related infections through Fe^{2+} chelating ability and supra-macromolecular re-arrangement and morphological changes in hydrous environments and biocompatibility [95]. The polymerization of the tertiary amine-containing cationic monomer with hydroxyl-tyrosol yielded copolymers with ideal amphiphilic balances, molar ratios, and functional groups (especially the aromatic rings) for biomedical device coatings. In other cases, polypropylene (PP) antioxidant random copolymers have been proven useful in hot water applications in the presence of nucleating agents. Even though the utility of nucleating agents was not demonstrated by Taresco et al. [95], Grabmann et al. [96] argued that the compounds contributed to the modification of the crystalline structure—a factor that translated to higher thermal stability, mechanical strength (toughness and creep); this was observed in β-nucleated PP-R grades. In contrast, non-nucleated grades with α-crystal structure did not exhibit improvements in thermal and mechanical properties.

The utility of antioxidant random copolymers transcends biomedical devices to encompass food packaging applications, but the suitability of the materials for such applications is offset by the rapid depletion of antioxidants and chemical changes induced by microwave heating. Shahidi [97] observed that the viability of polymers for food preservation depends on the ability of natural antioxidants, additives, and preservatives to prevent rancidity—a process that impacts the odor and sensory appeal of foods.

Alin and Hakkarainen [98] explored the suitability of propylene-ethylene copolymer (PP-C), propylene-ethylene random copolymer (PP-R), and polypropylene homopolymer (PP) in plastic packaging containers labeled as microwave safe. The microwave testing of the polymers established the following. First, excessive heating (and subsequent analysis using microwave-assisted extraction (MAE) and HPLC) triggered the migration of the antioxidant molecules, particularly Irgafos 168 and 100 from the propylene-ethylene copolymer (PP-C), propylene-ethylene random copolymer (PP-R), and polypropylene homopolymer (PP) [98]. The migration resulted in unequal distribution of the antioxidants diminishing the utility of the polymer packaging materials in food packaging. A notable observation was the influence of polypropylene materials on the migration rate of the antioxidants and the inverse relationship with crystallinity. Better migration resistance was noted in the polypropylene homo-polymer (PP) potentially due to higher crystallinity [98]; this pattern remained unchanged despite contact with fatty acid stimulants. The influence of acetic acid and ethanol/iso-octane concentration on the rate of migration is depicted in Table 11.

Table 11. Antioxidant migration under microwaving and heating conditions.

	Microwaving		Heating	
Sample	I168	I1010	I168	I1010
		90/10 Iso-octane/ethanol, 80 °C		
PPC	509	502	686 ± 31 a	726 ± 64 a
PPR	4735	1283	4955 ± 204 a	1342 ± 217 a
PP	615	190	748 ± 85 a	277 ± 45 a
		Iso-octane, 80 °C		
PPC	-	-	777	765
PPR	-	-	4909	1477
PP	-	-	780	429
		3% acetic acid, 100 °C		
PPC	b	9	b	b
		3% acetic acid, 120 °C		
PPC	b	9	b	b
		10% ethanol, 120 °C		
PPC	b	6	b	b

In light of the preliminary observations made by Alin and Hakkarainen [98], polypropylene homo-polymer (PP) is an ideal alternative compared to propylene-ethylene copolymer (PP-C), propylene-ethylene random copolymer (PP-R), considering that the migration of antioxidants promotes toxicity and emergence of undesirable flavors [99]. A fundamental constraint moving forward was the development of packaging materials that are suitable and resistant to antioxidant migration, regardless of the fatty acid concentration in food. The experimental data shows that the composition of the food items, exposure to heat, and microwave energy predicts the chemical composition of antioxidant polymers for food packaging.

8.1. Gallic Acid Grafted Polymers

Similar to lignin, polyphenol, and quercetin polymers, gallic acid grafted polymers are ideal for targeted drug delivery, packaging, and commercial, industrial applications. Cho et al. [100] documented successful synthesis and characterization of gallic acid-grafted-chitosans, which demonstrated superior antioxidant scavenging ability against various free radicals such as hydrogen peroxide (93% effectiveness) and 2,2-diphenyl-1-picrylhydrazyl (92% effectiveness) at low concentrations (50 μg/mL) [101]. The outcomes affirm the superiority of gallic acid grafted copolymers in free radical scavenging. In other studies, gallic acid grafted chitosan-casein phosphopeptide nanoparticles have demonstrated superior anticancer and antioxidant properties under simulated gastrointestinal conditions [102]. The observations made by alternatively, the incorporation of chitosan on resveratrol via free radical-induced grafting proven effective in the synthesis of resveratrol modified species [73]. Cho et al. [100] contrast with Curcio et al. [79], who documented successful free radical-induced grafting of gallic acid-chitosan and catechin-chitosan conjugates for antioxidant applications.

Considering that the in-vitro models were confined to the two free radicals (2,2-diphenyl-1-picrylhydrazyl, and H_2O_2, it remains unclear whether gallic acid-grafted-chitosans would retain the high antioxidant scavenging ability at higher concentrations. The outcomes documented by Cho et al. are in agreement with Spizzirri et al., who documented successful synthesis of antioxidant–protein conjugates (gallic acid (GA) and catechin (CT) attached on gelatin) via grafting reaction [103]. Similar to gallic acid grafted polymers, the Gallic Acid-Catechin-Gelatin copolymers exhibited superior scavenging ability against linoleic acid peroxide hydroxyl radicals and 2,2′-diphenyl-1-picrylhydrazyl.

8.2. Resveratrol Conjugated and Grafted Polymers

Resveratrol exhibits comparable antioxidant activity as quercetin, polyphenols, Vitamin C, and E, phenolic compounds, carotenoids, α-lipoic acid, melatonin, gallic acid, and

N-acetyl cysteine, captopril. However, the material is superior to other antioxidants based on its broad biological activity, including anticancer, anti-inflammatory and chemoprevention properties, and rapid metabolism within the body due to its shortened half-life [104]. The mechanical, biological, and chemical properties can be enhanced through modifications in the synthetic routes. For example, Siddalingappa et al. [104] reported the successful synthesis of Resveratrol-PEG ethers via alkali-mediated etherification. The plasma stability of the moieties was augmented by aggregation behavior and polymer hydrophobicity, and the antioxidant demonstrated ideal antioxidant behavior in mice models. Further customized changes were viable through carbodiimide coupling reaction-mediated surface and chemical modifications with α-methoxy poly (ethylene glycol)-co-polylactide succinyl ester resveratrol, MeO-PEG succinyl ester resveratrol, and α-methoxy-ω-carboxylic acid poly(ethylene glycol) succinylamide resveratrol [104].

9. Antioxidant-Polysaccharides Graft Polymers

Polysaccharide graft polymers play an integral role in biomedical (targeted drug delivery systems) and commercial engineering applications given their unique properties and wide range of sources (microbial and natural). In contrast to other precursors, polysaccharides are cheap and widely available [105]. Moreover, the materials have comparable properties as antioxidant terpene polymers in terms of biodegradation ability, water solubility, and non-toxicity. Grafted polysaccharides have attracted significant research attention based on a broad scope of application, chemical stability, efficiency, biocompatibility, tailored properties, and biodegradation ability. Recent advances in drug delivery research have demonstrated that the rate of in vivo drug delivery could be predicted by the extent of crosslinking and grafting [105]. The enhanced functionality is linked to chemical bonding between the grafted polymer chains and polymeric substrate. The free radicals are introduced onto the backbone structure through irradiation and chemical initiation.

Despite the promising material prospects, the polysaccharide graft polymers have not conclusively addressed challenges associated with traditional drug delivery systems, such as sustaining the desired drug dosage in tissues for an extended period (length of treatment). However, there are promising prospects that the uncontrolled drug release kinetics would be resolved through better absorption rates [105]. The progress made in targeted drug delivery by Pal and Das [105] is in line with Lemarchand et al. [106] research on polyester-polysaccharide nanoparticles for drug delivery systems. In particular, the nanoparticles graft polymers from poly (D, L-lactide) or poly(ε–caprolactone) (PCL) demonstrated ideal pharmacokinetic behavior linked to the synthesis route—emulsion solvent evaporation. The material exhibited superior stability with or without emulsions, and strong amphiphilic properties, which eliminated the need for surfactants.

The emulsion solvent evaporation technique's ability to influence the targeted drug delivery is comparable to RAFT (reversible addition-fragmentation chain transfer polymerization), reversible deactivation radical polymerization, and atom transfer radical polymerization; these two procedures were employed by Garcia-Valdez in the modification of polysaccharides to attain ideal control over the molecular weight distribution, and carefully control the macromolecular shapes, and sizes [107]. From an engineering perspective, the experimental outcomes reported following the adoption of different synthetic routes introduce unique challenges because the outcomes were based on laboratory-scale models; there are concerns that the procedures could lead to different outcomes in large experiments. The highlighted concerns could be resolved through advances in synthesis technologies, including lab-on-a-chip.

10. Poly (Trolox Ester) Polymers

Poly Trolox ester polymers are useful antioxidants in human cells based on their ability to relieve oxidative stress in injured cells and resolve iron particle toxicity at the nanoscale [107,108]. Even though there is a broad class of compounds that exhibits antioxidant behavior (including α-lipoic acid, N-acetyl cysteine, melatonin, gallic acid, captopril, taurine, catechin, and quercetin), the biological function of most is impaired by oxidative stress-linked biological incompatibility at the nanoscale [108]. Wattamwar et al. claimed that poly Trolox ester polymers were able to overcome these barriers by inhibiting the response of free radicals while retaining higher compatibility with medical devices. In contrast to quercetin, lignin, and polydopamine, poly Trolox ester polymers are a class of synthetic analogs of vitamin E (with superior hydrophilic, tunable particle sizes, antioxidant properties) [108]. The materials can be synthesized via a facile single emulsion technique; the nanoparticle concentration is varied over time to achieve the desired nanoparticle sizes.

Resolving iron particle toxicity is a priority in biomedical applications, given that the risk is elevated by the frequency of medical diagnosis and treatment (iron supplements for anemia and MRI contrast agents). Iron particle toxicity was mitigated through targeted delivery of functionalized poly Trolox ester polymers to the platelet endothelial cell adhesion molecules (PECAM-1) [101]. In theory, the modification of the pro-oxidant and antioxidant activity of poly Trolox ester polymers could be applied in virtually all high-oxidative-stress environments within the cells. However, in reality, this might not be feasible considering that poly (Trolox ester) nanoparticles' ability to induce/inhibit cellular oxidative stress depends on advanced programming of the molecules, based on knowledge of the molecular compounds being oxidized and the cumulative oxidative status of the cells [6]. If the listed variables are not known, the cellular function of poly (Trolox ester) nanoparticles might be impaired. Such complexities limit the clinical application of Trolox esters.

11. Polymerization of Vitamins

Reversible addition-fragmentation chain transfer (RAFT) is suitable for the Vitamin C and B2 mediated polymerization of synthetic monomers such as methacrylates, acrylates, and acrylamides under LED irradiation. The efficiency of the polymerization process is contingent on the optimization of the irrational wavelength and oxygen. Zhang et al. linked the high throughput in vitamin-mediated polymerization to the low reaction volume platform, molecular weight distribution, low reaction volumes and the concentrations of Vitamin B_2 and C [109]. The costs of the synthetic process can be offset by the inclusion of off-the-shelf vitamin supplements in place of reagent/analytical grade Vitamin C and B_2. The free radical-initiated chain polymerization of biomolecules has critical implications on disease prevalence, given that the progression of Alzheimer's disease and other neurodegenerative diseases can be linked to free radical damage of cells. On a positive note, this process can be offset by anti-oxygenic nutrients, such as vitamin-rich (C, E and co-enzyme Q) fruits and vegetables, which offer protective benefits against free radical damage [110]. The polymerization of compounds with complex architectures has been augmented by advances in technology, particularly photo-induced electron/energy transfer–reversible addition-fragmentation chain-transfer (PET-RAFT) polymerization.

The PET-RAFT process is remarkably superior relative to RAFT in isolation due to higher oxygen tolerance [111]. However, other issues remain unresolved, including the experimental testing of oxygen tolerant species. Most of the documented experiments were conducted in water—a factor that limits the utility of the process in the presence of organic solvents. The initiation of vitamin polymerization in the presence of organic solvents is critical in biological processes, given that vitamins and biological molecules comprise organic functional groups. The constraint was partially resolved by Gormley et al. [109], who documented successful polymerization of oxygen resistant compounds in organic solvents, dimethyl sulfoxide (DMSO). Nonetheless, it is of note that the initiation of the process requires a zinc tetraphenyl porphyrin (ZnTPP) catalyst. From an industrial and commercial perspective, the additional costs linked to catalysis pale in comparison relative to the additional benefits associated with successful PET-RAFT in organic solvents. The findings documented by Zhang et al. [110], Tappel et al. [111], and Gormley et al. [109] demonstrate the excellent potential of oxygen tolerant PET-RAFT and RAFT vitamin-based photo-initiation mechanism in biological applications.

12. Conclusions

The published literature has demonstrated the antioxidant properties of various naturally occurring compounds, such as quercetin, polyphenols, vitamins, and synthetic molecules including α-lipoic acid, N-acetyl cysteine, melatonin, gallic acid, lignin, captopril, taurine, and catechin; the free radical scavenging abilities predicted the potency of the antioxidants (hydrogen bonding within the ether O_2 atoms in peroxide radical) and the location of bulky tert-butyl functional groups. The extraction of antioxidants from agro-wastes generates high-quality phenolic antioxidants. On the downside, the benefits are offset by low yields and negative ecological effects, as well as synthesis-related trade-offs between functionality, chemical stability, and biocompatibility. Other constraints are related to the chemical properties of the products—polyphenols exhibit poor membrane permeability, rapid metabolism, and poor bioavailability and UV degradation, despite free radical grafting [73], step growth, free radical, and enzyme-catalyzed reactions. A vast majority of research in antioxidant materials so far has been focused on the entrapment or encapsulation of low-molecular-weight and enzymatic antioxidant compounds in polymersomes, micelles, or nanoparticles that often exhibit poor long-term stability. The structural incorporation of antioxidant moieties in polymers or direct polymerization while maintaining their intrinsic antioxidant properties has been attracting increasing interest from the research community. Polymerized antioxidants can be produced through grafting or backbone incorporation, and offer the potential advantages of protecting the antioxidant from inactivation, delivering continuous antioxidant protection as long as the polymer exists, as well as a relatively high mass of antioxidant payload.

Potential industrial applications of antioxidants in the food production industry are diverse, especially in the forms of blends with bio-polyether (PEO), bio-polyester (PLA), and commercial starch-based polymer (Mater-Bi) for photo-stabilization. The blended products exhibit better tensile strength, photostability, and elongation at break and modulus of elasticity. Active polymer packaging is becoming increasingly important as an emerging technology [112] that can significantly improve the quality and stability of food, eliminating direct addition of chemicals and the need to make significant changes in production processes. If antioxidant polymers are brought into packaging systems, major improvements in maintaining the stability of oxidation-sensitive food products can be made. Antioxidant polymers have been proven to be scavengers that function by reducing the presence of reactive oxygen species, which act as initiators of oxidation processes. In this review, we have seen that antioxidant polymers can be specifically designed and optimized for each specific product or application and in brief, they can be implemented by the food industry as new emerging sustainable polymeric materials.

Author Contributions: C.M. writing—original draft preparation, I.S.B. review and editing, T.B. review and editing. All authors have read and agreed to the published version of the manuscript.

Funding: This research received no external funding.

Institutional Review Board Statement: Not applicable.

Informed Consent Statement: Not applicable.

Data Availability Statement: The data presented in this study are available on request from the corresponding author.

Conflicts of Interest: The authors declare no conflict of interest.

References

1. Sawant, R.; Vaze, O.; D'Souza, G.; Rockwell, K.; Torchilin, V. Palmitoyl ascorbate-loaded polymeric micelles: Cancer cell targeting and cytotoxicity. *Pharm Res.* **2011**, *28*, 301–308. [CrossRef]
2. Augustyniak, A.; Bartosz, G.; Čipak, A.; Duburs, G.; Horáková, L.; Łuczaj, W.; Žarković, N. Natural and synthetic antioxidants: An updated overview. *Free Radic. Res.* **2010**, *44*, 1216–1262. [CrossRef] [PubMed]
3. Pryor, W.A.; Strickland, T.; Church, D.F. Comparison of the efficiencies of several natural and synthetic antioxidants in aqueous SDS [sodium dodecyl sulfate] micelle solutions. *J. Am. Chem. Soc.* **1988**, *110*, 2224–2229. [CrossRef]

4. Lourenço, S.C.; Moldão-Martins, M.; Alves, V.D. Antioxidants of Natural Plant Origins: From Sources to Food Industry Applications. *Molecules* **2019**, *24*, 4132. [CrossRef]
5. Flora, S.; Shrivastava, R.; Mittal, M. Chemistry and Pharmacological Properties of Some Natural and Synthetic Antioxidants for Heavy Metal Toxicity. *Curr. Med. Chem.* **2013**, *20*, 4540–4574. [CrossRef] [PubMed]
6. Wattamwar, P.P.; Hardas, S.S.; Butterfield, D.A.; Anderson, K.W.; Dziubla, T.D. Tuning of the pro-oxidant and antioxidant activity of trolox through the controlled release from biodegradable poly(trolox ester) polymers. *J. Biomed. Mater Res. Part A* **2011**, *99*, 184–191. [CrossRef]
7. Cateto, C.A.; Barreiro, M.F.; Rodrigues, A.E.; Belgacem, M.N. Kinetic study of the formation of lignin-based polyurethanes in bulk. *React. Funct. Polym.* **2011**, *71*, 863–869. [CrossRef]
8. Franco, P.; Incarnato, L.; De Marco, I. Supercritical CO_2 impregnation of α-tocopherol into PET/PP films for active packaging applications. *J. CO_2 Util.* **2019**, *34*, 266–273. [CrossRef]
9. Lukic, I.; Vulic, J.; Ivanovic, J. Antioxidant activity of PLA/PCL films loaded with thymol and/or carvacrol using $scCO_2$ for active food packaging. *Food Packag. Shelf Life* **2020**, *26*, 100578. [CrossRef]
10. Trucillo, P.; Campardelli, R.; Reverchon, E. Production of liposomes loaded with antioxidants using a supercritical CO_2 assisted process. *Powder Technol.* **2018**, *323*, 155–162. [CrossRef]
11. Ozkan, G.; Franco, P.; Capanoglu, E.; De Marco, I. PVP/flavonoid coprecipitation by supercritical antisolvent process. *Chem. Eng. Process. Process. Intensif.* **2019**, *146*, 107689. [CrossRef]
12. Bastante, C.C.; Cardoso, L.C.; Serrano, C.M.; de la Ossa, E.M. Supercritical impregnation of food packaging films to provide antioxidant properties. *J. Supercrit. Fluids* **2017**, *128*, 200–207. [CrossRef]
13. Grossman, A.; Vermerris, W. Lignin-based polymers and nanomaterials. *Curr. Opin. Biotechnol.* **2019**, *56*, 112–120. [CrossRef] [PubMed]
14. Ponomarenko, J.; Lauberts, M.; Dizhbite, T.; Lauberte, L.; Jurkjane, V.; Telysheva, G. Antioxidant activity of various lignins and lignin-related phenylpropanoid units with high and low molecular weight. *Holzforschung* **2015**, *69*, 795–805. [CrossRef]
15. Vinardell, M.P.; Mitjans, M. Lignins and Their Derivatives with Beneficial Effects on Human Health. *Int. J. Mol. Sci.* **2017**, *18*, 1219. [CrossRef]
16. Liu, Q.; Luo, L.; Zheng, L. Lignins: Biosynthesis and Biological Functions in Plants. *Int. J. Mol. Sci.* **2018**, *19*, 335. [CrossRef]
17. Abhilash, M.; Thomas, D. 15-Biopolymers for Biocomposites and Chemical Sensor Applications. In *Biopolymer Composites in Electronics*; Sadasivuni, K.K., Ponnamma, D., Kim, J., Cabibihan, J.-J., Eds.; AlMaadeed MABT-BC in Elsevier: Amsterdam, The Netherlands, 2017; pp. 405–435. [CrossRef]
18. Mena, P.; Calani, L.; Bruni, R.; Del Rio, D. Chapter 6-Bioactivation of High-Molecular-Weight Polyphenols by the Gut Microbiome. In *Diet-Microbe Interactions in the Gut*; Tuohy, K., Ed.; Academic Press: San Diego, CA, USA, 2015; pp. 73–101. [CrossRef]
19. Pizzorno, J.; Murray, M.T. *Textbook of Natural Medicine*, 4th ed.; Churchill Livingstone: London, UK, 2012; Available online: https://www.elsevier.com/books/textbook-of-natural-medicine/pizzorno/978-1-4377-2333-5 (accessed on 25 July 2021).
20. Gunawardena, D.; Govindaraghavan, S.; Münch, G. Anti-Inflammatory Properties of Cinnamon Polyphenols and their Monomeric Precursors. *Polyphen. Hum. Health Dis.* **2013**, *1*, 409–425. [CrossRef]
21. Gonçalves, C.; Dinis, T.; Batista, M.T. Antioxidant properties of proanthocyanidins of Uncaria tomentosa bark decoction: A mechanism for anti-inflammatory activity. *Phytochemistry* **2005**, *66*, 89–98. [CrossRef]
22. Chai, W.-M.; Ou-Yang, C.; Huang, Q.; Lin, M.-Z.; Wang, Y.-X.; Xu, K.-L.; Pang, D.D. Antityrosinase and antioxidant properties of mung bean seed proanthocyanidins: Novel insights into the inhibitory mechanism. *Food Chem.* **2018**, *260*, 27–36. [CrossRef]
23. Martinez-Avila, G.C.G.; Aguilera, A.F.; Saucedo, S.; Rojas, R.; Rodriguez, R.; Aguilar, C.N. Fruit Wastes Fermentation for Phenolic Antioxidants Production and Their Application in Manufacture of Edible Coatings and Films. *Crit. Rev. Food Sci. Nutr.* **2014**, *54*, 303–311. [CrossRef]
24. Iyer, K.A.; Zhang, L.; Torkelson, J.M. Direct Use of Natural Antioxidant-rich Agro-wastes as Thermal Stabilizer for Polymer: Processing and Recycling. *ACS Sustain. Chem. Eng.* **2016**, *4*, 881–889. [CrossRef]
25. Kricheldorf, H.R.; Kreiser-Saunders, I. Polymerizations of Lactide. *Polymer* **1994**, *35*, 4175–4180. [CrossRef]
26. Wang, C.E.; Stayton, P.S.; Pun, S.H.; Convertine, A.J. Polymer nanostructures synthesized by controlled living polymerization for tumor-targeted drug delivery. *J. Control. Release* **2015**, *219*, 345–354. [CrossRef]
27. Coulembier, O.; Degée, P.; Hedrick, J.L.; Dubois, P. From controlled ring-opening polymerization to biodegradable aliphatic polyester: Especially poly(β-malic acid) derivatives. *Prog. Polym. Sci.* **2006**, *31*, 723–747. [CrossRef]
28. Garbetta, A.D.; Antuono, I.; Melilli, M.G.; Sillitti, C.; Linsalata, V.; Scandurra, S. Inulin enriched durum wheat spaghetti: Effect of polymerization degree on technological and nutritional characteristics. *J. Funct. Foods* **2020**, *71*, 104004. [CrossRef]
29. Nanni, G.; Heredia-Guerrero, J.A.; Paul, U.C.; Dante, S.; Caputo, G.; Canale, C.; Athanassiou, A.; Fragouli, D.; Bayer, I.S. Poly (furfuryl alcohol)-polycaprolactone blends. *Polymers* **2019**, *11*, 1069. [CrossRef] [PubMed]
30. Kirschweng, B.; Tátraaljai, D.; Földes, E.; Pukánszky, B. Natural antioxidants as stabilizers for polymers. *Polym. Degrad. Stab.* **2017**, *145*, 25–40. [CrossRef]
31. Cirillo, G.; Iemma, F. *Antioxidant Polymers: Synthesis, Properties, and Applications*; John Wiley & Sons: Hoboken, NJ, USA, 2012.
32. Van Lith, R.; Ameer, G.A. Antioxidant Polymers as Biomaterial. In *Oxidative Stress Biomater*; Elsevier: Amsterdam, The Netherlands, 2016; pp. 251–296.

33. Quilez-Molina, A.I.; Heredia-Guerrero, J.A.; Armirotti, A.; Paul, U.C.; Athanassiou, A.; Bayer, I.S. Comparison of physicochemical, mechanical and antioxidant properties of polyvinyl alcohol films containing green tealeaves waste extracts and discarded balsamic vinegar. *Food Packag. Shelf Life* **2020**, *23*, 100445. [CrossRef]
34. Brocca, D.; Arvin, E.; Mosbæk, H. Identification of organic compounds migrating from polyethylene pipelines into drinking water. *Water Res.* **2002**, *36*, 3675–3680. [CrossRef]
35. Mahmood, Z.; Yameen, M.; Jahangeer, M.; Riaz, M.; Ghaffar, A.; Javid, I. *Lignin as Natural Antioxidant Capacity. Lignin-Trends Appl*; Intechopen: Budapest, Hungary, 2017.
36. Rahman, M.M.; Tsukamoto, J.; Rahman, M.; Yoneyama, A.; Mostafa, K.M. Lignin and its effects on litter decomposition in forest ecosystems. *Chem. Ecol.* **2013**, 540–553. [CrossRef]
37. Llerena, J.P.P.; Figueiredo, R.; Brito, M.S.; Kiyota, E.; Mayer, J.L.S.; Araujo, P. Deposition of lignin in four species of Saccharum. *Sci. Rep.* **2019**, *9*, 5877. [CrossRef]
38. Latif, N.H.A.; Rahim, A.A.; Brosse, N.; Hussin, M.H. The structural characterization and antioxidant properties of oil palm fronds lignin incorporated with p-hydroxyacetophenone. *Int. J. Biol. Macromol.* **2019**, *130*, 947–957. [CrossRef]
39. Kabir, A.S. Effects of Lignin as a Stabilizer or Antioxidant in Polyolefins. Electronic Thesis and Dissertation Repository. 2017, p. 4796. Available online: https://ir.lib.uwo.ca/etd/4796 (accessed on 25 July 2021).
40. Saito, T.; Perkins, J.H.; Jackson, D.C.; Trammel, N.E.; Hunt, M.A.; Naskar, A.K. Development of lignin-based polyurethane thermoplastics. *RSC Adv.* **2013**, *3*, 21832–21840. [CrossRef]
41. Cui, S.; Borgemenke, J.; Liu, Z.; Li, Y. Recent advances of "soft" bio-polycarbonate plastics from carbon dioxide and renewable bio-feedstocks via straightforward and innovative routes. *J. CO_2 Util.* **2019**, *34*, 40–52. [CrossRef]
42. Turner, D.A.; Williams, I.D.; Kemp, S. Resources, Conservation and Recycling Greenhouse gas emission factors for recycling of source-segregated waste materials. *Resour. Conserv. Recycl.* **2015**, *105*, 186–197. [CrossRef]
43. Zheng, J.; Suh, S. Strategies to reduce the global carbon footprint of plastics. *Nat. Clim. Chang.* **2019**, *9*. [CrossRef]
44. Nisticò, R. Aquatic-derived biomaterials for a sustainable future: A European opportunity. *Resources* **2017**, *6*, 65. [CrossRef]
45. Fernandez, J.G.; Ingber, D.E. Manufacturing of large-scale functional objects using biodegradable chitosan bioplastic. *Macromol. Mater. Eng.* **2014**, *299*, 932–938. [CrossRef]
46. Lackner, M. Bioplastics-Biobased plastics as renewable and/or biodegradable alternatives to petroplastics. In *Encyclopedia of Chemical Technology*; Othmer, K., Ed.; Wiley Publishing, Inc.: Hoboken, NJ, USA, 2015.
47. Sapei, L.; Padmawijaya, K.S.; Sijayanti, O.; Wardhana, P.J. The effect of banana starch concentration on the properties of chitosan-starch bioplastics. *Mater Sci.* **2015**, *7*, 101–105.
48. Liu, H.; Chung, H. Lignin-Based Polymers via Graft Copolymerization. *J. Polym. Sci. A Polym. Chem.* **2017**, *55*, 3515–3528. [CrossRef]
49. Boyer, C.; Bulmus, V.; Davis, T.P.; Ladmiral, V.; Liu, J.; Perrier, S. Bioapplications of RAFT polymerization. *Chem. Rev.* **2009**, *109*, 5402–5436. [CrossRef] [PubMed]
50. Lai, J.T.; Filla, D.; Shea, R. Functional Polymers from Novel Carboxyl-Terminated Trithiocarbonates as Highly Efficient RAFT Agents. *Macromolecules* **2002**, *35*, 6754–6756. [CrossRef]
51. da Silva, E.B.; Zabkova, M.; Araújo, J.; Cateto, C.; Barreiro, M.; Belgacem, M.; Rodrigues, A. An integrated process to produce vanillin and lignin-based polyurethanes from Kraft lignin. *Chem. Eng. Res. Des.* **2009**, *87*, 1276–1292. [CrossRef]
52. Qiu, Y.; Zhu, Z.; Miao, Y.; Zhang, P.; Jia, X.; Liu, Z.; Zhao, X. Polymerization of dopamine accompanying its coupling to induce self-assembly of block copolymer and application in drug delivery. *Polym. Chem.* **2020**, *11*, 2811–2821. [CrossRef]
53. Kwon, I.S.; Bettinger, C.J. Polydopamine nanostructures as biomaterials for medical applications. *J. Mater. Chem. B* **2018**, *6*, 6895–6903. [CrossRef]
54. Fichman, G.; Schneider, J. Dopamine Self-Polymerization as a Simple and Powerful Tool to Modulate the Viscoelastic Mechanical Properties of Peptide-Based Gels. *Molecules* **2021**, *26*, 1363. [CrossRef] [PubMed]
55. Barclay, T.G.; Hegab, H.M.; Clarke, S.R.; Ginic-Markovic, M. Versatile Surface Modification Using Polydopamine and Related Polycatecholamines: Chemistry, Structure, and Applications. *Adv. Mater. Interfaces* **2017**, *4*. [CrossRef]
56. Lakshminarayanan, R.; Madhavi, S.; Sim, C. Oxidative Polymerization of Dopamine: A High-Definition Multifunctional Coatings for Electrospun Nanofibers-An Overview. In *Dopamine-Health and Disease*; IntechOpen: London, UK, 2018. [CrossRef]
57. Du, X.; Li, L.; Behboodi-Sadabad, F.; Welle, A.; Li, J.; Heissler, S.; Zhang, H.; Plumeré, N.; Levkin, P.A. Bio-inspired strategy for controlled dopamine polymerization in basic solutions. *Polym. Chem.* **2017**, *8*, 2145–2151. [CrossRef]
58. Xiao, C.; Chu, X.; Yang, Y.; Li, X.; Zhang, X.; Chen, J. Hollow nitrogen-doped carbon microspheres pyrolyzed from self-polymerized dopamine and its application in simultaneous electrochemical determination of uric acid, ascorbic acid and dopamine. *Biosens. Bioelectron.* **2011**, *26*, 2934–2939. [CrossRef]
59. Kalimuthu, P.; John, A. Simultaneous determination of ascorbic acid, dopamine, uric acid and xanthine using a nanostructured polymer film modified electrode. *Talanta* **2010**, *80*, 1686–1691. [CrossRef]
60. Kohri, M.; Nannichi, Y.; Kohma, H.; Abe, D.; Kojima, T.; Taniguchi, T.; Kishikawa, K. Size control of polydopamine nodules formed on polystyrene particles during dopamine polymerization with carboxylic acid-containing compounds for the fabrication of raspberry-like particles. *Colloids Surf. A Physicochem. Eng. Asp.* **2014**, *449*, 114–120. [CrossRef]
61. Chen, T.-P.; Liu, T.; Su, T.-L.; Liang, J. Self-Polymerization of Dopamine in Acidic Environments without Oxygen. *Langmuir* **2017**, *33*, 5863–5871. [CrossRef] [PubMed]
62. Mrówczyński, R.; Markiewicz, R.; Liebscher, J. Chemistry of polydopamine analogues. *Polym. Int.* **2016**, *65*, 1288–1299. [CrossRef]

63. Aidoo, R.P.; Depypere, F.; Afoakwa, E.O.; Dewettinck, K. Industrial manufacture of sugar-free chocolates–Applicability of alternative sweeteners and carbohydrate polymers as raw materials in product development. *Trends Food Sci. Technol.* **2013**, *32*, 84–96. [CrossRef]
64. Guimarães, J.T.; Silva, E.K.; Lelis, A.L.R.C.; Cunha, R.; Freitas, M.Q.; Meireles, M.A.A.; da Cruz, A.G. Manufacturing a prebiotic whey beverage exploring the influence of degree of inulin polymerization. *Food Hydrocoll.* **2018**, *77*, 787–795. [CrossRef]
65. Karimi, R.; Azizi, M.H.; Ghasemlou, M.; Vaziri, M. Application of inulin in cheese as prebiotic, fat replacer and texturizer: A review. *Carbohydr. Polym.* **2015**, *119*, 85–100. [CrossRef]
66. Olkku, J.; Rha, C. Gelatinisation of starch and wheat flour starch—A review. *Food Chem.* **1978**, *3*, 293–317. [CrossRef]
67. Vervoort, L.; Mooter, G.V.D.; Augustijns, P.; Busson, R.; Toppet, S.; Kinget, R. Inulin hydrogels as carriers for colonic drug targeting: I. Synthesis and characterization of methacrylated inulin and hydrogel formation. *Pharm. Res.* **1997**, *14*, 1730–1737. [CrossRef]
68. Kumar, S.; Kesharwani, S.S.; Kuppast, B.; Bakkari, M.A.; Tummala, H. Pathogen-mimicking vaccine delivery system designed with a bioactive polymer (inulin acetate) for robust humoral and cellular immune responses. *J. Control. Release* **2017**, *261*, 263–274. [CrossRef]
69. Ito, H.; Wada, T.; Ohguchi, M.; Sugiyama, K.; Kiriyama, S.; Morita, T. The Degree of Polymerization of Inulin-Like Fructans Affects Cecal Mucin and Immunoglobulin A in Rats. *J. Food Sci.* **2008**, *73*, H36–H41. [CrossRef]
70. Boets, E.; Deroover, L.; Houben, E.; Vermeulen, K.; Gomand, S.V.; Delcour, J.A.; Verbeke, K. Quantification of In Vivo Colonic Short Chain Fatty Acid Production from Inulin. *Nutrients* **2015**, *7*, 8916–8929. [CrossRef]
71. Morrison, D.J.; Preston, T. Formation of short chain fatty acids by the gut microbiota and their impact on human metabolism. *Gut Microbes* **2016**, *7*, 189–200. [CrossRef]
72. Oliver, S.; Vittorio, O.; Cirillo, G.; Boyer, C. Enhancing the therapeutic effects of polyphenols with macromolecules. *Polym. Chem.* **2016**, *7*, 1529–1544. [CrossRef]
73. Curcio, M.; Puoci, F.; Iemma, F.; Parisi, O.I.; Cirillo, G.; Spizzirri, U.G.; Picci, N. Covalent Insertion of Antioxidant Molecules on Chitosan by a Free Radical Grafting Procedure. *J. Agric. Food Chem.* **2009**, *57*, 5933–5938. [CrossRef] [PubMed]
74. Uyama, H.; Kobayashi, S. Enzyme-catalyzed polymerization to functional polymers. *J. Mol. Catal. B Enzym.* **2002**, *19–20*, 117–127. [CrossRef]
75. Sun, X.; Bai, R.; Zhang, Y.; Wang, Q.; Fan, X.; Yuan, J.; Cui, L.; Wang, P. Laccase-Catalyzed Oxidative Polymerization of Phenolic Compounds. *Appl. Biochem. Biotechnol.* **2013**, *171*, 1673–1680. [CrossRef]
76. Hollmann, F.; Arends, I.W.C.E. Enzyme Initiated Radical Polymerizations. *Polymers* **2012**, *4*, 759–793. [CrossRef]
77. Puoci, F.; Morelli, C.; Cirillo, G.; Curcio, M.; Parisi, O.I.; Maris, P.; Sisci, D.; Picci, N. Anticancer activity of a quercetin-based polymer towards HeLa cancer cells. *Anticancer Res.* **2012**, *32*, 2843–2847. [PubMed]
78. Morici, E.; Arrigo, R.; Dintcheva, N.T. Quercetin as natural stabilizing agent for bio-polymer. *AIP Conf. Proc.* **2014**, *1599*, 314–317. [CrossRef]
79. Sunoqrot, S.; Al-Debsi, T.; Al-Shalabi, E.; Ibrahim, L.H.; Faruqu, F.N.; Walters, A.; Palgrave, R.G.; Al-Jamal, K.T. Bioinspired Polymerization of Quercetin to Produce a Curcumin-Loaded Nanomedicine with Potent Cytotoxicity and Cancer-Targeting Potential in Vivo. *ACS Biomater. Sci. Eng.* **2019**, *5*, 6036–6045. [CrossRef]
80. Xu, D.; Hu, M.-J.; Wang, Y.-Q.; Cui, Y.-L. Antioxidant Activities of Quercetin and Its Complexes for Medicinal Application. *Molecules* **2019**, *24*, 1123. [CrossRef]
81. Zhang, M.; Swarts, S.G.; Yin, L.; Liu, C.; Tian, Y.; Cao, Y.; Swarts, M.; Yang, S.; Zhang, S.B.; Zhang, K.; et al. Antioxidant Properties of Quercetin. *Adv. Exp. Med. Biol.* **2011**, *701*, 283–289. [CrossRef]
82. Sahiner, N. One step poly(quercetin) particle preparation as biocolloid and its characterization. *Colloids Surf. A Physicochem. Eng. Asp.* **2014**, *452*, 173–180. [CrossRef]
83. Pivec, T.; Kargl, R.; Maver, U.; Bračič, M.; Elschner, T.; Žagar, E.; Gradišnik, L.; Kleinschek, K.S. Chemical Structure–Antioxidant Activity Relationship of Water–Based Enzymatic Polymerized Rutin and Its Wound Healing Potential. *Polymers* **2019**, *11*, 1566. [CrossRef] [PubMed]
84. Ziyatdinova, G.; Kozlova, E.; Budnikov, H. Polyquercetin/MWNT-modified Electrode for the Determination of Natural Phenolic Antioxidants. *Electroanalysis* **2017**, *29*, 2610–2619. [CrossRef]
85. Ziyatdinova, G.K.; Kozlova, E.; Budnikov, H.C. Chronoamperometric evaluation of the antioxidant capacity of tea on a polyquercetin-modified electrode. *J. Anal. Chem.* **2017**, *72*, 382–389. [CrossRef]
86. Firdaus, M.; Espinosa, L.M.; De Meier, M.A.R. Terpene-Based Renewable Monomers and Polymers via Thiol À Ene Additions. *Macromolecules* **2011**, *44*, 7253–7262. [CrossRef]
87. Weems, A.C.; Chiaie, K.R.D.; Worch, J.; Stubbs, C.J.; Dove, A.P. Terpene- and terpenoid-based polymeric resins for stereolithography 3D printing. *Polym. Chem.* **2019**, *10*, 5959–5966. [CrossRef]
88. Winnacker, M.; Rieger, B. Recent Progress in Sustainable Polymers Obtained from Cyclic Terpenes: Synthesis, Properties, and Application Potential. *ChemSusChem* **2015**, *8*, 2455–2471. [CrossRef]
89. Singh, A.; Kamal, M. Synthesis and characterization of polylimonene: Polymer of an optically active terpene. *J. Appl. Polym. Sci.* **2012**, *125*, 1456–1459. [CrossRef]
90. Zhao, J.; Schlaad, H. Synthesis of Terpene-Based Polymers. In *Biosynthetic-Synthetic Polymer Conjugates*; Schlaad, H., Ed.; Springer: Berlin/Heidelberg, Germany, 2011.

91. Cox-Georgian, D.; Ramadoss, N.; Dona, C.; Basu, C. Therapeutic and Medicinal Uses of Terpenes. *Med. Plants* **2019**, *2019*, 333–359. [CrossRef]
92. Yang, Z.; Peng, H.; Wang, W.; Liu, T. Crystallization behavior of poly(ε-caprolactone)/layered double hydroxide nanocomposites. *J. Appl. Polym. Sci.* **2010**, *116*, 2658–2667. [CrossRef]
93. Jongedijk, E.; Cankar, K.; Buchhaupt, M.; Schrader, J.; Bouwmeester, H.; Beekwilder, J. Biotechnological production of limonene in microorganisms. *Appl. Microbiol. Biotechnol.* **2016**, *100*, 2927–2938. [CrossRef]
94. Coelho, F.M.; Vieira, R.P. Synthesis of Renewable Poly(limonene): A Kinetic Modeling Study to Improve the Polymerization. *Braz. Arch. Biol. Technol.* **2020**, *63*, 1–10. [CrossRef]
95. Taresco, V.; Crisante, F.; Francolini, I.; Martinelli, A.; D'Ilario, L.; Vitiani, L.R.; Buccarelli, M.; Pietrelli, L.; Piozzi, A. Antimicrobial and antioxidant amphiphilic random copolymers to address medical device-centered infections. *Acta Biomater.* **2015**, *22*, 131–140. [CrossRef] [PubMed]
96. Grabmann, M.K.; Wallner, G.M.; Maringer, L.; Buchberger, W.; Nitsche, D. Hot air aging behavior of polypropylene random copolymers. *J. Appl. Polym. Sci.* **2019**, *136*. [CrossRef]
97. Shahidi, F. *Handbook of Antioxidants for Food Preservation*; Elsevier B.V: Amsterdam, The Netherlands, 2015.
98. Alin, J.; Hakkarainen, M. Type of Polypropylene Material Significantly Influences the Migration of Antioxidants from Polymer Packaging to Food Simulants During Microwave Heating. *J. Appl. Polym. Sci.* **2010**, *118*, 1084–1093. [CrossRef]
99. Alin, J.; Hakkarainen, M. The significant effect of polypropylene material on the migration of antioxidants from food con-tainer to food simulants. In Proceedings of the 43rd IUPAC World Polymer Congress, Glasgow, UK, 11–16 July 2010.
100. Cho, Y.-S.; Kim, S.-K.; Ahn, C.-B.; Je, J.-Y. Preparation, characterization, and antioxidant properties of gallic acid-grafted-chitosans. *Carbohydr. Polym.* **2011**, *83*, 1617–1622. [CrossRef]
101. Hu, B.; Wang, Y.; Xie, M.; Hu, G.; Ma, F.; Zeng, X. Polymer nanoparticles composed with gallic acid grafted chitosan and bioactive peptides combined antioxidant, anticancer activities and improved delivery property for labile polyphenols. *J. Funct. Foods* **2015**, *15*, 593–603. [CrossRef]
102. Spizzirri, U.G.; Iemma, F.; Puoci, F.; Cirillo, G.; Curcio, M.; Parisi, O.I. Synthesis of antioxidant polymers by grafting of gallic acid and catechin on gelatin. *Biomacromolecules* **2009**, *10*, 1923–1930. [CrossRef]
103. Siddalingappa, B.; Benson, H.A.E.; Brown, D.H.; Batty, K.T.; Chen, Y. Stabilization of Resveratrol in Blood Circulation by Conjugation to mPEG and mPEG-PLA Polymers: Investigation of Conjugate Linker and Polymer Composition on Stability, Metabolism, Antioxidant Activity and Pharmacokinetic Profile. *PLoS ONE* **2015**, *10*, e0118824. [CrossRef] [PubMed]
104. Pal, S.; Das, R. Polysaccharide-Based Graft Copolymers for Biomedical Applications. *Polysacch. Based Graft Copolym.* **2013**, 325–345. [CrossRef]
105. LeMarchand, C.; Couvreur, P.; Besnard, M.; Costantini, D.; Gref, R. Novel Polyester-Polysaccharide Nanoparticles. *Pharm. Res.* **2003**, *20*, 1284–1292. [CrossRef]
106. Garcia-Valdez, O.; Champagne, P.; Cunningham, M. Graft modification of natural polysaccharides via reversible deactivation radical polymerization. *Prog. Polym. Sci.* **2018**, *76*, 151–173. [CrossRef]
107. Cochran, D.B.; Wattamwar, P.P.; Wydra, R.; Hilt, J.Z.; Anderson, K.W.; Eitel, R.E.; Dziubla, T.D. Suppressing iron oxide nanoparticle toxicity by vascular targeted antioxidant polymer nanoparticles. *Biomaterials* **2013**, *34*, 9615–9622. [CrossRef]
108. Wattamwar, P.P.; Mo, Y.; Wan, R.; Palli, R.; Zhang, Q.; Dziubla, T.D. Antioxidant Activity of Degradable Polymer Poly(trolox ester) to Suppress Oxidative Stress Injury in the Cells. *Adv. Funct. Mater.* **2010**, *20*, 147–154. [CrossRef]
109. Gormley, A.J.; Yeow, J.; Ng, G.; Conway, Ó.; Boyer, C.; Chapman, R. An Oxygen-Tolerant PET-RAFT Polymerization for Screening Structure–Activity Relationships. *Angew. Chem. Int. Ed.* **2018**, *57*, 1557–1562. [CrossRef]
110. Zhang, T.; Yeow, J.; Boyer, C. A cocktail of vitamins for aqueous RAFT polymerization in an open-to-air microtiter plate. *Polym. Chem.* **2019**, *10*, 4643–4654. [CrossRef]
111. Tappel, A.; Tappel, A. Oxidant free radical initiated chain polymerization of protein and other biomolecules and its relationship to diseases. *Med. Hypotheses* **2004**, *63*, 98–99. [CrossRef] [PubMed]
112. Tran, T.N.; Heredia-Guerrero, J.A.; Mai, B.T.; Ceseracciu, L.; Marini, L.; Athanassiou, A.; Bayer, I.S. Bioelastomers based on cocoa shell waste with antioxidant ability. *Adv. Sustain. Syst.* **2017**, *1*, 1700002. [CrossRef]

polymers MDPI

Article

An Innovative Computational Strategy to Optimize Different Furnish Compositions of Tissue Materials Using Micro/Nanofibrillated Cellulose and Biopolymer as Additives

Flávia P. Morais [1,*], Ana M. M. S. Carta [2], Maria E. Amaral [1] and Joana M. R. Curto [1,3,*]

1 Fiber Materials and Environmental Technologies Research Unit (FibEnTech-UBI), University of Beira Interior, Rua Marquês d'Ávila e Bolama, 6201-001 Covilhã, Portugal; mecca@ubi.pt
2 Forest and Paper Research Institute (RAIZ), R. José Estevão, 3800-783 Eixo, Portugal; ana.carta@thenavigatorcompany.com
3 Chemical Process Engineering and Forest Products Research Centre (CIEPQPF), University of Coimbra, R. Sílvio Lima, Polo II, 3004-531 Coimbra, Portugal
* Correspondence: flavia.morais@ubi.pt (F.P.M.); jmrc@ubi.pt (J.M.R.C.)

Abstract: The furnish management of tissue materials is fundamental to obtain maximum quality products with a minimum cost. The key fiber properties and fiber modification process steps have a significant influence on the structural and functional properties of tissue paper. In this work, two types of additives, a commercial biopolymer additive (CBA) that replaces the traditional cationic starch and micro/nanofibrillated cellulose (CMF), were investigated. Different formulations were prepared containing eucalyptus fibers and softwood fibers treated mechanically and enzymatically and both pulps with these two additives incorporated independently and simultaneously with drainage in the tissue process range. The use of these additives to reduce the percentage of softwood fibers on tissue furnish formulations was investigated. The results indicated that a maximum of tensile strength was obtained with a combination of both additives at the expense of softness and water absorbency. With a reduction of softwood fibers, the incorporation of additives increased the tensile strength and water absorbency with a slight decrease in HF softness compared with a typical industrial furnish. Additionally, a tissue computational simulator was also used to predict the influence of these additives on the final end-use properties. Both additives proved to be a suitable alternative to reduce softwood fibers in the production of tissue products, enhancing softness, strength and absorption properties.

Keywords: absorbency; commercial biopolymer additive (CBA); micro/nanofibrillated cellulose (CMF); softness; strength; tissue paper materials

Citation: Morais, F.P.; Carta, A.M.M.S.; Amaral, M.E.; Curto, J.M.R. An Innovative Computational Strategy to Optimize Different Furnish Compositions of Tissue Materials Using Micro/Nanofibrillated Cellulose and Biopolymer as Additives. *Polymers* **2021**, *13*, 2397. https://doi.org/10.3390/polym13152397

Academic Editors: José Miguel Ferri, Vicent Fombuena Borràs and Miguel Fernando Aldás Carrasco

Received: 26 May 2021
Accepted: 17 July 2021
Published: 21 July 2021

Publisher's Note: MDPI stays neutral with regard to jurisdictional claims in published maps and institutional affiliations.

1. Introduction

In recent years, there has been great growth potential in the tissue paper market. Factors such as the abundance of fibrous raw materials, the high demand for hygiene and health products and the evolution of socio-economic living standards have promoted this global growth. In addition, advances in manufacturing technologies for these materials have made the market increasingly competitive [1]. Tissue paper materials are increasingly used in consumer goods products such as napkins, facial tissues, paper towels, toilet papers, diapers and tissue masks. Tissue papers are produced to meet the desired properties for a specific application in the market such as a low basis weight, softness, strength and absorption [2]. To optimize the process, a great deal of effort been made to improve the properties of the raw materials used in the production of these papers [1–4]. The highest quality tissue products are produced with a blend of hardwood and softwood fiber pulps. The production of tissue papers with 100% eucalyptus fibers generally presents a higher softness and absorption; however, the strength is degraded during the creping and converting processes in the tissue manufacture [3,4]. Softwood fibers are introduced because these pulps provide strength to the papers and ensure the paper machine's runnability [5–8]. The fibers can also be subject to modification processes such as enzymatic and mechanical modifications. Refining is an important process in the development of functional tissue properties; however, it is associated with a large energy consumption [9,10]. The application of enzymes enhances not only the development of the tissue properties but also the reduction of energy costs associated with the refining process [11,12].

In addition to the modification of fiber pulps, chemical additives are used to modify properties or improve the production process for different materials. Different chemical additives used in the tissue paper industry are designed and applied according to different needs such as dry strength agents, wet strength agents, softeners and lotions [13,14]. Usually, these additives are cationic as the fibers carry negative charges and they are also more effective in much lower concentrations than anionic polymers. Nevertheless, these cationic additives normally adhere in excess to the paper, negatively affecting the creping process and the product quality [14]. The addition of strength agents can ensure that tissue products remain integrated when applied in wet conditions while the addition of debonding agents can improve the softness and bulk of these products [15–21]. However, reaching a balance between these two properties is a challenge because the conditions that maximize the softness are those that minimize strength properties. This balance can be achieved by selecting fibers, refining treatments and adding chemical polymers. Therefore, the tissue manufacturer must balance these factors in terms of cost and effect. The tissue paper industry provides the design, development and production of many chemical and polymeric additives [22].

The application of innovative and versatile additives such as biopolymers and micro/nanofibrillated cellulose (CMF), replacing synthetic additives currently used, is a sustainable strategy to be considered in order to produce high-quality tissue premium products with innovative features. Biopolymers as additives can be new natural and green approaches to new possibilities and advances in modifying or improving the final end-use tissue properties [23]. The same is true for CMF as an additive that has been extensively investigated for its application in tissue products due to its high specific surface area, excellent mechanical properties, flexibility, biocompatibility, biodegradability and lack of toxicity, among other properties [24–27]. The furnish formation is also affected by the CMF incorporation as an additive. CMF decreases porosity and air permeability when added to the tissue structures. CMF content and structural changes present a proportional behavior. These decreases are caused by the CMF bonding with the cellulose fibers in the 3D network structures, closing the porous structure. Furthermore, this reduction in porosity is also correlated with the increase in the structure's apparent density [28].

The presence of CMF can present an impact on the structure porosity because of the increasing number of hydrogen bonds present in this nanocellulose. The increase of these bonds is related to its increase in the surface/volume ratio due to the nanosized scale [28]. By filling the large pores created by the 3D fiber network, the CMF incorporation contributes to smaller pore dimensions and a narrower distribution of the structures, providing abundant pores with a uniform size and, consequently, a decrease in the properties of water absorption capacity and capillarity penetration [24,25]. The flexibility of CMF can be a filler for large pores in structures [25]. Although the mechanical properties of tissue products are improving, CMF also brings a difficulty to the dewatering of papermaking furnishes. The fibrillation degree influences the dewatering capacity of pulp suspensions. The drainage time increases depending on the CMF fibrillation degree. Particles resulting from fibrillation are more easily incorporated into the fiber network, partially closing the pores between the fibers and consequently limiting the wet network's ability to drain water. This drainage difficulty is also caused by the increased water retention capacity of CMF. However, this dewatering of pulp suspensions containing CMF can be improved with the use of retention aids and also with cationic starch-treated CMF with colloidal silica [28]. In addition to a negative impact on dewatering, the accumulation of CMF particles in the closed circuit of the tissue machine can increase the viscosity of the recirculating water, thus affecting processability. Therefore, there is a need to study the feasibility of using CMF to reduce the fiber content. Our study [24] confirmed that the use of 1 to 8% of CMF incorporated into tissue paper formulations, along with a reduction of hardwood and softwood fiber content, achieves a trade-off between the final end-use tissue properties and fiber blend costs, with drainability between 26 and 32 °SR (Schopper–Riegler degree). In this range of drainability, the furnish formulations are able to produce high-quality tissue papers without difficulties in tissue paper machine runnability and efficiency [2,24].

The application of these additives leads to the optimization of strength properties without compromising the softness and absorbency properties of these types of papers. These strategies can allow a reduction of softwood fibers in the production of tissue paper without affecting the strength performance and decrease the overall tissue manufacturing cost [24,27–30]. The aim of this work was to perform a comparative study between different tissue furnish compositions and the incorporation of two additives, a commercial tissue biopolymer additive (CBA) as a cationic starch substitute and CMF, and their effect on the properties of the final end-use tissue paper. For this purpose, the fiber and suspension drainability characterizations of different investigated samples were performed. Laboratory isotropic structures with a light basis weight of 20 g/m^2 were prepared and the softness, tensile strength and absorbency properties were evaluated.

2. Materials and Methods

2.1. Materials

An industrial, never-dried, bleached eucalyptus kraft pulp and an industrial, air-dried, bleached pinus kraft pulp were used in this work after disintegration according to ISO 5263-1. The additives used in this study were CBA and CMF. This commercial tissue additive, CBA, is a cationic polymeric solution specially designed as a replacement for cationic starch. CBA presents a viscosity of 500–900 cP, a pH of 2.1–4.0 and a density of 1.030 ± 0.040 g/cm^3 at 20 °C (data supplied by a commercial enterprise). CMF was supplied by a research institute and obtained by mechanical (refining and homogenization) and enzymatic (cellulase) treatments using an industrial bleached eucalyptus kraft pulp. This sample was characterized in our previous work, which can be used for more detailed information [24]. CMF (Figure 1) presents an intrinsic viscosity of 1577 mL/g (ISO 5351), a carboxyl group content of 8.4 ± 0.2 mmol/100 g (TAPPI T 237 om-93), a suspension drainability degree superior to 80 °SR (Schopper–Riegler) and 56% of particles below 1000 nm, representing 10% of particles below 200 nm [24].

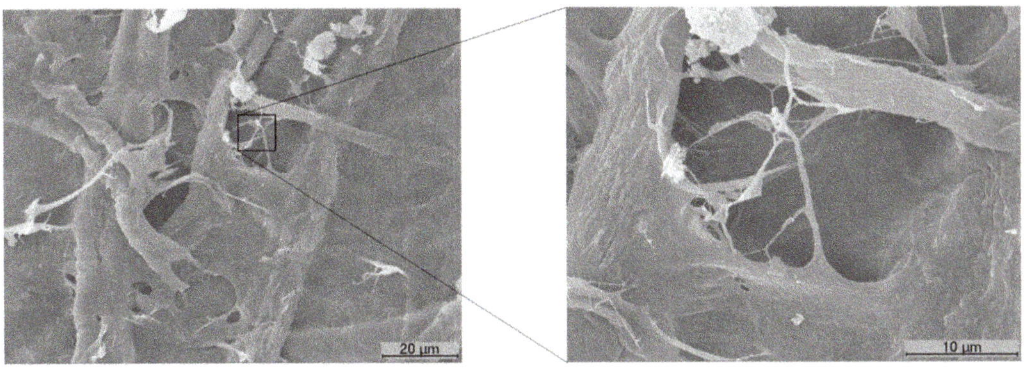

Figure 1. SEM image of the CMF additive, highlighting a high magnification image (3000×) where it is possible to verify the micro and nanofibrils of this sample.

2.2. Enzymatic and Mechanical Treatments

The softwood pulp was beaten in a PFI mill at 4000 revolutions under a refining intensity of 3.33 N/mm (ISO 5264-2) followed by an enzymatic treatment. The enzymatic treatment was performed according to the method described by Morais et al. [12]. The treatment was carried out at a consistency of 4% at pH 7 at 40 °C with continuous mechanical agitation to ensure an efficient mass transfer for 60 min and 30 mg of enzyme/g of pulp.

2.3. Tissue Formulations

Table 1 presents the various tissue formulations (investigated samples) carried out in this study. The different formulations were performed to study the effect of additives in the tissue structure properties and understand their implication in the reduction of softwood pulp usually used in this type of paper. The first formulation presents a similar combination of eucalyptus fibers and softwood fibers typically used in tissue mills. All of the samples presented similar drainability properties around 25 °SR, which is recommended for tissue paper products and means that drainage and processability are in the range for the tissue process and, consequently, processability is not negatively impacted [2].

Table 1. Tissue formulations used in the study.

	Eucalyptus Pulp	Softwood Pulp with Mechanical and Enzymatic Treatment	CBA (Commercial Biopolymer Additive)	CMF (Micro/Nanofibrillated Cellulose)
Formulation 1	75%	25%	-	-
Formulation 2	75%	25%	2%	-
Formulation 3	75%	25%	-	2%
Formulation 4	75%	25%	2%	2%
Formulation 5	90%	10%	-	-
Formulation 6	90%	10%	2%	-
Formulation 7	90%	10%	-	2%
Formulation 8	90%	10%	2%	2%

2.4. Preparation of Tissue Structures

The tissue structures were produced in a batch laboratory sheet former according to the method described by Morais et al. [6,7]. An adaptation to ISO 5269-1 was used, namely, the suppression of the pressing operation and the production of structures with a basis weight of 20 g/m^2. The tissue structures were prepared with the different formulations described in Table 1. Finally, all samples were conditioned at 23 ± 1 °C, with a relative humidity of 50 ± 2%, according to ISO 187.

2.5. Characterization of the Morphological, Drainability and Chemical Properties

The morphological properties of pulp fibers and formulations were evaluated automatically by an image analysis using a fiber analyzer, the MorFi equipment (TECHPAP, Grenoble, France).

In order to characterize the surfaces of the different formulations, the obtained tissue structures were coated with gold using a Sputter Quorum Q 15 OR ES (Laughton, East Sussex, UK) to be analyzed by scanning electron microscopy (SEM) using a Hitachi S-2700 model (Tokyo, Japan) with an accelerating voltage of 20 kV at different magnifications. ImageJ, an image analysis software, was also used to characterize the pore properties of the samples. The SEM images were processed and analyzed using a methodology with defined criteria for the stabilization of the average of the measured values [31].

The chemical functional groups were analyzed using Fourier transform infrared spectroscopy with attenuated total reflectance (FTIR-ATR) at a wavelength of 600–4000 cm^{-1} with a resolution of 4 cm^{-1}.

The drainability of pulp suspensions and formulations was evaluated by the °SR method (ISO 5267/1). Triplicate assays were performed for both methods.

2.6. Characterization of the Tissue Properties

2.6.1. Structural Properties

The structures produced were characterized in terms of basis weight (ISO 12625-6), thickness and bulk (ISO 12525-3). The basis weight was obtained by the quotient between the average mass of each structure and the respective area (0.02138 m^2). The thickness was obtained using a Frank-PTI micrometer. The bulk was obtained by the quotient between the thickness and the basis weight of the structures. Additionally, the porosity (apparent theoretical porosity) of the structure was also calculated according to the following equation:

$$Porosity\ (\%) = 100 \times \left(1 - \frac{\rho_{structures}}{\rho_{cellulose}}\right) \tag{1}$$

where $\rho_{cellulose}$ was the density of the cellulose (1.5 g/cm^3) [12] and $\rho_{structure}$ was the apparent density of the structures (g/cm^3).

2.6.2. Softness Properties

Softness was assessed using a Tissue Softness Analyzer, the TSA (Emtec). A two-step measurement was used. The first was a noise measurement where the real softness (TS7) was measured. The second step was a deformation measurement where stiffness, plasticity, elasticity and hysteresis were measured. The handfeel (HF) parameter value was calculated using both steps.

2.6.3. Strength Properties

Strength properties were determined according to ISO 12625-4 using the tensile index and calculated according to the following equation:

$$Tensile\ Index\ (\mathrm{Nm/g}) = \frac{\frac{break}{W}}{G} \tag{2}$$

where *break* was the breaking force (N), *W* was the sample width (50 mm) and *G* was the sample grammage or basis weight (g/m^2).

2.6.4. Absorbency Properties

The absorbency properties were determined by the methods of water absorbency capacity per unit of mass and Klemm capillary rise.

The immersion absorption method was used to determine the water absorption capacity per unit of mass of the samples. The assay was carried out according to an adaptation of ISO 12625-8. The sample was immersed for 30 s. At the end of that time, the sample was removed from the water and left to run on a support that guaranteed an amplitude of 30° for approximately 1 min. Finally, the amount of water that the sample was able to absorb during this time was determined. The assay was performed in triplicate for each sample.

The Klemm capillary rise method was measured through an adaptation of ISO 8787. The assay consisted of the water rising in the sample (50 mm wide) at a certain time. The capillary rise was recorded at 10, 20, 30, 60, 180, 300 and 600 s. Four readings were performed for each sample. All of the assays were performed in a controlled temperature and humidity laboratory.

2.7. Computational Studies

SimTissue, a tissue simulator, was developed specifically for furnish management and optimization by FibEnTech-UBI members, in the scope of Project InPaCTus. This tissue simulator was used to compare different end-use final tissue properties and to access them for the different formulations to those performed experimentally [24].

A 3D fiber-based computational simulator was used to model the tissue structures as planar random networks. This simulator is open-source software, and the code is available on GitHub (https://github.com/eduardotrincaoconceicao/voxelfiber, accessed on April 2021). A more detailed description of this 3D simulator can be found in Conceição et al. and Curto et al. [32,33]. The fibers are represented by a voxel chain and modeled according to fiber dimensions and properties such as the fiber length and width ratio, flexibility, fiber wall thickness and lumen dimensions [34]. The structural properties were obtained following this computational study and the resulting 3D computational structures. The computational studies were performed using MATLAB® (R 2020a, 9.8.0.1323502, MathWorks, Natick, MA, USA).

3. Results and Discussion

3.1. Characterization of Pulp Fibers and Tissue Formulations

In a first approach, the characterization of the raw materials used in the study was carried out. The eucalyptus pulp presented a length weighted by length of 0.729 ± 0.003 mm, a width of 19.1 ± 0.0 µm, a coarseness of 6.31 ± 0.04 mg/100 m and fine elements of 38.5 ± 0.4% in length. The softwood pulp presented a length weighted by length of 1.889 ± 0.052 mm, a width of 30.1 ± 0.2 µm, a coarseness of 20.52 ± 3.23 mg/100 m and fine elements of 36.6 ± 4.2% in length. In this study, only softwood pulp was subjected to fiber modification treatments, namely, refining and enzymatic treatments, as this eucalyptus pulp is already an industrial pulp suitable for the production of tissue papers. These treatments were carried out in order to improve fiber flexibility and inter-fiber bond properties with a greater strength three-dimensional matrix [35]. Analyzing the treatments separately, the refining effect was more pronounced than the enzymatic treatment. The results indicated that refining decreased the length weighted by length by 8% and coarseness by 30% and increased the fiber width by 7% and the fine content by 17% while the enzymatic treatment decreased the fiber length weighted by length by 2% and the coarseness by 18% and increased the fiber width by 23% and the fine content by 4%. The combination of these two treatments decreased the fiber length weighted by length by 10% and the coarseness by 26% and increased the fiber width by 7% and the fine content by 11% compared with the untreated pulp. Figure 2 presents an analysis of coarseness as a function of the length/width ratio (slenderness ratio) for the softwood samples. The results indicated that two groups can be considered with the fiber modification treatments applied. In the first one, the softwood fibers without treatment (63 of length/width ratio and 20.52 mg/100 m of coarseness) and with enzymatic treatment (61 of length/width ratio and 16.92 mg/100 m of coarseness) presented a higher slenderness ratio and higher coarseness. In the second group, the refined softwood fibers (54 of length/width ratio and 15.81 mg/100 m of coarseness) and with both treatments (52 of length/width ratio and 15.09 mg/100 m of coarseness) presented a lower slenderness ratio and lower coarseness. This suggests that the first group was more susceptible to presenting better tissue properties of softness and absorbency than the second group, which was more susceptible to presenting higher strength properties [7]. Additionally, the first group presented °SR between 12–14 and the second group presented 55–66 °SR. The mechanical refining and the combination of both treatments promoted a more efficient fiber fibrillation and consequently decreased the suspension drainability.

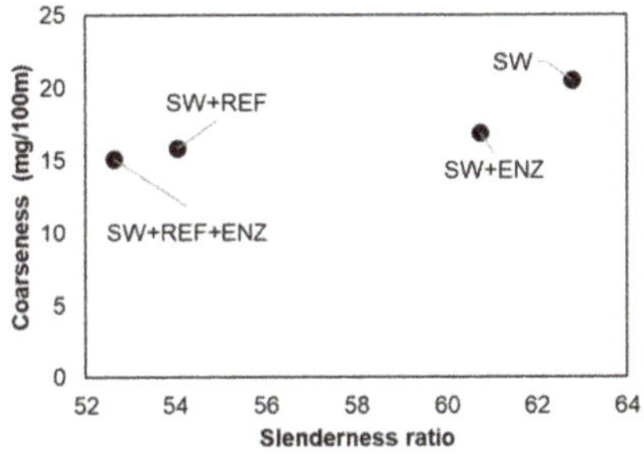

Figure 2. Coarseness as a function of the slenderness ratio of softwood fiber pulp (SW) with enzymatic treatment (SW + ENZ), with refining (SW + REF) and with the combination of both fiber modification treatments (SW + REF + ENZ).

The softwood pulp with the refining and enzymatic treatment was also assessed through SEM images (Figure 3a). This treatment made it possible to modify the properties of the softwood pulp, promoting inter-fiber bonds and the combination of key structural properties such as the pore dimension and distribution. The pore properties depend on the fiber properties and the mechanical and enzymatic processes to which they are subjected [36]. The results indicated that these treatments had an impact on the fiber dimension and flexibility, resulting in a structure with a surface porosity of 83 ± 1% and a pore diameter distribution of 35% between 2 and 10 μm and 65% between 10 and 27 μm (Figure 3b).

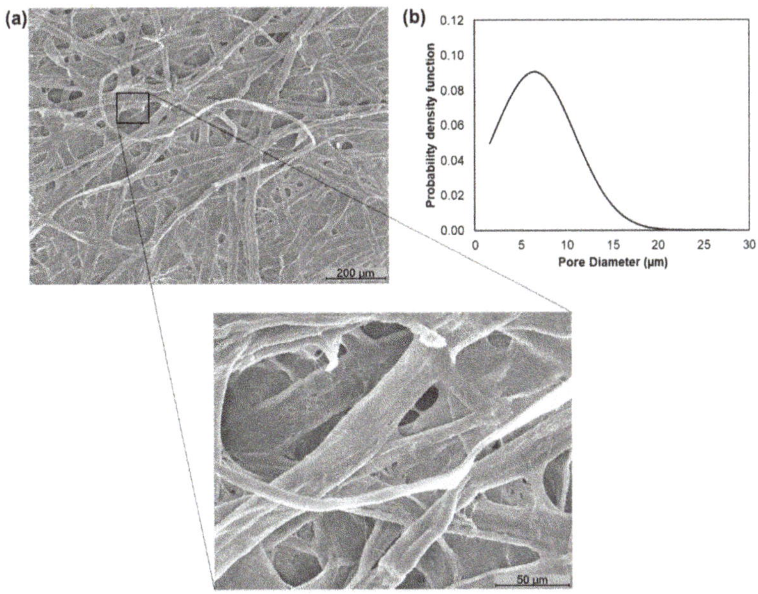

Figure 3. SEM image of the softwood pulp with refining and enzymatic treatment with a high magnification image (500×) (**a**) and its pore diameter distribution (**b**).

In a second approach, the characterization of the tissue formulations was carried out. As previously reported, this study compares two versatile additives: CBA and CMF. The MorFi analyzer is only able to evaluate the fiber suspension properties so the presence of the CBA does not affect its fibrous composition. The formulations with 75% eucalyptus pulp and 25% softwood pulp-treated and with incorporated additives presented a length weighted by length between 0.717 and 0.729 mm, a width between 19.4 and 19.9 µm, coarseness between 7.78 and 8.27 mg/100 m and fine elements between 32.7 and 35.7% in length. With a reduction of the softwood fibers, the formulations with 90% eucalyptus pulp and 10% softwood pulp-treated and with incorporated additives presented a length weighted by length between 0.719 and 0.722 mm, a width between 19.1 and 20.1 µm, coarseness between 6.77 and 7.61 mg/100 m and fine elements between 30.4 and 32.8% in length. Additionally, these formulations presented an °SR degree between 20 and 25 °SR. Although the softwood pulp presented low drainability, its combination with eucalyptus pulp and with different additives allowed the tissue formulations to present a range of °SRs suitable of producing premium tissue materials [2] without compromising the processability. Figure 4 presents the SEM image of the formulation 4 (75% eucalyptus pulp + 25% softwood pulp-treated + 2% CBA + 2% CMF) as well as its pore diameter distribution. The 3D structure matrix resulting from this formulation was presented as a multi-structured material with bonding between CMF fibrils and pulp fibers. The CBA addition also promoted a more closed structure with a surface porosity of 88 ± 1% and a pore diameter distribution of 40% between 1 and 10 µm and 60% between 10 and 22 µm (Figure 4b).

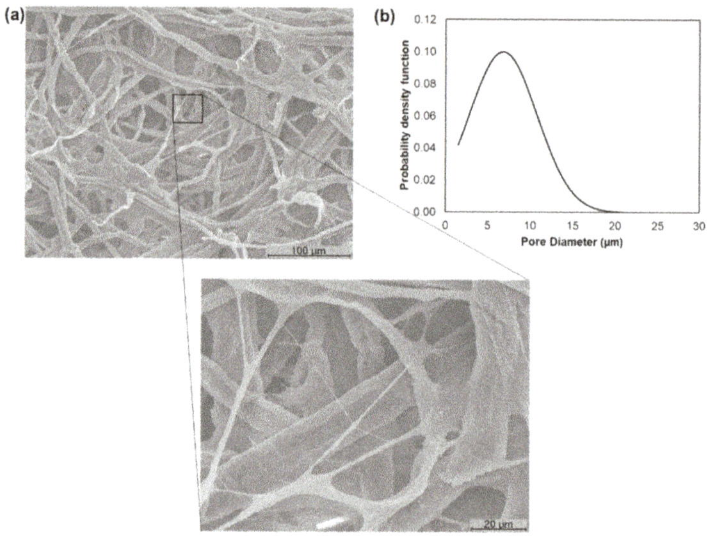

Figure 4. SEM image of tissue formulation 4 (75% eucalyptus pulp + 25% softwood pulp-treated + 2% CBA + 2% CMF) with a high magnification image (1000×) where is possible to verify the inter-fiber bonding between CMF and the structure in more detail (**a**) and its pore diameter distribution (**b**).

To complement this study, the FTIR-ATR technique allowed us to chemically characterize, identify and quantify the formulations with CBA and CMF, evaluating the existence of physical retentions in the structure without new chemical bonds between these additives and the cellulose eucalyptus and softwood fiber 3D matrix. Figure 5 shows the FTIR spectrum of formulation 1 (a typical furnish mixture at a tissue mill), formulation 2 (with CBA incorporation) and formulation 3 (with CMF incorporation). The typical absorption bands of hydroxyl groups between 3000 and 3500 cm^{-1} indicative of the –OH stretching of the intra- and inter-molecular interactions of hydrogen bonds were verified for the three formulations. A –OH band with a higher area was observed for the structures with CBA. The biggest differences in the FTIR spectrum were found in the bands between 2800 and 3000 cm^{-1}. This range corresponds with the =C–H stretching in the methyl groups of cellulose and hemicellulose. The band with the highest intensity in this range was observed for the formulation with CMF followed by the formulation with CBA and formulation 1. Additionally, the bands between 2200 and 2400 cm^{-1} correspond with the –OH asymmetrical stretching vibration of the carboxylic acid due to the enzymatic treatment applied to the materials [12]. The bands between 1500 to 1700 cm^{-1} were due to the C=O stretch of hemicelluloses. Other characteristics of cellulose were also observed between 800 and 1500 cm^{-1} such as the angular deformation of C-H and primary alcohol C-O bonds, the absorption band of C–O–C bonds, the β-glycosidic bonds between glucose units, C=H stretching, C=O stretch vibration in the syringyl ring and carboxylate ion group vibration.

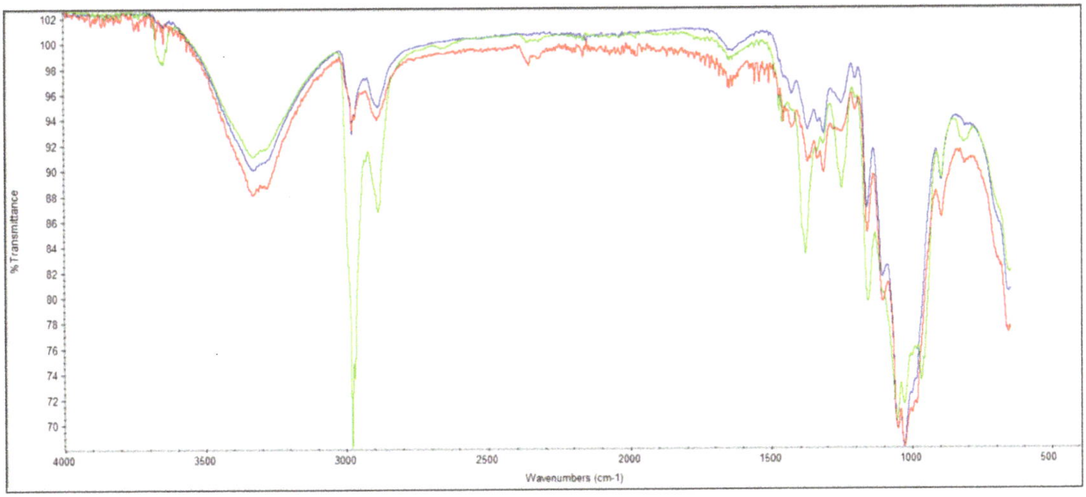

Figure 5. FTIR-ATR spectrum of formulation 1 (blue), formulation 2 (red) and formulation 3 (green).

3.2. *Characterization of the Properties of the Final End-Use Tissue*

3.2.1. Structural Properties

Bulk and porosity are essential parameters for the quality of tissue products that influence the softness and liquid absorption properties [6,7]. Figure 6a shows, for the different tissue formulations, the behavior of the structural properties. The addition of CBA in formulation 2 promoted a decrease in bulk and porosity by 1% and 0.1%, respectively, compared with formulation 1. The addition of CMF (formulation 3) showed a tissue structure with lower bulk (<4%) and porosity (<0.5%) compared with formulation 1. The combination of these versatile additives (formulation 4) also promoted an increase of hydrogen and inter-fiber bonds, inducing a decrease of 11% in bulk and 1% in porosity of the tissue structures in comparison with formulation 1. This densification effect can be observed in the cross-section of the structures, as shown in Figure 7. With the reduction of the reinforcement fibers by 15% in the formulations, bulk and porosity were increased. This reduction promoted an increase of 20% in bulk and 2% in porosity when comparing formulations 1 and 5. The same trend in formulations 1–4 was observed in formulations 5–8; i.e., the combination of both additives presented a higher influence on the structural properties followed by CMF and CBA. Formulations 6, 7 and 8 promoted a reduction in bulk by 4, 10 and 22%, respectively, and a reduction in porosity by 0.4, 1 and 3%, respectively, compared with the control (formulation 5). The addition of both CBA and CMF in formulations with a softwood fiber reduction (90% eucalyptus fiber and 10% softwood fiber) allowed the obtaining of structures with increased structural properties compared with typical industrial structures (75% of eucalyptus fiber and 25% softwood fiber). The results indicated that formulations 6 and 7 increased the structure bulk by 16% and 9%, respectively, compared with formulation 1 (Figure 8). However, the combination of these two additives promoted a 6% reduction in bulk compared with this formulation typically used industrially.

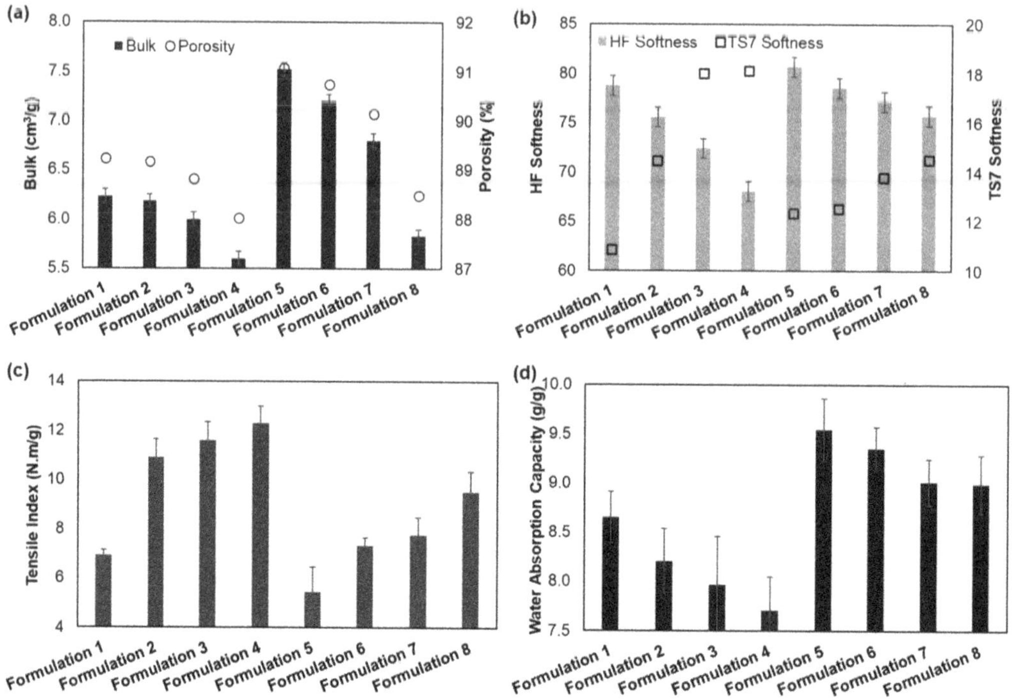

Figure 6. Properties of bulk and porosity (**a**), softness (**b**), tensile index (**c**) and water absorption capacity per unit of gram (**d**) of all tissue formulations.

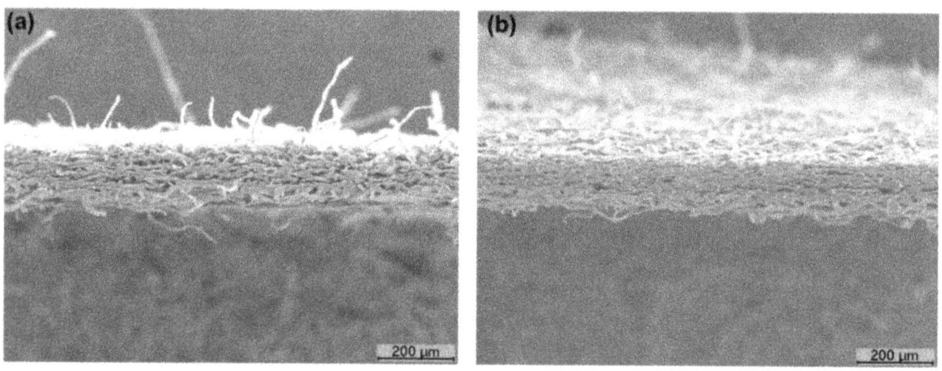

Figure 7. SEM images of the cross-section of the tissue structures prepared with formulation 1 (**a**) and formulation 4 (**b**).

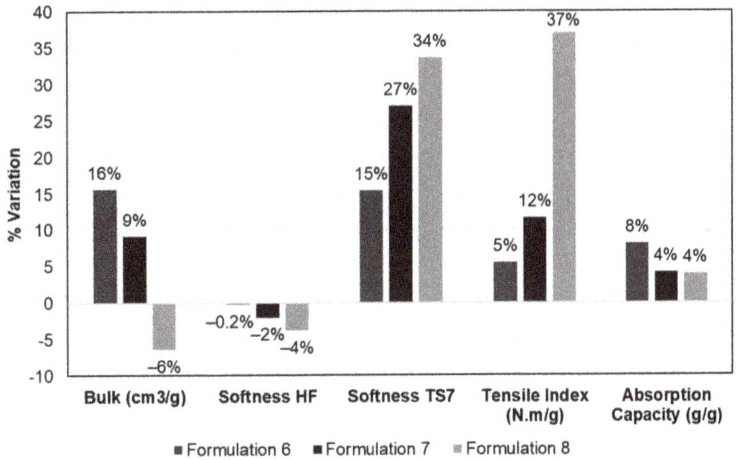

Figure 8. Percentage of variation in the properties of bulk, softness HF, softness TS7, tensile index and water absorption capacity of formulations 6, 7 and 8 compared with formulation 1 obtained with SimTissue.

3.2.2. Softness Properties

Softness is also a characteristic that is highly appreciated by consumers, resulting in the combination of structure flexibility, bulk and surface softness. Softness properties of the different formulations are shown in Figure 6b. The HF (handfeel) values decreased with the additives loaded in the paper 3D matrix. The increase in fiber flexibility, fine elements and, consequently, the number of inter-fiber bonds promoted a more closed and compact 3D matrix with less bulk, resulting in a structure with less softness. The addition of CBA in formulation 2 promoted a decrease in softness HF by 4% compared with formulation 1. The addition of CMF (formulation 3) showed a tissue structure with lower softness compared with formulation 1 with a decrease of 8%. The combination of these versatile additives (formulation 4) also promoted a decrease of 14% in the softness HF of the tissue structure compared with formulation 1. The reduction of the reinforcement fibers in the formulations increased softness properties by 2% compared with formulations 1 and 5. As with bulk properties, the same trend was also observed for softness properties. Formulations 6, 7 and 8 promoted a reduction in softness by 3%, 4% and 6%, respectively, compared with formulation 5. Regarding Figure 8, the softness HF values decreased by 0.2%, 2% and 4% with the incorporation of CBA, CMF and the combination of both, respectively. The production of structures with a low content of SW fibers and versatile additives promoted a small reduction in softness, improving strength properties, as reported below, contributing to the formation of a fibrous structure with less stiffness [37].

Softness TS7 is inversely related to softness HF values; therefore, it is influenced by the presence of fibers in the structure surface. TS7 values increased with the addition of additives in the structure (Figure 6b). This result indicated that the fibers in the structure presented higher bonds, causing structure densification and rigidity. The fibers in the z-direction were more adherent to the paper structure, reducing the number of free fibers on the tissue surface compared with the structures without additives (Figure 7). Compared with formulation 1, softness TS7 of the tissue structure increased by 33%, 65% and 67% with the incorporation of CBA, CMF and both, respectively (formulations 2–4). With the reduction of the reinforcement fibers in the formulations, this property also increased by 13% compared with formulations 1 and 5. Additionally, formulations 6, 7 and 8 promoted an increase in softness TS7 by 2%, 12% and 18%, respectively, compared with formulation 5. Compared with formulation 1 (75:25), softness TS7 increased by 15%, 27% and 34% with the incorporation of CBA, CMF and both, respectively, in formulations with a softwood fiber reduction (90:10) (Figure 8).

3.2.3. Strength Properties

The tensile index is an essential parameter not only for the product quality but also to ensure the paper machine runnability, which depends on the fiber intrinsic strength and the inter-fiber bonds. The results of the tensile index for each tissue formulation are shown in Figure 6c. The incorporation of additives promoted an increase in the tensile index of the tissue structures. Differences in strength improvements varied with the performance of CBA and CMF and became more noticeable with the combination of these additives. Formulation 3 with the addition of CMF showed improvements in the tensile index (+67% compared with formulation 1) related to formulation 2 with the addition of CBA (+57% compared with formulation 1). Formulation 4, with the incorporation of both additives, presented higher structure strength improvements by 77% compared with formulation 1. This result was corroborated by the results of the literature as the addition of CMF and chemical additives in the tissue structures improve strength properties of these products [13,25,27]. These additives promoted an increase in the surface area of the tissue structures due to the micro/nanofibrils and consequently formed stronger and more stable structures through inter-fiber interaction and hydrogen bonding [27]. This fact can also be verified by the densification of the tissue structures with the incorporation of both additives (Figure 7b).

Reinforcement fibers play a fundamental role in strength properties; therefore, the reduction of these fibers in formulation 5 promoted a 21% reduction of the tensile index compared with formulation 1. Additionally, formulations 6, 7 and 8 promoted an increase in the tensile index by 34%, 42% and 74%, respectively, compared with formulation 5. It is important to highlight that the strength gains obtained with the addition of CMF and CBA and the softwood reduction in the tissue structures were higher when compared with formulation 1. Despite the structures presenting a low basis weight and no pressing, these additives favored strength properties of these tissue structures. Additionally, as shown in Figure 8, the tensile index increased by 5%, 12% and 37% with the incorporation of CBA, CMF and both, respectively.

The commercial biopolymer additive used in this work was a cationic polymeric solution that is especially intended to be a substitute for cationic starch. Generally, cationic starch is used in the printing and writing paper industry with functions of a dry strength additive, emulsification of sizing agents, retention and drainage. This starch improves the formation, internal bonding, surface strength, tensile strength, writing and printing surface and energy consumption [38]. However, in the tissue paper industry, this starch has several inconveniences as the increase in strength properties promotes an abrupt reduction in softness properties. A solution found in the tissue industry was to design a practically constant biopolymer that would increase strength properties while preserving softness properties with a reduction of softwood fiber content. Furthermore, this additive allows the reduction of the contaminating organic load and costs as a result of reducing the use of cationic starch [38]. Therefore, the results of this work indicated that, compared with formulation 1, a typical industrial mixture can produce tissue structures containing 2% of the CBA and 10% of softwood fibers with 5% tensile index increases while maintaining the softness (Figure 8). This cationic polymeric solution also improved the inter-fiber bonding and CMF particle retention. This retention was proven by the 38% increase in strength properties with both additives (formulation 8) compared with formulation 1. The incorporation of CMF alone (formulation 7) only increased strength properties by 12% compared with formulation 1 (Figure 8).

In addition, an inverse relationship between the properties of softness and strength can also be observed in these formulations (Figure 9); i.e., strength improvements present a negative impact on the tissue structure softness. The formulations promoted an increase of 1.52 units of softness HF for each unit with a decreased tensile index ($y = -1.52x + 89.51$;

R^2 = 0.89). In our previous studies, this relationship was also observed in formulations with only eucalyptus fibers and without additives in which the softness HF increased by 2.4 units per unit of the decreased tensile index [6,7]. This can contribute to the reduction of costs associated with the use of softwood fibers in industrial furnishes as there is a balance between the properties of softness and strength in the tissue structures with the use of these versatile additives.

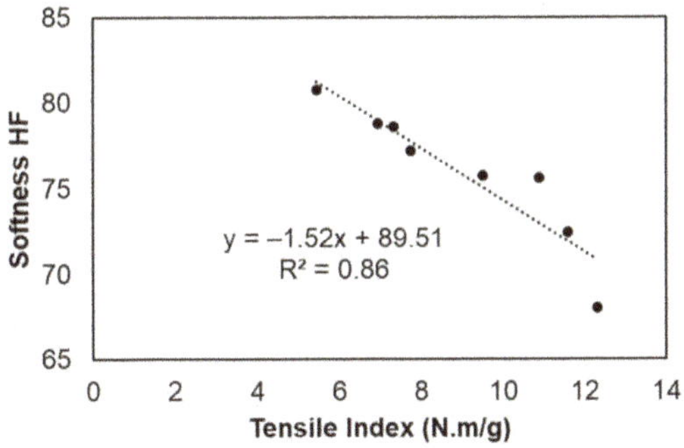

Figure 9. Correlation between the tensile index and softness HF of tissue formulations.

3.2.4. Absorbency Properties

Absorbency properties are essential in tissue products, mainly in paper towels and napkin s, depending on their fibrous structure. Tissue products with a more porous 3D matrix will present better water absorption as empty spaces for water interaction are more available [12]. Figure 6d presents the water absorption capacity of the tissue formulations in the study. Overall, the incorporation of CBA and CMF into the tissue formulations decreased the water absorption properties compared with the control. This result was also in line with the decrease in bulk and porosity properties (Figure 6a). This effect was observed from the SEM images in Figures 4a and 7b in which the formulations with these additives presented a less porous surface with a more compact cross-section. The addition of CBA and CMF (formulations 2 and 3) presented a decrease of 5% and 8% of this property, respectively, compared with formulation 1. The incorporation of both additives presented higher absorption decreases by 11% compared with formulation 1. The reduction of the reinforcement fibers in the formulations also increased the absorption properties by 10% compared with the formulations 1 and 5. Additionally, formulations 6, 7 and 8 promoted a reduction in absorption by 2%, 6% and 6%, respectively, compared with formulation 5. Compared with formulation 1, absorption increased by 8%, 4% and 4% with the incorporation of CBA, CMF and both, respectively, in formulations with a softwood fiber reduction (90:10) (Figure 8). In this work, we also found a good linear correlation (R^2 = 0.78) between the water absorption properties and the softness of the tissue structures made from these formulations (Figure 10a). The formulations promoted an increase of 5.39 units of softness HF for each unit of increased water absorption capacity ($y = -5.39x + 29.06$). The same inverse trend observed for the softness and strength properties was also verified for the absorbency and strength properties (Figure 10b). The tissue formulations promoted an increase of 3.37 units of water absorption capacity for each unit of decreased tensile index ($y = -3.37x + 38.19$; R^2 = 0.81).

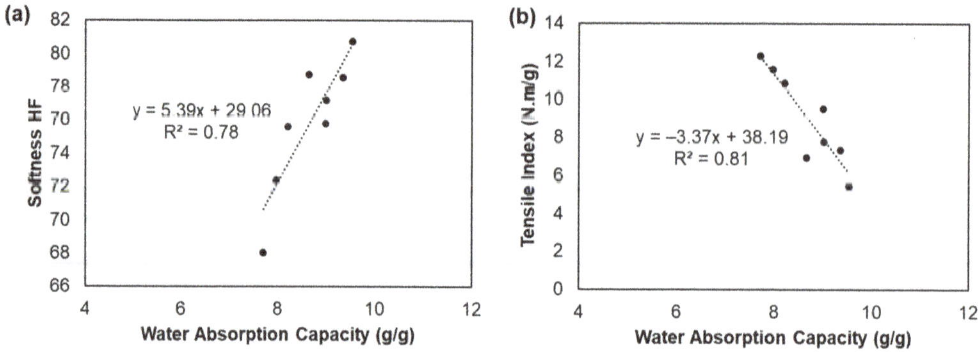

Figure 10. Correlation between the water absorption capacity and softness HF (**a**) and tensile index (**b**) of the tissue formulations.

The capillary rise of the structures depends on the adhesion and cohesion forces of the water molecules that ascend through the fiber walls or between the pores. Therefore, there must be a balance between the pore dimensions to promote the capillary rise efficiency. Inter-molecular bonds between fibers and water molecules in structures with higher pores can impair the capillary rise. On the other hand, in the case of channels that are too narrow, the water progression will be strangled by the limited space for capillary rise [39]. Figure 11 presents the Klemm capillary rise of the tissue formulations in this study. The same trend in the water absorption properties was observed in the capillary rise of tissue formulations. The formulations with a content of fewer softwood fibers presented better capillary ascents and the additives promoted the capillary rise decrease. The structures with a higher bulk and a more porous 3D matrix promoted a higher capillary rise. Comparatively of both additives, CBA promoted a better water affinity compared with CMF, corresponding with water capillary rise improvements.

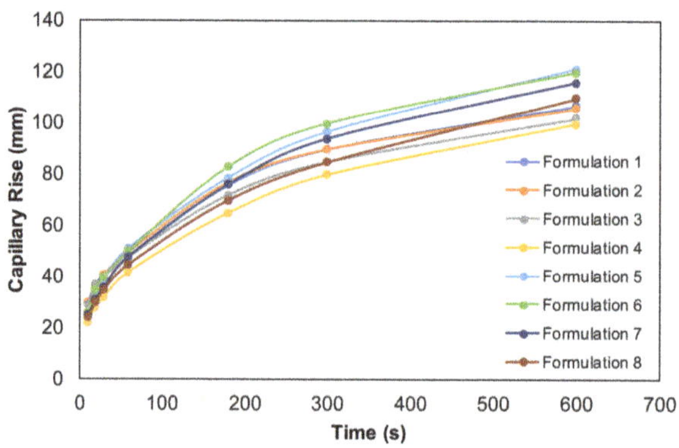

Figure 11. Klemm capillary rise method as a function of time for all tissue formulations.

These results indicated that a compromise could be found between the properties of softness, strength and absorbency with the incorporation of versatile additives in tissue formulations, reducing the incorporation of softwood fibers in an industrial furnish. This effect could promote a reduction in costs associated with the use of these fibers and also the mechanical and enzymatic refining applied to softwood fibers as the incorporation of these additives promoted structure strength improvements while maintaining the softness.

3.3. Computational Prediction of the CBA and CMF Performance in Formulations with 100% Eucalyptus Fibers

The high manufacturing costs associated with tissue paper materials are due to the market softwood pulp prices and the energy consumption associated with the fiber modification processes such as refining or beating. Finding strategies to reduce these processes is one of the focuses of the research on the management and optimization of the furnish and each type of tissue product. For this reason, we developed a tissue simulator, SimTissue, with the ability to predict the final end-use properties of different hypothetical scenarios of industrial interest. Our approach in this study was to estimate improvements in tissue properties with the incorporation of CBA and CMF, separately, in formulations with 100% eucalyptus fibers compared with formulations containing eucalyptus fibers and softwood fibers (Figure 12). Note that these simulations were carried out with a reinforcing fiber with high strength properties, according to a set of pulps studied previously [6,7]. It is important to present the simulations of these fiber mixtures as they represent the limits achievable in the tissue properties.

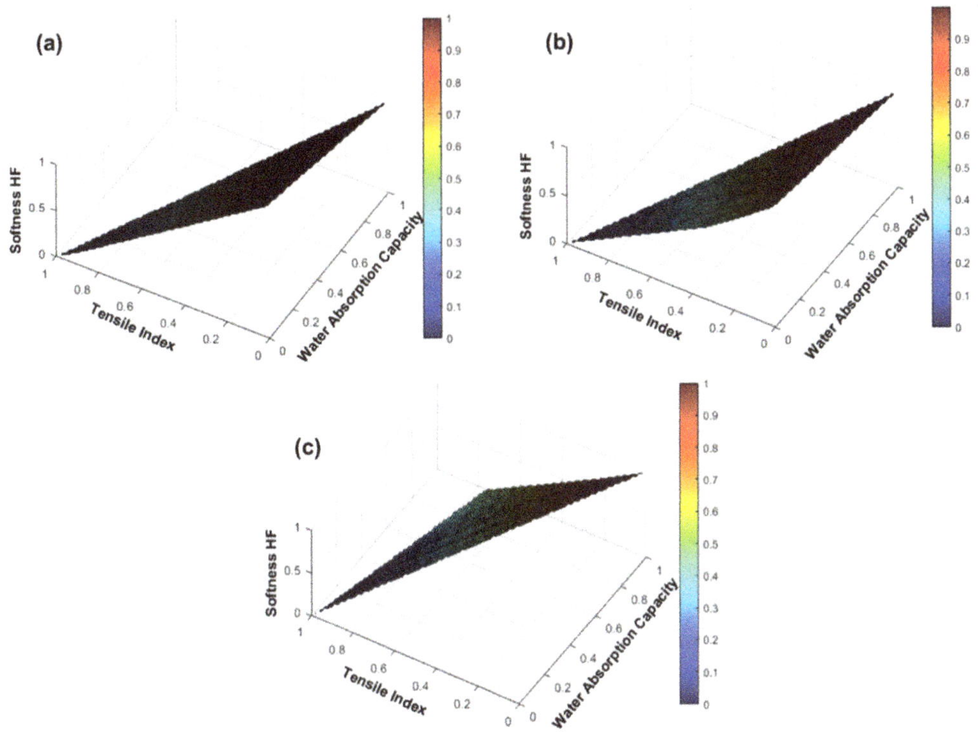

Figure 12. Evolution of the end-use tissue properties of softness HF, tensile index and water absorption capacity with the incorporation of CBA (**a**), CMF (**b**) and softwood fibers (**c**) in formulations with 100% eucalyptus fibers. All variables were normalized to present the same scale range.

As expected, the incorporation of CBA, CMF and softwood fibers promoted an improvement in strength properties with adverse effects on softness and absorbency. According to the tissue simulator, a formulation with 90% eucalyptus fibers and 10% softwood fibers promoted the production of a tissue product with a 68 softness HF, 15 Nm/g tensile index and 8 g/g water absorption capacity [24]. In order to replace this softwood fiber, the incorporation of 3% CBA allowed us to obtain improved tissue properties in the same range (72 softness HF, 15 Nm/g tensile index and 8 g/g water absorption capacity). The same trend was also verified for the incorporation of 14% CMF, presenting 70 softness HF, 15 Nm/g tensile index and 7 g/g water absorption capacity. However, in a previous work, we found that formulations with more than 10% CMF incorporation are not economically viable. The trade-off between the cost and the effectiveness of the formulations was achieved with the incorporation of CMF between 1 to 8%, presenting optimized properties at an industrial range [24].

This approach saves the consumption of softwood fibers in tissue production and is aimed at the use of biocompatible and biodegradable additives in the industrial production process. The simulator's predictive capacity allowed the optimization and development of innovative formulations, saving laboratory and industrial resources.

4. Conclusions

This work presented experimental and computational approaches to evaluate and predict the effects of the incorporation of CBA and CMF with eucalyptus and softwood fibers on the softness, strength and absorbency properties of tissue paper materials. Overall, the addition of both additives enhanced tensile strength properties at the expense of reduced softness and absorbency properties, followed by CMF and CBA. This cationic polymeric solution also improved the inter-fiber bonding and CMF particle retention. With the softwood fiber reduction in the formulations, the incorporation of additives promoted a strength and absorbency increase with a low softness decrease. It was possible to obtain a trade-off between the tissue end-use properties with CBA and CMF additions separately. With this work, it was possible to produce tissue furnish formulations using these additives with a reduction of the softwood fiber percentage in a drainage range of the tissue industrial process. This suggested that these additives presented a good potential to be introduced in the production of high-quality tissue materials such as tissue masks or premium napkins and facial papers. These biocompatible and biodegradable additives can promote the maximization of eucalyptus fibers in industrial tissue production, optimizing the process and associated manufacturing costs.

Author Contributions: Conceptualization, F.P.M. and J.M.R.C.; methodology, F.P.M. and J.M.R.C.; software, F.P.M. and J.M.R.C.; validation, F.P.M. and J.M.R.C.; formal analysis, F.P.M. and J.M.R.C.; investigation, F.P.M. and J.M.R.C.; resources, M.E.A., A.M.M.S.C. and J.M.R.C.; data curation, F.P.M. and J.M.R.C.; writing—original draft preparation, F.P.M. and J.M.R.C.; writing—review and editing, F.P.M., M.E.A., A.M.M.S.C. and J.M.R.C.; visualization, F.P.M. and J.M.R.C.; supervision, M.E.A. and J.M.R.C.; project administration, J.M.R.C.; funding acquisition, M.E.A., A.M.M.S.C. and J.M.R.C. All authors have read and agreed to the published version of the manuscript.

Funding: This research was supported by Project InPaCTus—Innovative Products and Technologies from eucalyptus, Project No 21 874 funded by Portugal 2020 through European Regional Development Fund (ERDF) in the frame of COMPETE 2020 no 246/AXIS II/2017.

Institutional Review Board Statement: Not applicable.

Informed Consent Statement: Not applicable.

Data Availability Statement: The raw/processed data required to reproduce these findings cannot be shared at this time due to legal or ethical reasons.

Acknowledgments: The authors are also very grateful for the support given by the Fiber Materials and Environmental Technologies Research Unit (FibEnTech-UBI) on the extent of the project reference UIDB/00195/2020.

Conflicts of Interest: The authors declare no conflict of interest.

References

1. Raunio, J.-P.; Löyttyniemi, T.; Ritala, R. Online quality evaluation of tissue paper structure on new generation tissue machines. *Nord. Pulp Pap. Res. J.* **2018**, *33*, 133–141. [CrossRef]
2. de Assis, T.; Reisinger, L.W.; Pal, L.; Pawlak, J.; Jameel, H.; Gonzalez, R.W. Understanding the Effect of Machine Technology and Cellulosic Fibers on Tissue Properties—A Review. *BioResources* **2018**, *13*, 4593–4629. [CrossRef]

3. Boudreau, J.; Germgård, U. Influence of various pulp properties on the adhesion between tissue paper and Yankee cylinder surface. *BioResources* **2014**, *9*, 2107–2114. [CrossRef]
4. de Assis, T.; Pawlak, J.; Pal, L.; Jameel, H.; Reisinger, L.W.; Kavalew, D.; Campbell, C.; Pawlowska, L.; Gonzalez, R.W. Comparison between uncreped and creped handsheets on tissue paper properties using a creping simulator unit. *Cellulose* **2020**, *27*, 5981–5999. [CrossRef]
5. Fišerová, M.; Gigac, J.; Stankovská, M.; Opálená, E. Influence of Bleached Softwood and Hardwood Kraft Pulps on Tissue Paper Properties. *Cellul. Chem. Technol.* **2019**, *53*, 469–477. [CrossRef]
6. Morais, F.P.; Bértolo, R.A.C.; Curto, J.M.R.; Amaral, M.E.C.C.; Carta, A.M.M.S.; Evtyugin, D.V. Comparative characterization of eucalyptus fibers and softwood fibers for tissue papers applications. *Mater. Lett. X* **2019**, *4*, 100028. [CrossRef]
7. Morais, F.P.; Bértolo, R.A.C.; Curto, J.M.R.; Amaral, M.E.C.C.; Carta, A.M.M.S.; Evtyugin, D.V. Characterization data of pulp fibres performance in tissue papers applications. *Data Brief* **2020**, *29*, 105253. [CrossRef]
8. Stankovská, M.; Gigac, J.; Fišerová, M.; Opálená, E. Blending impact of hardwood pulps with softwood pulp on tissue paper properties. *Wood Res.* **2020**, *65*, 447–458. [CrossRef]
9. Gigac, J.; Fišerová, M. Influence of pulp refining on tissue paper properties. *Tappi J.* **2008**, *7*, 27–32.
10. Chang, C.-H.; Yu, S.-T.; Perng, Y.-S. Effects of furnish and refining on properties of household paper. *Cellul. Chem. Technol.* **2018**, *52*, 433–440.
11. Demuner, B.J.; Junior, N.P.; Antunes, A.M.S. Technology prospecting on enzymes for the pulp and paper industry. *J. Technol. Manag. Innov.* **2011**, *6*, 148–158. [CrossRef]
12. Morais, F.P.; Carta, A.M.M.S.; Amaral, M.E.; Curto, J.M.R. Cellulose Fiber Enzymatic Modification to Improve the Softness, Strength, and Absorption Properties of Tissue Papers. *BioResources* **2021**, *16*, 846–861. [CrossRef]
13. Park, J.Y.; Melani, L.; Lee, H.; Kim, H.J. Effect of chemical additives on softness components of hygiene paper. *Nord. Pulp Pap. Res. J.* **2019**, *34*, 173–181. [CrossRef]
14. Valencia, C.; Valencia, Y.; Tovar, C.D.G. Synthesis and Application of a Cationic Polyamine as Yankee Dryer Coating Agent for the Tissue Paper-Making Process. *Polymers* **2020**, *12*, 173. [CrossRef]
15. Chan, E.; Woods, B.M.; Salaam, L.T.E. Soft Tissue Paper Having a Polyhydroxy Compound and Lotion Applied onto a Surface Thereof. U.S. Patent 7,972,475 B2, 5 July 2011.
16. Siqueira, E.J.; Salon, M.-C.B.; Belgacem, M.N.; Mauret, E. Carboxymethylcellulose (CMC) as a model compound of cellulose fibers and polyamideamine epichlorohydrin (PAE)-CMC interactions as a model of PAE-fibers interactions of PAE-based wet strength papers. *J. Appl. Polym. Sci.* **2015**, *132*, 42144. [CrossRef]
17. Adhikari, B.B.; Appadu, P.; Kislitsin, V.; Chae, M.; Choi, P.; Bressler, D.C. Enhancing the Adhesive Strength of a Plywood Adhesive Developed from Hydrolyzed Specified Risk Materials. *Polymers* **2016**, *8*, 285. [CrossRef]
18. Miller, J.H.; Sumnicht, D.W.; Oriaran, T.P.; Schuh, B.J.; Ramirez, A.J.; Lee, J.A. High Softness, High Durability Bath Tissues with Temporary Wet Strength. U.S. Patent 9,309,627 B2, 12 April 2016.
19. Miller, J.H.; Sumnicht, D.W.; Oriaran, T.P.; Schuh, B.J.; Ramirez, A.J.; Lee, J.A. Multi-Ply Bath Tissue with Temporary Wet Strength Resin and/or a Particular Lignin Content. U.S. Patent 9,879,382, 30 January 2018.
20. Chen, Z.; Zhang, L.; He, Z. Rethinking the determination of wet strength of paper. *BioResources* **2018**, *13*, 2184–2186. [CrossRef]
21. Zhang, Y.; Li, N.; Chen, Z.; Ding, C.; Zheng, Q.; Xu, J.; Meng, Q. Synthesis of High-Water-Resistance Lignin-Phenol Resin Adhesive with Furfural as a Crosslinking Agent. *Polymers* **2020**, *12*, 2805. [CrossRef]
22. Ginebreda, A.; Guilén, D.; Barceló, D.; Darbra, R.M. Additives in the Paper Industry. In *Global Risk-Based Management of Chemical Additives I: Production, Usage and Environmental Occurrence*, 1st ed.; Bilitewski, B., Darbra, R.M., Barceló, D., Eds.; Hdb Env Chem; Springer: Berlin/Heidelberg, Germany, 2012; pp. 11–34.
23. Shen, J.; Fatehi, P.; Ni, Y. Biopolymers for surface engineering of paper-based products. *Cellulose* **2014**, *21*, 3145–3160. [CrossRef]
24. Morais, F.P.; Carta, A.M.M.S.; Amaral, M.E.; Curto, J.M.R. Micro/nano-fibrillated cellulose (MFC/NFC) fibers as an additive to maximize eucalyptus fibers on tissue paper production. *Cellulose* **2021**, *28*, 6587–6605. [CrossRef]
25. Guan, M.; An, X.; Liu, H. Cellulose nanofiber (CNF) as a versatile filler for the preparation of bamboo pulp based tissue paper handsheets. *Cellulose* **2019**, *26*, 2613–2624. [CrossRef]
26. Patiño-Masó, J.; Serra-Parareda, F.; Tarrés, Q.; Mutjé, P.; Espinach, F.X.; Delgado-Aguilar, M. TEMPO-Oxidized Cellulose Nanofibers: A Potential Bio-Based Superabsorbent for Diaper Production. *Nanomaterials* **2019**, *9*, 1271. [CrossRef] [PubMed]
27. Zambrano, F.; Wang, Y.; Zwilling, J.D.; Venditti, R.; Jameel, H.; Rojas, O.; Gonzalez, R. Micro- and nanofibrillated cellulose from virgin and recycled fibers: A comparative study of its effects on the properties of hygiene tissue paper. *Carbohyd. Polym.* **2021**, *254*, 117430. [CrossRef] [PubMed]
28. Zambrano, F.; Starkey, H.; Wang, Y.; de Assis, C.A.; Venditti, R.; Pal, L.; Jameel, H.; Hubbe, M.A.; Rojas, O.J.; Gonzalez, R. Using Micro- and Nanofibrillated Cellulose as a Means to Reduce Weight of Paper Products: A Review. *BioResources* **2020**, *15*, 4553–4590. [CrossRef]
29. González, I.; Boufi, S.; Pèlach, M.A.; Alcalà, M.; Vilaseca, F.; Mutjé, P. Nanofibrillated cellulose as paper additive in Eucalyptus pulps. *BioResources* **2012**, *7*, 5167–5180. [CrossRef]
30. de Assis, T.C.A.; Iglesias, M.C.; Bilodeau, M.; Johnson, D.; Phillips, R.; Peresin, M.S.; Bilek, E.M.; Rojas, O.J.; Venditti, R.; Gonzalez, R. Cellulose micro- and nanofibrils (CMNF) manufacturing—financial and risk assessment. *Biofuels Bioprod. Biorefining* **2018**, *12*, 251–264. [CrossRef]

31. Hotaling, N.A.; Bharti, K.; Kriel, H.; Simon, C.G., Jr. DiameterJ: A Validated Open Source Nanofiber Diameter Measurement Tool. *Biomaterials* **2015**, *61*, 327–338. [CrossRef]
32. Conceição, E.L.T.; Curto, J.M.R.; Simões, R.M.S.; Portugal, A.T.G. Coding a simulation model of the 3D structure of paper. In *Computational Modeling of Objects Represented in Images. CompIMAGE 2010. Lecture Notes in Computer Science*, 1st ed., Barneva, R.P., Brimkov, V.E., Hauptman, H.A., Natal Jorge, R.M., Tavares, J.M.R.S., Eds.; Springer: Berlin/Heidelberg, Germany, 2010; Volume 6026, pp. 299–310.
33. Curto, J.M.R.; Conceição, E.L.T.; Portugal, A.T.G.; Simões, R.M.S. Three dimensional modeling of fibrous materials and experimental validation. *Mater. Werkst.* **2011**, *42*, 370–374. [CrossRef]
34. Morais, F.P.; Carta, A.M.M.S.; Amaral, M.E.; Curto, J.M.R. 3D Fiber Models to Simulate and Optimize Tissue Materials. *BioResources* **2020**, *15*, 8833–8848. [CrossRef]
35. de Assis, T.; Huang, S.; Driemeier, C.E.; Donohoe, B.S.; Kim, C.; Kim, S.H.; Gonzalez, R.; Jameel, H.; Park, S. Toward an understanding of the increase in enzymatic hydrolysis by mechanical refining. *Biotechnol. Biofuels* **2018**, *11*, 289. [CrossRef]
36. Martins, V.D.F.; Cerqueira, M.A.; Fuciños, P.; Garrido-Maestu, A.; Curto, J.M.R.; Pastrana, L.M. Active bi-layer cellulose-based films: Development and characterization. *Cellulose* **2018**, *25*, 6361–6375. [CrossRef]
37. Trepanier, R. Pulp fiber quality and the relationship with paper tissue properties. In Proceedings of the issue Conference & Expo 2017: The Power of TAPPI & RISI, Miami Beach, FL, USA, 3–6 October 2017; pp. 1–4.
38. Maurer, H.W. Starch in the Paper Industry. In *Starch—Chemistry and Technology. Food Science and Technology*, 3rd ed.; BeMiller, J., Whistler, R., Eds.; Academic Press: Cambridge, MA, USA, 2009; pp. 657–713.
39. Beuther, J.C.; Veith, M.W.; Zwick, K.J. Characterization of Absorbent Flow Rate in Towel and Tissue. *J. Eng. Fibers Fabr.* **2010**, *5*, 1–7. [CrossRef]

polymers MDPI

Article

Comparative Study of the Properties of Plasticized Polylactic Acid with Maleinized Hemp Seed Oil and a Novel Maleinized Brazil Nut Seed Oil

Aina Perez-Nakai [1], Alejandro Lerma-Canto [1], Ivan Dominguez-Candela [2], Daniel Garcia-Garcia [1], Jose Miguel Ferri [1] and Vicent Fombuena [1,*]

[1] Technological Institute of Materials (ITM), Universitat Politècnica de València (UPV), Plaza Ferrándiz y Carbonell 1, 03801 Alcoy, Spain; aipena@epsa.upv.es (A.P.-N.); allercan@epsa.upv.es (A.L.-C.); dagarga4@epsa.upv.es (D.G.-G.); joferaz@upvnet.upv.es (J.M.F.)

[2] Instituto de Seguridad Industrial, Radiofísica y Medioambiental (ISIRYM), Universitat Politècnica de València (UPV), Plaza Ferrándiz y Carbonell s/n, 03801 Alcoy, Spain; ivdocan@doctor.upv.es

* Correspondence: vifombor@upv.es

Abstract: In this study, for the first time, Brazil nut seed oil was chemically modified with maleic anhydride to obtain maleinized Brazil nut seed oil (MBNO). The same process was developed to obtain maleinized hemp seed oil (MHO). The use of MBNO and MHO was studied as bio-based plasticizers by incorporating them with different contents ranging from 0 to 10 phr in a polylactic acid (PLA) matrix. By means of mechanical, thermal and thermomechanical characterization techniques, the properties of the different formulations were studied to evaluate the plasticizing effect of the MBNO and MHO. With the addition of both plasticizers, a significant increase in ductile properties was observed, reaching an increase in elongation at break of 643% with 7.5 phr MBNO and 771% with 10 phr MHO compared to neat PLA. In addition, it has been observed that the mechanical resistant properties do not decrease, since the oils enhance the crystallization of PLA by increasing the free volume between its chains and counteracting the effect. Finally, a disintegration test was carried out under thermophilic conditions at 58 °C for 27 days, demonstrating that the incorporation of MHO and MBNO does not significantly affect the biodegradability of neat PLA.

Keywords: maleinized hemp seed oil; maleinized Brazil nut seed oil; bio-plasticizers; polylactic acid

Citation: Perez-Nakai, A.; Lerma-Canto, A.; Domingez-Candela, I.; Garcia-Garcia, D.; Ferri, J.M.; Fombuena, V. Comparative Study of the Properties of Plasticized Polylactic Acid with Maleinized Hemp Seed Oil and a Novel Maleinized Brazil Nut Seed Oil. *Polymers* **2021**, *13*, 2376. https://doi.org/10.3390/polym13142376

Academic Editors: Andreia F. Sousa, José Miguel Ferri, Vicent Fombuena Borràs and Miguel Fernando Aldás Carrasco

Received: 30 June 2021
Accepted: 16 July 2021
Published: 20 July 2021

1. Introduction

Since the end of the 20th century, environmental concerns and the development of sustainable strategies have been on the rise [1]. This concern carries over to the present day, where plastic represents a large part of human life, with up to 368 million tonnes of plastic produced globally in 2019 [2]. The problem arises mainly because of the negative impact on the ecosystems generated by the amount of plastic waste, since only 173 million tonnes are collected for recycling or landfill [3], but also by the exploitation of non-renewable natural resources, such as petroleum. The problem is magnified in the packaging industry, as it is the largest consumer of plastics, accounting for 39.6% of total plastic consumption in Europe [2]. In view of this situation, research is mainly focused on obtaining polymeric materials with formulations based on renewable resources or even with the property of being biodegradable. A wide variety of biopolymers or bio-based materials have now been obtained that contribute to this sustainable development.

Bio-based polymers are polymers that are made from biological substances, i.e., non-fossil materials [4]. These materials may or may not be biodegradable depending on their ability to be broken down by microorganisms [4]. Currently, some of the most promising biopolymers are extracted from biomass (starch, cellulose, protein, chitin, etc.), such as thermoplastic starches (TPSs), given the abundance of polysaccharides in the biomass. Another promising alternative are also those obtained by microbial production, such as polyhydroxyalkanoates (PHAs) and polyhydroxybutyrate (PHB) or polylactic acid (PLA) [5–7].

In 2020, bioplastics accounted for only 1.2 million tonnes, less than 1% of all plastics produced [3]. However, the market for these polymers has been growing for years and the trend is expected to increase, with production estimated to increase by 11.28% from 2019 to the end of 2025 [8]. Specifically, PLA is one of the biopolymers with the greatest potential for industrial use as a substitute for petroleum-based polymers [9,10]. This is due to its great potential due to the combination of its mechanical properties, easy processability and its price compared to other biopolymers [11,12]. PLA has a suitable thermal stability and resistance to being industrially processed by injection molding, welding, thermoforming or extrusion [13]. Moreover, as it is biodegradable, one of its most interesting applications is mainly in the packaging sector [14]. In terms of mechanical properties, this material is comparable to non-degradable polymers in the "commodity" range [14,15]. However, there is a major limitation of PLA, its brittleness, which is a major drawback in the packaging industry [16]. To alleviate this disadvantage, there are several proposals. One solution to improve the flexibility of this polymer is its blending with other polymers such as polyethylene glycol (PEG), thermoplastic starch (TPS) or polybutylene succinate-co-adipate (PBSA), among others. However, the lack of miscibility between the components makes it difficult to achieve this improvement in toughness [5,11,17,18]. For this reason, some authors proposed modified vegetable oils (MVOs) as compatibilizing agents or even as plasticizing additives [12,19]. The reason for this is that they are a renewable and sustainable substitute for synthetic modifiers [20,21], which are also respectful of human health due to their non-toxicity, as they do not generate the migration of substances such as Bisphenol A (BPA), as is the case with conventional epoxy resins used in consumer products [22].

Several techniques can be used for chemical modification, such as epoxidation [23,24], maleinization [5,10,25], acrylation or hydroxylation. The MVOs available on the market today are mainly epoxidized soybean oil (ESBO) [25] and epoxidized linseed oil (ELO) [26]. Authors such as Garcia-Garcia et al. or Chieng et al. [26,27] have reported the improvement of PLA stiffness properties with these oils. Additionally, other authors investigated other modified oils for PLA formulations such as epoxidized cottonseed oil (ECSO) [24] or epoxidized palm oil (EPO) [27]. Another commercial option is maleinized linseed oil, which has provided excellent properties to PLA, as reported Ferri et al. [28].

In this work, the chemical modification of process of oils carried out is the maleinization. Maleinization is a chemical process, usually carried out in a single step, which consists of incorporating maleic anhydride molecules into the triglycerides that make up vegetable oils when conjugated carbon–carbon double bonds are present. For this purpose, there are several methods that can be used, such as the so-called "ene" reaction, Diels-Alder addition and free radical copolymerization, the first being the most favorable and the one used in this work [5,29]. This procedure requires a temperature of about 200 °C to result in the addition of the anhydric groups at the allylic positions of the fatty acid, as shown in Figure 1.

Both Brazil nut (*Bertholletia excelsa*) and hemp seed (*Cannabis sativa* L.) are interesting as MVOs, since both have an interesting lipid profile for functionalization. The Brazil nut is a brown fruit generally cultivated in the Amazon [30]. Its oil is an interesting object of study, since it has between 60.8% and 72.5% lipids [30,31]. Among them, it has 75.6% unsaturated fatty acids (UFA) [30], and both monounsaturated (MUFA) and polyunsaturated (PUFA), which means a significant amount of double bonds that allow its functionalization by chemical processes such as maleinization. On the other hand, hemp seed oil is also attractive for this purpose, as it is high in linoleic acid (55.3%) and linolenic acid (20.3%) [32]. Both allow their oil to be extracted by cold pressing with a good yield and also have good oxidative stability [31,33]. Thereupon, this work evaluates the potential of MBNO and MHO as a bio-based plasticizer to improve the ductile properties of PLA related to its brittleness and the comparison of these with commercial maleinized linseed oil (MLO).

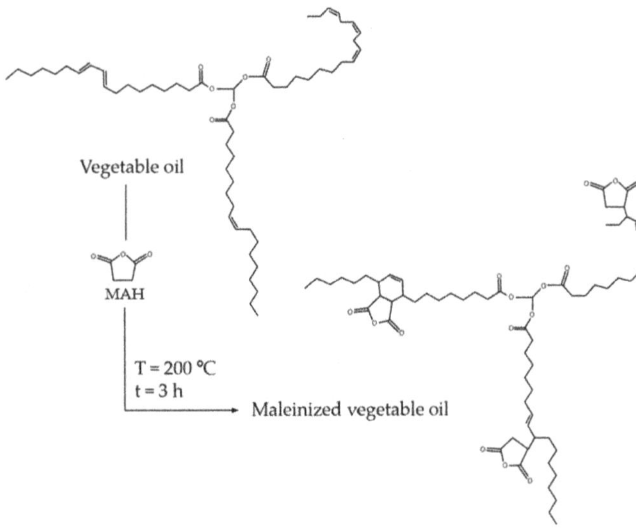

Figure 1. Schematic representation of the maleinization process of the triglyceride presents in vegetable oil.

2. Materials and Methods

2.1. Materials

The Brazil nuts used in this work were of Brazilian origin and were purchased from FrutoSeco (Alicante, Spain). On the other hand, the hemp seeds were obtained from a local market in Callosa de Segura (Alicante, Spain). The oil from both seeds was extracted by cold pressing with an extruder model DL-ZYJ05 purchased from Nanchang Dulong Industrial Company (Zucheng, China). The density of Brazil nut oil was 916 kg·m^{-3}, its iodine value (IV) between 99.5 and 102.6 g 100 g^{-1} and its acid value (AV) 0.2 mg·KOH g^{-1}. The density of hemp oil was 935 kg·m^{-3}, its IV is 163.8 g·100g^{-1} and its AV is 8 mg·KOH g^{-1}. For both cases, density, IV and AV values were obtained according to ISO 1675, ISO 3961 and ISO 660, respectively. For the maleinization of both oils, maleic anhydride supplied by Sigma Aldrich (Madrid, Spain), with a purity of 98% was used. The PLA used for mixing with the oils was a commercial-grade Ingeo Biopolymer 2003D from NatureWorks LLC (Minnetonka, MN, USA) in pellet form.

2.2. Maleinization of Brazil Nut (*Bertholletia excelsa*) and Hemp (*Cannabis sativa* L.) Oils

The same method was used for the maleinization of Brazil nut and hemp seed oil. For this purpose, a round flask with three necks with a capacity of 500 mL was used, where a stirrer was placed in the central neck and used at a speed of 300 rpm. A digital thermometer was placed in the second neck to measure the temperature during the process, and the extraction of samples and the introduction of maleic anhydride was carried out through the third neck. The maleinization ratio used was 2.4:1, following recommendations from previous works [5]. A total of 9 g of maleic anhydride per 100 g of each virgin oil was introduced. During the maleinization process, three temperature stages were used, at 180 °C, 200 °C and 220 °C, and samples were taken every 30 min. Initially, 300 g of oil was introduced into the round flask and when the first temperature of 180 °C was reached, 1/3 of the quantity of the total amount of maleic anhydride was added. This temperature was maintained for 1 h and increased to repeat the same at 200 °C and 220 °C. Finally, the mixture was allowed to cool to room temperature. The degree of maleinization of the oil was determined according to ISO 660:2009 using the following expression:

$$Acid\ Value = \frac{56.1 \times c \times V}{m} \quad (1)$$

where c is the exact concentration of the KOH standard solution used (mol·L^{-1}), V the volume of the KOH standard solution used (mL) and m the mass analyzed (g).

2.3. Manufacturing of PLA with Maleinized Vegetable Oils

Several PLA compositions were made with different phr of MBNO and MHO, which are summarized in Table 1, where the codes assigned to each of them are indicated. Prior to mixing, the PLA was dried for 24 h at 40 °C. The required amounts of PLA and oil were weighed and added, to be mixed manually. The mixtures were processed with a co-rotating twin-screw extruder (D = 30 mm; L/D = 20:1) from DUPRA (Alicante, Spain), at a rotation speed of 40 rpm with a temperature profile of 165 °C, 168.5 °C, 172.5 °C and 175 °C from the feed zone to die, respectively. The compositions obtained were cooled to room temperature, pelletized and dried at 40 °C for 24 h. Finally, each composition was shaped by injected molding from Mateu and Solé (Barcelona, Spain) at temperature profiles of 170 °C, 180 °C, 190 °C and 200 °C from the feed section to injection nozzle.

Table 1. Composition of plasticized PLA with different contents (phr) of maleinized vegetable oils and labeling of the formulations.

Reference	PLA	Maleinized Brazil Nut Oil—MBNO (phr)	Maleinized Hemp Seed Oil—MHO (phr)
PLA	100	0	0
PLA + 2.5 MBNO	100	2.5	0
PLA + 5 MBNO	100	5	0
PLA + 7.5 MBNO	100	7.5	0
PLA + 10 MBNO	100	10	0
PLA + 2.5 MHO	100	0	2.5
PLA + 5 MHO	100	0	5
PLA + 7.5 MHO	100	0	7.5
PLA + 10 MHO	100	0	10

2.4. Mechanical Characterization of PLA Formulations Plasticized with MBNO and MHO

Several standardized tensile, flexural, impact and hardness tests were carried out to determine the mechanical properties of the different PLA compositions manufactured with the different types and proportions of MVOs. The tensile and flexural tests were performed according to ISO 527 and ISO 178, respectively, with an Ibertest ELIB 30 universal testing machine from SAE Ibertest (Madrid, Spain). For the tensile test, a crosshead speed of 10 mm·min^{-1} was used with a 5 kN load cell. In the case of flexural test, the deformation rates were studied at 5 mm·min^{-1}. Both tests were performed on at least 5 different samples of each material and the average values of the elongation at break (%), tensile and flexural strength, and Young's modulus and flexural modulus were calculated. To determine the impact resistance by energy absorption, a 6 J Charpy pendulum from Metrotec S.A. (San Sebastian, Spain) was used. According to ISO 179, rectangular samples of 80 × 10 × 4 mm^3 without notches were employed. Finally, a Shore D hardness tester model 673-D from J. Bot S.A. (Barcelona, Spain) was used to determine the hardness according to ISO 868. A minimum of 5 samples were used and the results shown are the obtained average.

2.5. Thermal Analysis of PLA Formulations Plasticized with MBNO and MHO

The thermal properties of PLA and plasticized PLA compositions were determined by differential scanning calorimetry (DSC) and thermogravimetric analysis (TGA). The glass transition temperature (T_g), cold crystallization temperature (T_{cc}) and melting temperature

(T_m) were obtained by DSC with a Mettler Toledo DSC 821 (Schwerzenbach, Switzerland). The test conditions were under nitrogen atmosphere (flow rate 66 mL·min^{-1}) with a thermal program to remove thermal history consisting of a first heating from 30 °C to 200 °C at 10 °C·min^{-1}, followed by a cooling to 30 °C at 2 °C·min^{-1} and a last heating to 350 °C at 2 °C·min^{-1}. Thermal transitions were determined from the second heating. The percentage crystallinity of each material was calculated using the following equation:

$$X_c\ (\%) = \frac{\Delta H_m - \Delta H_c}{w \Delta H_m^o} \times 100 \quad (2)$$

where ΔH_m is the enthalpy of fusion, ΔH_c the cold crystallization enthalpy, w the mass fraction of the material and ΔH_m^o the enthalpy of fusion for a theoretical pure crystalline PLA structure, which was assumed to be 93 J g^{-1} [34].

TGA tests were carried out to determine the initial degradation temperature (T_0) and the maximum degradation temperature (T_{max}). The equipment used was a TGA/SDTA 851 from Mettler Toledo (Schwerzenbach, Switzerland) and the samples (7–10 mg) were tested under nitrogen atmosphere (flow rate 66 mL·min^{-1}) and heating from 30 °C to 700 °C at 20 °C·min^{-1}. T_0 was determined at 5% mass loss, while T_{max} was calculated from the first derivate of the TGA curves (DTG).

2.6. Thermomechanical Characterization of PLA Formulations Plasticized with MBNO and MHO

To analyze the changes in storage modulus (G') and damping factor, a dynamic thermo-mechanical analysis (DMTA) was performed with an oscillatory rheometer AR G2 from TA Instruments (New Castle, DE, EEUU), equipped with a clamp attachment for solid samples. The 40 × 10 × 4 mm^3 rectangular samples were tested in torsion mode and subjected to a thermal program from 30 °C to 130 °C with a heating rate of 2 °C·min^{-1}. The frequency was set at 1 Hz and the maximum deformation (γ) was 0.1%.

2.7. Degree of Disintegration under Composting Conditions of PLA Formulations Plasticized with MBNO and MHO

The degree of disintegration of each compound was carried out according to the recommendations of the ISO 20200 standard. For this purpose, 7 different square samples for each compound of dimensions 25 × 25 × 1 mm^3 were buried in controlled soil. Previously, samples were dried at 40 °C for 24 h. To evaluate the disintegration process, samples were unburied on days 3, 7, 14, 17, 21, 24 and 27, washed with distilled water, dried at 45 °C for 24 h and weighed in an analytical balance. The degree of disintegration was calculated with the following equation given in ISO 20200:

$$D = \frac{m_i - m_r}{m_i} \times 100 \quad (3)$$

where m_i is the initial mass of the sample before the test and m_r is the weight of the sample extracted from compost soil on different days after drying. Photographs were also taken of each sample recovered on the days indicated to also qualitatively assess the disintegration process.

3. Results

3.1. Synthesis of Maleinized Hemp Seed Oil and Brazil Nut Oil

The evolution of the AV values along the three temperature stages of 180 °C, 200 °C and 220 °C used throughout the maleinization process of BNO and HO has been plotted in Figure 2. Initially, the AV of BNO is 0.20 mg KOH g^{-1} and 8 mg KOH g^{-1} for the HO. In both cases, it can be observed how these values increase significantly and analogously after the first hour at 180 °C to approximately 47 mg KOH g^{-1} and 50 mg KOH g^{-1} for the BNO and HO, respectively, indicating that maleinization is occurring. After two hours, at the second temperature of 200 °C, another significant increment can be observed up to values close to 90 mg KOH g^{-1} in the case of BNO and up to 80 mg KOH g^{-1} in the case of HO. Finally, at 220 °C, after three hours, there is a drastic increase in the AV up to 105 mg KOH g^{-1} in HO and even up to 130 mg KOH g^{-1} in BNO. In the two oils, the AV follows a very similar trend in which a clear increase is observed as the temperature increases. This is due to the fact that at temperatures close to 200 °C, the "ene" reaction is favored, in which maleic anhydride can easily bind to an allylic position of the unsaturated fatty acid [35]. As the temperature of 220 °C is maintained over time, the acidity index tends to stabilize, in the case of HO at 105 mg KOH g^{-1} and at 130 mg KOH g^{-1} for BNO. Both values are in complete agreement with the AV of commercial MLO reported by Quiles-Carrillo et al. [36], which indicates that it ranges between 105 and 130 mg KOH g^{-1}, suggesting that both the MBNO and MHO developed have similar characteristics to one of the few maleinized oils available on the market today.

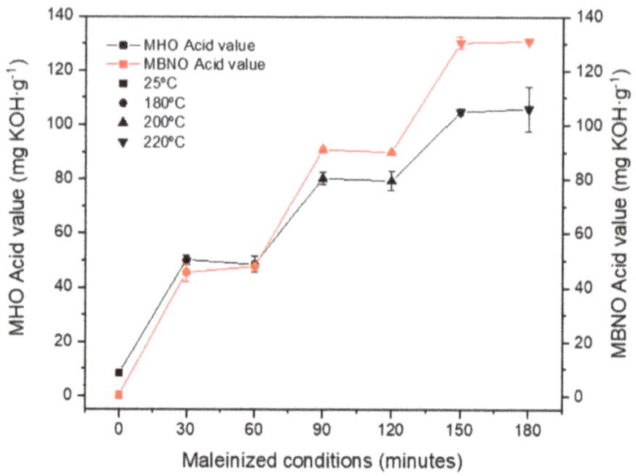

Figure 2. Effect of temperature and time on the efficiency of the maleinization of HO and BNO with maleic anhydride.

Another consequence observed as a result of the maleinization process is an important change in the color of the oils. Figure 3 shows a progressive color change of BNO, in which, at 180 °C, the color is light yellow, similar to the color of the original oil, and changes to a reddish color as maleinization occurs. The same occurs with HO. Ernzen et al. [37] reported a similar change during the maleinization process of soybean oil, in which the oil changed from a yellow to an orange–reddish color.

(a) (b) (c)

Figure 3. Influence of the reaction temperature and time on the color during maleinization of BNO, (**a**) 180 °C–0 min, (**b**) 200 °C–120 min and (**c**) 220 °C–180 min.

3.2. *Effect of MBNO and MHO on Mechanical Properties of Plasticized PLA Formulations*

The results obtained from the mechanical properties of the different PLA formulations with both MBNO and MHO show a significant reduction in PLA stiffness, indicating that both oils are effective as renewable PLA plasticizers. Figure 4 shows the tensile mechanical properties of PLA with MBNO and MHO. Unplasticized PLA has a Young's modulus close to 3000 MPa (2977 ± 21 MPa) and a tensile strength of 35.8 ± 7.3 MPa, relatively high values in the thermoplastic commodity range. However, its toughness is low because it shows an elongation at break of 7.4 ± 7%. By incorporating MBNO into the PLA matrix, an increase in elongation is observed from 2.5 phr, reaching the maximum value of 52% ± 3.0% at 7.5 phr, which represents an increase of 643% with respect to neat PLA. From this concentration of MBNO, a decrease in elongation occurs, probably due to plasticizer saturation, a phenomenon that some authors call an anti-plasticization effect [38]. On the other hand, the effect of MHO on the elongation at break is observed from 7.5 phr MHO where it increases drastically up to 42% ± 6% and even increases up to 61% ± 3.0% with the addition of 10 phr MHO, i.e., 771% more than neat PLA. These results obtained are close to the results reported by Ferri et al. [28], in which an elongation of 78.4% was obtained by incorporating commercial MLO into PLA. In Figure 5, it is possible to observe the changes in the appearance of the different tensile specimens after the test. Figure 5a shows specimens made with different MBNO contents. It can be seen how from 7.5 phr of MBNO content, the specimen has a lower final elongation. Figure 5b shows the specimens with different MHO contents after the tensile test. The main difference with respect to MBNO is observed in the PLA + 10% MHO sample, whose elongation is higher than that obtained with 7.5 phr MHO. This increase in elongation at break is due to the enhancement of molecular mobility, which is explained by several plasticization theories. The first of these, the lubricity theory, holds that the plasticizer functions as a molecular lubricant of the polymer. On the other hand, gel theory, applied to amorphous thermoplastics such as the grade of PLA used in this work, suggests that the plasticizer molecules are placed between the polymer chains and weaken the interactions between them. Finally, the free volume theory argues that the plasticizer increases the free volume and thus decreases the interactions between the polymer chains [39].

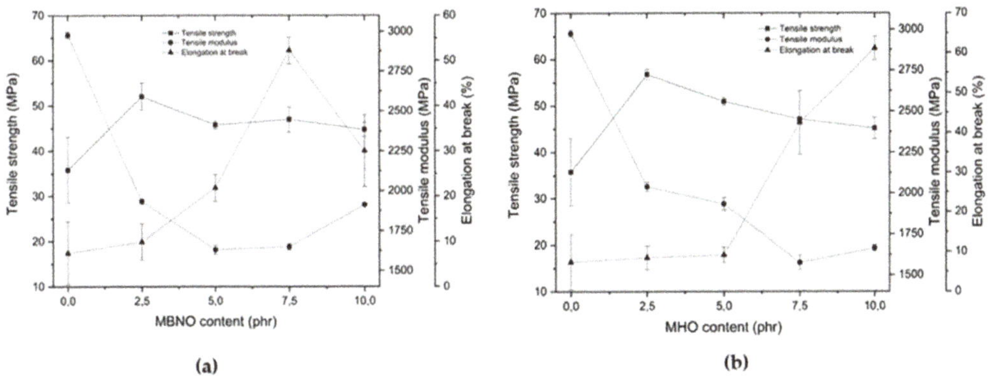

Figure 4. Plot of evolution of tensile properties of PLA formulations plasticized with various contents of MBNO (**a**) and MHO (**b**).

In addition, with the incorporation of both oils, an increase in tensile strength is observed with 2.5 phr, with respect to neat PLA, and then a decreasing trend as the percentage of oil incorporation increases. Several authors report a similar effect, but with lower strength values than PLA as an oil content increases. Garcia-Garcia et al. [40] reported a decrease in strength from 46.5 MPa to 42.2 MPa with the addition of 10% epoxidized karanja oil—EKO. In the case of MBNO and MHO, by incorporating between 2.5 and 10 phr, higher strengths were obtained than those of neat PLA, counterbalanced, in turn, by an improvement in its elongation. This phenomenon can be explained by the fact that MHO and MBNO, by improving the mobility between PLA chains, also facilitate crystallization [41], which at the same time generates an increase in its stiffness. Finally, Young's modulus follows a decreasing trend in both cases, obtaining the lowest value with the incorporation of 5 phr MBNO of up to 45.2% and with 7.5 phr MHO of up to 47%. From these contents in phr, the modulus tends to increase as the elongation tends to decrease. This decreasing evolution is in line with the results reported by authors such as Carbonell-Verdu et al. [24], who with the incorporation of ECSO to PLA obtained a reduction in the tensile modulus of 47.5% with respect to neat PLA. The high decrease in tensile modulus obtained with MBNO and MHO is related to the decrease in molecular interactions that facilitate movement.

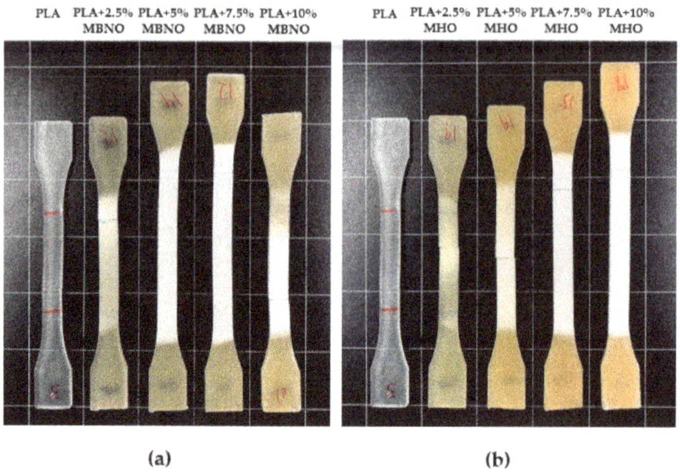

Figure 5. Plot of evolution of tensile properties of PLA formulations plasticized with various contents of MBNO (**a**) and MHO (**b**).

In parallel, Figure 6 shows a decrease in strength and flexural modulus as MBNO and MHO content increases. In the case of MBNO, a maximum reduction in flexural strength of 37.7% and flexural modulus of 20% is obtained with 10 phr and 39.6% and 18.4%, respectively, with 10 phr MHO content. However, as with the tensile properties, it is at 5 phr MBNO and 7.5 phr MHO that the decrease in strength and modulus tends to stabilize, suggesting plasticizer saturation. Ferri et al. [28] reported a decrease in the flexural strength of PLA with commercial MLO of up to 24%, and also a saturation around 15 phr MLO. The antiplasticizing effect of polymers depends, above all, on the molecular weight and the concentration of the diluent, so that in each formulation it is produced with a specific percentage of plasticizer [42].

On the other hand, Figure 7 summarizes the energy absorbed by Charpy's impact test and the Shore D hardness of neat PLA and formulations. The energy absorbed by Charpy's impact test is closely related to the toughness of the material; therefore, it is also representative for evaluating the effectiveness of MBNO and MHO as a plasticizer. Due to its brittleness, neat PLA has a relatively low absorbed energy (around 35.5 kJ·m^{-2}). By adding plasticizers, an increase in absorbed energy is observed, in the case of MBNO up to 20% higher and in the case of MHO up to 46% higher. The energy absorbed by a material depends both on the deformation capacity linked to the ductility properties and on the breaking strength, which is related to the mechanical properties [28]; therefore, the results are in full agreement with the obtained values of elongation, modulus and breaking strength. As for hardness, the value decreases progressively as more plasticizer is incorporated, although it is not as strongly visualized as with other properties. With MBNO the Shore D hardness decreased from 80 to 71.7 at 7.5 phr MBNO, and with 10 phr MHO to 72.9. Other studies reported a similar decrease from 75.6 to 59.6 with the addition of 22.5% commercial ELO to PLA/hazelnut shell flour (HSF) blends [43].

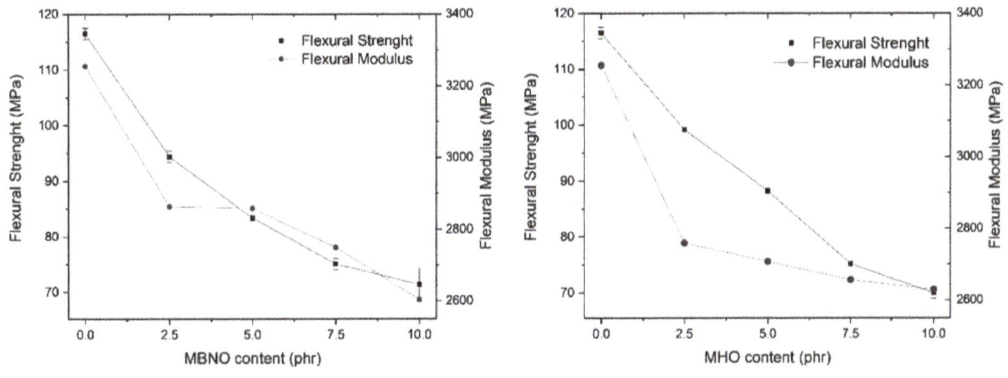

(a) (b)

Figure 6. Plot of evolution of flexural properties of PLA formulations plasticized with various contents of MBNO (**a**) and MHO (**b**).

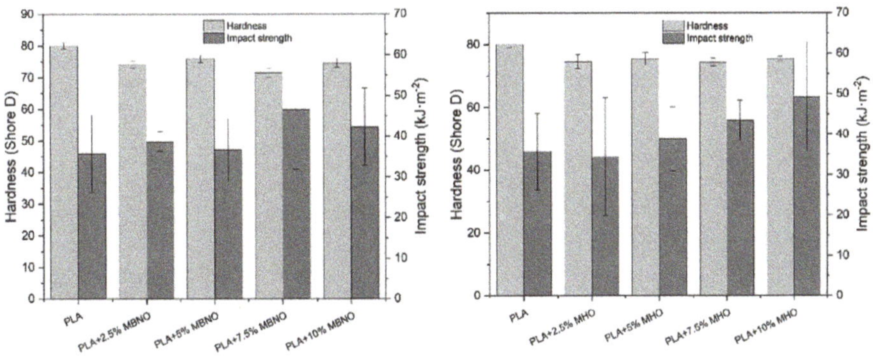

(a) (b)

Figure 7. Graphic representation of impact strength and hardness (Shore D) of PLA formulations plasticized with various contents of MBNO (**a**) and MHO (**b**).

3.3. *Effect of MBNO and MHO on Thermal Properties of Plasticized PLA Formulations*

Thermal stability of PLA and plasticized formulations with different content of MBNO and MHO was performed by thermogravimetry analysis (TGA). Figure 8 show the weight loss versus temperature curves for each sample (TG) and their corresponding first derivative curves (DGT). Table 2 shows some characteristic thermal parameters of the thermograms, such as the degradation onset temperature ($T_{5\%}$), which indicates the temperature at which a 5% weight loss occurs, and the maximum degradation temperature (T_{max}), which corresponds to the peak of the first derivative curve. Neat PLA possesses good thermal stability with a $T_{5\%}$ of 347.9 °C and a T_{max} of 386 °C. As can be seen, adding MBNO or MHO to PLA causes a slight decrease in $T_{5\%}$ as they are incorporated in higher quantities. In this case, the addition of 10 phr MNBO results in a $T_{5\%}$ reduction of 4.5 °C, while with 10 phr MHO, the reduction is 10.1 °C. Regarding the maximum degradation temperature, it can be observed that both plasticizers lead to a decrease in temperature, obtaining the lowest value at 10 phr, with a reduction of 16.4 °C for both plasticizers. Similar behavior was observed by Garcia-Campo et al. [44] for PLA/PHB/PCL blends compatibilized with epoxidized soybean oil (ELO). The authors observed how the addition of ELO into the blend resulted in a reduction in $T_{5\%}$ by 19.1 °C and T_{max} by 20 °C.

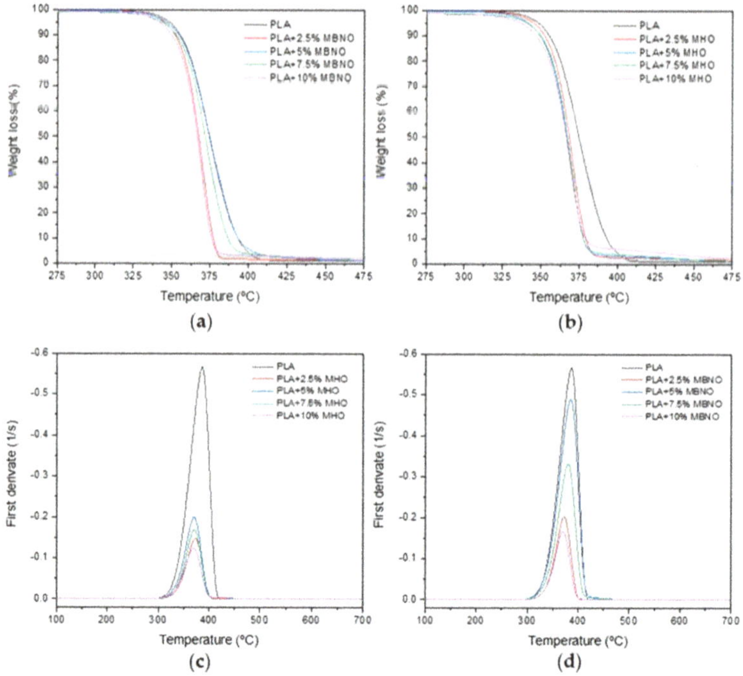

Figure 8. TGA (**a**,**b**) and DTGA (**c**,**d**) of unplasticized PLA and PLA plasticized with different content of MBNO and MHO.

Table 2. Summary of the TGA and DSC thermal parameters of PLA unplasticized and plasticized with different amounts of MBNO and MHO content.

Samples	TGA Parameters		DSC Parameters					
	$T_{5\%}$ (°C) [a]	T_{max} (°C)	T_g (°C)	T_{cc} (°C)	ΔH_c (Jg^{-1})	T_m (°C)	AH_m (Jg^{-1})	X_{PLA} (%)
PLA	347.9	386.2	61.0	123.64	1.2	150.7	3.2	1.7
PLA + 2.5%MBNO	346.6	372.1	58.0	100.8	24.9	153.3	26.8	2.1
PLA + 5%MBNO	346.5	384.6	57.9	120.8	9.2	150.3	15.0	5.8
PLA + 7.5%MBNO	344.5	379.3	58.7	123.7	3.3	151.2	10.0	6.8
PLA + 10%MBNO	343.4	369.6	57.8	123.5	2.6	151.0	3.2	7.3
PLA + 2.5%MHO	346.9	373.1	57.8	103.1	21.0	153.3	22.6	1.6
PLA + 5% MHO	342.9	370.6	56.8	101.8	20.8	153.6	28.4	7.5
PLA + 7.5% MHO	342.2	370.6	57.3	113.9	13.5	150.5	19.5	6.0
PLA + 10% MHO	337.8	369.6	56.9	109.9	9.7	149.2	20.3	10.6

[a] $T_{5\%}$, calculated at 5% mass loss.

Table 2 shows the main thermal properties obtained by DSC. On the other hand, Figure 9 shows the plot representation of the dynamic curve of DSC obtained with the unplasticized PLA and PLA plasticized with different contents of MBNO and MHO. As can be seen, both maleinized oils have a direct effect on some thermal properties of PLA. Unplasticized PLA has a glass transition temperature (T_g) located at 61 °C, a cold crystallization temperature (T_{cc}) at 123.6 °C and a melting temperature located at 150.7 °C. Both plasticizers, MBNO and MHO, decrease the T_g and T_{cc} of PLA, which is indicative of an increase in the mobility of the polymer chains at lower temperatures, evidencing the plasticizing effect of both maleinized oils [45]. The same evolution was reported by Dominguez-Candela et al. [46], who employed epoxidized chia oil (ECO) in PLA. On the other hand, no significant variations were observed in T_m, showing that after the addition of the plasticizers, the T_m remained at values around 150–153 °C for all the samples. As it is possible to observe in Figure 9, the addition of lower contents of maleinized oils provides two small melting temperature peaks, which are associated to the formation of two regions with different crystallinities, a result of the compatibilization process between the PLA matrix and the plasticizer. The addition of higher oil contents broadens these two peaks, producing an overlap between them. Regarding crystallinity, it is observed that PLA without plasticizing has a crystallinity of 1.7%, whereas as MBNO or MHO is added, the crystallinity increases progressively. As can be seen in Table 2, with a 10 phr MBNO, the crystallinity of PLA increases to 7.3%, while with the 10 phr MHO, it increases to 10.6%. A similar increase was reported by Carbonell-Verdu et al. [5], who observed that the addition of 10 phr commercial MLO and MCSO to PLA increased its crystallinity up to 11.6% and 19.1%, respectively. The increase in crystallinity with the addition of plasticizers, in this case MBNO and MHO, is caused by the facilitation of the laminar rearrangement of the amorphous zones of PLA, which is caused by the increased mobility of the polymer chains [47].

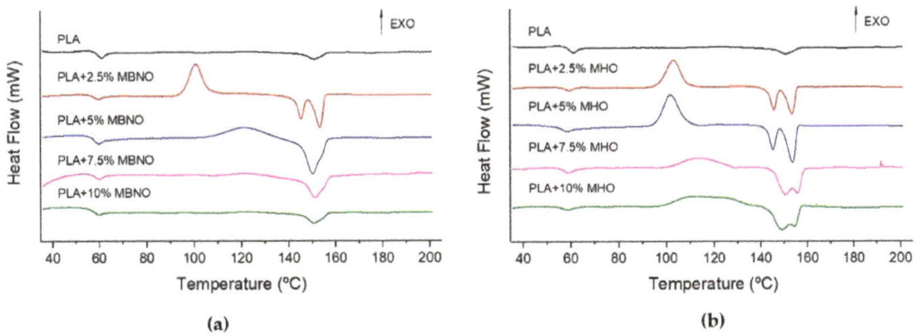

Figure 9. Dynamic DSC curve of PLA unplasticized and plasticized with different amounts of MBNO (**a**) and MHO content (**b**).

3.4. Effect of MBNO and MHO on Thermomechanical Properties of Plasticized PLA Formulations

The evolution of the storage modulus (G′) as a function of temperature is shown in Figure 10a,b. Neat PLA, at room temperature, has a storage modulus above 1000 MPa and this remains constant up to 60 °C. With the incorporation of MBNO and MHO in the PLA matrix, no significant changes are observed in terms of G′ at room temperature, since in all of them the values remain between 900 and 1100 MPa, although it is observed that with the incorporation of the plasticizers, the G′ tends to decrease slightly, showing its plasticizer effect. This is followed by a drop in G′ of up to three orders of magnitude, a change that is related to T_g. In this case, a significant difference can be observed between the plasticized samples and neat PLA. As can be seen after the addition of 10 phr MBNO, the T_g decreases from 60 °C to 48 °C and to almost 50 °C with 7.5 phr MHO. The almost 10 °C shift in T_g with both plasticizers is due, as described above, to the fact that the plasticizer generates an increase in the free volume, which improves the mobility of the chain due to a lower interaction between them, being able to move with lower energy [45]. Neat PLA above 75 °C behaves like an elastomer up to 95–100 °C. Above this temperature, the energetic conditions favor and promote crystallization, improving the elastic behavior and leading to an increase in the storage modulus up to 60 MPa. Here, again, differences are observed when incorporating the plasticizers into PLA. As can be seen, this temperature decreases by around 10 °C in all the MBNO-plasticized formulations and by around 5 °C in the MHO-plasticized samples with respect to neat PLA. This decrease in temperature is due to the higher mobility of the chains, which allows their rearrangement into a packed structure with lower energy [48].

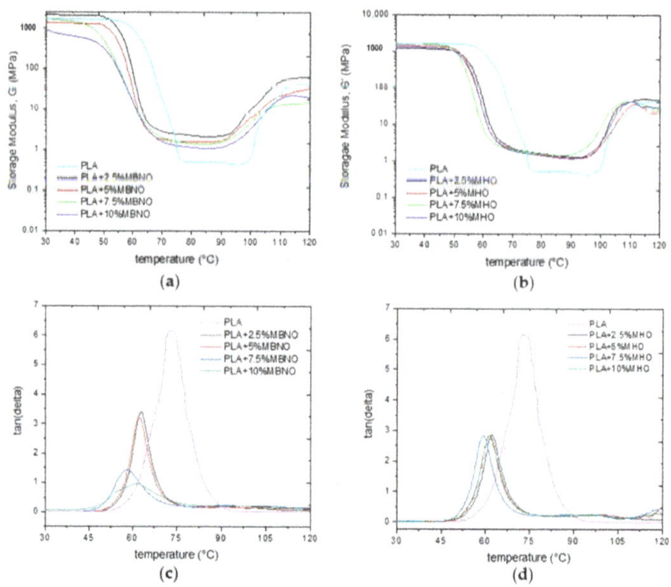

Figure 10. Storage modulus (**a**,**b**) and damping factor (**c**,**d**) of unplasticized PLA and PLA plasticized with different content of MBNO and MHO as a function of temperature.

The damping factor (tan δ), which represents the energy lost due to viscous behavior relative to the energy stored due to elastic behavior, is shown in Figure 10c,d. The peak tan δ, is a way to obtain the T_g value of the materials and, as can be seen, while that of neat PLA is around 73 °C, in the formulations plasticized with MBNO and MHO, this temperature decreases to 58 °C and 60 °C for samples plasticized with 7.5 phr MBNO and MHO, respectively. Santos et al. [49] observed a similar decrease in T_g with the addition to 10 wt.% and 20 wt.% of oligoesters obtained from sunflower oil biodiesel in PLA. The authors observed that the T_g of PLA decreased from 62 °C to 52 °C and 44 °C, respectively.

3.5. Disintegration under Composting Condition of PLA Formulations

The disintegration process of PLA formulations with MBNO and MHO in compost soil is shown visually in Figures 11 and 12, respectively. Initially, the samples were translucent in all formulations; however, after 3 days of incubation, a change in visual appearance to opaque was observed. This may be due to several factors. One cause can be attributed to the 50% relative humidity test condition, since possible hydrolytic degradation due to water absorption affects the refractive index [50]. On the other hand, taking into account the fact that the test was carried out under thermophilic conditions at 58 °C, the opacity may also be due to crystallization, since this temperature is close to the T_g obtained in the thermal study by DSC. As can be seen in Figure 13, where the weight loss with respect to the initial mass at different periods of incubation for two PLA formulations with MBNO (a) and MHO (b) after 7 days was buried, the samples began to lose mass, which led to increased embrittlement, which is observed in the images with the appearance of cracks. However, it was not until day 14 that significant weight loss and inconsistency of the samples was observed. In the case of neat PLA, a faster degradation than in the plasticized PLA was observed, since on day 17 it had already exceeded 90% mass loss, the degree of disintegration determined by the ISO 20200 standard for considering a material to be disintegrable. In the case of plasticized PLA, in the formulations with 2.5 phr and 5 phr of both MBNO and MHO, 90% mass loss was reached at day 27, while with the 7.5 phr and 10 phr formulations, it was reached at day 24. Although the difference in the disintegration time of PLA formulations with plasticizer is small, and in all cases, the time is longer than with neat PLA, and a slight increase in the disintegration rate is observed when more plasticizer is added. This delay in disintegration with respect to neat PLA when introducing plasticizer is due to the fact that the PLA grade used in this work is very amorphous, as can be seen in the thermal analysis; when introducing plasticizer, crystallinity increases, making it difficult for microorganisms to act in the degradation, which act faster in amorphous domains [51,52]. A similar trend was reported by Balart et al. [53], who observed an increase in disintegration time upon the incorporation of ELO into the PLA matrix. Therefore, in view of the results, as a conclusion of the disintegration study, it has been demonstrated that the maleinized oils slightly retard the disintegration of PLA samples. However, PLA compounds plasticized with MBNO and MHO can be considered equally biodegradable by composting.

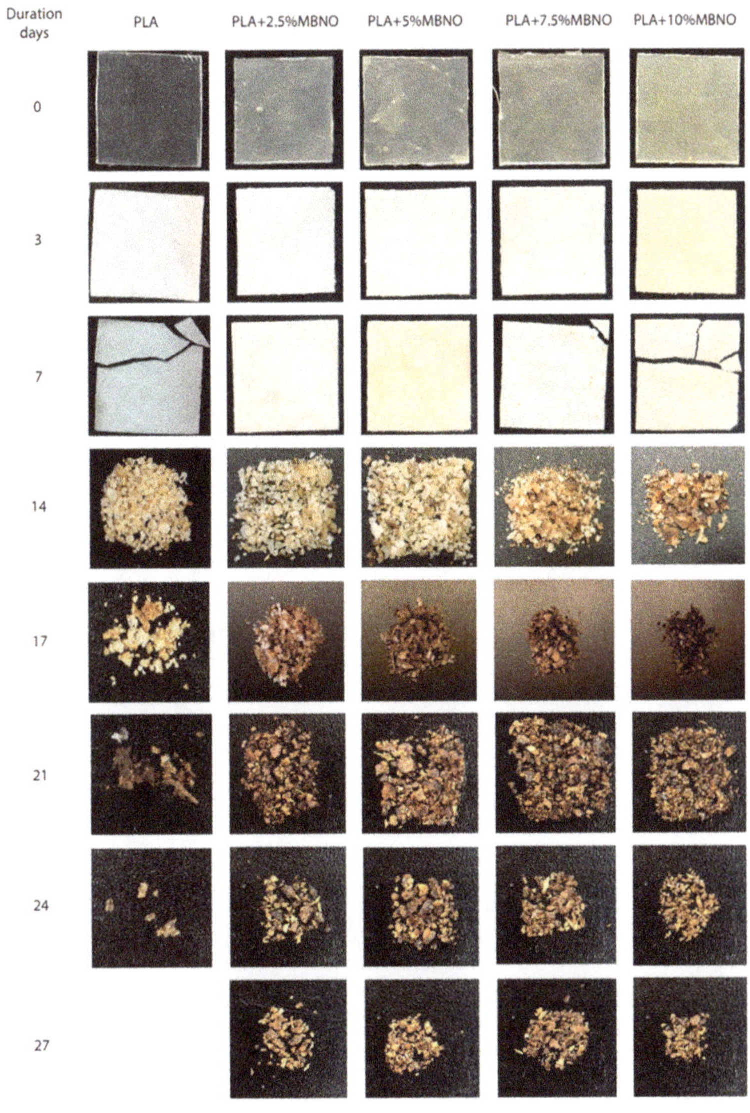

Figure 11. Visual appearance of disintegration under composting conditions of PLA and PLA plasticized with different contents of MBNO.

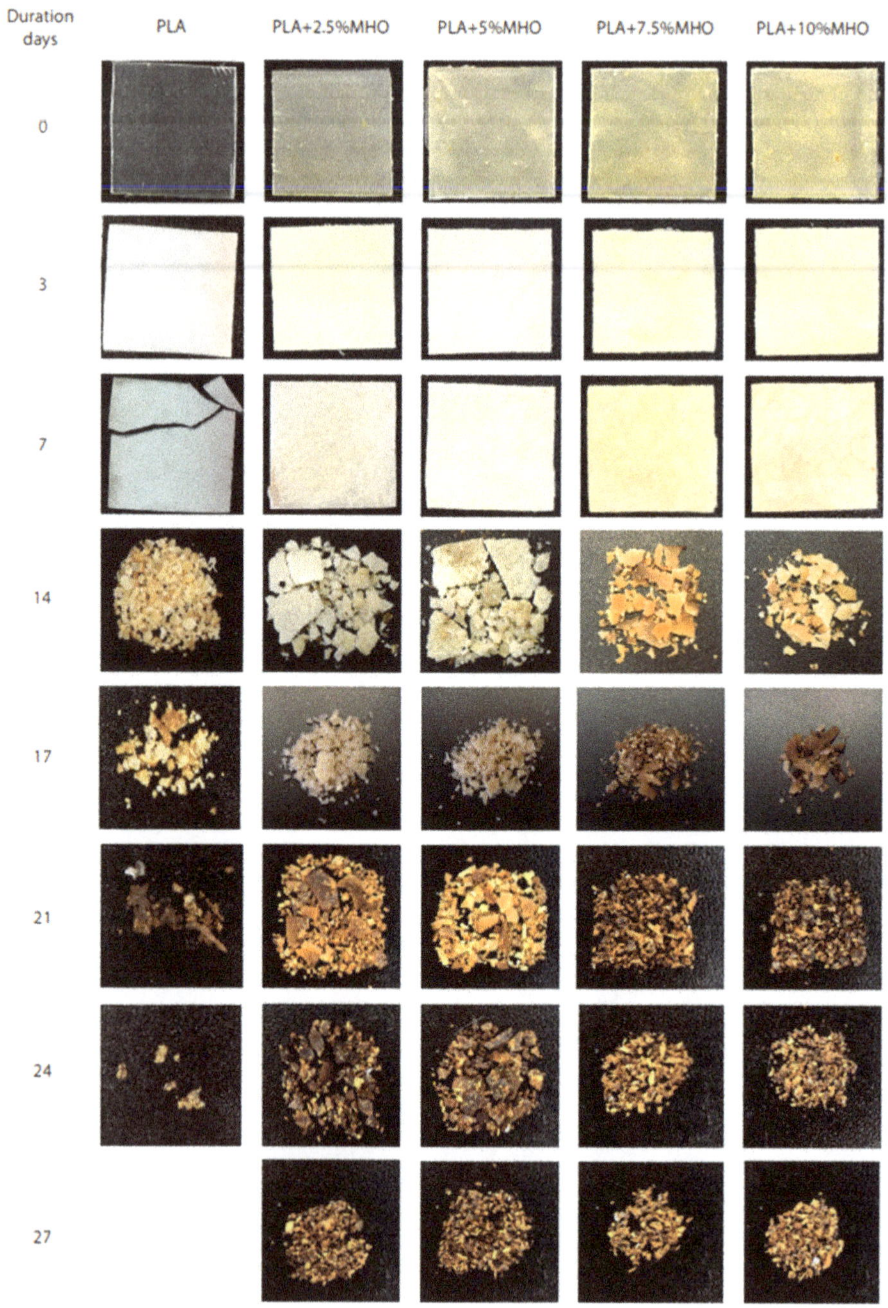

Figure 12. Visual appearance of disintegration under composting conditions of PLA and PLA plasticized with different contents of MHO.

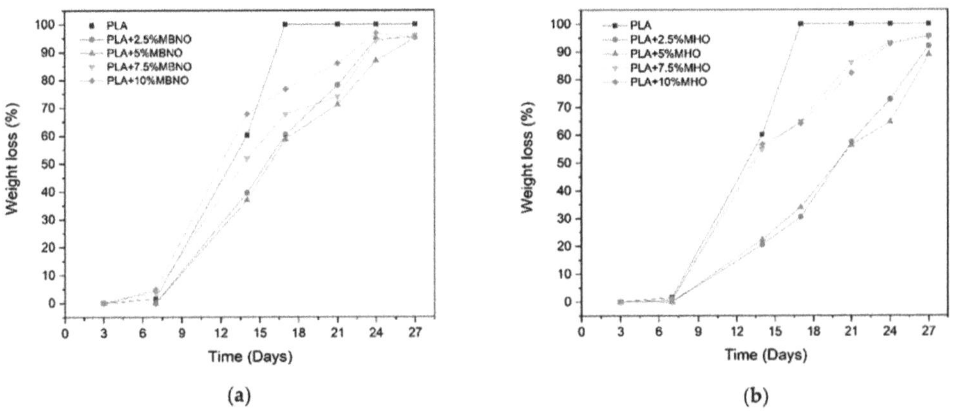

Figure 13. Weight loss recorded during disintegration test of PLA formulations with MBNO (**a**) and MHO (**b**).

4. Conclusions

This research work has developed for the first time a maleinized Brazil nut oil (MBNO). This, as well as a maleinized hemp oil (MHO), were introduced into the PLA matrix to study and analyze their effect as bio-based plasticizers. In addition, the results obtained show MBNO and MHO provided a similar performance compared with a commercially available maleinized linseed oil (MLO), demonstrating its potential. For example, the elongation at break of neat PLA is 7.4%, which is quite low given its brittleness. With the addition of MBNO and MHO, an improvement of 643% and 771%, respectively, was achieved. On the other hand, with values higher than 7.5 phr MBNO, a decrease in the elongation at break was observed due to the anti-plasticizing effect produced by saturation in the mixture. In terms of absorbed impact energy, a 20% higher value was obtained with the addition of 7.5 phr MBNO and 46% higher with 10 phr MHO compared to unplasticized PLA. In addition, mechanical properties such as tensile strength do not decrease, as both maleinized oils provide an improvement in the mobility of PLA chains and an increase in free volume, increasing the degree of crystallinity and thus counteracting the plasticizing effect that would decrease these properties. Thermal parameters, such as T_g, show a decreasing influence with the presence of these bio-plasticizers developed, but not drastically. Finally, the disintegration test under composting conditions showed that the addition of MBNO and MHO, although slightly delaying the process, did not lead to a loss of the biodegradability of PLA. Therefore, both MBNO and MHO are shown as potential bio-plasticizers of organic origin to increase the ductility of PLA without affecting its mechanical properties and not affecting its biodegradability, which makes them two bio-plasticizers of interest for industrial applications.

Author Contributions: Conceptualization, A.P.-N. and V.F.; methodology, A.P.-N., A.L.-C. and I.D.-C.; validation, D.G.-G., J.M.F. and V.F.; formal analysis, A.P.-N., A.L.-C. and I.D.-C.; investigation, A.P.-N. and A.L.-C.; resources, D.G.-G. and V.F.; data curation, D.G.-G.; writing—original draft preparation, A.P.-N., J.M.F. and D.G.-G.; writing—review and editing, V.F.; visualization, D.G.-G.; supervision, J.M.F. and V.F.; project administration, V.F.; funding acquisition, J.M.F., D.G.-G. and V.F. All authors have read and agreed to the published version of the manuscript.

Funding: I.D.-C. wants to thank Universitat Politècnica de València for his FPI grant (PAID-2019-SP20190013) and Generalitat Valenciana (GVA) for his FPI grant (ACIF/2020/233). J.M.F. thanks the postdoc contract (APOSTD/2019/122) Generalitat Valenciana (2019–2021).

Institutional Review Board Statement: Not applicable.

Informed Consent Statement: Not applicable.

Acknowledgments: The authors want to thank the Cátedra FACSA-FOVASA of Water, Waste and Circular Economy for their support, which promotes and supports training, dissemination, innovation, social responsibility and entrepreneurship activities in the field of circular economy and has kindly provided the necessary chemical reagents.

Conflicts of Interest: The authors declare no conflict of interest.

References

1. *Report of the World Commission on Environment and Development: Our common future*; Development and International Co-operation: Environment, Oxford University Press: London, UK, 1987; Volume 10, p. 300.
2. *An Analysis of European Plastics Production, Demand and Waste Data*; Plastics Europe: Brussels, Belgium, 2020; p. 64.
3. Rai, P.; Mehrotra, S.; Priya, S.; Gnansounou, E.; Sharma, S. Recent advances in the sustainable design and applications of biodegradable polymers. *Bioresour. Technol.* **2021**, *325*, 124739. [CrossRef]
4. Pasaoglu, M.E.; Koyuncu, I. Substitution of Petroleum-Based Polymeric Materials Used in the Electrospinning Process with Nanocellulose: A Review and Future Outlook. *Chemosphere* **2020**, *269*, 128710. [CrossRef]
5. Carbonell-Verdu, A.; Garcia-Garcia, D.; Dominici, F.; Torre, L.; Sanchez-Nacher, L.; Balart, R. PLA films with improved flexibility properties by using maleinized cottonseed oil. *Eur. Polym. J.* **2017**, *91*, 248–259. [CrossRef]
6. Mittal, V.; Akhtar, T.; Matsko, N. Engineering. Mechanical, thermal, rheological and morphological properties of binary and ternary blends of PLA, TPS and PCL. *Macromol. Mater. Eng.* **2015**, *300*, 423–435. [CrossRef]
7. Chivrac, F.; Pollet, E.; Averous, L. Progress in nano-biocomposites based on polysaccharides and nanoclays. *Mater. Sci. Eng. R Rep.* **2009**, *67*, 1–17. [CrossRef]
8. *Biodegradable Plastics Market Is Expected to Grow USD 8971.32 Million by the End of 2025*; Valuates Reports; PR Newswire: Bangalore, India, 2020.
9. Yu, Y.; Cheng, Y.; Ren, J.; Cao, E.; Fu, X.; Guo, W. Plasticizing effect of poly(ethylene glycol)s with different molecular weights in poly(lactic acid)/starch blends. *J. Appl. Polym. Sci.* **2015**, *132*, 132. [CrossRef]
10. Arrieta, M.P.; Fortunati, E.; Dominici, F.; Rayón, E.; López, J.; Kenny, J.M. PLA-PHB/cellulose based films: Mechanical, barrier and disintegration properties. *Polym. Degrad. Stab.* **2014**, *107*, 139–149. [CrossRef]
11. Fortunati, E.; Armentano, I.; Iannoni, A.; Kenny, J.M. Development and thermal behaviour of ternary PLA matrix composites. *Polym. Degrad. Stab.* **2010**, *95*, 2200–2206. [CrossRef]
12. Ferri, J.M.; Garcia-Garcia, D.; Sánchez-Nacher, L.; Fenollar, O.; Balart, R. The effect of maleinized linseed oil (MLO) on mechanical performance of poly (lactic acid)-thermoplastic starch (PLA-TPS) blends. *Carbohydr. Polym.* **2016**, *147*, 60–68. [CrossRef] [PubMed]
13. Carrasco, F.; Gámez-Pérez, J.; Santana, O.; Maspoch, M. Processing of poly (lactic acid)/organomontmorillonite nanocomposites: Microstructure, thermal stability and kinetics of the thermal decomposition. *Chem. Eng. J.* **2011**, *178*, 451–460. [CrossRef]
14. Nampoothiri, K.M.; Nair, N.R.; John, R. An overview of the recent developments in polylactide (PLA) research. *Bioresour. Technol.* **2010**, *101*, 8493–8501. [CrossRef]
15. Wee, Y.-J.; Kim, J.-N.; Ryu, H.-W. Biotechnological production of lactic acid and its recent applications. *Biotechnology* **2006**, *44*, 163–172.
16. Chieng, B.W.; Ibrahim, N.A.; Yunus, W.M.Z.W.; Hussein, M.Z. Poly (lactic acid)/poly (ethylene glycol) polymer nanocomposites: Effects of graphene nanoplatelets. *Polymers* **2014**, *6*, 93–104. [CrossRef]
17. Müller, P.; Bere, J.; Fekete, E.; Móczó, J.; Nagy, B.; Kállay, M.; Gyarmati, B.; Pukánszky, B. Interactions, structure and properties in PLA/plasticized starch blends. *Polymers* **2016**, *103*, 9–18. [CrossRef]
18. Malwela, T.; Ray, S.S. Enzymatic degradation behavior of nanoclay reinforced biodegradable PLA/PBSA blend composites. *Int. J. Biol. Macromol.* **2015**, *77*, 131–142. [CrossRef] [PubMed]
19. Ghorui, S.; Bandyopadhyay, N.; Ray, D.; Sengupta, S.; Kar, T. Industrial Crops and Products. Use of maleated castor oil as biomodifier in unsaturated polyester resin/fly ash composites. *Ind. Crop. Prod.* **2011**, *34*, 893–899. [CrossRef]
20. Meier, M.A.; Metzger, J.O.; Schubert, U.S.J. Plant oil renewable resources as green alternatives in polymer science. *Chem. Soc. Rev.* **2007**, *36*, 1788–1802. [CrossRef] [PubMed]
21. Maisonneuve, L.; Lebarbé, T.; Grau, E.; Cramail, H.J. Structure—Properties relationship of fatty acid-based thermoplastics as synthetic polymer mimics. *Polym. Chem.* **2013**, *4*, 5472–5517. [CrossRef]
22. Liao, C.; Liu, F.; Alomirah, H.; Loi, V.D.; Mohd, M.A.; Moon, H.-B.; Nakata, H.; Kannan, K.J.E.s. Bisphenol S in urine from the United States and seven Asian countries: Occurrence and human exposures. *Technology* **2012**, *46*, 6860–6866. [CrossRef]
23. Janković, M.R.; Govedarica, O.M.; Sinadinović-Fišer, S.V.J. The epoxidation of linseed oil with in situ formed peracetic acid: A model with included influence of the oil fatty acid composition. *Ind. Crop. Prod.* **2020**, *143*, 111881. [CrossRef]
24. Carbonell-Verdu, A.; Samper, M.D.; Garcia-Garcia, D.; Sanchez-Nacher, L.; Balart, R.J. Plasticization effect of epoxidized cottonseed oil (ECSO) on poly (lactic acid). *Ind. Crop. Prod.* **2017**, *104*, 278–286. [CrossRef]
25. Stanzione, J., III; La Scala, J. Sustainable polymers and polymer science: Dedicated to the life and work of Richard P. Wool. *J. Appl. Polym. Sci.* **2016**, *133*. [CrossRef]
26. Garcia-Garcia, D.; Ferri, J.M.; Montanes, N.; Lopez-Martinez, J.; Balart, R.J. Plasticization effects of epoxidized vegetable oils on mechanical properties of poly (3-hydroxybutyrate). *Polym. Int.* **2016**, *65*, 1157–1164. [CrossRef]

27. Chieng, B.W.; Ibrahim, N.A.; Then, Y.Y.; Loo, Y.Y.J. Epoxidized vegetable oils plasticized poly (lactic acid) biocomposites: Mechanical, thermal and morphology properties. *Molecules* **2014**, *19*, 16024–16038. [CrossRef]
28. Ferri, J.M.; Garcia-Garcia, D.; Montanes, N.; Fenollar, O.; Balart, R.J. The effect of maleinized linseed oil as biobased plasticizer in poly (lactic acid)-based formulations. *Polym. Int.* **2017**, *66*, 882–891. [CrossRef]
29. Quiles-Carrillo, L.; Blanes-Martínez, M.; Montanes, N.; Fenollar, O.; Torres-Giner, S.; Balart, R.J. Reactive toughening of injection-molded polylactide pieces using maleinized hemp seed oil. *Eur. Polym. J.* **2018**, *98*, 402–410. [CrossRef]
30. Chunhieng, T.; Goli, T.; Piombo, G.; Pioch, D.; Brochier, J.; Montet, D. Recent analysis of the composition of Brazil nut Bertholletia excelsa. *Amaz. Bois Trop.* **2004**, *280*.
31. Cornelio-Santiago, H.P.; Bodini, R.B.; de Oliveira, A.L.J. Potential of Oilseeds Native to Amazon and Brazilian Cerrado Biomes: Benefits, Chemical and Functional Properties, and Extraction Methods. *J. Am. Oil Chem. Soc.* **2021**, *98*, 3–20. [CrossRef]
32. Mikulcová, V.; Kašpárková, V.; Humpolíček, P.; Buňková, L.J. Formulation, characterization and properties of hemp seed oil and its emulsions. *Molecules* **2017**, *22*, 700. [CrossRef]
33. Abuzaytoun, R.; Shahidi, F.J. Oxidative stability of flax and hemp oils. *J. Am. Oil Chem. Soc.* **2006**, *83*, 855–861. [CrossRef]
34. Ferri, J.M.; Fenollar, O.; Jorda-Vilaplana, A.; García-Sanoguera, D.; Balart, R.J. Effect of miscibility on mechanical and thermal properties of poly (lactic acid)/polycaprolactone blends. *Polym. Int.* **2016**, *65*, 453–463. [CrossRef]
35. Eren, T.; Küsefoğlu, S.H.; Wool, R.J. Polymerization of maleic anhydride–modified plant oils with polyols. *J. Appl. Polym. Sci.* **2003**, *90*, 197–202. [CrossRef]
36. Quiles-Carrillo, L.; Montanes, N.; Sammon, C.; Balart, R.; Torres-Giner, S.J. Compatibilization of highly sustainable polylactide/almond shell flour composites by reactive extrusion with maleinized linseed oil. *Ind. Crop. Prod.* **2018**, *111*, 878–888. [CrossRef]
37. Ernzen, J.R.; Bondan, F.; Luvison, C.; Henrique Wanke, C.; De Nardi Martins, J.; Fiorio, R.; Bianchi, O.J. Structure and properties relationship of melt reacted polyamide 6/malenized soybean oil. *J. Appl. Polym. Sci.* **2016**, *133*, 133. [CrossRef]
38. Mikus, P.-Y.; Alix, S.; Soulestin, J.; Lacrampe, M.; Krawczak, P.; Coqueret, X.; Dole, P.J. Deformation mechanisms of plasticized starch materials. *Carbohydr. Polym.* **2014**, *114*, 450–457. [CrossRef]
39. Daniels, P.H.J. Brief overview of theories of PVC plasticization and methods used to evaluate PVC-plasticizer interaction. *J. Vinyl Addit. Technol.* **2009**, *15*, 219–223. [CrossRef]
40. Garcia-Garcia, D.; Carbonell-Verdu, A.; Arrieta, M.; López-Martínez, J.; Samper, M.J. Stability. Improvement of PLA film ductility by plasticization with epoxidized karanja oil. *Polym. Degrad. Stab.* **2020**, *179*, 109259. [CrossRef]
41. Cui, L.; Zhang, R.; Wang, Y.; Zhang, C.; Guo, Y.J. Effect of plasticizer poly (ethylene glycol) on the crystallization properties of stereocomplex-type poly (lactide acid). *Wuhan Univ. J. Nat. Sci.* **2017**, *22*, 420–428. [CrossRef]
42. Moraru, C.; Lee, T.C.; Karwe, M.; Kokini, J.J. Plasticizing and Antiplasticizing Effects of Water and Polyols on a Meat-Starch Extruded Matri. *J. Food Sci.* **2002**, *67*, 3396–3401. [CrossRef]
43. Balart, J.; Fombuena, V.; Fenollar, O.; Boronat, T.; Sánchez-Nacher, L.J. Processing and characterization of high environmental efficiency composites based on PLA and hazelnut shell flour (HSF) with biobased plasticizers derived from epoxidized linseed oil (ELO). *Compos. Part B Eng.* **2016**, *86*, 168–177. [CrossRef]
44. Garcia-Campo, M.J.; Quiles-Carrillo, L.; Masia, J.; Reig-Pérez, M.J.; Montanes, N.; Balart, R.J. Environmentally friendly compatibilizers from soybean oil for ternary blends of poly (lactic acid)-PLA, poly (ε-caprolactone)-PCL and poly (3-hydroxybutyrate)-PHB. *Materials* **2017**, *10*, 1339. [CrossRef]
45. Dobircau, L.; Delpouve, N.; Herbinet, R.; Domenek, S.; Le Pluart, L.; Delbreilh, L.; Ducruet, V.; Dargent, E.J. Molecular mobility and physical ageing of plasticized poly (lactide). *Polym. Eng. Sci.* **2015**, *55*, 858–865. [CrossRef]
46. Dominguez-Candela, I.; Ferri, J.M.; Cardona, S.C.; Lora, J.; Fombuena, V.J. Dual Plasticizer/Thermal Stabilizer Effect of Epoxidized Chia Seed Oil (Salvia hispanica L.) to Improve Ductility and Thermal Properties of Poly (Lactic Acid). *Ploymers* **2021**, *13*, 1283. [CrossRef] [PubMed]
47. Xiong, Z.; Li, C.; Ma, S.; Feng, J.; Yang, Y.; Zhang, R.; Zhu, J.J. The properties of poly (lactic acid)/starch blends with a functionalized plant oil: Tung oil anhydride. *Carbohydr. Polym.* **2013**, *95*, 77–84. [CrossRef] [PubMed]
48. Silverajah, V.; Ibrahim, N.A.; Zainuddin, N.; Yunus, W.M.Z.W.; Hassan, H.A.J. Mechanical, thermal and morphological properties of poly (lactic acid)/epoxidized palm olein blend. *Molecules* **2012**, *17*, 11729–11747. [CrossRef]
49. Santos, E.F.; Oliveira, R.V.; Reiznautt, Q.B.; Samios, D.; Nachtigall, S.M.J. Sunflower-oil biodiesel-oligoesters/polylactide blends: Plasticizing effect and ageing. *Polym. Test.* **2014**, *39*, 23–29. [CrossRef]
50. Fukushima, K.; Tabuani, D.; Abbate, C.; Arena, M.; Ferreri, L.J. Effect of sepiolite on the biodegradation of poly (lactic acid) and polycaprolactone. *Polym. Degrad. Stab.* **2010**, *95*, 2049–2056. [CrossRef]
51. Ray, S.S.; Yamada, K.; Ogami, A.; Okamoto, M.; Ueda, K.J. New polylactide/layered silicate nanocomposite: Nanoscale control over multiple properties. *Macromol. Rapid Commun.* **2002**, *23*, 943–947. [CrossRef]
52. Tokiwa, Y.; Calabia, B.P.J. Biodegradability and biodegradation of poly (lactide). *Appl. Microbiol. Biotechnol.* **2006**, *72*, 244–251. [CrossRef]
53. Balart, J.; Montanes, N.; Fombuena, V.; Boronat, T.; Sánchez-Nacher, L.J. Disintegration in compost conditions and water uptake of green composites from poly (lactic acid) and hazelnut shell flour. *J. Polym. Environ.* **2018**, *26*, 701–715. [CrossRef]

polymers

MDPI

Article

Effects of Poloxamer Content and Storage Time of Biodegradable Starch-Chitosan Films on Its Thermal, Structural, Mechanical, and Morphological Properties

Abril Fonseca-García [1,2], Carolina Caicedo [3], Enrique Javier Jiménez-Regalado [1], Graciela Morales [1] and Rocio Yaneli Aguirre-Loredo [1,2,*]

1 Centro de Investigación en Química Aplicada (CIQA), Blvd. Enrique Reyna Hermosillo 140, Saltillo, Coahuila 25294, Mexico; abril.fonseca@ciqa.edu.mx (A.F.-G.); enrique.jimenez@ciqa.edu.mx (E.J.J.-R.); graciela.morales@ciqa.edu.mx (G.M.)
2 CONACYT-CIQA, Blvd. Enrique Reyna Hermosillo 140, Saltillo, Coahuila 25294, Mexico
3 Grupo de Investigación en Química y Biotecnología (QUIBIO), Facultad de Ciencias Básicas, Universidad Santiago de Cali, Pampalinda, Santiago de Cali 760035, Colombia; carolina.caicedo03@usc.edu.co
* Correspondence: yaneli.aguirre@ciqa.edu.mx

Abstract: Biodegradable packaging prepared from starch is an alternative to fossil-based plastic packaging. However, the properties of starch packaging do not comply with the necessary physicochemical properties to preserve food. Hence, in a previous study, we reported the preparation of a composite polymer material based on starch-chitosan-pluronic F127 that was found to be an adequate alternative packaging material. In this study, we modified the physicochemical properties of this material by storing it for 16 months under ambient conditions. The results indicate that the incorporation of pluronic F127 in the blend polymer can help avoid the retrogradation of starch. Moreover, at higher concentrations of pluronic F127, wettability is reduced. Finally, after storage, the materials exhibited surface modification, which is related to a color change and an increase in solubility, as well as a slight increase in stiffness.

Keywords: biodegradable packaging; biopolymer; starch; storage conditions; thermal properties; physicochemical properties

Citation: Fonseca-García, A.; Caicedo, C.; Jiménez-Regalado, E.J.; Morales, G.; Aguirre-Loredo, R.Y. Effects of Poloxamer Content and Storage Time of Biodegradable Starch-Chitosan Films on Its Thermal, Structural, Mechanical, and Morphological Properties. *Polymers* **2021**, *13*, 2341. https://doi.org/10.3390/polym13142341

Academic Editors: José Miguel Ferri, Vicent Fombuena Borràs and Miguel Fernando Aldás Carrasco

Received: 25 June 2021
Accepted: 10 July 2021
Published: 17 July 2021

1. Introduction

Synthetic-origin plastics have shaped various areas of our life, bringing with them numerous new and improved applications. However, their advantages, the environmental pollution resulting from their excessive use, slight or prolonged degradation, and inadequate waste disposal are critical. Studies have been focused on the development of polymeric materials of biological and biodegradable origin using biopolymers such as starch, cellulose, chitosan, and proteins that do not contribute to the accumulation of CO_2 at the end of their life cycles but instead promote sustainable development in the face of any emerging ecological crisis [1–4].

Starch is an excellent choice for developing biodegradable packaging materials, owing to its abundance in nature and higher economical viability. As a nontoxic polymer, it can be safely used for food packaging applications; moreover, it does not impart flavor to the packaged food. Therefore, there is no risk of affecting the organoleptic properties of the packed foods [5].

However, biodegradable packaging based on starch and other biopolymers, such as chitosan or gelatin, are susceptible to hydration and thus do not resist high relative humidity conditions, direct contact with liquid water, or foods with high moisture content [6–8]. Therefore, through previous research, mixtures of starch with other biopolymers have been developed with better barrier properties against moisture [9–12]. A mixture of cornstarch, chitosan, and different proportions of a poloxamer, known as pluronic F127, was developed [11]. The study evaluated the effects of the pluronic concentration and its mechanical and thermal properties, permeability to water vapor, and solubility in water of its starch chitosan films. The presence of the poloxamer considerably improved the moisture resistance of biodegradable packaging materials when made using starch in a concentration range of 1−5%. The packaging materials are less susceptible to water with a higher the concentration of poloxamer [11].

Storage time and conditions also significantly affect the properties of starch-based packaging materials due to the retrogradation exhibited by starch. Starch is composed of amylose and amylopectin. Amylose is a molecule with a mainly linear structure, whereas amylopectin is a highly branched molecule. The amylose-to-amylopectin ratio in corn starch granules is reported to be approximately 25/75 [13]. The proportion, size, and molecular organization of amylose and amylopectin, as well as the concentration of solids, significantly affects the retrogradation rate of materials made using starch [14,15]. The retrogradation of native starch mainly occurs due to the rearrangement of the amylose chains [16]. After storing for a long time, the amorphous molecules of the gelatinized starch tend to recrystallize to again form the ordered structures of double-helix crystallites. Thus, in gelatinized starch, both components, amylose and amylopectin, crystallize to form double helices generated from the external branches of amylopectin or amylose molecules [14,17].

The results obtained in a previous study were promising, suggesting this material, made using starch-chitosan-poloxamer, to be a viable alternative for producing packaging materials [11]. However, determining the behavior of this material during storage period is important.

This study hypothesized that new poloxamer-containing starch-chitosan formulations could maintain their structural integrity for several months, significantly delaying the starch retrogradation process that typically occurs in short periods. Therefore, in this study, the thermal, structural, mechanical, and morphological behaviors of corn starch-chitosan and poloxamer F127 stored for zero and sixteen months at room temperature were evaluated using differential scanning calorimetry (DSC), X-ray, FTIR-attenuated total reflection (ATR), tensile stress, wettability by the contact angle, color, and scanning electron microscopy (SEM) techniques.

2. Materials and Methods

2.1. Materials

The following biopolymers were used: corn starch from KMC (30% amylose; Brande, Denmark); shrimp shell chitosan (practical grade, deacetylation degree $\geq$ 75%; Sigma-Aldrich, Saint Louis, MO, USA); pluronic F127 (70% ethylene oxide; Sigma-Aldrich, Saint Louis, MO, USA). Glacial acetic acid (Productos Químicos Monterrey SA, Monterrey, Mexico) and glycerol (reagent grade; Meyer, Mexico City, Mexico) were also used.

2.2. Film Preparation

The methodology proposed by Fonseca-García, Jiménez-Regalado [11] was followed for producing biodegradable films. The polymers were used in various concentrations: starch 5% (*w/w*), chitosan 1% (*w/w*), and poloxamer (0%, 1%, 3%, and 5% *w/w* of biopolymers). The polymer solutions were mixed in a ratio of 60:20:20, and glycerol was added as a plasticizer at 25% (*w* of glycerol/*w* of total polymers). Fifty milliliters of each polymer dispersion was poured into a plastic mold and dried at 60 °C (Duo-Vac Oven, Lab-Line Instruments, Inc., Melrose Park, IL, USA) for four hours. The dried films were peeled off and stored in airtight storage bags for 16 months in the dark and at ambient conditions with a relative humidity (RH) of 32.73% and temperature of 25 °C.

2.3. Color

Color was determined using a portable spectrophotometer (Mini Scan EZ 4500L, HunterLab, Reston, VA, USA) based on the CIELab scale with the coordinates L^*, a^*, and b^* [18]. The films were placed on a white copy paper used as standard with color coordinate values of $L^* = 91.76$, $a^* = 2.14$, and $b^* = -10.64$.

2.4. Contact Angle

The contact angle was determined using the video goniometer (Rame-Hart instrument co., Succasunna, NJ, USA) to determine the hydrophobicity of the films [10]. A 5-µL drop of deionized water was placed on the opaquest side of a film sample (the side in contact with the mold) with dimensions of 3 cm × 1.5 cm, using a micropipette, and six measurements were recorded. Subsequently, with a program created using LabVIEW 2014 software, the image was captured, and angles formed by the water droplets on the film were calculated using ImageJ software.

2.5. Scanning Electron Microscopy

To analyze the morphology of the packaging films, SEM (JCM-6000, JEOL, Tokyo, Japan) was performed with a voltage of 10 kV. The films were coated with a gold-palladium layer (DESK II coating system, Denton Vacuum Inc., Moorestown, NJ, USA) [11].

2.6. Fourier Transform Infrared Spectroscopy (FTIR)

The vibrational modes of principal functional groups present in the packaging film were analyzed by FTIR spectrum using a Nicolet iS50 FTIR (Thermo Fisher Scientific, Madison, WI, USA) by ATR technique [10].

2.7. X-ray Diffraction (XRD) Analysis

The XRD of biodegradable films was determined using an X-ray diffractometer (D500, Siemens Aktiengesellschaft, Munich, Germany) with a voltage of 35 kV, current of 25 mA, and Cu Kα radiation of 1.5406 Å over a Bragg angular range (2θ) of 5–35° [11].

2.8. Differential Scanning Calorimetry

DSC was used to measure the gelatinization peak (T_P), melting temperatures (T_{m1} and T_{m2}), and the melting enthalpy (ΔH_{m2}). The analyses were performed at a temperature range of 25–300 °C under a nitrogen atmosphere (50 cm^3 min^{-1}) using a thermogravimetric analyzer (DSC 2 STAR System, Mettler-Toledo AG, Schwerzenbach, Switzerland), according to the standard ASTM D3418-15 [19].

2.9. Water Solubility

To determine the susceptibility of biodegradable films in regards to contact with liquid water at room temperature, the methodology proposed by Fonseca-García, Jiménez-Regalado [11] was followed without modifications.

2.10. Mechanical Properties

The tensile strength and percentage of elongation at break were determined using the methodology proposed by Fonseca-García, Jiménez-Regalado [11]. The experimental data regarding the mechanical properties were analyzed using OriginPro 8.5.0 SR1 software (OriginLab Corporation, Northampton, MA, USA) through an analysis of variance and the Tukey test with a significance level of $p < 0.05$.

3. Results and Discussion

3.1. Appearance and Color

All biodegradable films developed in this study appeared to be homogeneous. Figure 1 shows the starch-chitosan composite films with different proportions of the poloxamer pluronic F127 stored for 0 and 16 months; the images were captured on a black background for contrast. All films presented a translucent white color, as seen in Figure 1, and exhibited no significant visible changes during the 16-month storage process. They did not become brittle or more brittle to handle with time in storage. No visibly considerable color change was noted in the films when compared to the initial time (0) with the samples stored for 16 months. However, when evaluating the film color using a colorimeter, it was observed that the samples did change their coloration after 16 months of storage when compared to the initial time, as indicated by the values of the L^*, a^*, and b^* color parameters in Table 1.

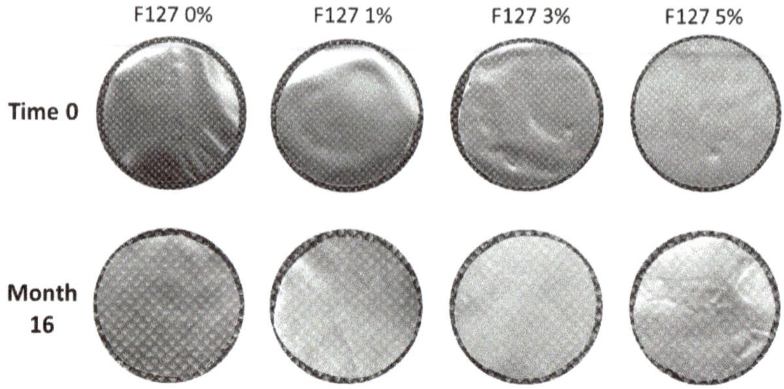

Figure 1. Biodegradable films based on corn starch and chitosan with pluronic F127 at 0%, 1%, 3%, and 5%; images were captured in the initial storage time (time 0) and after 16 months of storage at room temperature; the photos of the films at 16 months were taken at higher magnification.

Table 1. Effect of poloxamer content and storage time on water solubility, color, contact angle, and mechanical properties of corn starch-chitosan biodegradable films.

Content of F127	Water Solubility (%S)	Tensile Strength (MPa)	Elongation at Break (%)	Hunter Color Values [1]			Contact Angle (°)
				L^*	a^*	b^*	Θ
Time 0 months							
F127 0%	42.58 §	4.24 a,§	149.05 c,§	93.74	2.29	−11.50	99.93
F127 1%	7.81 §	6.49 a,b,§	53.30 a,§	93.52	1.87	−9.31	33.22
F127 3%	4.02 §	4.96 a,§	95.20 b,§	94.22	2.31	−11.56	43.88
F127 5%	3.34 §	3.68 a,§	84.96 b,§	94.53	2.10	−10.54	46.35
Time 16 months							
F127 0%	24.40	5.18 b	68.53 c	90.88	1.53	−8.05	67.66
F127 1%	17.66	7.01 c	60.60 c	90.81	1.47	−7.75	31.89
F127 3%	24.17	5.96 b	30.67 b	91.09	1.41	−7.79	46.09
F127 5%	56.98	3.27 a	12.40 a	89.53	1.30	−6.02	50.74

Mean values of at least three replicates. § Experimental data and significant difference taken from Fonseca-García, Jiménez-Regalado [11]. [1] Hunter color parameters, L^*: color coordinate, lightness (0 = black, 100 = white); a^*: color coordinate, greenness and redness (+ = red, − = green); b^*: color coordinate, blueness to yellowness (+ = yellow, − = blue). L^* standard= 91.76, a^* standard = 2.14, and b^* standard = −10.64. a,b,c: values with different letter in a same column denote significant difference.

3.2. *Morphology by SEM*

Figure 2 shows a comparison of the films with different amounts of poloxamer evaluated at time 0 and 16 months of storage. The evaluation of the initial time (zero) showed that in the biodegradable films formulated from starch and chitosan, the presence of the poloxamer significantly improved the morphology of the surface of these materials. In the film formulated without F127, an irregular surface was observed without cracks. This appearance was gradually reduced until a smoother surface was obtained with an increase in the content of F127.

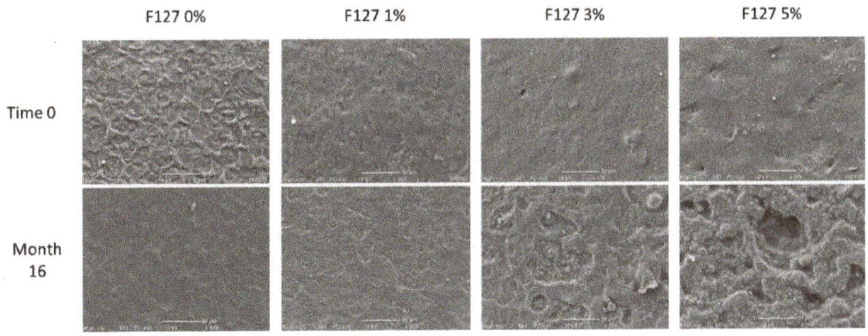

Figure 2. SEM images for the surface (500×) of composite films of corn starch-chitosan with pluronic F127 at ratios of 0%, 1%, 3%, and 5% stored for 0 and 16 months.

A significant change was observed in the morphology of all the composite biodegradable films after 16 months of storage under ambient conditions; however, the most noticeable or drastic changes were observed in the materials containing F127 at concentrations of 3% and 5% (Figure 2). A less homogeneous morphology was observed in these films, which may be due to a change in the organizational structure of the starch chains. Garalde, Thipmanee [20] reported a similar morphology in thermoplastic starch/poly (butylene adipate-co-terephthalate) (PBAT) films, which occurred when the starch was selectively removed from the composite film.

3.3. Contact Angle

The wettability of the biodegradable films was evaluated in terms of the contact angle; Table 1 shows the contact angle values of the starch-chitosan-poloxamer (F127) films. In films F127 0% and F127 1%, after 16 months of storage, the hydrophilicity increased as the contact angle decreased from 99.93° to 67.66° and 33.22° to 31.89°, respectively. In the case of F127 3% and F127 5%, the hydrophilicity decreased as the contact angle increased from 43.88° to 46.09° and 46.25° to 50.74°, respectively. The results suggest that pluronic F127 has a considerable effect on the wettability of biodegradable starch-chitosan films.

3.4. FTIR

Figure 3 shows the spectra of the films stored for different durations. After 16 months of storage, the FTIR-ATR analysis showed modifications in the absorption in the region of 1300 cm^{-1} to 900 cm^{-1} for all materials (F127 0%, F127 1%, F127 3%, and F127 5%); the bands in this region are sensitive to the gelation of starch because of their association with C–O stretching of the ring, linkages (C–O–C), and COH groups. Notably, the band at 1017 cm^{-1} is reported as sensitive to amorphous starch, which is constant in the spectra after 16 months of storage [21–23]. Casu and Reggiani [24] reported that the band at 3300 cm^{-1}, attenuated after 16 months of storage, can be assigned to the O-H stretching of the groups in amorphous amylose. Moreover, the molecule of water showed that the absorption bands at 3300 and 1646 cm^{-1} are associated with OH stretching and deformation vibrations, respectively [24]. In addition, the absorption band at 2883 cm^{-1} was associated with C-H bond stretching, which did not show a modification in the four materials after 16 months [25]. Finally, another important phenomenon identified in the spectra of all films after storage was the disappearing of the absorption band at 1150 cm^{-1}, which was associated with C-O stretching of C–O–C in glycosidic linkage; this result suggests the depolymerization of the starch molecule [26].

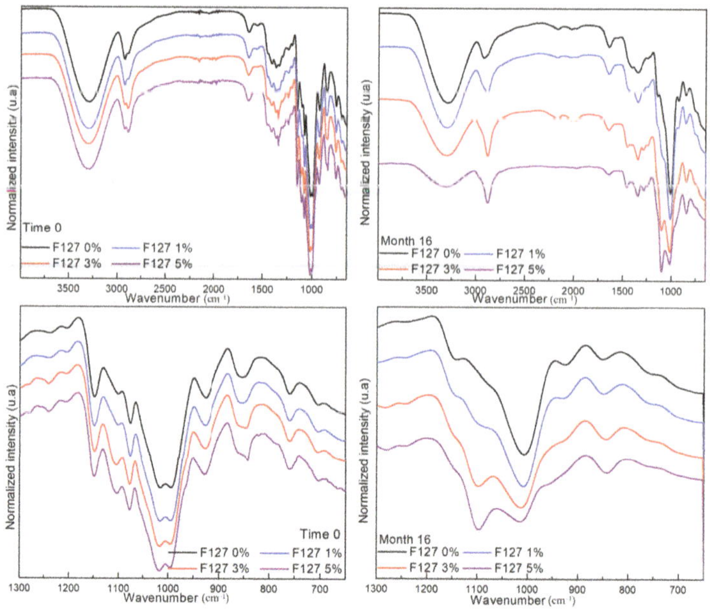

Figure 3. FTIR spectra of biodegradable films of corn starch-chitosan with pluronic F127 at ratios of 0%, 1%, 3%, and 5% stored for 0 and 16 months.

3.5. X-ray Diffraction

To determine if there was retrogradation or modification in the conformation of the structural matrix of the biodegradable films, Figure 4 shows the XRD patterns of biodegradable films F127 0%, F127 1%, F127 3%, and F127 5%. The patterns show that there was a higher atomic ordering in the biodegradable film in F127 0% after 16 months of storage. Peaks at 16°, 19.65°, and 22° were defined after storage; these peaks were related to the crystalline starch, which indicates the retrogradation of starch in this film. However, in films F127 1%, F127 3%, and F127 5%, there was no increase in the atomic ordering after 16 months of storage, as peaks related to starch could not be identified. In F127 3% and F127 5% films, there were two peaks at 19° and 23°, which were less intense after 16 months; these peaks were associated with pluronic 127 [11]. The X-ray patterns indicate that pluronic F127 avoided retrogradation after the storage period of 16 months. Mina Hernandez [27] reported a significant increase in the rearrangement of the polymer chains of mixtures of thermoplastic starch and polycaprolactone during short storage periods (5 and 26 days). Furthermore, the author found that the retrogradation of starch-based polymers occurs more rapidly when the materials are stored in conditions of high RH, which also negatively impacts their mechanical performance.

3.6. Thermal Behavior by DSC

The results of the thermal behavior of the biodegradable films are presented in Table 2. DSC experiments have shown that the melting temperature (T_m) of the neat corn starch film decreases with a significant increase in enthalpy (ΔH_m), indicating a gain in the ordering of the polymeric structures involved. In contrast, chitosan films exhibited an increase of 15 °C after 16 months of storage. The enthalpy values for this biopolymer decreased as a function of the evaluated time. Notably, the mixture of starch and chitosan (F127 0%) exhibited a thermal behavior in which starch transitions predominated. The incorporation of the poloxamer in the starch-chitosan matrix ensured gelatinization, with the absence of the band at ~62 °C (T_P, gelatinization peak). A decrease was observed in the melting temperature (56.1 °C) corresponding to neat F127. The increase in poloxamer content resulted in an increase in the melting temperature and a decrease in enthalpy after 16 months, with respect to its nonfunctionalized homologous (F127 0%). This allowed for demonstrating the plasticizing effect generated by the poloxamer due to the destabilization of the crystalline regions. The fusion of intra- and inter-molecular double helices and the partial recovery of the crystalline structure of amylopectin (Figure S1, Supplementary Material) can be observed [28]. The crystals formed during the starch retrogradation process are less orderly and homogeneous than native starch; this is reflected in their lower melting temperatures. This effect is similar to that observed in other starches, such as sago [29]. The DSC parameters of starch in biodegradable films are summarized in Table 2, and this result is in accordance with that obtained from the XRD analysis.

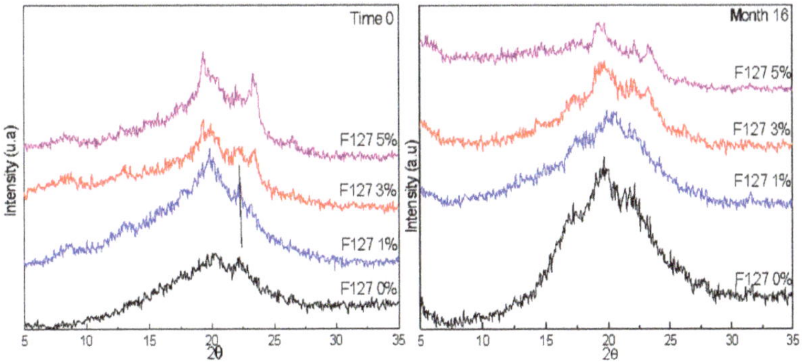

Figure 4. X-ray diffraction patterns of biodegradable films of corn starch-chitosan with pluronic F127 at ratios of 0%, 1%, 3%, and 5% stored for 0 and 16 months.

Table 2. DSC parameters of biodegradable films of corn starch-chitosan with pluronic F127 at ratios of 0%, 1%, 3%, and 5% stored for 0 and 16 months.

Film Sample	T_p (°C)	T_{m1} (°C)		T_{m2} (°C)		ΔH_{m2} (J g^{-1})	
	Month 0	Month 0	Month 16	Month 0	Month 16	Month 0	Month 16
Corn starch	63.13	–	–	141.03	117.75	6.0	18.7
Chitosan	–	–	–	104.68	119.47	28.0	19.6
F127		56.10	56.10				
F127 0%	62.06	–	–	138.96	106.63	6.4	10.3
F127 1%	–	45.52	51.52	137.63	112.94	8.4	9.1
F127 3%	–	51.95	51.68	–	108.05	–	6.3
F127 5%	–	47.84	51.46	114.36	115.22	8.5	4.0

– negligible.

3.7. Water Solubility (WS)

The WS of biodegradable materials is a crucial parameter because it is important that they can decompose in both terrestrial and aquatic environments after being discarded, where they can potentially contaminate the environment on several occasions. A significant increase in the water solubility capacity of the starch-chitosan-based materials was observed (Table 1) at all poloxamer contents (1%, 3%, and 5%) after 16 months of storage with water solubility values of 18%, 24%, and 57%, respectively, compared to the materials evaluated at time zero [11]. Shaker, Elbadawy [30] found that by using different types of poloxamers, such as pluronic F127 and F68, the degree of solubility and the dissolution rate of drugs improved, thus facilitating their usage in pharmaceutical production.

3.8. Mechanical Properties

The mechanical performance of corn starch and chitosan films with various poloxamer concentrations was significantly affected with storage time, as shown in Table 1. A slight increase in the stiffness of the materials was observed, presenting an increase in the tensile stress values. In comparison, the percentage of elongation of the materials decreased when the content of poloxamer was >3%. This mechanical behavior may be associated with two changes in the films. First, SEM micrographs showed that when the poloxamer content was >3% in the films, a higher superficial modification was observed. Second, based on the FTIR spectra of the films, the starch contained in these materials was presumed to depolymerize after storage for 16 months. Furthermore, hydrogen bonds are the predominant ones in these biodegradable films; however, this type of bond is known to be weak, and the tenacity is affected in this case. This mechanical property is diminished by lower values of elongation, without resistance being sacrificed.

4. Conclusions

Storage time can significantly modify the physicochemical, morphological, and structural properties of biodegradable films comprising starch, chitosan, and poloxamer. This information is important to ascertain the state of packaging materials of biodegradable origin when stored at room temperature. Based on physicochemical characterization, there is evidence suggesting that poloxamer avoids the retrogradation of cornstarch. The biodegradable films showed a slight modification in their surface morphology after 16 months, similar to erosion, and this phenomenon can modify the color, WS, mechanical behavior, and wettability of films. Lower melting temperatures and enthalpies in the poloxamer films, as well as storage for 16 months, indicate that the retrogradation of amylopectin only partially recovers the crystalline structure of the native starch. The use in the pharmaceutical, medical, and packaging industry of a biodegradable material such as the one developed in this study may be possible, owing to the favorable properties observed in its mechanical performance and acceptable solubility in liquid water, which can be sufficiently solubilized in a short period of time with the presence of the poloxamer.

Supplementary Materials: The following are available online at https://www.mdpi.com/article/10.3390/polym13142341/s1, Figure S1: DSC thermograms of biodegradable films of corn starch-chitosan with pluronic F127 at ratios of 0%, 1%, 3%, and 5% and stored for 0 and 16 months.

Author Contributions: Conceptualization, A.F.-G., E.J.J.-R. and R.Y.A.-L.; methodology, A.F.-G. and R.Y.A.-L.; formal analysis, A.F.-G., C.C. and R.Y.A.-L.; investigation, A.F.-G., E.J.J.-R. and R.Y.A.-L.; visualization, A.F.-G., E.J.J.-R. and R.Y.A.-L.; supervision, E.J.J.-R. and R.Y.A.-L.; project administration, E.J.J.-R. and R.Y.A.-L.; writing—original draft preparation, A.F.-G., C.C., E.J.J.-R., G.M. and R.Y.A.-L.; writing—review and editing, A.F.-G., C.C., E.J.J.-R., G.M. and R.Y.A.-L. All authors have read and agreed to the published version of the manuscript.

Funding: This research was funded by Centro de Investigación en Química Aplicada (CIQA) under internal projects 6521 and 6610, and by Dirección General de Investigaciones (DGI) of Universidad Santiago de Cali under call no. 01-2021.

Institutional Review Board Statement: Not applicable.

Informed Consent Statement: Not applicable.

Data Availability Statement: The data presented in this study are available on request from the corresponding author.

Acknowledgments: Thank you to M.C. Maria Guadalupe Mendez-Padilla and M.C. Myrna Salinas Hernández, M.C. Blanca Huerta Martinez, and LNMG-CIQA for DSC, DRX, and SEM characterizations, respectively. Fonseca-García and Aguirre-Loredo thank CONACYT for their nominations as a researchers assigned to CIQA.

Conflicts of Interest: The authors declare no conflict of interest.

References

1. Gómez-Aldapa, C.A.; Velazquez, G.; Gutierrez, M.C.; Castro-Rosas, J.; Jiménez-Regalado, E.J.; Aguirre-Loredo, R.Y. Characterization of Functional Properties of Biodegradable Films Based on Starches from Different Botanical Sources. *Starch Stärke* **2020**, *72*, 1900282. [CrossRef]
2. Wang, H.; Ding, F.; Ma, L.; Zhang, Y. Edible films from chitosan-gelatin: Physical properties and food packaging application. *Food Biosci.* **2021**, *40*, 100871. [CrossRef]
3. Yadav, M.; Behera, K.; Chang, Y.-H.; Chiu, F.-C. Cellulose Nanocrystal Reinforced Chitosan Based UV Barrier Composite Films for Sustainable Packaging. *Polymers* **2020**, *12*, 202. [CrossRef]
4. Abera, G.; Woldeyes, B.; Demash, H.D.; Miyake, G. The effect of plasticizers on thermoplastic starch films developed from the indigenous Ethiopian tuber crop Anchote (Coccinia abyssinica) starch. *Int. J. Biol. Macromol.* **2020**, *155*, 581–587. [CrossRef]
5. Samsudin, H.; Hani, N.M. Chapter 8—Use of Starch in Food Packaging. In *Starch-Based Materials in Food Packaging*; Villar, M.A., Barbosa, S.E., García, M.A., Castillo, L.A., López, O.V., Eds.; Academic Press: Cambridge, MA, USA, 2017; pp. 229–256.
6. Aguirre-Loredo, R.Y.; Rodriguez-Hernandez, A.I.; Morales-Sánchez, E.; Gomez-Aldapa, C.; Velazquez, G. Effect of equilibrium moisture content on barrier, mechanical and thermal properties of chitosan films. *Food Chem.* **2016**, *196*, 560–566. [CrossRef]
7. Beigmohammadi, F.; Barzoki, Z.M.; Shabaniyan, M. Rye Flour and Cellulose Reinforced Starch Biocomposite: A Green Approach to Improve Water Vapor Permeability and Mechanical Properties. *Starch Stärke* **2020**, *72*, 1900169. [CrossRef]
8. Meng, W.; Shi, J.; Zhang, X.; Lian, H.; Wang, Q.; Peng, Y. Effects of peanut shell and skin extracts on the antioxidant ability, physical and structure properties of starch-chitosan active packaging films. *Int. J. Biol. Macromol.* **2020**, *152*, 137–146. [CrossRef] [PubMed]
9. Aguirre-Loredo, R.Y.; Rodríguez-Hernández, A.I.; Chavarría-Hernández, N. Physical properties of emulsified films based on chitosan and oleic acid. *CyTA J. Food* **2014**, *12*, 305–312. [CrossRef]
10. Caicedo, C.; Loredo, R.Y.A.; García, A.F.; Ossa, O.H.; Arce, A.V.; Pulgarin, H.L.C.; Torres, Y. Ávila Rheological, Thermal, Superficial, and Morphological Properties of Thermoplastic Achira Starch Modified with Lactic Acid and Oleic Acid. *Molecules* **2019**, *24*, 4433. [CrossRef]
11. Fonseca-García, A.; Jiménez-Regalado, E.J.; Aguirre-Loredo, R.Y. Preparation of a novel biodegradable packaging film based on corn starch-chitosan and poloxamers. *Carbohydr. Polym.* **2021**, *251*, 117009. [CrossRef] [PubMed]
12. Gómez-Aldapa, C.A.; Velazquez, G.; Gutiérrez, M.C.; Rangel-Vargas, E.; Castro-Rosas, J.; Aguirre-Loredo, R.Y. Effect of polyvinyl alcohol on the physicochemical properties of biodegradable starch films. *Mater. Chem. Phys.* **2020**, *239*, 122027. [CrossRef]
13. Panaitescu, D.M.; Frone, A.N.; Ghiurea, M.; Chiulan, I. Influence of storage conditions on starch/PVA films containing cellulose nanofibers. *Ind. Crop. Prod.* **2015**, *70*, 170–177. [CrossRef]
14. Roman, L.; Yee, J.; Hayes, A.; Hamaker, B.R.; Bertoft, E.; Martinez, M.M. On the role of the internal chain length distribution of amylopectins during retrogradation: Double helix lateral aggregation and slow digestibility. *Carbohydr. Polym.* **2020**, *246*, 116633. [CrossRef] [PubMed]
15. Yu, S.; Ma, Y.; Sun, D.-W. Impact of amylose content on starch retrogradation and texture of cooked milled rice during storage. *J. Cereal Sci.* **2009**, *50*, 139–144. [CrossRef]
16. Leyva, J.D.H.; Alonso, L.; Rueda-Enciso, J.; Yee-Madeira, H.; Bello-Perez, L.; Alvarez-Ramirez, J. Morphological, physicochemical and functional characteristics of starch from Marantha ruiziana Koern. *LWT* **2017**, *83*, 150–156. [CrossRef]
17. Eliasson, A.C. *10—Gelatinization and Retrogradation of Starch in Foods and Its Implications for Food Quality, in Chemical Deterioration and Physical Instability of Food and Beverages*; Skibsted, L.H., Risbo, J., Andersen, M.L., Eds.; Woodhead Publishing: Cambridge, UK, 2010; pp. 296–323.
18. Gómez-Aldapa, C.A.; Díaz-Cruz, C.A.; Castro-Rosas, J.; Jiménez-Regalado, E.J.; Velazquez, G.; Gutiérrez, M.C.; Aguirre-Loredo, R.Y. Development of Antimicrobial Biodegradable Films Based on Corn Starch with Aqueous Extract of *Hibiscus sabdariffa* L. *Starch Stärke* **2021**, *73*, 2000096. [CrossRef]
19. ASTM. *D3418. Standard Test Method for Transition Temperatures and Enthalpies of Fusion and Crystallization of Polymers by Differential Scanning Calorimetry*; ASTM International: Philadelphia, PA, USA, 2015.
20. Garalde, R.A.; Thipmanee, R.; Jariyasakoolroj, P.; Sane, A. The effects of blend ratio and storage time on thermoplastic starch/poly(butylene adipate-co-terephthalate) films. *Heliyon* **2019**, *5*, e01251. [CrossRef]
21. Rindlava, Å.; Hulleman, S.H.; Gatenholma, P. Formation of starch films with varying crystallinity. *Carbohydr. Polym.* **1997**, *34*, 25–30. [CrossRef]
22. Rubens, P.; Snauwaert, J.; Heremans, K.; Stute, R. In situ observation of pressure-induced gelation of starches studied with FTIR in the diamond anvil cell. *Carbohydr. Polym.* **1999**, *39*, 231–235. [CrossRef]
23. Van Soest, J.J.G.; De Wit, D.; Tournois, H.; Vliegenthart, J.F.G. Retrogradation of Potato Starch as Studied by Fourier Transform Infrared Spectroscopy. *Starch Stärke* **1994**, *46*, 453–457. [CrossRef]
24. Casu, B.; Reggiani, M. Infrared spectra of amylose and its oligomers. In *Journal of Polymer Science Part C: Polymer Symposia*; Wiley Subscription Services, Inc., A Wiley Company: New York, NY, USA, 1964.
25. Kahvand, F.; Fasihi, M. Plasticizing and anti-plasticizing effects of polyvinyl alcohol in blend with thermoplastic starch. *Int. J. Biol. Macromol.* **2019**, *140*, 775–781. [CrossRef] [PubMed]
26. Pawlak, A.; Mucha, M. Thermogravimetric and FTIR studies of chitosan blends. *Thermochim. Acta* **2003**, *396*, 153–166. [CrossRef]

27. Hernandez, J.H.M. Effect of the Incorporation of Polycaprolactone (PCL) on the Retrogradation of Binary Blends with Cassava Thermoplastic Starch (TPS). *Polymers* **2020**, *13*, 38. [CrossRef]
28. Vamadevan, V.; Bertoft, E. Impact of different structural types of amylopectin on retrogradation. *Food Hydrocoll.* **2018**, *80*, 88–96. [CrossRef]
29. Li, C.; Hamaker, B.R. Effects of different storage temperatures on the intra- and intermolecular retrogradation and digestibility of sago starch. *Int. J. Biol. Macromol.* **2021**, *182*, 65–71. [CrossRef] [PubMed]
30. Shaker, M.A.; Elbadawy, H.M.; Shaker, M.A. Improved solubility, dissolution, and oral bioavailability for atorvastatin-Pluronic® solid dispersions. *Int. J. Pharm.* **2020**, *574*, 118891. [CrossRef]

polymers MDPI

Article

Prediction of the Miscibility of PBAT/PLA Blends

Shen Su [1,2]

[1] Department of Circular and Bio-Based Plastics, Fraunhofer UMSICHT, Institute for Environment, Safety and Energy Technology, Osterfelder Str. 3, 46047 Oberhausen, Germany; shen.su@umsicht.fraunhofer.de or shen.su@ruhr-uni-bochum.de; Tel.: +49-208-8598-1422

[2] Department of Mechanical Engineering, Ruhr-University Bochum, Universitaetsstr. 150, 44780 Bochum, Germany

Abstract: Designing polymer structures and polymer blends opens opportunities to improve the performance of plastics. Blending poly(butylene adipate-co-terephthalate) (PBAT) and polylactide (PLA) is a cost-effective approach to achieve a new sustainable material with complementary properties. This study aimed to predict the theoretical miscibility of PBAT/PLA blends at the molecular level. First, the basic properties and the structure of PBAT and PLA are introduced, respectively. Second, using the group contribution methods of van Krevelen and Hoy, the Hansen and Hildebrand solubility parameters of PBAT and PLA were calculated, and the effect of the molar ratio of the monomers in PBAT on the miscibility with PLA was predicted. Third, the dependence of the molecular weight on the blend miscibility was simulated using the solubility parameters and Flory–Huggins theory. Next, the glass transition temperature of miscible PBAT/PLA blends, estimated using the Fox equation, is shown graphically. According to the prediction and simulation, the blends with a number-average molecular weight of 30 kg/mol for each component were thermodynamically miscible at 296 K and 463 K with the possibility of spinodal decomposition at 296 K and 30% volume fraction of PBAT. This study contributes to the strategic synthesis of PBAT and the development of miscible PBAT/PLA blends.

Keywords: Flory–Huggins; free energy of mixing; glass transition temperature; group contribution; molecular weight; miscibility prediction; PBAT; PLA; simulation; solubility parameter

Citation: Su, S. Prediction of the Miscibility of PBAT/PLA Blends. *Polymers* **2021**, *13*, 2339. https://doi.org/10.3390/polym13142339

Academic Editors: José Miguel Ferri, Vicent Fombuena Borràs and Miguel Fernando Aldás Carrasco

Received: 20 June 2021
Accepted: 14 July 2021
Published: 16 July 2021

1. Introduction

Physical blending is an economical approach to achieve complementary material properties without developing a new polymer. Poly(butylene adipate-co-terephthalate) (PBAT) and polylactide (PLA) currently account for 13.5 and 18.8 percent of global bioplastic production capacity, respectively [1]. Blends made of commercially available PBAT and PLA tend to macrophase-separate and exhibit two glass transition temperatures (T_g) [2–4]. Therefore, unmodified PBAT/PLA blends have poor miscibility.

Experimentally, PBAT/PLA blends are often prepared by melt blending. At an elevated temperature for a sufficiently long time, transesterification can take place between the two polyesters, resulting in enhanced miscibility [5]. Two criteria often used to evaluate the blend miscibility [6–8] include (1) phase morphology: whether it is homogeneous down to the molecular level; (2) glass transition temperature: whether a binary blend exhibits a single T_g. The morphological investigation depends on the measuring points of the samples in scanning electron microscopy (SEM). A blend may not be truly miscible at the molecular level, even though it shows a homogeneous phase if examined on sufficiently large length scales [6]. Thermal analysis using differential scanning calorimetry (DSC) can only determine the miscibility of polymer blends with well-separated T_g values [9]. Furthermore, different heating rates can lead to a variation in T_g values to some extent. Therefore, both the blend preparation and the analysis can practically affect the actual blend miscibility.

To exclude the experimental influences, a theoretical study of the intrinsic miscibility is necessary. The concept of group contributions has been applied to calculate the solubility parameters and the blend miscibility [10]. According to the literature research, Dil et al. indicate that PBAT and PLA have the Hildebrand solubility parameters (HiSP) of 22.2 $MPa^{1/2}$ and 21.9 $MPa^{1/2}$, respectively [11]. Ding et al. report that the δ (PBAT) and δ(PLA) are 21.9 $MPa^{1/2}$ and 20.7 $MPa^{1/2}$, respectively [12]. As a result, the difference of solubility parameters δ(PBAT-PLA) is 0.3 or 1.2 $MPa^{1/2}$. Materials with similar HiSP have a high affinity with each other [13]. However, the PBAT/PLA blend miscibility is still difficult to determine directly from the difference in solubility parameters. One reason for this is the irregular structure of PBAT molecules that consist of two randomly arranged monomers (BA and BT). Their molar ratio can differ due to the synthesis and degradation. If PBAT were degraded, the resulting lower molecular PBAT-chains might have different ratios of BA and BT. Another reason is that the intermolecular interaction depends on the molecular weights of both polymers, blend ratio, and temperature [14]. To the author's best knowledge, only Park et al. [15] have studied the miscibility of blends (PETG/PLA) with some similarity in molecular structure to PBAT/PLA blends from the theoretical aspect. Therefore, PBAT/PLA blends need more investigation regarding the solubility parameters and miscibility from the molecular level and the thermodynamic aspect. A good understanding of the intrinsic miscibility is not only of academic interest but also contributes to designing miscible PBAT/PLA blends without adding expensive additives.

This study aimed to explore the dependence of the PBAT structure and the molecular weights of PBAT and PLA with different weight ratios on the blend miscibility. First, group contributions theories were used to estimate the effect of the molar ratio of the two monomers in PBAT on the solubility parameter and the difference of solubility parameter between both polymers. Second, the Flory–Huggins model was used to establish phase diagrams and spinodal curves. Furthermore, the composition-dependent glass transition of miscible PBAT/PLA blends was estimated. The first novelty was to find a correlation between the molar ratios of the two monomers in PABT and the solubility parameter between PBAT and PLA. The second novelty was to simulate the phase diagrams and spinodal curves of PBAT/PLA blends, taking into account parameters including temperature, molecular weights, and the component ratios, as well as the calculated solubility parameters.

2. Materials

Poly(butylene adipate-co-terephthalate) (PBAT) (Figure 1) is a fully biodegradable polyester with two types of dimers: BT and BA. One dimer is the rigid section consisting of 1,4-butanediol and terephthalic acid monomers. The other dimer is the flexible section consisting of 1,4-butanediol and adipic acid monomers [16], resulting in high flexibility and high ductility [4]. The largest manufacturer, BASF (Ludwigshafen, Germany), produces PBAT under the brand name Ecoflex® (e.g., Ecoflex F Blend C1200). This polymer possesses a density of 1.26 g/mol, a number average molecular weight (M_n) of 52.1 kg/mol, and a polydispersity index of 2.0.

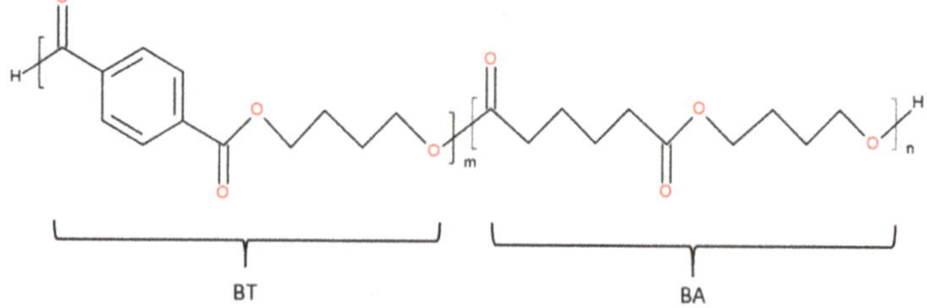

Figure 1. Chemical structure of PBAT.

Poly(lactic acid) or polylactide (PLA) is a biodegradable and renewable aliphatic polyester [17]. Poly(lactic acid) is produced by direct polycondensation of lactic acid. Polylactide can be produced by ring-opening polymerization of cyclic lactide. PLA possesses two optical active and crystallizable isomeric forms: PDLA and PLLA (Figure 2). PLA exhibits many favorable features such as high modulus of elasticity, high strength and high transparency (in the amorphous state), and good processability [18]. The largest manufacturer, NatureWorks (Minnetonka, MN, USA), produces PLA under the trade name Ingeo^{TM} (for example, Ingeo^{TM} Biopolymer 2003D). This polymer has a density of 1.24 g/mol, a number average molecular weight (M_n) of 127.0 kg/mol, a polydispersity index of 1.6, and a D-isomer content of approximately 4.4%.

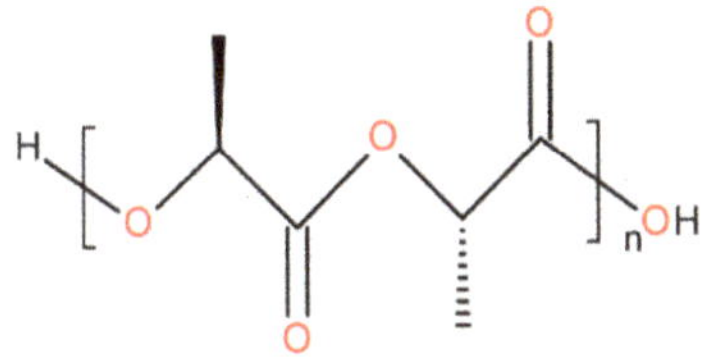

Figure 2. Chemical structure of PLLA.

3. Prediction of Solubility Parameters

The basis of a molecular group contribution method is additivity. This assumes that the physical property of a polymer is calculable by the additive contributions from the individual structural and functional groups in its repeating unit. Using a group contribution method enables us to establish a correlation between the chemical structure of polymers and their interaction. The term "solubility parameter" quantifies the intermolecular interaction.

Coleman et al. has reported a group contribution method to calculate the one-dimensional Hildebrand solubility parameter (HiSP) for polymers with T_g values below room temperature [19]. However, PLA possesses a glass transition temperature higher than room temperature (approx. 60 °C). Therefore, Coleman's method was inappropriate in this miscibility study of PBAT/PLA blends. Group contribution methods developed by van Krevelen [10] and Hoy [20] were used to estimate the three-dimensional HSP and one-dimensional HiSP for polymers. Compared with the real system, the group contribution methods neglect the possible reactions between the blend components, such as transesterification while melt blending, which may change the molecular structures.

In this section, these two different group contribution methods were applied to calculate the solubility parameters of PBAT and PLA. As mentioned in the Introduction, PBAT can consist of different ratios of its monomers (BA and BT). Furthermore, the sequence in the chain may change due to, e.g., polymer degradation. Considering these variations, the estimation of its solubility parameter was performed using three different assumptions:

(1) PBAT (alternating): alternating arrangement with a molar ratio of 1/1.
(2) (PBT): only monomers of BT constitute the polymer.
(3) (PBA): only monomers of BA constitute the polymer.

3.1. Solubility Parameter Calculation According to van Krevelen's Method

The group contributions F_{di}, F_{pi}, and E_{hi} [10] are applied to calculate HSP, including dispersion interactions (δ_d), polar interactions (δ_p), and hydrogen bond interactions (δ_h), as well as the total Hildebrand solubility parameters (δ_t or HiSP). The following equations are used for the calculation:

$$\delta_d = \frac{\sum F_{di}}{V} \quad (1)$$

$$\delta_p = \frac{\sqrt{\sum F_{pi}^2}}{V} \quad (2)$$

$$\delta_h = \sqrt{\frac{\sum E_{hi}}{V}} \quad (3)$$

$$\delta_t = \sqrt{\delta_d^2 + \delta_p^2 + \delta_h^2} \quad (4)$$

where F_{di} represents the group contributions of type i to the dispersion component F_d of the molar attraction constant; F_{pi} represents the group contributions to the polar component F_p; E_{hi} is the hydrogen-bonding energy per structural group i; V is the molar volume. Details of the calculation are described in the Supplementary Material on the sheet "van Krevelen". The calculated HSP and HiSP using van Krevelen's method are listed (Table 1).

Table 1. Calculated solubility parameters using van Krevelen's method.

Polymer Type	δ_d [MPa$^{1/2}$]	δ_p [MPa$^{1/2}$]	δ_h [MPa$^{1/2}$]	HiSP [MPa$^{1/2}$]
alternating PBAT	18.21	5.89	9.17	21.22
PBT	19.63	6.19	9.37	22.62
PBA	14.99	5.00	8.45	17.92
PLA	15.33	8.44	10.98	20.66

Due to the random arrangement of the copolymers BT and BA and possible different molar ratios, the solubility parameters for each segment is different. In the case of a BT-rich PBAT (molar ratio of BT $\geq$ 50%), the HiSP of PBAT was between 21.22 and 22.62, which are the values of alternating PBAT with BT/BA = 1/1 and PBT. In the case of a BA-rich PBAT (molar ratio of BA $\geq$ 50%), the HiSP of PBAT was between 17.92 and 21.22, the values of PBA and alternating PBAT with BA/BT =1/1. The calculated HiSP of PBAT (alternating) and PLA was 21.22 [MPa$^{1/2}$] and 20.66 [MPa$^{1/2}$], respectively, so the difference of total solubility parameter $\Delta\delta$ between them was 0.56 [MPa$^{1/2}$]. Since the HiSP of PLA was in the range of BA-rich PBAT (Figure 3), varying the molar ratio of the monomers in PBAT may lead to a smaller difference in solubility parameters and an increase in the intermolecular interaction between PBAT and PLA.

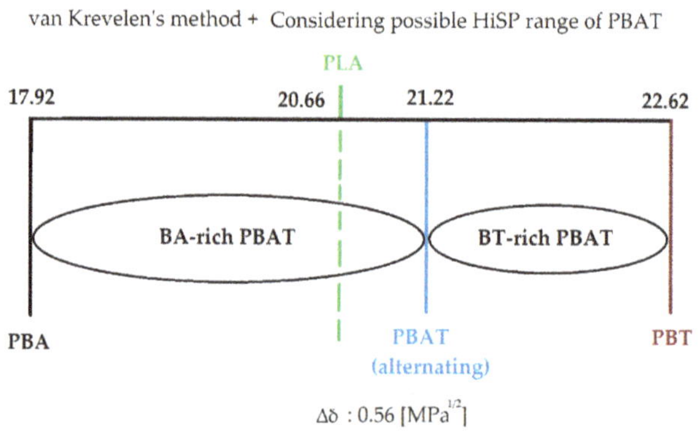

Figure 3. Values and ranges of calculated solubility parameters using van Krevelen's method.

3.2. Solubility Parameter Calculation According to Hoy's Method

The methods of van Krevelen and Hoy, based on the same basic assumption of additivity, have the same order of accuracy. These two are of the same order of accuracy for predicting the solubility parameters [10]. However, van Krevelen and Hoy have applied different ways of calculation. Van Krevelen's method first predicts the partial solubility parameters, which are then used to calculate the total solubility parameter [10,21]. In contrast, Hoy's method first determines the total solubility parameter using the molar attraction functions and auxiliary equations. From the total, the three partial ones are calculated using additive molar functions and expressions for the components [20,22]. The equations used in Hoy's method are below.

$$\alpha(P) = 777\Delta_T(P)/V \quad (5)$$

$$n = 0.5/\Delta_T(P) \quad (6)$$

$$\delta_t = (F_i + \frac{B}{\bar{n}})/V \quad (7)$$

$$\delta_P = \delta_t(\frac{1}{\alpha(P)}\frac{F_P}{F_t + B/\bar{n}})^{1/2} \quad (8)$$

$$\delta_h = \delta_t[(\alpha(P) - 1/\alpha(P))]^{1/2} \quad (9)$$

$$\delta_d = (\delta_t^2 - \delta_P^2 - \delta_h^2)^{1/2} \quad (10)$$

where $\alpha(P)$ is the molecular aggregation number of a polymer; $\Delta_T(P)$ is the Lydersen correction for polymers derived by Hoy; V is the molar volume; B is a base value of 277; F_t is the molar attraction function; F_P is the polar component of molar attraction function; n is the molecular aggregation number. For explanations of variables used before, see Section 3.1 "van Krevelen's method".

The three-dimensional HSP and the overall one-dimensional HiSP were calculated (Table 2). (For details of the calculation, see the Supplementary Material on the sheet "Hoy").

Table 2. Calculated solubility parameters using Hoy's method.

Polymer Type	δ_d [MPa$^{1/2}$]	δ_p [MPa$^{1/2}$]	δ_h [MPa$^{1/2}$]	HiSP [MPa$^{1/2}$]
alternating PBAT	15.94	11.43	9.35	21.73
PBT	15.64	12.82	11.32	23.18
PBA	16.13	10.44	7.98	20.80
PLA	14.02	12.73	9.77	21.31

From the table above, it can be seen that BA-rich PBAT had a HiSP in the range of 20.8–21.73, while BT-rich PBAT possessed a HiSP in the range of 21.73 to 23.18 [MPa$^{1/2}$]. The calculated HiSP for PBAT and PLA was 21.73 [MPa$^{1/2}$] and 21.31 [MPa$^{1/2}$], respectively, indicating a solubility parameter difference of 0.42 [MPa$^{1/2}$]. Furthermore, Hoy's method shows the same tendency as van Krevelen's method: that the HiSP of PLA was in the range of the values of BA-rich PBAT (Figure 4).

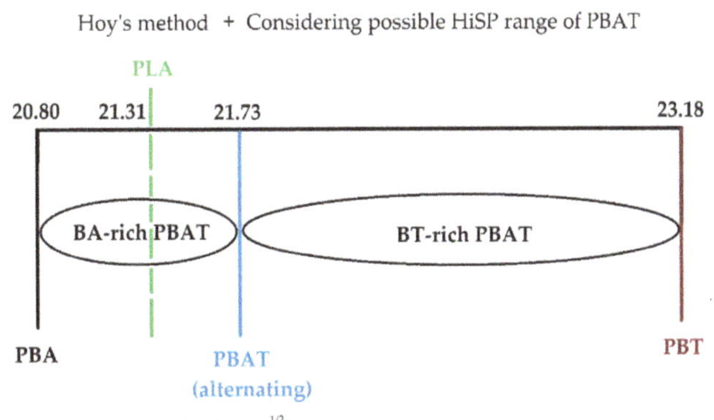

Figure 4. Values and ranges of calculated solubility parameters using Hoy's method.

According to the two group contribution methods, the mean value of the difference in calculated total solubility parameters between alternating PBAT and PLA was 0.49 MPa$^{1/2}$.

The two different methods show consistency in the total solubility parameter and that PLA was in the range of the one of BA-rich PBAT. Furthermore, the same tendency of the solubility parameter can be seen in both methods:

δ(PBA) < δ(PLA) < δ(alternating PBAT) < δ(PBT)

4. Simulation of the Blend Miscibility

Based on the solubility parameters, the miscibility of PBAT/PLA blends can be further studied from the aspect of thermodynamics. The thermodynamic criterion of solubility of two dissimilar components is described by the equation:

$$\Delta G_M = \Delta H_M - T\Delta S_M \quad (11)$$

where ΔG_m is the free energy of mixing; the ΔH_M is the enthalpy of mixing (heat of mixing); T is the absolute temperature; ΔS_M is the entropy of mixing.

A negative value of ΔG_m is generally required to obtain a miscible system. For low molecular weight materials, an increasing temperature generally results in an increase in miscibility as the $T\Delta S_M$ term increases, thus driving G_M to be more negative [8,23]. However, both PBAT and PLA are high molecular weight molecules, implying that the negative contribution from the $T\Delta S_M$ term is small. An equation based on the Flory–Huggins theory correlates the miscibility of a polymer blend with several parameters [24].

$$\frac{\Delta G_M}{RTV} = \frac{\phi_1}{V_1} ln\phi_1 + \frac{\phi_2}{V_2} ln\phi_2 + \frac{\phi_1\phi_2}{RT}(\delta_1 - \delta_2)^2 \quad (12)$$

where ΔG_M is the free enthalpy of mixing; R the gas constant of 8.31 kg·m^2/s^2·mol·K; T the absolute temperature; V the volume of the system; ϕ_1 and ϕ_2 the volume fraction of component, respectively; δ_1 and δ_2 are the solubility parameters; V_1 and V_2 are the molar volumes, respectively. Since the densities of PBAT and PLA are very close, for simplicity, we assumed that both polymers had the same density of 1.25 g/mol. Based on this assumption, for PBAT/PLA blends, the volume fraction is equal to the mass fraction.

The phase diagram (Figure 5) offers a simulation of the PBAT/PLA blend miscibility with assumptions at room temperature (296 K, approximately 23 °C), which is important for the blend preparation by solution blending. PBAT structure was alternating; the HiSP difference was 0.49 MPa$^{1/2}$; the density of both polymers was 1.25 kg/mol; the blend had the molecular weights: M_n52/127, M_n52/60, M_n52/30, and M_n30/30. As an example, M_n52/127 implies M_n(PBAT) is 52 kg/mol and M_n(PLA) is 127 kg/mol, respectively. Details of the calculation are in the Supplementary Material on the sheet "Flory–Huggins".

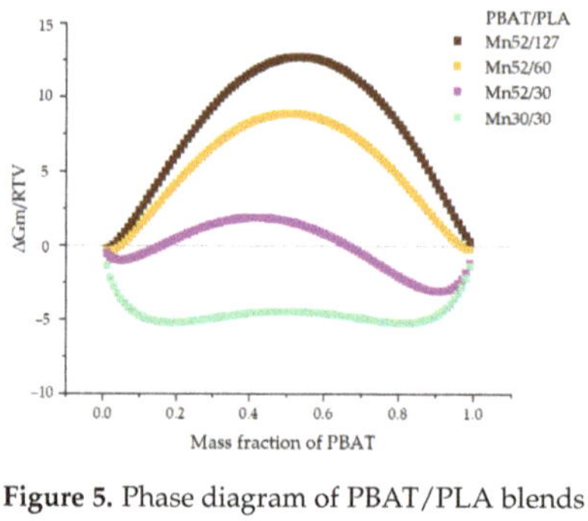

Figure 5. Phase diagram of PBAT/PLA blends with various molecular weights at 296 K.

From the phase diagram, the curve of the PBAT/PLA blends M_n52/127 with the respective molecular weight of 52 and 127 kg/mol is above the zero line, implying blend immiscibility made of the commercially available polymers. A similar trend is shown by the curve M_n52/60, representing the blend made of the original PBAT and PLA with molecular weights of 60 kg/mol. When the molecular weight of PLA was reduced to 30 kg/mol, M_n52/30 showed an interesting curve: in the region of low PBAT content (<15 wt.%), the value of free energy of mixing was slightly negative; this was followed by the curve with a PBAT mass fraction of 15–70 wt.%. Starting with a PBAT content of approximately 70 wt.%, the curve was in the negative region again. The PBAT/PLA blends with the molecular weight of 30 kg/mol (M_n30/30) showed negative values of the free energy of mixing in the whole range. However, it can be seen that in the middle range (40–60 wt.% PBAT), this curve exhibited a slight increase, while the curve shows even lower negative values at the PBAT/PLA ratio of about 20/80 or 80/20. Generally, a lower negative value would lead to a higher probability of miscibility.

Since PBAT/PLA blends are often melt-blended at an elevated temperature, the blend miscibility at 463 K, approx. 190 °C, was simulated (Figure 6).

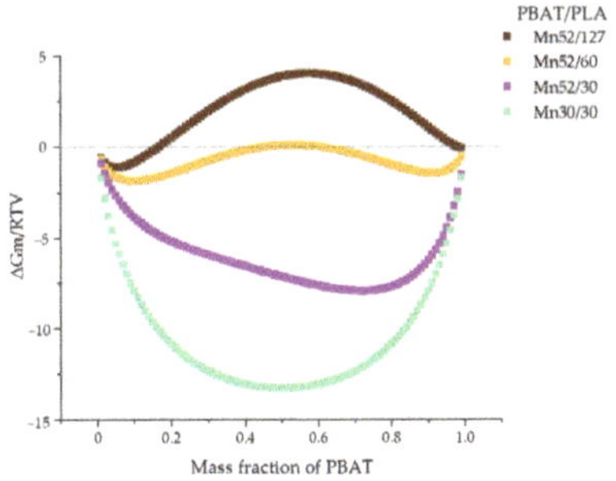

Figure 6. Phase diagram of PBAT/PLA blends with various molecular weights at 463 K.

Even at this higher temperature, M_n52/127 (PBAT/PLA blends consisting of original commercial polymers) showed a curve mostly close to or above the zero line, implying poor miscibility. With the decreased molecular weights of PLA, the curve of M_n52/60 displays a maximum slightly above the zero line (when PBAT content is approx. 50–60 wt.%) and two minima slightly below the zero line (when PBAT content is around 10 wt.% or 90 wt.%), indicating a relatively unstable state of miscibility. A small fluctuation of the temperature or composition may change the theoretical miscibility. When the molecular weights of PLA were 30 kg/mol, M_n52/30 blends showed negative values of the free energy of mixing in the whole range, indicating the blend should be miscible at 463 K. The blends with an M_n of 30 kg/mol for each component demonstrated even lower values, indicating thermodynamically miscibility in the molten state at 463 K.

It is concluded that the mixing behavior depends strongly on the molecular weights of both components, their ratios, and the temperature. As shown in both phase diagrams, the miscibility of PBAT/PLA generally increases with decreasing molecular weights. At 296 K, the blends M_n52/127, M_n52/60, and M_n52/30 each showed a maximum positive value of ΔG_M in the middle range (PBAT content: approx. 50 wt.%) in the curves, indicating poor miscibility of these blends, especially at the blend ratio of 50/50. Generally, in addition to the negative value of ΔG_M, the second derivative of ΔG_M with respect to the volume fraction of the second blend components was a necessary condition of the blend miscibility. At 296 K, the blends M_n30/30 showed negative values of ΔG_M in the whole range of compositions while displaying two minima (at the ratio of about 10/90 and 90/10) and one maximum (at the ratio of 50/50). According to the second derivative of ΔG_M (eq.12 = 0, at 30% volume fraction of PBAT), spinodal decomposition can occur at 296 K due to the negative value of free energy of mixing but beginning of the convex curve of the spinodal (Figure 7).

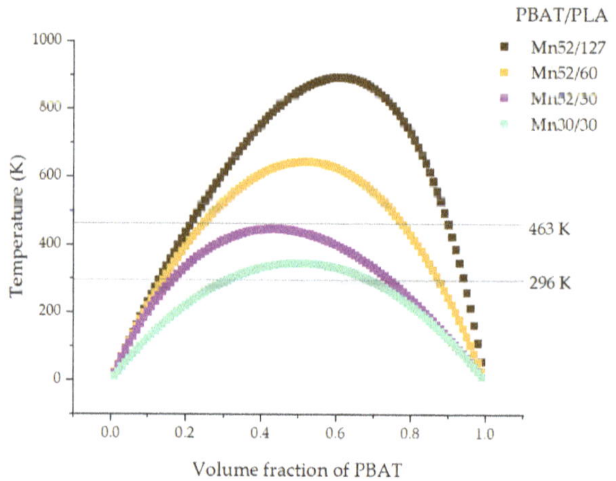

Figure 7. Spinodal curves for PBAT/PLA blends.

At 463 K (e.g., for melt-blending), the curve of the blends $M_n30/30$ shows negative values in the entire range of compositions indicating good miscibility. Even the blends $M_n52/30$ would be miscible in the molten state while melt-blending at 463 K. After that, if the PBAT/PLA blends $M_n52/30$ were stored at a lower temperature between −28 °C and 61 °C (corresponding to the T_g values of both polymers) for a sufficiently long time, the PBAT chains would behave like rubber. For this reason, the blends $M_n52/30$ could change the miscibility from miscible in the molten state to partially miscible or immiscible at a lower temperature after a long enough time. However, the PBAT/PLA with an M_n of 30 kg/mol for each component should be miscible both at 296 K and 463 K, according to the simulation.

5. Estimation of the Glass Transition Temperature

T_g can reflect whether a polymer blend is completely miscible. A fully miscible PBAT/PLA blend exhibits a single T_g. Generally, the T_g value of binary miscible PBAT/PLA blends is predictable based on the Fox equation:

$$\frac{1}{T_g(\text{Blend})} = \frac{m_{\text{PBAT}}}{T_g(\text{PBAT})} + \frac{1 - m_{\text{PBAT}}}{T_g(\text{PLA})} \tag{13}$$

where T_g(Blend) is the predicted glass transition temperature of the miscible PBAT/PLA blend; m_{PBAT} is the mass (weight) fraction of PBAT; T_g(PBAT) and T_g(PLA) are the glass transition temperature of neat PBAT and neat PLA, respectively.

The value of T_g(PBAT) and T_g(PLA) was −28.3 °C and 61.6 °C, respectively [4]. For details of the T_g(Blend) calculation, see the Supplementary Material on the sheet "Fox". The composition-dependent T_g of miscible PBAT/PLA blends is shown graphically (Figure 8).

T_g(Blend) tends to decrease with an increasing mass fraction of PBAT. The glass transition temperature is slightly below 40 °C for PBAT/PLA (20/80). This value is about 10 °C for PBAT/PLA (50/50). Moreover, this value decreases to approximately −15 °C when the ratio of PBAT/PLA is 80/20. To the author's best knowledge, the glass transition temperatures have not been studied for fully miscible PBAT/PLA blends without compatibilizers. Unmodified PBAT/PLA blends with a wide range of ratios (0/100, 10/90, … 90/10, 100/0) have been reported to have two almost unchanged glass transition temperatures at about −30 °C and 61 °C, which correspond to the T_g of neat PBAT and PLA in the DSC [4].

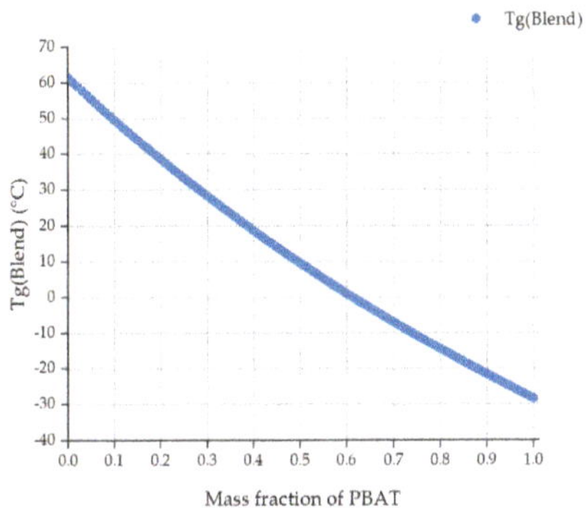

Figure 8. Glass transition temperature of miscible PBAT/PLA blends.

6. Conclusions

In this study, the blend miscibility of PBAT/PLA blends was predicted. The solubility parameters calculated using the group contribution methods of van Krevelen and Hoy had a mean difference of 0.49 $MPa^{1/2}$ between alternating PBAT and PLA. To a certain extent, a higher affinity would be possible between the two polymers when the monomers BA and BT reach a molar ratio exceeding 1 to 1. In this way, the structural optimization of PBAT will fundamentally improve the solubility of PBAT and PLA. Furthermore, a simulation of the miscibility of PBAT/PLA blends was established by using the calculated HiSP and different parameters. According to the simulation, the state of a PBAT/PLA blend can vary from immiscible to miscible, depending strongly on the molecular weights and weight ratio of both polymers at a constant temperature. Generally, the higher the molecular weights, the lower the predicted probability of the blend miscibility. Another tendency is that the higher the temperature, the higher the probability of the blend miscibility. The blends M_n52/30 displayed negative values in the whole range of compositions at 463 K. If M_n52/30 were melt-blended and then stored at a temperature above the T_g of PBAT (−28 °C, approx. 245 K) for enough long time, the miscibility could change from miscible to partially miscible or immiscible, due to the mobility of PBAT chains. The blends M_n30/30 showed negative values of ΔG_M both at 296 K and 463 K, according to the simulation. However, spinodal decomposition of M_n30/30 can appear at 296 K (at 30% volume fraction of PBAT) due to the negative value of ΔG_M and the curvature of the spinodal. Moreover, the glass transition temperature of miscible PBAT/PLA blends was calculated using the Fox equation. A single T_g would show at about 40 °C, 10 °C, and −15 °C for PBAT/PLA blends with the composition of (20/80), (50/50), and (80/20), respectively. This study gives the theoretical prediction of the miscibility for PBAT/PLA blends. The next scientific challenge will be the experimental discovery of to what extent the theoretical prediction is consistent with the practical results, especially the molecular weight-dependent miscibility of PBAT and PLA.

Supplementary Materials: The following are available online at https://www.mdpi.com/article/10.3390/polym13142339/s1, Excel table with the sheets of "vanKrevelen", "Hoy", "Flory–Huggins" and "Fox".

Funding: This research received no external funding.

Institutional Review Board Statement: Not applicable.

Informed Consent Statement: Not applicable.

Data Availability Statement: For details of the data used in this paper, see the supplementary material.

Acknowledgments: The author gives sincere thanks to Rodion Kopitzky (Fraunhofer UMSICHT) for discussions.

Conflicts of Interest: The authors declare no conflict of interest.

References

1. European Bioplastics e.V. Bioplastics Market Data. Available online: https://www.european-bioplastics.org/market/ (accessed on 24 May 2021).
2. Liu, T.-Y.; Lin, W.-C.; Yang, M.-C.; Chen, S.-Y. Miscibility, thermal characterization and crystallization of poly(l-lactide) and poly(tetramethylene adipate-co-terephthalate) blend membranes. *Polymer* **2005**, *46*, 12586–12594. [CrossRef]
3. Farsetti, S.; Cioni, B.; Lazzeri, A. Physico-Mechanical Properties of Biodegradable Rubber Toughened Polymers. *Macromol. Symp.* **2011**, *301*, 82–89. [CrossRef]
4. Su, S.; Duhme, M.; Kopitzky, R. Uncompatibilized PBAT/PLA Blends: Manufacturability, Miscibility and Properties. *Materials* **2020**, *13*, 4897. [CrossRef] [PubMed]
5. Kim, Y.J.; Park, O.O. Miscibility and Biodegradability of Poly(Butylene Succinate)/Poly(Butylene Terephthalate) Blends. *J. Polym. Environ.* **1999**, *7*, 53–66. [CrossRef]
6. Arrighi, V.; Cowie, J.M.; Fuhrmann, S.; Youssef, A. Miscibility Criterion in Polymer Blends and its Determination. In *Encyclopedia of polymer blends. Volume 1, Fundamentals*; Isayev, A.I., Ed.; Wiley-VCH: Weinheim, Germany, 2010; pp. 153–198, ISBN 9783527805204.
7. Imre, B.; Renner, K.; Pukanszky, B. Interactions, structure and properties in poly(lactic acid)/thermoplastic polymer blends. *Express Polym. Lett.* **2014**, *8*, 2–14. [CrossRef]
8. Hamad, K.; Kaseem, M.; Ayyoob, M.; Joo, J.; Deri, F. Polylactic acid blends: The future of green, light and tough. *Prog. Polym. Sci.* **2018**, *85*, 83–127. [CrossRef]
9. Fekete, E.; Földes, E.; Pukánszky, B. Effect of molecular interactions on the miscibility and structure of polymer blends. *Eur. Polym. J.* **2005**, *41*, 727–736. [CrossRef]
10. van Krevelen, D.W.; Nijenhuis, K.t. *Properties of Polymers: Their Correlation with Chemical Structure; Their Numerical Estimation and Prediction from Additive Group Contributions/D.W. van Krevelen*, 4th, completely rev. ed.; Revised by K. Te Nijenhuis; Elsevier B.V.: Amsterdam, The Netherlands; Boston, MA, USA, 2009; ISBN 9780080915104.
11. Dil, E.J.; Carreau, P.J.; Favis, B.D. Morphology, miscibility and continuity development in poly(lactic acid)/poly(butylene adipate-co-terephthalate) blends. *Polymer* **2015**, *68*, 202–212. [CrossRef]
12. Ding, Y.; Feng, W.; Huang, D.; Lu, B.; Wang, P.; Wang, G.; Ji, J. Compatibilization of immiscible PLA-based biodegradable polymer blends using amphiphilic di-block copolymers. *Eur. Polym. J.* **2019**, *118*, 45–52. [CrossRef]
13. Hansen, C.M. *Hansen Solubility Parameters: A User's Handbook*, 2nd ed.; CRC Press: Boca Raton, FL, USA, 2007; ISBN 9781420006834.
14. Nedoma, A.J.; Robertson, M.L.; Wanakule, N.S.; Balsara, N.P. Measurements of the Composition and Molecular Weight Dependence of the Flory−Huggins Interaction Parameter. *Macromolecules* **2008**, *41*, 5773–5779. [CrossRef]
15. Park, J.Y.; Hwang, S.Y.; Yoon, W.J.; Yoo, E.S.; Im, S.S. Compatibility and physical properties of poly(lactic acid)/poly(ethylene terephthalate glycol) blends. *Macromol. Res.* **2012**, *20*, 1300–1306. [CrossRef]
16. Kijchavengkul, T.; Auras, R.; Rubino, M.; Selke, S.; Ngouajio, M.; Fernandez, R.T. Biodegradation and hydrolysis rate of aliphatic aromatic polyester. *Polym. Degrad. Stab.* **2010**, *95*, 2641–2647. [CrossRef]
17. Farah, S.; Anderson, D.G.; Langer, R. Physical and mechanical properties of PLA, and their functions in widespread applications—A comprehensive review. *Adv. Drug Deliv. Rev.* **2016**, *107*, 367–392. [CrossRef] [PubMed]
18. Su, S.; Kopitzky, R.; Tolga, S.; Kabasci, S. Polylactide (PLA) and Its Blends with Poly(butylene succinate) (PBS): A Brief Review. *Polymers* **2019**, *11*, 1193. [CrossRef] [PubMed]
19. Coleman, M.M.; Serman, C.J.; Bhagwagar, D.E.; Painter, P.C. A practical guide to polymer miscibility. *Polymer* **1990**, *31*, 1187–1203. [CrossRef]
20. Hoy, K.L. Solubility Parameter as a Design Parameter for Water Borne Polymers and Coatings. *J. Coat. Fabr.* **1989**, *19*, 53–67. [CrossRef]
21. Stefanis, E.; Panayiotou, C. Prediction of Hansen Solubility Parameters with a New Group-Contribution Method. *Int. J. Thermophys.* **2008**, *29*, 568–585. [CrossRef]
22. Kitak, T.; Dumičić, A.; Planinšek, O.; Šibanc, R.; Srčič, S. Determination of Solubility Parameters of Ibuprofen and Ibuprofen Lysinate. *Molecules* **2015**, *20*, 21549–21568. [CrossRef] [PubMed]
23. Robeson, L.M. *Polymer Blends: A Comprehensive Review*; Carl Hanser Verlag: Munich, Germany; Cincinnati, OH, USA, 2007; ISBN 1569904081.
24. Jost, V.; Kopitzky, R. Blending of Polyhydroxybutyrate-co-valerate with Polylactic Acid for Packaging Applications—Reflections on Miscibility and Effects on the Mechanical and Barrier Properties. *Chem. Biochem. Eng. Q.* **2015**, *29*, 221–246. [CrossRef]

polymers

Article

Rheological Characterization and Modeling of Thermally Unstable Poly(3-hydroxybutyrate-co-3-hydroxyvalerate) (PHBV)

Silvia Lajewski *, Annika Mauch, Kalman Geiger and Christian Bonten

Institut für Kunststofftechnik, University of Stuttgart, Pfaffenwaldring 32, 70569 Stuttgart, Germany; a_mauch@t-online.de (A.M.); kalman.geiger@ikt.uni-stuttgart.de (K.G.); christian.bonten@ikt.uni-stuttgart.de (C.B.)

* Correspondence: silvia.lajewski@ikt.uni-stuttgart.de; Tel.: +49-711-685-62831

Abstract: Presently, almost every industry uses conventional plastics. Its production from petroleum and extensive plastic pollution cause environmental problems. More sustainable alternatives to plastics include bioplastics such as poly(3-hydroxybutyrate-co-3-hydroxyvalerate) (PHBV), which is produced by bacteria and is biodegradable even in seawater. High temperature sensitivity as well as massive thermal degradation cause difficulties during the processing of PHBV. The aim of this work is to create a detailed rheological characterization and master curves to gain deeper knowledge about the material and its processing parameters. The rheological characterization was performed with frequency sweeps in the range of 0.1 rad/s to 628 rad/s and time sweeps over 300 s. Creating master curves at the reference temperature of 180 °C with the software IRIS delivers Carreau and Arrhenius parameters. These parameters allow for a calculation of the master curves for all other temperatures by means of the temperature shift factor. Moreover, the rheological measurements reveal a minimum rheological measurement temperature of 178 °C and a surprisingly high activation energy of 241.8 kJ/mol.

Keywords: bioplastics; PHBV; rheological characterization; plastics processing

Citation: Lajewski, S.; Mauch, A.; Geiger, K.; Bonten, C. Rheological Characterization and Modeling of Thermally Unstable Poly(3-hydroxybutyrate-co-3-hydroxyvalerate) (PHBV). *Polymers* **2021**, *13*, 2294. https://doi.org/10.3390/polym13142294

Academic Editors: José Miguel Ferri, Vicent Fombuena Borràs and Miguel Fernando Aldás Carrasco

Received: 28 June 2021
Accepted: 8 July 2021
Published: 13 July 2021

1. Introduction

Presently, plastics are used for every imaginable application. Because of the industry's focus on the benefits of the processing and application of plastics, important aspects of plastics have been neglected. In addition to the problematic use of finite resources such as petroleum, masses of plastic waste and the creation of plastic patches in oceans produce a major environmental threat [1]. In order to solve these problems without losing the positive properties of conventional plastics, the industry is in search of better alternatives. A very up-to-date and detailed overview of current research trends in biobased and biodegradable polymers can be found in [2]. A promising biopolymer that can make a small but important contribution along the way in the fight against the ocean garbage patches is poly(3-hydroxybutyrate-co-3-hydroxyvalerate) (PHBV).

PHBV belongs to the group of polyhydroxyalkanoates (PHA) and is a copolymer of polyhydroxybutyrate (PHB) [3]. The linear molecular structure of PHBV consists of a linear carbonyl chain with randomly distributed methyl and ethyl side-chains [4]. PHBV is a biodegradable, biocompatible, hydrophobic, and non-toxic polymer [5,6]. It is produced, for instance, by the Gram-negative bacterium Ralstonia eutropha as an energy and carbon store. The production takes place under a lack of oxygen, nitrogen, and phosphor as well as a surplus of carbon [7]. To produce the characteristic side-chains of PHBV, propionic acid and valeric acid are added [8]. The brittle and stiff thermoplastic polymer shows viscoelastic properties [9]. In addition to the application of PHBV for films and packaging, there is also research on applications in medical technology. However, due to high production costs as well as its brittle and temperature-sensitive behavior, PHBV is not currently established in industry and PHA only accounted for 1.7% of the global production capacities of bioplastics in 2020 [10].

Depending on the composition, the melting point of PHBV varies between 50 °C and 180 °C [11]. Temperatures above the melting point cause unstable behavior of PHBV. Thus, the influence of high temperatures leads to a strong degradation of the polymer chains. The thermal degradation processes with α-deprotonation, β-elimination, and random chain scission proceed via several intermediates. At the end of thermal degradation, propylene, acetaldehyde, ketene, acetone, and carbon dioxide occur [12,13].

Viscosity curves are suitable tools for the analysis of the rheological properties of PHBV. These measurements are carried out in the molten state and can therefore accurately reflect the processing behavior of the plastics considered in the application. Even though PHBV has so far been used mainly in blends, the rheological characterization of the pure material is also important in order to obtain a statement of the expected material behavior. Only in this way can the pros and cons for the use of this biopolymer be assessed.

The viscosity η is plotted double-logarithmically over the angular frequency ω or the shear rate $\dot{\gamma}$ (cf. Figure 1). A strong linear decrease at high frequencies is called the pseudo-plastic flow area or shear thinning behavior. At low frequencies, there is a constant viscosity called zero shear viscosity η_0. Between the areas of zero shear viscosity and shear thinning, a transition occurs. The measured shear viscosity function of polymers with linear molecular structure (PS, PE-HD, PC, or even PHBV) is precisely approached with the Carreau model. The Carreau model, shown in Equation (1), calculates the viscosity η as a function of the shear rate $\dot{\gamma}$ with three parameters: the zero shear viscosity A_0, the reciprocal transient shear rate A_1, and the gradient of the pseudo-plastic flow area A_2.

$$\eta(\dot{\gamma}) = \frac{A_0}{(1 + A_1\dot{\gamma})^{A_2}} \quad (1)$$

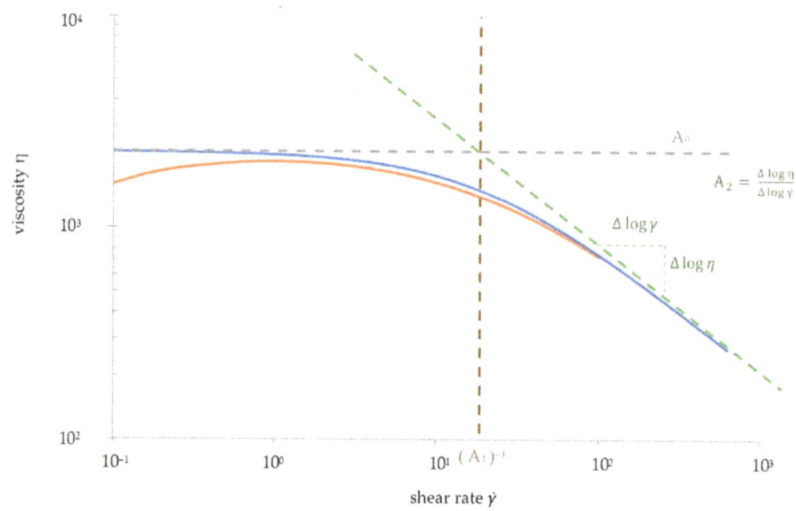

Figure 1. Viscosity curve of thermoplastics. The blue curve is an approximated viscosity curve derived from the Carreau model, and the orange curve is from the rheological measurement of PHBV.

Because of the temperature sensitivity of thermoplastics, the viscosity curve shifts with the change of temperature. At higher temperatures, the viscosity curve shifts by

−45° [14,15]. According to Equation (2), the temperature shift can be calculated in the Carreau model (3) with the temperature shift factor a_T:

$$\eta(\dot{\gamma}) = \frac{a_T \cdot A_0}{(1 + a_T \cdot A_1 \dot{\gamma})^{A_2}} \tag{2}$$

The nondimensional temperature shift factor a_T results from the Arrhenius model as the best approach for the temperature-dependent viscosity function of semi-crystalline polymers like PHBV and can be calculated following Equation (3) [15–17]:

$$a_T = \exp\left(\frac{E_0}{R}\left(\frac{1}{T} - \frac{1}{T_0}\right)\right) \tag{3}$$

To plot viscosity curves, rheological measurements are necessary. Important values of a rheological measurement are the storage modulus G′ and the loss modulus G″. G′ contains the reversible elastic energy, whereas G″ represents the irreversible viscous energy component.

The complex viscosity can be used for an easier analysis of rheological measurements. As shown in Equation (4), the complex viscosity is calculated via the ratio of shear stress τ and the shear rate $\dot{\gamma}$ or the angular frequency ω [18].

$$\eta^* = \frac{\tau(t)}{\dot{\gamma}(t)} = \frac{\tau_0 \sin(\omega t + \delta)}{\gamma_0 \omega \cos(\omega t)} \tag{4}$$

The empirical approach of Cox–Merz [19] in Equation (5) specifies the composition between the viscosity η and the amount of the complex viscosity η^*. At low frequencies, the viscosity corresponds to the amount of the complex viscosity if the shear rate takes the same value as the angular frequency [16].

$$\eta(\dot{\gamma}) = |\eta^*(\omega)| \text{ if } \dot{\gamma} = \omega \tag{5}$$

Thus, the graph of the complex viscosity and angular frequency can be linked with the graph of the viscosity and the shear rate.

In a preliminary work, viscosity curves of PHBV were plotted from rheological measurements. A frequency sweep (FS) from 100 rad/s to 0.1 rad/s was performed to obtain the viscosity curves. The viscosity curves did not fulfill the expected and discussed model curves. As shown in Figure 1, the viscosity decreases at frequencies lower than 10 rad/s where a constant zero shear viscosity is expected. This decrease can be explained with thermal degradation.

Due to the extreme thermal degradation already occurring just above the melting temperature, it is difficult to accurately predict viscosity curves at temperatures below or above the measurement temperature. Moreover, the reasonable range of temperature and frequency as well as the zero shear viscosity at a certain temperature are unknown. However, for further rheological measurements, applications, and processing of PHBV, it is necessary to gain this knowledge. Therefore, the aim of this work is to determine important rheological parameters of PHBV. Additionally, we try to better understand the thermal degradation of PHBV by comparing and analyzing measured and calculated viscosity curves.

2. Materials and Methods

2.1. Material

PHBV pellets from TianAn Biopolymer, Ningbo, China, of type ENMAT Y1000P were used for all measurements. According to the manufacturer, this type has a valerate content of only 1–2%. The molar mass was described in the literature as 485,000 g/mol [20]. The pellets were dried overnight in a vacuum drier from Binder GmbH, Tuttlingen, Germany, at 40 °C.

2.2. Methods

Rheological measurements were performed on the computer-controlled plate–plate rheometer Discovery HR-2 from TA Instruments, New Castle, Great Britain. A plate–plate rheometer measures torque with a rotatable upper plate. Using torque, angular frequency, and geometry data, the complex viscosity and angular frequency can be calculated [14,21]. Additional data from the rheological measurement are the storage and loss modulus.

At the beginning of the measurement, the pellets were placed at the lower plate and molded for three minutes. After three minutes, the gap of 1 mm was approached and the excess mass was removed. Three minutes later, the measurement was started as a frequency sweep or a time sweep. In the case of the frequency sweeps, the measurement was made from high to low frequencies. The parameters for the rheological measurement are presented in Table 1.

Table 1. Measurement parameters of the rheological measurements.

	Frequency Sweep	Time Sweep
preparation/molding time in min	6	6
plate diameter in mm	25	25
gap in mm	1	1
temperature in °C	177–190	180–190
angular frequency in rad/s	628.0–1.0 628.0–0.1	0.1 1.0 10.0 100.0
time in s	approx. 102	300
strain in %	5	5

The software IRIS Rheo-Hub 2020 from IRIS Development LLC, Amherst, MA, USA was used to evaluate the measured data. The horizontal temperature shift is calculated with the "t-T-Shift mode" function. For the viscosity shift with the Carreau model, the function "Viscosity Fit -> 2nd Carreau" is used. All shifting parameters and data can be exported.

3. Results and Discussion

3.1. Preliminary Measurement for Error Analysis

In order to analyze the fluctuations at several rheological measurements, a time sweep for 300 s at a temperature of 180 °C and a frequency of 1 rad/s was set five times. Figure 2 shows the semi-logarithmical plot of the complex viscosity over time.

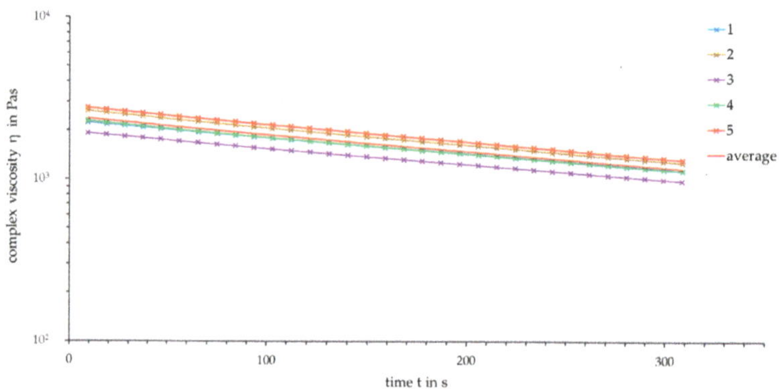

Figure 2. Error analysis, time sweep (ϑ = 180 °C, ω = 1 rad/s).

High deviations between the measurements occurred, even though all measurements were set under the same conditions. At the beginning of the measurements, the standard deviation amounted to 12.55% and decreased to 10.27% after five minutes. The decrease of the standard deviation can be explained by the decreasing viscosity and longer measurement time. Further measurements at 190 °C show a lower standard deviation between 10.39% and 7.94%. Thus, the high standard deviation can be explained not only by the material properties of PHBV but also by the rheological measurement process, which is susceptible to inaccuracies in handling. For the following discussions and analysis, the standard deviation is given as 12.5%.

3.2. Minimum Rheological Measurement Temperature

Thermal degradation has a huge impact on the viscosity curves of PHBV. Therefore, it is helpful to work with low measurement temperatures to minimize degradation effects. In order to investigate the impact of temperature on PHBV and to find the minimum measurement temperature, frequency sweeps were performed at different temperatures. Each frequency sweep at a certain temperature was measured three times, and the average was taken. For a clear illustration, error bars are omitted in Figure 3.

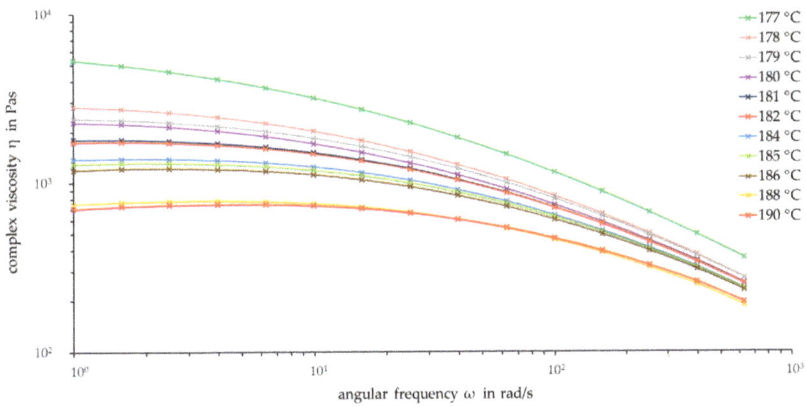

Figure 3. Frequency sweeps at different temperatures (ϑ = 177–190 °C, ω = 1–628 rad/s).

As can be seen, a higher measurement temperature resulted in a lower complex viscosity. Despite the high standard deviation, the viscosity curves are perfectly ordered according to temperature. The viscosity curves for 188 °C and 190 °C are visibly lower than the other curves. With the decrease in the complex viscosity between 10 rad/s and 1 rad/s, the thermal degradation of PHBV for high temperatures is illustrated. In the temperature range between 186 °C and 178 °C, the slopes of the viscosity curves above 100 rad/s are comparable. This can be explained by the assumption that in an ideal case the viscosity curves shift with the temperature by $-45°$. Consequently, the slope of -1, which is equal to a shift of $-45°$, shows a temperature invariance in the form of the viscosity curve. In Figure 3, the viscosity curves with different temperatures approach this temperature invariance with their behavior in the pseudo-plastic flow area. At lower frequencies, between 10 rad/s and 1 rad/s, the viscosity curves differentiate further. A zero shear viscosity has not yet been reached yet, especially at lower temperatures.

The viscosity curve at 177 °C is remarkably higher than the curve at 178 °C. Due to the low temperature of 177 °C, the PHBV pellets cannot be completely melted with the given rheological measurement setup. Other measurements with longer melting times up to eight minutes showed the same effect. As is common with thermoplastic polymers, PHBV does not have a precise melting point due to the entangled polymer chains but rather a melting range where the polymer plastifies. Further research with differential scanning calorimetry (DSC) confirms this behavior: PHBV shows a pronounced melting peak with a peak temperature of 177.4 °C. Only complete melting of the sample material allows meaningful rheological measurements. Therefore, the minimum suitable temperature for rheological measurements of the PHBV pellets is 178 °C, which is an important parameter for its later processing.

It is worth mentioning that even a change in the measurement temperature of only 1 °C effects visible changes of the complex viscosity. This behavior indicates a high temperature dependency on the viscosity of PHBV and is a challenge in the processing and application of PHBV.

3.3. Master Curve at the Reference Temperature of 180 °C

With the measurement data from temperatures between 178 °C and 181 °C, it is possible to calculate a master curve for the reference temperature of 180 °C via IRIS. This master curve delivers the Carreau parameters A_0 = 2297.3 Pas, A_1 = 0.054 s, and A_2 = 0.605, as well as the Arrhenius parameter E_0/R = 29,077.3 K, which includes the activation energy E_0 = 241,763.2 J/mol. This value is unusually high. In the literature, data of 25–80 kJ/mol can be found for polymer melts in general [15].

After calculating the temperature shift factors a_T with Equation (3), all master curves for other temperatures can be calculated by applying Equation (2). Figure 4 shows the calculated Carreau model as well as the IRIS viscosity fit and the measurement data for the reference temperature 180 °C.

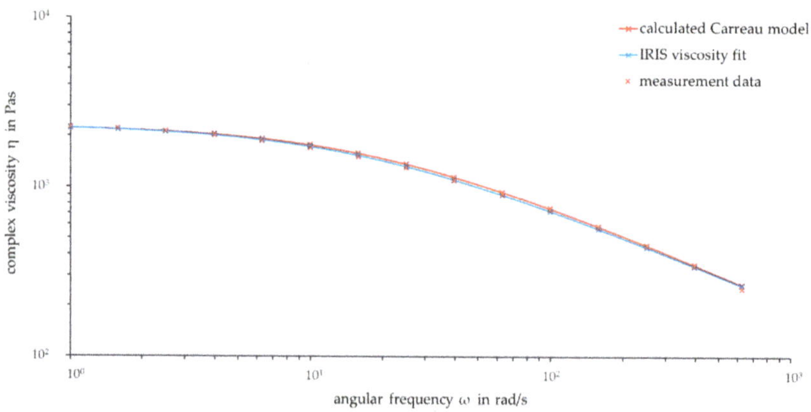

Figure 4. Frequency sweep: Calculated Carreau model, IRIS viscosity fit, and measurement data from reference temperature (ϑ = 180 °C, ω = 1–628 rad/s).

The viscosity curve of the calculated Carreau model is similar to the IRIS viscosity fit. As the temperatures of the two curves correspond to each other, this behavior was expected. Between a frequency of 10 rad/s and 250 rad/s, the viscosity curve of the calculated Carreau model is slightly higher than the curve fitted with IRIS.

It is interesting to compare the calculated values at 180 °C with the values of a time sweep at 180 °C (not shown here). The Carreau model does not consider thermal degradation explicitly, so the calculated Carreau model values are considered ideal. Comparing the start data with the data of the time sweep after the same time as the frequency sweep underlines the temporal and thermal degradation of PHBV. These degradation processes intensify with increasing time and decreasing frequency.

At a frequency of 100 rad/s, the calculated Carreau data are similar to the data of the time sweep after the same time. This could be an indicator that thermal degradation does not considerably affect the Carreau model because the master curve is based on data including thermal degradation. With decreasing frequency and increasing time, the conformity decreases due to increasing external influences. Especially for low frequencies, thermal degradation can be considered as having a minor impact on the calculated Carreau values.

Overall, the comparisons show that degradation processes occur at the reference temperature of 180 °C. The long preparation time of 6 min before a rheological measurement can start has to be considered. The discussed thermal degradation processes start before the start of the actual measurement. Actual processing steps, such as extrusion and injection molding, often need shorter times in the range of only a few minutes. For this reason, there should be deeper investigations into if and how the calculated data of the complex viscosity can be used for industrial processing.

3.4. *Calculated Master Curves from the Parameters at the Reference Temperature 180 °C*

In the following graphs, error bars for the calculated Carreau model curves are shown for easier analysis. The error bars are based on the above-mentioned standard deviation of 12.5% since the calculated Carreau parameters are based on measurement data including errors.

First, the viscosity curves from measurement temperatures below 180 °C are investigated. Figure 5 shows the double-logarithmically plotted complex viscosity over the frequency at the temperature of 178 °C.

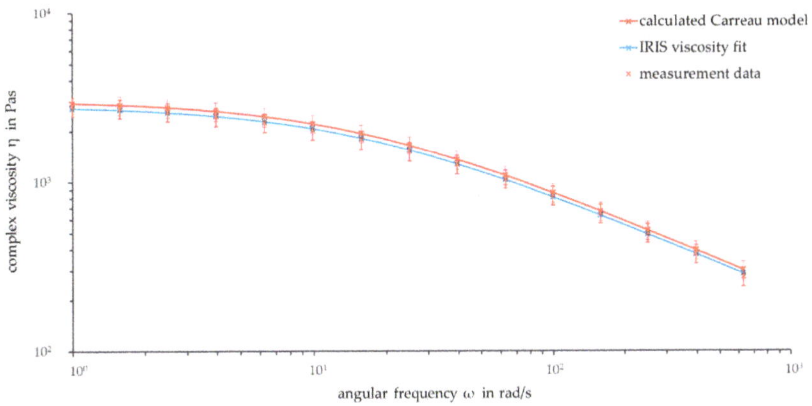

Figure 5. Frequency sweep: Calculated Carreau model, IRIS viscosity fit, and measurement data (ϑ = 178 °C, ω = 1–628 rad/s).

The calculated viscosity curve and the measured and fitted curve show a similar course, with the data of the calculated Carreau model curve being slightly higher than the measurement data. Compared to the data at 180 °C, the viscosity curves are shifted to higher viscosity values. This is due to the fact that the movement of the polymer chains slows down with a decrease in temperature, so the viscosity increases. As a result, the temperature shift factor increases for temperatures below the reference temperature.

If the measurement temperature exceeds the melting point, a stronger degradation of the PHBV chains occurs and results in lower complex viscosities. For temperatures above 182 °C, the complex viscosity curve of the calculated Carreau model is below the measurement data. It is notable that the difference between calculated and measured values increases with higher frequencies. This is interesting, because the deviation increases at lower frequencies for temperatures below 180 °C. The reason for increasing differences at higher frequencies can be found in the transition area of the viscosity curve. Especially at measurement temperatures above 185 °C, the superposition of degradation and viscosity change in the transition area begins after a shorter time and therefore at higher frequencies. Therefore, the transition area of the calculated Carreau model overlaps with a higher transition angular frequency. The Carreau model overestimates the temperature-dependent behavior of the complex viscosity at high temperatures. Higher values for the calculated viscosity were expected, because of the minor influence of degradation on the Carreau model.

The viscosity curves at 190 °C presented in Figure 6 show further details. A temperature of 190 °C is the first temperature where no overlap within the standard deviations of calculated and measured data are found and where the viscosity curves can be clearly separated. A possible reason for this is that at 190 °C, the material reaches a state in which thermal degradation has a strong influence on the PHBV chains. In general, the Carreau model underestimates the viscosity curves at temperatures above the reference temperature. On the one hand, this is explainable, because a higher complex viscosity of the Carreau model without thermal degradation was expected. On the other hand, this could be a sign that thermal degradation of PHBV chains is not as high as assumed.

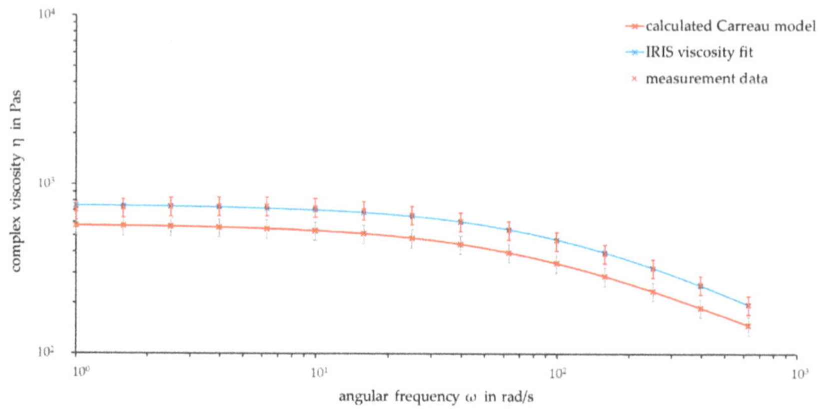

Figure 6. Frequency sweep: Calculated Carreau model, IRIS viscosity fit, and measurement data (ϑ = 190 °C, ω = 1–628 rad/s).

For temperatures below and near the reference temperature, the Carreau model overestimates the viscosity. All in all, the Carreau model calculates and illustrates the viscosity curve satisfyingly. Since the calculated viscosity curves up to a measurement temperature of 182 °C are higher than the measured data, an investigation into whether this is due to the proximity to the reference temperature or to the proximity to the melting temperature should be performed.

One possibility to further analyze the zero shear viscosity and the validity range of the viscosity curves is to extend the range of the curves from 0.1 rad/s to 628 rad/s. The analysis is only useful up to the frequencies where the calculated viscosity curve intersects with the measured viscosity curve (for high temperatures) or where the calculated viscosity curve deviates from the measured viscosity curves (for lower temperatures). Figure 7 shows the frequency sweep from 628 rad/s to 0.1 rad/s at 190 °C.

The measured viscosity curve crosses the calculated viscosity curve at 1 rad/s. For lower frequencies, the measured viscosity curve decreases while the calculated viscosity curve stays constant. Thus, the deviation between both viscosity curves increases with increasing angular frequency. At 0.25 rad/s, the error bars of the viscosity curves with the standard deviation intersect for the last time. As a result, the viscosity curve at 190 °C probably reaches zero shear viscosity between 1 rad/s and 0.25 rad/s.

The measured and calculated viscosity curves at a measurement temperature of 180 °C deviate with decreasing frequency. There is no intersection for the viscosity curves because the measured viscosity curve is lower than the calculated viscosity curve for the whole measurement. From 628 rad/s to 0.4 rad/s, there is an intersection of the error bars. Thus, the validity area at a measurement temperature of 180 °C starts at 0.4 rad/s.

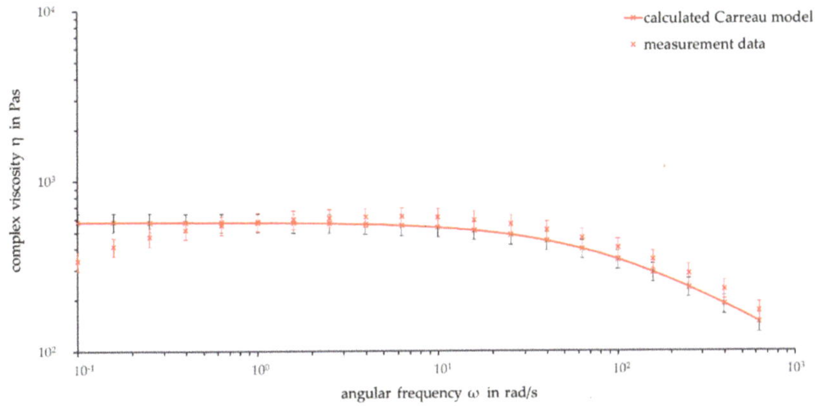

Figure 7. Frequency sweep: Calculated Carreau model and measurement data (ϑ = 190 °C, ω = 0.1–628 rad/s).

4. Conclusions

The performed rheological measurements of PHBV are in the linear viscosity area and have a standard deviation of 12.5%. The minimum rheological measurement temperature of PHBV is 178 °C. Using the IRIS software, a master curve at the reference temperature of 180 °C based on the measurement data from 178 °C to 181 °C was calculated. This master curve delivers Carreau and Arrhenius parameters that can be used to calculate the master curves for other temperatures.

Master curves for measurement temperatures below the reference temperature show higher calculated values than the measurement data. At measurement temperatures above 182 °C, the calculated viscosity data are lower than the measurement data. Thus, the Carreau model underestimates the viscosity of PHBV at higher temperatures. The error bars of the master curves until 185 °C intersect with the error bars from the measurement data, suggesting that the master curves from 178 °C to 185 °C can be used for analysis. Due to the extension of the frequency area to 0.1 rad/s, the zero shear viscosity for 190 °C can be set at 1.0 rad/s to 576.5 Pas. Analogously, the zero shear viscosity for 180 °C can be assumed at 0.4 rad/s as 1732.7 Pas.

The investigated procedure to create master curves with the Carreau model results in viable curves. To minimize the difference between calculated and measured data, other calculation models should be investigated in the future. For example, the calculation of the temperature shift factor with the Vogel-Fulcher and the WLF models should be investigated in further projects. The molar mass also has a great influence on the rheological properties of PHBV. An analogous investigation of different PHBV types would provide further highly interesting insights into the degradation behavior of PHBV.

The rheological characterization underlines the high temperature sensitivity of PHBV. A temperature shift as small as 1 °C shows visible differences in the viscosity curves. This is further confirmed by the unusually high flow activation energy E_0 = 241.8 kJ/mol and leaves only a small temperature window for processes and applications. More detailed investigations will allow better characterization of PHBV and allow for applications of the biodegradable plastic PHBV in the future.

Author Contributions: Project administration, S.L.; methodology, K.G.; data curation and investigation, A.M.; writing, A.M. and S.L.; review and editing, C.B. and K.G.; supervision, C.B. All authors have read and agreed to the published version of the manuscript.

Funding: The project on which this report is based was funded by the German Federal Ministry of Food and Agriculture under the funding code 22039518. The responsibility for the content of this publication lies with the authors.

Institutional Review Board Statement: Not applicable.

Informed Consent Statement: Not applicable.

Data Availability Statement: Not applicable.

Acknowledgments: We would like to thank our colleagues Tobias Schaible for assistance with the evaluation of the rheological measurements and Jochen Wellekötter for the linguistic revision. Furthermore, we thank the Fachagentur für Nachwachsende Rohstoffe (FNR) for funding within the framework of the "Renewable Resources" program.

Conflicts of Interest: The authors declare no conflict of interest.

References

1. Egger, M.; Sulu-Gambari, F.; Lebreton, L. First evidence of plastic fallout from the North Pacific Garbage Patch. *Sci. Rep.* **2020**, *10*, 7495. [CrossRef] [PubMed]
2. Vaishnav, A. *Microbial Polymers: Applications and Ecological Perspectives*; Springer Nature: Basingstoke, UK, 2021.
3. Zhao, H.; Cui, Z.; Wang, X.; Turng, L.-S.; Peng, X. Processing and Characterization of Solid and Microcellular Poly(Lactic Acid)/Polyhydroxybutyrate-Valerate (PLA/PHBV) Blends and PLA/PHBV/Clay Nanocomposites. *Compos. Part B Eng.* **2013**, *51*, 79–91. [CrossRef]
4. Rivera-Briso, A.L.; Serrano-Aroca, Á. Poly(3-Hydroxybutyrate-*co*-3-Hydroxyvalerate): Enhancement Strategies for Advanced Applications. *Polymers* **2018**, *10*, 732. [CrossRef] [PubMed]
5. Laycock, B.; Halley, P.; Pratt, S.; Werker, A.; Lant, P. The Chemomechanical Properties of Microbial Polyhydroxyalkanoates. *Prog. Polym. Sci.* **2013**, *38*, 536–583. [CrossRef]
6. Avella, M.; Martuscelli, E.; Raimo, M. Properties of Blends and Composites Based on Poly(3-Hydroxy)Butyrate (PHB) and Poly(3-Hydroxybutyrate-Hydroxyvalerate) (PHBV) Copolymers. *J. Mater. Sci.* **2000**, *35*, 523–545. [CrossRef]
7. Süss, R.; Leitner, V.; Paulik, C.; Putz, R.; Kamm, B. Quantifizierungsmethoden für die Produktion von PHB mittels Halomonas Halophila. *BIOspektrum* **2019**, *25*, 226–228. [CrossRef]
8. Pötter, M.; Madkour, M.H.; Mayer, F.; Steinbüchel, A. Regulation of Phasin Expression and Polyhydroxyalkanoate (PHA) Granule Formation in Ralstonia Eutropha H16. *Microbiology* **2002**, *148*, 2413–2426. [CrossRef] [PubMed]
9. Sudesh, K.; Abe, H.; Doi, Y. Synthesis, Structure and Properties of Polyhydroxyalkanoates: Biological Polyesters. *Prog. Polym. Sci.* **2000**, *25*, 1503–1555. [CrossRef]
10. European Bioplastics, Nova-Institute. Global Production Capacities of Bioplastics 2020. Available online: https://www.european-bioplastics.org/market/ (accessed on 21 January 2021).
11. Choi, H.J.; Kim, J.; Jhon, M.S. Viscoelastic Characterization of Biodegradable Poly(3-Hydroxybutyrate-Co-3-Hydroxyvalerate)1. *Polymer* **1999**, *40*, 4135–4138. [CrossRef]
12. Liu, Q.-S.; Zhu, M.-F.; Wu, W.-H.; Qin, Z.-Y. Reducing the Formation of Six-Membered Ring Ester during Thermal Degradation of Biodegradable PHBV to Enhance Its Thermal Stability. *Polym. Degrad. Stab.* **2009**, *94*, 18–24. [CrossRef]
13. Xiang, H.; Wen, X.; Miu, X.; Li, Y.; Zhou, Z.; Zhu, M. Thermal Depolymerization Mechanisms of Poly(3-Hydroxybutyrate-Co-3-Hydroxyvalerate). *Prog. Nat. Sci. Mater. Int.* **2016**, *26*, 58–64. [CrossRef]
14. Bonten, C. *Kunststofftechnik, 3. Auflage*; Hanser: München, Germany, 2020.
15. Schröder, T. *Rheologie der Kunststoffe*; Hanser: München, Germany, 2018.
16. Osswald, T.A.; Rudolph, N. *Polymer Rheology*; Hanser Publications: Cincinnati, OH, USA, 2015.
17. Morrison, F.A. *Understanding Rheology*; Topics in chemical engineering; Oxford University Press: New York, NY, USA, 2001.
18. Mours, M.; Winter, H.H. *IRIS Rheo-Hub Handbook*; Iris Development LLC: New Orleans, LA, USA, 2007.
19. Cox, W.P.; Merz, E.H. Correlation of Dynamic and Steady Flow Viscosities. *J. Polym. Sci.* **1958**, *28*, 619–622. [CrossRef]
20. McDonough, K.; Itrich, N.; Casteel, K.; Menzies, J.; Williams, T.; Krivos, K.; Price, J. Assessing the biodegradability of microparticles disposed down the drain. *Chemosphere* **2017**, *175*, 452–458. [CrossRef] [PubMed]
21. Mezger, T.G. *Das Rheologie Handbuch*; Vincentz Network: Hannover, Germany, 2019. [CrossRef]

polymers

MDPI

Review

Macro and Micro Routes to High Performance Bioplastics: Bioplastic Biodegradability and Mechanical and Barrier Properties

Olivia A. Attallah [1], Marija Mojicevic [1,*], Eduardo Lanzagorta Garcia [1], Muhammad Azeem [1], Yuanyuan Chen [1], Shumayl Asmawi [2] and Margaret Brenan Fournet [1]

1 Materials Research Institute, Athlone Institute of Technology, N37 HD68 Athlone, Ireland; oadly@ait.ie (O.A.A.); e.lgarcia@research.ait.ie (E.L.G.); m.azeem@research.ait.ie (M.A.); yuanyuanchen@ait.ie (Y.C.); mfournet@ait.ie (M.B.F.)

2 Fundamental and Applied Science Department, Universiti Teknologi PETRONAS, Bandar Seri Iskandar 32610, Perak Darul Ridzuan, Malaysia; muhammad_24531@utp.edu.my

* Correspondence: mmojicevic@ait.ie

Abstract: On a score sheet for plastics, bioplastics have a medium score for combined mechanical performance and a high score for biodegradability with respect to counterpart petroleum-based plastics. Analysis quickly confirms that endeavours to increase the mechanical performance score for bioplastics would be far more achievable than delivering adequate biodegradability for the recalcitrant plastics, while preserving their impressive mechanical performances. Key architectural features of both bioplastics and petroleum-based plastics, namely, molecular weight (M_w) and crystallinity, which underpin mechanical performance, typically have an inversely dependent relationship with biodegradability. In the case of bioplastics, both macro and micro strategies with dual positive correlation on mechanical and biodegradability performance, are available to address this dilemma. Regarding the macro approach, processing using selected fillers, plasticisers and compatibilisers have been shown to enhance both targeted mechanical properties and biodegradability within bioplastics. Whereas, regarding the micro approach, a whole host of bio and chemical synthetic routes are uniquely available, to produce improved bioplastics. In this review, the main characteristics of bioplastics in terms of mechanical and barrier performances, as well as biodegradability, have been assessed—identifying both macro and micro routes promoting favourable bioplastics' production, processability and performance.

Keywords: biomaterials; biodegradation; bioplastics; mechanical performance; barrier performance; processability

Citation: Attallah, O.A.; Mojicevic, M.; Garcia, E.L.; Azeem, M.; Chen, Y.; Asmawi, S.; Brenan Fournet, M. Macro and Micro Routes to High Performance Bioplastics: Bioplastic Biodegradability and Mechanical and Barrier Properties. *Polymers* **2021**, *13*, 2155. https://doi.org/10.3390/polym13132155

Academic Editors: José Miguel Ferri, Vicent Fombuena Borràs and Miguel Fernando Aldás Carrasco

Received: 29 May 2021
Accepted: 25 June 2021
Published: 30 June 2021

1. Introduction

Pervasive plastics are leaving an indelible imprint on our planet. As high performance and energy-saving materials, plastics are ubiquitous and central to socio-economic advancement. Current mainstay plastics are processed from fossil fuel resources, with production requirements expected to double over the next 20 years. After use, these recalcitrant plastics are contributing to waste stockpiles and alarming pollution. Recycling technologies, which primarily include mechanical and thermochemical approaches, does not meet the efficiency levels required to safeguard the planet and adequately revalorise plastics as new products. The current linear economic model of resource mining, use and discarding, is now widely recognised as unsustainable. A circular approach, where resources are repurposed cyclically, akin to biological lifecycles, is essential in achieving a sustainable socio-economic ecosystem.

Nature readily operates elegant and efficient regenerative cycles for natural polymers and end of life bio-based materials. Such biodegradation and bio-regeneration processes involve microbial, enzymatic and biocatalytic activities for depolymerisation and repolymerisation. Petroleum-based plastics, with their smooth surface topographies, extensive hydrophobic chains and lack of bio-accessible organic chemical groups, are strongly bio-inert and largely incompatible with bioprocessing, leading to their persistence over century timescales within land and water environments. Biomass provides a wealth of renewable and bio-waste resources for bioplastics synthesis. Many of these bio-based plastics, encompass capacities for biodegradation and bioprocessing with high performance features akin to petroleum-based plastics. The realisation of bioplastics that exhibit a complete set of mechanical and biodegradability, hold the promise of delivering material of ecologically sustainable, low carbon footprint circularity.

Bioplastics to date, however, have not achieved wide acceptability by the industry. Incompatibility with existing sorting infrastructures and high temperature mechanical recycling implemented for fossil-based plastics, along with raised production costs, are limiting factors. Technical shortcomings, such as brittleness, lower gas barrier functions and processing performances, have also played a role in keeping current market penetration levels in the region at just 2%. Combining high performance for consumer applications and continuous low carbon closed loop regeneration within plastics poses considerable challenges. At a fundamental structural level, polymeric features associated with good mechanical and fluid barrier properties are typically prohibitive to biodegradability. Petroleum-based plastics achieve the required degrees of high mechanical strength combined with flexibility and strong liquid and gas barrier properties by packing their sleek chemically structured chains into signature crystalline and amorphous regional arrangements. The tight alignment of chemically simple chains at high degrees of crystallinity also renders these plastics largely incompatible with biodegradation processes that require bioactivities, including enzymatic hydrolysis. Bioplastics, in contrast, by the very fact that they are generated from bio-based resources, are inherently more complex with more elaborate chemical structures. This provides both a means to progress their mechanical performance properties and provides amenability to bioactivity with higher levels of hydrolysable groups available for post use biodegradation and biodepolymerisation. To date, equivalent results are readily achievable, and in cases, results outperform particular mechanical properties for polylactic acid (PLA) and polyhydroxyalkanoates (PHA) bioplastics compared with conventional fossil-based thermoplastics. The potential to address performance limitations by a combination of bottom up and top-down approaches using considered chemical structure modifications and blending and composite formations, holds the promise of framing a new generation of bioplastics that encompass sustainability with performance.

In this review, the sustainability/performance triangulation between the biodegradability, mechanical and barrier properties of bioplastics is discussed. Approaches to overcoming the gap between industrially required mechanical and barrier performances and biodegradability are overviewed and related to the potential to build a new generation of high-performance sustainable plastics.

2. Bioplastic Production

Biopolymers can be obtained directly from biomass, as in the case of proteins and polysaccharides and synthetic biopolymers, such as PLA. Biodegradable polymers, including polycaprolactone (PCL), polyglycolic acid (PGA) and polybutylene succinate-co-adipate (PBSA), are primarily synthesised from petrochemicals. Microbial fermentation of biopolymers, including PHA and bacterial cellulose (BC), operates under relatively benign low energy conditions, and hence, is a highly favourable sustainable production route. Various microorganisms can accumulate PHAs as storage materials when cultivated under different nutrient and environmental conditions. This ability allows their survival under stressful conditions. The number and size of the PHA granules, the monomer composition, macromolecular structure and physico-chemical properties vary, depending on the producing microorganisms, the feedstock supplied and the operation conditions [1–3]. On the other hand, some bacteria can produce BC. This exopolysaccharide is a naturally occurring, chemically pure, free of hemicellulose, lignin and pectin, which is why BC purification is an easy process demanding low energy consumption. Nevertheless, production yields are very low, cultivation times are extensive, and thickness of layers is limited, which are major drawbacks in the conventional BC production process, affecting the range of possible applications [4]. For the industrial production of bioplastics, three limitations are very important. These include requirements of specialised growth conditions, expensive precursors, and high recovery costs. Building an increased body of knowledge on producing microbes' metabolism, biosynthetic pathways and their regulation is essential in overcoming these limitations [1,5].

3. Biosynthesised Plastics

In recent years, PHA polymers have emerged as one of the most promising biodegradable materials. Unlike commonly used fossil-based plastics or PLA, which requires the additional step of lactic acid polymerisation, PHAs are the product of bacterial metabolism and have a function of cytoplasmic inclusions. These materials, thanks to the over 150 monomer units, can have a variety of polymer properties that can compete with commonly used plastics, such as polyethene or polypropylene [6]. Monomer composition is connected to the substrate specificity of PHA synthase, hence the type of derived PHA is highly connected to the microorganism used for its production. For example, biosynthesis of short-chain length PHAs (SCL-PHAs) consisting of poly-3-hydroxybutyrate P(3HB) homopolymers is a three-step process regulated by 3-ketothiolase, acetoacetyl-CoA reductase and the SCL PHA synthase [7,8]. Bacteria, such as *Aeromonas caviae* and *Pseudomonas stutzeri*, are proven producers of PHA synthases with wide substrate specificity. Moreover, these enzymes have been identified after recombinant expression in *Ralstonia eutropha*, previously PHA negative. These PHA synthases can produce copolymers of SCL- and medium-chain length PHAs (MCL-PHA) [9]. It has been reported that combinatorial mutations in *P. aeruginosa*, *P. oleovorans*, and *P. putida* resulted in their ability to synthesise PHAs with 3-hydroxybutyrate (3HB), 3-hydroxyhexanoate (3HHx), 3-hydroxyocatanoate (3HO), 3-hydroxydecanoate (3HD), or 3-hydroxydodecanoate (3HDD) monomers [10–13]. These structures are presented in Table 1. The development of recombinant strains and using genetic engineering techniques can lead to improved mechanical and thermal characteristics of PHA materials. These properties are highly dependent on various factors during upstream and downstream processes. Modification of PHAs chemical structure, such as the introduction of functional groups or producing blends and copolymers, can also affect the quality of these materials [14]. Traditional chemical synthesis techniques have been successfully used for creating block copolymers with PHA materials. For example, block copolymers, including PHB blocks balancing with other materials (poly(6-hydroxyhexanoate), poly(3-hydroxyoctanoate), monomethoxy-terminated poly(ethylene glycol) (mPEG), and poly(ethylene glycol) (PEG)), have been reported. These structures are presented in Table 2 [8,15].

Table 1. Chemical structures of monomers described as units of PHA copolymer producing strains.

3-Hidroxyacids	Structure
butyric (3HB)	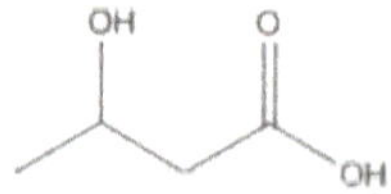
hexanoic (3HHx)	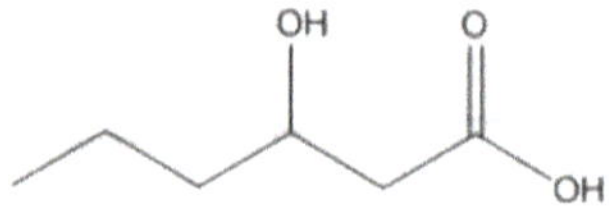
octanoic (3HO)	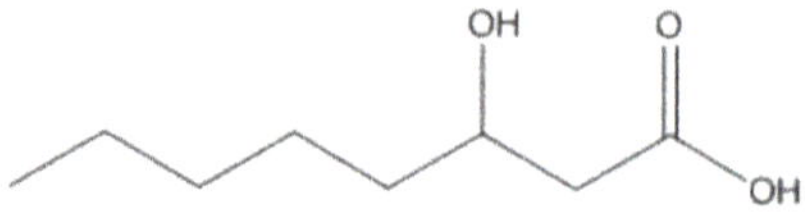
decanoic (3HD)	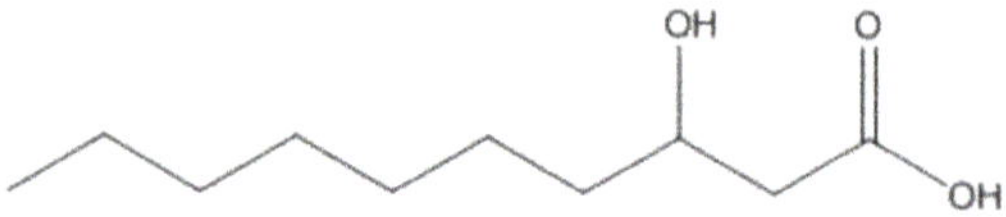
dodecanoic (3HDD)	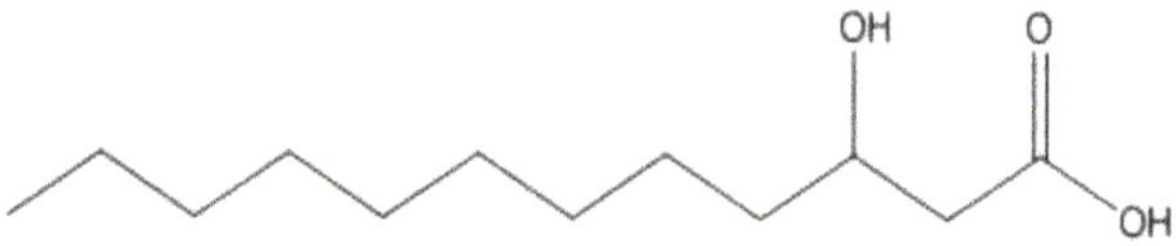

Table 2. Chemical structures of polymers commonly found as building blocks in PHA related block copolymers.

Polymer	Structure
poly(6-hydroxyhexanoate)	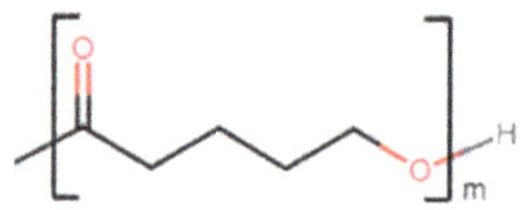
poly(3-hydroxyoctanoate)	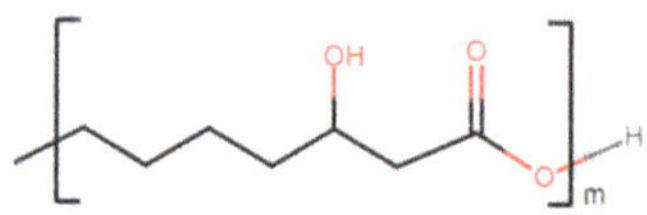
monomethoxy-terminated poly(ethylene glycol) (mPEG)	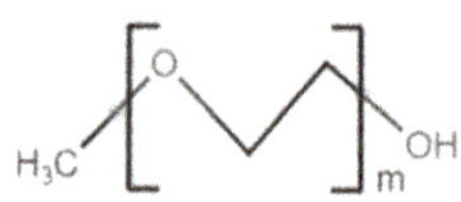
poly(ethylene glycol)	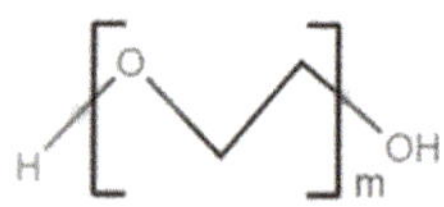

PHAs can also be derived from various substrates, including industrial waste streams [16], food waste [17], supplemented solid biodiesel waste, plant oils [7]. Besides these, seaweeds were found to be great feedstock for PHAs productions [18]. These materials can be produced using different strategies, including batch, fed-batch and continuous processes, and various conditions could be used for their conduction. Batch cultivations are easy and simple to operate, but production yields are very low. On the other hand, fed-batch cultivation can provide higher product and cell concentrations (no substrate inhibition) [19].

Continuous cultivation are also considered to be a viable strategy, providing required concentrations of limiting substrates, such as fatty acids and their derivatives [20]. Nevertheless, large scale application is still impractical with this type of production. Another strategy used for PHA production is solid state fermentation (SSF)—microbial cultivation on solid support made of appropriate substrates. SSF can be performed with inexpensive cultivation media, such as substrates, based on agro-industrial residues. This strategy provides disposing of waste, while valuable compounds are being produced at the same time [21]. There are few additional advantages of SSF over submerged fermentation: Easier aeration, higher substrate concentration, as well as reduced downstream processing steps. However, keeping conditions constant during the process is the biggest drawback to the market-ready production of PHAs using this strategy [22].

Bacteria from the genus *Komagataeibacter* (former *Gluconacetobacter*) synthesises another interesting, biosynthesised bioplastic—bacterial cellulose (BC). The BC production is a multi-step, strictly regulated process that involves regulatory proteins, few enzymes and catalytic complexes. The formation of 1,4-β-glucan chains is the first step in BC synthesis. This process, together with chains' assembly and crystallisation, occurs intracellularly. The second step involves extracting cellulose chains from the cells and their assembly in fibrils [4,23]. The yield and properties of the resulting material highly depend on the used bacterial strain and conditions of the conducted process, including medium composition, aeration, shaking, etc. Morphological and physical properties of the resulting material are in correlation to the cultivation broth composition. BC production can be improved using genetically modified producing strains, isolating novel strains with the ability to produce BC, and investigating process parameters and their significance. Florea et al. isolated *Komagataeibacter rhaeticus* strain able to produce a high yield of BC, while growing in low nitrogen level medium [24]. In order to increase BC production in limited oxygen conditions, genetically engineered strains have been developed [25]. Hungund et al. reported using ethyl methanesulfonate and ultraviolet radiation for *Gluconacetobacter xylinus* NCIM 2526 strain improvement that resulted in a significantly higher yield of BC (30%) [26].

Different strategies can be used to produce BC, including aerated submerged cultivation, static culture and airlift bioreactors [27]. Effects of these conditions on cellulose mechanical properties are presented in Figure 1.

Tensile strength, polymerisation degree and crystallinity index are highly influenced by BC structure. Produced in static conditions, BC forms pellicles on the surface of the medium. This gelatinous membrane can vary in thickness up to few centimetres depending on substrates' availability. Produced like this, BC has a significantly higher crystallinity index and tensile strength than the BC produced in agitated culture. Supply of air in agitated cultures will result in pellets with a higher crystallinity index. Crucial factors in BC production are the design of the reactor and proper control of conditions and process. These parameters highly affect the yield and quality of the resulting material. No matter which strategy is used, pH, oxygen supply, and temperature are essential conditions that should be carefully monitored. These parameters are mostly strain-dependent, but it was shown that the optimal pH value for cell growth and BC production is usually between 4.0 and 7.023, while the optimal temperature is in the range 28–30 °C [28].

Despite the great mechanical properties of the resulting material, static culture cannot provide uniformity of the cultivation broth; hence, cells are not equally exposed to nutrients and the thickness of the BC layer can be uneven. Additionally, productivity achieved in this strategy is very low, and it demands an extensive period of cultivation [29,30]. To improve the productivity of this process, the fed-batch strategy was developed, and a constant BC production rate was achieved for 30 days [31]. Submerged cultivations with agitation provide uniformity of nutrients, especially oxygen, resulting in higher yields in comparison to the static cultivations and making the production process cost-effective. Products of agitated fermentation depend on applied agitation speed and may include various forms of cellulose: Spheres, pellets, fibrous suspension [32]. As previously mentioned, higher productivity is the biggest advantage of submerged compared to static cultivation, but drawbacks, such as products' shape consistency and limited mechanical properties, are issued to overcome [33]. Another problem that can limit BC yield in aerated cultivation is a synthesis of gluconic acid. Due to the high agitation rates and hydrostatic stresses, the production of the secondary, protective metabolites is favoured over cellulose. Nevertheless, submerged cultivation of BC was implemented in different types of bioreactors, such as airlift, stirred tank and rotating disk. Production of this material on a large scale is still an issue, and designing new or improving existing equipment for this purpose is an important subject of research [34].

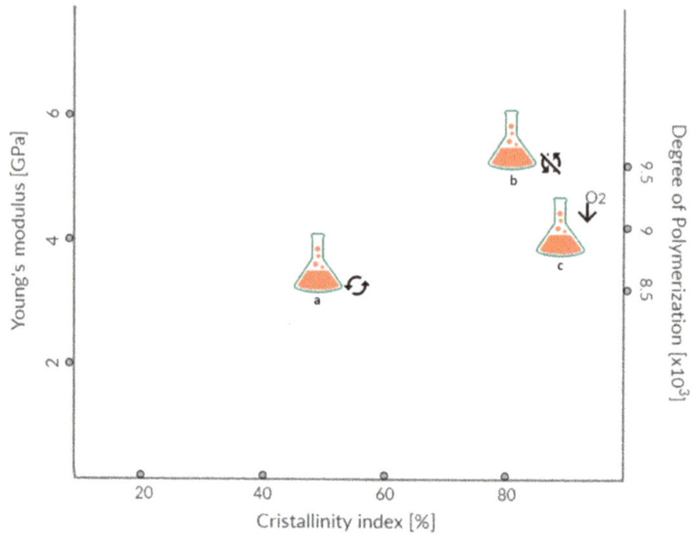

Figure 1. Young's modulus, crystallinity index and degree of polymerisation of BC depending on cultivation conditions: a—with shaking, b—without shaking, c—with additional oxygen supply.

3.1. Bioplastic Mechanical Performance

Bio-based polyesters (PLA, and PHAs) exhibit similar mechanical properties and can even exceed conventional plastic performances. Figure 2 demonstrates the Maximum Tensile Strength (MPa) and maximum Tensile Elongation (%) of bioplastics compared to petroleum-derived plastics. PLA is one of the most prominent bioplastics in terms of global consumption. It possesses several desirable properties, such as biocompatibility, biodegradability, composability and low toxicity to humans. The mechanical properties of PLA are greatly affected by the degree of PLA crystallinity. PLA derived from 93%, or more L-lactic acid can be semi-crystalline, while it is strictly amorphous when derived from 50–93% L-lactic acid. Thus, high tensile strength can be observed in films of high L-lactide content. Tensile strength and impact resistance are also influenced by the degree of crosslinking and the annealing of L-PLA, which increases the stereoregularity of the chain [35]. Comparison of mechanical properties between poly(98% L-lactide) and poly(94% L-lactide) showed a slightly greater elongation at yield for 98% than 94% L-lactide. However, poly(94% L-lactide) has an elongation at the break seven times greater than poly(98% L-lactide), indicating more plastic behaviour with 94% of L-lactide [35].

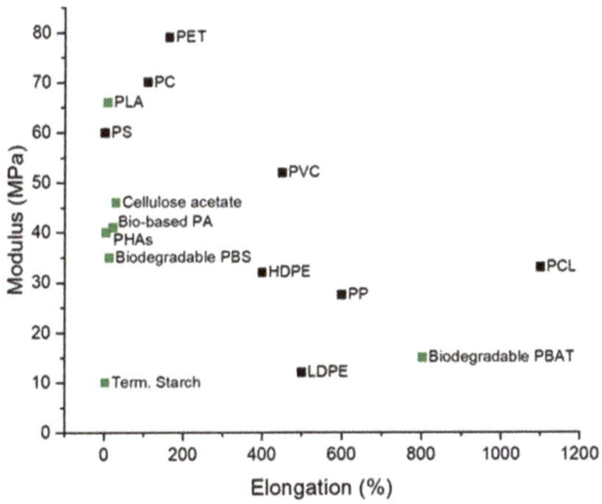

Figure 2. Maximum Tensile Strength (MPa) and Maximum Tensile Elongation (%) of bioplastics compared to petroleum-derived plastics, Data from Ref. [36].

For racemic mixtures, a study by Chen et al. demonstrated that polymerisation of 50% D-Lactide and 50% L-Lactide usually results in forming an amorphous polymer of poly (DL-lactide) [37].

As a packaging material, PLA offers high stiffness (greater than polyethene terephtalate (PET) and polystyrene (PS)), good clarity (similar to PET), relatively low processing temperatures, excellent resistance to fats and grease, and good breathability suitable for fruits and vegetable storage. Such characteristics make PLA a potential candidate to replace PS, polyethene (PE) and polypropylene (PP) in the fabrication of disposable cups, salad boxes and cold food packaging [38]. Nevertheless, PLA brittleness with less than 10% elongation at the break renders it unsuitable as a pure material for applications that require plastic deformations at higher stress levels [39]. Additionally, PLA's poor gas moisture permeability performance make it unsuitable for many beverage bottle applications [38].

On the other hand, PHAs gained considerable interest as a green alternative to petrochemically derived plastics, as they are biocompatible, biodegradable and synthesised from renewable resources [40,41]. PHB is the only polymer from the PHAs family to be produced in large quantities. This material is considered an aliphatic polyester with a linear polymer chain, composed of monomers of 3-hydroxybutyrate with a chromophoric carbonyl group. Being a member of the PHAs, PHB is also characterised by having a methyl (CH_3) as an alkyl replacing group, which provides it with a hydrophobic charge. The regularity of the polymerised PHB chain has a direct influence on its degree of crystallinity that, in turn, is influenced by the synthesis route used. Isotactic PHB, which has chiral carbon in absolute configuration R, is obtained through bacterial fermentation, while syndiotactic PHB is synthesised through a synthetic route from monomers with setting R and S. As isotactic PHB presents a more regular structure, it will allow a higher crystallinity than syndiotactic [42]. Favourable PHB properties in terms of melting point, strength, modulus and barrier properties promotes it as a substitute for PP, low density polyethene (LDPE), polyvinyl chloride (PVC) and PET in packaging applications. Differences in chemical structures between PLA, PHB and previously mentioned commonly used plastics are shown in Figure 3.

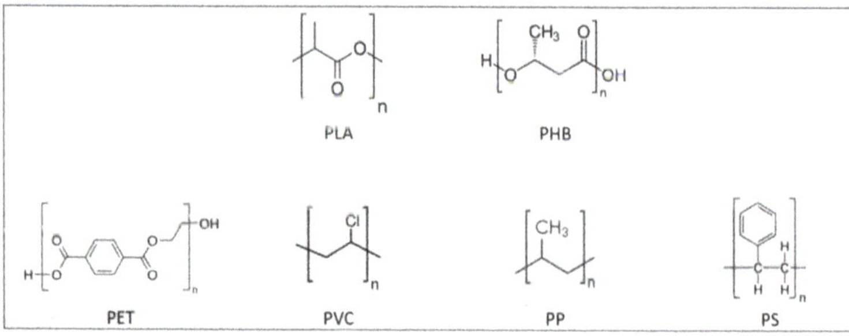

Figure 3. Chemical structures of biodegradable: Polylactic acid (PLA) and polyhydroxybutyrate PHB); and nonbiodegradable polymers: Polyethene terephatalate (PET), polyvinyl chloride (PVC), polypropylene (PP), polystyrene (PS).

Nevertheless, as a bioplastic, PHB has drawbacks, such as being brittle, hard and thermally unstable, making it challenging to use for applications like injection moulding in food industries [43,44]. As a pure material, PHB is highly crystalline (around 80%), resulting in the previously mentioned brittle nature and low elongations. The brittle nature of PHB is associated with a secondary crystallisation of the amorphous phase at ambient temperature. Another important issue is the glass temperature (T_g) of PHB. The T_g is close to room temperature resulting in secondary crystallisation taking place during storage, which, combined with a low nucleation density feature, leads to large spherulite formations which can grow over long durations leading to inter-spherulitic cracks [43]. Generally, spherulites are formed when PHB is crystallised from the melt, with band spacing between them depending on the crystallisation temperature [45]. Cracks are always present within spherulites in melt-crystallised PHB, and subsequent growth of the cracks leads to failure of the polymer. Two distinct types of crack exist in PHB spherulites, which can run either radially or circumferentially within the spherulites. Radial cracks occur more frequently in films crystallised at lower temperatures, while circumferential cracks occur when PHB is crystallised at high temperatures [46]. Another problem with PHB processing is the narrow processing window. The melting temperature of PHB is around 180 °C, therefore, processing temperature should be at least 190 °C. However, thermal degradation at this point happens rapidly, drastically reducing the acceptable residence time in the processing equipment to a few minutes only [47].

However, notwithstanding the limitations of PLA and PHAs, these bioplastic polymers have the potential to be fine-tuned to extend their application range comparable to fossil-based thermoplastics.

3.2. Bioplastic Barrier Performance

Inadequate fluid and gas barrier properties strongly impedes the utilisation of biopolymers in applications, such as the food-packaging sector. The established utilisation of PET and polyolefin family polymers in packaging applications is due to their combination of low cost, transparency, good barrier (to oxygen and water vapours), and mechanical characteristics. The oxygen permeability of PET (0.04 barrer) is much stronger than that of the recently developed biopolymers [48]. Among previously mentioned biopolymers, the most widely used—PHB stand apart having significant gas barrier properties (0.01 barrer), which are comparable to benchmark polymers, such as PET [49]. However, PHB's brittle mechanical nature precludes its suitability for food packaging applications. PLA, while having tunable mechanical properties, has oxygen permeability levels in the region of 0.26 barrer, restricting its use within several food-packaging applications [50]. As an approach to overcoming these barrier limitations, the addition of fillers to block the gas and moisture molecular pathways through the polymers, on a nano-micro scale level, is an attractive option [51].

An ideal filler should have a high surface area, aspect ratio and suitable chemical compatibility to provide enhanced mechanical and gas barrier properties at low filler content. The filler geometrical characteristics are an important factor in reducing gas permeability. The higher the aspect ratio, the greater the surface activity, which leads to an increase in mechanical properties, as well as gas barrier properties of the polymer matrix.

In terms of the orientation of fillers inside the polymer matrix, the Nielson model is commonly used. This is an ideal case where the orientation of the filler is perpendicular to the direction of diffusion and is generally not readily achievable. The modified version of the Nielson model is proposed by Bharadwaj by introducing the orientation parameter (S) [52]. If S = 0, designates perfect orientation and the Bharadwaj model will be reduced to the Nielson model, and maximum permeability reduction will be observed as indicated in Figure 4. Different models based on volume fraction and aspect ratio have also been developed to compare theoretical data with experimentally investigated gas barrier results [53].

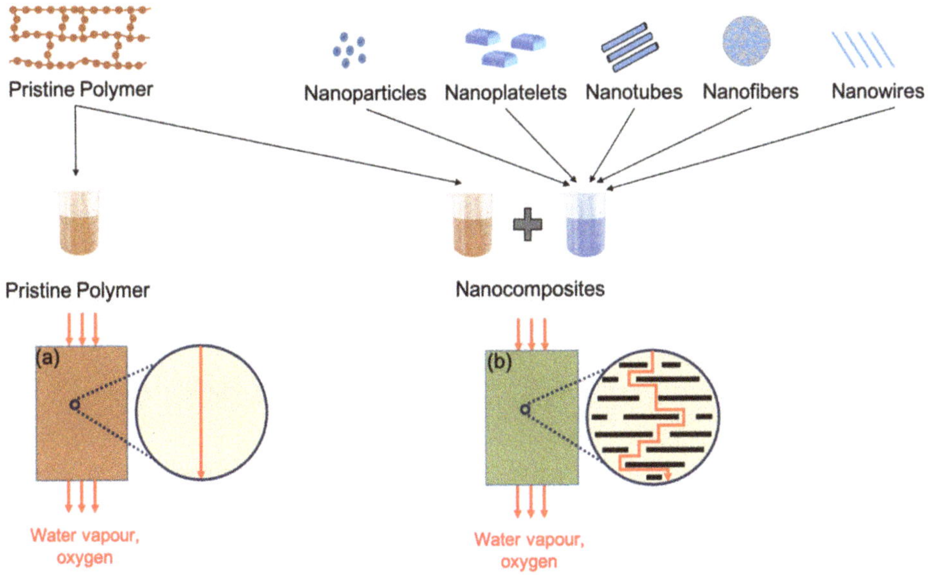

Figure 4. Effect of filler addition on gas barrier properties of nanocomposites: (**a**) Poor barrier properties in pristine polymer, due to direct diffusion pathways for gas molecules, (**b**) improved barrier properties in nanocomposites due to longer diffusion pathways.

Another aspect to be considered while studying the bioplastics' barrier properties is the dispersion and interchain compatibility of fillers within the bioplastics' matrices. Well exfoliated nano-fillers in polymer matrices give optimal reinforcement and contribute to other material performance characteristics [51]. A major issue with the dispersion of fillers in polymers is their hydrophilic nature, which causes inefficient compatibility with the hydrophobic polymer phase. Therefore, treatments are adopted to promote better interactions and good dispersion between the polymer phase and the fillers [54]. Moreover, the structural characteristics of the filler define the contribution imparted to the mechanical and gas barrier properties of polymers. The structural format of fillers dramatically impacts the gas barrier properties of their host polymers [55]. This is due to the higher crystallinity, which increases the effective path of diffusion and impedes the passage of gas molecules through the polymers rendering them suitable for packaging applications.

Figure 5 demonstrates the effect of fillers addition on the barrier properties, especially the oxygen (O_2) permeability of commonly used petroleum-based plastics and biopolymers. The data is compiled by converting oxygen permeability values from different units into a single unit (barrer). The addition of smaller amounts of fillers in biopolymers has drastically reduced the permeability of oxygen, fulfilling the criteria of ideal gas barrier material (LDPE, PET, and high density polyethene (HDPE)). Among biopolymers, PHAs showed more hindrance to the passage of gas molecules in their pristine polymers as compared to mentioned petroleum-based polymers. Functionalised graphene oxide (Gr-O) proved to be the best filler. The impressive reduction of oxygen permeability by Gr-O could be related to the strong interfacial adhesion between Gr-O and PHA polymer matrix [56].

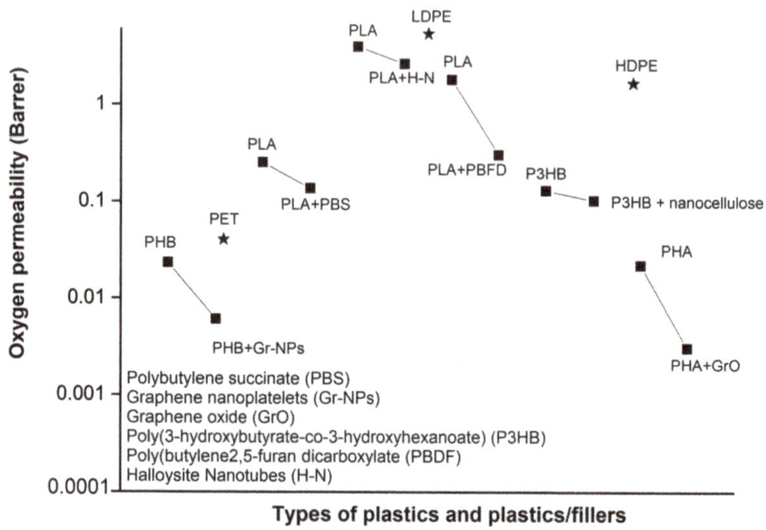

Figure 5. Effect of nanofillers on oxygen permeability of various biopolymers.

3.3. *Bioplastic Processing and Formulation*

Blending and composite formation is an established route for achieving improved technical and processing performance within polymer engineering. Blending PHAs, in particular, PHB with other polymers, offers opportunities to improve processability by lowering the processing temperature and reducing the brittle nature of these biopolymers. The physical, chemical and molecular architectural aspects dictate the enhancement of the polymer blend achievable through compounding techniques and introduced additives. Plasticiser additives can improve polymer viscosity and improve chain mobility during processing. Thermal stabilising additives can be used to eliminate premature degradation of polymers during processing, such as antioxidants, which guard against the presence of oxygen in the processing environment. Compatibiliser additives can improve miscibility between polymers by inducing flexible physical dipole-dipole interactions, or hydrogen bonding [57]. Nanocomposite additives, such as nanocrystals and nanofibres, can significantly improve the mechanical strength and gas barrier properties of the polymers if they are well dispersed in the biopolymer matrices. Natural fibres as an example of nanocomposite have been recently introduced as the main component in fibre reinforced biopolymer composites [58]. The intermolecular hydrogen bonds connecting the polymer chains of natural fibres provide a linear crystalline structure with a tensile strength reaching 15 GPa [59]. Such great strength is also accompanied by other advantages, as low cost, abundance, biodegradability, easy recyclability and fabrication of low weight composite materials [58]. All these properties made natural fibres perfect candidates as fillers in biopolymer composites and can compete with glass or carbon fibres. Accordingly, several studies were performed to evaluate the effect of incorporating natural fibres in biopolymer composites to improve the composite's mechanical and barrier properties. Among natural fibres, cellulose [60–66], hemp [67–73], kenaf [74–78] and flax [79–88] were the most studied ones. It is worth mentioning that nanocomposites of natural fibres or crystals added as fillers without plasticiser or compatibiliser, results in their poor dispersion and decrease the quality of polymer composite. Alternatively, plasticisers of hydrophilic nature, when mixed with biopolymers or their blends, tend to increase the wettability and O_2 permeability and deteriorate the barrier properties of the polymer composite.

Thus, to obtain better performance of nanocomposites with plasticisers or compatibilisers in biopolymer composites, both additives should be used together. The addition of nanocomposites with plasticisers improves the interfacial adhesion between the nanocomposites and the polymer matrix, allowing better dispersion and consequently provides a more tortuous path for gas and water and increase the barrier properties. Other approaches were introduced to improve the dispersion of the nanocomposites of natural fibres or crystals in polymer matrices. These include physical and chemical treatments of the nanocomposites before mixing with biopolymers. Bio-based coatings were also applied to natural fibres reinforced biocomposites as a means of inducing the hydrophobicity, and thus, improve the barrier properties of the biocomposite.

3.3.1. Blends and Composites

As previously mentioned, the processability and formability of PHB represent a drawback in industrial applications; blending with PLA provides a potential route to facilitating its introduction in the market, while also improving PLA properties at the same time. Several studies on PLA-PHB blends have been conducted in recent years, with the results typically showing a slightly higher Young's modulus than neat PHB and neat PLA [89]. Blends with PHB content of 50% or higher have shown lower values of tensile stress and elongation at the break in comparison to pure PLA [90]. However, above 60% PLA content has been reported to increase elongation at the break by up to 12%, with values even comparable to typical thermoplastics achieved on the addition of plasticisers [57,89,91–93]. Furthermore, the PLA-PHB 75:25 blend demonstrated higher mechanical performance than neat PLA, and greater impact resistance than the homopolymers on their own [89,94]. Jandas et al. reported the incorporation of PHB within PLA matrix in different ratios resulting in intermediate properties for the blends. The ductility of PLA increased consistently as PHB concentrations increased from 10 to 30 wt%. The maximum increase in percentage elongation was observed in the 70:30 ratio, suggesting some degree of molecular interaction between the macromolecules of PLA and PHB within the blend (Figure 6). However, tensile modulus and tensile strength were considerably decreased in the case of the blends, compared to pure PLA, as well as a corresponding decrease in stiffness. Blends prepared at 70:30 ratio were used for trial with compatibilisers and preparation of blend composites, due to the optimum elongation at the break and impact strength exhibited [57].

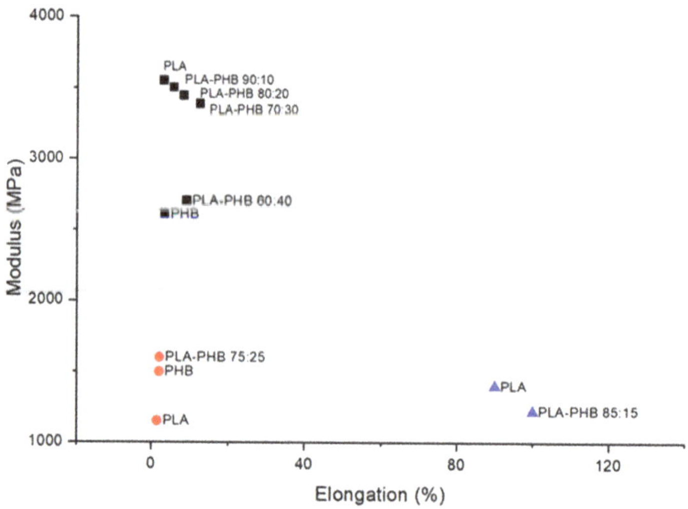

Figure 6. Young's Modulus (MPa) vs. elongation at the break (%) of PLA, PHB and their blends, reported by Jandas et al. (black squares), Data from Ref. [57], Armentano et al. (blue triangles), Data from Ref. [91] and Arrieta et al. (red circles), Data from Ref.[94].

3.3.2. Compatibilisers and Plasticisers

Facilitation of processability and improved flexibility in PLA/PHB blends is achievable using plasticisers [94]. Plasticisers are available as cost-effective, readily available materials on the market and are also generally of natural origin and include: Oxypropylated glycerin (or laprol), glycerol, glycerol triacetate, 4-nonylphenol, 4,40-dihydroxydiphenylmethane, acetyl tributyl citrate, salicylic ester, acetylsalicylic acid ester, soybean oil, epoxidised soybean oil, dibutyl phthalate, triethyl citrate, dioctyl phthalate, dioctyl sebacate, acetyl tributyl citrate, di-2-ethylhexylphthalate, tri(ethylene glycol)-bis(2-ethylhexanoate), triacetine, and fatty alcohols with or without glycerol fatty esters. Blends of PEG (2–5%) and PHB produced by solvent casting technique has been demonstrated to increase the elongation at the break by up to about four times compared with the original neat PHB. This behaviour was attributed to a plasticising effect of PEG, which acts to weaken the intermolecular forces between the adjacent polymer chains. The changes in free volume reduced the melting temperatures of the system and are also associated with an accompanying reduction in tensile strength [95].

As shown in Figure 7, compatibilisers, such as maleic anhydride, have been applied to PLA and PLA/PHB blends to impart additional flexibility and improve the blend's young's modulus by the induction of flexible physical interactions, including dipole-dipole or hydrogen bonding [57]. For industrial purposes, considerable attention needs to be paid to the selection of suitable plasticisers or compatibilisers as most tend to negatively impact other mechanical properties, such as lowering barrier properties of bioplastic blends, which can restrict their use in packaging applications [89].

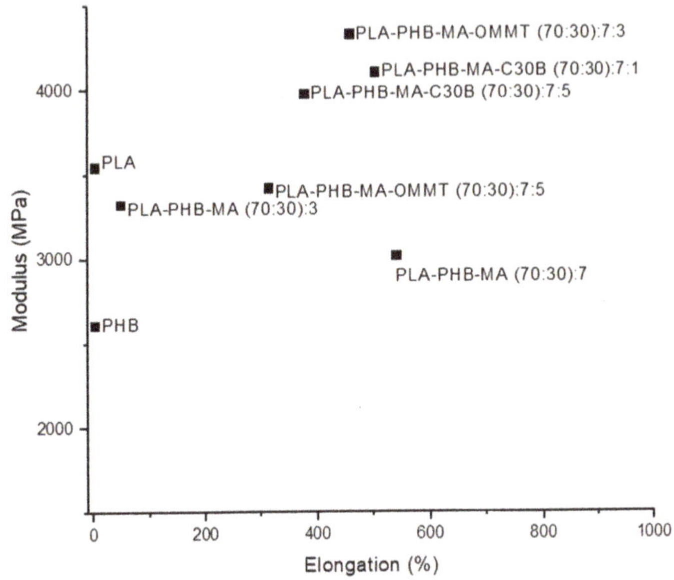

Figure 7. Young's Modulus (MPa) vs. Elongation (%) of PLA, PHB and their blends using MA as a compatibiliser, together with OMMT and C30B nanoclays, Data from Ref. [57].

3.3.3. Natural Fillers

The use of naturally sourced fibre-based fillers, with their considerable tensile strength and high sustainability, presents a key route to facilitating polymer circularity. The hydrophilic nature of bare natural fibres with no surface treatment requires an address, due to the incompatibility with hydrophobic biopolymer. Biopolymer composites reinforced with untreated natural fibres typically exhibit non-uniform fibre dispersion with interfaces promoting crack formation. These material defects can lead to premature mechanical failure of the composites [58,59,96]. These defects are attributable to the natural fibre hydrophilicity, which hinders proper mixing with the biopolymer hydrophobic matrix causing poor fibre/matrix interfacial bonding [59]. While high susceptibility of moisture absorption by the hydrophilic natural fibres can support the growth of fungi and bacteria and deteriorate the physical and mechanical properties of the bio-composites [58], this fact can be a highly useful feature that can be availed of post use for achieving biocyclability.

Different approaches are available to enhance the interfacial adhesion between the fibres and matrix, resulting in the fabrication of biocomposites with better mechanical and barrier properties. Among these solutions is the surface modification of natural fibres. Surface modification can result in increasing the sites of reaction, offering new functionality to the fibre surface and enhancing surface roughness by removing impurities [59]. Such modification will lead eventually to the improvement in mechanical properties and reduction of the water absorption of the fabricated biocomposite. The techniques applied to modify the natural surface fibres before inclusion in biopolymer matrices include physical, chemical, biological treatments and their combinations. Some examples of physical treatment techniques are calendaring, stretching, hybrid yarns production and thermal treatments; while chemical techniques include, alkali swelling, silane modifications, graft copolymerisation, and treatment with isocyanate, mercerisation [97].

A representative selection of the recently applied surface modifications for natural fibres reinforced biocomposites are given in Table 3.

Table 3. Surface treatment of natural fibres reinforced bio-composites.

Process	Natural Fibre Used	Biopolymer Matrix	Outcomes			Ref.
			Interfacial Adhesion	Mechanical Properties	Barrier Properties	
Treatment with compatibilisers; Polyglycerol polyglycidyl ether (SR-4GL), Trimethylol propane polyglycidyl ether (SR-TMP), and (Polyglycerol polypropyleneoxide (SC-P1000)	Cellulose fibres	PLA	Improved interfacial adhesion between fibres and PLA and		Inhibited degradation of the PLA matrix	[65]
STEFAC TM 8170, surfactant modification	Cellulose fibres	PLA/PHB		Enhanced mechanical performance	Improved water resistance, reduced oxygen and UV-light transmission, as well as appropriate disintegration in compost	[94]
Alkali treatment	Kenaf fibre	PHB		Reduction in the crystallinity of PHB (up to 6% reduction), making it more ductile, and improvement of the flexural modulus by up to 11%.		[75]
Silane treatment	Flax fibre	PLA	Improvement to fibre/matrix adhesion with 2% *w/w* silane content, yet further improvement of the fibre-matrix interface can be partially resolved by silane/alkali treatment combination.	Improved mechanical properties		[80]
Alkali treatment	Flax fibres	PLA		T_g values of fabricated bio-composites were lowered by 10 °C for 10% NaOH treatment and 15 °C for 30% NaOH treatment		[82]
Treatment with ethylene plasma	Flax fibres	PHB	Improved interfacial adhesion strength in the bio-composite		Improved thermal resistance	[88]

Combinations of natural materials with petroleum-based plastics is an option that is also under development. While facilitating the high mechanical performance, considerations are required when using this approach as further dilemma's may be posed regarding factors, such as continued resource depletion dependencies and degradation and biodegradation pathways, which potentially lead to increased microplastics production.

3.3.4. Bio-Coatings

The application of bio-based coatings to biocomposites and natural fibre reinforced biocomposites is a promising approach proposed to overcome the significant water uptake propensities of natural fibres and increase the moisture resistance of bioplastics for fluid barrier property application requirements. Exposure to long term environmental/hygroscopic ageing necessitates the induction of a higher level of hydrophobicity in chemically modified natural fibre reinforced bio-composites and bioplastics in general. Introducing bio-based coatings to natural fibres reinforced biocomposites, ensures the environmentally friendly and biodegradable nature of fibres. Besides, bio-based coatings are obtained from renewable resources and have superior hydrophobic characteristics [62]. For instance, polyurethane (PU) coatings were first introduced as bio-based coating resins in the 1950s [98]. PU coatings mainly provide their composites with high solvent resistance, hydrolytic stability, resistance to acid–base conditions and weather-ability [62]. Currently, most of the industrially produced PUs are petroleum-based polyols. Thus, renewable resources, such as vegetable oil [99–101], canola oil [102], soybean oil [103–105] and castor oil [106–108], are thoroughly investigated and were able to produce PU coating of competing properties to that of petroleum-based ones.

Polyfurfuryl alcohol (PFA) is another attractive type of bio-based coatings that can be used for barrier property application requirements. PFA has a low cost manufacturing process and can be obtained from natural resources as the agricultural residue of wheat, birch wood, hazelnut shells, corn, rice hulls, oat and sugar cane [62]. In addition, PFA's hydrophobicity, great heat distortion temperature and resistance to chemical erosion make it an excellent candidate as a coating material for bioplastics and natural fibres reinforced biocomposites.

Despite such potentials of PU and PFA as bio-based coatings, there is almost no work presented in the literature on the application of PU and PFA as coatings for natural fibre reinforced composites or biocomposites, except for a recent study done by Mokhothu et al. [109]. This study proposed using PU and PFA as bio-based coatings to composites containing flame-retardant treated natural fibres (flax) and phenolic resin. For three days, uncoated and coated samples were subjected to 90 °C and 90% relative humidity. Analysis was performed to the relative moisture content and mechanical properties and compared with the commercially available water-resistant product (FIRESHELL® (F1E)). Concerning the mechanical properties, PFA coated samples showed the highest modulus value (1.93 GPa) after being subjected to environmental conditioning with respect to uncoated (1.59 GPa); PU (1.05 GPa) and F1E (0.98 GPa) coated composites. Besides, the PFA and PU coated samples showed high stress at the break and a decreased elongation at the break in comparison to F1E coated ones. The moisture content of the conditioned PFA and PU coated composites was significantly reduced by 75% and 30%, respectively, when compared to uncoated and F1E coated composites [109].

4. Bioplastics Biodegradability

There is an important distinction between degradable polymers and biodegradable polymers. Degradable polymers are defined as polymers that can be depolymerised or recycled under controlled conditions and processes. According to the American Society for Testing and Materials (ASTM) definition, biodegradable polymers are polymers that can undergo decomposition into carbon dioxide, methane, water, inorganic compounds, or biomass, which can be measured by standardised tests, in a specified period, reflecting available disposal conditions (ASTM standard D6813). The mechanism of biodegradation is that the molecular weight of biodegradable polymers reduced, due to hydrolysis and oxidation, followed by breaking down into natural elements, such as water and carbon dioxide, via microorganisms. Aliphatic polyesters are the most economically viable biodegradable polymers, with PHB being among the most mechanically promising. The structural changes to the polymer molecules can be described in terms of three main categories of actions or mechanisms, namely, chain depolymerisation, random chain scission, and substituent reactions [110]. As defined by the IUPAC, depolymerisation is the process of converting a polymer into a monomer or a mixture of monomers/oligomers. Therefore, chain depolymerisation means the chain reaction is responding to the transformation of the macromolecular polymer chain into its constituent micromolecular monomers. Random scission is defined in the IUPAC Gold Book as a chemical reaction resulting in the breaking of skeletal bonds. It is also defined as a degradation mechanism that assumes a random cleavage of bonds along the macromolecular polymer chains [111]. This leads to the production of fragments that steadily decrease in length, which may eventually be small enough to allow for the removal of micromolecules. Substituent reactions refer to the kinetic reactions carried out by the constituent monomers of a polymer chain, which differ between polymers. According to Ghosh (1990), each kind of substituent has a characteristic chemical nature and reactivity [112]. However, as substituent reactions can only be observed at relatively low temperatures, substituent reactions only assume prominence when initiated and accomplished at temperatures lower than those of the breaking temperature of main chain bonds of a polymer.

Several factors affect the degradability of a polymer. In general, the surface conditions, the first-order structures, and the high order structures of a polymer play a major role in determining the rate of degradation. Surface conditions, such as hydrophilicity and surface area, directly correspond with the overall degradability of a polymer. Additionally, external environmental factors, such as humidity and temperature, also affect the overall degradability. Humidity introduces water molecules to a polymer and may result in a hydrolysis process, depending on the susceptibility or hydrophobicity of the polymer. Furthermore, the crystallinity of a polymer is also proportional to the degradability of a polymer, so that the lower the degree of crystallinity, the higher the degradability of the polymer. According to Tokiwa et al. (2009), this can be attributed to the fact that enzymes generally interact with the amorphous regions within a polymer, which are loosely packed together as compared to the crystalline regions. Moreover, the melting temperature (T_m) of polyesters greatly affects their enzymatic degradation. This is evident from the fact that aliphatic polyesters and polycarbonates with low Tm have a greater biodegradability than aliphatic polyurethanes and polyamides, which have higher Tm [113]. This is due to the large melting enthalpy change values of the latter, which can be attributed to the presence of hydrogen bonds among the polymer chains. The introduction of heat into a polymer matrix generally weakens the intermolecular bonds, resulting in an increased rate of degradation. With biodegradation specifically, the microbial species introduced to the polymer, directly correlates with the level of microbial activity, which in turn determines the rate of degradation of a biodegradable polymer [114]. The degree of microbial activity is also heavily influenced by nutrient and oxygen content in the biodegradation environment.

As described earlier, selected microorganisms can produce and storing PHAs. The ability to synthesise these molecules does not imply the capacity to also degrade them, in the case where extracellular hydrolases capable of converting polymers are also expressed [115,116]. Under nutrient-limited conditions, degradation occurs when the limitation is removed. Currently, six hundred PHA depolymerases from various microorganisms have been identified and categorised within eight families [117]. The degradation of these polymers is affected by many factors, such as type of enzyme, temperature, moisture, and nutrients composition [118]. Degradation rates of PHAs are also related to the microbial population density. It was shown that during degradation of P(3HB-co-3HV) copolymer, microbes at first attach to the polymer and then begin secreting degrading enzymes. Although PHA producing/degrading microbes usually express high specificity towards

P(3HB), many microbes have been identified with wide substrate specificity. *Xanthomonas* spp., for example, has the ability to produce enzymes for PHAs with aromatic side chain degradation and can also degrade P(3HB), P(3HO), and poly-3-hydroxy-5-phenylvalerate (P(3HPV)) [119]. The type of polymer also plays an important role in degradability. In addition to the presence of side chains, length and composition are also significant factors. Manna et al. report that homopolymers have higher degradation rates in comparison with copolymers of PHAs [120]. Other studies have demonstrated opposing results [118], which may be explained because, in these cases, the experiments were conducted in natural environments where previously mentioned factors (nutrients, moisture, temperature etc.) were non-controllable. Kusaka et al. showed that PHAs degradation ability is negatively correlated to the M_w and crystallinity [121]. The format and shape of the polymer material is also a significant factor for PHAs degradation, with thin films degrading faster than thicker films. Soil and climatic conditions are further factors that can affect the PHA degradation rate [28,118]. Boyandin et al. examined PHA films degradation response and reported that humid and the hot Vietnamese climate facilitated degradation of PHA [122].

Additives, such as fillers, are another factor that can affect the biodegradability of the bioplastic in which they are added, as demonstrated in Figures 8 and 9. There is no general guarantee that the addition of fillers will enhance or inhibit biodegradability as the effect of fillers on a polymer is mainly dependent on its chemical and physical aspects, such as size, geometry, surface area, and the surface energy of its particles [123] (Murphy, 2001). These aspects directly affect the overall degradation ability of a polymer. In general, the effect of additives on the biodegradability of a polymer is largely dependent on the properties of the additives, such as hydrophobicity and amenability to bacterial growth on the surface.

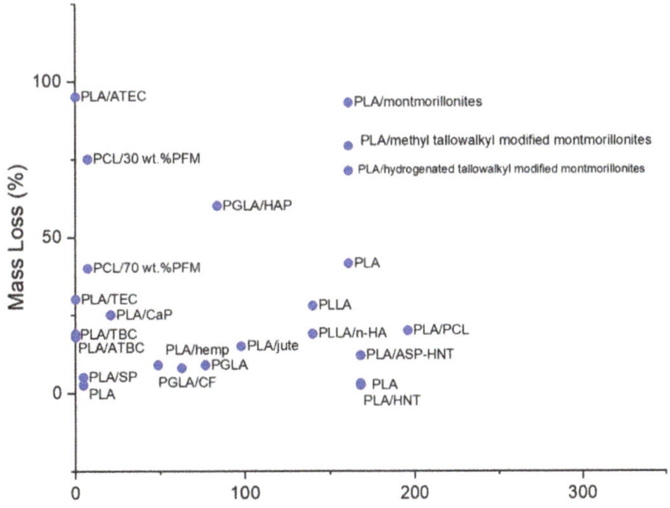

Figure 8. Mass Loss percentage of composites in various hydrolytic degradation environments.

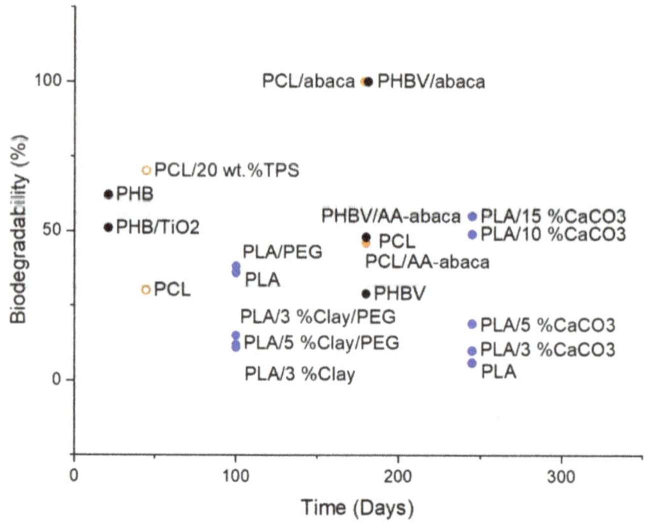

Figure 9. Biodegradability of PLA, PCL and PHB with the addition of various fillers and plasticisers in various soil burial degradation environments.

Aframehr et al. report a study on the effect of calcium carbonate ($CaCO_3$) in soil burial biodegradation where the $CaCO_3$ fillers, act to increase the biodegradability of PLA. The weight loss of $CaCO_3$ nanocomposites is approximately two times higher than other nanocomposites, with a weight loss of around 55% for PLA/15% $CaCO_3$, 49% for PLA/10% $CaCO_3$, 19% for PLA/5% $CaCO_3$, 10% for PLA/3% $CaCO_3$, and around 6% for neat PLA after a soil exposure time of 35 weeks [124]. A study by Teramoto et al. investigated the effect of treated and untreated abaca fibre filler on the biodegradability of poly(3-hydroxybutyrate-co-3-hydroxyvalerate) (PHBV) after being subjected to a soil-burial environment for a duration of 180 days. Neat PHBV exhibited the least biodegradability at only around 29% weight loss, followed by PHBV/AA-abaca at around 48% weight loss, and lastly, PHBV/untreated abaca with the highest biodegradability, which can be seen in the high degree of fragmentation after 60 days [125].

Altaee et al. conducted a study on the biodegradation of PHB and titanium oxide (PHB-TiO_2) composites in a soil burial environment with pH 7.30 and a humidity of 80% at 30 °C and found that PHB-TiO_2 exhibits a lower weight loss of only ~51% after three weeks as compared to ~62% weight loss of neat PHB through the same duration [126]. Paul et al. (2005) studied the degradation of nanocomposites of PLA with unmodified and organo-modified montmorillonites and found that montmorillonites filler enhances hydrolytic degradation. PLA with unmodified montmorillonites exhibited the greatest decrease in M_w after 23 weeks of hydrolytic degradation with a 93.1% loss in M_w. These results were followed by PLA with montmorillonites treated with bis-(2-hydroxyethyl) methyl tallowalkyl ammonium cations and PLA with montmorillonites treated with dimethyl-2-ethylhexyl (hydrogenated tallowalkyl) ammonium cations at 79.2% and 71.2% M_w loss, respectively. In comparison, unfilled PLA is found to only have a 41.6% M_w decrease compared to its initial value [127]. Chen et al. found that PLA with halloysite nanotubes (HNTs) as filler has a greater rate of hydrolytic degradation as compared to neat PLA as shown from the mass reduction of 3.1% for PLA/HNT as compared to that of neat PLA at only 2.6% in an in vitro environment in SBF at 37 °C by the end of the 24th week of degradation. It was also reported that PLA with HNTs surface treated with 3-aminopropyltriethoxysilane (ASP) has an even greater hydrolytic degradation, due to better interfacial adhesion between PLA and HNTs, which is evident from the mass reduction of 12.1% [128]. A study by Navarro et al. (2005) investigated the effect of the addition of calcium phosphate (CaP) glass to PLA on its hydrolytic degradability. During the first three weeks of the degradation, PLA/CaP composites experienced a greater weight loss than neat PLA, but an increase of weight of the PLA/CaP composite was reported after three weeks. This may be credited to forming hydrated calcium phosphate precipitate on the composite. The maximum weight loss exhibited by week 3 is 25%, and a final weight loss percentage of about 22% on week 6. In a comparison, the weight loss of neat PLA is around 1% from week 3 through week 5 [129]. Moreover, a study by Huang et al. (2013) investigated the hydrolytic degradability of poly(L-lactic acid) (PLLA)/nanohydroxyapatite (n-HA) and found that the rate of degradation of PLLA/n-HA composite was slower than neat PLLA. This is evident from the weight loss of only around 19% for PLLA/n-HA composite and about 28% for neat PLLA after 20 weeks in a PBS environment with an initial pH of around 7.4 [130]. A study by Valapa et al. (2016) investigated the hydrolytic degradation behaviour of sucrose palmitate (SP) reinforced PLA (PLA-SP) nanocomposites in acidic (pH 2), basic (pH 12), and neutral (pH 7) hydrolytic degradation environments and found that the rate of degradation is increased with the addition of sucrose palmitate. This can be seen in the mass loss percentage of about 5.1% for PLA-SP as compared to only around 2.6% mass loss percentage of neat PLA in a pH 7 degradation solution at 35 °C after 115 h [131].

Alternatively, the alterations proposed by the addition of plasticisers and compatibilisers to bioplastics directly affect their degradability since plasticisers and compatibilisers decrease the glass transition temperature (T_g) of the polymers they are blended with, as shown in Figures 8 and 9 [132]. Like fillers, there is no general guarantee that the addition of plasticisers will positively or negatively impact biodegradation as it is highly dependent on the properties of the plasticisers used. Some of the more common forms of plasticisers include citrate esters and phthalates (phthalate, isophthalate, terephthalate), the latter being biodegradable as degradation by microorganisms is considered as the most effective method of degradation for phthalates plasticisers [133]. A study by Labrecque et al. (1997) researched using triethyl citrate (TEC), tributyl citrate (TBC), acetyl triethyl citrate (ATEC) and acetyl tributyl citrate (ATBC) as plasticisers for PLA and their effect on enzyme-catalysed hydrolytic degradation. The study found that all citrate esters enhanced the degradability of PLA. At a concentration of 20%, ATEC plasticised PLA has the highest weight loss of around 95%, followed by neat PLA at around 48%, and a decreased rate of degradation in TEC, TBC, and ATBC with weight losses of around 30%, 19%, and 18%, respectively, after a degradation period of 6 h [134]. A 2009 study by Ozkoc and Kemaloglu found that the addition of PEG and clay plasticisers to PLA decreases the rate of biodegradation after exposure to the composting environment for 100 days. This is evident from the weight loss percentage values of about 15%, 12%, and 11% for PLA/3%Clay/PEG, PLA/5%Clay/PEG, and PLA/3%Clay, respectively, in comparison to that of neat PLA with a 36% weight loss. However, the weight loss percentage of PLA/PEG shows a slight increase as compared to neat PLA with a value of around 38% weight loss [135].

Careful selection and monitoring of additives and composite formation, hence, has the potential for exploitation both to enhance the target mechanical properties and simultaneously promote biodegradation.

5. Conclusions

The fundamental many-faceted aspects of biopolymer architectures afford greater versatility for configuration and exploitation compared with the main ubiquitous recalcitrant synthetic petroleum-based plastics. This is a pivotal trait that is available for harnessing to enable bioplastics to meet both the high mechanical and barrier performances of their petroleum-based counterparts together with full sustainable biodegradability and circularity. Their very bio-nature means bioplastics are inherently more elaborate at a structural level. This increases accessibility to bio-interactions for enzymatic biodegradation, biodepolymerisation and biorepolymerisation, as well as supporting routes to improving mechanical and barrier performances. While several bioplastics have already achieved certain mechanical and barrier performance criteria, which are equivalent and even exceed those of corresponding petroleum-based plastics, courses of action for resolving the remaining limitations are becoming increasingly accessible.

Here, avenues for the advancement of the performance of bioplastics with respect to mechanical and barrier properties alongside biodegradability are discussed. The key architectural features properties are M_w and crystallinity, which typically exhibit an inversely dependent relationship with mechanical performance and biodegradability for both bioplastics and petroleum-based plastics. Increasing M_w and crystallinity are generally associated with higher mechanical performance and decreased biodegradability as lower crystallinity corresponds to looser chain packing, facilitating enzyme access.

Both macro and micro strategies have the potential for dual positive correlation on the mechanical and biodegradability performances of bioplastics. Regarding the macro approach, new possibilities are afforded, such as harnessing intricate bioplastic structures and formats, which include microfibrillar frameworks and combination with advanced methods, such as in situ polymerisation. Blending and compounding with additives as selected fillers, plastisisers and compatibilisers are also being demonstrated for improved mechanical features, without decreasing biodegradability.

Whereas, regarding the micro approach, expanding PHAs families and large numbers for possible monomer units, present the potential to engineer biodegradable plastics with equivalent target petroleum plastic performances without associated environmental pollution. Further routes include metabolic pathway alteration, design of high specificity substrates, intricate copolymer and block copolymer and genetic modification to produce strains to achieve next-generation multifunctional biopolymers.

In these respects, bioplastic polymers, in contrast to Petroleum-based plastics, have not yet been tailored or even adequately explored to establish their capacities for current and future applications. Given the range and diversity of options available for bioplastics development, there are excellent prospects to extend their application range on a comparable scale to fossil-based thermoplastics and beyond. Further innovations can be expected as the knowledge and new capacities for the manipulation of biopolymers advances, and spawns outputs in related and novice disciplines. The realisation of high performance plastics, without recalcitrance, pollution or resource depletion and switching to regenerative low carbon circularity, has the potential to both safeguard and promote future prosperity for the planet and its inhabitants.

Author Contributions: O.A.A. Conceptualization, Data curation, Investigation, Writing—original draft; E.L.G. Investigation, Data curation, Formal analysis; Writing—original draft; M.A. Investigation, Data curation; S.A. Investigation, Data curation; Y.C. Data curation, Formal analysis; Writing—original draft; M.B.F. Supervision, Validation, Writing—original draft; M.M. Investigation, Visualization, Writing—original draft, Writing—review & editing. All authors have read and agreed to the published version of the manuscript.

Funding: This project has received funding from the European Union's Horizon 2020 research and innovation programme under grant agreement No 870292 (BioICEP).

Institutional Review Board Statement: Not applicable.

Informed Consent Statement: Not applicable.

Data Availability Statement: The data presented in this study are available on request from the corresponding author.

Acknowledgments: This project has received funding from the European Union's Horizon 2020 research and innovation programme under grant agreement No 870292 (BioICEP) and is supported by the National Natural Science Foundation of China (grant numbers: Institute of Microbiology, Chinese Academy of Sciences: 31961133016, Beijing Institute of Technology: 31961133015, Shandong University: 31961133014).

Conflicts of Interest: The authors declare no conflict of interest.

References

1. Madison, L.L.; Huisman, G.W. Metabolic Engineering of Poly(3-Hydroxyalkanoates): From DNA to Plastic. *Microbiol. Mol. Biol. Rev.* **1999**, *63*, 21–53. [CrossRef] [PubMed]
2. Koller, M.; Maršálek, L.; Dias, M.M.D.S.; Braunegg, G. Producing microbial polyhydroxyalkanoate (PHA) biopolyesters in a sustainable manner. *New Biotechnol.* **2017**, *37*, 24–38. [CrossRef] [PubMed]
3. Koller, M. A Review on Established and Emerging Fermentation Schemes for Microbial Production of Polyhydroxy-alkanoate (PHA) Biopolyesters. *Fermentation* **2018**, *4*, 30. [CrossRef]
4. Rajwade, J.M.; Paknikar, K.; Kumbhar, J.V. Applications of bacterial cellulose and its composites in biomedicine. *Appl. Microbiol. Biotechnol.* **2015**, *99*, 2491–2511. [CrossRef]
5. Sudesh, K.; Abe, H.; Doi, Y. Synthesis, structure and properties of polyhydroxyalkanoates: Biological polyesters. *Prog. Polym. Sci.* **2000**, *25*, 1503–1555. [CrossRef]
6. Muneer, F.; Rasul, I.; Azeem, F.; Siddique, M.H.; Zubair, M.; Nadeem, H. Microbial Polyhydroxyalkanoates (PHAs): Efficient Replacement of Synthetic Polymers. *J. Polym. Environ.* **2020**, *28*, 2301–2323. [CrossRef]
7. Ciesielski, S.; Mozejko-Ciesielska, J.; Pisutpaisal, N. Plant oils as promising substrates for polyhydroxyalkanoates pro-duction. *J. Clean. Prod.* **2015**, *106*, 408–421. [CrossRef]
8. Shuai, X.-T.; Jedlinski, Z.; Luo, Q.; Farhod, N. Synthesis of novel block copolymers of poly(3-hydroxybutyric acid) with poly(ethylene glycol) through anionic polymerisation. *Chin. J. Polym. Sci.* **2000**, *18*, 19–23.
9. Fu, J.; Sharma, P.; Spicer, V.; Krokhin, O.V.; Zhang, X.; Fristensky, B.; Wilkins, J.A.; Cicek, N.; Sparling, R.; Levin, D.B. Effects of impurities in biodiesel-derived glycerol on growth and expression of heavy metal ion homeostasis genes and gene products in Pseudomonas putida LS46. *Appl. Microbiol. Biotechnol.* **2015**, *99*, 5583–5592. [CrossRef]
10. Noda, I.; Lindsey, S.B.; Caraway, D. Nodax™ Class PHA Copolymers: Their Properties and Applications. In *Beneficial Microorganisms in Food and Nutraceuticals*; Springer Science and Business Media LLC: Berlin/Heidelberg, Germany, 2010; pp. 237–255.
11. Kunioka, M.; Doi, Y. Thermal degradation of microbial copolyesters: Poly(3-hydroxybutyrate-co-3-hydroxyvalerate) and poly(3-hydroxybutyrate-co-4-hydroxybutyrate). *Macromolecules* **1990**, *23*, 1933–1936. [CrossRef]
12. Avella, M.; Martuscelli, E.; Raimo, M. Review Properties of blends and composites based on poly(3-hydroxy)butyrate (PHB) and poly(3-hydroxybutyrate-hydroxyvalerate) (PHBV) copolymers. *J. Mater. Sci.* **2000**, *35*, 523–545. [CrossRef]
13. Qiu, Y.-Z.; Han, J.; Guo, J.-J.; Chen, G.-Q. Production of Poly(3-hydroxybutyrate-co-3-hydroxyhexanoate) from Gluconate and Glucose by Recombinant Aeromonas hydrophila and Pseudomonas putida. *Biotechnol. Lett.* **2005**, *27*, 1381–1386. [CrossRef]
14. McAdam, B.; Fournet, M.B.; McDonald, P.; Mojicevic, M. Production of Polyhydroxybutyrate (PHB) and Factors Impacting Its Chemical and Mechanical Characteristics. *Polymers* **2020**, *12*, 2908. [CrossRef]
15. Ravenelle, F.; Marchessault, R.H. One-Step Synthesis of Amphiphilic Diblock Copolymers from Bacterial Poly([R]-3-hydroxybutyric acid). *Biomacromolecules* **2002**, *3*, 1057–1064. [CrossRef]
16. Amaro, T.M.M.M.; Rosa, D.; Comi, G.; Iacumin, L. Prospects for the Use of Whey for Polyhydroxyalkanoate (PHA) Pro-duction. *Front. Microbiol.* **2019**, *10*, 992. [CrossRef]
17. Colombo, B.; Favini, F.; Scaglia, B.; Sciarria, T.P.; D'Imporzano, G.; Pognani, M.; Alekseeva, A.; Eisele, G.; Cosentino, C.; Adani, F. Enhanced polyhydroxyalkanoate (PHA) production from the organic fraction of municipal solid waste by using mixed microbial culture. *Biotechnol. Biofuels* **2017**, *10*, 1–15. [CrossRef] [PubMed]
18. Rajendran, N.; Sharanya, P.; Sneha Raj, M.; Ruth Angeeleena, B.; Rajam, C. Seaweeds can be a new source for bio-plastics. *J. Pharm. Res.* **2012**, *5*, 1476–1479.
19. Chee, J.Y.; Tan, Y.; Samian, R.; Sudesh, K. Isolation and Characterization of a Burkholderia sp. USM (JCM15050) Capable of Producing Polyhydroxyalkanoate (PHA) from Triglycerides, Fatty Acids and Glycerols. *J. Polym. Environ.* **2010**, *18*, 584–592. [CrossRef]
20. Koller, M.; Salerno, A.; Dias, M.; Reiterer, A.; Braunegg, G. Modern biotechnological polymer synthesis: A review. *Food Technol. Biotechnol.* **2010**, *48*, 255–269.
21. Pandey, A. Solid-state fermentation. *Biochem. Eng. J.* **2003**, *13*, 81–84. [CrossRef]
22. Sindhu, R.; Pandey, A.; Binod, P. Solid-state fermentation for the production of Poly(hydroxyalkanoates). *Chem. Biochem. Eng. Q.* **2015**, *29*, 173–181. [CrossRef]
23. Reiniati, I.; Hrymak, A.N.; Margaritis, A. Recent developments in the production and applications of bacterial cellulose fibers and nanocrystals. *Crit. Rev. Biotechnol.* **2017**, *37*, 510–524. [CrossRef]
24. Florea, M.; Hagemann, H.; Santosa, G.; Abbott, J.; Micklem, C.; Spencer-Milnes, X.; Garcia, L.D.A.; Paschou, D.; Lazenbatt, C.; Kong, D.; et al. Engineering control of bacterial cellulose production using a genetic toolkit and a new cellulose-producing strain. *Proc. Natl. Acad. Sci. USA* **2016**, *113*, E3431–E3440. [CrossRef] [PubMed]
25. Liu, M.; Li, S.; Xie, Y.; Jia, S.; Hou, Y.; Zou, Y.; Zhong, C. Enhanced bacterial cellulose production by Gluconacetobacter xylinus via expression of Vitreoscilla hemoglobin and oxygen tension regulation. *Appl. Microbiol. Biotechnol.* **2018**, *102*, 1155–1165. [CrossRef]
26. Hungund, B.; Gupta, S. Strain improvement of Gluconacetobacter xylinus NCIM 2526 for bacterial cellulose production. *Afr. J. Biotechnol.* **2010**, *9*, 32.
27. Neves, N.M.; Reis, R.L. (Eds.) *Biomaterials from Nature for Advanced Devices and Therapies*; Wiley: Hoboken, NJ, USA, 2016; pp. 384–399.

28. Volova, T.G.; Prudnikova, S.; Sukovatyi, A.G.; Shishatskaya, E. Production and properties of bacterial cellulose by the strain Komagataeibacter xylinus B-12068. *Appl. Microbiol. Biotechnol.* **2018**, *102*, 7417–7428. [CrossRef] [PubMed]
29. Lin, S.-P.; Calvar, I.L.; Catchmark, J.M.; Liu, J.-R.; Demirci, A.; Cheng, K.-C. Biosynthesis, production and applications of bacterial cellulose. *Cellulose* **2013**, *20*, 2191–2219. [CrossRef]
30. Shah, N.; Ul-Islam, M.; Khattak, W.A.; Park, J.K. Overview of bacterial cellulose composites: A multipurpose advanced material. *Carbohydr. Polym.* **2013**, *98*, 1585–1598. [CrossRef] [PubMed]
31. Hsieh, J.-T.; Wang, M.-J.; Lai, J.-T.; Liu, H.-S. A novel static cultivation of bacterial cellulose production by intermittent feeding strategy. *J. Taiwan Inst. Chem. Eng.* **2016**, *63*, 46–51. [CrossRef]
32. Ul-Islam, M.; Khan, S.; Ullah, M.W.; Park, J.K. Bacterial cellulose composites: Synthetic strategies and multiple applications in bio-medical and electro-conductive fields. *Biotechnol. J.* **2015**, *10*, 1847–1861. [CrossRef] [PubMed]
33. Singhsa, P.; Narain, R.; Manuspiya, H. Physical structure variations of bacterial cellulose produced by different Komaga-taeibacter xylinus strains and carbon sources in static and agitated conditions. *Cellulose* **2018**, *25*, 1571–1581. [CrossRef]
34. Parte, F.G.B.; Santoso, S.P.; Chou, C.-C.; Verma, V.; Wang, H.-T.; Ismadji, S.; Cheng, K.-C. Current progress on the production, modification, and applications of bacterial cellulose. *Crit. Rev. Biotechnol.* **2020**, *40*, 397–414. [CrossRef] [PubMed]
35. Auras, R.; Harte, B.; Selke, S. An Overview of Polylactides as Packaging Materials. *Macromol. Biosci.* **2004**, *4*, 835–864. [CrossRef]
36. Zhao, X.; Cornish, K.; Vodovotz, Y. Narrowing the Gap for Bioplastic Use in Food Packaging: An Update. *Environ. Sci. Technol.* **2020**, *54*, 4712–4732. [CrossRef]
37. Chen, S.-Q.; Lopez-Sanchez, P.; Wang, D.; Mikkelsen, D.; Gidley, M.J. Mechanical properties of bacterial cellulose synthesised by diverse strains of the genus Komagataeibacter. *Food Hydrocoll.* **2018**, *81*, 87–95. [CrossRef]
38. Jem, K.J.; Tan, B. The development and challenges of poly (lactic acid) and poly (glycolic acid). *Adv. Ind. Eng. Polym. Res.* **2020**, *3*, 60–70. [CrossRef]
39. Farah, S.; Anderson, D.G.; Langer, R. Physical and mechanical properties of PLA, and their functions in widespread applications—A comprehensive review. *Adv. Drug Deliv. Rev.* **2016**, *107*, 367–392. [CrossRef] [PubMed]
40. Visakh, P.M. CHAPTER 1. Polyhydroxyalkanoates (PHAs), their Blends, Composites and Nanocomposites: State of the Art, New Challenges and Opportunities. In *Polyhydroxyalkanoate (PHA) Based Blends, Composites and Nanocomposites*; Green Chemistry Series; Royal Society of Chemistry: Cambridge, UK; 2014; pp. 1–17. [CrossRef]
41. Goh, L.-K.; Purama, R.K.; Sudesh, K. Enhancement of Stress Tolerance in the Polyhydroxyalkanoate Producers without Mobilization of the Accumulated Granules. *Appl. Biochem. Biotechnol.* **2014**, *172*, 1585–1598. [CrossRef]
42. Dos Santos, A.J.; Valentina, L.V.O.D.; Schulz, A.A.H.; Duarte, M.A.T. From Obtaining to Degradation of PHB:Material Properties. Part I. *Ing. Cienc.* **2017**, *13*, 269–298. [CrossRef]
43. El-Hadi, A.M.; Schnabel, R.; Straube, E.; Müller, G.; Henning, S. Correlation between degree of crystallinity, morphology, glass temperature, mechanical properties and biodegradation of poly (3-hydroxyalkanoate) PHAs and their blends. *Polym. Test.* **2002**, *21*, 665–674. [CrossRef]
44. Smith, M.K.M.; Paleri, D.M.; Abdelwahab, M.; Mielewski, D.F.; Misra, M.; Mohanty, A.K. Sustainable composites from poly(3-hydroxybutyrate) (PHB) bioplastic and agave natural fibre. *Green Chem.* **2020**, *22*, 3906–3916. [CrossRef]
45. Barham, P.J.; Keller, A.; Otun, E.L.; Wills, H.H.; Holmes, P.A. Crystallization and morphology of a bacterial thermoplastic: Poly-3-hyd roxybutyrate. *J. Mater. Sci.* **1984**, *19*. [CrossRef]
46. Barham, P.J.; Keller, A. The relationship between microstructure and mode of fracture in polyhydroxybutyrate. *J. Polym. Sci. Part B Polym. Phys.* **1986**, *24*, 69–77. [CrossRef]
47. Janigová, I.; Lacík, I.; Chodák, I. Thermal degradation of plasticized poly(3-hydroxybutyrate) investigated by DSC. *Polym. Degrad. Stab.* **2002**, *77*, 35–41. [CrossRef]
48. Shim, S.H.; Kim, K.T.; Lee, J.U.; Jo, W.H. Facile Method to Functionalize Graphene Oxide and Its Application to Poly(ethylene terephthalate)/Graphene Composite. *ACS Appl. Mater. Interfaces* **2012**, *4*, 4184–4191. [CrossRef] [PubMed]
49. Papadopoulou, E.L.; Basnett, P.; Paul, U.C.; Marras, S.; Ceseracciu, L.; Roy, I.; Athanassiou, A. Green Composites of Poly(3-hydroxybutyrate) Containing Graphene Nanoplatelets with Desirable Electrical Conductivity and Oxygen Barrier Properties. *ACS Omega* **2019**, *4*, 19746–19755. [CrossRef] [PubMed]
50. Dhar, P.; Gaur, S.S.; Soundararajan, N.; Gupta, A.; Bhasney, S.M.; Milli, M.; Kumar, A.; Katiyar, V. Reactive Extrusion of Polylactic Acid/Cellulose Nanocrystal Films for Food Packaging Applications: Influence of Filler Type on Thermomechanical, Rheological, and Barrier Properties. *Ind. Eng. Chem. Res.* **2017**, *56*, 4718–4735. [CrossRef]
51. Azeem, M.; Jan, R.; Farrukh, S.; Hussain, A. Improving gas barrier properties with boron nitride nanosheets in poly-mer-composites. *Results Phys.* **2019**, *12*, 1535–1541. [CrossRef]
52. Bharadwaj, R.K. Modeling the Barrier Properties of Polymer-Layered Silicate Nanocomposites. *Macromolecules* **2001**, *34*, 9189–9192. [CrossRef]
53. Moggridge, G.; Lape, N.K.; Yang, C.; Cussler, E. Barrier films using flakes and reactive additives. *Prog. Org. Coat.* **2003**, *46*, 231–240. [CrossRef]
54. Cui, Z.; Martinez, A.P.; Adamson, D.H. PMMA functionalized boron nitride sheets as nanofillers. *Nanoscale* **2015**, *7*, 10193–10197. [CrossRef]

55. Gitari, B.; Chang, B.P.; Misra, M.; Navabi, A.; Mohanty, A.K. A comparative study on the mechanical, thermal, and water barrier properties of PLA nanocomposite films prepared with bacterial nanocellulose and cellulose nanofibrils. *BioResources* **2019**, *14*, 1867–1889.
56. Xu, P.; Yang, W.; Niu, D.; Yu, M.; Du, M.; Dong, W.; Chen, M.; Lemstra, P.J.; Ma, P. Multifunctional and robust polyhydroxyalkanoate nanocomposites with superior gas barrier, heat resistant and inherent antibacterial performances. *Chem. Eng. J.* **2020**, *382*, 122864. [CrossRef]
57. Jandas, P.J.; Mohanty, S.; Nayak, S.K. Morphology and Thermal Properties of Renewable Resource-Based Polymer Blend Nanocomposites Influenced by a Reactive Compatibilizer. *ACS Sustain. Chem. Eng.* **2013**, *2*, 377–386. [CrossRef]
58. Kick, T.; Grethe, T.; Mahltig, B. A Natural Based Method for Hydrophobic Treatment of Natural Fiber Material. *Acta Chim. Slov.* **2017**, *64*, 373–380. [CrossRef] [PubMed]
59. Lu, N.; Oza, S.; Tajabadi, M.G. Surface Modification of Natural Fibers for Reinforcement in Polymeric Composites. In *Surface Modification of Biopolymers*; John Wiley & Sons, Inc.: Hoboken, NJ, USA, 2015; pp. 224–237.
60. Zhijiang, C.; Yi, X.; Haizheng, Y.; Jia, J.; Liu, Y. Poly(hydroxybutyrate)/cellulose acetate blend nanofiber scaffolds: Prepa-ration, characterization and cytocompatibility. *Mater. Sci. Eng. C* **2016**, *58*, 757–767. [CrossRef]
61. Panaitescu, D.M.; Nicolae, C.A.; Gabor, A.R.; Trusca, R. Thermal and mechanical properties of poly(3-hydroxybutyrate) reinforced with cellulose fibers from wood waste. *Ind. Crop. Prod.* **2020**, *145*, 112071. [CrossRef]
62. Mokhothu, T.H.; John, M.J. Review on hygroscopic aging of cellulose fibres and their biocomposites. *Carbohydr. Polym.* **2015**, *131*, 337–354. [CrossRef]
63. Joffe, R.; Pupure, L.; Berthold, F.; Varna, J. Micro-structure and Mechanical Properties in PLA Reinforced with Cellulosic Fiber Sheets Made by Wet Forming Method. In Proceedings of the 8th International Conference on Composites Testing and Model Identification (CompTest 2017), Leuven, Belgium, 5–7 April 2017.
64. Masmoudi, F.; Bessadok, A.; Dammak, M.; Jaziri, M.; Ammar, E. Biodegradable packaging materials conception based on starch and polylactic acid (PLA) reinforced with cellulose. *Environ. Sci. Pollut. Res.* **2016**, *23*, 20904–20914. [CrossRef]
65. Kyutoku, H.; Maeda, N.; Sakamoto, H.; Nishimura, H.; Yamada, K. Effect of surface treatment of cellulose fiber (CF) on durability of PLA/CF bio-composites. *Carbohydr. Polym.* **2019**, *203*, 95–102. [CrossRef]
66. Piekarska, K.; Sowinski, P.; Piorkowska, E.; Haque, M.M.U.; Pracella, M. Structure and properties of hybrid PLA nano-composites with inorganic nanofillers and cellulose fibers. *Compos. Part A Appl. Sci. Manuf.* **2016**, *82*, 34–41. [CrossRef]
67. Hu, R.; Lim, J.-K. Fabrication and Mechanical Properties of Completely Biodegradable Hemp Fiber Reinforced Polylactic Acid Composites. *J. Compos. Mater.* **2007**, *41*, 1655–1669. [CrossRef]
68. Pommet, M.; Juntaro, J.; Heng, J.; Mantalaris, A.; Lee, A.; Wilson, K.; Kalinka, G.; Shaffer, M.S.P.; Bismarck, A. Surface Modification of Natural Fibers Using Bacteria: Depositing Bacterial Cellulose onto Natural Fibers To Create Hierarchical Fiber Reinforced Nanocomposites. *Biomacromolecules* **2008**, *9*, 1643–1651. [CrossRef]
69. Xu, Z.; Yang, L.; Ni, Q.; Ruan, F.; Wang, H. Fabrication of high-performance green hemp/polylactic acid fibre composites. *J. Eng. Fibers Fabr.* **2019**, *14*. [CrossRef]
70. Michel, A.; Billington, S. Characterization of poly-hydroxybutyrate films and hemp fiber reinforced composites exposed to accelerated weathering. *Polym. Degrad. Stab.* **2012**, *97*, 870–878. [CrossRef]
71. Keller, A. Compounding and mechanical properties of biodegradable hemp fibre composites. *Compos. Sci. Technol.* **2003**, *63*, 1307–1316. [CrossRef]
72. Sawpan, M.A.; Pickering, K.L.; Fernyhough, A. Characterisation of hemp fibre reinforced Poly(Lactic Acid) composites. *Int. J. Mater. Prod. Technol.* **2009**, *36*, 229. [CrossRef]
73. Smoca, A.; Kukle, S.; Zelca, Z. Properties of Hemp Fibres Reinforced PLA Composites. *Key Eng. Mater.* **2019**, *800*, 205–209. [CrossRef]
74. Avella, M.; Buzõarovska, A.; Errico, M.E.; Gentile, G.; Grozdanov, A.; Bogoeva-Gaceva, G. Poly(3-hydroxybutyrate-co-3-hydroxyvalerate)-based biocomposites reinforced with kenaf fibers. *J. Appl. Polym. Sci.* **2007**, *104*, 3192–3200. [CrossRef]
75. Hassan, A.; Isa, M.R.M.; Ishak, Z.A.M. Improving thermal and mechanical properties of injection moulded Kenaf Fibre-reinforced Polyhydroxy-butyrate composites through fibre surface treatment. *BioResources* **2019**, *14*, 3101–3116.
76. Buzarovska, A.; Bogoeva-Gaceva, G.; Grozdanov, A.; Avella, M. Crystallization behavior of polyhydroxybutyrate in model composites with kenaf fibers. *J. Appl. Polym. Sci.* **2006**, *102*, 804–809. [CrossRef]
77. Graupner, N.; Müssig, J. A comparison of the mechanical characteristics of kenaf and lyocell fibre reinforced poly(lactic acid) (PLA) and poly(3-hydroxybutyrate) (PHB) composites. *Compos. Part A Appl. Sci. Manuf.* **2011**, *42*, 2010–2019. [CrossRef]
78. Kuciel, S.; Liber-Knec, A. Biocomposites based on PHB filled with wood or kenaf fibers. *Polimery* **2011**, *56*, 218–223. [CrossRef]
79. Bayart, M.; Gauvin, F.; Foruzanmehr, M.R.; Elkoun, S.; Robert, M. Mechanical and moisture absorption characterization of PLA composites reinforced with nano-coated flax fibers. *Fibers Polym.* **2017**, *18*, 1288–1295. [CrossRef]
80. Georgiopoulos, P.; Kontou, E.; Georgousis, G. Effect of silane treatment loading on the flexural properties of PLA/flax unidirectional composites. *Compos. Commun.* **2018**, *10*, 6–10. [CrossRef]
81. Karsli, N.G.; Aytac, A. Properties of alkali treated short flax fiber reinforced poly(lactic acid)/polycarbonate composites. *Fibers Polym.* **2014**, *15*, 2607–2612. [CrossRef]
82. Aydın, M.; Tozlu, H.; Kemaloglu, S.; Aytac, A.; Ozkoc, G. Effects of Alkali Treatment on the Properties of Short Flax Fi-ber-Poly(Lactic Acid) Eco-Composites. *J. Polym. Environ.* **2011**, *19*, 11–17. [CrossRef]

83. Shanks, R.A.; Hodzic, A.; Ridderhof, D. Composites of poly(lactic acid) with flax fibers modified by interstitial polymeri-zation. *J. Appl. Polym. Sci.* **2006**, *99*, 2305–2313. [CrossRef]
84. Foruzanmehr, M.; Vuillaume, P.; Elkoun, S.; Robert, M. Physical and mechanical properties of PLA composites reinforced by TiO^2 grafted flax fibers. *Mater. Des.* **2016**, *106*, 295–304. [CrossRef]
85. Ventura, H.; Claramunt, J.; Rodríguez-Pérez, M.; Ardanuy, M. Effects of hydrothermal aging on the water uptake and tensile properties of PHB/flax fabric biocomposites. *Polym. Degrad. Stab.* **2017**, *142*, 129–138. [CrossRef]
86. Chilali, A.; Assarar, M.; Zouari, W.; Kebir, H.; Ayad, R. Mechanical characterization and damage events of flax fabric-reinforced biopolymer composites. *Polym. Polym. Compos.* **2019**, *28*, 631–644. [CrossRef]
87. Rytlewski, P.; Stepczyńska, M.; Moraczewski, K.; Malinowski, R.; Karasiewicz, T.; Sikorska, W.; Żenkiewicz, M. Flax fibers reinforced polycaprolactone modified by triallyl isocyanurate and electron radiation. *Polym. Compos.* **2019**, *40*, 481–488. [CrossRef]
88. Lee, S.G.; Choi, S.-S.; Park, W.H.; Cho, D. Characterization of surface modified flax fibers and their biocomposites with PHB. *Macromol. Symp.* **2003**, *197*, 089–100. [CrossRef]
89. Arrieta, M.P.; Samper, M.D.; Aldas, M.; López, J. On the Use of PLA-PHB Blends for Sustainable Food Packaging Applications. *Materials* **2017**, *10*, 1008. [CrossRef]
90. Zhang, M.; Thomas, N.L. Blending polylactic acid with polyhydroxybutyrate: The effect on thermal, mechanical, and biodegradation properties. *Adv. Polym. Technol.* **2011**, *30*, 67–79. [CrossRef]
91. Armentano, I.; Fortunati, E.; Burgos, N.; Dominici, F.; Luzi, F.; Fiori, S.; Jimenez, A.; Yoon, K.; Ahn, J.; Kang, S.; et al. Processing and characterization of plasticized PLA/PHB blends for biodegradable multiphase systems. *Express Polym. Lett.* **2015**, *9*, 583–596. [CrossRef]
92. Armentano, I.; Fortunati, E.; Burgos, N.; Dominici, F.; Luzi, F.; Fiori, S.; Jiménez, A.; Yoon, K.; Ahn, J.; Kang, S.; et al. Bio-based PLA_PHB plasticized blend films: Processing and structural characterization. *LWT* **2015**, *64*, 980–988. [CrossRef]
93. Abdelwahab, M.; Flynn, A.; Chiou, B.-S.; Imam, S.; Orts, W.; Chiellini, E. Thermal, mechanical and morphological characterization of plasticized PLA–PHB blends. *Polym. Degrad. Stab.* **2012**, *97*, 1822–1828. [CrossRef]
94. Arrieta, M.P.; Samper, M.D.; Lopez, J.; Jiménez, A. Combined Effect of Poly(hydroxybutyrate) and Plasticizers on Polylactic acid Properties for Film Intended for Food Packaging. *J. Polym. Environ.* **2014**, *22*, 460–470. [CrossRef]
95. Rodrigues, J.A.F.R.; Parra, D.F.; Lugão, A.B. Crystallization on films of PHB/PEG blends. *J. Therm. Anal. Calorim.* **2005**, *79*, 379–381. [CrossRef]
96. Ali, A.; Shaker, K.; Nawab, Y.; Jabbar, M.; Hussain, T.; Militky, J.; Baheti, V. Hydrophobic treatment of natural fibers and their composites—A review. *J. Ind. Text.* **2018**, *47*, 2153–2183. [CrossRef]
97. Lalit, R.; Mayank, P.; Ankur, K. Natural Fibers and Biopolymers Characterization: A Future Potential Composite Material. *Stroj. Cas.* **2018**, *68*, 33–50. [CrossRef]
98. Awasthi, S.; Agarwal, D. Influence of cycloaliphatic compounds on the properties of polyurethane coatings. *J. Coat. Technol. Res.* **2007**, *4*, 67–73. [CrossRef]
99. Alagi, P.; Choi, Y.J.; Hong, S.C. Preparation of vegetable oil-based polyols with controlled hydroxyl functionalities for thermoplastic polyurethane. *Eur. Polym. J.* **2016**, *78*, 46–60. [CrossRef]
100. Miao, S.; Wang, P.; Su, Z.; Zhang, S. Vegetable-oil-based polymers as future polymeric biomaterials. *Acta Biomater.* **2014**, *10*, 1692–1704. [CrossRef] [PubMed]
101. Deka, H.; Karak, N. Bio-based hyperbranched polyurethanes for surface coating applications. *Prog. Org. Coat.* **2009**, *66*, 192–198. [CrossRef]
102. Kong, X.; Liu, G.; Curtis, J.M. Novel polyurethane produced from canola oil based poly(ether ester) polyols: Synthesis, characterization and properties. *Eur. Polym. J.* **2012**, *48*, 2097–2106. [CrossRef]
103. Tan, S.; Abraham, T.; Ference, D.; Macosko, C.W. Rigid polyurethane foams from a soybean oil-based Polyol. *Polymer* **2011**, *52*, 2840–2846. [CrossRef]
104. Ismail, E.A.; Motawie, A.; Sadek, E.; Ismail, E.A.; Motawie, A.; Sadek, E. Synthesis and characterization of polyurethane coatings based on soybean oil–polyester polyols. *Egypt. J. Pet.* **2011**, *20*, 1–8. [CrossRef]
105. Datta, J.; Głowińska, E. Effect of hydroxylated soybean oil and bio-based propanediol on the structure and thermal properties of synthesized bio-polyurethanes. *Ind. Crop. Prod.* **2014**, *61*, 84–91. [CrossRef]
106. Fu, C.; Zheng, Z.; Yang, Z.; Chen, Y.; Shen, L. A fully bio-based waterborne polyurethane dispersion from vegetable oils: From synthesis of precursors by thiol-ene reaction to study of final material. *Prog. Org. Coat.* **2014**, *77*, 53–60. [CrossRef]
107. Thakur, S.; Karak, N. Castor oil-based hyperbranched polyurethanes as advanced surface coating materials. *Prog. Org. Coat.* **2013**, *76*, 157–164. [CrossRef]
108. Gurunathan, T.; Mohanty, S.; Nayak, S.K. Isocyanate terminated castor oil-based polyurethane prepolymer: Synthesis and characterization. *Prog. Org. Coat.* **2015**, *80*, 39–48. [CrossRef]
109. Mokhothu, T.H.; John, M.J. Bio-based coatings for reducing water sorption in natural fibre reinforced composites. *Sci. Rep.* **2017**, *7*, 1–8. [CrossRef]
110. Van Krevelen, D.W.; Te Nijenhuis, K. *Properties of Polymers: Their Correlation with Chemical Structure; Their Numerical Estimation and Prediction from Additive Group Contributions*; Elsevier: Amsterdam, The Netherlands, 2009.
111. Sanchez, P.; A Perez-Maqueda, L.; Perejon, A.; Criado, J.M. A new model for the kinetic analysis of thermal degradation of polymers driven by random scission. *Polym. Degrad. Stab.* **2010**, *95*, 733–739. [CrossRef]

112. Ghosh, P. *Polymer Science and Technology;* Tata McGraw-Hill Education: New York, NY, USA, 1990.
113. Tokiwa, Y.; Calabia, B.P.; Ugwu, C.U.; Aiba, S. Biodegradability of Plastics. *Int. J. Mol. Sci.* **2009**, *10*, 3722–3742. [CrossRef]
114. Bernard, M. Industrial Potential of Polyhydroxyalkanoate Bioplastic: A Brief Review. *USURJ: Univ. Sask. Undergrad. Res. J.* **2014**, *1*. [CrossRef]
115. Muhammadi, S.; Afzal, M.; Hameed, S. Bacterial polyhydroxyalkanoates-eco-friendly next generation plastic: Production, biocompatibility, biodegradation, physical properties and applications. *Green Chem. Lett. Rev.* **2015**, *8*, 56–77. [CrossRef]
116. Ahmed, T.; Shahid, M.; Azeem, F.; Rasul, I.; Shah, A.A.; Noman, M.; Hameed, A.; Manzoor, N.; Manzoor, I.; Muhammad, S. Biodegradation of plastics: Current scenario and future prospects for environmental safety. *Environ. Sci. Pollut. Res.* **2018**, *25*, 7287–7298. [CrossRef] [PubMed]
117. Knoll, M.; Hamm, T.M.; Wagner, F.; Martinez, V.; Pleiss, J. The PHA Depolymerase Engineering Database: A systematic analysis tool for the diverse family of polyhydroxyalkanoate (PHA) depolymerases. *BMC Bioinform.* **2009**, *10*, 89. [CrossRef]
118. Mergaert, J.; Webb, A.; Anderson, C.; Wouters, A.; Swings, J. Microbial degradation of poly(3-hydroxybutyrate) and poly(3-hydroxybutyrate-co-3-hydroxyvalerate) in soils. *Appl. Environ. Microbiol.* **1993**, *59*, 3233–3238. [CrossRef]
119. Jendrossek, D.; Handrick, R. Microbial Degradation of Polyhydroxyalkanoates. *Annu. Rev. Microbiol.* **2002**, *56*, 403–432. [CrossRef]
120. Manna, A.; Paul, A. Degradation of microbial polyester poly(3-hydroxybutyrate) in environmental samples and in culture. *Biodegradation* **2000**, *11*, 323–329. [CrossRef] [PubMed]
121. Kusaka, S.; Iwata, T.; Doi, Y. Properties and biodegradability of ultra-high-molecular-weight poly[(R)-3-hydroxybutyrate] produced by a recombinant Escherichia coli. *Int. J. Biol. Macromol.* **1999**, *25*, 87–94. [CrossRef]
122. Boyandin, A.N.; Prudnikova, S.; Filipenko, M.L.; Khrapov, E.A.; Vasil'Ev, A.D.; Volova, T. Biodegradation of polyhydroxyalkanoates by soil microbial communities of different structures and detection of PHA degrading microorganisms. *Appl. Biochem. Microbiol.* **2011**, *48*, 28–36. [CrossRef]
123. Murphy, J. Chapter 4—Modifying Specific Properties: Mechanical Properties—Fillers. In *Additives for Plastics Handbook*, 2nd ed.; Murphy, J., Ed.; Elsevier Science: Amsterdam, The Netherlands, 2001; Volume 1, pp. 19–35.
124. Aframehr, W.M.; Molki, B.; Heidarian, P.; Behzad, T.; Sadeghi, M.; Bagheri, R. Effect of calcium carbonate nanoparticles on barrier properties and biodegradability of polylactic acid. *Fibers Polym.* **2017**, *18*, 2041–2048. [CrossRef]
125. Teramoto, N.; Urata, K.; Ozawa, K.; Shibata, M. Biodegradation of aliphatic polyester composites reinforced by abaca fiber. *Polym. Degrad. Stab.* **2004**, *86*, 401–409. [CrossRef]
126. Altaee, N.; El-Hiti, G.A.; Fahdil, A.; Sudesh, K.; Yousif, E. Biodegradation of different formulations of polyhydroxybutyrate films in soil. *SpringerPlus* **2016**, *5*, 1–12. [CrossRef]
127. Paul, M.-A.; Delcourt, C.; Alexandre, M.; Degée, P.; Monteverde, F.; Dubois, P. Polylactide/montmorillonite nanocomposites: Study of the hydrolytic degradation. *Polym. Degrad. Stab.* **2005**, *87*, 535–542. [CrossRef]
128. Chen, Y.; Murphy, A.; Scholz, D.; Geever, L.M.; Lyons, J.G.; Devine, D.M. Surface-modified halloysite nanotubes reinforced poly(lactic acid) for use in biodegradable coronary stents. *J. Appl. Polym. Sci.* **2018**, *135*, 46521. [CrossRef]
129. Navarro, M.; Ginebra, M.; Planell, J.; Barrias, C.; Barbosa, M. In vitro degradation behavior of a novel bioresorbable composite material based on PLA and a soluble CaP glass. *Acta Biomater.* **2005**, *1*, 411–419. [CrossRef]
130. Huang, J.; Xiong, J.; Liu, J.; Zhu, W.; Wang, D. Investigation of the in vitro degradation of a novel polylactide/nanohydroxyapatite composite for artificial bone. *J. Nanomater.* **2013**, *2013*, 3. [CrossRef]
131. Valapa, R.B.; Pugazhenthi, G.; Katiyar, V. Hydrolytic degradation behaviour of sucrose palmitate reinforced poly(lactic acid) nanocomposites. *Int. J. Biol. Macromol.* **2016**, *89*, 70–80. [CrossRef] [PubMed]
132. Jamarani, R.; Erythropel, H.C.; Nicell, J.A.; Leask, R.L.; Marić, M. How Green is Your Plasticizer? *Polymers* **2018**, *10*, 834. [CrossRef]
133. Boll, M.; Geiger, R.; Junghare, M.; Schink, B. Microbial degradation of phthalates: Biochemistry and environmental impli-cations. *Environ. Microbiol. Rep.* **2020**, *12*, 3–15. [CrossRef] [PubMed]
134. Labrecque, L.V.; Kumar, R.A.; Gross, R.A.; McCarthy, S.P. Citrate esters as plasticizers for poly(lactic acid). *J. Appl. Polym. Sci.* **1997**, *66*, 1507–1513. [CrossRef]
135. Ozkoc, G.; Kemaloglu, S. Morphology, biodegradability, mechanical, and thermal properties of nanocomposite films based on PLA and plasticized PLA. *J. Appl. Polym. Sci.* **2009**, *114*, 2481–2487. [CrossRef]

polymers

MDPI

Article

Nanocomposite Materials Based on TMDCs WS_2 Modified Poly(L-Lactic Acid)/Poly(Vinylidene Fluoride) Polymer Blends

Mohammed Naffakh

Escuela Técnica Superior de Ingenieros Industriales, Universidad Politécnica de Madrid (ETSII-UPM), José Gutiérrez Abascal 2, 28006 Madrid, Spain; mohammed.naffakh@upm.es

Abstract: Novel multifunctional biopolymer blend nanocomposites composed of poly(vinylidene fluoride)(PVDF) and tungsten disulfide nanotubes (INT-WS_2) that are layered transition metal dichalcogenides (TMDCs) were easily prepared by applying an economical, scalable, and versatile melt processing route. Furthermore, their synergistic effect to enhance the properties of poly(L-lactic acid) (PLLA) matrix was investigated. From morphological analysis, it was shown that the incorporation of 1D (INT)-WS_2 into the immiscible PLLA/PVDF mixtures (weight ratios: 80/20, 60/40, 40/60, and 20/80) led to an improvement in the dispersibility of the PVDF phase, a reduction in its average domain size, and consequently a larger interfacial area. In addition, the nanoparticles INT-WS_2 can act as effective nucleating agents and reinforcing fillers in PLLA/PVDF blends, and as such, greatly improve their thermal and dynamic-mechanical properties. The improvements are more pronounced in the ternary blend nanocomposites with the lowest PVDF content, likely due to a synergistic effect of both highly crystalline PVDF and 1D-TMDCs nano-additives on the matrix performance. Considering the promising properties of the developed materials, the inexpensive synthetic process, and the extraordinary properties of environmentally friendly and biocompatibe 1D-TMDCs WS_2, this work may open up opportunities to produce new PLLA/PVDF hybrid nanocomposites that show great potential for biomedical applications.

Keywords: TMDCs-WS_2; PLLA; PVDF; nanomaterials; morphology; crystallization; dynamic-mechanical properties

Citation: Naffakh, M. Nanocomposite Materials Based on TMDCs WS_2 Modified Poly(L-Lactic Acid)/Poly(Vinylidene Fluoride) Polymer Blends. *Polymers* **2021**, *13*, 2179. https://doi.org/10.3390/polym13132179

Academic Editors: José Miguel Ferri, Vicent Fombuena Borràs and Miguel Fernando Aldás Carrasco

Received: 14 June 2021
Accepted: 28 June 2021
Published: 30 June 2021

1. Introduction

The biobased, biodegradable aliphatic polyester poly(L-lactic acid) (PLLA) that is derived from natural resources, such as corn and sugar beet, is a highly versatile polymer and a promising alternative to petroleum-derived polymers in many applications as a result of its good biodegradability, renewability, reasonably good mechanical properties, and processability [1–3]. Additionally, it is known to be an excellent base polymer for biomedical applications, including drug delivery systems (DDS), sutures, and clips due to the fact that it has good biocompatibility and that its degradation products are benign to the human body. Moreover, unlike other biopolymers like, poly(hydroxyalkanoates) (PHA), poly(ethylene glycol) (PEG), and poly(ε-caprolactone) (PCL), PLLA can be processed using many different methods, such as extrusion, film casting, fiber spinning, and blow molding, due to its higher thermal processability [4]. However, in spite of its many beneficial attributes, historically, its commercial viability has been limited by poor stability during long melt molding and processing cycles, poor production efficiency, and overall high costs. Since the cost and brittleness of PLLA are quite high, it is not economically feasible to use it alone for day-to-day use as a packaging material without blending. Thus, blends of PLLA with several synthetic and biopolymers have been prepared in an effort to enhance the properties of PLLA. PLLA blends with poly(vinylidene fluoride) (PVDF), poly(butylenes succinate adipate) (PBSA), poly(ethylene glycol) (PEG), poly(methyl methacrylate) (PMMA), polypropylene (PP), polyethylene (PE), poly(ethylene oxide) (PEO), and poly(butylenes adipate-co-terephthalate) (PBAT) have been reported to improve the properties, such as toughness, modulus, impact strength, crystallization behavior, and thermal stability, compared to the neat polymer [5–9]. Notwithstanding, successful enhancement in the nucleation and crystallization behavior has been reported for PLLA by immiscible blending with PVDF via epitaxial and interface-assisted nucleation [5]. By using the classical method of fiber extrusion, without any special spinnerets, fibrous structures were obtained consisting of a polylactic acid (PLA) matrix filled with PVDF micro- and nanofibers [10]. This kind of fibers, as a hybrid system, can be successfully used for producing implants and prostheses. PVDF film was widely used in the filtration of protein because of its hydrophobicity being able to reduce surface fouling [11]. In addition, PVDF was also reported as a scaffold for cell culture because the piezoelectric properties can enhance cellular adherence and proliferation [12,13]. Previous studies also found that PLLA can facilitate the α- to β-phase transition of PVDF under eletrospinning and uniaxial stretching [14,15]. Electrospun PLA/PVDF mats exhibited higher cell proliferation for L929 fibroblasts than both PLA and PVDF mats [14]. However, to attain PLA blends with good general properties, typically, some sort of compatibilization strategy is required. For more detailed information on PLA property modification, the recent review by Zheng et al. [16] is recommended, which includes the use of copolymer, reactive polymer, nanoparticle, and low molecular weight chemical addition, as well as interfacial compatibilization, exchange reactions, and dynamic vulcanization. The conventional approach to compatibilizing polymer blends is via the use of copolymers as it is an efficient means to achieve good blend compatibility. However, commercial unavailability of specific copolymers and the fact that they must be synthesized prior to blending is one of its drawbacks [16]. Recent nanotechnology advances have been applied to PLLA-based polymers, resulting in improved chemical, mechanical, and biological properties. Advanced nanocomposite materials were obtained by filling the polymer matrix with both synthetic and natural nanoparticles [17–20]. The addition of CNTs to polymer composite structures with natural fiber has opened a new era of polymer composites for various structural applications. As polymer matrix reinforcements, different types of CNTs with specific and unique functional groups interact with hydroxyl groups in natural fiber cellulose chains, thus modifying the natural fiber surface [17]. The emerging halloysite-based bionanocomposites are used in applications such as biomedicine, packaging, corrosion protection, and restoration of cultural heritages [19]. In this way, Lisuzzo et al. demonstrated that the chitosan coating of halloysite nanotubes driven by electrostatic interactions can be considered a suitable strategy to obtain drug delivery systems with tunable properties [20]. Of particular interest is the use of layered transition metal dicalcogenide (TMDC) nanostructures, such as tungsten disulfide (WS_2) and molybdenum disulfide (MoS_2), which are broadband semiconductors with multidimensional structural anisotropy, 0D (IF), 1D (INT), and 2D [21–26]. These environmentally friendly and biocompatible TMDCs nanoparticles also have demonstratable processing, performance, design, and cost advantages over nanoclays, CNTs, etc. when manufacturing advanced biopolymer nanocomposites poly(3-hydroxybutyrate) (PHB), Bio-PNC INT-WS_2, poly(propylene fumarate) (PPF), poly(ether ether ketone) (PEEK), PLLA, etc. [27–30]. Furthermore, the cytotoxicity of INT-WS_2 is comparable to regular environmental particulate matter and much lower than other nanoparticles, like silica or carbon black [31,32]. Moreover, blending nanoparticles with an immiscible mixture (PLLA-blend) expands the possible routes for compatibilization, and unlike polymeric compatibilizers, nanoparticles are not specific to the nature of the immiscible components of the mixture and are easily incorporated by mixing. Further, the addition of nanoparticles can significantly improve the materials' properties, combining the attributes of the base polymer blend with the characteristics of the nanoparticle [16,33].

The object of the current research is to demonstrate the advantage of using INT-WS_2 as a suitable nano-reinforcement to improve the performance of promising PLLA/PVDF polymer blends. New melt-processable nanocomposites were prepared via a scalable, versatile, and cheap procedure, without the addition of compatibilizers or modifiers. In particular, the effect of INT-WS_2 on the morphology, thermal, processability, and mechanical properties of the resulting PLLA/PVDF/INT-WS_2 nanocomposites is considered.

2. Experimental Section

2.1. Materials and Processing

Poly(L-lactic acid) (PLLA) and poly(vinylidene fluoride) (PVDF) were purchased from Goodfellow Ltd. (Huntingdon, UK). Multiwall WS_2 1D nanotubes (INT-WS_2) with diameters of 30–150 nm and lengths of 1–20 nm were obtained from NanoMaterials Ltd. (Yavne, Israel) (see Section 3.1). Both the blends and nanocomposites were prepared following the same procedure: each mixture of PVDF and PLLA, with or without INT-WS_2, was dispersed in a small volume of ethanol (HPLC grade, Sigma-Aldrich Química SL, Madrid, Spain) and homogenized by mechanical stirring and bath ultrasonication for approximately 15 min. Subsequently, the dispersion was partially dried in vacuum at 60 °C under a pressure of about 70 mbar for 24 h. The PLLA/PVDF blends are designated as 80/20, 60/40, 40/60, and 20/80, where the numbers indicate the weight percentages of PLLA and PVDF, respectively. In the case of PLLA/PVDF/INT-WS_2 nanocomposites, the INT-WS_2 fraction was 0.5 wt.% of the total composite weight and the ratio of PLLA and PVDF was the same as in the binary blends (80/20-INT(79.6/19.9/0.5), 60/40-INT(59.7/39.8/0.5), 40/60-INT (39.8/59.7/0.5), and 20/80-INT(19.9/79.6/0.5)). In a previous study [30], the crystallization behavior and mechanical properties of PLLA filled with different amounts of INT–WS_2 (0.1 wt.%, 0.5 wt.%, and 1.0 wt.%) were investigated, and it was found that 0.5 wt.% loading led to the highest property improvements. Therefore, 0.5 wt.% INT–WS_2 was chosen as an optimum concentration to prepare the ternary PLLA/PVDF/INT–WS_2 nanocomposites. For the sake of comparison, reference samples of PLLA/INT-WS_2 (0.5 wt.%) (PLLA-INT) and PVDF/INT-WS_2 (0.5 wt.%) (PVDF-INT) nanocomposites were also prepared in the same way. The melt-mixing of the resulting dispersions (~6 g) was performed using a micro-extruder (Thermo-Haake Minilab system) operating at 190 °C with a rotor speed of 100 rpm for 10 min. Then, the samples were pressed into films of 0.5 mm thickness in a hot press system using two heating/cooling plates.

2.2. Characterization Studies

The morphology of the samples was characterized using an ultra-high field-emission scanning microscopy (FESEM) (SU8000, Hitachi Co., Tokyo, Japan). Cryogenically fractured surfaces from film specimens were coated with a ~5 nm Au/Pd layer to avoid charging during electron irradiation.

Wide-angle x-ray diffraction (WAXS) analysis were carried out using a Bruker D8 Advance diffractometer (Bruker AXS GmbH, Karlsruhe, Germany) and a Ni-filtered CuKα radiation source. Diffractograms were recorded at a 0.2 °/s scan speed and a resolution of 40 points/degree over the 2θ region of 5 to 35 °C. Diffractograms were also recorded at a scanning rate of approximately 7 °C/min to understand the dynamic crystallization and melting behavior of the samples. This was performed by initially holding them at a temperature of 190 °C for 5 min to erase any thermal history, and then cooling from 190 to 30 °C, followed by reheating them over the same temperature range and rate.

Thermogravimetric analysis (TGA) was performed using a TA Instruments Q50 thermobalance (Waters Cromatografía, S.A., Cerdanyola del Vallès, Spain) under nitrogen gas (flow rate = 60 mL/min) at 10 °C/min, over a temperature range of 100–800 °C.

Differential scanning calorimetry (DSC) was performed on a Perkin Elmer DSC7/7700 differential scanning calorimeter (Perkin-Elmer España SL, Madrid, Spain), calibrated with indium (T_m = 156.6 °C, ΔH_m = 28.45 kJ/kg) and zinc (T_m = 419.47 °C, ΔH_m = 108.37 kJ/kg) under the flow of nitrogen gas (25 mL/min). The samples were first heated to 220 °C and held at the same temperature for 5 min to erase the thermal history. Then, the crystallization of the samples was carried out by cooling from 220 to 40 °C, followed by heating cycles at 10 °C/min over the interval of temperatures between 40 and 220 °C. The crystallization/melting enthalpy of PLLA in the blend nanocomposites was determined by considering the weight fraction of PLLA in the nanocomposites.

Dynamic mechanical analysis (DMA) was performed on rectangular shaped samples using a Mettler DMA 861 device (Mettler-Toledo, Greifensee, Switzerland), at three frequencies of 0.1, 1, and 10 Hz in the tensile mode. An 8 N oscillating dynamic force using an amplitude of 17 µm at fixed frequency was adopted. The relaxation spectra were recorded over the temperature range −100 °C to 150 °C, at a heating rate of 3 °C/min.

3. Results

3.1. Morphology

Tailoring phase morphology of immiscible blends through the addition of nanoparticles is a universally accepted strategy of forming compatible nanocomposites polymer blends, and typically results in improved upon physical properties. The reason for this is that the added nanoparticles typically locate at the interface of the polymer domains and act as interfacial modifiers, strengthening the interfacial adhesion. Nanoparticles can induce the formation of fine dispersed phase particles from coalescence during melt processing, thus stabilizing the fine morphology and therefore maintaining the properties of the blend.

The cryogenically fractured surface morphologies using SEM of the PLLA/PVDF blends can be seen in Figure 1. For the 80/20 PLLA/PVDF mixture, a distinct two-phase morphology was noted, with the PVDF phase dispersed evenly within the PLLA matrix. Furthermore, the mean diameter of the domains augmented with increasing PVDF content. When the PLLA content < 60 wt.%, the phase morphology was reversed and the PLLA phase became dispersed in the PVDF. For the ternary hybrid PLLA/PVDF/INT nanocomposites and individual PLLA/INT and PVDF/INT nanocomposites (data are not shown), it was found that the INTs were uniformly dispersed at the nanoscale without evidence of aggregates or agglomerates (see arrows pointing to individual INT-WS_2 tubes in the images), verifying the effectiveness of the melt extrusion conditions (Figure 2). The INT-WS_2 nanoparticles also improved the compatibility of the two phases, as demonstrated by an important reduction in the size of the dispersed PVDF domain. This can be attributed to the formed morphological structure in which INT-WS_2 was mainly dispersed in the PLLA matrix and at the PLLA/PVDF interface.

To achieve similar morphology, other authors employed more elaborate methodologies. For example, carboxyl functionalized multi-walled carbon nanotubes (MWCNTs) were used by Wu et al. to improve the compatibility of immiscible PLA/PCL mixtures [34]. In the case of PLA and PBS, Chen et al. employed double functionalized organoclay (TFC) to increase phase compatibility [35]. They also found that the concentration of TFC had a significant effect on the blend morphology, though when the TFC content < 0.5 wt.%, the PBS domain did not change in size and it was almost exclusively found within the PLA regions. Another nanoparticle that has been used to make immiscible PLA mixtures compatible is silica, as reported by Odent et al. in a mixture of PLA with a gummy copolyester based on 3-caprolactone, P[CL-*co*-LA] [36]. It was found that P[CL-*co*-LA] with spherical nodules dispersed regularly in the PLA matrix in a blank mixture containing 10% by weight of P[CL-*co*-LA], while surface treated (5 wt.% hexamethyldisilazane) spherical nodules disappeared. On the other hand, the compatibility of hydroxyl functionalized and poly(3-caprolactone)-*b*-poly(L-lactide) diblock copolymer grafted polyhedral oligomeric silsesquioxane (POSS-OH and POSS-PCL-*b*-PLLA, respectively) on PLA/PCL blends were analyzed by Monticelli et al. [37]. In both cases, the adhesion between PLA and PCL increased, and with POSS-PCL-*b*-PLLA, a nearly homogeneous microstructure was formed.

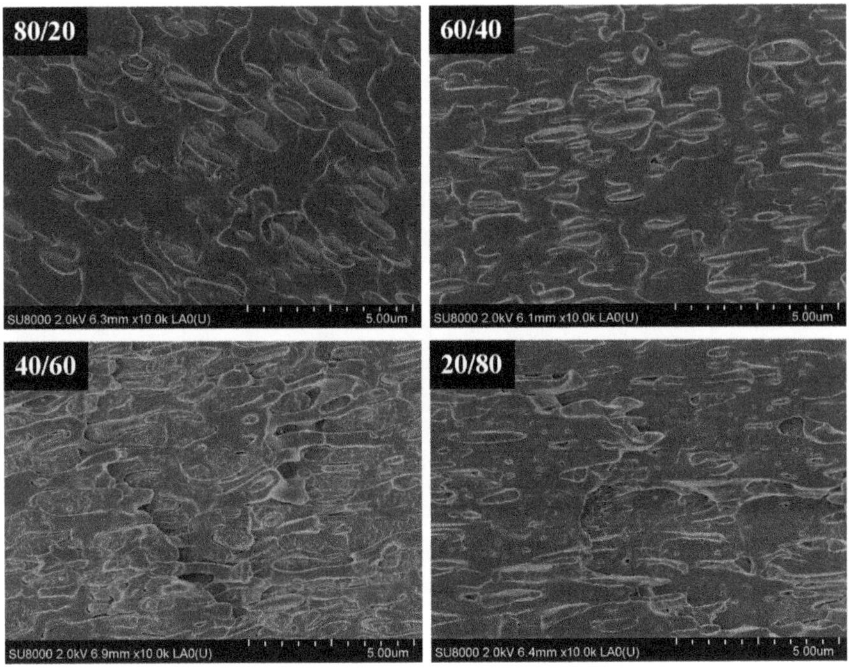

Figure 1. High-resolution SEM image for PLLA/INT-WS_2, PVDF/INT-WS_2, binary PLLA/PVDF, and ternary PLLA/PVDF/INT-WS_2 hybrid nanocomposites.

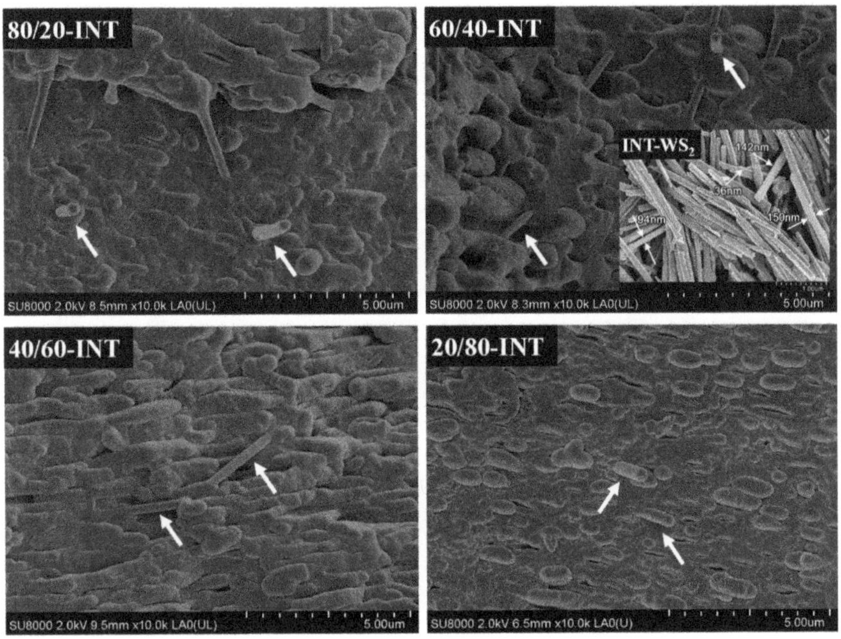

Figure 2. High-resolution SEM image for ternary PLLA/PVDF/INT-WS_2 hybrid nanocomposites; the inset shows the nanotube dimensions (thicknesses) of as received INT-WS_2.

3.2. Thermal Stability

The thermal stability conditions the processing limits of this type of system, and as such, the study of the effect of blend composition on the degradation temperatures is of utmost importance. Figure 3 shows the variation of the integral and differential thermogravimetric curves, TGA and DTG, of PLLA/PVDF blends as a function of the blend composition. The initial weight loss of PLLA started at around 326 °C, and in the case of PVDF, at a temperature about 100 °C higher. The differential curves show a single peak for each polymer which implies that the degradation processes occurred in a single step. The values of the characteristic degradation temperatures T_i (temperature for 2% weight loss), T_{10} (temperature for 10% weight loss), T_{max} (temperature corresponding to the maximum rate of weight loss), and R_{max} (rate of maximum decomposition) are summarized in Table 1.

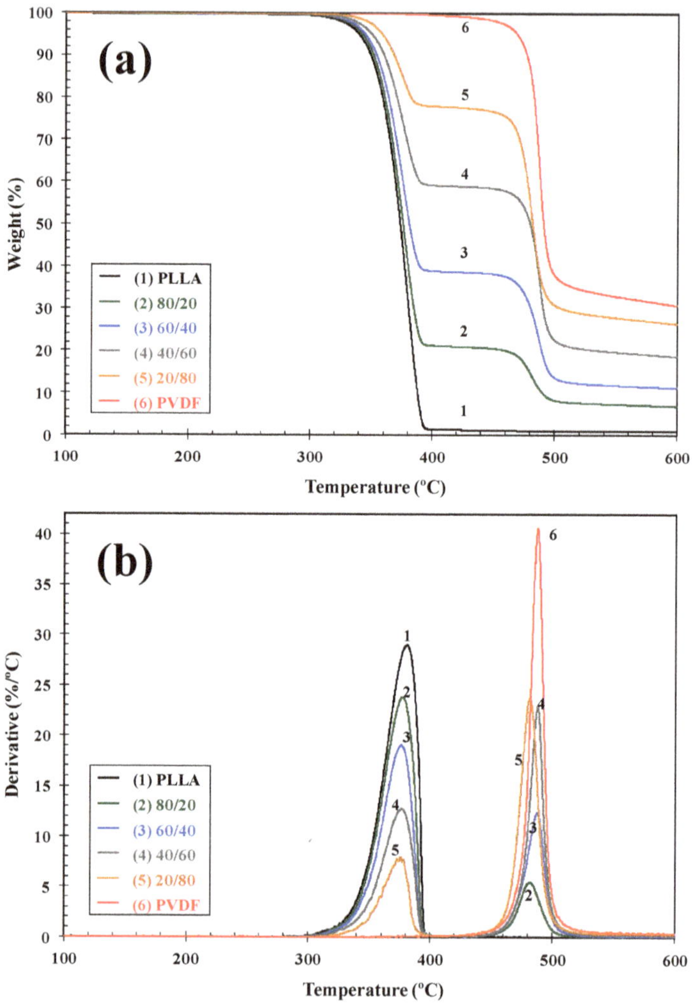

Figure 3. (**a**) Thermogravimetric (TGA) and (**b**) derivative thermogravimetric (DTG) curves of PLLA/PVDF blends.

Table 1. TGA parameters of different PLLA/PVDF/INT-WS_2 blend nanocomposites based on PLLA, PVDF, and INT-WS_2.

Material	T_i (°C)	T_{10} (°C)	T_{max} (°C)	R_{max} (%/°C)
PLLA	325.6	347.0	381.0	29.01
80/20	327.4	351.9	378.2/480.7	23.85/05.34
60/40	332.8	353.7	376.4/488.0	19.00/12.21
40/60	338.3	360.0	377.3/488.0	12.75/22.97
20/80	346.5	368.2	375.5/481.6	8.02/23.73
PVDF	420.9	453.5	478.4	40.34
PLLA-INT	325.8	349.0	377.4	29.39
80/20-INT	327.4	351.9	378.0/489.2	22.61/05.58
60/40-INT	330.2	352.0	377.1/485.3	18.39/09.34
40/60-INT	336.4	358.3	374.8/482.0	13.27/14.96
20/80-INT	344.8	367.1	371.9/476.5	6.95/23.86
PVDF-INT	396.6	437.0	454.5	17.68

T_i: temperature for 2% weight loss; T_{10}: temperature for 10% weight loss; T_{max}: temperature corresponding to the maximum rate of weight loss; and R_{max}: rate of maximum decomposition.

It is clear that the presence of PVDF influenced the degradation behavior of PLLA and vice versa. The addition of PVDF to PLLA hardly increased the thermal stability of the PLLA biopolymer (T_i and T_{10}), and the thermal degradation of both components took place via different mechanisms, since two well-defined decomposition stages can be observed. Although the change in T_{max} values was small (Figure 4a), the change in the corresponding values of maximum rate of decomposition (R_{max}) is representative of the variation in the composition of the blends, Figure 4b. In particular, the R_{max} of PLLA/PVDF blends clearly fell dramatically with respect to the values observed of neat PLLA and PVDF. Analogously, the addition of 0.5 wt.% INT-WS_2 to the PLLA/PVDF blends slightly affected the thermal stability (Table 1), without apparent perturbation of the mechanisms of degradation of the blend components, as can be deduced from the presence of the degradation maxima corresponding to PVDF and PLLA (Figure 5).

3.3. Crystallization Behavior

DSC was used to study the crystallization behavior of the biopolymers and it was found that PVDF, despite having a similar melting point to PLLA, crystallizes at a faster rate and has a higher crystallization temperature [5]. It can be understood therefore that the PVDF will separate from the melt and crystallize first during cooling. Any modification to the PLLA/PVDF domain interface would affect this process of phase separation, crystallization rate, and hence, crystal morphology.

Dynamic DSC cooling scans at a rate of 10 °C/min of the prepared PLLA/PVDF blends (Figure 6a) show that as the ratio of PVDF increased, the PLLA crystallization exotherms shifted to higher temperatures and had a higher enthalpy. It should be noted that the crystallization exotherm of pure PLLA with a crystallization enthalpy of only 8.0 J/g can be difficult to detect. For comparison, the samples containing only 0.5 wt.% INT-WS_2 are shown in Figure 6b, where it is seen that the crystallization exotherms shifted toward higher temperatures and the enthalpy increased, and in the case of neat PLLA, up to 41.6 J/g. To be able to compare this change in T_c vs. PLLA concentration for the PLLA/PVDF blends and those containing INT-WS_2 nanoparticles, Figure 7a is presented. Regarding the blends, a strong increase in the crystallization temperature of PLLA was found upon increasing PVDF content, from 92.2 °C for neat PLLA to 133.1 °C for raw PVDF. A noticeable increase up to 111.3 °C was already found with the incorporation of only 20 wt.% PVDF, but no significant change in T_c was observed with a further increase of PVDF content. The T_c of PLLA increased dramatically after blending with PVDF, indicating that PVDF accelerated the melt-crystallization of PLLA. In the same way, the presence of INT-WS_2 caused an increase in the crystallization temperature, both in the neat polymers and in the blends, the rise being higher than 25 °C for the neat PLLA and about 2 °C for the neat PVDF (see Table 2). With regard to the nanocomposites, this aspect was also observed, though it seems to be less dependent on the concentration of the polymers. As such, it would suggest that the nanofillers provoked nucleation in both polymeric components, with the effect being more pronounced for PLLA. In contrast, 2D-WS_2 nanosheets have been shown to slow down the crystallization rate of PLLA [25]. Such differences suggest that the nanoparticle shape plays a fundamental role in PLLA crystallization. This discrepancy is likely related to several factors, including the nanofiller geometry, its surface energy, roughness, and crystalline structure as well as on the filler ability to form the critical nucleus [17,22,23]. It should be noticed that, in the case of the 60/40-INT nanocomposite, a double crystallization exotherm was found (Figure 6b), with T_c values of 114.1 °C and 134.5 °C, which is likely related to the presence of two distinct macrophases, one containing the majority of the 0.5 wt.% INT-WS_2 and the other encompassing very little. This tendency was also observed for the 40/60-INT and 20/80-INT nanocomposite samples.

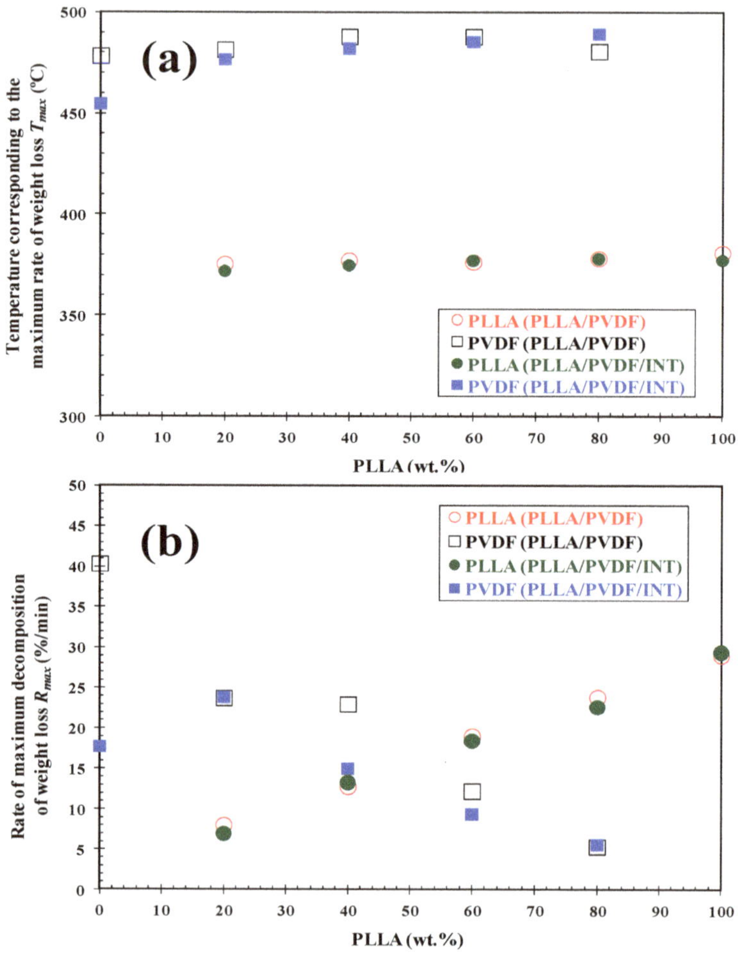

Figure 4. (**a**) Variation of the temperature and (**b**) rate of maximum decomposition (T_{max}/R_{max}) of PLLA and PVDF in the binary PLLA/PVDF and ternary PLLA/PVDF/INT-WS2 hybrid nanocomposites with composition.

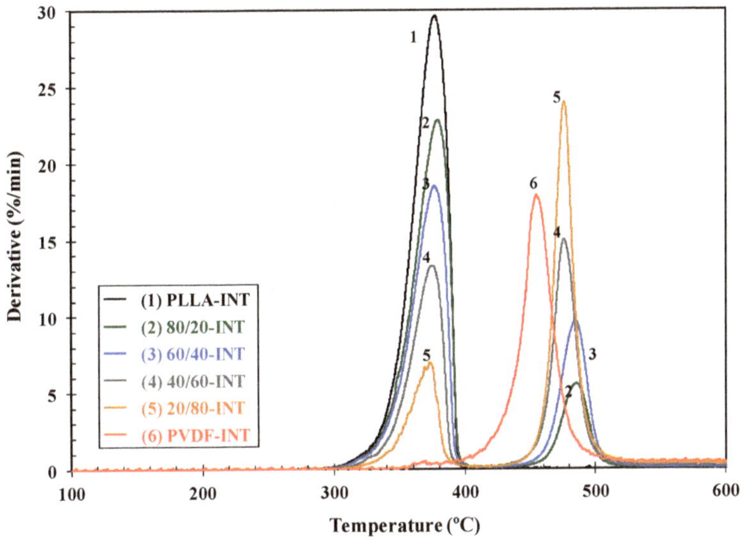

Figure 5. DTG curves of PLLA/PVDF/INT-WS_2 nanocomposites.

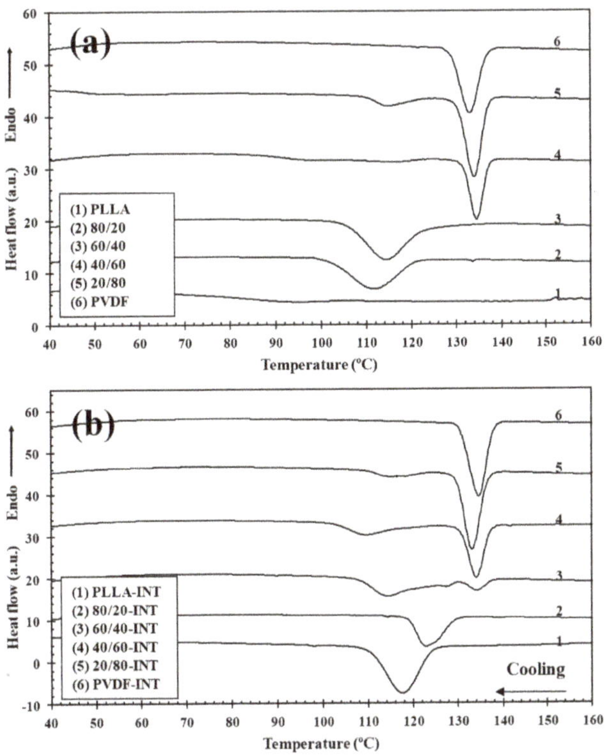

Figure 6. DSC thermograms of the non-isothermal crystallization of (**a**) binary PLLA/PVDF and (**b**) ternary PLLA/PVDF/INT-WS_2 hybrid nanocomposites obtained during cooling from the melt to room temperature at 10 °C/min.

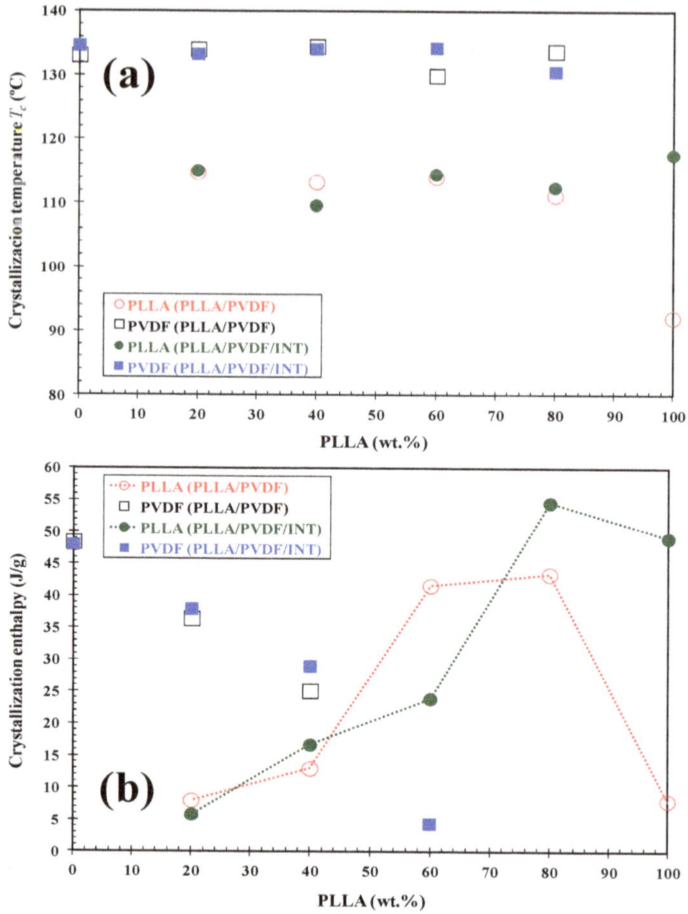

Figure 7. (**a**) Variation of the crystallization temperature (T_c) and (**b**) crystallization enthalpy (ΔH_c) of PLLA and PVDF in the binary PLLA/PVDF and ternary PLLA/PVDF/INT-WS_2 hybrid nanocomposites with composition.

Table 2. DSC parameters of different PLLA/PVDF/INT-WS_2 blend nanocomposites based on PLLA, PVDF, and INT-WS_2.

Material	$T_{c,\ PLLA}$ (°C)	$\Delta H_{c,\ PLLA}$ (J/g)	$T_{c,\ PVDF}$ (°C)	$\Delta H_{c,\ PVDF}$ (J/g)	$T_{cc,\ PLLA}$ (°C)	ΔH_{cc} (J/g)	T_m (°C)	ΔH_m (J/g)
PLLA	92.2	8.0	-	-	104.2	27.9	166.1	47.4
80/20	111.3	43.4	133.8	-	-	-	161.0/166.9	53.0
60/40	114.1	41.6	130.0	-	-	-	164.5/139.9	56.0
40/60	113.3	13.0	134.5	25.1	-	-	169.0	45.4
20/80	114.8	8.0	134.0	36.4	-	-	166.0/168.7	46.8
PVDF	-	-	133.1	48.4	-	-	166.2/169.3	48.9
PLLA-INT	117.6	49.1	-	-	-	-	163.9/168.3	52.2
80/20-INT	112.5	54.6	130.6	-	-	-	164.0/169.4	49.2
60/40-INT	114.5	23.9	134.3	4.3	-	-	164.2/169.2	54.4
40/60-INT	109.6	16.7	134.1	29.0	-	-	166.4/170.2	44.4
20/80-INT	115.0	5.8	133.3	37.9	-	-	166.4/169.0	48.4
PVDF-INT	-	-	134.6	48.0	-	-	169.8	47.2

T_c: crystallization temperature; ΔH_c: crystallization entalphy; T_{cc}: cold-crystallization temperature; ΔH_{cc}: cold-crystallization entalphy; T_m: melting temperature; ΔH_m: melting entalphy.

The previously mentioned nucleation effect that led to the increase in crystallization temperature is highly important, particularly when evaluating the crystallization enthalpy tendency. This can be seen in Figure 7b where the variation of the crystallization enthalpy (ΔH_c) versus the PLLA wt.% of the blends and the nanocomposites is presented, with the parameters of crystallization detailed in Table 2. For the materials without nanoparticle addition, the value of ΔH_c reduced from 48.4 J/g for PVDF, that is 47% crystalline ($\Delta H_{100\ PVDF}$ = 103 J/g for perfect crystals [38]) to 8.0 J/g for PLLA, with 8.6% crystallinity ($\Delta H_{100\ PLLA}$ = 93 J/g [39]). However, as can be seen, the increased presence of PVDF provoked a significant increment in the ΔH_c of PLLA due to it helping to speed up PLLA's overall crystallization rate. This is reflected in both an increase the temperature of crystallization from melt and enthalpy of crystallization and cold-crystallization suppression (see next section). Using a similar PLLA/PVDF blend, the crystal nucleation was also investigated by Pan et al., where they found that transcrystallization of one polymer type can occur on the crystalline surface of the other polymer [5]. Based on this, it is suggested that the PVDF presence in the immiscible blends promoted PLLA crystallization via two routes, interface-assisted and heterogeneous epitaxial nucleation. The interface between the two phases reduce the surface free energy, facilitating crystal nuclei to form via heterogeneous nucleation. In addition to PVDF crystallization, phase separation can bring about the molecular ordering, alignment, and/or orientation of PLLA chains at the PLLA/PVDF domain interface via interdiffusion, further aiding crystal embryo development [5]. Regarding the samples containing INT-WS_2, ΔH_c was observed to increase dramatically as a result of the nucleating effect of the nanoparticles, reaching 54.6 J/g (58.7% crystallinity) for 80/20-INT, the dual-additive system was more important for nucleation than of PVDF alone. Figure 8 illustrates the WAXS profiles of neat PLLA, 80/20, and 80/20-INT recorded during cooling from the melt to room temperature. The characteristic peaks of α-form of PLLA in WAXS patterns (see the following part) appeared as soon as the material attained an appreciable degree of crystallinity. The appearance of these peaks relates well to the crystallization temperature calculated from DSC curves. These results also indicate that the presence of INT-WS_2 accelerated the crystallization rate of PLLA in the PLLA/PVDF-INT blend nanocomposites. This led to the appearance of the characteristic of the crystalline diffraction of PLLA at higher temperature. Nucleating effects due to the presence of nanofillers have previously been reported for PLLA filled with inorganic nanotubes, nano-calcium carbonate, nano-zinc citrate, graphene oxide and fullerenes (C_{60}), nanoclay, and carbon nanotubes [30]. In particular, it was shown that INT-WS_2 exhibited much more prominent nucleation activity on the crystallization of PLLA than other specific nucleating agents or nano-sized fillers.

3.4. Melting Behavior

The melting of semicrystalline thermoplastics is a very complex process significantly influenced by the crystallization conditions. In some circumstances, two peculiarities are observed in the DSC heating scans of semi-crystalline PLLA [40,41]. One is the emergence of a small exothermic peak just before the melting peak and the other is the occurrence of a double melting peak, which has usually been interpreted in terms of a pre-exiting morphology and/or reorganization [30,40]. These two phenomena can be well explained by taking into consideration the crystallization conditions in parallel with the α and α′ crystal formation requirements [41]. When PLLA is crystallized at temperatures corresponding to α crystal formation, the small exotherm appearing just before the single melting peak is due to the transformation of disordered α′ crystals to the ordered α-form. On the other hand, a double melting behavior appears when the crystallization temperature is situated in the region of simultaneous α′ and α type formation. For high crystallization temperatures, only α crystals are produced leading to a single melting peak.

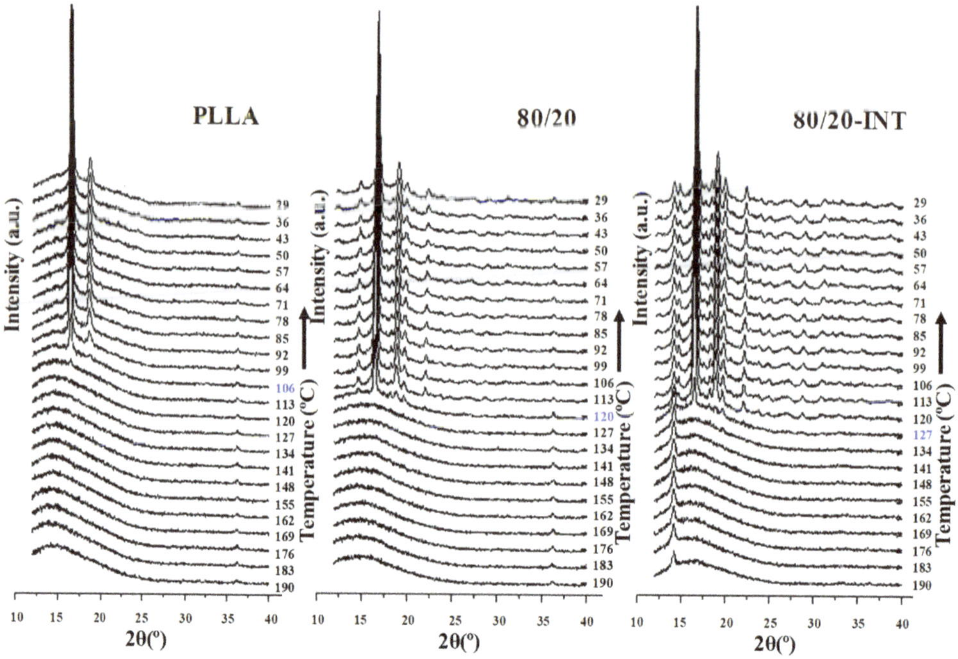

Figure 8. WAXS diffractograms of the dynamic crystallization of PLLA, PLLA/PVDF (80/20), and PLLA/PVDF/INT-WS_2 (80/20-INT) nanocomposites.

Figure 9 compares the heating thermograms after cooling from the melt of binary (PLLA/PVDF) and ternary (PLLA/PVDF/INT) hybrid nanocomposites with those of neat PLLA and PVDF. Because the melting peaks of PLLA and PVDF converge in the temperature range of 155−175 °C, the effects of blending on the melting behavior of PLLA cannot be easily identified. It is seen that PLLA presented a maximum endotherm at 166.1 °C of melt after the described exothermic cold-crystallization process. From Figure 10 of the WAXS diffractograms, it can be observed the PLLA sample did not show polymorphism when crystallized under the same conditions as those used for the DSC, with only the main (200)/(110) diffraction peak at 16.7° clearly visible, relating to the PLLA α-form [5,30]. This was similar for the high PLLA content blends, that also principally exhibited the PLLA α-phase characteristic diffraction. In the case of PVDF and high content PVDF blends, diffraction planes of (100), (020), (110), and (021) relating to 2θ = 17.7°, 18.5°, and 20.0° diffraction peaks, respectively were seen, and are typical of the PVDF α-phase [42,43]. In the case when both polymers or polymer blends with nanofiller (INT-WS_2 [30]) were present, all representative diffraction peaks were observed. From Figure 9b, it can be seen that INT-WS_2 addition to the mixed PLLA/PVDF blends significantly affected the melting behavior due to the suppression of the cold-crystallization processes as a result of heterogeneous nucleation and the consequences of this as noted earlier [30]. It is also important to note that the ΔH_m values of the binary (PLLA/PVDF) and ternary (PLLA/PVDF/INT) hybrid nanocomposites were higher than those of neat PLLA and PVDF, which was more distinct for the PVDF-rich blends. This is ascribed to the positive effects of both PVDF and INT-WS_2 on PLLA crystallization.

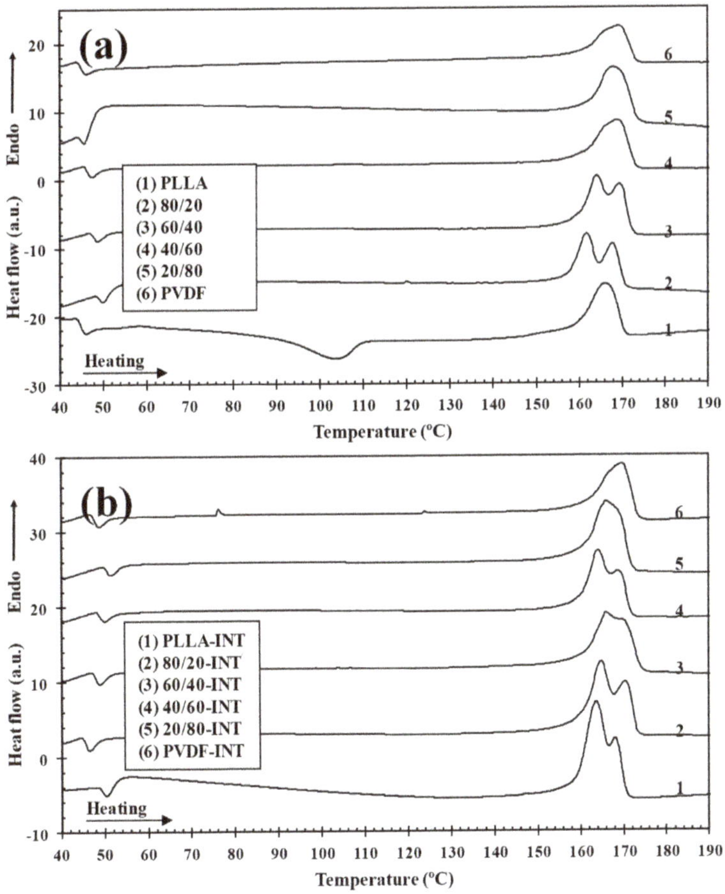

Figure 9. DSC thermograms of melting of (**a**) binary PLLA/PVDF and (**b**) ternary PLLA/PVDF/INT-WS_2 hybrid nanocomposites obtained during heating at 10 °C/min after cooling from the melt to room temperature at 10 °C/min.

3.5. *Dynamic Mechanical Analysis*

DMA is a technique widely used to characterize a material's physical properties, such as its glass transition, but is also sensitive to other relaxation processes, making it particularly relevant to assess the impact the addition of the nanofiller has on these events. For example, the storage modulus (E') and loss tangent (*tan δ*) curves for pure PLLA, PVDF, and their binary and ternary nanocomposite blends, prepared by quenching from the melt state, are shown as a function of temperature (see Figures 11 and 12, respectively). It can be seen that between −70 to −20 and 40 to 70 °C, the E' of the blends decreased sharply as they passed through the glass transition regions of PVDF and PLLA, respectively. After this, in the temperature ranging from 80 to 140 °C, the E' of the blends rose slightly due to the PLLA component cold crystallizing (Figure 11a). Below the T_g of PVDF as its composition increased in the blend, E' also improved according to the rule of mixtures. However, between the T_g of the two polymers, E' was seen to decrease with increasing PVDF content, as at this point the PVDF had transitioned from a leathery to rubbery state. The T_g and E' (at 25 °C) values for all the samples are presented in Table 3, where it is seen that with increasing PLLA content, E' increased due to it having a higher modulus than

PVDF at this temperature. The addition of the INT-WS_2 nanofiller to the samples resulted in them all having higher E' throughout the complete testing temperature range. This is related to the nanofillers' nucleating properties on the polymers and it enhancing their stiffness (Figure 11b), as can be seen, for example, comparing PLLA/PVDF (80/20) with the sample with 0.5 wt.% INT-WS_2, where the room temperature modulus increased 27% from 2274 MPa.

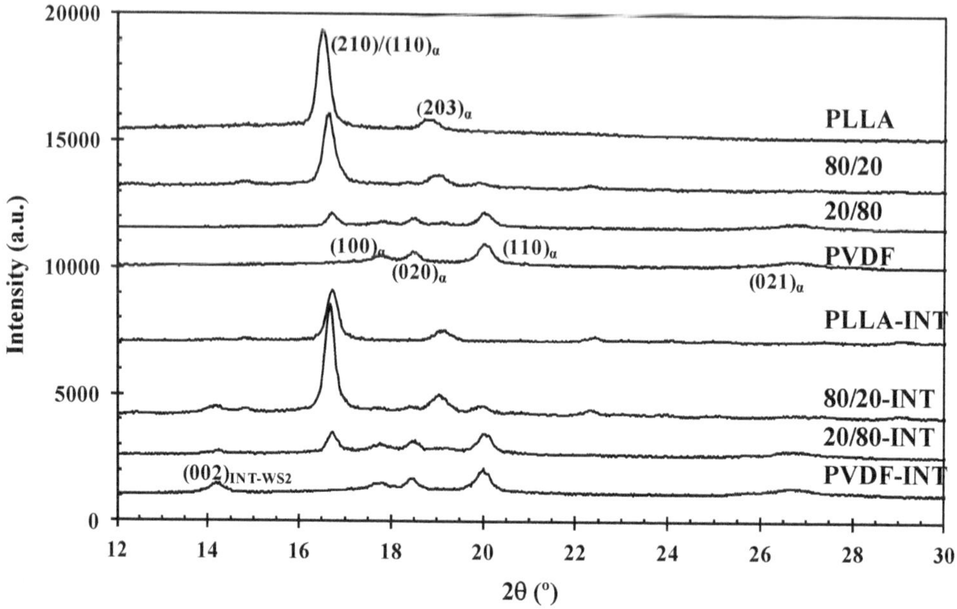

Figure 10. WAXS diffractograms of PLLA, PVDF, binary PLLA/PVDF, and ternary PLLA/PVDF/INT-WS_2 hybrid nanocomposites obtained at room temperature after dynamic crystallization from the melt.

Table 3. DMA parameters of different PLLA/PVDF/INT-WS_2 blend nanocomposites based on PLLA, PVDF, and INT-WS_2.

Material	$E'_{25°C}$ (GPa)	$T_{g, PVDF}$ (°C)	$T_{g, PLLA}$ (°C)
PLLA	3127	-	55
80/20	2274	-	54
60/40	2055	−36	57
40/60	1985	−37	54
20/80	1776	−38	50
PVDF	1560	−37	-
PLLA-INT	3640	-	54
80/20-INT	2886	−37	53
60/40-INT	2415	−35	53
40/60-INT	2293	−39	56
20/80-INT	2154	−37	-
PVDF-INT	1923	−37	-

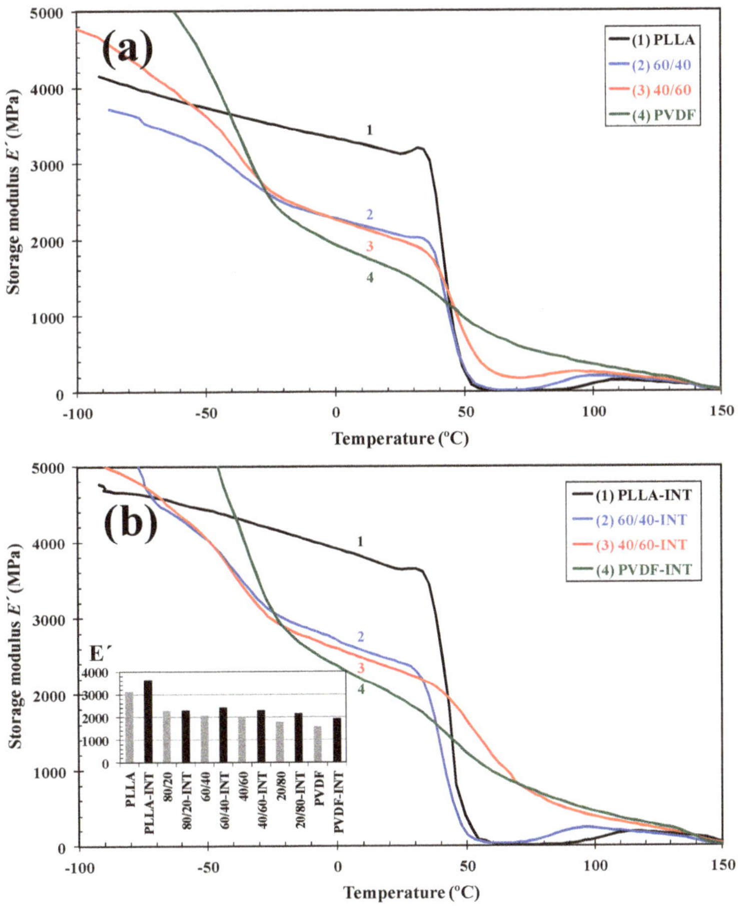

Figure 11. Evolution of the storage modulus (E') as a function of temperature for indicated (**a**) binary PLLA/PVDF and (**b**) ternary PLLA/PVDF/INT-WS_2 hybrid nanocomposites obtained in the tensile mode at 1 Hz; inset is the room temperature values of storage modulus (E') obtained for all binary and ternary hybrid nanocomposites.

From Figure 12a of the *tan δ* curves, it can be seen that the immiscible blends all presented individual T_g transitions at temperatures characteristic of the pure PVDF and PLLA polymers, though in the case of PLLA, the T_g peak decreased slightly and broadened, possibly related to its partial miscibility with PVDF at the interface between the two polymers or its nuclei at this location also increasing chain mobility. The mobility of the PLLA chain segments were improved with INT-WS_2 addition as can be seen with an increase in the *tan δ* peak of 60/40-INT (Figure 12b) compared to the same blend without nanofiller. The position (glass transition temperature) and height of the *tan δ* peak are associated with segmental mobility. The decrease in T_g means the enhancement of chain segment mobility and the increase in the height of the *tan δ* peak reveals the rise in segmental mobility. Both the decrease in T_g and the increase in the height of *tan δ* indicate that the blend nanocomposites had high mobility of chain segments.

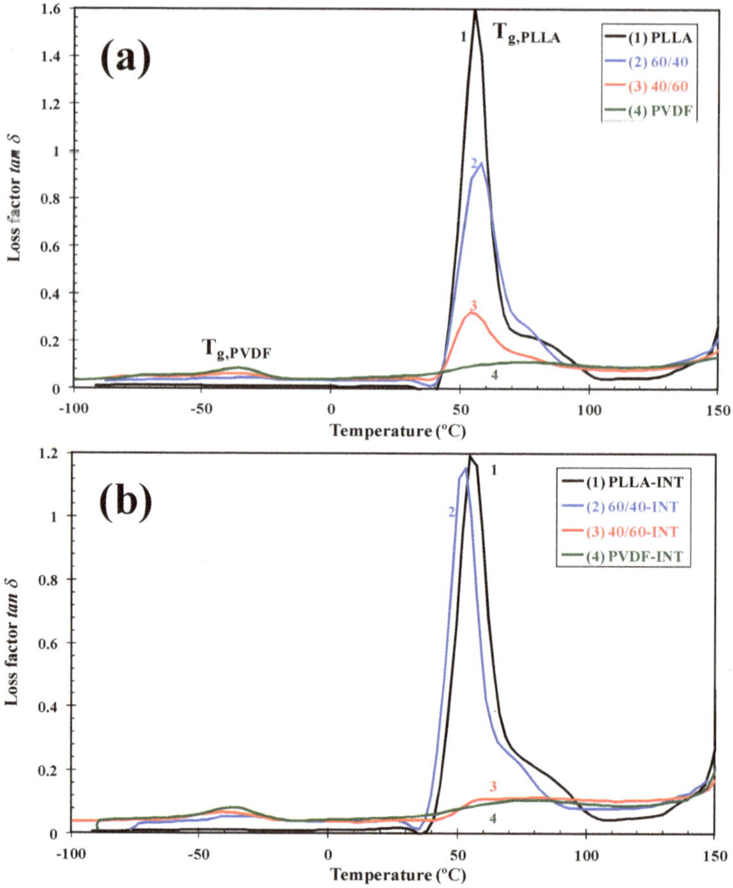

Figure 12. Evolution of the loss factor (*tan δ*) as a function of temperature for indicated (**a**) binary PLLA/PVDF and (**b**) ternary PLLA/PVDF/INT-WS_2 hybrid nanocomposites obtained in the tensile mode at 1 Hz.

4. Conclusions

This investigation provides evidence of the successful preparation and performance of new multifunctional biopolymer blend nanocomposites containing poly(L-lactic acid) (PLLA), poly(vinylidene fluoride)(PVDF), and 1D-TMDCs WS_2. The addition of PVDF and inorganic nanotubes was found to be efficient as an alternative route to produce advanced PLLA/PVDF blend nanocomposites processed via the widely used melt processing. The dispersion of the nanofiller from SEM was observed to be dispersed well and helped modify the blend interface morphology. The PVDF temperature of crystallization was higher and its rate faster than PLLA despite them having comparable melting points. Its effect on PLLA thermal stability was minimal, though it did increase the rate of PLLA crystallization. The incorporated nanofiller INT-WS_2 had a nucleating effect on both PLLA and PDVF, though it was more prominent on PLLA and PLLA-rich blends. This effect also led to higher enthalpies of crystallization and cold-crystallization omission, but with the final crystalline structure maintained. Finally, the nanocomposite blends with a high PLLA content demonstrated significant mechanical improvements compared to blends without filler, with the possible outlook of these materials being used as medical implants. The use of PLLA and PVDF ensured biocompatibility of the composite, and the presence of inorganic nanotubes additionally provided it with mechanical strength and processability.

Funding: This research was funded by FEDER/Ministerio de Ciencia, Innovación y Universidades-Agencia Estatal de Investigación/MAT2017-84691-P and Subprograma Estatal de Incorporación/IED2019-001134-I.

Institutional Review Board Statement: Not applicable.

Informed Consent Statement: Not applicable.

Data Availability Statement: The data presented in this study are available on request from the corresponding author.

Acknowledgments: The author would also like to acknowledge the assistance and support of Marían A. Gómez (ICTP-CSIC) for the realization of the projects and David Heras for the preparation of the samples.

Conflicts of Interest: The author declares no conflict of interest.

References

1. Drumright, R.E.; Gruber, P.R.; Henton, D.E. Polylactic acid technology. *Adv. Mater.* **2000**, *12*, 1841–1846. [CrossRef]
2. Sinclair, R.G. The case for the polylactic acid as a commodity packaging plastic. *Pure Appl. Chem. A* **1996**, *33*, 585–597. [CrossRef]
3. Lim, L.T.; Auras, R.; Rubino, M. Processing technologies for poly(lactic acid). *Prog. Polym. Sci.* **2008**, *33*, 820–852. [CrossRef]
4. Rhim, J.W.; Mohanty, A.K.; Singh, S.P.; Ng, P.K.W. Effect of the processing methods on the performance of polylactide films: Thermocompression versus solvent casting. *J. Appl. Polym. Sci.* **2006**, *101*, 3736–3742. [CrossRef]
5. Pan, P.; Shan, G.; Bao, Y. Enhanced nucleation and crystallization of poly(l-lactic acid) by immiscible blending with poly(vinylidene fluoride). *Ind. Eng. Chem. Res.* **2014**, *53*, 3148–3156. [CrossRef]
6. Anderson, K.S.; Hillmyer, M.A. The influence of block copolymer microstructure on the toughness of compatibilized polylactide/polyethylene blends. *Polymer* **2004**, *45*, 8809–8823. [CrossRef]
7. Girija, B.G.; Sailaja, R.R.N.; Madras, G. Thermal degradation and mechanical properties of PET blends. *Polymer* **2005**, *90*, 147–153. [CrossRef]
8. Reddy, N.; Nama, D.; Yang, Y. Polylactic acid/polypropylene polyblend fibers for better resistance to degradation. *Polym. Degrad. Stab.* **2008**, *93*, 233–241. [CrossRef]
9. Rasal, R.M.; Janrkor, A.V.; Hirt, D.E. Poly(lactic acid) modifications. *Prog. Polym. Sci.* **2010**, *35*, 338–356. [CrossRef]
10. Fryczkowski, R.; Fryczkowska, B.; Binias, W.; Janicki, J. Morphology of fibrous composites of PLA and PVDF. *Compos. Sci. Technol.* **2013**, *89*, 186–193. [CrossRef]
11. Mazzei, R.; Smolko, E.; Tadey, D.; Gizzi, L. Radiation grafting of NIPAAm on PVDF nuclear track membranes. *Nucl. Instrum. Methods B* **2000**, *170*, 419–426. [CrossRef]
12. Zheng, J.; He, A.; Li, J.; Han, C.C. Polymorphism control of poly(vinylidene fluoride) through electrospinning. *Macromol. Rapid Commun.* **2007**, *28*, 2159–2162. [CrossRef]
13. Rodrigues, M.T.; Gomes, M.E.; Mano, J.F.; Reis, R.L. PVDF membranes induce cellular proliferation and differentiation in static and dynamic conditions. *Mater. Sci. Forum* **2008**, *587*, 72–76. [CrossRef]
14. Chen, H.C.; Tsai, C.H.; Yang, M.C. Mechanical properties and biocompatibility of electrospun polylactide/poly(vinylidene fluoride) mats. *J. Polym. Res.* **2011**, *18*, 319–327. [CrossRef]
15. Xie, Q.; Ke, K.; Jiang, W.R.; Yang, W.; Liu, Z.Y.; Xie, B.H.; Yang, M.B. Role of poly(lactic acid) in the phase transition of poly(vinylidene fluoride) under uniaxial stretching. *J. Appl. Polym. Sci.* **2013**, *129*, 1686–1696. [CrossRef]
16. Zeng, J.B.; Li, K.A.; Du, A.K. Compatibilization strategies in poly(lactic acid)-based blends. *RSC Adv.* **2015**, *5*, 32546–32565. [CrossRef]
17. Mohd Nurazzi, N.; Asyraf, M.R.M.; Khalina, A.; Abdullah, N.; Sabaruddin, F.A.; Kamarudin, S.H.; Ahmad, S.; Mahat, A.M.; Lee, C.L.; Aisyah, H.A.; et al. Fabrication, Functionalization, and application of carbon nanotube-reinforced polymer composite: An Overview. *Polymers* **2021**, *13*, 1047. [CrossRef]
18. Yu, C.; Hu, K.; Yang, Q.; Wang, D.; Zhang, W.; Chen, G.; Kapyelata, C. Analysis of the storage stability property of carbon nanotube/recycled polyethylene-modified asphalt using molecular dynamics simulations. *Polymers* **2021**, *13*, 1658. [CrossRef] [PubMed]
19. Bertolino, V.; Cavallaro, G.; Milioto, S.; Lazzara, G. Polysaccharides/halloysite nanotubes for smart bionanocomposite materials. *Carbohydr. Polym.* **2020**, *245*, 116502. [CrossRef] [PubMed]
20. Lisuzzo, L.; Cavallaro, G.; Milioto, S.; Lazzara, G. Halloysite nanotubes coated by chitosan for the controlled release of khellin. *Polymers* **2020**, *12*, 1766. [CrossRef]
21. Tenne, R.; Redlich, M. Recent progress in the research of inorganic fullerene-like nanoparticles and inorganic nanotubes. *Chem. Soc. Rev.* **2010**, *39*, 1423–1434. [CrossRef]

22. Naffakh, M.; Díez-Pascual, A.M.; Marco, C.; Ellis, G.; Gómez-Fatou, M.A. Opportunities and challenges in the use of inorganic fullerene-like nanoparticles to produce advanced polymer nanocomposites. *Prog. Polym. Sci.* **2013**, *38*, 1163–1231. [CrossRef]
23. Naffakh, M.; Díez-Pascual, A.M. Thermoplastic polymer nanocomposites based on inorganic fullerene-like nanoparticles and inorganic nanotubes. *Inorganics* **2014**, *2*, 291–312. [CrossRef]
24. Rao, C.N.R.; Maitra, U.; Waghmare, U.V. Extraordinary attributes of 2-dimensional MoS_2 nanosheets. *Chem. Phys. Lett.* **2014**, *609*, 172–183. [CrossRef]
25. Naffakh, M.; Fernández, M.; Shuttleworth, P.S.; Garcia, A.M.; Moreno, D.A. Nanocomposite materials with poly(l-lactic acid) and transition-metal dichalcogenide nanosheets 2D-TMDCs WS_2. *Polymers* **2020**, *12*, 2699. [CrossRef] [PubMed]
26. Zak, A.; Sallacan Ecker, L.; Fleischer, N.; Tenne, R. Large-scale synthesis of WS_2 multiwall nanotubes and their dispersion, an Update. *Sens. Transducers. J.* **2011**, *12*, 1–10.
27. Lalwani, G.; Henslee, A.M.; Farshid, B.; Parmar, P.; Lin, L.; Qin, Y.X.; Kasper, F.K.; Mikos, A.G.; Sitharaman, B. Tungsten disulfide nanotubes reinforced biodegradable polymers for bone tissue engineering. *Acta Biomater.* **2013**, *9*, 8365–8373. [CrossRef]
28. Naffakh, M.; Marco, C.; Ellis, G.; Cohen, S.R.; Laikhtman, A.; Rapoport, L.; Zak, A. Novel poly(3-hydroxybutyrate) nanocomposites containing WS_2 inorganic nanotubes with improved thermal, mechanical and tribological properties. *Mater. Chem. Phys.* **2014**, *147*, 273–284. [CrossRef]
29. Naffakh, M.; Díez-Pascual, A.M. Nanocomposite biomaterials based on poly(etherether-ketone) (PEEK) and WS_2 inorganic nanotubes. *J. Mater. Chem. B* **2014**, *2*, 4509–4520. [CrossRef]
30. Naffakh, M.; Marco, C.; Ellis, G. Development of novel melt-processable biopolymer nanocomposites based on poly(L-lactic acid) and WS_2 inorganic nanotubes. *CrystEngComm* **2013**, *16*, 5062–5072. [CrossRef]
31. Pardo, M.; Shuster-Meiseles, T.; Levin-Zaidman, S.; Rudich, A.; Rudich, Y. Low cytotoxicity of inorganic nanotubes and fullerene-like nanostructures in human bronchial epithelial cells: Relation to inflammatory gene induction and antioxidant response. *Environ. Sci. Technol.* **2014**, *48*, 3457–3466. [CrossRef] [PubMed]
32. Goldman, E.B.; Zak, A.; Tenne, R.; Kartvelishvily, E.; Levin-Zaidman, S.; Neumann, Y.; Stiubea-Cohen, R.; Palmon, A.; Hovav, A.H.; Aframian, D.J. Biocompatibility of tungsten disulfide inorganic nanotubes and fullerene-like nanoparticles with salivary gland cells. *Tissue Eng. Part A* **2015**, *21*, 1013–1023. [CrossRef]
33. Naffakh, M.; Diez-Pascual, A.M.; Marco, C. Polymer blend nanocomposites based on poly(L-lactic acid), polypropylene and WS_2 inorganic nanotubes. *RSC Adv.* **2016**, *6*, 40033–40044. [CrossRef]
34. Wu, D.; Zhang, Y.; Zhang, M.; Yu, W. Selective Localization of multiwalled carbon nanotubes in poly(ε-caprolactone)/polylactide blend. *Biomacromolecules* **2009**, *10*, 417–424. [CrossRef] [PubMed]
35. Chen, G.X.; Chen, H.S.; Kim, E.S.; Yoon, J.S. Compatibilization-like effect of reactive organoclay on the poly(l-lactide)/poly(butylene succinate) blends. *Polymer* **2005**, *46*, 11829–11836. [CrossRef]
36. Odent, J.; Habibi, Y.; Raquez, J.M.; Dubois, P. Ultra-tough polylactide-based materials synergistically designed in the presence of rubbery e-caprolactone-based copolyester and silica nanoparticles. *Compos. Sci. Technol.* **2013**, *84*, 86–91. [CrossRef]
37. Monticelli, O.; Calabrese, M.; Gardella, L.; Fina, A.; Gioffredi, E. Silsesquioxanes: Novel compatibilizing agents for tuning the microstructure and properties of PLA/PCL immiscible blends. *Eur. Polym. J.* **2014**, *58*, 69–78. [CrossRef]
38. Mead, W.T.; Zacharidas, A.E.; Shimada, T.; Porter, R.S. Solid state extrusion of poly(vinylidene fluoride)—1: Ram and hydrostatic extrusion. *Macromolecules* **1979**, *12*, 473–478. [CrossRef]
39. Fischer, E.W.; Sterzel, H.J.; Wegner, G. Investigation of the structure of solution grown crystals of lactide copolymers by means of chemical reactions. *Kolloid-Zeitschrift und Zeitschrift für Polymere* **1973**, *251*, 980–990. [CrossRef]
40. Saeidou, S.; Huneault, M.A.; Li, H.; Park, C.B. Poly(lactic acid) crystallization. *Prog. Polym. Sci.* **2012**, *37*, 1657–1677. [CrossRef]
41. Pan, P.J.; Kai, W.H.; Zhu, B.; Dong, T.; Inoue, Y. Polymorphic transition in disordered poly(l-lactide) crystals induced by annealing at elevated temperatures. *Macromolecules* **2008**, *41*, 4296–4304. [CrossRef]
42. Hasegawa, R.; Takahashi, Y.; Chatani, Y.; Tadokoro, H. Crystal Structures of three crystalline forms of poly(vinyldene fluoride). *Polym. J.* **1972**, *3*, 600–610. [CrossRef]
43. Naegele, D.; Yoon, D.Y.; Broadhurst, M.G. Formation of a new crystal form (α_p) of poly(vinylidene fluoride) under electric field. *Macromolecules* **1978**, *11*, 1297–1298. [CrossRef]

polymers

MDPI

Article

Role of Organically-Modified Zn-Ti Layered Double Hydroxides in Poly(Butylene Succinate-Co-Adipate) Composites: Enhanced Material Properties and Photodegradation Protection

Jie-Mao Wang, Hao Wang, Erh-Chiang Chen, Yun-Ju Chen and Tzong-Ming Wu *

Department of Materials Science and Engineering, National Chung Hsing University, 250 Kuo Kuang Road, Taichung 402, Taiwan; D9866023@mail.nchu.edu.tw (J.-M.W.); eddie115923@icloud.com (H.W.); erchiang.chen@gmail.com (E.-C.C.); bitcoin2243@gmail.com (Y.-J.C.)
* Correspondence: tmwu@dragon.nchu.edu.tw

Abstract: In this research, the effects of Zn-Ti layered double hydroxide (Zn-Ti LDH) as a UV-protection additive, which was added to the poly(butylene succinate-co-adipate) (PBSA) matrix, were investigated. Stearic acid was used to increase the hydrophobicity of Zn-Ti LDH via ion-exchange method. Transmission electron microscopy images of PBSA composites showed that modified Zn-Ti LDH (m-LDH) well-dispersed in the polymer matrix. Due to the effect of heterogeneous nucleation, the crystallization temperature of the composite increased to 52.9 °C, and the accompanying crystallinity increased to 31.0% with the addition of 1 wt% m-LDH. The additional m-LDH into PBSA copolymer matrix significantly enhanced the storage modulus, as compared to pure PBSA. Gel permeation chromatography and Fourier transform infrared spectroscopy analysis confirmed that the addition of m-LDH can reduce the photodegradation of PBSA.

Keywords: photodegradation; biodegradable polymer; poly(butylene succinate-co-adipate); Zn-Ti LDH

Citation: Wang, J.-M.; Wang, H.; Chen, E.-C.; Chen, Y.-J.; Wu, T.-M. Role of Organically-Modified Zn-Ti Layered Double Hydroxides in Poly(Butylene Succinate-Co-Adipate) Composites: Enhanced Material Properties and Photodegradation Protection. *Polymers* **2021**, *13*, 2181. https://doi.org/10.3390/polym13132181

Academic Editors: José Miguel Ferri, Vicent Fombuena Borràs and Miguel Fernando Aldás Carrasco

Received: 16 June 2021
Accepted: 29 June 2021
Published: 30 June 2021

1. Introduction

Using biodegradable polymers as substitutes for common plastics has attracted immense interest in reducing the environmental impact of plastic waste [1,2]. Nanomaterial additives is one of the effective methods to enhance the mechanical, thermal, and crystallization properties of biodegradable polymers [3,4]. Many studies have been conducted on the effect of nanomaterials on the degradation behavior of biodegradable polymers in the soil, compost, or simulated environments in the laboratory [5–7]. However, the impacts of light, water, and temperature on the degradability of polymers in the natural environment require further investigation. These factors usually deteriorate the properties of biodegradable polymers and affect their lifetime. For polymeric materials that are often exposed to outdoors, sunlight is the main cause of photodegradation and performance loss [8,9]. The photodegradation behavior of several biodegradable polymers, including poly(l-lactide) (PLA), poly(butylene adipate-co-terephthalate) (PBAT), and poly(butylene succinate-co-adipate) (PBSA) have been reported [10–12]. The photodegradation feature of the polymer itself and the influence of nanomaterials as fillers have received extensive attention to increase the lifetime of biodegradable polymers under sunlight exposure. Chen et al. prepared a nanocomposite by PBAT and clay, and showed that inorganic particles can absorb or reflect photon energy and reduce the intensity of the light, inhibiting the photodegradation of the polymer [13]. Zhang et al. discussed the effect of ZnO on the photodegradation of PBSA [14] and showed that ZnO hinders the photodegradation of PBSA, but does not significantly change the photodegradation mechanism of PBSA. The ultraviolet light with high energy UV-C (wavelengths 100–280 nm) is absorbed by the earth's atmosphere while allowing the transmission of UV-B (280–320 nm) and UV-A (320–400 nm) [15]. Therefore, UV-B and UV-A contribute to the photodegradation of most polymers that are used outdoors. To reduce the photodegradation of biodegradable polymers, the UV-B and UV-A absorption capacity of nanomaterials should be further investigated [9,16].

The layered double hydroxide (LDH) conventionally prepared by bivalent and trivalent cations has been well-known for its ability to affect mechanical, crystallization, thermal, and biodegradation properties of biodegradable polymers [17,18]. The general formula for these LDHs is $(M^{2+}{}_{1-x}M^{3+}{}_{x}(OH)_2)(A^{n-})_{x/n} \cdot mH_2O$, where M^{2+} and M^{3+} are divalent (Mg^{2+}, Zn^{2+}, Cu^{2+}) and trivalent (Al^{3+}, Cr^{3+}, Fe^{3+}) cations, and A^{n-} represents interlamellar anions [19]. The distance between two adjacent layers, which depends mainly on the nature of the interlayer species and their electrostatic interactions with the main layers, can be adjusted by introducing anionic compounds into the interlayer by ion exchange to replace its native anions [20]. The reports also pointed out that by modifying the LDH with a hydrophobic aliphatic carbon chain, the compatibility between originally hydrophilic LDH and the hydrophobic polymer can be improved [21]. The eco-friendliness and biocompatibility of LDHs have been demonstrated [22]. Zn-Ti LDH, which consists of bivalent (Zn^{2+}) and tetravalent (Ti^{4+}) cations, was developed by Saber et al. [23]. Compared to LDHs with other metal ions (e.g., Mg-Al LDH and Zn-Al LDH), Zn-Ti LDH can provide better protection in broadband UV [24]. Wang et al. also found that Zn-Ti LDH is a safe UV-shielding material as it has lower photocatalytic activity than TiO_2 and ZnO [15]. Ekambaram et al. also gave same discussion and indicated its ability to shield UV radiation [25]. In the study of Naseem et al., Zn-Ti LDH as a UV-absorbing nanomaterial provided a method for protecting polypropylene from UV-vis degradation [26].

PBSA is an aliphatic biodegradable copolyester, which synthesized via polycondensation of 1,4-butanediol in the presence of succinic and adipic acids [27]. It is worth noting that 1,4-butanediol and succinic acid can not only be extracted from oil, but also via fermentation [28]. The photodegradation reaction of PBSA produces carboxyl end groups (C=O) and chain scission. At the same time, the peak of C=O in the FTIR spectrum can be used to study the evolution of photodegradation. [14,29]. Due to the appropriate degradation rate, thermal stability, mechanical property, and good processability, practical applications of PBSA can be found in mulch films, where the photodegradation stability is a crucial property. The characteristics of Zn-Ti LDH, such as excellent biocompatibility, broadband UV protection, and lower photocatalytic activity, makes it suitable as a UV-protection additive in the PBSA matrix [14,29]. Reviewing past literature regarding PBSA nanocomposites, many studies focus on the addition of nanomaterial and its modification to achieve a better dispersion in the PBSA matrix [5,17,18]. Therefore, to improve the dispersion of Zn-Ti LDHs in PBSA, biocompatible and nontoxic stearic acid (SA) was used to modify in this study [30]. The changes in the physical properties of PBSA composites at various photodegradation periods were investigated. The crystallization, rheology, and thermal and mechanical properties of the PBSA composites were evaluated for their practical applications.

2. Materials and Methods

2.1. Materials

Commercial PBSA was purchased from Mitsubishi Chemical Co. (Tokyo, Japan), under the trade name BioPBS™ FD92PM. Zinc nitrate hexahydrate ($Zn(NO_3)_2{\cdot}6H_2O$), titanium tetrachloride ($TiCl_4$), urea (CH_4N_2O), sodium hydroxide (NaOH), and stearic acid (SA, $C_{18}H_{38}O_2$) were acquired from Sigma-Aldrich (St. Louis, MO, USA). All chemicals were used without further purification.

2.2. Synthesis and Modification of m-LDH

Zn-Ti LDH was synthesized by the co-precipitation method. $TiCl_4$ (0.44 mL), $Zn(NO_3)_2{\cdot}6H_2O$ (4.76 g) and urea (6.0 g) were dissolved in deionized water, the mixture was stirring at 95 °C for 48 h under nitrogen atmosphere. After preparation, the obtained insoluble (i.e., Zn-Ti LDHs) was washed by deionized water and ethanol, eventually dried at 80 °C in vacuum 24 h. The SA modification of Zn-Ti LDH was used by ion-exchange method. SA (0.2 M) and Zn-Ti LDH (0.5 g) were added into the deionized water. The solution controlled at pH = 10 to 11 by drop 0.1 M aqueous NaOH and stirring at 80 °C for 24 h under nitrogen atmosphere. The produced m-LDH was washed by deionized water and ethanol several times, then dried at 50 °C in vacuum 24 h.

2.3. PBSA/LDH Composites Preparation

PBSA was dissolved in dichloromethane (Sigma-Aldrich) for 2 h. Different amount of m-LDH were added into dichloromethane and ultrasonicated for 10 mins to reach dispersions. After that, the m-LDH solution was added into PBSA solution slowly and stirring for 1 h. The mix solution was solvent casting onto a glass petri dish at room temperature for 24 h then dried in vacuum at 50 °C for 48 h. The obtained composites are identified as PBSA/m-LDH-x, where x is the wt% of m-LDH in composite. For further characterization, samples were hot-pressed to plate at 120 °C then cooling at room temperature.

2.4. Artificial Photodegradation Test

The samples were irradiated under artificial conditions with an artificial light source (Philips CLEO HPA 400S, Amsterdam, Netherlands), which has a mainly radiation between 300 and 400 nm. The sample dimension is 10 mm × 10 mm × 1 mm (L × W × T) prepared by hot-press. The temperature of sample surface is about 45 °C. The relative humidity of the environment is about 50%.

2.5. Characterization

The TEM images of PBSA composites were performed using JEM-2010 (JEOL, Tokyo, Japan). The scanning electron microscopy (SEM) images of PBSA composites were performed using JSM-6700F (JEOL, Tokyo, Japan). The contact angle was carried out on FTA200 (First Ten Angstroms, Portsmouth, VA, USA). FTIR experiments of Zn-Ti LDH, SA, and m-LDH were carried out on a spectrometer (Spectrum One, Perkin-Elmer, Waltham, MA, USA) in the range from 450 to 4000 cm^{-1}. The PBSA and its composites were performed in attenuated total reflection (ATR) mode. The UV absorption spectrum was acquired by a U-3900 UV–vis spectrophotometer (Hitachi, Tokyo, Japan) in the range of wavelength from 250 to 400 nm. X-ray diffractometer (Bruker D8, Karlsruhe, Germany) equipped with a Ni-filtered Cu Kα radiation was used for the experiments of wide-angle X-ray diffraction (WAXD). The measurements of WAXD were carried out in the range of 2θ = 1–40° at scanning rate of 1°/min. The thermal degradation of specimens was operated using thermal gravimetric analyzer (TGA 2950, TA Instruments, New Castle, DE, USA). The experiment was carried out from room temperature to 600 °C under atmospheric environment at a heating rate of 10 °C/min. The crystallization behavior was carried out by a Pyris Diamond DSC (Perkin-Elmer, Waltham, MA, USA) and all experiments were performed under nitrogen environment. All specimens were heated to the designed temperatures (T_{ds}), which were about 40 °C higher than the melting temperatures of neat PBSA, at a rate of 10 °C/min and held for 5 min. Then, they were cooled to −30 °C at a rate of 10 °C/min, which was called 1st cooling. Finally the samples were heated to T_{ds} at a rate of 10 °C/min, which was called 2nd heating. The crystallinity degree (X_c) of neat PBSA and the composites was obtained according to enthalpy of the melting peak (ΔH_f) of 2nd heating. The storage modulus (E′) was evaluated by DMA8000 (Perkin-Elmer, MA, USA) from −80 to 40 °C at 2 °C/min heating rate and 1 Hz constant frequency. Sample size is 20 mm × 10 mm × 0.5 mm (L × W × T). The experiment was performed in atmospheric environment. Molecular weights of the samples were determined by GPC (LC-4000, JASCO, Tokyo, Japan) with a refractive index detector (RI-4030, JASCO, Tokyo, Japan) calibrated with standard polystyrene. Dichloromethane was used as the mobile phase with a 1 mL/min flow rate.

3. Results and Discussion

3.1. Characterization of m-LDH

The FTIR spectra of Zn-Ti LDH, m-LDH, and SA are shown in Figure 1a. A broad absorption band between 3200 and 3400 cm^{-1} of Zn-Ti LDH indicates the stretching mode of hydroxyl groups and physisorbed water in the interlayer of LDH. The bands at 1508, 1388, and 1047 cm^{-1} are attributed to the interlayer CO_3^{2-}. In addition, the bands below 1000 cm^{-1} are metal-oxygen (MO, O-M-O, or M-O-M) signals [15,23,24]. After the chemical modification with SA, the absorption bands at 2920, 2850, and 1462 cm^{-1} are related to the CH_2 asymmetric vibrations, symmetric vibrations, and scissor mode of SA, respectively [31]. Notably, the bands attributed to the CO_3^{2-} of Zn-Ti LDH and the band at 1701 cm^{-1} attributed to the COOH group of SA disappeared or became weaker in the spectrum of m-LDH. At the same time, the clear absorption band at 1596 cm^{-1} of m-LDH is due to COO^- stretching vibrations, indicating that the SA was converted to stearate anion (from $H_3C(CH_2)_{16}COOH$ to $H_3C(CH_2)_{16}COO^-$) [23]. Figure 1b,c give the photographs of water droplets on the surface of Zn-Ti LDH and m-LDH, respectively. The wettability of Zn-Ti LDH changed with SA modification. Compared to Zn-Ti LDH, a larger water contact angle of m-LDH means lower hydrophilicity, which may result in a better dispersion in the PBSA matrix.

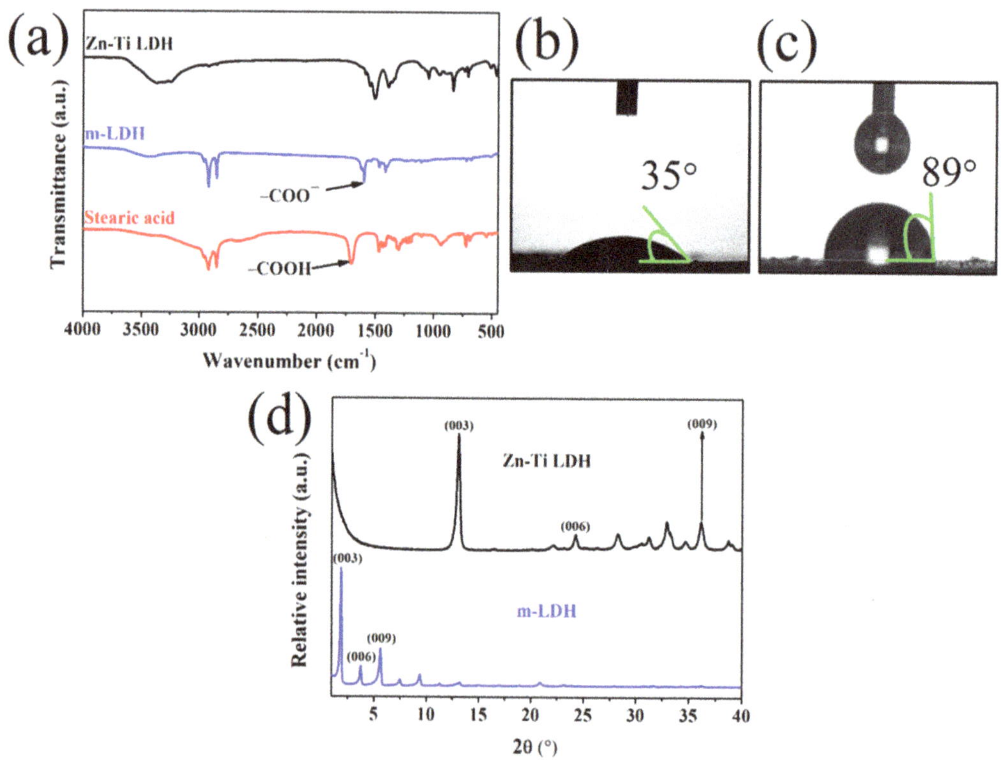

Figure 1. (**a**) The FTIR result of Zn-Ti LDH, m-LDH, and SA; the photographs of water contact angle on the surface of (**b**) Zn-Ti LDH and (**c**) m-LDH; (**d**) XRD result of Zn-Ti LDH and m-LDH.

Figure 1d shows the XRD patterns of Zn-Ti LDH and m-LDH. The diffraction peaks of Zn-Ti LDH at $2\theta = 13.05°, 24.33°, 28.23°, 32.90°, 33.28°, 36.14°$, and $38.70°$ correspond to the crystal planes of (003), (006), (012), (100), (101), (009), and (015), respectively [23,32]. The d-spacing of (003), (006), and (009) calculated by Bragg's law are 0.68, 0.37, and 0.25 nm, respectively. After chemical modification using SA, the diffraction peaks of (003), (006), and (009) planes of LDH shifted to smaller angles at 2θ = 1.88°, 3.73°, and 5.64°, respectively. The d-spacing corresponding to (003), (006), and (009) planes of m-LDH increases to 4.69, 2.37, and 1.57 nm, respectively. The chemical modification can expand the interlayer distance of LDH, indicating that the ion exchange of SA was successful and consistent with the above FTIR results. In addition, the chain length of the stearate anion is about 2.25 nm [23]. From the increase in d-spacing from 0.68 nm of ZN-Ti LDH to 4.69 nm of m-LDH, we deduce that the intercalation of SA formed bilayer structures with an inclined angle of 63°.

The TGA curves of Zn-Ti LDH, m-LDH, and SA performed in the air environment are shown in Figure 2a. The 10% weight loss and Char yield weight percent at 600 °C of Zn-Ti LDH, m-LDH, and SA are shown in Table A1. The main thermal decomposition of Zn-Ti LDH occurs from 200 to 300 °C, which corresponded to the evaporation of carbonate anions [23]. After chemical modification using SA, there is no rapidly weight drop as same as Zn-Ti LDH at the temperature below 300 °C, and the residual weight of m-LDH is 46.9% at 600 °C. Combining the analysis results of FTIR, XRD and TGA, it can be seen that SA had been inserted into the Zn-Ti LDH laminates by ion exchange method. It can be seen from the change of contact angle that SA was also adsorbed on the surface of m-LDH.

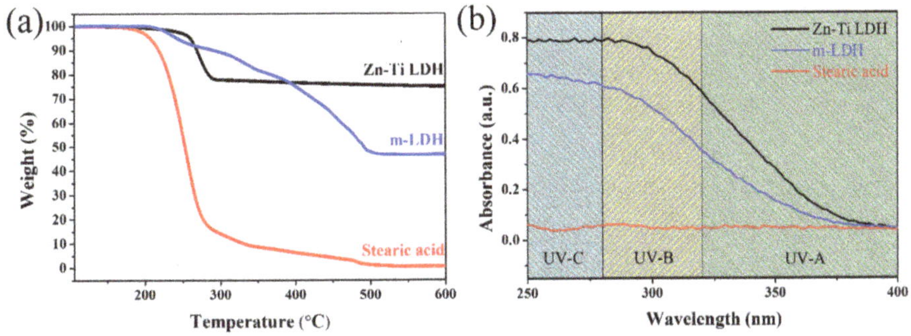

Figure 2. (**a**) TGA and (**b**) UV-vis result of Zn-Ti LDH, m-LDH, and SA.

The UV–vis absorbance spectrum of Zn-Ti LDH, SA, and m-LDH are shown in Figure 2b. An absorption signal from 250 to 370 nm is observed for Zn-Ti LDH, which includes whole UV-B and most of the UV-A irradiation range. Simultaneously, no significant UV-absorbing character of SA is observed from 250 to 400 nm. After modification, the absorption intensity of m-LDH is lower than Zn-Ti LDH. However, the m-LDH still shows a significant absorption signal from 250 to 370 nm, which indicates a good application potential in UV irradiation protecting.

3.2. *Characterization of PBSA and Its Composites*

Figure 3a shows the TEM images of PBSA/m-LDH-5. It can be seen that the stacking layers of m-LDH dispersed in PBSA matrix. Figure 3b shows the XRD diffraction patterns of PBSA and its composites. The XRD experimental data of PBSA shows three diffraction peaks at 2θ = 19.4°, 21.7°, and 22.4°, corresponding to (020), (021), and (110) crystallographic planes of monoclinic PBS, respectively [17,18]. The XRD results show that the addition of m-LDH did not change the crystal structure of PBSA. In addition, by increasing m-LDH content, the peak positions of (003), (006), and (009) planes for m-LDH remained unchanged, and the intensities of these peaks increased. These results indicate that the initial layer stacking structure of m-LDH is still present in PBSA.

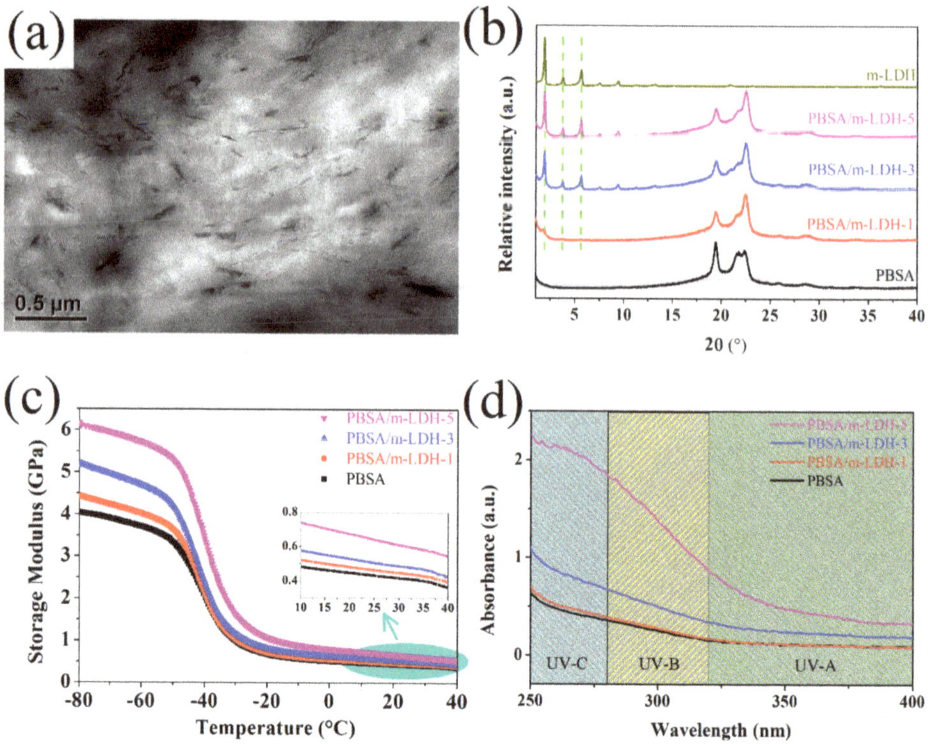

Figure 3. TEM image of (**a**) PBSA/m-LDH-5 and (**b**) XRD, (**c**) DMA, and (**d**) UV-vis result of PBSA and its composites.

The mechanical properties, represented by the storage modulus (E′), of the PBSA and its composites evaluated by DMA are shown in Figure 3c. The phenomenon of E′ dropping rapidly at approximately −40 °C contributes to the occurrence of glass transition of PBSA [27]. The effect of m-LDH on the mechanical properties of PBSA can be seen from E′ at the temperature below T_g. The value of E′ at −80 °C increases significantly with increasing filler content. Detailed data are presented in Table 1. The enhancement of mechanical properties for the PBSA composites may be attributed to the reinforcement effect of the additional stiffness of LDH [5]. In addition, the PBSA/m-LDH-5 composites exhibit the highest mechanical properties at 25 °C, as shown in Figure 3c and Table 1.

Table 1. Result of DMA, and DSC analysis for PBSA and its composites.

Sample Name	E′ (Gpa)	E′ (Gpa)	T_c (°C)	T_m (°C)	ΔH_f (J/g)	X_c (%)
PBSA	4.04	0.43	44.3	87.1	32.1	27.4
PBSA/m-LDH-1	4.42	0.46	52.9	74.0/87.4	36.3	31
PBSA/m-LDH-3	5.19	0.5	50.4	74.3/86.9	34.6	29.5
PBSA/m-LDH-5	6.16	0.63	48.8	73.3/87.1	33.5	28.6

E'_{-80} and E'_{25} : storage modulus at −80 and 25 °C, respectively, measurement by DMA. T_c: crystallization temperature during 1st cooling trace, measurement by DSC. T_m: crystalline melting temperature during 2nd heating trace, measurement by DSC. X_c: crystallinity, which obtained by the following equation, $Xc(\%) = \Delta H_f / \left[(1-\varphi)\Delta H_f^0\right] \times 100\%$, where $\Delta H_f^0 = 117.2\ \mathrm{Jg^{-1}}$ for PBSA with 72% BS group ratio, and φ is the weight fraction of the filler in the composites.

The UV–vis absorbance spectrum of PBSA and its composites are shown in Figure 3d. A relatively weak absorption of PBSA was observed in the range from 250 to 320 nm, which is attributed to the absorption of the carbonyl group [14]. By adding m-LDH into the PBSA polymer matrix, the composites show an improvement in UV-absorbing character in the whole UV-B and UV-A region. The absorption intensity of PBSA composites increases with increasing m-LDH content. This result indicates that excellent UV absorption property of m-LDH can be used as a UV protecting additive for PBSA.

The effect of the chemical modification of nanofiller on polymer composite has been reported by Zhang et al., which shows an increasing T_c with modified nanofiller but a decreasing T_c by an unmodified nanofiller [33]. In this study, the crystallization behavior of PBSA and its composites were analyzed using 1st cooling and 2nd heating of DSC measurement. The crystallization temperature (T_c) of PBSA/m-LDH-1 shown in Figure 4a is higher than that of PBSA. With increasing m-LDH content, T_c gradually decreased but was still higher than that of PBSA. The detailed data are presented in Table 1. The increase in T_c is due to the heterogeneous nucleation caused by m-LDH. Further, increasing m-LDH content may hinder the chain motion of PBSA during crystallization, thus, leading to a decrease in T_c [3,20].

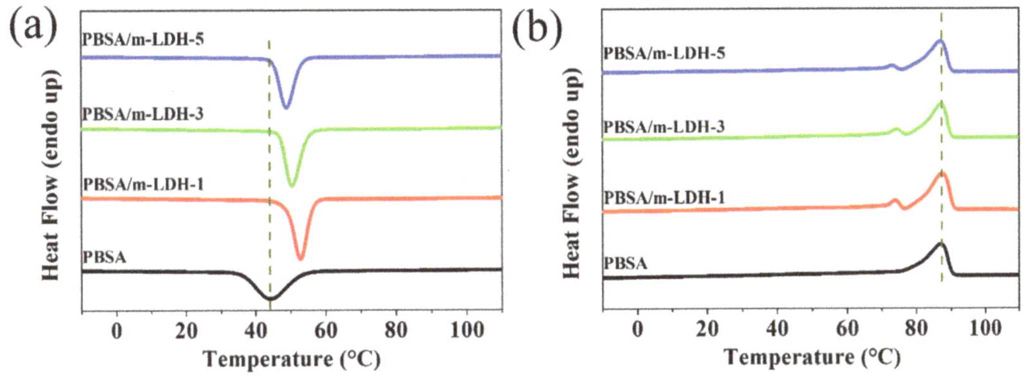

Figure 4. (**a**) DSC 1st cooling and (**b**) 2nd heating curves of PBSA and its composites.

Figure 4b shows the melting behavior in the 2nd heating of PBSA and its composites. The melting temperatures (T_m) of PBSA and its composites are almost the same in all samples, showing that the melting behavior of the PBSA crystallite was not affected by adding fillers. The detailed data are presented in Table 1. An obvious difference observed in the composites with m-LDH revealed a small melting peak appearing before the main peak of samples. These small melting peaks might be attributed to the melt–recrystallization–remelt phenomenon of the PBSA crystallite. The crystallinity of polymer can be calculated by dividing the observed ΔH_f by the theoretical value (ΔH_f^0) for perfectly (100%) crystalline polymer. The theoretical ΔH_f^0 for polybutylene succinate (PBS) and polybutylene adipate (PBA) are 110.3 and 135.0 J/g, respectively. The theoretical ΔH_f^0 for PBSA can be calculated via the basis of the butylene succinate (BS)/butylene adipate (BA) group contribution method [34]. In this study, the PBSA material has 72% BS composition (determined via ^{1}H nuclear magnetic resonance, shown in Figure A1). The crystallinity of all samples are presented in Table 1. By adding m-LDH into PBSA, the crystallinity increases from 27.4% for PBSA to 31.0, 29.5, and 28.6% for PBSA/m-LDH-1, PBSA/m-LDH-3, and PBSA/m-LDH-5, respectively. The increase in crystallinity might be due to the effect of m-LDH via heterogeneous nucleation. Compared to PBSA, the higher crystallization temperature of PBSA/m-LDH-1 leads to better chain motion during crystal growth at previous cooling process. Higher m-LDH content hinders chain motion during crystallization, which leads to the crystallinity decrease with increasing m-LDH content. At the same time, the lower crystallization temperature compared to PBSA/m-LDH-1 is not conducive to chain motion. Higher m-LDH content hinders chain motion during crystallization, which leads to a decrease in crystallinity with increasing m-LDH content.

3.3. Characterization of PBSA/m-LDH Composites after Irradiation

The photodegradation caused by UV could induce the polymer chain scission, which supports the change in molecular weight with increasing irradiation time [14]. Figure 5a shows the change in number average molecular weight (Mn) for PBSA and PBSA/m-LDH composites after a period of irradiation time. These results demonstrate that the photodegradation causes a remarkable reduction in molecular weight in all samples, but the addition of higher m-LDH content could reduce the degradation rates. The result indicates that m-LDH can play a significant role in the photodegradation protection. In addition, the molecular weight of each sample drops sharply in the first week. To understand the difference in the degradation rate of samples on this relatively short irradiation time, the FTIR analysis was applied. As a result of hydroxyl end group oxidation and main chain scission from photolysis at ester linkages, the terminal carboxyl groups are generated. Thus, as irradiation progresses, an increase in C=O peaks is observed in the FTIR spectrum. Therefore, the carbonyl index is defined as [14,29]:

$$\left(\frac{A^{t}_{C=O}}{A^{t}_{C-H}}\right)/\left(\frac{A^{t0}_{C=O}}{A^{t0}_{C-H}}\right) \quad (1)$$

where $A^{t0}_{C=O}$ and $A^{t}_{C=O}$ are the intensity of carboxyl groups at 1712 cm^{-1} before and after irradiation, respectively; A^{t0}_{C-H} and A^{t}_{C-H} are assigned to the C–H stretching peak at 2858 cm^{-1}, which is used as the reference for calculating the value of the carbonyl index, before and after irradiation, respectively [14,28]. Therefore, the higher carbonyl index indicates the poor photostability of materials. The evolution of C=O at 1712 cm^{-1} of PBSA and PBSA/m-LDH composites are shown in Figure A2. Figure 5b shows the carbonyl index of the corresponding samples. The increase of m-LDH content could remarkably reduce the carbonyl index in different irradiation time, which indicates decreased photodegradation of PBSA.

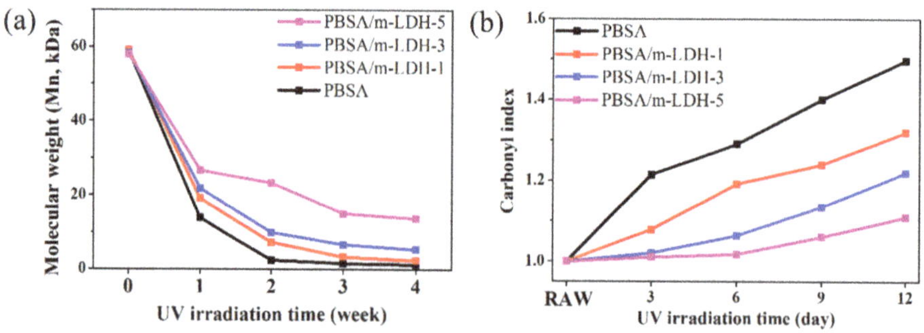

Figure 5. (**a**) Molecular weight (Mn) and (**b**) carbonyl index of PBSA and PBSA/m-LDH composites after different irradiation time.

The result of samples after artificial photodegradation test shows the m-LDH is an effective nanomaterial to reduce the photodegradation of PBSA. Based on past literature, Zn-Ti LDH has been shown to have UV absorption ability. This study also showed that Zn-Ti LDH keeps this feature after SA modification. The photodegradation of polymer starts from the surface and then develops along the depth [14]. In addition, destruction via photodegradation can induce the entry of oxygen and promote further degradation. For pure PBSA, UV light could enter the inside of the material without additional hindrance, causing the above degradation reaction. For PBSA/m-LDH composites, UV light can be absorbed by m-LDH, which might decreased the photo intensity enter the inside of the material, causing the less degradation reaction. The morphologies of PBSA and PBSA/m-LDH-5 after different UV irradiation time are shown in Figure 6. Prior to irradiation, both of them exhibited a smooth surface with no significant defects. After irradiation, the morphology became rough and was characterized with cracks. After 4 weeks, PBSA showed a rougher surface, indicating a stronger photodegradation behavior than PBSA/m-LDH-5.

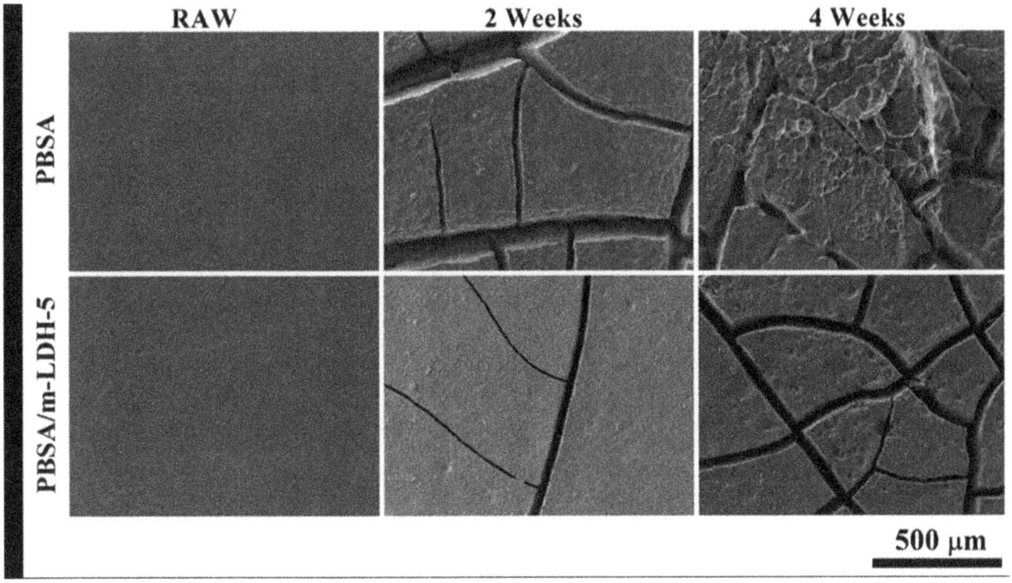

Figure 6. SEM images of surface morphology of PBSA and PBSA/m-LDH-5 after different irradiation time.

4. Conclusions

The hydrophobic UV absorption nanofiller was prepared by SA-modified Zn-Ti LDH. The dispersed m-LDH in PBSA matrix shows its stacking structure via WAXD analysis. While not affecting the crystal structure of PBSA, both the crystallization temperature and crystallinity of PBSA/m-LDH composites were increased due to the addition of m-LDH with heterogeneous nucleation. The composite with 1 wt% m-LDH content had the highest T_c and crystallinity, which are 52.9 °C and 31.0%, respectively. In addition, the mechanical properties of the composite were enhanced with increasing m-LDH content. By adding 5 wt% m-LDH into PBSA, the storage modulus at −80 °C increased from 4.04 GPa of neat PBSA matrix to 6.16 GPa of composites. The m-LDH decreased the photodegradation behavior of PBSA, which was confirmed by GPC, FTIR, and SEM analysis. This may attribute to m-LDH absorbs the partial incident light entering the composite and reduces the effect of UV radiation. Based on the fact that both Zn-Ti LDH, SA, and PBSA are biocompatible and have low toxicity, the biofriendly composites with photodegradation-resistant properties could be deployed for suitable applications, such as biodegradable mulching films.

Author Contributions: Conceptualization, J.-M.W. and T.-M.W.; methodology, T.-M.W.; software, E.-C.C. and J.-M.W.; validation, H.W. and J.-M.W.; formal analysis, J.-M.W. and H.W.; investigation, Y.-J.C. and E.-C.C.; resources, E.-C.C.; data curation, Y.-J.C. and J.-M.W.; writing—original draft preparation, Y.-J.C. and J.-M.W.; writing—review and editing, T.-M.W.; visualization, T.-M.W.; supervision, T.-M.W.; project administration, T.-M.W.; funding acquisition, T.-M.W. All authors have read and agreed to the published version of the manuscript.

Funding: Not applicable.

Institutional Review Board Statement: Not applicable.

Informed Consent Statement: Not applicable.

Data Availability Statement: Not applicable.

Acknowledgments: The financial support of this work is provided by the Ministry of Science and Technology (MOST) under Grand MOST 109-2221-E-005-069-NY3 and the Ministry of Education under the project of Innovation and Development Center of Sustainable Agriculture (IDCSA). The authors would like to thank the Instrument Center of National Chung Hsing University for help with measurements of Transmission electron microscopy.

Conflicts of Interest: The authors declare no conflict of interest. The funders had no role in the design of the study; in the collection, analyses, or interpretation of data; in the writing of the manuscript, or in the decision to publish the results.

Appendix A

Table A1. Results of TGA analysis for Zn-Ti LDH, m-LDH and stearic acid.

Sample Name	10 wt% Loss Temperature (°C)	Char Yield at 600 °C (wt%)
Zn-Ti LDH	269.2	75.0
m-LDH	292.0	46.9
stearic acid	218.7	0.5

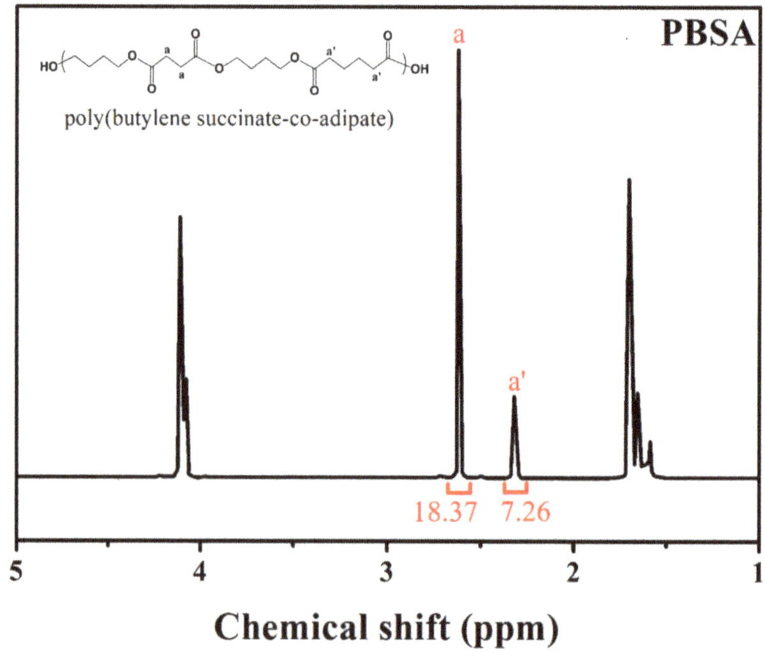

Figure A1. ^{1}H nuclear magnetic resonance (H-NMR) spectra of PBSA.

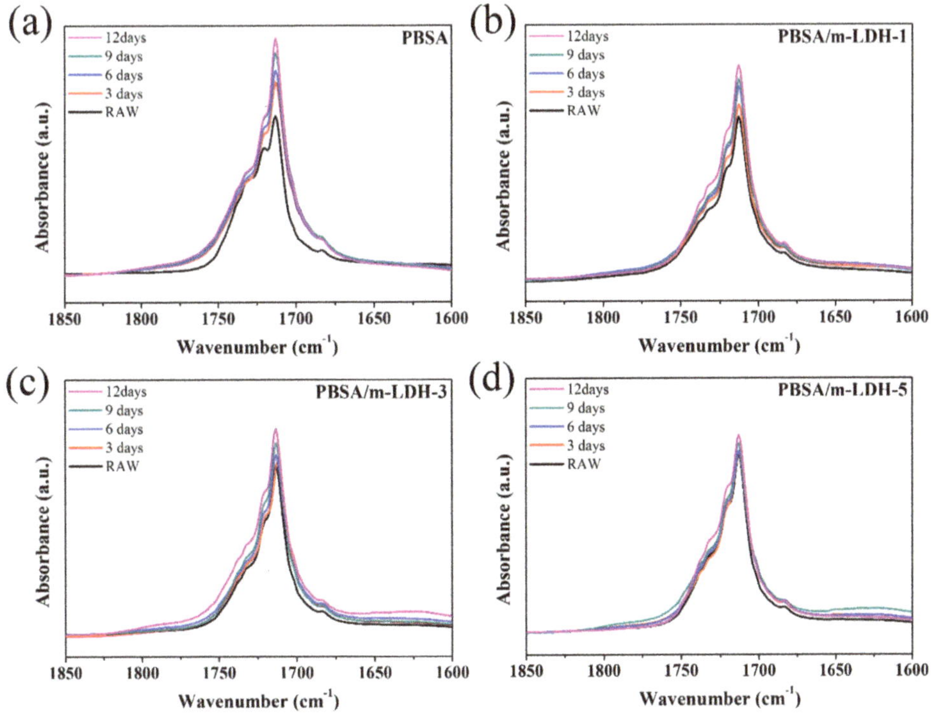

Figure A2. FTIR result of PBSA and its composites after different irradiation time. (**a**) PBSA; (**b**) PBSA-m-LDHD-1; (**c**) PBSA-m-LDHD-3; (**d**) PBSA-m-LDHD-5.

References

1. Rocca-Smith, J.R.; Pasquarelli, R.; Lagorce-Tachon, A.; Rousseau, J.; Fontaine, S.; Aguié-Béghin, V.; Debeaufort, F.; Karbowiak, T. Toward Sustainable PLA-Based Multilayer Complexes with Improved Barrier Properties. *ACS Sustain. Chem. Eng.* **2019**, *7*, 3759–3771. [CrossRef]
2. Xiong, S.-J.; Pang, B.; Zhou, S.-J.; Li, M.-K.; Yang, S.; Wang, Y.-Y.; Shi, Q.; Wang, S.-F.; Yuan, T.-Q.; Sun, R.-C. Economically Competitive Biodegradable PBAT/Lignin Composites: Effect of Lignin Methylation and Compatibilizer. *ACS Sustain. Chem. Eng.* **2020**, *8*, 5338–5346. [CrossRef]
3. Wang, J.-M.; Ding, S.-J.; Wu, T.-M. Rheology, crystallization behavior, and mechanical properties of poly(butylene succinate-co-terephthalate)/cellulose nanocrystal composites. *Polym. Test.* **2020**, *87*, 106551. [CrossRef]
4. Wang, H.-T.; Wang, J.-M.; Wu, T.-M. Synthesis and characterization of biodegradable aliphatic–aromatic nanocomposites fabricated using maleic acid-grafted poly[(butylene adipate)-co-terephthalate] and organically modified layered zinc phenylphosphonate. *Polym. Int.* **2019**, *68*, 1531–1537. [CrossRef]
5. Kuo, D.-L.; Wu, T.-M. Crystallization Behavior and Morphology of Hexadecylamine-Modified Layered Zinc Phenylphosphonate and Poly(Butylene Succinate-co-Adipate) Composites with Controllable Biodegradation Rates. *J. Polym. Environ.* **2019**, *27*, 10–18. [CrossRef]
6. Phua, Y.J.; Lau, N.S.; Sudesh, K.; Chow, W.S.; Mohd Ishak, Z.A. Biodegradability studies of poly(butylene succinate)/organo-montmorillonite nanocomposites under controlled compost soil conditions: Effects of clay loading and compatibiliser. *Polym. Degrad. Stab.* **2012**, *97*, 1345–1354. [CrossRef]
7. Wang, H.; Wei, D.; Zheng, A.; Xiao, H. Soil burial biodegradation of antimicrobial biodegradable PBAT films. *Polym. Degrad. Stab.* **2015**, *116*, 14–22. [CrossRef]
8. Souza, P.M.S.; Morales, A.R.; Sanchez, E.M.S.; Mei, L.H.I. Study of PBAT Photostabilization with Ultraviolet Absorber in Combination with Hindered Amine Light Stabilizer and Vitamin E, Aiming Mulching Film Application. *J. Polym. Environ.* **2018**, *26*, 3422–3436. [CrossRef]
9. Kijchavengkul, T.; Auras, R.; Rubino, M.; Selke, S.; Ngouajio, M.; Fernandez, R.T. Formulation selection of aliphatic aromatic biodegradable polyester film exposed to UV/solar radiation. *Polym. Degrad. Stab.* **2011**, *96*, 1919–1926. [CrossRef]
10. Santonja-Blasco, L.; Ribes-Greus, A.; Alamo, R.G. Comparative thermal, biological and photodegradation kinetics of polylactide and effect on crystallization rates. *Polym. Degrad. Stab.* **2013**, *98*, 771–784. [CrossRef]

11. Maurer-Jones, M.A.; Monzo, E.M. Quantifying Photochemical Transformations of Poly(butylene adipate-co-terephthalate) Films. *ACS Appl. Polym. Mater.* **2021**, *3*, 1003–1011. [CrossRef]
12. Puchalski, M.; Szparaga, G.; Biela, T.; Gutowska, A.; Sztajnowski, S.; Krucińska, I. Molecular and Supramolecular Changes in Polybutylene Succinate (PBS) and Polybutylene Succinate Adipate (PBSA) Copolymer during Degradation in Various Environmental Conditions. *Polymers* **2018**, *10*, 251. [CrossRef] [PubMed]
13. Chen, J.-H.; Yang, M.-C. Preparation and characterization of nanocomposite of maleated poly(butylene adipate-co-terephthalate) with organoclay. *Mater. Sci. Eng. C* **2015**, *46*, 301–308. [CrossRef] [PubMed]
14. Zhang, Y.; Xu, J.; Guo, B. Photodegradation behavior of poly(butylene succinate-co-butylene adipate)/ZnO nanocomposites. *Colloids Surf. A Physicochem. Eng. Asp.* **2016**, *489*, 173–181. [CrossRef]
15. Wang, X.-R.; Li, Y.; Tang, L.-P.; Gan, W.; Zhou, W.; Zhao, Y.-F.; Bai, D.-S. Fabrication of Zn-Ti layered double hydroxide by varying cationic ratio of Ti4+ and its application as UV absorbent. *Chin. Chem. Lett.* **2017**, *28*, 394–399. [CrossRef]
16. Morro, A.; Catalina, F.; Sanchez-León, E.; Abrusci, C. Photodegradation and Biodegradation Under Thermophile Conditions of Mulching Films Based on Poly(Butylene Adipate-co-Terephthalate) and Its Blend with Poly(Lactic Acid). *J. Polym. Environ.* **2019**, *27*, 352–363. [CrossRef]
17. Chen, Y.-A.; Tsai, G.-S.; Chen, E.-C.; Wu, T.-M. Thermal degradation behaviors and biodegradability of novel nanocomposites based on various poly [(butylene succinate)-co-adipate] and modified layered double hydroxides. *J. Taiwan Inst. Chem. Eng.* **2017**, *77*, 263–270. [CrossRef]
18. Chen, Y.-A.; Tsai, G.-S.; Chen, E.-C.; Wu, T.-M. Crystallization behaviors and microstructures of poly (butylene succinate-co-adipate)/modified layered double hydroxide nanocomposites. *J. Mater. Sci.* **2016**, *51*, 4021–4030. [CrossRef]
19. Nyambo, C.; Songtipya, P.; Manias, E.; Jimenez-Gasco, M.M.; Wilkie, C.A. Effect of MgAl-layered double hydroxide exchanged with linear alkyl carboxylates on fire-retardancy of PMMA and PS. *J. Mater. Chem.* **2008**, *18*, 4827–4838. [CrossRef]
20. Yang, J.-H.; Zhang, W.; Ryu, H.; Lee, J.-H.; Park, D.-H.; Choi, J.Y.; Vinu, A.; Elzatahry, A.A.; Choy, J.-H. Influence of anionic surface modifiers on the thermal stability and mechanical properties of layered double hydroxide/polypropylene nanocomposites. *J. Mater. Chem. A* **2015**, *3*, 22730–22738. [CrossRef]
21. Quispe-Dominguez, R.; Naseem, S.; Leuteritz, A.; Kuehnert, I. Synthesis and characterization of MgAl-DBS LDH/PLA composite by sonication-assisted masterbatch (SAM) melt mixing method. *RSC Adv.* **2019**, *9*, 658–667. [CrossRef]
22. Zhang, Z.; Hua, Z.; Lang, J.; Song, Y.; Zhang, Q.; Han, Q.; Fan, H.; Gao, M.; Li, X.; Yang, J. Eco-friendly nanostructured Zn–Al layered double hydroxide photocatalysts with enhanced photocatalytic activity. *CrystEngComm* **2019**, *21*, 4607–4619. [CrossRef]
23. Saber, O.; Tagaya, H. New Layered Double Hydroxide, Zn–Ti LDH : Preparation and Intercalation Reactions. *J. Incl. Phenom. Macrocycl. Chem.* **2003**, *45*, 107–115. [CrossRef]
24. Li, Y.; Tang, L.-P.; Zhou, W.; Wang, X.-R. Fabrication of intercalated p-aminobenzoic acid into Zn-Ti layered double hydroxide and its application as UV absorbent. *Chin. Chem. Lett.* **2016**, *27*, 1495–1499. [CrossRef]
25. Egambaram, O.P.; Pillai, S.K.; Lategan, M.; Ray, S.S. Nanostructured Zn-Ti layered double hydroxides with reduced photocatalytic activity for sunscreen application. *J. Nanoparticle Res.* **2019**, *21*, 53. [CrossRef]
26. Naseem, S.; Lonkar, S.P.; Leuteritz, A.; Labuschagné, F.J.W.J. Different transition metal combinations of LDH systems and their organic modifications as UV protecting materials for polypropylene (PP). *RSC Adv.* **2018**, *8*, 29789–29796. [CrossRef]
27. Tserki, V.; Matzinos, P.; Pavlidou, E.; Vachliotis, D.; Panayiotou, C. Biodegradable aliphatic polyesters. *Part I. Properties and biodegradation of poly (butylene succinate-co-butylene adipate). Polym. Degrad. Stab.* **2006**, *91*, 367–376.
28. Rafiqah, S.A.; Khalina, A.; Harmaen, A.S.; Tawakkal, I.A.; Zaman, K.; Asim, M.; Nurrazi, M.N.; Lee, C.H. A Review on Properties and Application of Bio-Based Poly(Butylene Succinate). *Polymers* **2021**, *13*, 1436. [CrossRef] [PubMed]
29. Cai, L.; Qi, Z.; Xu, J.; Guo, B.; Huang, Z. Study on the Photodegradation Stability of Poly(butylene Succinate-co-butylene Adipate)/TiO_2 Nanocomposites. *J. Chem.* **2019**, *2019*, 5036019. [CrossRef]
30. Hu, F.-Q.; Zhao, M.-D.; Yuan, H.; You, J.; Du, Y.-Z.; Zeng, S. A novel chitosan oligosaccharide–stearic acid micelles for gene delivery: Properties and in vitro transfection studies. *Int. J. Pharm.* **2006**, *315*, 158–166. [CrossRef]
31. Chen, C.; Ruengkajorn, K.; Buffet, J.-C.; O'Hare, D. Water adsorbancy of high surface area layered double hydroxides (AMO-LDHs). *RSC Adv.* **2018**, *8*, 34650–34655. [CrossRef]
32. Djaballah, R.; Bentouami, A.; Benhamou, A.; Boury, B.; Elandaloussi, E.H. The use of Zn-Ti layered double hydroxide interlayer spacing property for low-loading drug and low-dose therapy. *Synthesis, characterization and release kinetics study. J. Alloy. Compd.* **2018**, *739*, 559–567. [CrossRef]
33. Zhang, X.; Zhang, Y. Poly(butylene succinate-co-butylene adipate)/cellulose nanocrystal composites modified with phthalic anhydride. *Carbohydr. Polym.* **2015**, *134*, 52–59. [CrossRef] [PubMed]
34. Nikolic, M.S.; Djonlagic, J. Synthesis and characterization of biodegradable poly(butylene succinate-co-butylene adipate)s. *Polym. Degrad. Stab.* **2001**, *74*, 263–270. [CrossRef]

polymers

MDPI

Article

Gum Rosin as a Size Control Agent of Poly(Butylene Adipate-Co-Terephthalate) (PBAT) Domains to Increase the Toughness of Packaging Formulations Based on Polylactic Acid (PLA)

Miguel Aldas [1,2,*], José Miguel Ferri [1,*], Dana Luca Motoc [3], Laura Peponi [4], Marina Patricia Arrieta [5,6] and Juan López-Martínez [1]

1 Instituto de Tecnología de Materiales (ITM), Universitat Politècnica de València (UPV), 03801 Alcoy, Spain; jlopezm@mcm.upv.es
2 Departamento de Ciencia de Alimentos y Biotecnología, Facultad de Ingeniería Química y Agroindustria, Escuela Politécnica Nacional, 170517 Quito, Ecuador
3 Department of Automotive and Transport Engineering, Transilvania University of Brasov, Eroilor Av., 500036 Brasov, Romania; danaluca@unitbv.ro
4 Instituto de Ciencia y Tecnología de Polímeros (ICTP-CSIC), C/Juan de la Cierva 3, 28006 Madrid, Spain; lpeponi@ictp.csic.es
5 Departamento de Ingeniería Química Industrial y del Medio Ambiente, Escuela Técnica Superior de Ingenieros Industriales, Universidad Politécnica de Madrid (ETSII-UPM), Calle José Gutiérrez Abascal 2, 28006 Madrid, Spain; m.arrieta@upm.es
6 Grupo de Investigación: Polímeros, Caracterización y Aplicaciones (POLCA), 28006 Madrid, Spain
* Correspondence: miguel.aldas@epn.edu.ec (M.A.); joferaz@upvnet.upv.es (J.M.F.); Tel.: +593-999-736-444 (M.A.); +34-699-495-982 (J.M.F.)

Citation: Aldas, M.; Ferri, J.M.; Motoc, D.L.; Peponi, L.; Arrieta, M.P.; López-Martínez, J. Gum Rosin as a Size Control Agent of Poly(Butylene Adipate-Co-Terephthalate) (PBAT) Domains to Increase the Toughness of Packaging Formulations Based on Polylactic Acid (PLA). *Polymers* **2021**, *13*, 1913. https://doi.org/10.3390/polym13121913

Academic Editor: Łukasz Klapiszewski

Received: 11 May 2021
Accepted: 7 June 2021
Published: 8 June 2021

Abstract: Gum rosin (GR) was used as a natural additive to improve the compatibility between polylactic acid, PLA, and poly(butylene adipate-co-terephthalate, PBAT, blended with 20 wt.% of PBAT (PLA/PBAT). The PBAT was used as a soft component to increase the ductility of PLA and its fracture toughness. The coalescence of the PBAT domains was possible due to the plasticization effect of the GR component. These domains contributed to increasing the toughness of the final material due to the variation and control of the PBAT domains' size and consequently, reducing the stress concentration points. The GR was used in contents of 5, 10, 15, and 20 phr. Consequently, the flexural properties were improved and the impact resistance increased up to 80% in PLA/PBAT_15GR with respect to the PLA/PBAT formulation. Field emission scanning electron microscope (FESEM) images allowed observing that the size of PBAT domains of 2–3 μm was optimal to reduce the impact stress. Differential scanning calorimetry (DSC) analysis showed a reduction of up to 8 °C on the PLA melting temperature and up to 5.3 °C of the PLA glass transition temperature in the PLA/PBAT_20GR formulation, which indicates an improvement in the processability of PLA. Finally, transparent films with improved oxygen barrier performance and increased hydrophobicity were obtained suggesting the potential interest of these blends for the food packaging industry.

Keywords: polylactic acid; PLA; poly(butylene adipate-co-terephthalate; PBAT; gum rosin; biodegradable polymers; barrier properties

1. Introduction

Polylactic acid (PLA) is one of the most widely consumed biodegradable and compostable polymers. In the field of medicine, it is used for its excellent compatibility with the human body [1]. However, its low toughness makes it necessary to modify it by incorporating additives to obtain a more ductile material, with better barrier properties, higher hydrophobicity, and higher stability to temperature and external agents (UV, humidity, etc.). Its modification allows greater applicability, there being several industrial sectors in which it can be used, among them agriculture [2], packaging [3,4], medical fields [5], 3D printing [6], textile fibers [7], and composites [8–10].

The PLA processing method with the more favorable industrial viability is its physical mixing or blending [11,12]. The modification of biodegradable polymers through physical blending with another biodegradable polymer shows many advantages since it offers the opportunity to create a new material with desired properties. Moreover, it is relatively simple and cost-effective to blend polymeric materials in the melt state, based on available processing technologies commonly used at the industrial level (i.e., extrusion, injection molding, film-forming, etc.) [13]. Many biodegradable polymer formulations have been obtained by blending polymeric matrices to modulate their mechanical, thermal, rheological, and morphological behavior. The literature refers to blends of PLA with other polymers or copolymers such as poly(ε-caprolactone) (PCL) [14], thermoplastic starch (TPS) [15], poly(butylene succinate) (PBS) [16], poly(butylene succinate-co-adipate) (PBSA) [17], poly(glycolic acid) (PGA) [18], poly(hydroxybutyrate) (PHB) [12,19], poly(hydroxybutyrate-co-valerate) (PHBV) [20], and poly(butylene adipate-co-terephthalate) (PBAT) [21], among others. Among these, PBAT has gained interest in the development of PLA/PBAT blends intended for film manufacturing due to its high flexibility [22] and its inherent biodegradable character [23]. However, in most of the scientific works reported up to now, poor miscibility or total immiscibility between components of the blend was observed and the expected synergism to improve the overall properties was not achieved.

Currently, additives can be used to increase the miscibility between polymers through various modification mechanisms that seek to increase either the interaction between the different polymeric phases through compatibilization, the plasticization of one of the components to increase the free volume, which facilitates the miscibility of a second component, or interaction through free radicals (reactive mixing). The components of the blend, together with the reactive agent, undergo a reaction and molecular chemical change that influence the mechanical and thermal properties. Wang et al. formulated PLA-based blends with different PBAT contents, additivated with 0.75% of Joncryl ADR 4368. As a result, the compatibility of both polymers was increased. Specifically, the elongation at break was increased by 18% and the impact absorption energy went from 4.85 kJ/m^2 for the PLA/20PBAT to 5.21 kJ/m^2 for the formulation 80PLA-20PBAT_0.75Joncryl [24]. Arruda et al. also used Joncryl ADR-4368 on PLA-PBAT blends and demonstrated its compatibilizing effect. The formulation 60PLA-40PBAT-0.6Joncryl increased the elongation at break by 1200% (compared to the percentage of elongation of neat PLA). Such a compatibilizer acts as a crosslinking and/or branching agent for both polymers, providing higher strength and Young's modulus to each of the polymers separately, although it increases the ductility of all the studied formulations [25]. Wu et al. added 0.3, 0.5, and 0.75 wt.% of 2,5-bis (tert-butyl peroxy) -2,5-dimethyl hexane (Luperox 101), observing a crosslinking reaction and some interaction between the components (PBAT and PBS) of the ternary PLA-based blend. By adding 0.3 phr of Luperox 101 to the PLA/20PBAT/20PBS formulation, it was possible to increase by up to 10 times the impact energy absorbed, with respect to the polymeric matrix without additives, considerably reducing the glass transition temperature, T_g, of PLA and increasing the elongation at break [26]. On the other hand, some natural plasticizers such as vegetable oils (VO), obtained from seeds, are presented as an effective alternative, sometimes acting as compatibilizers [27,28]. Their use in the packaging sector is of interest due to the high resistance of plasticizers to migration in food contact conditions [29,30], being able to increase the solubility of the blend components and therefore, act as a compatibilizer in polymeric blends. These can be chemically modified allowing greater interaction with the polymeric chains thanks to the added functional groups, reacting with polar groups such as hydroxyl, carboxylic, etc. [31,32]. According to Bocque et al., an efficient plasticizer has to be able to increase the molecular free volume and be endowed with ester groups (reactive functional groups that provide cohesion) and aromatic groups that increase its compatibilizing effect [33]. In this sense, Carbonell-Verdú et al. compatibilized PLA blended with 20 wt.% of PBAT by using cottonseed oil-based derivatives [22]. They observed that the low miscibility of PLA/PBAT could be improved by compatibilization with epoxidized cottonseed oil (ECSO) and maleinized cottonseed oil (MCSO). Moreover, both additives were able to considerably increase the elongation at break of the PLA/PBAT blend without compromising mechanical strength.

On the other hand, colophony or gum rosin (GR) and its derivatives have gained interest in the field of polymeric materials as highly versatile and multifunctional natural additives, both with synthetic plastic matrices and with biodegradable matrices [34–36]. For example, Arrieta et al. used a gum rosin ester as a natural viscosity increasing agent in a blend based on polyvinyl chloride (PVC) plasticized with epoxidized linseed oil (ELO). In that study, the gum rosin derivative showed good compatibility with the PVC synthetic matrix. Furthermore, it was verified that a composition between 40 and 50 phr of rosin ester present in PVC can increase the viscosity of the blends up to 10 times [37]. On the other hand, in previous work, the effect of GR and two gum rosin esters on the properties of a commercial blend of TPS, PBAT, and PCL was studied, and the versatility of the resin and its derivatives were verified. Furthermore, the GR acted as a plasticizer, and on the other hand, the gum rosin esters provided a solubilizing and compatibilizing effect of the biodegradable blends. This behavior influenced the properties of each of the studied formulations, especially in those with 15 wt.% of GR where the processing temperature was reduced by up to 50 °C and the toughness increased up to 500%, compared to the neat polymeric matrix [38]. Finally, when the interaction between gum rosin and gum rosin derivatives with Mater-Bi type bioplastic was studied through microscopic techniques, the improvement of the miscibility of the components and the solubility effect conferred to the PBAT phase thanks to the compatibilizing effect of the GR and its derivatives were confirmed [39,40].

The main objective of this work is to compatibilize PLA/PBAT-based blends with GR and to study the plasticizing effect conferred by the GR on the PLA/PBAT binary blends, focusing on the use of a natural resin as the main novelty of the present work. The processability aspects, as well as the main mechanical properties of the formulations, the thermal stability, and the barrier properties were assessed. In addition, the microstructure was studied by field emission scanning electron microscopy (FESEM) which allowed to see the interface of both polymers and to evaluate the plasticizing effect of GR on the toughness of the PLA/PBAT blend.

2. Materials and Methods

2.1. Materials

The poly(lactic acid) (PLA) used was Ingeo^{TM} Biopolymer, commercial-grade 6201D, supplied by NatureWorks LLC (Minnetonka, MN, USA). The pellet's density was 1.24 g/cm^3, the melt flow index was 15–30 g/10 min measured at 210 °C, and it contained 2% of D-lactic acid. The poly(butylene adipate-co-terephthalate) (PBAT) was a commercial-grade Biocosafe^{TM} 2003 F, supplied by Xinfu Pharmaceutical Co. Ltd. (Zhejiang, China) and it was characterized by a density of 1.25 g/cm^3 and a melt flow index < 6 g/10 min at 190 °C. As the additive, gum rosin (GR), supplied by Sigma-Aldrich (Mostoles, Spain), was used.

2.2. Blends Preparation

Blends containing PLA, PBAT, and GR were prepared with different compositions, which are summarized in Table 1. The percentage of PBAT in the PLA matrix was fixed at 20 wt.% based on results reported in the literature [22]. At this percentage, PLA-20%PBAT blends have given interesting results in terms of toughness. To study the effect of the resin in the PLA/PBAT blends, the content of gum rosin added was set from 0 to 20 phr (parts of GR per hundred parts of PLA/PBAT blend). Neat PLA, neat PBAT, and PBAT with 10 phr of GR were also prepared to compare their properties with the previous blends.

Table 1. Composition and coding of PLA/PBAT/GR studied blends.

Code	PLA wt.%	PBAT wt.%	GR phr
PLA	100	-	-
PLA/PBAT	80	20	-
PLA/PBAT_5GR	80	20	5
PLA/PBAT_10GR	80	20	10
PLA/PBAT_15GR	80	20	15
PLA/PBAT_20GR	80	20	20
PBAT	-	100	-
PBAT_10GR	-	100	10

All materials were dried at 50 °C for 48 h in an air circulation oven before processing. After, the formulations were premixed in a zipper bag. To obtain the final materials, the procedure was followed as described: (1) extrusion of the material formulations in a twin-screw extruder (Dupra S.L, Castalla, Spain), L/D ratio of 25, with a temperature profile from 185 to 140 °C (from die to hopper) at 50 rpm; (2) milling into pellets; and (3) injection molding in an injection molding machine (Sprinter-11, Erinca S,L., Barcelona, Spain), with a temperature profile from 185 to 175 °C, to obtain test specimens. The test specimens were standard rectangular specimens (80 × 10 × 4 mm) and standard tensile specimens "1BA" (length ≥ 75 mm, width 10 mm, and thickness ≥ 2 mm) according to ISO 527 [41]. The films of each formulation were further obtained by the solvent casting method, using chloroform as solvent. In particular, 20 g of each formulation was dissolved in 100 mL of chloroform and heated at 30 °C under vigorous stirring for 1 h. The obtained solutions were cast onto a 20-cm diameter mold and the films were obtained after complete solvent evaporation at room temperature after 24 h.

2.3. *Material Characterization*

2.3.1. Colorimetric Properties and Visual Appearance Evaluation

The color properties of each formulation were determined in the CIEL*a*b* color space using a Konica CM-3600d ColorFlex-Diff2, HunterLab, Hunter Associates Laboratory, Inc, (Reston, VA, USA), using the standard rectangular specimens as the sample. The instrument was calibrated with a white standard tile and the color coordinates: L* (lightness), a* (red-green) and b* (yellow-blue), and yellowness index (YI) were measured. Measurements were carried out in quintuplicate at random positions over the sample surface. For the visual appearance evaluation, film samples were placed on a background image, to evaluate their transparency and the change in tonality due to the incorporation of GR. All samples were photographed in the same ambient light conditions and the evaluated films presented a uniform and constant thickness of 200 μm.

2.3.2. Mechanical Characterization

Tensile and flexural tests were carried out in a universal testing machine ELIB 30 from S.A.E. Ibertest (Madrid, Spain) at room temperature, according to ISO 527 [42] and ISO 178 [43], respectively. Both tests were performed with a loading cell of 5 kN and a test speed of 10 mm/min, using five samples from each formulation in each test. Moreover, the typical stress-strain curve for each formulation was plotted from one representative curve, which showed the average behavior of the formulation. Furthermore, the toughness of the materials was calculated with the area under this curve using the OriginPro2015 program. Five values of toughness were assessed for each formulation and the average and standard deviation values are reported.

Impact absorbed energy measurements were carried out with a Charpy pendulum machine from Metrotec S.A. (San Sebastian, Spain), using a 6 J pendulum under the ISO 179 [44]. Five specimens were tested, and the mean and standard deviation are reported.

The hardness of PLA/PBAT/GR formulations was measured using a Shore D durometer, Model 673-D, from Instrument J.Bot S.A. (Barcelona, Spain), following the guidelines of ISO 868 [45]. Twenty measurements were taken from aleatory parts of the samples, and the mean and standard deviation are reported as hardness values.

The heat deflection temperature (HDT) was determined by the A method according to ISO 75 [46], which recommends a load of 1.8 MPa and a heating rate of 120 °C/h. In addition, the Vicat softening temperature (VST) was assessed using the ISO 306 [47], method B (with a load of 50 N and a heating rate of 50 °C/h). Both tests were carried out in a VICAT/HDT station DEFLEX 687-A2 from Metrotec SA (San Sebastián, Spain). For each property, three specimens were tested, and the mean and standard deviation are reported.

2.3.3. Microstructural Characterization

A field emission scanning electron microscope (FESEM) model Zeiss Ultra, from Oxford Instruments (Abingdon, UK) was used to obtain micrographs from the cryofractured surfaces of the rectangular samples. The acceleration voltage was set to 2 kV. Previously, the samples were coated with an ultrathin platinum layer in a vacuum using a coater, model MED020 (Leica, Leica Microsystems). Micrographs of PLA with GR have not been reported, since there is literature that shows the effect of gum rosin on the PLA matrix [48].

2.3.4. Thermal and Thermomechanical Characterization

The thermal properties of the PLA/PBAT/GR formulations were obtained by using two analyses, differential scanning calorimetry (DSC) (Mettler-Toledo model 821, Schwerzenbach, Switzerland) and thermogravimetric analysis (TGA) (Linseis, model TGA 1000, Linseis Messgeraete GmbH, Selb, Germany). In the programming of the DSC system, the thermal cycles were carried out, under a nitrogen atmosphere, with a heating and cooling rate of 10 °C/min. The purpose of the first heating was to eliminate the thermal history and was carried out from 25 °C to 200 °C. Later it was cooled down to −50 °C and finally, the second heating was carried out to 250 °C. The degree of crystallinity (X_c), obtained from the DSC thermograms, was calculated using Equation (1).

$$X_c\ (\%) = 100 \times \frac{\Delta H_m - \Delta H_{cc}}{\Delta H_m(100\%)} \times \frac{1}{W_{PLA}}, \quad (1)$$

where ΔH_m and ΔH_{cc} are the melting and the cold crystallization enthalpies, respectively, ΔH_m (100%) is the calculated melting enthalpy of purely crystalline PLA (93 J/g) [49]. W_{PLA} is the weight fraction of PLA in the formulation.

The mass loss obtained by TGA was carried out at a heating rate of 10 °C/min, measured in the range of 30 °C to 800 °C, under a nitrogen atmosphere and with a flow rate of 30 mL/min. The $T_{5\%}$ was taken at the temperature where the 5% of mass loss was reached. The T_{max}, the temperature where the degradation rate was maximum, was obtained at the peak of the first derivative of the TGA curve (DTG curves). Additionally, dynamic mechanical thermal analysis (DMTA) was performed in torsion mode, from −90 °C to 120 °C at a heating rate of 2 °C/min, a frequency of 1 Hz, and 0.1% of maximum deformation. The test was done on rectangular samples sizing 40 × 10 × 4 mm^3 in an oscillatory rheometer AR G2 from TA Instruments (New Castle, DE, USA) equipped with a special clamp system for solid samples.

2.3.5. Oxygen Permeability Measurements of PLA/PBAT/GR Formulations

Oxygen transition rate (OTR) values were obtained using an oxygen permeation analyzer from Systech Instruments-Model 8500 (Metrotec S.A, Lezo, Spain) at a pressure of 2.5 atm and room temperature. Films were introduced in the diffusion chamber. Pure oxygen (99.9% purity) flowed through the upper half of the sample chamber and nitrogen flowed through the lower half of the chamber. It dragged the oxygen flowing through the film and was measured by an oxygen detector. To obtain an average value of OTR per film thickness (OTR.e), three measurements were made. Thickness was measured precisely at

25 °C using a Digimatic Micrometer Series 293 MDC-Lite (Mitutoyo, Kanagawa, Japan) with an error of 0.001 mm. Twenty readings were taken at random positions over the 14 cm diameter circle films.

2.3.6. Static Water Contact Angle Measurements of PLA/PBAT/GR Formulations

The wettability was measured by water contact angle at room temperature using an Easy Drop Standard goniometer FM140 (KRÜSS GmbH, Hamburg, Germany). The equipment was provided with a camera and analyzer software (Drop Shape Analysis SW21; DSA1 from KRÜSS GmbH, Hamburg, Germany). Ten contact angles were measured randomly using distiller water as contact liquid on the surface film with a microsyringe. Five measurements were carried out for each drop and the average value was calculated.

2.3.7. Statistical Analysis

The statistical analysis was performed to establish the effect of the GR content on the properties of the PLA/PBAT matrix. The significant differences in all the properties were statistically assessed at 95% confidence level according to Tukey's test using a one-way analysis of variance (ANOVA) by means of OriginPro2018 software (OriginLab, Northampton, MA, USA).

3. Results and Discussion

3.1. *Visual Appearance and Color Properties*

One of the most important requirements of packaging materials for consumers' acceptance is seeing the packed food through the packaging material. Figure 1 shows the visual appearance of the obtained films. It is possible to observe the high transparency of the films suggesting their potential application in the food packaging field. The addition of 20 wt.% of PBAT showed some loss in transparency. The further incorporation of GR also led to a partial decrease in the high transparency of PLA, which was more marked with the increasing amount of GR. In the case of the neat PBAT formulation, it showed less transparency than neat PLA and the incorporation of GR was practically imperceptible on the transparency of the sample (PBAT_10GR).

The film color properties were determined in the CIEL*a*b* space and the results are summarized in Table 2. PLA showed the highest lightness (L*), in good accordance with the visual appearance of the films. Meanwhile, the lowest lightness was observed for the PLA/PBAT blend with the higher amount of GR (PLA/PBAT_20GR). The a* values (which correspond to red-green coloration), although significant ($p < 0.05$), did not highly change its values in any of the studied formulations. In contrast, the b* coordinate and the yellowness index (YI) significantly ($p < 0.05$) and considerably increased with the resin content, since the blend became more yellow due to the inherent characteristics of the GR.

Table 2. Color change of the studied formulations.

Formulation	Color Change			
	L*	a*	b*	YI
PLA	41.2 ± 0.7^{a}	-1.3 ± 0.4^{a}	-2.7 ± 0.6^{a}	-12.3 ± 2.4^{a}
PLA/PBAT	87.5 ± 0.3^{b}	-0.7 ± 0.1^{b}	2.1 ± 0.3^{b}	3.8 ± 0.6^{b}
PLA/PBAT_5GR	86.2 ± 0.6^{c}	-1.7 ± 0.1^{c}	8.0 ± 0.8^{c}	14.5 ± 1.5^{c}
PLA/PBAT_10GR	82.7 ± 0.9^{d}	-1.4 ± 0.1^{a}	17.7 ± 1.8^{d}	33.5 ± 3.4^{d}
PLA/PBAT_15GR	82.4 ± 0.6^{d}	-0.9 ± 0.3^{b}	20.7 ± 1.2^{e}	39.1 ± 2.1^{e}
PLA/PBAT_20GR	79.5 ± 0.8^{e}	-0.1 ± 0.2^{d}	28.2 ± 1.1^{f}	53.1 ± 2.1^{f}
PBAT	83.7 ± 0.7^{f}	-0.6 ± 0.2^{b}	6.3 ± 0.5^{g}	12.5 ± 1.1^{c}
PBAT_10GR	76.8 ± 0.5^{g}	1.3 ± 0.3^{e}	20.9 ± 0.7^{e}	43.4 ± 1.4^{g}

[a–g] Different letters within the same property show statistically significant differences between formulations ($p < 0.05$).

Figure 1. The visual appearance of PLA, PBAT, PBAT_10GR, PLA/PBAT, and PLA/PBAT with 5, 10, 15, and 20 phr GR resin films.

3.2. *Microstructural Characterization*

Figure 2 shows the effect of the GR resin on the partially miscible PLA/PBAT blends, as well as the effect on the neat PBAT, taken as reference. Neat PLA (Figure 2a) showed a flat surface with small prominences, characteristic of a brittle break of the material under cryofracture conditions. The blend of PLA with 20% PBAT (PLA/PBAT, Figure 2b) showed a smoother surface with PBAT domains sizing less than 0.5 μm, a characteristic size that shows that the components have partial miscibility, although it is very poor [22]. The incorporation of a 5 phr of GR (Figure 2c) showed significant differences in the morphology of the cryofracture surface. Specifically, the PBAT domains were larger (between 0.5–1.5 μm), with an average size of 1 μm. Moreover, PBAT domains turned from presenting irregular shapes (Figure 2b) to almost perfect spherical shapes. This is indicative of the loss of affinity and miscibility between PLA and PBAT when adding GR. However, with a 5 phr of GR, there was still some interaction between the PLA and the PBAT matrices since the PBAT domains broke through the plane of the fracture (in other words, the PBAT spheres broke through the crack of the fracture). This means that the PLA-PBAT interaction was greater than the cohesion forces of PBAT. Moreover, it should be highlighted that, although higher, the PBAT domains showed good adhesion with the PLA matrix at the interface.

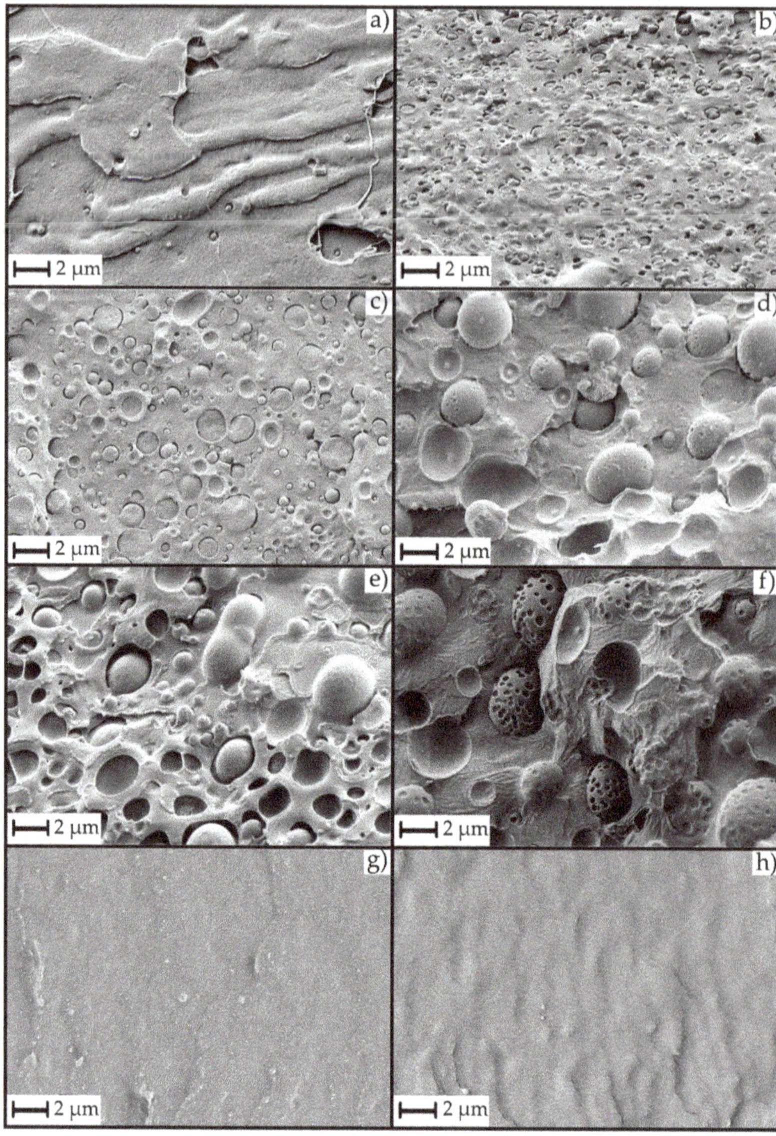

Figure 2. FESEM images at 5000 X of studied materials: (**a**) PLA, (**b**) PLA/PBAT, (**c**) PLA/PBAT_5GR, (**d**) PLA/PBAT_10GR, (**e**) PLA/PBAT_15GR, (**f**) PLA/PBAT_20GR, (**g**) PBAT, and (**h**) PBAT_10GR.

This type of fracture did not persist with higher concentrations of GR. Figure 2d,e show the PLA/PBAT formulation with 10 and 15 phr of GR, respectively. In these images, the larger size of the PBAT domains (1.5–4 μm) can be observed. Although there were slight differences in PBAT domain sizes, their fractures were not similar to Figure 2c. In fact, in these formulations, the PBAT domains were not broken and showed complete PBAT spheres with small (nanoscale) domains of GR. The non-breakage of the PBAT spheres was due to the lower miscibility and interaction between the PLA and the PBAT domains, which is due to the phobic effect that exists between the PLA matrix and the GR resin [48]. This lack of interaction generated points of zero interaction around the PBAT spheres that prevented their breakage, although it improved the impact energy absorption, probably due to the PBAT domains still showing good adhesion with the PLA matrix at the interface, since there was not a gap between both polymeric matrices. Finally, Figure 2f shows the formulation with a 20 phr of GR, where the PBAT domains had an approximate average size of 4–5 μm and exhibited signs of GR saturation within them. These nanodomains, which were less evident in the other formulations, generated an important reduction in the PLA/PBAT interactions and resulted in a decrease of the impact energy absorption. Therefore, the formulations with PBAT domain sizes of 2–3 μm were those that improved the impact energy absorption or the toughness.

Additionally, Figure 2g,h show the cryofractured surface of neat PBAT and PBAT with 10 phr of GR. It is possible to verify, along with the mechanical properties, that GR acted as a plasticizer for the PBAT. There were no appreciable differences in the morphology of the cryofractured surfaces of both materials since PBAT is a soft material that became even more softer by the addition of GR, demonstrating the plasticization effect of the GR.

3.3. Mechanical Properties of the PLA/PBAT/GR Formulations

Table 3 shows the main values of the mechanical properties, maximum tensile and flexural strength, Young's moduli, elongation at break, impact absorption energy (Charpy), Shore D hardness, and the HDT temperature of each obtained formulation, as well as the neat PLA as a reference.

It was observed that the tensile strength of the PLA/PBAT formulations significantly decreased ($p < 0.05$) by 6.3%, 17%, 23.2%, and 29.1% (compared to PLA/PBAT) when adding 5, 10, 15, and 20 phr of GR, respectively. However, the toughness (calculated as the area under the stress-strain curve) of the formulation with a 5 phr of GR (PLA/PBAT_5GR) significantly increased ($p < 0.05$) compared to the base formulation PLA/PBAT by 40%, as shown in Figure 3. In Figure 3, it is observed that as the content of GR increased, the toughness (energy per unit volume) of the formulations significantly decreased ($p < 0.05$). In contrast, the drastic increase in the toughness of PBAT by adding 10 phr of GR should be noted. This fact explains that GR acts as a PBAT plasticizer since it increased its elongation and decreased its tensile strength.

On the other hand, it was observed that when adding 5 and 10 phr of GR, Young's modulus (Table 3) did not significantly decrease ($p > 0.05$) compared to PLA/PBAT. However, from contents higher than 10 phr of GR, where GR saturates notoriously in the PBAT domains (as shown by FESEM in Figure 2), the saturations behaved as reinforcements, increasing Young's modulus mean values. Specifically, Young's modulus of the formulations with a 15 phr GR showed a reduction of less than 10% relative to the reference formulation (PLA/PBAT). For contents of 20 phr GR, it showed an increase of the value, being 6.6% higher than the reference. The tendency of the mean values suggests that at low contents, GR acts as a plasticizer of the PBAT component, making the PBAT domains less compatible with the PLA matrix (because PLA is incompatible with GR). At contents higher than the saturation point (at 10 phr of GR), some nano-scale domains were generated. These domains acted as reinforcement, increasing the mean value of the Young's modulus of the materials. It should be noted that GR resin has higher affinity and therefore higher compatibility with PBAT, confirming its ability to plasticize it.

Table 3. Mechanical properties of PLA/PBAT blends with different contents of gum rosin (GR) as additive.

Formulation \ Property	Tensile Strength (MPa)	Young's Modulus (MPa)	Elongation at Break (%)	Flexural Strength (MPa)	Flexural Modulus (MPa)	Charpy Impact Energy (KJ/m^2)	Hardness (Shore D)	HDT Temperature (°C)
PLA	65.1 ± 1.7 ^{a}	2100 ± 250 ^{a}	6.4 ± 1.6 ^{a}	108.8 ± 8.8 ^{a}	3170 ± 150 ^{a,b}	* 34.6 ± 2.8 ^{a}	77 ± 1 ^{a}	58.0 ± 0.8 ^{a}
PLA/PBAT	50.5 ± 0.5 ^{b}	1680 ± 200 ^{b}	16.4 ± 1.2 ^{b}	74.9 ± 8.6 ^{b}	2720 ± 130 ^{a}	5.1 ± 1.4 ^{b}	71 ± 1 ^{b}	57.8 ± 0.6 ^{a}
PLA/PBAT _5GR	47.3 ± 1.2 ^{b}	1440 ± 200 ^{b}	7.3 ± 1.4 ^{a}	67.2 ± 0.8 ^{b}	2510 ± 30 ^{a}	8.3 ± 1.2 ^{b,c}	72 ± 1 ^{c}	56.6 ± 0.6 ^{a,b}
PLA/PBAT _10GR	41.9 ± 0.4 ^{c}	1430 ± 100 ^{b}	5.2 ± 0.8 ^{a,c}	48.0 ± 7.8 ^{c}	2530 ± 180 ^{a}	9.3 ± 0.7 ^{c}	71 ± 1 ^{b}	55.2 ± 0.4 ^{b,c}
PLA/PBAT _15GR	38.8 ± 2.8 ^{c}	1510 ± 90 ^{b}	3.8 ± 0.4 ^{c,d}	29.7 ± 1.1 ^{d}	3400 ± 190 ^{b}	10.3 ± 1.3 ^{c}	74 ± 1 ^{d}	54.8 ± 0.8 ^{b,c}
PLA/PBAT _20GR	35.8 ± 4.5 ^{c}	1790 ± 220 ^{a,b}	1.7 ±0.4 ^{d}	28.3 ± 3.3 ^{d}	3020 ± 180 ^{a,b}	6.9 ± 0.3 ^{b,c}	75 ± 1 ^{d}	53.8 ± 0.8 ^{c}
PBAT	13.6 ± 1.4 ^{d}	110 ± 40 ^{c}	487 ± 70 ^{e}	6.8 ± 0.5 ^{e}	80 ± 10 ^{c}	No break	41 ± 1 ^{e}	36.8 ± 0.4 ^{d}
PBAT_10GR	14.9 ± 1.7 ^{d}	80 ± 10 ^{c}	720 ± 15 ^{f}	7.2 ± 0.8 ^{e}	60 ± 10 ^{c}	No break	38 ± 1 ^{f}	35.6 ± 0.2 ^{d}

* PLA sample tested in specimens without notch. $^{a–f}$ Different letters within the same property show statistically significant differences between formulations ($p < 0.05$).

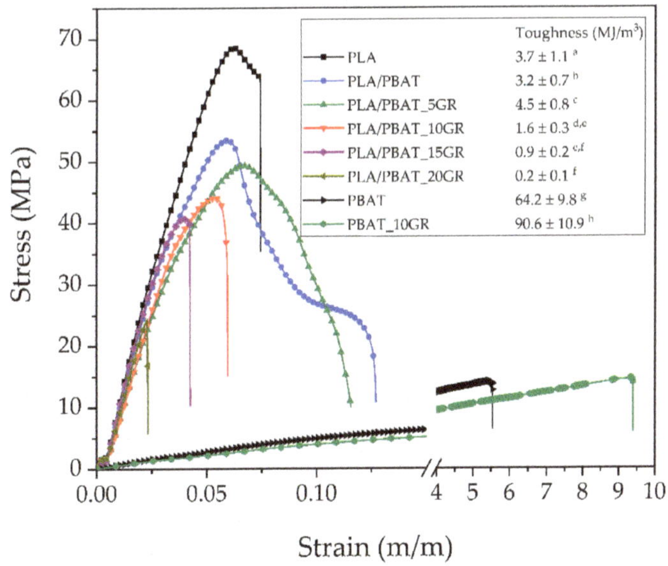

Figure 3. The toughness of PLA, PBAT, PBAT_10GR, PLA/PBAT, and PLA/PBAT with 5, 10, 15, and 20 phr of GR resin. [a–h] Different letters show statistically significant differences between formulations ($p < 0.05$).

Table 3 also shows the mechanical properties of PBAT and PBAT_10GR, the behavior of which can be considered as an indicator of the compatibility between the PBAT and the GR. The significant reduction ($p < 0.05$) in Shore D hardness, which went from 41 to 38 and the reduction in Young's modulus, both in the flexion and in the tensile test, of the PBAT_10GR formulation compared to neat PBAT, show that the GR exerted a plasticizing effect on the PBAT matrix. Moreover, a significant increase ($p < 0.05$) of 47.6% in elongation was observed, which reinforces the idea of the plasticizing effect. However, the elongation at break of the PLA/PBAT formulations with different GR contents suffered a significant reduction ($p < 0.05$) of 55.5%, 68.3%, 76.8%, and 88.0%, compared to the PLA/PBAT formulation, when adding 5, 10, 15, and 20 phr of GR, respectively. It is well known that the miscibility of PBAT and PLA is partial with some compatibility [22,24,50]. Nevertheless, a phenomenon of coalescence of the PBAT domains was generated with the addition of GR to the PLA/PBAT formulation, creating larger domains and reducing the interaction between PLA and PBAT. This behavior can be explained since the PLA does not assimilate the GR in its matrix, which remains isolated. This phenomenon makes the GR assimilation by the PBAT domains easier, and therefore, the size of these domains increased with the GR content.

In contrast, a significant increase ($p < 0.05$) in the impact energy absorption was observed (Table 3) when incorporating 10 and 15 phr GR into the PLA/PBAT formulation, managing to increase it by 75.5 and 79.2%, respectively. For contents higher than 15 phr GR, the impact energy absorption mean values began to decrease due to saturation of the GR on the PBAT domains, generating a phase separation between the PLA/PBAT matrix and the GR saturations. This effect was corroborated by FESEM analysis (Figure 2f). According to the literature, when the domains of the ductile and dispersed material in a rigid polymeric matrix have a size of 2–5 μm, the energy absorption is maximum. This phenomenon has been demonstrated in a large number of materials, for example in the synthesis of HIPS [51], where polybutadiene forms domains into the PS matrix. At smaller or larger domain sizes, the energy absorption value decreases again. Therefore, the stress concentration generated in materials with poor interaction will depend on the size of the domains of the minority component, in this case, the PBAT-GR system. In the present study, the toughness of materials, especially at impact, could be controlled and improved thanks to the phobicity between PLA and GR and the affinity between PBAT and GR. If the GR resin showed a good affinity with both polymers (PLA and PBAT), the toughness modification would depend exclusively on the composition of the polymeric components and their interaction.

Finally, the hardness (Shore D) of the studied blends significantly varied ($p < 0.05$) with the incorporation of GR. This property significantly increased for 15 and 20 phr of GR contents. The plasticizing effect of the GR resin on the neat PBAT significantly reduced the hardness (Table 3). The HDT significantly decreased ($p < 0.05$) with increasing GR content, going from 57.8 °C for PLA/PBAT to 53.8 °C for the formulation with a 20 phr of GR. Therefore, the processability of these materials improved as the GR content increased.

3.4. Thermal and Thermomechanical Properties of the PLA/PBAT/GR Formulations

The calorimetric curves of the PLA, PBAT, PBAT_10GR, PLA/PBAT, and PLA/PBAT formulations containing 5, 10, 15, and 20 phr of GR formulation were obtained by differential scanning calorimetry (DSC analysis) and are reported in Figure 4. In addition, Table 4 shows the main thermal transitions such as the glass transition temperature (T_g), cold crystallization temperature (T_{cc}), the meting (ΔH_m) and crystallization (ΔH_{cc}) enthalpies, and the degree of crystallinity (X_c).

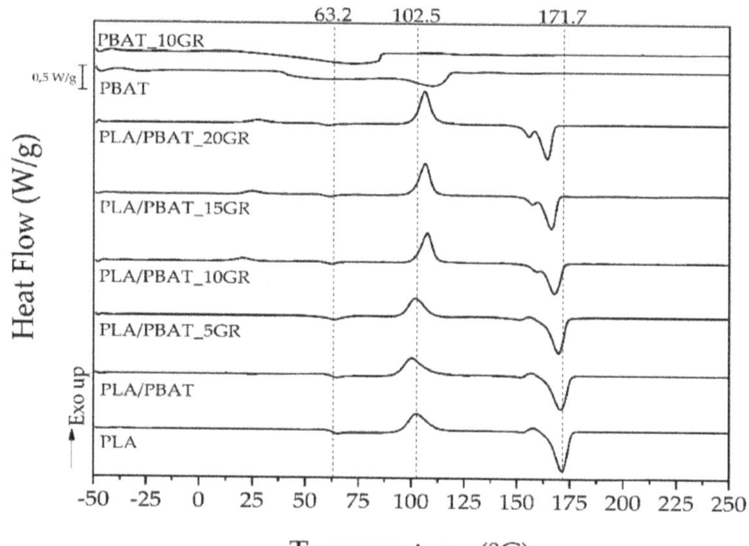

Figure 4. DSC curves of PLA, PBAT, PBAT_10GR, and the studied PLA/PBAT formulations with different content of GR.

The T_g related to the PLA component of the blend tended to decrease both when adding PBAT (1 °C lower for the PLA/PBAT formulation, due to the partial miscibility between PLA and PBAT) and when adding GR resin. Neat PLA had a T_g value of 63.2 °C, while this value significantly dropped ($p < 0.05$) to 57.9 °C for the T_g of the PLA/PBAT formulation with a 20 phr of GR. This decrease in the T_g of PLA is due to the saturation of GR, which acts as a lubricant, facilitating the movement of the chains. However, the PBAT_10GR had a significantly higher T_g than the T_g of the neat PBAT, going from −25.9 °C to −20.8 °C. This increment in the T_g when GR was additivated to the PBAT could be due to an increase in PBAT crystallinity due to the presence of the GR. It is important to mention that the T_g values of PBAT and PBAT_10GR formulations were obtained by DMA analysis since this transition is not easily observable by DSC.

Table 4. Thermal properties of studied formulations and neat matrixes materials.

Formulation	T_{gPBAT} * (°C)	T_{gPLA} (°C)	T_{ccPLA} (°C)	ΔH_{ccPLA} (J/g)	T_{mPLA} (°C)	ΔH_{mPLA} (J/g)	$X_{c\ PLA}$ (%)
PLA	-	63.2 ± 1.2 [a]	102.5 ± 0.8 [a]	26.0 ± 1.5 [a]	171.7 ± 0.9 [a]	−32.8 ± 1.3 [a]	7.4 ± 0.9 [a]
PLA/PBAT	−33.5 ± 1.1 [a]	62.3 ± 1.6 [a,b]	100.8 ± 0.8 [a]	23.9 ± 1.3 [a]	170.3 ± 1.1 [a]	−30.5 ± 1.6 [a,b]	8.9 ± 0.3 [a,b]
PLA/PBAT_5GR	−20.6 ± 0.5 [b]	61.8 ± 1.3 [a,b]	101.8 ± 1.5 [a]	21.6 ± 1.5 [b]	169.3 ± 1.3 [a,b]	−28.8 ± 1.9 [b]	9.5 ± 0.8 [a,b]
PLA/PBAT_10GR	−21.3 ± 1.1 [b,c]	60.9 ± 0.9 [a,b,c]	107.4 ± 0.6 [b]	23.9 ± 1.0 [a,b]	167.5 ± 1.2 [b,c]	−31.7 ± 1.1 [a]	11.6 ± 1.1 [b]
PLA/PBAT_15GR	−23.1 ± 0.9 [c,d]	59.2 ± 0.5 [b,c]	106.5 ± 1.3 [b]	23.3 ± 0.6 [a,b]	165.9 ± 0.8 [c,d]	−29.9 ± 0.9 [a,b]	10.4 ± 0.7 [a,b]
PLA/PBAT_20GR	−24.2 ± 1.1 [d]	57.9 ± 1.0 [c]	106.0 ± 1.0 [b]	25.1 ± 1.2 [a]	163.9 ± 0.9 [d]	−30.0 ± 1.0 [a,b]	8.2 ± 1.0 [a,b]
					T_{mPBAT} (°C)	ΔH_{mPBAT} (J/g)	
PBAT	−25.9 ± 0.7 [d]	–	–	–	110.8 ± 0.9 [e]	−21.4 ± 0.8 [c]	
PBAT_10GR	−20.8 ± 0.7 [b]	–	–	–	79.7 ± 1.2 [f]	−19.4 ± 1.2 [c]	

* T_g PBAT determined by DMA analysis explained below. [a–f] Different letters show statistically significant differences between formulations ($p < 0.05$).

The same effect was observed with the T_m of PLA and PBAT when adding GR, 171.7 °C and 110.8 °C being the temperatures for neat PLA and neat PBAT, respectively. A significant reduction to 163.9 ° C for the PLA/PBAT formulation with a 20 phr of GR and up to 79.7 °C for the PBAT_10GR formulation was achieved. This decrease confirmed the plasticizing effect of the GR resin on the PBAT and the lubricating effect exerted by the saturated GR on the PLA matrix.

Table 4 also shows the degree of crystallinity (X_c) of the PLA fraction in the blends when adding PBAT and GR. A slight increase (not statistically significant, $p > 0.05$) in X_c was observed when adding PBAT, since the microdomains of PBAT act as a nucleating agent due to the interaction between PLA and PBAT. When adding 10 phr of GR, a statistically ($p < 0.05$) higher increase was observed, reaching X_c values of 11.6%. However, as the GR content increased above 10 phr, X_c decreased (reaching similar values of neat PLA) since the PBAT domains coalesced thanks to the GR and fewer nucleation points were generated from the PLA spherulites.

By DMA technique, the T_g and the storage modulus "G" from torsional tests were obtained and the results are reported in Figure 5. Despite the incorporation of PBAT and GR into the PLA matrix, the values of G′ (Figure 5a) did not suffer a big change at lower temperatures, which is in good agreement with the Young's modulus trend discussed in the mechanical characterization (Table 3). It was observed that the incorporation of 20% of PBAT generated a partially miscible blend, since T_g changed from 70.2 °C for neat PLA to 67.1 °C for the PLA/PBAT, as shown by the δ peaks of Figure 5b. Al-Itry et al. obtained a lower decrease (only 1 °C) in T_g when incorporating 20% PBAT [52]. A higher reduction in T_g was observed when adding GR resin. Specifically, a value of 65.5 °C was obtained for the formulation with 5 phr of GR and 64.8 °C for the formulation with 10 phr of GR. This behavior demonstrates the plasticizing effect of GR on the PBAT domains and the lubricant effect on the PLA matrix. The values of T_g obtained by DMA were slightly higher than those obtained by DSC but with the same trend.

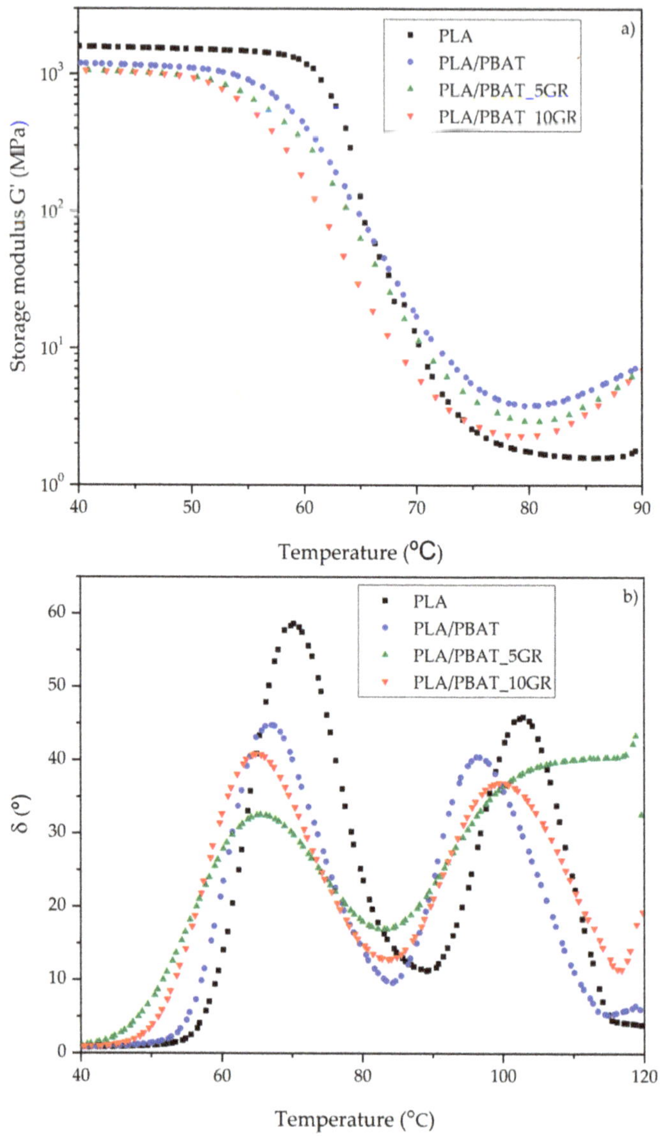

Figure 5. (**a**) Storage modulus (G′) and (**b**) loss factor (δ) of PLA and PLA/PBAT as references, and representatives PLA/PBAT formulations with 5 and 10 phr of GR.

Figure 6 shows the results of the thermogravimetric analysis (TGA). The $T_{5\%}$ reflects that the addition of PBAT to the PLA matrix (PLA/PBAT) did not significantly modify ($p > 0.05$) the thermal stability of neat PLA. This shows that the interactions between PLA and PBAT were poor. The T_{max} of PLA/PBAT showed a lower but not significantly different value than neat PLA ($p > 0.05$). The effect of adding GR was to obtain significantly lower $T_{5\%}$ values, which were due to the partial degradation of the GR component [38]. At higher contents of GR, the saturation of the resin directly affected the thermal stability since the components were not interacting with the PLA matrix due to the weak interaction of the PBAT domains (as also shown by FESEM).

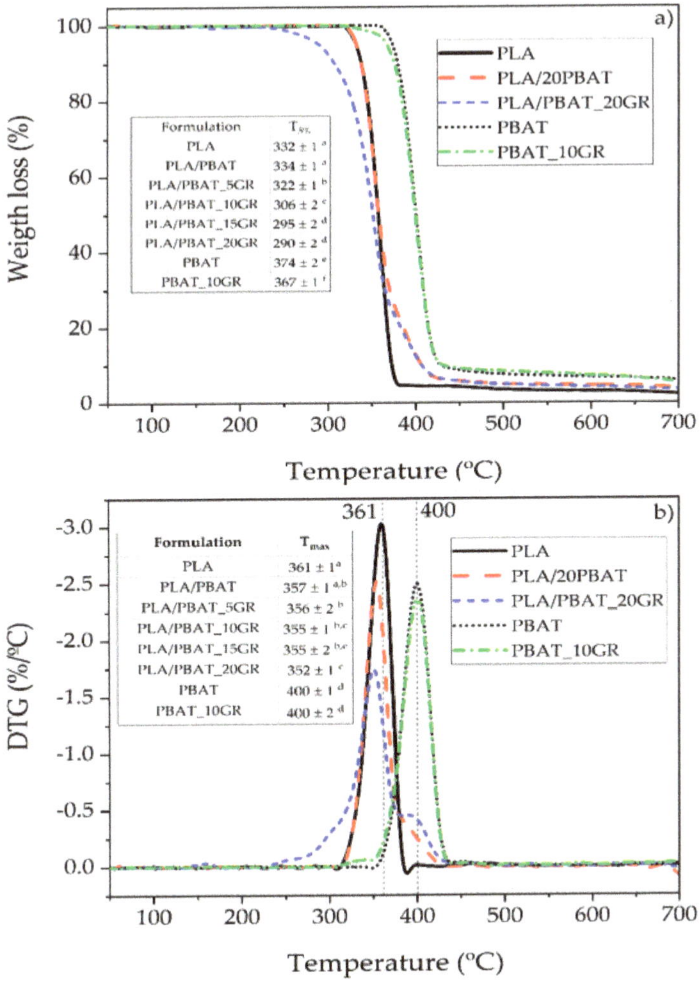

Figure 6. (**a**) TGA and (**b**) DTG curves of PLA and PBAT matrixes, and studied representative blends: PLA/PBAT, PLA/PBAT_20GR, and PBAT_10GR. [a–f] Different letters within the same property show statistically significant differences between formulations ($p < 0.05$).

3.5. Oxygen Permeability of Films of PLA/PBAT/GR Formulations

Since these materials are intended for food packaging applications, the barrier performance against oxygen is very relevant to protect the foodstuffs from oxidation processes. Thus, the oxygen transmission rate was measured and Table 5 summarizes the OTR*e values of the obtained formulations. PLA showed significantly better oxygen barrier performance than PBAT ($p < 0.05$). The incorporation of 20 wt.% of PBAT into the formulation led to a significant increment of the oxygen permeation of PLA (increment of 72%) reaching values close to that of PBAT. This behavior was due to the low miscibility of PLA and PBAT in this formulation, which allowed oxygen diffusion through the porous/defects present in the film. The incorporation of 5 phr of GR into the PLA/PBAT blend significantly reduced the oxygen permeability by 19.5%, mainly due to the homogeneous dispersion of GR into the PLA/PBAT blend matrix. This dispersion contributed to a reduction of the defects observed in PLA/PBAT as well as to the increased crystallinity of the formulation, which led to a better oxygen barrier performance as was already observed in the PLA-based blends [19]. Higher amounts of GR, 10 phr, led to the best result, showing a significant reduction of 35% in the oxygen permeability ($p < 0.05$) with respect of PLA/PBAT formulation and being closer to that of PLA (12% higher than PLA), in good agreement with the highest crystallinity observed in this formulation. Those formulations with higher contents of GR, 15 and 20 phr, resulted in a worse oxygen barrier than PLA/PBAT_10GR but were still better than PLA/PBAT. The saturation effect of GR into the PBAT domains generated pores/defects that allowed oxygen diffusion through the film. Nevertheless, the good adhesion between the increased PBAT domains due to the GR presence with the PLA matrix at the interface still allowed to obtain better barrier performance than PLA/PBAT. The oxygen transmission results obtained here were higher than traditional petrochemical plastics widely used in the packaging sector, such as EVOH (OTR*e < 4 $cm^3 mm/m^2/day$) [53] or PET (OTR*e < 3 cm^3 $mm/m^2/day$), but substantially lower than that of LDPE (OTR*e between 160 and 240 $cm^3 mm/m^2/day$) [54]. Thus, the materials developed here could be used in several food packaging applications as a potential alternative to some packaging materials made of conventional plastics (i.e., polyolefins) directly as films or in more complex formulations (i.e., multilayers systems).

Table 5. Oxygen permeability measurements of PLA, PLA/PBAT, and PLA/PBAT with 5, 10, 15, and 20 phr of GR resin.

Formulation	OTR*e ($cm^3 \cdot mm/m^2/day$)	Wettability (°)
PLA	43.8 ± 2.2 [a]	67.2 ± 2.1 [a]
PLA/PBAT	75.2 ± 0.7 [b]	77.6 ± 1.6 [b]
PLA/PBAT_5GR	60.5 ± 3.7 [c,d,e]	74.4 ± 2.6 [c]
PLA/PBAT_10GR	49.1 ± 1.7 [a,d]	74.5 ± 1.4 [c]
PLA/PBAT_15GR	53.4 ± 2.6 [c,d]	69.5 ± 4.3 [d]
PLA/PBAT_20GR	57.6 ± 1.0 [d]	67.1 ± 3.1 [a]
PBAT	79.2 ± 1.9 [b]	74.1 ± 3.0 [c]
PBAT_10GR	66.0 ± 4.2 [e]	82.2 ± 2.9 [e]

[a–f] Different letters within the same property show statistically significant differences between formulations ($p < 0.05$).

3.6. Wettability Performance of Films

Another important issue in films for food packaging is their protection against humidity. Thus, the water contact angle was measured to get information regarding the surface hydrophilicity/hydrophobicity of the materials. Table 5 shows the water contact angle measurements. PLA presented a WCA value of 67.2°, similar to that obtained by Carbonell-Verdú et al. [22], while PBAT showed a significantly higher WCA value of 74.1°, being more hydrophobic. The PLA/PBAT blend showed even significantly higher values than those of PBAT, probably due to the high roughness of this formulation since the wettability is strongly dependent not only on the surface chemical properties but also on the surface topography [55]. The incorporation of GR into the PLA/PBAT blend produced a significant decrease ($p < 0.05$) of the WCA, particularly evident in those formulations with a high amount of GR (15 and 20 phr), increasing the hydrophilicity of the surface. Meanwhile, the incorporation of 5 and 10 phr led to a reduction of 3° of the WCA of PLA/PBAT matrix. An excess of GR, such as in the case of PLA/PBAT_15GR and PLA/PBAT_20GR, produced an increase in the size of PBAT domains. However, the WCA of these formulations were significantly lower than that of PBAT. These behaviors could be ascribed to the lower dispersion of PBAT domains into the polymeric matrix as well as the loss of the interfacial tension between PLA and PBAT [56].

4. Conclusions

In this work, PLA-PBAT blends were melt-blended and compatibilized through the incorporation of gum rosin. An improvement in both tensile and impact toughness was observed when adding gum rosin (GR) to the formulation composed of a PLA matrix with 20 wt.% of PBAT as a 'soft' component. Such increment is due to the coalescent effect of the PBAT domains into the PLA matrix due to the plasticizing effect of GR. The flexural modulus was also improved and the tensile strength increased by 80% compared to the PLA/PBAT formulation. Morphologically, it was observed that the size of the PBAT domains of 2–3 μm was optimal to reduce stress concentrations in impact conditions. Concerning neat PLA, a significant reduction of up to 8 °C of the melting temperature and up to 5.3 °C of the glass transition temperature was observed, which denotes an improvement of the processability of PLA in the blends containing PBAT and GR. Regarding the application of the obtained blends as films for packaging, PLA-PBAT-GR films were transparent with luminescence (L*) values very close to neat PLA; therefore, all the films obtained presented a very good visual appearance for the intended use. Moreover, improved barrier properties were observed as a reduction in OTR*e of up to 35% compared to the PLA/PBAT blend film and additionally showing an increase in hydrophobicity, as an increase in the water contact angle value from 67.2° for PLA to 74.5° for the PLA/PBAT_10GR film formulation was observed. Finally, the obtained results show that GR can be used as a dispersed phase size control agent to improve the toughness of PLA/PBAT formulations. The results obtained demonstrate the potential of the PLA/PBAT/GR films to be produced at an industrial level and further used in the food packaging field to replace traditional non-biodegradable plastics.

Author Contributions: Conceptualization, M.A., J.M.F., M.P.A., and J.L.-M.; data curation, M.A., J.M.F., D.L.M., L.P., M.P.A., and J.L.-M.; formal analysis, M.A., J.M.F., M.P.A., and J.L.-M.; funding acquisition, J.L.-M.; investigation, M.A., J.M.F., and M.P.A.; methodology, M.A. and J.M.F.; project administration, M.P.A. and J.L.-M.; resources, J.L.-M.; supervision, D.L.M., L.P., M.P.A., and J.L.-M.; validation, M.A., J.M.F., D.L.M., L.P., M.P.A., and J.L.-M.; visualization, M.A., J.M.F., D.L.M., L.P., M.P.A., and J.L.-M.; writing—original draft, M.A., J.M.F., and M.P.A.; writing—review and editing, M.A., J.M.F., D.L.M., L.P., and M.P.A. All authors have read and agreed to the published version of the manuscript.

Funding: This research was funded by the Spanish Ministry of Economy and Competitiveness (MINECO), project: PROMADEPCOL (MAT2017-84909-C2-2-R).

Institutional Review Board Statement: Not applicable.

Informed Consent Statement: Not applicable.

Data Availability Statement: The data presented in this study are available on request from the corresponding author.

Acknowledgments: M.A. thanks Secretaría de Educación Superior, Ciencia, Tecnología e Innovación (SENESCYT) and Escuela Politécnica Nacional (EPN). J.M.F. thanks the postdoc contract (APOSTD/2019/122) Generalitat Valenciana (2019-2021).

Conflicts of Interest: The authors declare no conflict of interest.

References

1. Nova-Institut GmbH. Market Study and Database on Bio-Based Polymers in the World: Capacities, Production and Applications: Status Quo and Trends towards 2020. July 2013. Available online: http://bio-based.eu/top-downloads/ (accessed on 15 February 2021).
2. Castro-Aguirre, E.; Iñiguez-Franco, F.; Samsudin, H.; Fang, X.; Auras, R. Poly(lactic acid)—Mass production, processing, industrial applications, and end of life. *Adv. Drug Deliv. Rev.* **2016**, *107*, 333–366. [CrossRef]
3. Arrieta, M.; Fortunati, E.; Dominici, F.; Rayón, E.; López, J.; Kenny, J.M. PLA-PHB/cellulose based films: Mechanical, barrier and disintegration properties. *Polym. Degrad. Stab.* **2014**, *107*, 139–149. [CrossRef]
4. Siracusa, V.; Rocculi, P.; Romani, S.; Rosa, M.D. Biodegradable polymers for food packaging: A review. *Trends Food Sci. Technol.* **2008**, *19*, 634–643. [CrossRef]

5. Aydin, E.; Planell, J.A.; Hasirci, V. Hydroxyapatite nanorod-reinforced biodegradable poly(l-lactic acid) composites for bone plate applications. *J. Mater. Sci. Mater. Med.* **2011**, *22*, 2413–2427. [CrossRef]
6. Martinez, J.I.; Carrillo, L.Q.; Lascano, D.; Bou, S.F.; Vitoria, T.B. Effect of infill parameters on mechanical properties in additive manufacturing. *DYNA Ing. E Ind.* **2020**, *95*, 412–417. [CrossRef]
7. Baig, G.A.; Carr, C.M. Surface and Structural Damage to PLA Fibres during Textile Pretreatments. *Fibres Text. East. Eur.* **2016**, *24*, 52–58. [CrossRef]
8. Pawlak, F.; Aldas, M.; López Martínez, J.; Samper, M.D. Effect of Different Compatibilizers on Injection-Molded Green Fiber-Reinforced Polymers Based on Poly(lactic acid)-Maleinized Linseed Oil System and Sheep Wool. *Polymers* **2019**, *11*, 1514. [CrossRef]
9. Carrasco, M.F.A.; Rouault, N.J.; Azor, J.M.F.; Martínez, J.L.; Madrigal, M.D.S. A new bio-based fibre-reinforced polymer obtained from sheep wool short fibres and PLA. *Green Mater.* **2020**, *8*, 79–91. [CrossRef]
10. Pawlak, F.; Aldas, M.; Parres, F.; López-Martínez, J.; Arrieta, M. Silane-Functionalized Sheep Wool Fibers from Dairy Industry Waste for the Development of Plasticized PLA Composites with Maleinized Linseed Oil for Injection-Molded Parts. *Polymers* **2020**, *12*, 2523. [CrossRef]
11. Lim, L.-T.; Auras, R.; Rubino, M. Processing technologies for poly(lactic acid). *Prog. Polym. Sci.* **2008**, *33*, 820–852. [CrossRef]
12. Arrieta, M.P.; Samper, M.D.; Aldas, M.; López, J. On the Use of PLA-PHB Blends for Sustainable Food Packaging Applications. *Materials* **2017**, *10*, 1008. [CrossRef]
13. Arrieta, M.; Fortunati, E.; Burgos, N.; Peltzer, M.; López, J.; Peponi, L. Nanocellulose-Based Polymeric Blends for Food Packaging Applications. In *Multifunctional Polymeric Nanocomposites Based on Cellulosic Reinforcements*; Elsevier BV: Amsterdam, The Netherlands, 2016; pp. 205–252.
14. Ferri, J.M.; Fenollar, O.; Jorda-Vilaplana, A.; García-Sanoguera, D.; Balart, R. Effect of Miscibility on Mechanical and Thermal Properties of Poly (Lactic Acid)/Polycaprolactone Blends. *Polym. Int.* **2016**, *65*, 453–463. [CrossRef]
15. Teixeira, E.D.M.; Curvelo, A.A.; Corrêa, A.C.; Marconcini, J.M.; Glenn, G.M.; Mattoso, L.H. Properties of thermoplastic starch from cassava bagasse and cassava starch and their blends with poly (lactic acid). *Ind. Crop. Prod.* **2012**, *37*, 61–68. [CrossRef]
16. Elwathig, H.; You, W.; He, J.; Yu, M. Dynamic Mechanical Properties and Thermal Stability of Poly(Lactic Acid) and Poly(Butylene Succinate) Blends Composites. *J. Fiber Bioeng. Inform.* **2013**, *6*, 85–94.
17. Xi, Y.; Huan, X.; Karin, O.; Minna, H. Poly(Lactide)-g-Poly(Butylene Succinate-Co-Adipate) with High Crystallization Capacity and Migration Resistance. *Materials* **2016**, *9*, 313. [CrossRef]
18. Takayama, T.; Daigaku, Y.; Ito, H.; Takamori, H. Mechanical properties of bio-absorbable PLA/PGA fiber-reinforced composites. *J. Mech. Sci. Technol.* **2014**, *28*, 4151–4154. [CrossRef]
19. Arrieta, M.P.; Samper, M.D.; Lopez, J.; Jiménez, A. Combined Effect of Poly(hydroxybutyrate) and Plasticizers on Polylactic acid Properties for Film Intended for Food Packaging. *J. Polym. Environ.* **2014**, *22*, 460–470. [CrossRef]
20. Muniyasamy, S.; Ofosu, O.; John, M.J.; Anandjiwala, R.D. Mineralization of Poly(lactic acid) (PLA), Poly(3-hydroxybutyrate-co-valerate) (PHBV) and PLA/PHBV Blend in Compost and Soil Environments. *J. Renew. Mater.* **2016**, *4*, 133–145. [CrossRef]
21. Quero, E.; Müller, A.J.; Signori, F.; Coltelli, M.-B.; Bronco, S. Isothermal Cold-Crystallization of PLA/PBAT Blends With and Without the Addition of Acetyl Tributyl Citrate. *Macromol. Chem. Phys.* **2012**, *213*, 36–48. [CrossRef]
22. Carbonell-Verdu, A.; Ferri, J.M.; Dominici, F.; Boronat, T.; Sánchez-Nacher, L.; Balart, R.; Torre, L. Manufacturing and compatibilization of PLA/PBAT binary blends by cottonseed oil-based derivatives. *Express Polym. Lett.* **2018**, *12*, 808–823. [CrossRef]
23. Del Campo, A.; de Lucas-Gil, E.; Rubio-Marcos, F.; Arrieta, M.P.; Fernández-García, M.; Fernández, J.F.; Muñoz-Bonilla, A. Accelerated disintegration of compostable Ecovio polymer by using ZnO particles as filler. *Polym. Degrad. Stab.* **2021**, *185*, 109501. [CrossRef]
24. Wang, X.; Peng, S.; Chen, H.; Yu, X.; Zhao, X. Mechanical properties, rheological behaviors, and phase morphologies of high-toughness PLA/PBAT blends by in-situ reactive compatibilization. *Compos. Part B Eng.* **2019**, *173*, 107028. [CrossRef]
25. Arruda, L.C.; Magaton, M.; Bretas, R.E.S.; Ueki, M.M. Influence of chain extender on mechanical, thermal and morphological properties of blown films of PLA/PBAT blends. *Polym. Test.* **2015**, *43*, 27–37. [CrossRef]
26. Wu, F.; Misra, M.; Mohanty, A.K. Super Toughened Poly(lactic acid)-Based Ternary Blends via Enhancing Interfacial Compatibility. *ACS Omega* **2019**, *4*, 1955–1968. [CrossRef] [PubMed]
27. Lligadas, G.; Ronda, J.C.; Galià, M.; Cádiz, V. Renewable polymeric materials from vegetable oils: A perspective. *Mater. Today* **2013**, *16*, 337–343. [CrossRef]
28. Carbonell-Verdu, A.; Garcia-Sanoguera, D.; Jordá-Vilaplana, A.; Sanchez-Nacher, L.; Balart, R. A New Biobased Plasticizer for Poly(Vinyl Chloride) Based on Epoxidized Cottonseed Oil. *J. Appl. Polym. Sci.* **2016**, *133*. [CrossRef]
29. Carbonell-Verdu, A.; Samper, M.D.; Garcia-Garcia, D.; Sanchez-Nacher, L.; Balart, R. Plasticization effect of epoxidized cottonseed oil (ECSO) on poly(lactic acid). *Ind. Crop. Prod.* **2017**, *104*, 278–286. [CrossRef]
30. Garcia-Garcia, D.; Carbonell-Verdu, A.; Arrieta, M.P.; López-Martínez, J.; Samper, M. Improvement of PLA film ductility by plasticization with epoxidized karanja oil. *Polym. Degrad. Stab.* **2020**, *179*, 109259. [CrossRef]
31. Ferri, J.M.; Garcia-Garcia, D.; Montanes, N.; Fenollar, O.; Balart, R. The effect of maleinized linseed oil as biobased plasticizer in poly(lactic acid)-based formulations. *Polym. Int.* **2017**, *66*, 882–891. [CrossRef]
32. Carbonell-Verdu, A.; Garcia-Garcia, D.; Dominici, F.; Torre, L.; Sanchez-Nacher, L.; Balart, R. PLA films with improved flexibility properties by using maleinized cottonseed oil. *Eur. Polym. J.* **2017**, *91*, 248–259. [CrossRef]

33. Bocqué, M.; Voirin, C.; Lapinte, V.; Caillol, S.; Robin, J.J. Petro-based and bio-based plasticizers: Chemical structures to plasticizing properties. *J. Polym. Sci. Part A Polym. Chem.* **2016**, *54*, 11–33. [CrossRef]
34. Pavon, C.; Aldas, M.; De La Rosa-Ramírez, H.; López-Martínez, J.; Arrieta, M.P. Improvement of PBAT Processability and Mechanical Performance by Blending with Pine Resin Derivatives for Injection Moulding Rigid Packaging with Enhanced Hydrophobicity. *Polymers* **2020**, *12*, 2891. [CrossRef]
35. Pavon, C.; Aldas, M.; López-Martínez, J.; Ferrándiz, S. New Materials for 3D-Printing Based on Polycaprolactone with Gum Rosin and Beeswax as Additives. *Polymers* **2020**, *12*, 334. [CrossRef] [PubMed]
36. Aldas, M.; Pavon, C.; López-Martínez, J.; Arrieta, M.P. Pine Resin Derivatives as Sustainable Additives to Improve the Mechanical and Thermal Properties of Injected Moulded Thermoplastic Starch. *Appl. Sci.* **2020**, *10*, 2561. [CrossRef]
37. Arrieta, M.P.; Samper, M.D.; Jiménez-López, M.; Aldas, M.; Lopez, J. Combined effect of linseed oil and gum rosin as natural additives for PVC. *Ind. Crop. Prod.* **2017**, *99*, 196–204. [CrossRef]
38. Aldas, M.; Ferri, J.M.; Lopez-Martinez, J.; Samper, M.D.; Arrieta, M.P. Effect of pine resin derivatives on the structural, thermal, and mechanical properties of Mater-Bi type bioplastic. *J. Appl. Polym. Sci.* **2020**, *137*, 48236. [CrossRef]
39. Aldas, M.; Rayón, E.; López-Martínez, J.; Arrieta, M.P. A Deeper Microscopic Study of the Interaction between Gum Rosin Derivatives and a Mater-Bi Type Bioplastic. *Polymers* **2020**, *12*, 226. [CrossRef]
40. Aldas, M.; Pavon, C.; Ferri, J.; Arrieta, M.; López-Martínez, J. Films Based on Mater-Bi® Compatibilized with Pine Resin Derivatives: Optical, Barrier, and Disintegration Properties. *Polymers* **2021**, *13*, 1506. [CrossRef] [PubMed]
41. International Standards Organization. *ISO 527-2. Plastics—Determination of Tensile Properties—Part 2: Test Conditions for Moulding and Extrusion Plastics*; International Standards Organization: Geneva, Switzerland, 2012.
42. International Standards Organization. *ISO 527-1:2012—Plastics—Determination of Tensile Properties—Part 1: General Principles*; International Standards Organization: Geneva, Switzerland, 2012.
43. International Standards Organization. *ISO 178:2010—Plastics—Determination of Flexural Properties*; International Standards Organization: Geneva, Switzerland, 2010.
44. International Standards Organization. *ISO 179-1:2010—Plastics—Determination of Charpy Impact Properties/ Part 1: Non-Instrumented Impact Test*; International Standards Organization: Geneva, Switzerland, 2010.
45. International Standards Organization. *ISO 868:2003—Plastics and Ebonite—Determination of Indentation Hardness by Means of a Durometer (Shore Hardness)*; International Standards Organization: Geneva, Switzerland, 2003.
46. International Standards Organization. *ISO 75-1:2013—Plastics—Determination of Temperature of Deflection under Load—Part 1: General Test Method*; International Standards Organization: Geneva, Switzerland, 2013.
47. International Standards Organization. *ISO 306:2013—Plastics—Thermoplastic Materials—Determination of Vicat Softening Temperature (VST)*; International Standards Organization: Geneva, Switzerland, 2013.
48. De La Rosa-Ramírez, H.; Aldas, M.; Ferri, J.M.; López-Martínez, J.; Samper, M.D. Modification of Poly (Lactic Acid) through the Incorporation of Gum Rosin and Gum Rosin Derivative: Mechanical Performance and Hydrophobicity. *J. Appl. Polym. Sci.* **2020**, *137*, 49346. [CrossRef]
49. Turner, J.F.; Riga, A.; O'Connor, A.; Zhang, J.; Collis, J. Characterization of drawn and undrawn poly-L-lactide films by differential scanning calorimetry. *J. Therm. Anal. Calorim.* **2004**, *75*, 257–268. [CrossRef]
50. Abdelwahab, M.A.; Taylor, S.; Misra, M.; Mohanty, A.K. Thermo-mechanical characterization of bioblends from polylactide and poly(butylene adipate-co-terephthalate) and lignin. *Macromol. Mater. Eng.* **2015**, *300*, 299–311. [CrossRef]
51. Ubaldo Alarcón, A. *Polimerización In Situ de Poliestireno de Alto Impacto En Presencia de Nanopartículas de Sílice y Micropartículas de Mg(OH)2 y Su Influencia Sobre La Cinética de Polimerización, La Morfología de Las Fase Elastomérica y La Retardancia a La Flama En Los Nan*; Consejo Nacional de Ciencia y Tecnología: Mexico City, Mexico, 2017.
52. Al-Itry, R.; Lamnawar, K.; Maazouz, A.; Billon, N.; Combeaud, C. Effect of the simultaneous biaxial stretching on the structural and mechanical properties of PLA, PBAT and their blends at rubbery state. *Eur. Polym. J.* **2015**, *68*, 288–301. [CrossRef]
53. Rahnama, M.; Oromiehie, A.; Ahmadi, S.; Ghasemi, I. Effect of different blend compositions on properties of low-density polyethylene/ethylene vinyl alcohol/clay toward high oxygen barrier nanocomposite films. *Polym. Sci. Ser. A* **2017**, *59*, 533–542. [CrossRef]
54. Arrieta, M.P.; Garrido, L.; Faba, S.; Guarda, A.; Galotto, M.J.; De Dicastillo, C.L. Cucumis metuliferus Fruit Extract Loaded Acetate Cellulose Coatings for Antioxidant Active Packaging. *Polymers* **2020**, *12*, 1248. [CrossRef]
55. Iglesias-Montes, M.L.; Luzi, F.; Dominici, F.; Torre, L.; Manfredi, L.B.; Cyras, V.P.; Puglia, D. Migration and Degradation in Composting Environment of Active Polylactic Acid Bilayer Nanocomposites Films: Combined Role of Umbelliferone, Lignin and Cellulose Nanostructures. *Polymers* **2021**, *13*, 282. [CrossRef] [PubMed]
56. Zolali, A.M.; Favis, B.D. Partial to complete wetting transitions in immiscible ternary blends with PLA: The influence of interfacial confinement. *Soft Matter* **2017**, *13*, 2844–2856. [CrossRef]

polymers

Article

Dual Plasticizer/Thermal Stabilizer Effect of Epoxidized Chia Seed Oil (*Salvia hispanica* L.) to Improve Ductility and Thermal Properties of Poly(Lactic Acid)

Ivan Dominguez-Candela [1], Jose Miguel Ferri [2], Salvador Cayetano Cardona [1], Jaime Lora [1] and Vicent Fombuena [2,*]

[1] Instituto de Seguridad Industrial, Radiofísica y Medioambiental (ISIRYM) Universitat Politècnica de València (UPV), Plaza Ferrándiz y Carbonell s/n, 03801 Alcoy, Spain; ivdocan@doctor.upv.es (I.D.-C.); scardona@iqn.upv.es (S.C.C.); jlora@iqn.upv.es (J.L.)

[2] Technological Institute of Materials (ITM), Universitat Politècnica de València (UPV), Plaza Ferrándiz y Carbonell 1, 03801 Alcoy, Spain; joferaz@upvnet.upv.es

* Correspondence: vifombor@upv.es

Abstract: The use of a new bio-based plasticizer derived from epoxidized chia seed oil (ECO) was applied in a poly(lactic acid) (PLA) matrix. ECO was used due to its high epoxy content (6.7%), which led to an improved chemical interaction with PLA. Melt extrusion was used to plasticize PLA with different ECO content in the 0–10 wt.% range. Mechanical, morphological, and thermal characterization was carried out to evaluate the effect of ECO percentage. Besides, disintegration and migration tests were studied to assess the future application in packaging industry. Ductile properties improve by 700% in elongation at break with 10 wt.% ECO content. Field emission scanning electron microscopy (FESEM) showed a phase separation with ECO content equal or higher than 7.5 wt.%. Thermal stabilization was improved 14 °C as ECO content increased. All plasticized PLA was disintegrated under composting conditions, not observing a delay up to 5 wt.% ECO. Migration tests pointed out a very low migration, less than 0.11 wt.%, which is to interest to the packaging industry.

Keywords: PLA; epoxidized chia seed oil (ECO); plasticizers; migration; disintegration

Citation: Dominguez-Candela, I.; Ferri, J.M.; Cardona, S.C.; Lora, J.; Fombuena, V. Dual Plasticizer/Thermal Stabilizer Effect of Epoxidized Chia Seed Oil (*Salvia hispanica* L.) to Improve Ductility and Thermal Properties of Poly(Lactic Acid). *Polymers* **2021**, *13*, 1283. https://doi.org/10.3390/polym13081283

Academic Editor: Beom Soo Kim

Received: 20 March 2021
Accepted: 12 April 2021
Published: 14 April 2021

1. Introduction

Currently, a global plastics production of 368 million tons was recorded in 2019, an increment of 2.5% from 2018. Conventional polymers such a polyethylene (PE), polypropylene (PP), polyvinyl chloride (PVC), poly(ethylene terephthalate) (PET), polystyrene (PS), and polyamide (PA), represent approximately 70% of plastic demand in Europe. With regard to the packing industry, which represents around 40% of total demand, these polymers are leading the plastic demand [1]. The majority are non-biodegradables as well as manufactured by petrochemical industries (non-renewable resources) [2]. The food packaging industry generates a large volume of waste due to it short-lifespan and its recycling is often limited to those not contaminated with food products. According to Plastics Europe 2020, about 39.5% of post-consumer waste was recycled, while 18.5% (3.2 million tons) ended up in landfills [1]. These non-recycled plastics need to be managed to avoid the presence in seas, lakes, and rivers which threatens the environment [3,4].

Concerning biodegradable polymers, their presence is increasing in the food packing industry. Several biodegradable polymers such as poly(lactic acid) (PLA), thermoplastic polyurethane (TPU), and polyhydroxyalkanoates (PHAs) have been applied as new alternatives [5–8]. The most employed polymer is PLA (about 10.9%), which is obtained by fermentation of polysaccharides or sugar extracted from potato, sugarcane, corn, etc., thus obtained by renewable resources [9]. PLA is currently manufactured for common applications such as salad cups, lamination films, drinking cups, containers, etc. [10].

Biodegradation of PLA is produced by hydrolysis, resulting in harmless and non-toxic substances [11,12]. Nowadays, PLA is considered economically competitive and its properties such as good processability, high transparency, water solubility resistance, biodegradability, recyclability, etc., make it suitable for good packaging [13,14]. Besides, the energy saved in production is around 22–55% with regard to petroleum-based polymers, thus contributing to a decrease in environmental impact [15]. PLA is characterized by its high tensile modulus, although some drawbacks such as brittle nature with elongation at break of less than 9%, a narrow processing window, poor thermal stability, etc., are detected [16,17]. Several methodologies to improve ductile properties have been carried out successfully using copolymers, blends, or plasticizers in a PLA matrix [18–20].

The plasticizer market is increasing its annual demand in the polymeric industries. Phthalates (PTs) are a common plasticizer and additive to provide transparency, flexibility, and durability properties to a polymer matrix, commonly found in food packaging, medical equipment, building materials, toys, etc. [21,22]. The annual world production of PTs as plasticizer is approximately 80% [21]. However, studies show a migration phenomenon from polymer matrix to element in contact causing health and environment impact. It is well known that exposure of PTs to human lives produces endocrine damage, and reproduction and carcinogenic effects [23]. According to the European Union and other organizations, specific PTs such as diisobutyl phthalate (DIBP) or diethylhexyl phthalate (DEHP) among others, are banned for contents above 0.1 wt.% [24]. Among other alternatives, poly(ethylene glycol) (PEG), polyethylene oxide (PEO), and adipates are widely studied in a PLA matrix, obtaining an excellent improvement of ductile properties [25–27]. However, petrochemical-based plasticizers are questioned for toxicity and therefore there is a continuous attempt to obtain bio-based plasticizers [28].

Vegetable oils (VOs) are an interesting route to achieve renewable plasticizers for three main reasons: they are widely available, have a low toxicity, and are biodegradable. Two reactive sites are identified in fatty acids of vegetable oils to bring compatibility with a polymer matrix: double bonds and ester groups [29]. To increase this compatibility, vegetable oil can be epoxidized, which consists of introducing epoxy groups (oxirane ring) in double bonds. Several studies reported the use of epoxidized vegetable oils in PLA matrix, thus obtaining a bio-based polymer with high performance as a plasticizer. Some epoxidized oils such as epoxidized linseed oil (ELO) and epoxidized soybean oil (ESBO) are available commercially at competitive prices. Several studies using epoxidized oil with a non-elevated number of oxirane groups have been reported. The study performed by Qiong Xu et al. reported an improvement of elongation at break from 3.98 to 6.5% using 9 wt.% of ESBO [30]. Further percentage led to a decrease in ductile properties due to plasticizer saturation. Garcia-Garcia et al. found an increment of elongation at break from 7.8 to 15% with 5 wt.% of epoxidized Karanja oil [31]. In respect to impact energy, an evident improvement of 32% was obtained, confirming an effective plasticization. More interesting results were found by Balart et al., who reported an increment of 450% of elongation at break with respect to neat PLA employing ELO with 8% of epoxy groups [32].

Chia seed oil (CO) is a promising VO characterized by its high iodine value (above 190 g I_2/100 g oil) [33], becoming suitable to be epoxidized. Epoxidized chia seed oil has not been previously applied in a polymer matrix and could present an elevated epoxy content, improving the compatibility between PLA.

As different authors have reported that the interaction between PLA and epoxidized chia seed oil (ECO) could occur between the carbonyl group, from ester linkage, present in the PLA main chain and the epoxy group of ECO. This reaction mechanism was proposed by Al-Mulla et al. [34]. Although the interaction mechanism is not very strong, the terminal location of the hydroxyl groups in PLA gives them a high availability to react with the epoxy groups [30]. Thus, based on these previous studies, Figure 1 shows the chemical interaction between PLA and ECO. This new bio-based plasticizer could be an alternative to ELO and ESBO plasticizers. For this reason, the aim of this work is studying the potential of epoxidized chia seed oil as a new bio-based plasticizer for PLA to be used in the packaging sector. Mechanical and thermal properties were tested to evaluate the effect of different percentage of ECO in a PLA matrix. A migration test was evaluated as an important property in the packaging sector. Finally, a disintegration test was carried out to evaluate the effect of ECO in polymer degradation.

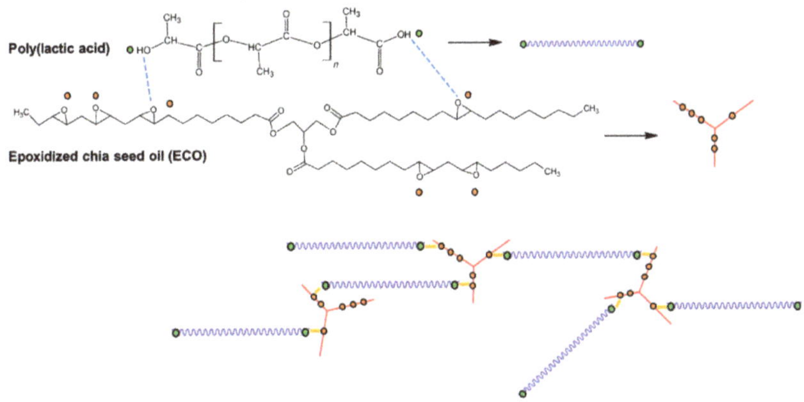

Figure 1. Schematic representation of chemical interactions between PLA and ECO.

2. Materials and Methods

2.1. Materials and Sample Preparation

Poly(lactic acid) (PLA) with a commercial grade 2003D was supplied by NatureWorks LLC (Minnetonka, MN, USA). Its density was 1.24 g·cm^{-3}, with an approximate molecular weight of 120,000 g·mol^{-1}. The base of the plasticizer was chia oil extracted through cold pressing extraction using a CZR-309 press machine (Changyouxin Trading Co., Zhucheng, China) from edible chia seed (*Salvia hispanica*, L.) supplied by Frutoseco (Bigastro, Alicante, Spain). The oil was characterized by a density of 0.928 g cm^{-3} at 25 °C and iodine value of 197. Epoxidation process was carried out in situ with hydrogen peroxide (30% *v/v*), acetic acid (99.7%), and sulfuric acid (97%) supplied by Sigma Aldrich (Sigma Aldrich, Madrid, Spain). Epoxidized chia seed oil (ECO) presented an epoxy content of 6.71%, equivalent to 238 EEW (Equivalent Epoxy Weight). This value was obtained following the guidelines of the ASTM D1652–97. ECO presented an iodine value of 25 and density of 1.026 g·cm^{-3} at 25 °C, which made it suitable to be used as plasticizer.

In the first stage, PLA pellets were dried at 60 °C for 24 h to remove moisture. After this, PLA pellets and ECO were weighed according to the compositions indicated in Table 1. Percentages of ECO were selected, taking into account published articles that consider that more than 10% of oil shows signs of plasticizer saturation [35–37]. All the five compositions selected were extruded at constant speed (40 rpm) using a twin-screw co-rotating extruder from DUPRA S.L (Castalla, Alicante, Spain) with the following temperature profiles: 162.5 (feeding zone), 165, 170, and 175 °C (die). After extrusion, samples were air-cooled to room temperature, pelletized, and dried again at 60 °C for 24 h for further processing by injection. Each composition was shaped by injected molding using a Meteor 270/75 from Mateu & Solé (Barcelona, Spain) at temperature profiles of 170, 180, 190, and 200 °C from feed section to injection nozzle. Cavity filling time was 1 s and the cooling time was set to 10 s.

Table 1. Composition of ECO plasticized PLA materials and labelling.

Reference	Parts by Weight (wt.%)	
	PLA	ECO
PLA	100	0
PLA_2.5%ECO	97.5	2.5
PLA_5%ECO	95	5
PLA_7.5%ECO	92.5	7.5
PLA_10%ECO	90	10

2.2. Mechanical and Thermal Characterization

The effect of ECO on mechanical properties was studied using standard tensile, impact test, and hardness. Tensile properties were obtained in a universal test machine ELIB 30 from S.A.E Ibertest (Madrid, Spain). Five different samples were tested using 5 kN load cell and a crosshead speed of 10 mm·min^{-1} according to ISO 527. Furthermore, an axial extensometer from S.A.E Ibertest (Madrid, Spain) was used to obtain tensile modulus with high accuracy. The impact strength was tested in a 1 J Charpy pendulum from Metrotec S.A. (San Sebastián, Gipuzkoa, Spain), as is indicated in ISO 179. Shore D hardness was carried out following the guidelines of the ISO 868 using a durometer 673-D from J. Bot S.A (Barcelona, Spain). In both tests, a minimum of 5 samples were used and the results shown are the obtained average.

Thermomechanical properties were assessed by using standard Vicat softening temperature (VST) by using a load of 5 kg and a heating rate of 50 °C·h^{-1} according to ISO 306. Moreover, heat deflection temperature (HDT) was obtained following the guidelines of ISO 75 with a load of 296 g and a heating rate of 120 °C·h^{-1}. Both values were obtained in a Vicat-HDT station mod. DEFLEX 687-A2 from Metrotec S.A. (San Sebastián, Gipuzkoa, Spain). At least five different specimens for each composition were tested and average values were calculated.

Additionally, storage modulus (G′) and damping factor (tan δ) were evaluated in torsion mode by dynamic mechanical thermal analysis (DMTA) in an oscillatory rheometer AR-G2 from TA instrument (New Castle, DE, USA). Shaped samples (4 × 10 × 40 mm^3) were evaluated with a dynamic program from 30 to 110 °C using a heating rate of 2 °C·min^{-1}. The maximum deformation was set to 0.1% with a constant frequency of 1 Hz.

Thermal transitions of PLA plasticized with different contents of ECO were obtained by differential scanning calorimetry (DSC) in a DSC mod. 821 from Mettler-Toledo Inc. (Schwerzenbach, Switzerland). Samples with an average weight of 6–8 mg were subjected to the following temperature program: 1st heating program from 25 to 300 °C at 10 °C·min^{-1} to remove thermal history, 2nd cooling program from 300 to 25 °C at 10 °C·min^{-1}, and 3rd heating program from 25 to 300 °C at 10 °C·min^{-1}. All thermal cycles were performed in a nitrogen atmosphere with a flow rate of 66 mL·min^{-1}. The percentage of crystallinity of different PLA formulations with ECO was determined by Equation (1):

$$X_c\ (\%) = \left[\frac{\Delta H_m - \Delta H_{cc}}{\Delta H_{m(100\%)}}\right] \cdot \frac{1}{w_{sample}} \times 100 \quad (1)$$

where ΔH_{cc} and ΔH_m represent cold crystallization and melting enthalpies (J·g^{-1}), respectively. The weight amount of PLA is represented by w_{sample} (g). Theoretical melting of PLA 100% crystalline ($\Delta H_{m(100\%)}$) was 93 J·g^{-1}, as is reported [10].

Thermogravimetric analysis was carried out in a TGA/SDTA 851 thermobalance from Mettler-Toledo Inc (Schwerzenbach, Switzerland). A heating ramp from 30 to 700 °C at constant heating rate of 10 °C·min^{-1} and constant flow rate of nitrogen (66 mL·min^{-1}) were set to evaluate samples with average weight of 10 mg. The temperature when a 5% weight loss has been reached and the maximum degradation were obtained in order to evaluate thermal stability of the different samples of PLA plasticized with ECO.

2.3. Morphology Characterization

Fractured surface from impact test was observed by field emission scanning electron microscopy (FESEM) model ZEISS ULTRA 55 from Oxford Instruments (Oxfordshire, UK). Previously fractured surfaces were coated with a thin metallic layer (Au-Pd alloy) employing a sputter coater EM MED020 from Leica Microsystems (Wetzlar, Germany) to avoid electrical charging. All samples were observed using an acceleration voltage of 2 kV.

2.4. Disintegration under Composting Conditions

Disintegration test was conducted in aerobic conditions according to ISO 20200 at temperature of 58 °C and a relative humidity of 55% using a synthetic compost reactor (300 × 200 × 100 mm^3). Sample sizing of 25 × 25 × 1 mm^3 were placed in a carrier bag and buried in controlled soil. Previously, all films manufactured with neat and plasticized PLA with ECO were dried at 40 °C over 24 h. Seven different samples of each formulation were employed for the disintegration process under composting conditions. Each control day (3, 7, 14, 17, 21, 24, and 28 days), a different sample of each formulation was unburied while the rest remained in the process. The removed samples were washed with distilled water and dried 24 h before to be weighed in an analytical balance. All tests were carried out in triplicate to ensure reliability. Average disintegration percentage of extracted samples was calculated using Equation (2).

$$W_l\ (\%) = \frac{w_0 - w}{w_0} \cdot 100 \tag{2}$$

where w_0 is referred to initial dry weight of the sample and w is the weight of the sample extracted from compost soil on different days after drying. Furthermore, optical images were taken to record the progression of disintegration along time.

2.5. Migration of Plasticizer by Solvent Extraction Test

Migration test was studied by solvent extraction using n-hexane solvent as is indicated by several reports [36,38,39]. Samples of PLA and PLA plasticized with ECO were immersed in n-hexane solvent at different temperatures (30, 40, 50, and 60 °C) over 8 h in an air circulating oven mod. Selecta 2001245 by JP Selecta S.A. (Barcelona, Spain). Before and after experiments, all samples were dried at 40 °C for 24 h to ensure the absence of solvent. The weight loss of plasticizer (WL_p) obtained by migration test was calculated using Equation (3).

$$WL_p\ (\%) = \frac{w_b - w_a}{w_b} \cdot 100 \tag{3}$$

where w_b is the weight of samples before experiment and w_a is the weight after migration test.

3. Results and Discussion

3.1. Effect of ECO in PLA on the Mechanical Properties

One of the main disadvantages of PLA is its low ductile property, which gives it a characteristic brittleness. As is shown in Figure 2, neat PLA used in the present study reaches an elongation at break of 8% and a tensile strength higher than 45 MPa. The resulting tensile modulus was higher than 3100 MPa, as plotted in Figure 3. The plasticization effect was clearly observed in the different formulations of PLA with ECO. For example, the addition of 2.5 wt.% of ECO (PLA_2.5%ECO) provided higher ductile properties, with an elongation at break around 18%. Consequently, mechanical properties like tensile strength and tensile modulus slightly decreased (40.9 and 3040 MPa, respectively), which was attributed to the elastomeric and toughening effect of ECO plasticizer. In Figure 2, it was possible to observe how the slope of the stress–strain curve became lower as the ECO content in the samples increased. This decrease resulted in a lower tensile modulus. At the same time, the elongation at break increased notably, reaching values of 64.5% for samples with 10 wt.% of ECO. As a consequence of this variation in mechanical and ductile properties, the toughness of the samples was clearly improved. This increase represents a 700% rise compared to neat PLA. Therefore, the presence of epoxy groups in ECO interacts with hydroxyl groups present in PLA, decreasing intermolecular forces and, consequently, increasing its ductile properties [30]. These results are in accordance with previous studies on the use of epoxidized vegetable oils like PLA plasticizer. Yu-Qiong et al. reported an increase of 123% employing epoxidized palm oil [30]. This lower increase in elongation values was due to the lower content of epoxy groups in palm oil (3.23%) with respect to chia oil (6.71%), which is one of the main parameters to be taken into account. For example, other authors have reported similar increases of 700% or even higher using epoxidized vegetable oils with content of epoxy groups around 5.8% [36,40,41]. Thus, stronger interactions occurred as epoxy content increased due to the increased presence of reactive groups Therefore, in view of the results obtained, it seems that once epoxy group values of 5.8% or higher were reached, a saturation effect was shown, not reaching further increases in the elongation at break. Where the use of ECO with PLA appeared to have a direct effect was on the attenuating effect of the sharp drop in mechanical properties. For example, Chieng et al. reported that using palm oil in 10 wt.% proportion reduces the tensile strength of the neat PLA by almost 50%. However, this decrease was only 21% when ECO was employed [35]. Therefore, the higher oxirane oxygen content in ECO than, for example in palm oil, can interact more strongly with PLA chains leading to intense polymer-plasticizer interactions allowing to maintain mechanical and ductile balanced properties. As a consequence of these interactions between PLA and ECO, it was remarkable that samples with 7.5 and 10 wt.% ECO content provided constant values in tensile mechanical properties. On the other hand, this attenuation of decrease in mechanical properties also suggested, as other cited authors have pointed out, that a percentage higher than 10 wt.% produces a negative effect on ductile properties due to the excess of plasticizer and a possible phase separation. Some authors like Sanyang et al. and Rizzuto et al. remarked that once all free volume is full of plasticizer, a decrease of elongation at break occurs due to free volume reduction [42,43]. Similar results were reported by Emad et al., who observed a decrease of elongation at break above of 9 wt.% of epoxidized palm oil [44]. Therefore, PLA_10%ECO obtained the highest elongation at break (64.3%), together with a reduction of 6.7% in tensile modulus (2893 MPa) and 21% in tensile strength (33.4 MPa) in respect to initial value (PLA neat).

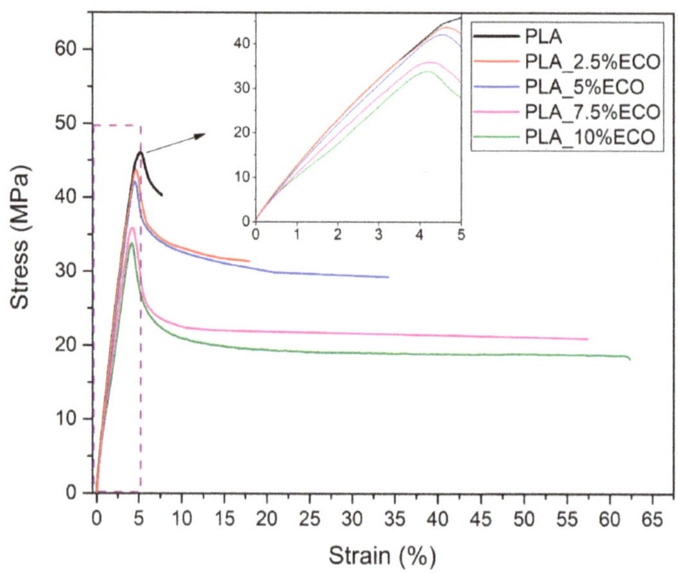

Figure 2. Plot evolution of characteristic stress-strain curves of PLA with different epoxidized chia seed oil (ECO) contents.

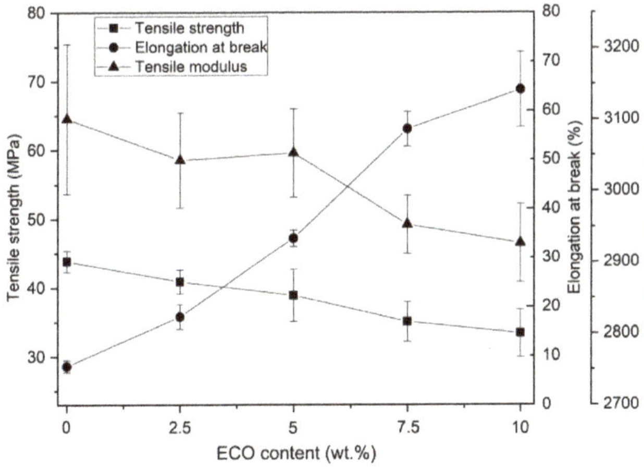

Figure 3. Plot evolution of tensile mechanical properties of PLA with different epoxidized chia seed oil (ECO) contents.

Other techniques that provide substantial information about the plasticizing effect of ECO on PLA are the Charpy impact test and Shore D hardness. As shown previously, the addition of ECO provides a plasticizing effect. As a result, tensile strength and tensile modulus decrease slightly with increment of ECO, contrary to the elongation at break. Impact absorbed energy is a parameter related to the ductile properties and toughness. Therefore, as expected initially, PLA is a brittle polymer, which has a low impact absorption capacity (37.1 $J \cdot m^{-2}$), but adding ECO improves absorbed impact energy due to elastomeric and toughing effect (Figure 4). It is possible to observe how an increase from 2.5 to 10 wt.% of ECO provides a gradual gain in impact absorption capacity. Sample ECO_10%ECO obtained the highest value (68.3 $J \cdot m^{-2}$), which represented an increase of 85% with respect to non-plasticized PLA. These results were in concordance with Carbonell-Verdu et al., who evaluated the plasticization effect on PLA using a 7.5 wt.% of epoxidized cottonseed oil, obtaining an increase on the impact of absorbed energy of 18% [45]. Again, the greater content of epoxy groups in ECO (6.7%) compared to cottonseed oil (5.8%), provided higher ductile properties. On the other hand, the plasticizing effect of ECO inversely affected the hardness of the samples. As the ECO content increased, the hardness decreased. However, in the same way that the tensile strength analysis showed a stabilization in the decrease of the properties of ECO_7.5%ECO and ECO_10%ECO samples, the difference between the hardness of neat PLA and ECO_10%ECO sample was only less than 6%.

Regarding the morphological changes produced by the incorporation of ECO to PLA, Figure 5a showed a brittle morphology with smooth surface and very low plastic deformation characteristic of neat PLA. With increasing ECO content, remarkable changes can be observed. Figure 5b, which represents PLA_2.5%ECO, showed a slight change in surface roughness. PLA_5%ECO, Figure 5c, shows a rougher surface as well as presence of filaments thus indicating an increase in ductility as a consequence of plasticization. So, an increase in roughness and a higher filament density was observed with increasing ECO. However, a higher presence of plasticizer (equal or more than 7.5 wt.%) began to display some spherical voids due to the plasticizer saturation, Figure 5d,e. A similar finding was reported by Ferri et al., who observed spherical voids above 5 wt.% of maleinized linseed oil as a plasticizer for PLA [46]. Phase separation was produced, and a worse miscibility occurred [31]. Finally, in Figure 5f, the presence of high density of filaments and voids in sample PLA_10%ECO, indicating a plasticizer saturation, can be observed better (2500×).

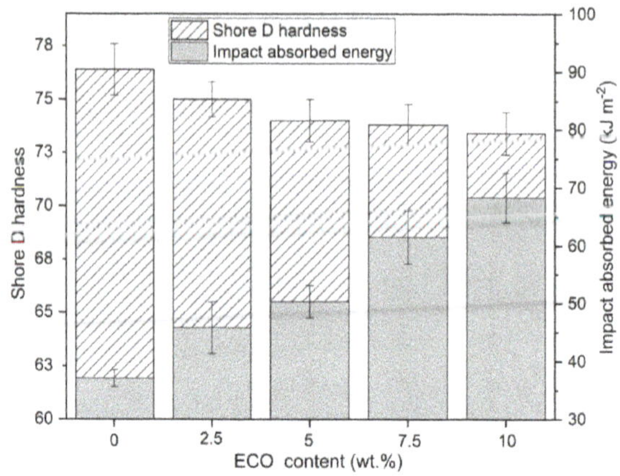

Figure 4. Plot evolution of Shore D hardness and impact absorbed energy of PLA with different epoxidized chia seed oil (ECO) content.

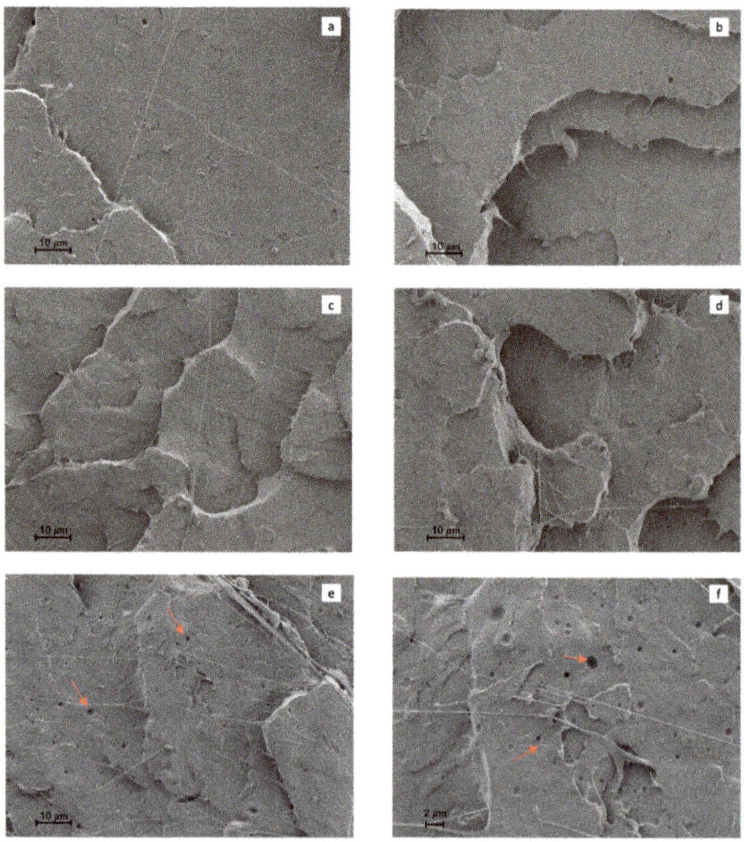

Figure 5. Fracture surface morphology of Charpy test at 1000× by field emission scanning electron microscopy (FESEM): (**a**) neat PLA; (**b**) PLA_2.5%ECO; (**c**) PLA_5%ECO; (**d**) PLA_7.5%ECO; (**e**) PLA_10%ECO; and (**f**) PLA_10%ECO at 2500×.

3.2. Effect of ECO in PLA on the Thermomechanical Properties of PLA

Storage modulus (G′) and damping factor (tan δ) were assessed by dynamic mechanical response. In Figure 6, the viscoelastic behavior of PLA_ECO formulations were exposed. Two characteristic changes in the storage modulus could be distinguished. The first change, between 50 and 70 °C, was the drop of storage modulus, which was related to glass transition temperature (T_g) at around 60 °C, as reported by Yong et al. [47]. The second change, between 80 and 100 °C, was recognized as the beginning of cold crystallization process. The addition of ECO to PLA resulted in a loss of storage modulus at lower temperatures. This was due to the plasticizing effect that ECO exerts on the PLA matrix, which increases the free volume between PLA chains before a saturation effect, decreasing the interaction between them [48]. In addition, at room temperature, neat PLA showed a storage modulus value of 1300 MPa, while the plasticized PLA formulations showed a decrease of this modulus up to 1000 MPa as a consequence of the plasticization effect. On the other hand, the beginning of cold crystallization decreased as ECO content increased, obtaining a shift from 87 up to 84 °C for plasticized PLA. This effect was due to plasticizer enabling the rearrangement in packed structure under lower energetic conditions.

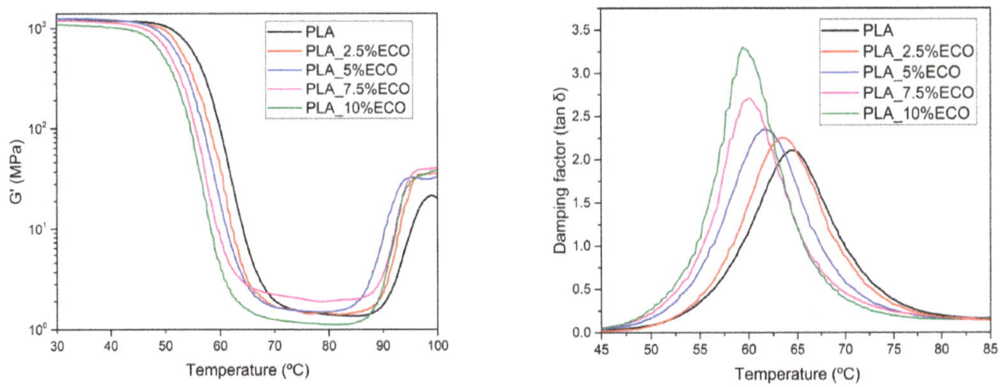

(**a**) (**b**)

Figure 6. Plot evolution of dynamic mechanical thermal analysis (DMTA) of PLA with different ECO contents: (**a**) storage modulus (G′); (**b**) damping factor (tan δ).

The temperature of the tan δ peak was a great manner to obtain an accurate value of T_g which was moved to lower temperatures as ECO content increased. Specifically, the T_g values were reduced from 64.2 (PLA neat) up to 61.9 and 59.5 °C for PLA_5%ECO and PLA_10%ECO, respectively. Regarding tan delta peak magnitude, it is related to their molecular mobility. As Chieng et al. reported, the addition of plasticizer content led to increase the intensity of tan delta due to higher molecular mobility caused by the plasticizing effect [35]. Then, a significant increase in respect to neat PLA in the magnitude of the damping factor was observed with higher ECO content. On the other hand, Silverajah et al. observed that increasing the content of epoxidized palm oil plasticizer in PLA led to decrease the temperature of tan δ peak [49].

Table 2 shows a summary of values obtained for Vicat softening temperature (VST) and heat deflection temperature (HDT) of neat PLA and PLA plasticized with ECO. These thermomechanical parameters are directly related to mechanical resistant properties. For this reason, the decreasing trend was the same that had been observed previously in tensile strength and modulus. Regarding VST, a remarkable decrease was detected when ECO content increased. Neat PLA had a VST of 56.6 °C, while samples with 7.5 and 10 wt.% ECO, respectively, reduced this value 4.4 °C. A similar trend was observed in HDT values, obtaining a difference of 2.6 °C between neat PLA and PLA_10%ECO sample. The addition of ECO to PLA samples in different percentages facilitated the sliding of the polymeric chains, being also favored by the slight increase in temperature that takes place in the VST and HDT tests. The ECO molecules decreased the intermolecular attraction forces between the PLA and the macromolecules, obtaining the plasticization of the materials [50].

Table 2. Vicat softening temperature (VST) and heat deflection temperature (HDT) of PLA with different epoxidized chia seed oil (ECO) content.

Reference	VST (°C)	HDT (°C)
PLA	56.6 ± 1.5	52.8 ± 0.5
PLA_2.5%ECO	55 ± 1.2	51.8 ± 0.6
PLA_5%ECO	53.2 ± 1.3	50.4 ± 0.4
PLA_7.5%ECO	52.2 ± 1.1	50.4 ± 0.6
PLA_10%ECO	52.2 ± 1.4	50.2 ± 0.4

3.3. *Effect of ECO in PLA on the Thermal Properties of PLA*

The addition of ECO led to a change in the main thermal transitions of neat PLA. Neat PLA presented a T_g at 62 °C, cold crystallization temperature (T_{cc}) at 119.4 °C, and melt temperature T_m at 150 °C, as can be seen in Table 3. As percentage of ECO increased in PLA matrix, a remarkable decrease of T_g values (62 °C from neat PLA up to 56.8 °C for PLA_10%ECO) was observed. It was seen that PLA_7.5%ECO and PLA_10%ECO showed similar T_g values (56.3 and 56.8 °C, respectively), demonstrating the plasticizer saturation effect. Additionally, the T_{cc} values presented a slight decrease with respect to neat PLA, obtaining in all plasticizer formulations lower values than neat PLA. Melt temperature (T_m) presented also a slight decrease with the incorporation of ECO, concluding that the presence of plasticizer also affected the melt temperature due to the increase of chain mobility. The same evolution was reported by Garcia-Garcia et al., who employed epoxidized Karanja oil in PLA [31]. Addition of ECO allowed to increase the free volume of polymer chains and thus provide better movement at lower temperatures [48]. The changes in the degree of crystallinity (X_c) also indicated the chain motions of PLA. It was clearly observed that X_c of PLA increased with the rise of ECO content, confirming that plasticizer enables the mobility of chains to form stable crystallites at lower energy conditions. Specifically, above 5 wt.% ECO, more evident changes were observed, where the highest crystallinity was found at PLA_10%ECO with 11.5%. This value was almost 67% higher than neat PLA, indicating the enhancement of chain mobility.

Table 3. Main thermal parameters of PLA plasticized with different ECO contents obtained using DSC.

Reference	T_g (°C) [1]	T_{cc} (°C) [2]	ΔH_{cc} ($J \cdot g^{-1}$) [3]	T_m (°C) [4]	ΔH_m ($J \cdot g^{-1}$) [5]	X_c (%) [6]
PLA	62.0	119.4	8.00	150.0	15.0	7.5
PLA_2.5%ECO	60.0	117.3	12.5	149.7	19.6	7.9
PLA_5%ECO	59.2	117.7	10.9	149.0	18.4	8.5
PLA_7.5%ECO	56.3	118.0	11.6	148.2	20.3	10.1
PLA_10%ECO	56.8	118.5	8.4	148.4	18.0	11.5

[1] Glass transition temperature; [2] Cold crystallization temperature; [3] Cold crystallization enthalpy; [4] Melt temperature; [5] Melt enthalpy; [6] Degree of crystallization.

Thermogravimetric analysis (TGA) was assessed for PLA formulations with different contents of ECO. Temperature at 5% weight loss ($T_{5\%}$) and temperature at maximum degradation (T_{max}) showed an important increase in thermal stabilization. CO presented higher thermal stability at elevated temperatures than neat PLA, obtaining in $T_{5\%}$ around of 320 °C and T_{max} around of 425 °C, as was reported by Timilsena et al. [33]. On the other hand, in order to study the influence of the epoxidation process in thermal stability of the vegetable oil, the authors compared an epoxidized linseed oil [51] and virgin linseed oil [52], obtaining very similar results. Linseed oil is characterized by an epoxy content very similar to ECO. For this reason, thermal stability of CO could be considered the same as ECO.

Thus, as can be seen in Figure 7a, higher contents of ECO provided an improvement of thermal stabilization with respect to neat PLA. Regarding weight loss derivate (Figure 7b), it was observed that samples up to 5 wt.% ECO showed an increase of $T_{5\%}$, while higher contents caused a decrease, possibly due to the first evidence of plasticizer saturation [31]. On the other hand, T_{max} of neat PLA was 390.4 °C and increased up to 404.9 °C for PLA formulations with a 2.5–5% ECO, showing a slight decrease when the saturation effect was beginning. Typical plasticizers employed in PLA as citrate esters or ATBC reduce thermal stability of PLA when their content increases [53,54]. However, in this report, different trends were observed in $T_{5\%}$ and T_{max}. Addition of ECO produced an evident delay of degradation temperature, as was reported by Garcia-Garcia et al., who used epoxidized oils as plasticizers [20]. The reason for this behavior was due to the presence of epoxy groups of ECO that allowed the scavenging of acid groups, obtaining a better thermal stabilization [55].

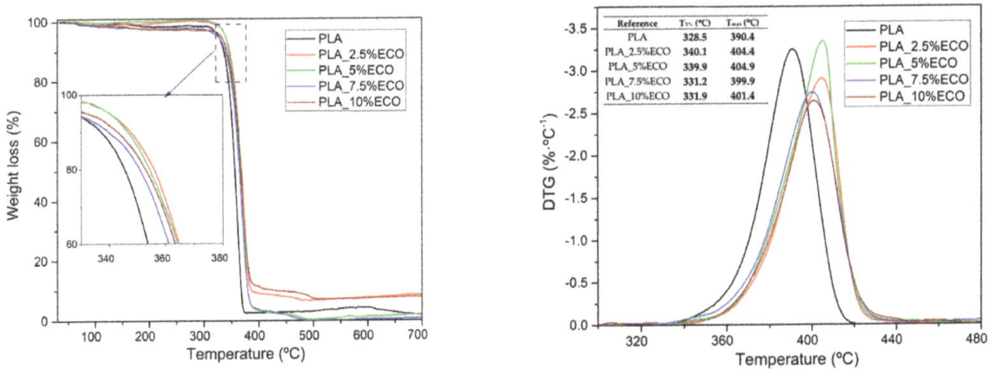

(a) (b)

Figure 7. Thermal parameters of degradation of PLA with different epoxidized chia seed oil (ECO) contents. (**a**) Weight loss; (**b**) derivative thermogravimetry. $T_{5\%}$ is temperature at 5% weight loss; T_{max} is temperature at maximum degradation.

3.4. Disintegration under Composting Conditions

Disintegration process under compost soil of neat and plasticized PLA is shown in both Figures 8 and 9, in which the visual appearance and the weight loss in respect to the initial mass, respectively, are plotted. After 3 days of incubation, film samples changed their visual appearance from translucid to opaque due to increased crystallinity and possible water absorption. It is important to remark that this experiment was conducted at thermophile conditions, with constant temperature of 58 °C and 50% relative humidity. The proximity to the T_g, as was studied by DSC, can induce an increment of chain mobility and thus the crystallization that causes the increasing opacity [31]. After 7 days buried in controlled compost soil, neat PLA and the sample with less amount of ECO, started the embrittlement process and slight weight loss, as plotted in Figure 9. However, samples with higher amounts of ECO did not show any signs of disintegration. After 14 days buried, visual changes and weight loss were evident in all samples. Neat PLA disintegrates faster than plasticized PLA with ECO. Neat PLA obtained the highest weight loss with 60.2%, while PLA_10%ECO lost 35% of its initial weight. Above 21 days buried in controlled conditions, samples showed a physical inconsistency and disintegration. According to ISO 20200, a disintegrable material was considered when the degree of disintegration achieved 90%. Neat PLA and samples with 2.5 and 5 wt.% of ECO achieved more than 90% of weight loss in respect to their initial value. However, samples with 7.5 and 10 wt. % needed 3 more days to reach this value, indicating a delay of the disintegration process provided by an increasing amount of ECO. These results are in concordance with Balart et al., who observed a reduction of disintegration capacity by increasing the epoxidized linseed oil content in PLA matrix [56]. This delay was directly related with the fact that samples plasticized with ECO possess a higher degree of crystallinity as reported in Table 3. Biodegradability is usually done by lipases, proteases, and esterase secreted from microorganism in the soil compost [57] and this microorganism acts faster in amorphous domains [58,59]. Thus, addition of ECO leads to a delay in the disintegrations process but, in general terms, PLA films developed with ECO can be considered as biodegradable according to ISO 20200.

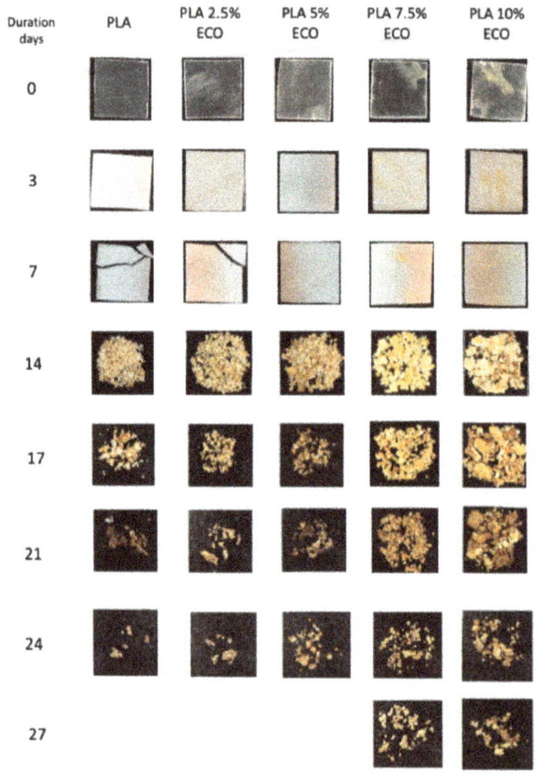

Figure 8. Visual appearance of disintegration under composting conditions.

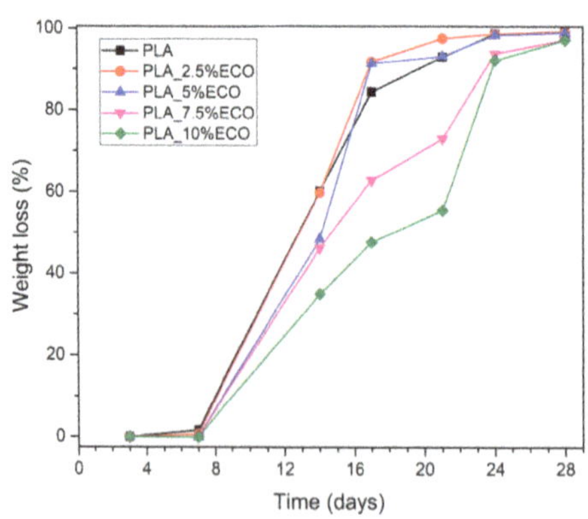

Figure 9. Weight loss recorded during disintegration test.

3.5. Migration of ECO by Solvent Extraction Test

PLA is considered a safe polymer for food contact applications [60]. For this reason, the development of a new plasticizer, as ECO, needs to be studied due to plasticizer migration being an important drawback [61,62]. Plasticizer migration is defined as the capacity of transferring plasticizer molecules from the surface of the matrix to the contact medium [63] and the assay developed by solvent extraction is a quite aggressive test to provide information about the potential use at industrial scale. In Figure 10, the percentage of migration of plasticizer from PLA matrix using *n*-hexane as dissolvent at different temperatures is plotted. Neat PLA, taken as control sample, presented a similar value below 0.02% between 30 and 60 °C due to absence of plasticizer. Regarding plasticized PLA with ECO, an increment of percentage of migration is appreciated as temperature increases achieving a highest migration of 0.108% with PLA_7.5%ECO and PLA_10%ECO at 60 °C. These results were lower than those reported by Carbonell-Verdu et al., who employed epoxidized cottonseed oil, obtaining values up to 0.12% [45]. A higher content of epoxy groups in ECO provided stronger interactions with the hydroxyl groups of PLA matrix and, as a consequence, lower migration levels in respect to other plasticizers with less reactive groups, indicating a correct functionality to be employed at an industrial scale. On the other hand, the high molecular weight characterizing the vegetable oils (around 900 g·mol^{-1}) could be another positive effect to minimize the migration level, compared with industrial plasticizers with lower molecular weight, as for example tributyl citrate plasticizers (350–400 g·mol^{-1}) [48].

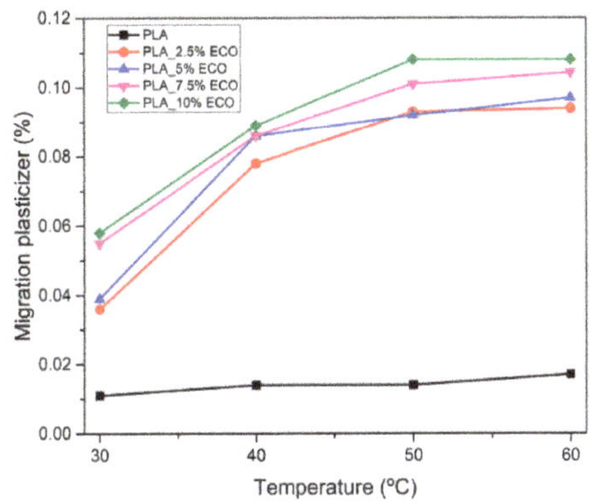

Figure 10. Migration of epoxidized chia seed oil (ECO) plasticizer in PLA matrix by n-hexane solvent extraction.

4. Conclusions

Epoxidized chia seed oil (ECO) was applied in a PLA matrix to evaluate its effect as a new bio-based plasticizer. The low elongation at break of neat PLA (8%) was improved up to values of 64.5% using 10 wt.% ECO, obtaining an improvement of 700%. Morphological images showed spherical voids at equal or higher percentage of 7.5 wt.% ECO in PLA matrix, indicating the beginning of plasticizer saturation. With regard to absorbed impact energy, an almost twice as high value was obtained with 10 wt.% ECO in respect to neat PLA, and, as a consequence, a decrease of hardness. Then, ECO led to enhanced chain mobility of PLA, which induced an increase in free volume and a reduction in intermolecular forces. This lubrication of chains also led to reduce the glass temperature around 4.7 °C with 7.5–10 wt.% ECO and slightly the cold crystallization temperature. Thermal stability was highly improved up to 14.0 °C in respect to neat PLA. The disintegration ability of PLA up to 5 wt.% ECO content was not affected, meanwhile at 7.5–10 wt.%, it was slightly delayed, being considered, in general terms, biodegradable formulations. Finally, very low migration of plasticizer was detected in migration test by using *n*-hexane. The maximum migration recorded was 0.108% with 10 wt.% ECO at 60 °C, while lower migration was obtained (<0.06%) at 30 °C. Therefore, ECO is a promising bio-based plasticizer with potential to be applied in the packaging industry.

Author Contributions: Conceptualization, I.D.-C. and V.F.; methodology, I.D.-C.; validation, I.D.-C., J.M.F.; formal analysis, S.C.C. and J.M.F.; investigation, I.D.-C.; resources, V.F and J.L.; data curation, S.C.C.; writing—original draft preparation, I.D.-C. and V.F.; writing—review and editing, V.F. and J.M.F.; visualization, J.L.; supervision, J.L. and V.F.; project administration, V.F. All authors have read and agreed to the published version of the manuscript.

Funding: I.D.-C. wants to thank Universitat Politècnica de València for his FPI grant (PAID-2019-SP20190013) and Generalitat Valenciana (GVA) for his FPI grant (ACIF/2020/233). J.M.F. thanks the postdoc contract (APOSTD/2019/122) Generalitat Valenciana (2019–2021).

Institutional Review Board Statement: Not applicable.

Informed Consent Statement: Not applicable.

Acknowledgments: Authors want to thank the support of the Cátedra FACSA-FOVASA of Water, Waste and Circular Economy, which promotes and supports training, dissemination, innovation, social responsibility, and entrepreneurship activities in the field of circular economy and has kindly provided the necessary chemical reagents.

Conflicts of Interest: The authors declare no conflict of interest.

References

1. Europe, P. Plastics—The Facts An analysis of European plastics production, demand and waste. Plastics Europe, Brussels. Available online: https://www.plasticseurope.org/download_file/force/4261/181 (accessed on 15 March 2021).
2. Muthuraj, R.; Misra, M.; Mohanty, A.K. Biodegradable compatibilized polymer blends for packaging applications: A literature review. *J. Appl. Polym. Sci.* **2018**, *135*, 45726. [CrossRef]
3. Yonkos, L.T.; Friedel, E.A.; Perez-Reyes, A.C.; Ghosal, S.; Arthur, C.D. Microplastics in Four Estuarine Rivers in the Chesapeake Bay, U.S.A. *Environ. Sci. Technol.* **2014**, *48*, 14195–14202. [CrossRef] [PubMed]
4. Song, Y.K.; Hong, S.H.; Jang, M.; Han, G.M.; Rani, M.; Lee, J.; Shim, W.J. A comparison of microscopic and spectroscopic identification methods for analysis of microplastics in environmental samples. *Mar. Pollut. Bull.* **2015**, *93*, 202–209. [CrossRef]
5. Râpă, M.; Miteluţ, A.C.; Tănase, E.E.; Grosu, E.; Popescu, P.; Popa, M.E.; Rosnes, J.T.; Sivertsvik, M.; Darie-Niţă, R.N.; Vasile, C. Influence of chitosan on mechanical, thermal, barrier and antimicrobial properties of PLA-biocomposites for food packaging. *Compos. Part. B Eng.* **2016**, *102*, 112–121. [CrossRef]
6. Muller, J.; González-Martínez, C.; Chiralt, A. Combination of poly (lactic) acid and starch for biodegradable food packaging. *Materials* **2017**, *10*, 952. [CrossRef] [PubMed]
7. Ivonkovic, A.; Zeljko, K.; Talic, S.; Lasic, M. Biodegradable packaging in the food industry. *J. Food Saf. Food Qual.* **2017**, *68*, 26–38.
8. Musioł, M.; Sikorska, W.; Adamus, G.; Janeczek, H.; Kowalczuk, M.; Rydz, J. (Bio)degradable polymers as a potential material for food packaging: Studies on the (bio)degradation process of PLA/(R,S)-PHB rigid foils under industrial composting conditions. *Eur. Food Res. Technol.* **2016**, *242*, 815–823. [CrossRef]
9. Murariu, M.; Dubois, P. PLA composites: From production to properties. *Adv. Drug Deliv. Rev.* **2016**, *107*, 17–46. [CrossRef]
10. Vink, E.T.; Rábago, K.R.; Glassner, D.A.; Gruber, P.R. Applications of life cycle assessment to NatureWorks™ polylactide (PLA) production. *Polym. Degrad. Stab.* **2003**, *80*, 403–419. [CrossRef]
11. Gross, R.A.; Kalra, B. Biodegradable polymers for the environment. *Science* **2002**, *297*, 803–807. [CrossRef]
12. Castro-Aguirre, E.; Iñiguez-Franco, F.; Samsudin, H.; Fang, X.; Auras, R. Poly(lactic acid)—Mass production, processing, industrial applications, and end of life. *Adv. Drug Deliv. Rev.* **2016**, *107*, 333–366. [CrossRef] [PubMed]
13. Leja, K.; Lewandowicz, G. Polymer biodegradation and biodegradable polymers-A review. *Pol. J. Environ. Stud.* **2010**, *19*, 255–266.
14. Malathi, A.; Santhosh, K.; Nidoni, U. Recent trends of biodegradable polymer: Biodegradable films for food packaging and application of nanotechnology in biodegradable food packaging. *Curr. Trends Technol. Sci.* **2014**, *3*, 73–79.
15. Rasal, R.M.; Janorkar, A.V.; Hirt, D.E. Poly (lactic acid) modifications. *Prog. Polym. Sci.* **2010**, *35*, 338–356. [CrossRef]
16. Farah, S.; Anderson, D.G.; Langer, R. Physical and mechanical properties of PLA, and their functions in widespread applications—A comprehensive review. *Adv. Drug Deliv. Rev.* **2016**, *107*, 367–392. [CrossRef] [PubMed]
17. Shayan, M.; Azizi, H.; Ghasemi, I.; Karrabi, M. Influence of modified starch and nanoclay particles on crystallization and thermal degradation prop-erties of cross-linked poly (lactic acid). *J. Polym. Res.* **2019**, *26*, 1–12. [CrossRef]
18. Lemmouchi, Y.; Murariu, M.; Dos Santos, A.M.; Amass, A.J.; Schacht, E.; Dubois, P. Plasticization of poly(lactide) with blends of tributyl citrate and low molecular weight poly(d,l-lactide)-b-poly(ethylene glycol) copolymers. *Eur. Polym. J.* **2009**, *45*, 2839–2848. [CrossRef]
19. Kowalczyk, M.; Pluta, M.; Piorkowska, E.; Krasnikova, N. Plasticization of polylactide with block copolymers of ethylene glycol and propylene glycol. *J. Appl. Polym. Sci.* **2012**, *125*, 4292–4301. [CrossRef]

20. Garcia-Garcia, D.; Fenollar, O.; Fombuena, V.; Lopez-Martinez, J.; Balart, R. Improvement of Mechanical Ductile Properties of Poly(3-hydroxybutyrate) by Using Vegetable Oil Derivatives. *Macromol. Mater. Eng.* **2017**, *302*, 1600330. [CrossRef]
21. Yang, J.; Li, Y.; Wang, Y.; Ruan, J.; Zhang, J.; Sun, C. Recent advances in analysis of phthalate esters in foods. *TRAC Trends Anal. Chem.* **2015**, *72*, 10–26. [CrossRef]
22. Khosravi, K.; Price, G.W. Determination of phthalates in soils and biosolids using accelerated solvent extraction coupled with SPE cleanup and GC–MS quantification. *Microchem. J.* **2015**, *121*, 205–212. [CrossRef]
23. Liu, X.; Sun, Z.; Chen, G.; Zhang, W.; Cai, Y.; Kong, R.; Wang, X.; Suo, Y.; You, J. Determination of phthalate esters in environmental water by magnetic Zeolitic Imidazolate Framework-8 solid-phase extraction coupled with high-performance liquid chromatography. *J. Chromatogr. A* **2015**, *1409*, 46–52. [CrossRef] [PubMed]
24. Stevens, M. *Polymer Chemistry: An Introduction*; Oxford University Press: New York, NY, USA, 1999.
25. Chieng, B.W.; Ibrahim, N.A.; Yunus, W.M.Z.W.; Hussein, M.Z. Plasticized poly(lactic acid) with low molecular weight poly(ethylene glycol): Mechanical, thermal, and morphology properties. *J. Appl. Polym. Sci.* **2013**, *130*, 4576–4580. [CrossRef]
26. Tsou, C.-H.; Gao, C.; De Guzman, M.; Wu, D.-Y.; Hung, W.-S.; Yuan, L.; Suen, M.-C.; Yeh, J.-T. Preparation and characterization of poly(lactic acid) with adipate ester added as a plasticizer. *Polym. Polym. Compos.* **2018**, *26*, 446–453. [CrossRef]
27. Shirai, M.; Grossmann, M.; Mali, S.; Yamashita, F.; Garcia, P.; Müller, C. Development of biodegradable flexible films of starch and poly(lactic acid) plasticized with adipate or citrate esters. *Carbohydr. Polym.* **2013**, *92*, 19–22. [CrossRef] [PubMed]
28. Bocqué, M.; Voirin, C.; Lapinte, V.; Caillol, S.; Robin, J.-J. Petro-based and bio-based plasticizers: Chemical structures to plasticizing properties. *J. Polym. Sci. Part. A Polym. Chem.* **2016**, *54*, 11–33. [CrossRef]
29. Vieira, M.G.A.; da Silva, M.A.; dos Santos, L.O.; Beppu, M.M. Natural-based plasticizers and biopolymer films: A review. *Eur. Polym. J.* **2011**, *47*, 254–263. [CrossRef]
30. Xu, Y.-Q.; Qu, J.-P. Mechanical and rheological properties of epoxidized soybean oil plasticized poly(lactic acid). *J. Appl. Polym. Sci.* **2009**, *112*, 3185–3191. [CrossRef]
31. Garcia-Garcia, D.; Carbonell-Verdu, A.; Arrieta, M.; López-Martínez, J.; Samper, M. Improvement of PLA film ductility by plasticization with epoxidized karanja oil. *Polym. Degrad. Stab.* **2020**, *179*, 109259. [CrossRef]
32. Balart, J.; Fombuena, V.; Fenollar, O.; Boronat, T.; Sánchez-Nacher, L. Processing and characterization of high environmental efficiency composites based on PLA and hazelnut shell flour (HSF) with biobased plasticizers derived from epoxidized linseed oil (ELO). *Compos. Part. B Eng.* **2016**, *86*, 168–177. [CrossRef]
33. Timilsena, Y.P.; Vongsvivut, J.; Adhikari, R.; Adhikari, B. Physicochemical and thermal characteristics of Australian chia seed oil. *Food Chem.* **2017**, *228*, 394–402. [CrossRef]
34. Al-Mulla, E.A.J.; Yunus, W.M.Z.W.; Ibrahim, N.A.B.; Rahman, M.Z.A. Properties of epoxidized palm oil plasticized polytlactic acid. *J. Mater. Sci.* **2010**, *45*, 1942–1946. [CrossRef]
35. Chieng, B.W.; Ibrahim, N.A.; Then, Y.Y.; Loo, Y.Y. Epoxidized Vegetable Oils Plasticized Poly(lactic acid) Biocomposites: Mechanical, Thermal and Morphology Properties. *Molecules* **2014**, *19*, 16024–16038. [CrossRef] [PubMed]
36. Carbonell-Verdu, A.; Samper, M.D.; Garcia-Garcia, D.; Sanchez-Nacher, L.; Balart, R. Plasticization effect of epoxidized cottonseed oil (ECSO) on poly(lactic acid). *Ind. Crop. Prod.* **2017**, *104*, 278–286. [CrossRef]
37. Garcia-Garcia, D.; Ferri, J.M.; Montanes, N.; Lopez-Martinez, J.; Balart, R. Plasticization effects of epoxidized vegetable oils on mechanical properties of poly(3-hydroxybutyrate). *Polym. Int.* **2016**, *65*, 1157–1164. [CrossRef]
38. Fenollar, O.; Sanchez-Nacher, L.; Garcia-Sanoguera, D.; López, J.; Balart, R. The effect of the curing time and temperature on final properties of flexible PVC with an epoxidized fatty acid ester as natural-based plasticizer. *J. Mater. Sci.* **2009**, *44*, 3702–3711. [CrossRef]
39. Carbonell-Verdu, A.; Garcia-Sanoguera, D.; Jordá-Vilaplana, A.; Sanchez-Nacher, L.; Balart, R. A new biobased plasticizer for poly(vinyl chloride) based on epoxidized cottonseed oil. *J. Appl. Polym. Sci.* **2016**, *133*, 43642/1–43642/10. [CrossRef]
40. Thuy, N.T.; Nam, B.X.; Duc, V.M. Study to improve the properties of polylactic acid by epoxidized crude rubber seed oil. *Vietnam. J. Chem.* **2019**, *57*, 735–740. [CrossRef]
41. Thakur, S.; Cisneros-Lopez, E.O.; Pin, J.-M.; Misra, M.; Mohanty, A.K. Green Toughness Modifier from Downstream Corn Oil in Improving Poly(lactic acid) Performance. *ACS Appl. Polym. Mater.* **2019**, *1*, 3396–3406. [CrossRef]
42. Rizzuto, M.; Mugica, A.; Zubitur, M.; Caretti, D.; Müller, A.J. Plasticization and anti-plasticization effects caused by poly(lactide-ran-caprolactone) addition to double crystalline poly(l-lactide)/poly(ε-caprolactone) blends. *CrystEngComm* **2016**, *18*, 2014–2023. [CrossRef]
43. Sanyang, M.L.; Sapuan, S.M.; Jawaid, M.; Ishak, M.R.; Sahari, J. Effect of sugar Palm-derived cellulose reinforcement on the mechanical and water barrier properties of sugar palm starch biocomposite films. *Polymers* **2015**, *7*, 1106–1124. [CrossRef]
44. Al-Mulla, E.A.J.; Ibrahim, N.A.B.; Shameli, K.; Ahmad, M.B.; Yunus, W.M.Z.W. Effect of epoxidized palm oil on the mechanical and morphological properties of a PLA–PCL blend. *Res. Chem. Intermed.* **2014**, *40*, 689–698. [CrossRef]
45. Carbonell-Verdu, A.; Boronat, T.; Quiles-Carrillo, L.; Fenollar, O.; Dominici, F.; Torre, L. Valorization of Cotton Industry Byproducts in Green Composites with Polylactide. *J. Polym. Environ.* **2020**, *28*, 2039–2053. [CrossRef]
46. Ferri, J.M.; Garcia-Garcia, D.; Montanes, N.; Fenollar, O.; Balart, R. The effect of maleinized linseed oil as biobased plasticizer in poly(lactic acid)-based formulations. *Polym. Int.* **2017**, *66*, 882–891. [CrossRef]
47. Yong, A.X.; Sims, G.D.; Gnaniah, S.J.; Ogin, S.L.; Smith, P.A. Heating rate effects on thermal analysis measurement of T g in composite materials. Advanced Manufacturing. *Polym. Compos. Sci.* **2017**, *3*, 43–51.

48. Dobircau, L.; Delpouve, N.; Herbinet, R.; Domenek, S.; Le Pluart, L.; Delbreilh, L.; Ducruet, V.; Dargent, E. Molecular mobility and physical ageing of plasticized poly(lactide). *Polym. Eng. Sci.* **2015**, *55*, 858–865. [CrossRef]
49. Silverajah, V.S.G.; Ibrahim, N.A.; Yunus, W.M.Z.W.; Abu Hassan, H.; Woei, C.B. A Comparative Study on the Mechanical, Thermal and Morphological Characterization of Poly(lactic acid)/Epoxidized Palm Oil Blend. *Int. J. Mol. Sci.* **2012**, *13*, 5878–5898. [CrossRef]
50. Alam, J.; Alam, M.; Raja, M.; Abduljaleel, Z.; Dass, L.A. MWCNTs-Reinforced Epoxidized Linseed Oil Plasticized Polylactic Acid Nanocomposite and Its Electroactive Shape Memory Behaviour. *Int. J. Mol. Sci.* **2014**, *15*, 19924–19937. [CrossRef]
51. Jebrane, M.; Cai, S.; Panov, D.; Yang, X.; Terziev, N. Synthesis and characterization of new vinyl acetate grafting onto epoxidized linseed oil in aqueous media. *J. Appl. Polym. Sci.* **2015**, *132*. [CrossRef]
52. Suryanarayana, C.; Rao, K.C.; Kumar, D. Preparation and characterization of microcapsules containing linseed oil and its use in self-healing coatings. *Prog. Org. Coat.* **2008**, *63*, 72–78. [CrossRef]
53. Arrieta, M.P.; Samper, M.D.; López, J.; Jiménez, A. Combined Effect of Poly(hydroxybutyrate) and Plasticizers on Polylactic acid Properties for Film Intended for Food Packaging. *J. Polym. Environ.* **2014**, *22*, 460–470. [CrossRef]
54. Arrieta, M.; de Dicastillo, C.L.; Garrido, L.; Roa, K.; Galotto, M. Electrospun PVA fibers loaded with antioxidant fillers extracted from Durvillaea antarctica algae and their effect on plasticized PLA bionanocomposites. *Eur. Polym. J.* **2018**, *103*, 145–157. [CrossRef]
55. Torres-Giner, S.; Montanes, N.; Fenollar, O.; García-Sanoguera, D.; Balart, R. Development and optimization of renewable vinyl plastisol/wood flour composites exposed to ultraviolet radiation. *Mater. Des.* **2016**, *108*, 648–658. [CrossRef]
56. Balart, J.F.; Montanes, N.; Fombuena, V.; Boronat, T.; Sánchez-Nacher, L. Disintegration in Compost Conditions and Water Uptake of Green Composites from Poly(Lactic Acid) and Hazelnut Shell Flour. *J. Polym. Environ.* **2017**, *26*, 701–715. [CrossRef]
57. Tokiwa, Y.; Calabia, B.P. Biodegradability and biodegradation of poly(lactide). *Appl. Microbiol. Biotechnol.* **2006**, *72*, 244–251. [CrossRef]
58. Gil-Castell, O.; Badia, J.; Kittikorn, T.; Strömberg, E.; Martínez-Felipe, A.; Ek, M.; Karlsson, S.; Ribes-Greus, A. Hydrothermal ageing of polylactide/sisal biocomposites. Studies of water absorption behaviour and Physico-Chemical performance. *Polym. Degrad. Stab.* **2014**, *108*, 212–222. [CrossRef]
59. Ray, S.S.; Yamada, K.; Ogami, A.; Okamoto, M.; Ueda, K. New Polylactide/Layered Silicate Nanocomposite: Nanoscale Control Over Multiple Properties. *Macromol. Rapid Commun.* **2002**, *23*, 943–947. [CrossRef]
60. Conn, R.; Kolstad, J.; Borzelleca, J.; Dixler, D.; Filer, L.; Ladu, B.; Pariza, M. Safety assessment of polylactide (PLA) for use as a food-contact polymer. *Food Chem. Toxicol.* **1995**, *33*, 273–283. [CrossRef]
61. Till, D.; Reid, R.; Schwartz, P.; Sidman, K.; Valentine, J.; Whelan, R. Plasticizer migration from polyvinyl chloride film to solvents and foods. *Food Chem. Toxicol.* **1982**, *20*, 95–104. [CrossRef]
62. Wang, X.; Song, M.; Liu, S.; Wu, S.; Thu, A.M. Analysis of phthalate plasticizer migration from PVDC packaging materials to food simulants using molecular dynamics simulations and artificial neural network. *Food Chem.* **2020**, *317*, 126465. [CrossRef] [PubMed]
63. Hakkarainen, M. Migration of Monomeric and Polymeric PVC Plasticizers. *Chromatogr. Sustain. Polym. Mater.* **2008**, *211*, 159–185. [CrossRef]

polymers

Article

Recyclability Analysis of Starch Thermoplastic/Almond Shell Biocomposite

Ana Ibáñez-García [1,*], Asunción Martínez-García [1] and Santiago Ferrándiz-Bou [2]

1 AIJU, Technological Institute for Children's Products & Leisure, 03440 Ibi, Alicante, Spain; sunymartinez@aiju.es
2 Technological Institute of Materials (ITM), Universitat Politècnica de València (UPV), Plaza Ferrándiz y Carbonell 1, 03801 Alcoy, Alicante, Spain; sferrand@mcm.upv.es
* Correspondence: anaibanyez@aiju.es; Tel.: +34-965-554-475

Abstract: This article is focused on studying the effect of the reprocessing cycles on the mechanical, thermal, and aesthetic properties of a biocomposite. This process is based on starch thermoplastic polymer (TPS) filled with 20 wt% almond shell powder (ASP) and epoxidized linseed oil (ELO) as a compatibilizing additive. To do so, the biocomposite was prepared in a twin-screw extruder, molded by injection, and characterized in terms of its mechanical, thermal, and visual properties (according to CieLab) and the melt flow index (MFI). The analyses carried out were tensile, flexural, Charpy impact tests, differential scanning calorimetry (DSC), thermogravimetric analysis (TGA). The effects of the reprocessing were also studied for the biodegradable unfilled TPS polymer. The results showed that TPS and TPS/ASP biocomposite suffer changes progressively on the properties studied after each reprocessing cycle. Furthermore, it was observed that the addition of ASP intensified these effects regarding TPS. However, in spite of the progressive degradation in both cases, it is technically feasible to reprocess the material at least three times without needing to incorporate virgin material.

Keywords: almond shell; starch thermoplastic polymer; biodegradable; biocomposite; recycling; reprocessing cycle; injection molding; natural filler

Citation: Ibáñez-García, A.; Martínez-García, A.; Ferrándiz-Bou, S. Recyclability Analysis of Starch Thermoplastic/Almond Shell Biocomposite. *Polymers* **2021**, *13*, 1159. https://doi.org/10.3390/polym13071159

Academic Editor: Elisabetta Ranucci

Received: 28 February 2021
Accepted: 1 April 2021
Published: 5 April 2021

Publisher's Note: MDPI stays neutral with regard to jurisdictional claims in published maps and institutional affiliations.

1. Introduction

Recently, there has been a considerable interest in the development of biobased polymers to decrease dependency on petroleum-based polymers due to environmental concerns. Biopolymers derived from renewable resources have a wide range of applications in different industries due to their specific characteristics. As a result of researches and advances in R&D, the properties of these materials have been improved, and this ushered in new markets ranging from packaging, food service, consumer electronics, automotive, agriculture/horticulture, and toys to textiles. Although packaging, either rigid or flexible, remains the largest field of application for these materials, with almost 53 percent (1.14 million tons) of the total bioplastics market in 2019 [1]. Despite this interest, there are serious doubts about the behavior of these plastics during processing. One of them is the possibility of reprocessing the material generated by defective production or cut parts (sprues, distribution channels, etc.).

Traditionally, the reprocessing of the thermoplastic materials led to a significant deterioration of the material properties. This deterioration was the result of the decrease in molecular weight caused by the breakage of the polymeric chains that occurs when the material is subjected to a high shear process. On the other hand, loss of mechanical and thermal properties as well as discolorations are some of the most common degradation problems in polymers as a result of the recycling process. Recycled polymers must possess a set of minimum performance characteristics to meet specific requirements after the recycling process. Therefore, it is of great interest to know the impact of successive reprocessing cycles and conditions on the physical, mechanical, and thermal properties of polymers.

Biodegradable materials, such as polylactic acid (PLA), polyhydroxyalkanoate (PHA), starch-based polymer (TPS), are thermoplastic materials and, therefore, they can be reprocessed. However, these materials are much more sensitive to thermo-mechanical degradation [2].

Mechanical recycling has been studied in biodegradable polyesters, such as PLA, for several reasons. On the one hand, because plastic waste is generated during the transformation process [3]. On the other hand, to extend the useful life of the material, the possibility of reusing post-consumer waste has been studied [4].

The reprocessing of PLA in different transformation processes has also been studied. One of the processes evaluated was the continuous extrusion up to ten times. The tensile properties of PLA do not depend significantly on the number of extrusion cycles, while the impact strength and viscosity clearly decrease by increasing extrusion cycles [4]. On the contrary, when the material is processed by injection molding, the values of tensile strength, breaking strain, hardness, and viscosity decreased progressively from the first processing [3]. This loss of mechanical properties is due to the polymer chain breaking during processing, which results in a significant decrease in molecular weight [2]. Another study that additionally corroborates the loss of mechanical properties with the number of injection cycles showed changes in the color of injected parts, which intensifies as the processing cycles increase [5]. Furthermore, the effect of drying a PLA/PBAT (polybutylene adipate terephthalate) blend before reprocessing has also been studied recently. Rheological characterization showed that when the sample was processed after drying, less degradation was observed, as hydrolysis degradation of the components was minimized. However, regardless of whether the mixture was dried or not, after the fifth extrusion process, the mechanical properties did not decrease significantly [6].

Unlike PLA, few studies are available on the mechanical recycling of other pure biopolymers, such as PHA or TPS. PHA can be recycled mechanically, assuming the loss of molecular weight and mechanical properties [2]. As for TPS, it was found through a comparative study that the reprocessing capacity depends on the chemical structure of the polymer. While one of the references, Mater-Bi TF01U/095R, could be reprocessed up to ten times, the other one, Mater-Bi YI014U/C, was not viable due to the large loss of mechanical properties of the material from the first reprocessing cycle [7].

When it comes to the mechanical recycling of biocomposites, they have also been studied. In general, it is expected that biopolymers reinforced with natural fibers are less likely to be mechanically recycled than the material itself, as biocomposites are more sensitive to thermo-mechanical degradation [2].

TPS is one of the most promising polymers to replace petroleum-based ones. TPS is based on different starches (wheat, corn, yucca, potato, etc.) [8]. Chemically, starch is constituted of two polysaccharides: amylose and amylopectin. It is partially crystalline and biodegradable in different media: water, soil, and compost. Depending on their ratio, mechanical properties can vary to a great extent. TPS is easy to process but very sensitive to thermal and mechanical degradation in the process and moisture. The density is higher than the most conventional thermoplastic polymers, between 1.2 and 1.5 g/cm^3 [9]. The mechanical properties of TPS are generally inferior to those of petrochemical plastics, so it is common to find commercial grades blended with other degradable polymers (PCL (polycaprolactone), PVOH (Polyvinyl alcohol), PLA, PHA) to improve their mechanical properties without affecting their biodegradability.

Our previous work focused on TPS/ASP biocomposites, with a constant ASP content of 20 wt%, and several epoxidized vegetable oils (EVOs) were added at 5, 10, and 20 parts per hundred resin (phr) of TPS/ASP biocomposites, as additives. Said oils were epoxidized soybean oil (ESBO), epoxidized linseed oil (ELO), and epoxidized corn oil (ECO). The purpose of the work was to study the effect on the properties of injected biodegradable parts. The most optimal performance was attained for biocomposites with ELO or ESBO between 10 and 20 phr.

Taking as a reference one of the above-mentioned optimal formulations, the main objective of this present study is to study the influence of reprocessing cycles on the properties of TPS and TPS/ASP biocomposite in terms of their mechanical and thermal properties and changes in morphology, visual appearance, and melt flow index.

This study shows the degradation of TPS and TPS/ASP and establishes possible strategies for the reuse of the discarded parts generated during the manufacturing process with these materials.

2. Materials and Methods

2.1. Materials

For this study, a commercially available starch-based polymer, Mater-Bi® EI51N0 of Novamont, was studied. This bio-based and biodegradable polymer has a melt flow index (MFI) of 19 g/10 min (190 °C/2.16 kg) and a density of 1240 kg/m^3 (data provided by Novamont). This grade was selected for its properties, which are quite similar to polypropylene (PP), of high use in injection molding. Table 1 shows some properties of the as-received material.

Table 1. Properties of Mater-Bi® EI51N0 extract to datasheet supplied by Novamont.

Property	Test	Value
Melting temperature (°C)	ASTM-D3418	167
Tensile strength at break (MPa)	ASTM-D638	39
Elongation at break (%)	ASTM-D638	2.5
Young Modulus (MPa)	ASTM-D638	2200

Almond shell powder (ASP) with particle size between 0.05 and 0.125 mm provided by Hermen Europe, S.L. (Spain) was used for this study (Figure 1a). It consisted of a mixture of different almond shell varieties. Figure 1b shows that the predominant particle size range was about 0.08–0.125 mm. Table 2 shows the content of fixed carbon, volatile matter, humidity, ash, and the chemical composition of the main components of the used almond shell: hemicellulose, lignin, and cellulose, determined by thermogravimetric analysis (TGA) in previous work [10].

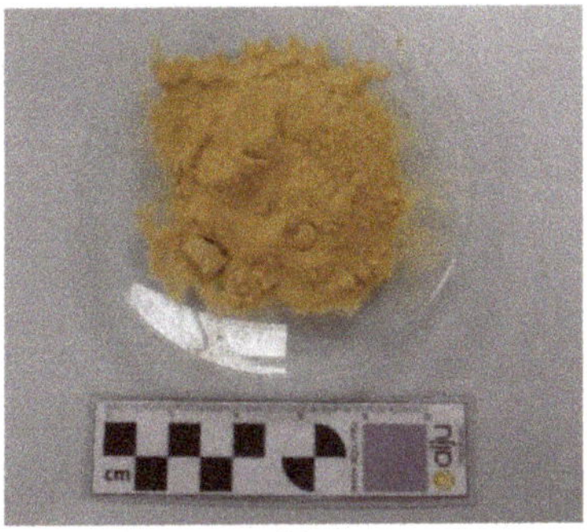

(a)

(b)

Figure 1. (**a**) Almond shell powder (ASP) powder, (**b**) Scanning electron microscope (SEM) image of ASP (×200 magnification and a scale marker of 100 μm).

Table 2. Content of moisture, volatile matter, fixed carbon ash, and composition of the used almond shell powder (ASP).

	Moisture (%)	Volatile Matter (%)	Fixed Carbon (%)	Ash(%)	Hemicellulose (%)	Cellulose (%)	Lignin (%)
ASP	7.4	64.5	19.4	8.7	37	27	36

The compatibilization of starch polymer and ASP was carried out with epoxidized linseed oil (ELO), supplied by Traquisa S.L. (Spain). Some of its properties are shown in Table 3.

Table 3. Properties of epoxidized linseed oil (ELO) datasheet supplied by Traquisa.

Property	Value
Density at 20 °C (kg/m^3)	1050–1060
Viscosity at 25 °C (cP)	8–12
Acid value (mg KOH/g)	$\leq$1
Iodine content	$\leq$5
Oxygen epoxide (%)	$\geq$8

2.2. *Experimental Procedure*

2.2.1. Preparation of Composites

Before processing, TPS and ASP were dried separately for 4 h at 60 °C and 24 h at 105 °C, respectively, in a SLW 115STD INOX air-circulating oven from POL-EKO (Wodzisław Śląski, Poland) to minimize its moisture and to avoid hydrolytic reactions [11].

A composite of starch-based polymer with ASP was developed using a BERSTORFF ZE26 × 44D-BASIC co-rotating twin-screw extruder (26:44 L/D). Before feeding the material into the extruder, a manual pre-mixing of the different components, Mater-Bi® EI51N0/ASP/ELO, was carried out and fed into the extruder through the main hopper. The ratio of TPS/ASP was set at 20 wt% since this content has shown balanced mechanical properties and appealing aesthetics in a previous study dealing with TPS biocomposites [12]. ELO was added at 10 parts per hundred resin (phr) of biocomposite. The temperature profile was set as follows: 130-185-185-185-185 °C (from feeding zone to die). The rotating speed was 80 rpm. The extruded material was finally pelletized using an air knife.

2.2.2. Injection Molding

Testing samples were molded using an injection molding machine Ergotech 110–430h/310V from DEMAG (Demag, Germany). The injection conditions used to develop recycled TPS and biocomposite TPS/AASP test specimens are listed in Table 4. After each injection cycle, the material was milled and dried at 60 °C for 6 h. Then, TPS was processed a total of four cycles and TPS/ASP biocomposite six cycles. The difference in the number of processing cycles between TPS and TPS/ASP was due to the poor quality of the specimens after TPS-4. TPS-5 presented injection defects, such as shrinkage. The nomenclature of recycled TPS and TPS/ASP biocomposite is given in Table 5. Specimens were conditioned at a temperature of 23 °C and relative humidity of 50% for at least 16 h before testing.

Table 4. Injection condition of starch-based polymer and biocomposite.

Parameters	TPS	TPS/ASP
Barrel profile (°C)	210-190-170-140-50	200-190-170-140-50
Mold temperature (°C)	35	35
Injection speed (rpm)	60	60
Pack pressure (bar)	400	400
Pack time (s)	15	15
Back pressure (bar)	50	50
Cooling time (s)	40	40

Table 5. Nomenclature of recycled biocomposite.

Injection Cycle Number	Recycling Order	Designation of As-Received Material	Designation of Biocomposites
1	0	TPS-1	TPS/ASP-1
2	1	TPS-2	TPS/ASP-2
3	2	TPS-3	TPS/ASP-3
4	3	TPS-4	TPS/ASP-4
5	4	—	TPS/ASP-5
6	5	—	TPS/ASP-6

2.2.3. Color Measurements

The influence of the reprocessing cycles on the color of the TPS and TPS/SP biocomposite specimens was recorded with a CR-200 Chroma Meter from KONICA MINOLTA (Tokyo, Japan). Moreover, the color indexes (L*, a* and b*) were measured according to the following criteria: L* = 0, darkness; L* = 100, lightness; + a* = red, − a* = Green and + b* = yellow, − b* = blue. From these coordinates, it was possible to determine the color difference associated with this space. The color variation, ΔE^*_{ab}, was obtained by the following Equation (1) and compared with the color coordinates of the neat TPS (TPS-1) and TPS/ASP (TPS/ASP-1).

$$\Delta E^*_{ab} = \sqrt{(\Delta L^*)^2 + (\Delta a^*)^2 + (\Delta b^*)^2} \quad (1)$$

The color change was assessed according to experimentally verified data [13]: Unnoticeable ($\Delta E^*_{ab} < 1$), only an experienced observer can notice the difference ($1 < \Delta E^*_{ab} < 2$), an unexperienced observer notices the difference ($2 < \Delta E^*_{ab} < 3.5$), there is a clear noticeable difference ($3.5 < \Delta E^*_{ab} < 5$), and the observer notices two different colours ($\Delta E^*_{ab} > 5$).

2.2.4. Mechanical Properties

The mechanical properties of TPS and TPS/ASP biocomposites were determined to study the capacity of their reprocessability.

Tensile and flexural tests were performed using an 6025 universal testing machine from INSTRON (Canton, Massachusetts, USA) with 5 kN power sensors. The tensile test was performed according to ISO 527 standard, using a crosshead speed of 1 mm/min for Young's Modulus determination and 5 mm/min for tensile and elongation at break determination. The extensometer used was MTS 634.11F-54. Recorded values included ultimate tensile strength (UTS), Young's modulus, and strain at break. A total of 5 specimens from each material were tested using standardized samples 1A (dogbone).

Impact tests were performed with a Resil 5.5 impact testing device (CEAST RESILIMPACTOR) with a 1 Joule hammer. Then, test samples were cut and tested according to ISO 179 standard. A total of 10 sample tests from each material was tested.

2.2.5. Scanning Electron Microscopy (SEM)

The impact fracture surface obtained after a Charpy impact test was analyzed using a Jeol JSM-840 SEM system. Samples were gold-coated before analysis, and the energy of the electron beam was 20 kV.

2.2.6. Thermal Analysis

The main thermal degradation parameters of biocomposites, initial degradation temperature (T_{onset}), and maximum mass loss rate temperature (Tmax) were studied by TGA using a TA Instrument Q500 thermogravimetric analyzer from TA INSTRUMENTS (Delaware (New Castle, DE, USA)). Then, those samples with an average weight between 8 and 10 mg were placed in standard platinum crucibles of 70 µL. In this case, biocomposites were subjected to the following temperature program: from 30 to 600 °C under nitrogen (N_2)

atmosphere at a rate of 10 °C/min, and from 600 to 1000 °C under oxygen (O_2) atmosphere at a rate of 10 °C/min with a purge gas flow of 10 mL/min.

Later on, thermal transitions of developed biocomposites were studied by differential scanning calorimetry (DSC) in a DSC Q200 calorimeter from TA INSTRUMENTS (New Castle, DE, USA). Then, those samples with an average weight of 8–10 mg were placed in standard aluminum crucibles and subjected to a three-step program that consisted of an initial heating cycle from 0 °C to 250 °C at a rate of 10 °C/min to remove thermal history, followed by cooling to −20 °C at a rate of 5 °C/min, and a second heating cycle to 350 °C at a rate of 10 °C/min. All the tests were run in an N_2 atmosphere with a constant flow of 50 mL/min. Finally, thermal transitions were determined: temperature (T_c) and enthalpy (ΔH_c) of crystallization after the cooling cycle and temperature (T_f) and enthalpy of fusion (ΔH_f) after the second heating.

2.2.7. Melt Flow Index

The determination of the melt flow index (MFI) was carried out according to ISO 1133 standard with a MFI TWELVINDEX equipment from ASTFAAR (Milan, Italy) at 190 °C and 2.16 kg. The cut time between two consecutive measurements was 15 s.

3. Results

3.1. Visual Aspect

The appearance of the injected specimens of TPS and TPS/ASP after the different reprocessing cycles are gathered in Figure 2a,b, respectively. Table 6 summarizes the color indexes (L*, a* and b*), and the color variation measured by ΔE^*_{ab}, with respect to TPS and TPS/ASP biocomposite with only the first processing cycle (TPS-1 and TPS/ASP-1, respectively).

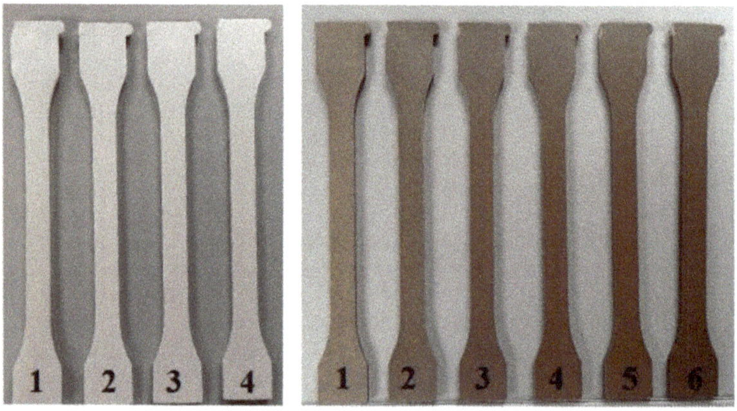

(a) (b)

Figure 2. Visual aspect of the injected specimens subjected to different reprocessing cycles. (**a**) Starch thermoplastic polymer (TPS) (**b**) TPS/ASP biocomposite.

The reprocessing cycles had more effect on the change in color of TPS/ASP biocomposite than in the unfilled TPS. As expected, the L* value decreased (white to black) in both cases. a* and b* coordinates changed progressively after each reprocessing cycle. Moreover, after the second processing cycle, the obtained ΔE^*_{ab} value of TPS was 1.00 (TPS-2). After the third reprocessing cycle, the value slightly increased to 1.91 (TPS-3); only an experienced observer could notice the difference in color [13]. However, in the case of TPS/ASP, after the second processing cycle, the value obtained of ΔE^*_{ab} was 3.96 (TPS/ASP-2), which resulted in samples whose change in color could be noticed by an observer ($3.5 < \Delta E^*_{ab} < 5$). The color variation clearly showed an increasing tendency in reprocessing, as expected due to degradation [13]. The ΔE^*_{ab} was more remarkable and noticeable for TPS/ASP biocomposite.

Table 6. Color parameters (L^*, a^*, b^* and ΔE^*_{ab} injection-molded samples of starch thermoplastic polymer (TPS) and TPS/ASP biocomposite subjected to different reprocessing cycles.

Samples	L^*	a^*	b^*	ΔE^*_{ab}
TPS-1	92.50 ± 0.48	0.72 ± 0.19	4.69 ± 0.16	—
TPS-2	91.76 ± 0.26	0.81 ± 0.05	4.69 ± 0.16	1.00
TPS-3	90.85 ± 0.16	0.65 ± 0.05	4.98 ± 0.16	1.91
TPS-4	89.88 ± 0.27	0.51 ± 0.03	5.28 ± 0.16	2.91
TPS/ASP-1	57.35 ± 0.74	6.61 ± 0.40	10.77 ± 0.18	—
TPS/ASP-2	53.95 ± 0.41	5.79 ± 0.10	8.92 ± 0.45	3.96
TPS/ASP-3	52.70 ± 0.44	5.96 ± 0.06	9.18 ± 0.19	4.96
TPS/ASP-4	50.25 ± 0.21	5.92 ± 0.03	9.25 ± 0.21	7.29
TPS/ASP-5	49.52 ± 0.63	5.82 ± 0.08	9.56 ± 0.37	7.96
TPS/ASP-6	47.90 ± 0.72	5.98 ± 0.15	9.92 ± 0.18	9.51

3.2. Thermal Properties

Table 7 shows the main thermal degradation parameters, initial degradation temperature (T_{onset}), and maximum mass loss rate temperature (T_{max1} and T_{max2}) obtained by TGA. Figures 3 and 4 show the TGA curves of TPS and TPS/ASP biocomposite, respectively, obtained after different reprocessing cycles. Furthermore, all samples presented a two-step process because the thermogravimetric analysis was carried out under an N_2 and O_2 atmosphere. When it comes to the comparative TGA curves of TPS samples obtained after different reprocessing cycles, those suggested no significant changes in the thermal degradation parameters since the curves overlapped. TPS-1 presented moderate thermal stability with T_{onset}, and T_{max1} of 326.1 °C and 362.1, respectively. It is noticeable that, after the fourth injection cycle, the values remained almost invariable. This indicated that the reprocessing cycles did not have a significant effect on thermal degradation at the current processing temperatures. Regarding the TGA thermograms of TPS/ASP, the addition of ASP reduced the thermal stability of the biocomposite because almond shells degrade faster than polymer matrix [10,12], and T_{onset} and T_{max1} moved towards a lower temperature. T_{max2} remained practically unchanged. However, in a previous study, which is in the process of publication, it was found that the addition of 10 phr ELO had a positive effect on the thermal stability of this type of biocomposite, increasing T_{onset} and T_{max1} by 6 °C: TPS/ASP-1 biocomposite presented a T_{onset} of 315.0 °C and T_{max1} of 346.0 °C. After the second injection cycle, T_{onset} decreased by 9 °C and continued to decrease progressively after each reprocessing cycle. Nevertheless, the T_{max1} remained almost invariable until the fifth injection cycle (340.6 °C).

Table 7. Thermal properties of the injection-molded samples of TPS and TPS/ASP biocomposite subjected to different reprocessing cycles obtained by thermogravimetric analysis (TGA).

Samples	T_{onset} (°C)	T_{Max1} (°C)	T_{Max2} (°C)	Residual Weight (%)
TPS-1	326.1 ± 0.4	362.1 ± 0.3	405.6 ± 1.7	1.42 ± 0.04
TPS-2	327.9 ± 0.1	360.9 ± 0.2	409.6 ± 2.7	1.45 ± 0.02
TPS-3	325.9 ± 0.3	360.7 ± 1.2	409.3 ± 2.0	1.66 ± 0.30
TPS-4	325.3 ± 1.6	360.2 ± 1.4	409.5 ± 0.6	1.42 ± 0.03
TPS/ASP-1	315.0 ± 3.4	346.0 ± 0.5	406.9 ± 1.0	1.28 ± 0.04
TPS/ASP-2	305.9 ± 1.2	345.4 ± 1.4	407.6 ± 1.0	1.37 ± 0.06
TPS/ASP-3	303.9 ± 0.9	345.4 ± 0.5	406.9 ± 1.0	1.23 ± 0.01
TPS/ASP-4	299.4 ± 0.3	345.0 ± 0.9	406.5 ± 0.5	1.31 ± 0.06
TPS/ASP-5	300.4 ± 1.1	340.6 ± 2.2	407.6 ± 0.6	1.26 ± 0.22
TPS/ASP-6	298.8 ± 0.9	333.9 ± 0.2	405.7 ± 1.6	1.24 ± 0.09

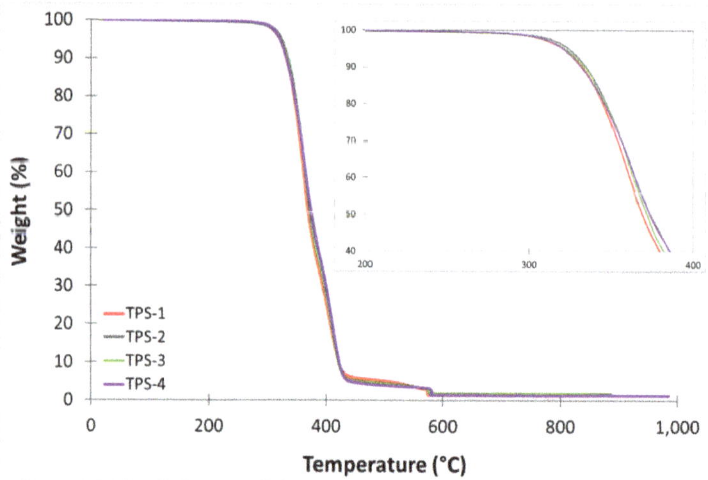

(a)

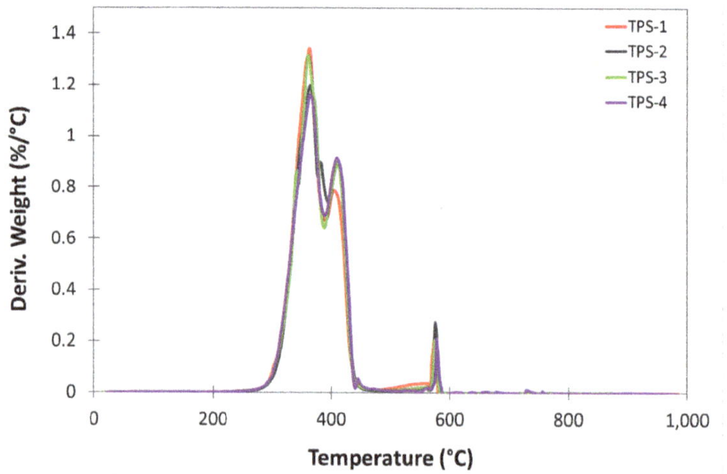

(b)

Figure 3. (**a**) Thermogravimetric analysis (TGA) thermograms corresponding to different reprocessing cycles of TPS (**b**) first derivative (DTG) curves.

To obtain the main thermal transition of the material a differential scanning calorimetry was used. Figures 5 and 6 show the DSC curves achieved from the cooling and second heating scans of TPS and TPS/ASP biocomposite samples, respectively, obtained after different reprocessing cycles. In addition, Table 8 summarizes the most relevant thermal parameters obtained from the cooling and second heating of the samples subjected to different reprocessing cycles. Bastioli et al. [14] described Novamont's starch-based technology as a process that could destroy amylose and amylopectin from starch. Thus, the main endothermic (T_{m1} y T_{m2}) and exothermic (T_c) peaks were related to the melting and the cold crystallization of the crystalline structure of the synthetic biodegradable polymer presented in Mater-Bi, respectively. After the first injection run (TPS-1), TPS presented a bimodal endothermic peak (Figure 5b). The first one, smaller, at around 161.4 °C (T_{m1}), and the second one, more pronounced, at 168.5 °C (T_{m2}). These are associated with the fusion of the crystalline structure as a result of chain scission caused by different resistance to heat. The TPS-1 thermogram corresponding to the cooling scan presented a unique main exothermic peak around 109.5 °C (Figure 5a). As can be observed, after different reprocessing cycles, the thermal transitions of TPS, such as T_c, T_m, and their respective enthalpy, did not present significant changes. The addition of 20 wt% produced a slight reduction in all parameters. After the first injection run, TPS/ASP-1 also presented a bimodal endothermic peak. The melting point temperature, T_{m1}, decreased from 161.4 °C to 157.1 °C and the T_{m2} from 168.5 °C to 166.4 °C. Regarding the normalized melting enthalpy (ΔH_m), it decreased from 23.8 to 17.2 J/g. These results indicated that the addition of almond shell to the starch-based polymer decreases the crystallization of the molecular chains [12]. After the second injection cycle (TPS/ASP-2), T_{m1} and T_{m2} decreased 2–3 °C compared to TPS/ASP-1 and continued to decrease progressively after each reprocessing cycle until 144.6 °C and 162.7 °C (Figure 6b), respectively. The decrease in the values may be attributed to the higher mobility of the polymer chains as a result of the reduction in the molecular weight during the recycling process. In addition, it was noticed that the crystallization temperature (T_c) slightly decreased upon adding ASP, thus indicating that the presence of ASP made crystallization start later, compared to as-received TPS. Furthermore, it was noticed that the crystallization temperature and enthalpy of TPS/ASP decreased progressively after each reprocessing cycle (Figure 6b). This indicated a loss of crystalline structure and a major difficulty in the crystallization process (i.e., starting at a lower temperature) as the number of processing cycles was increased [15].

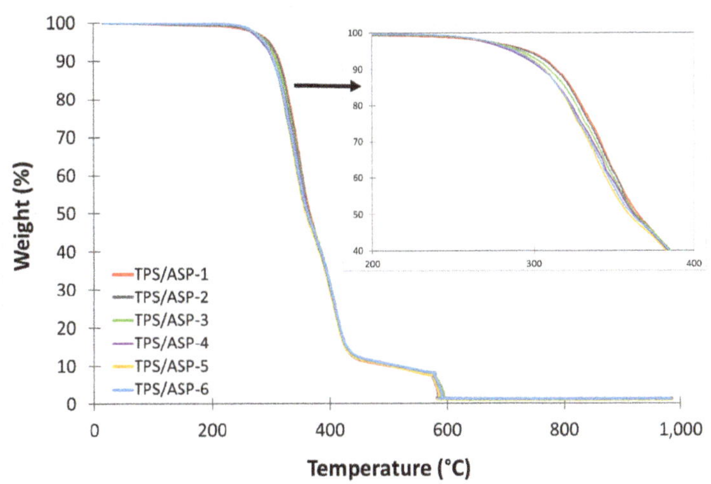

(a)

Figure 4. *Cont.*

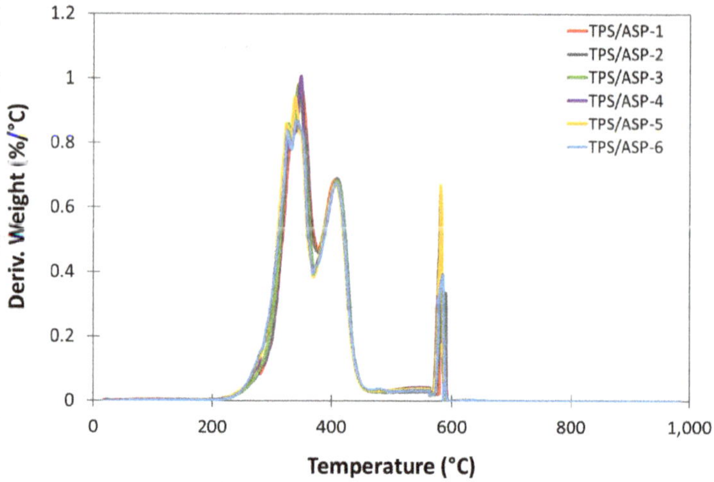

(b)

Figure 4. (**a**) TGA thermograms corresponding to different reprocessing cycles of TPS/ASP biocomposite (**b**) first derivative (DTG) curves.

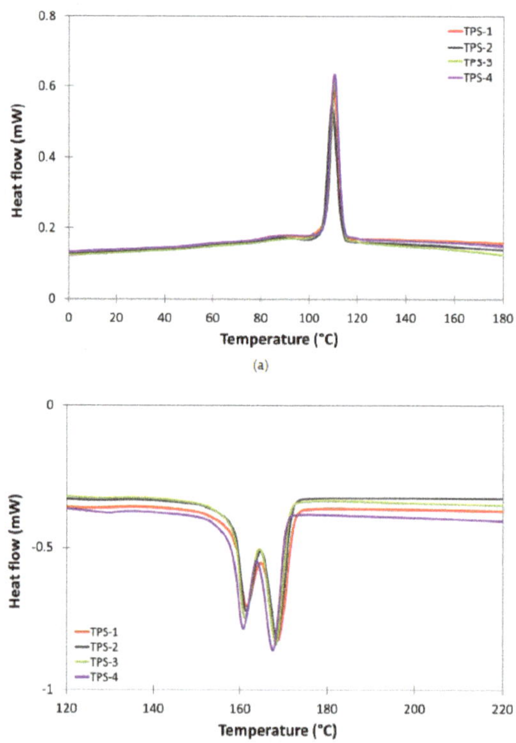

(a)

(b)

Figure 5. Comparative plot of differential scanning calorimetry (DSC) curves of TPS after different reprocessing cycles: (**a**) cooling cycle, (**b**) second heating cycle.

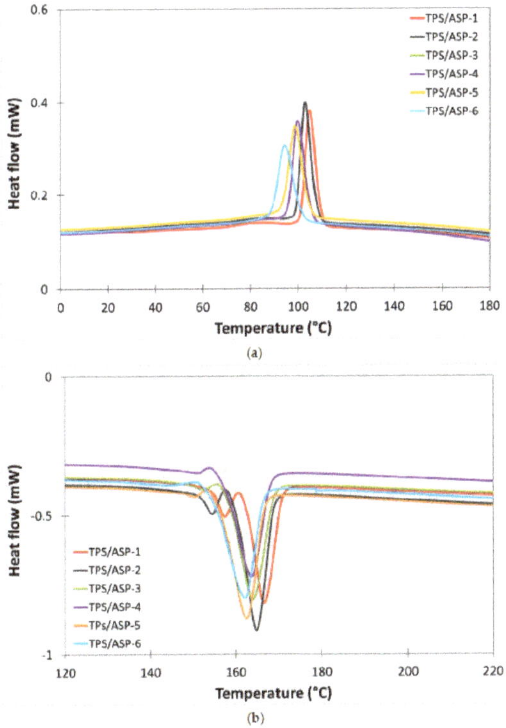

Figure 6. Comparative plot of DSC curves of TPS/ASP after different reprocessing cycles: (**a**) cooling cycle, (**b**) second heating cycle.

Table 8. Thermal properties of the injection-molded samples of TPS and TPS/ASP biocomposite subjected to different reprocessing cycles obtained by differential scanning calorimetry (DSC).

Samples	T_c (°C)	ΔH_c (J/g^{-1})	T_{m1} (°C)	T_{m2} (°C)	ΔH_f (J/g^{-1})
TPS-1	109.5 ± 0.2	22.6 ± 1.1	161.4 ± 0.4	168.5 ± 0.1	23.8 ± 1.6
TPS-2	110.3 ± 0.3	23.4 ± 0.7	161.3 ± 0.2	168.4 ± 0.1	24.9 ± 0.6
TPS-3	110.4 ± 0.1	23.1 ± 0.7	161.1 ± 0.1	168.2 ± 0.1	25.2 ± 0.6
TPS-4	110.4 ± 0.3	23.3 ± 0.3	160.8 ± 0.2	167.6 ± 0.1	24.0 ± 1.0
TPS/ASP-1	104.6 ± 0.2	16.0 ± 1.2	157.1 ± 0.4	166.4 ± 0.4	17.2 ± 1.4
TPS/ASP-2	102.3 ± 0.8	15.9 ± 0.1	153.9 ± 0.5	164.6 ± 0.4	18.0 ± 0.2
TPS/ASP-3	100.5 ± 0.5	15.1 ± 0.2	152.7 ± 0.5	164.0 ± 0.1	17.6 ± 0.8
TPS/ASP-4	98.6 ± 1.7	14.2 ± 1.7	150.8 ± 0.4	163.1 ± 0.8	18.2 ± 0.4
TPS/ASP-5	98.4 ± 0.2	14.9 ± 0.9	150.8 ± 0.3	162.6 ± 0.1	18.1 ± 0.3
TPS/ASP-6	94.0 ± 0.4	13.5 ± 1.5	144.6 ± 0.3	162.7 ± 0.7	17.1 ± 1.3

3.3. Mechanical Properties

Table 9 shows a summary of the tensile properties of both TPS and TPS/ASP biocomposite after each reprocessing cycle. The TPS-1 presented Young's modulus of 1658 MPa and tensile strength of 39.4 MPa. As can be seen in Table 9, the number of reprocessing cycles influenced the tensile properties, decreasing the values progressively. In particular, after the third processing cycle, Young's modulus significantly decreased to 1184 MPa. Regarding the tensile and elongation at break, changes were more noticeable after the fourth processing. That indicated that the material was less ductile.

Table 9. Mechanical properties, Young's modulus (E), tensile strength (σ_M), elongation at tensile strength (ε_M), tensile strength at break (σ_R), elongation at break (ε_R), of TPS and TPS/ASP biocomposite subjected to different reprocessing cycles.

Samples	Young's Modulus (MPa)	σ_M (MPa)	ε_M (%)	σ_R (MPa)	ε_R (%)
TPS-1	1650 ± 32	39.4 ± 0.7	2.8 ± 0.1	17.5 ± 0.7	13.0 ± 1.6
TPS-2	1570 ± 78	36.5 ± 0.5	2.7 ± 0.1	16.9 ± 2.7	10.7 ± 1.2
TPS-3	1184 ± 79	36.6 ± 0.7	3.1 ± 0.1	17.2 ± 0.5	10.7 ± 1.2
TPS-4	1108 ± 58	34.2 ± 0.5	3.1 ± 0.1	33.4 ± 1.5	3.3 ± 0.1
TPS/ASP-1	1050 ± 33	16.8 ± 0.2	2.2 ± 0.1	11.7 ± 0.4	7.7 ± 2.3
TPS/ASP-2	1040 ± 40	16.6 ± 0.4	2.1 ± 0.1	11.2 ± 0.5	6.8 ± 0.7
TPS/ASP-3	1057 ± 83	16.8 ± 0.2	2.1 ± 0.1	13.2 ± 0.2	4.3 ± 0.5
TPS/ASP-4	1066 ± 17	16.7 ± 0.1	2.3 ± 0.1	16.7 ± 0.1	2.3 ± 0.1
TPS/ASP-5	1028 ± 41	15.1 ± 0.4	1.8 ± 0.1	15.1 ± 0.4	1.8 ± 0.1
TPS/ASP-6	1052 ± 47	13.4 ± 0.9	1.5 ± 0.1	13.4 ± 0.9	1.5 ± 0.1

In a previous study, the addition of 20 wt% of ASP produced an increment in Young's Modulus and a reduction in the rest of the tensile parameters (tensile and elongation at maximum strength, tensile, and elongation at break) [12]. However, in this work, as can be observed in Table 8, Young's modulus of TPS/ASP biocomposite was lower than the unfilled TPS. The addition of ELO increased the flexibility of the biocomposite, reaching values of Young's modulus even lower than those of the as-received TPS. Thus, the tensile and elongation of the TPS/ASP biocomposite decreased due to its progressive lack of capacity to sustain deformation. When it comes to the effect of reprocessing TPS/ASP biocomposite in tensile properties, it did not have a significant effect on Young's modulus, which indicated similar rigidity. The rest of the parameters (tensile and elongation strength/break) kept similar values to TPS/ASP-1 until the fourth reprocessing.

Figure 7 shows the evolution of impact strength in both the reprocessed TPS and TPS/ASP biocomposite. The impact tests of TPS were carried out with notched samples because the material presented a high strength and did not break, even after the applied reprocessing cycles. The as-received TPS (TPS-1) presented an impact strength of 6.0 kJ/m^2. This value barely decreased after the second injection cycle (TPS-2). After the third and fourth injection cycle, a slight reduction in energy absorption capacity was attained, reaching values around 4.5 kJ/m^2. The reprocessing cycles did not have a significant effect on the impact strength. This behavior was observed in other grades of starch-based polymer (Solanyl, [16]). The addition of 20 wt% of ASP drastically affected the impact strength. While the TPS samples needed notching to do the test, the TPS/ASP biocomposite did not. The TPS/ASP biocomposite showed an impact strength of 14.13 kJ/m^2. After the second reprocessing cycle, the impact strength decreased until 12.26 kJ/m^2. Additionally, an evident reduction in energy absorption capacity was attained when increasing the reprocessing cycles, reaching values down to 9.70 kJ/m^2 (TPS/ASP-6). This behavior was observed in PLA systems with cellulose fibers [17–19]. Finally, this loss in the impact strength can be attributed to the polymer degradation occurring during each thermo-mechanical recycle.

The fracture of the surface after impact was studied by SEM analysis. TPS surface after the first injection cycle (Figure 8a) showed less roughness than after the fourth injection cycle (TPS/ASP-4) (Figure 8b). This increase in the surface roughness can be directly attributed to the degradation of the TPS matrix. SEM micrographs of the impact fracture surfaces of TPS/ASP biocomposite gave qualitative information about the dispersion of almond shell particles. The fracture surfaces of TPS/ASP after the first injection cycle (Figure 8c) showed that the almond shell was homogeneously distributed in the polymer matrix. As with the TPS surface, it could be observed that the surface roughness and the imperfections on the impact fracture surface increased after the fourth injection cycle (TPS/ASP-4) (Figure 8d).

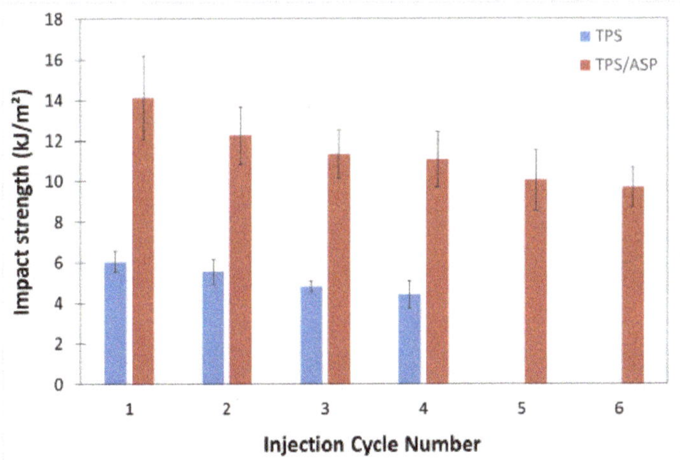

Figure 7. Charpy Impact strength corresponding to injection-molded samples of TPS and TPS/ASP biocomposite subjected to different reprocessing cycles.

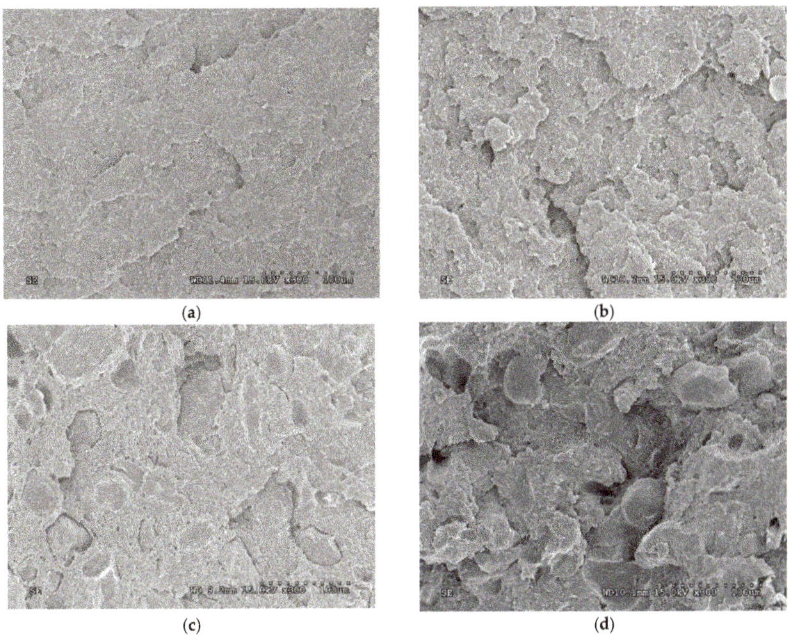

Figure 8. SEM micrographs of the impact fracture surfaces of the composites: (**a**) TPS after the first injection cycle; (**b**) TPS after the fourth injection cycle; (**c**) TPS/ASP biocomposite after the first injection cycle; (**d**) TPS/ASP biocomposite after the fourth injection cycle.

3.4. Melt Flow Index

It is a well-known fact that polymer chain breaking due to polymer degradation results in reduced molecular weight and increased flowability with temperature [20]. MFI is a simple test that can potentially offer detailed information about the degradation a polymer material has undergone [5,20]. Figure 9 shows the evolution of the MFI values after different reprocessing cycles. As can be seen, the MFI suffered a drastic increment in the value in both TPS and TPS/ASP biocomposite from the second injection process (TPS-2 and TPS/ASP-2). This effect was observed in other grades of commercial Mater-Bi and Mater-Bi YI014U/C, in which MFI increased from 9 to 16 g/10 min after the second processing [7]. TPS/ASP-1 biocomposite presented an MFI value higher than TPS-1, probably because TPS/ASP was previously subjected to a thermal process to obtain the pellets. However, despite the evident degradation of the material according to the MFI values obtained, this was not reflected in the decrease in the mechanical and thermal properties of the material.

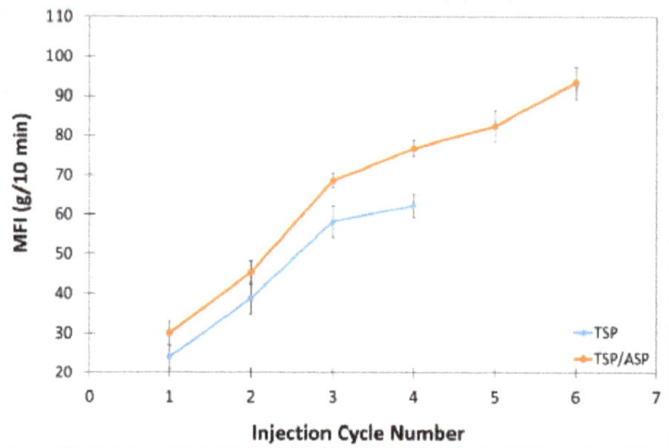

Figure 9. Variation of the melt flow index (MFI) after different reprocessing cycles.

4. Conclusions

This work describes the influence of the recycling, by injection molding reprocessing, of TPS and TPS/ASP biocomposite samples. To do so, the visual aspect, mechanical and thermal properties, and melt flow index were studied. TPS was processed four times and TPS/ASP biocomposite six times.

Experimental results revealed that reprocessing cycles had more effect on the TPS/ASP biocomposite. The change in color of TPS was 2.9 after all four cycles, meaning that only an experienced observer could notice the difference in color. However, after only two injection cycles, the change in color of TPS/ASP biocomposite was higher than unfilled TPS, and an observer could notice the change in color ($\Delta E^*_{ab} > 5$).

Regarding the thermal properties, TGA and DSC analyses showed that after four injection cycles of TPS (TPS-4), the values remained almost invariable compared to TPS-1. This indicated that the reprocessing cycles did not have a significant effect on thermal degradation and main thermal transitions. On the other hand, in the case of TPS/ASP, in the second injection cycle, a significant reduction in T_{onset} (9 °C) was produced and to a lesser extent in the melting temperature (T_{m1} and T_{m2} decreased in 2–3 °C respect to TPS/ASP-1). The values continued to decrease progressively after each reprocessing cycle.

Moreover, it was observed that the reprocessing cycles had an influence on the mechanical properties of the TPS after the third processing cycle. In particular, Young's modulus was reduced, indicating a loss of rigidity. Furthermore, after the fourth processing (TPS-4), the material lost ductility. In regard to the TPS/ASP biocomposite, the reprocessing did not have significant effects on tensile properties. The results maintained similar values to TPS/ASP-1 until the sixth cycle. As for impact strength, a slight reduction in energy absorption capacity was obtained after the second processing cycle (TPS-2), and no significant variations on the impact strength were observed with more cycles. However, TPS/ASP biocomposite showed an evident reduction in energy absorption capacity after the second processing cycle and continued when increasing reprocessing cycles. This confirmed the negative effect of biocomposite reprocessing on toughness.

Finally, the MFI showed a noticeable increase in both TPS and TPS/ASP biocomposites. Despite this, it was not reflected in such a drastic decrease in the mechanical and thermal properties of the material.

As a general conclusion, this study revealed that the TPS/ASP biocomposite is more sensitive than TPS to recycling regarding mechanical and thermal properties and visual aspects. From an industrial point of view, TPS could be reprocessed at least four times, as shown in this work, without the need of adding a virgin material. The TPS/ASP biocomposite could be recycled up to two or three cycles, but the impact strength has to be taken into account if it is a critical property for the product considered as it is the most affected property.

Author Contributions: Conceptualization was devised by A.M.-G. and S.F.-B.; methodology, validation, formal analysis was carried out by A.I.-G.; investigation, resources, data curation, and writing—original draft preparation was performed by A.I.-G.; writing—review and editing, A.I.-G.; supervision, A.M.-G. and S.F.-B.; project administration, A.M.-G.; All authors have read and agreed to the published version of the manuscript.

Funding: This research was funded by GVA. GVA provided a funding line to boost AIJU's capacity and activity to develop excellence in R&D, with the aim to share the results of the research carried out and facilitate and promote the transfer of knowledge to companies within the Valencian Community.

Institutional Review Board Statement: Not applicable.

Informed Consent Statement: Not applicable.

Data Availability Statement: The data presented in this study are available on request from the author.

Conflicts of Interest: The authors declare no conflict of interest.

References

1. European Bioplastics. Available online: https://www.european-bioplastics.org/about-us/members-membership/members-list/ (accessed on 3 September 2020).
2. Soroudi, A.; Jakubowicz, I. Recycling of bioplastics, their blends and biocomposites: A review. *Eur. Polym. J.* **2013**, *49*, 2839–2858. [CrossRef]
3. Pillin, I.; Montrelay, N.; Bourmaud, A.; Grohens, Y. Effect of thermo-mechanical cycles on the physico-chemical properties of poly(lactic acid). *Polym. Degrad. Stab.* **2008**, *93*, 321–328. [CrossRef]
4. Żenkiewicz, M.; Richert, J.; Rytlewski, P.; Moraczewski, K.; Stepczyńska, M.; Karasiewicz, T. Characterisation of multi-extruded poly(lactic acid). *Polym. Test.* **2009**, *28*, 412–418. [CrossRef]
5. Agüero, A.; Morcillo, M.D.C.; Quiles-Carrillo, L.; Balart, R.; Boronat, T.; Lascano, D.; Torres-Giner, S.; Fenollar, O. Study of the Influence of the Reprocessing Cycles on the Final Properties of Polylactide Pieces Obtained by Injection Molding. *Polymers* **2019**, *11*, 1908. [CrossRef] [PubMed]
6. Paolo, F.; Mantia, L.; Botta, L.; Mistretta, M.C.; Di Fiore, A.; Titone, V. Recycling of a Biodegradable Polymer Blend. *Polymers* **2020**, *12*, 2297.
7. Lopez, J.P.; Gironès, J.; Méndez, J.A.; Puig, J.; Pèlach, M. Àngels Recycling Ability of Biodegradable Matrices and Their Cellulose-Reinforced Composites in a Plastic Recycling Stream. *J. Polym. Environ.* **2011**, *20*, 96–103. [CrossRef]
8. Nafchi, A.M.; Moradpour, M.; Saeidi, M.; Alias, A.K. Thermoplastic starches: Properties, challenges, and prospects. *Starch Stärke* **2013**, *65*, 61–72. [CrossRef]
9. Debiagi, F.; Mello, L.R.P.; Mali, S. *Thermoplastic Starch-Based Blends: Processing, Structural, and Final Properties*; Elsevier Inc.: Amsterdam, The Netherlands, 2017; pp. 153–186. [CrossRef]
10. García, A.I.; García, A.M.; Bou, S.F. Study of the Influence of the Almond Shell Variety on the Mechanical Properties of Starch-Based Polymer Biocomposites. *Polymers* **2020**, *12*, 2049. [CrossRef] [PubMed]
11. Borchani, K.E.; Carrot, C.; Jaziri, M. Biocomposites of Alfa fibers dispersed in the Mater-Bi®type bioplastic: Morphology, mechanical and thermal properties. *Compos. Part A Appl. Sci. Manuf.* **2015**, *78*, 371–379. [CrossRef]
12. Ibáñez-García, A.; Martínez-García, A.; Ferrándiz-Bou, S. Influence of Almond Shell Content and Particle Size on Mechanical Properties of Starch-Based Biocomposites. *Waste Biomass Valorization* **2021**, 1–14. [CrossRef]
13. Mokrzycki, W.; Tatol, M. Color Difference Delta E—A Survey Colour Difference Δ E—A Survey Faculty of Mathematics and Informatics. *Mach. Graph. Vis.* **2011**, *20*, 383–411.
14. Bastioli, C. *Handbook of Biodegradable Polymers*; Taylor & Francis: Oxfordshire, UK; Rapra Technology: Shrewsbury, UK, 2005.

15. Morreale, M.; Liga, A.; Mistretta, M.C.; Ascione, L.; La Mantia, F.P. Mechanical, Thermomechanical and Reprocessing Behavior of Green Composites from Biodegradable Polymer and Wood Flour. *Materials* **2015**, *8*, 7536–7548. [CrossRef] [PubMed]
16. Shojaeiarani, J.; Bajwa, D.S.; Rehovsky, C.; Bajwa, S.G.; Vahidi, G. Deterioration in the Physico-Mechanical and Thermal Properties of Biopolymers Due to Reprocessing. *Polymers* **2019**, *11*, 58. [CrossRef] [PubMed]
17. Chaitanya, S.; Singh, I.; Song, J.I. Recyclability analysis of PLA/Sisal fiber biocomposites. *Compos. Part B Eng.* **2019**, *173*, 106895. [CrossRef]
18. Graupner, N.; Albrecht, K.; Ziegmann, G.; Enzler, H.; Muessig, J. Influence of reprocessing on fibre length distribution, tensile strength and impact strength of injection moulded cellulose fibre-reinforced polylactide (PLA) composites. *Express Polym. Lett.* **2016**, *10*, 647–663. [CrossRef]
19. Bourmaud, A.; Åkesson, D.; Beaugrand, J.; Le Duigou, A.; Skrifvars, M.; Baley, C. Recycling of L-Poly-(lactide)-Poly-(butylene-succinate)-flax biocomposite. *Polym. Degrad. Stab.* **2016**, *128*, 77–88. [CrossRef]
20. Běhálek, L.; Dobránsky, J.; Pollák, M.; Borůvka, M.; Brdlík, P. Application of Physical Methods for the Detection of a Thermally Degraded Recycled Material in Plastic Parts Made of Polypropylene Copolymer. *Materials* **2021**, *14*, 552. [CrossRef] [PubMed]

polymers

MDPI

Article

Biocompatibility and Pharmacological Effects of Innovative Systems for Prolonged Drug Release Containing Dexketoprofen in Rats

Liliana Mititelu-Tartau [1,†], Maria Bogdan [2,*], Daniela Angelica Pricop [3,*], Beatrice Rozalina Buca [1,†], Loredana Hilitanu [1,†], Ana-Maria Pauna [1,†], Lorena Anda Dijmarescu [4,†] and Eliza Gratiela Popa [5]

1 Department of Pharmacology, Faculty of Medicine, "Grigore T. Popa" University of Medicine and Pharmacy, 700115 Iasi, Romania; liliana.tartau@umfiasi.ro (L.M.-T.);beatrice-rozalina.buca@umfiasi.ro (B.R.B.); ln.rusu@yahoo.com (L.H.); ana-maria-raluca-d-pauna@umfiasi.ro (A.-M.P.)

2 Department of Pharmacology, Faculty of Pharmacy, University of Medicine and Pharmacy, 200349 Craiova, Romania

3 Department of Physics, Faculty of Physics, "Al. I. Cuza" University, 700506 Iasi, Romania

4 Department of Obstetrics-Gynecology, Faculty of Medicine, University of Medicine and Pharmacy, 200349 Craiova, Romania; lorenadijmarescu@yahoo.com

5 Department of Pharmaceutical Technology, Faculty of Pharmacy, "Grigore T. Popa" University of Medicine and Pharmacy, 700115 Iasi, Romania; eliza.popa@umfiasi.ro

* Correspondence: bogdanfmaria81@yahoo.com (M.B.); daniela.a.pricop@gmail.com (D.A.P.)

† These authors contributed equally to the study.

Citation: Mititelu-Tartau, L.; Bogdan, M.; Pricop, D.A.; Buca, B.R.; Hilitanu, L.; Pauna, A.-M.; Dijmarescu, L.A.; Popa, E.G. Biocompatibility and Pharmacological Effects of Innovative Systems for Prolonged Drug Release Containing Dexketoprofen in Rats. *Polymers* **2021**, *13*, 1010. https://doi.org/10.3390/polym13071010

Academic Editors: José Miguel Ferri, Vicent Fombuena Borràs and Miguel Fernando Aldás Carrasco

Received: 26 February 2021
Accepted: 22 March 2021
Published: 25 March 2021

Abstract: The present study reports on the in vivo biocompatibility investigation and evaluation of the effects of liposomes containing dexketoprofen in somatic sensitivity in rats. Method: The liposomes were prepared by entrapping dexketoprofen in vesicular systems stabilized with chitosan. The in vivo biocompatibility was evaluated after oral administration in white Wistar rats: Group I (DW): distilled water 0.3 mL/100 g body weight; Group II (DEX): dexketoprofen 10 mg/kg body weight (kbw); Group III (nano-DEX): liposomes containing dexketoprofen 10 mg/kbw. Blood samples were collected from caudal lateral vein one day and seven days after the substance administration, to assess the eventual hematological, biochemical, and immunological changes. The investigation of somatic pain reactivity was performed using the hot plate test, to count the latency time response evoked by the thermal paws' noxious stimulation. Results: Original liposomes entrapping dexketoprofen, with mean size of 680 nm and good stability, were designed. Laboratory analysis indicated no substantial variances between the three treated groups. The treatment with liposomes containing dexketoprofen resulted in a prolongation of the latency time response, statistically significant in the interval between 90 min and 10 h, in the hot plate test. Conclusions: The use of liposomes with dexketoprofen proved a good in vivo biocompatibility in rats and prolonged analgesic effects in the hot plate test.

Keywords: liposomes; dexketoprofen; biocompatibility; hot plate test; rats

1. Introduction

Nanomedicine offers highly valuable research and practical application tools in the medical field, for improving the current methods of prevention, diagnosis, and targeted therapy in various pathological conditions, from simple inflammatory to neoplastic diseases [1]. Nano-sized carrier systems exhibit new or improved physical, chemical, and biological characteristics, and the obtained innovative compounds having the same dimensions as biological structures can interact more quickly at the bio-molecular level, both on the surface and inside the cell [2].

Contemporary papers highlight many medical applications of nanotechnology in the pharmaceutical field, given the design of new nanoparticulate systems for the transport and release of active substances, as well as in the field of regenerative medicine (nano-robots and devices used in cell regeneration), in preventive medicine, diagnosis, and treatment of diseases [3]. Researchers have historically tried to incorporate different agents into nano-systems: thus, analgesic–antipyretic agents, non-steroidal anti-inflammatory drugs, or analgesics belonging to the opioid group were loaded into nanoparticles through various methods [4–9].

Dexketoprofen (DEX) is a dextro-enantiomer of the non-steroidal anti-inflammatory drug ketoprofen, acting through the inhibition of the prostaglandin biosynthesis by blocking both cyclooxygenases (COX-1 and COX-2) [10–12]. The water-soluble salt of this non-steroidal anti-inflammatory drug, the (+)-(S)-2-(3-benzoylphenyl) propionic acid tromethamine derivative (dexketoprofen tromethamine), possesses analgesic, antipyretic, and anti-inflammatory properties, and is used in the treatment of acute and chronic pain in medical, as well as in post-surgery, conditions [13–15].

The pharmacodynamic effects of DEX are manifested 30 min after oral administration (when the maximum plasma concentration is reached), with a half-life of 4–6 h [16]. It has a high capacity to bind the plasma proteins, and the apparent volume of distribution is less than 0.25 L/kg. The liver plays a central role in DEX biotransformation, and it is excreted urinarily, with a terminal half-life around 1.65 h [17]. Due to its short half-life and need to be administered frequently over 24 h, much research has been conducted aiming to obtain prolonged-release compounds. Moreover, the design of such formulations loading this non-steroidal anti-inflammatory agent aims to reduce its adverse effects [18].

Chitosan is obtained by the partial de-acetylation of chitin. It has various properties, due to different values of molecular weight, de-acetylation degree, and sequencing pattern (random or block distribution of de-acetylated residues along the main chain) [19]. Chitosan has a high content of primary amines, which induce a cationic polyelectrolyte behavior, interacting easily with cellular membranes, but also with the lipid bilayer of the vesicles [20]. Positive charging of chitosan-coated lipid vesicles is possible due to pH-dependent protonation/deprotonation processes [21]. At low pH values, chitosan is a water-soluble cationic polyelectrolyte, due to protonated amino groups; at high values of pH, the polymer becomes water-insoluble, losing its charge [22]. In the latter case, the electrostatic repulsion of chitosan is low, facilitating the formation of interpolymer bonds (liquid crystal domains or network bonding) [23]. Thus, fibers, films or hydrogels are formed, depending on the preset conditions for the initiation of transition from the soluble state to insoluble state; this transition takes place usually at pH values between 6 and 6.5, which is a convenient interval for biological applications.

A variety of nanoparticles consist of DEX-coated zinc oxide quantum dots, which demonstrated good penetrability of the drug through rats' skin after transdermal application [24]. In order to improve the pharmacokinetic properties and to decrease adverse effects, nanoparticles containing DEX entrapped in montmorillonite (a natural layered structure of phyllosilicate) were prepared, which proved to deliver the active substance in the body after oral use in acute postoperative pain [25].

Different solid formulations based on lipids (such as triglycerides, fatty acids, steroids, waxes), surfactant and water have been prepared, using the ultrasonication technique, in order to obtain DEX-loaded nano-systems for the sustained release of the drug in different diseases accompanied with pain [26]. A new design of analgesic-loaded nano-systems consists of the encapsulation of DEX-trometamol in nanovesicles based on the polysaccharide chitosan, using a spray-drying method [27].

The aim of our study was to obtain original formulations of DEX-loaded nanoparticle systems with chitosan and phosphatidylcholine, and also the experimental investigation of the biocompatibility and the pharmacological effects of nanovesicles containing DEX in rats.

2. Materials and Methods

2.1. Substances

L-α-phosphatidylcholine (Egg Yolk PC 99% TLC purity), dexketoprofen tromethamine (99% purity, mean molecular weight M_w = 375.42 g/mol) and chitosan (M_w = 310.0 g/mol, 3.26 polydispersity index, 79.7%*N*-de-acetylation degree) were purchased from Vanson

Chemicals, Redmond, WA, USA. Chitosan was dissolved in 0.5% (*v*/*v*) acetic acid solution with 99.5% purity purchased from the Chemical Company. The homogeneous 0.5% (*w*/*v*) chitosan solution was prepared by stirring at room temperature for 24 h and subjected to degassing by centrifugation at 1500 rpm for 30 min.

2.2. Procedure of Preparation the DEX-Loaded Liposomes Stabilized with Chitosan

The liposomes were designed by entrapping DEX inside L-α-phosphatidylcholine lipid vesicles, which were obtained by dissolving 25 mg of lipid in 1 mL chloroform. Through evaporation, a dry lipid film was achieved, which was hydrated by adding 50 mL of DEX stock solution in deionized water at 10 mg/L. During this process, the solution was moderately ultrasonically stirred (10% amplitude mode at 20 kHz ± 500 Hz standard frequency, Sonoplus Bandeline set-up), at room temperature, for 10 min, in order to disrupt the monolayer structure. To transform the multilamellar into unilamellar vesicles, the lipid solution was sonicated. The chitosan solution 0.5% (*w*/*w*) dissolved in 0.5% (*v*/*v*) acetic acid [28,29] was used to coat the vesicles, Thereafter, the colloidal solution was dialyzed at room temperature for 10 h (using a Sigma D6191-25EA dialysis tubing cellulose membrane with pore size of 12,000 Da MWCO, Sigma Aldrich Chemical Co, Schnelldorf, Germany), to remove its acidity.

For the preparation of lipid vesicles, 15 mg of lipid were dissolved in 1 mL chloroform; through solvent evaporation, a lipid dry film was obtained; the lipid layers were hydrated with 30 mL DEX solution (10 mg/L) in deionized water. The mixture was put to magnetically stirred for 2 h for complete dissolution. A total of 19.2 mL chitosan 0.5% (*w*/*w*) was added to the blurry vesicle suspension; in order to transform the multilamellar vesicles to unilamellar vesicles, the suspension was ultrasonicated for 20 min (amplitude modulus 10% at 20 kHz ± 500 Hz standard frequency, Sonoplus Bandeline configuration). Through ultrasonication, a suspension with a transparent aspect was obtained, with a pH value of 4 [28,29]. Given that stabilizing the vesicles with chitosan led to an acid pH, the suspension was further dialyzed for 10 h, up to a pH value of 6.7.

Due to its protonated amino groups (in acid to neutral solutions), which are incompatible with the stability of the bilayer hydrophobic core, the polymer chitosan coats the surface of the vesicles. Its physicochemical parameters, such as intra- and intermolecular hydrogen bonding and cationic transformation in acidic environment, make this polymer very attractive and useful for the development of suitable drug delivery formulations [30–33].

The anionic polyelectrolyte properties of chitosan provide interaction with anionic molecules of the vesicles' bilayer, as well as with the cells' membranes. In acid-to-neutral solutions, the amino groups of chitosan become protonated through bonding to intra- and intermolecular hydrogen of the vesicle bilayer. Formation of a stable positively charged layer around the unilamellar vesicles is related to the saturation absorption of chitosan molecules. These phenomena lead to obtaining confined, small-sized vesicles, with a morphology close to spherical. From this perspective, this polymer is very attractive and useful for designing modified release dosage forms [30–33].

2.3. Analysis of the Prepared Liposomes

The DEX vesicles were analyzed for size distribution and Zeta-potential profile with a Malvern Zetasizer Nano ZS ZEN-3500 device (Worcestershire, UK). The Zetasizer evaluation of sizes was performed on unit samples and on 1:10 dilution for the Zeta potential.

The liposome shapes were examined in differential interference contrast (DIC) optical microscopy, and size distribution was assessed by the visual comparison of micrographs with a standard scale measurement, using a Nikon Eclipse Ti-E Inverted Microscope equipped with an NIS Elements Basic Research imaging software (NIS E-Br).

The pH values of the obtained solutions were measured using a Sartorius Professional PP-50 pH Meter.

The in vitro release kinetics of DEX from nano-DEX, from DEX solution, was investigated over 12 h using the regular dialysis method. This technique is based on physical separation of the samples, offering the possibility of determining the drug molar concentration at periodic moments of time. Ten milliliters of the tested solutions were placed into a cellulose acetate dialysis bag (MW cut off 12–14 KDa, Sigma Aldrich Chemical Co., Schnelldorf, Germany), representing the inner dissolution compartment. Then, the bag was introduced into a large glass with 200 mL of the release medium (consisting of phosphate buffered saline, pH 7.4, Sigma Aldrich Chemical Co., Schnelldorf, Germany), under continuous magnetic stirring of 100 rpm at 37 ± 0.5 °C.

The molar concentration of DEX was determined using a UV–Vis spectrophotometer (Hewlett Packard 8453, equipped with an HP Chem-Station software for assessing the absorbance between 259 and 281.6 nm), by extracting 2 mL of the release medium at certain intervals of time: 15, 30, 45, 60, 90, and 120 min, 3, 4, 6, 8, 10, and 12 h, respectively. After each determination, the same volume of medium was replaced in the dissolution compartment.

The dissolution profile of DEX released from nano-DEX was compared with those assessed for DEX solution (1 mg/mL), identical to that existent in 10 mL nano-DEX, using the same UV–VIS method.

2.4. *The In Vivo Biocompatibility Evaluation of Chitosan-Coated Liposomes Entrapping DEX*

The tested substances were administered in rats, in order to assess the in vivo biocompatibility and to estimate the effects on the somatic nociceptive reactivity.

Three groups of six rats (weighting 200–250 g each, 10 weeks old), with equal distribution of the sexes were used in the experiment. The animals receiving the tested substances orally (using an eso-gastric tube), were as follows:

1. Group I (DW): distilled water 0.3 mL/100 g body weight;
2. Group II (DEX): dexketoprofen 10 mg/kg body weight (kbw);
3. Group III (nano-DEX): liposomes containing dexketoprofen 10 mg/kbw.

White, specific-pathogen-free (SPF), healthy Wistar rats were used in the experiment; the rats had no genetic modifications and were provided by the National Medical-Military Institute for Research and Development "Cantacuzino", Baneasa Station, Bucuresti, Romania.

The animal groups were randomly set, and no inclusion-exclusion criteria of animals were used in the experiment.

The biocompatibility of liposomes containing DEX was estimated by evaluating their effects on the hematological, biochemical, and immunological parameters after the administration in rats. The tested drugs were administered in a single dose in the first day of the experiment, and one day and seven days after substances administration, blood samples were collected from the lateral caudal vein, to evaluate the white blood count (polymorphonuclear neutrophils, PMN; lymphocytes, Ly; eosinophils, E; monocytes, M; basophils, B), the liver enzyme activity (aspartate transaminase, AST; alanine aminotransferase, ALT; lactate dehydrogenase, LDH), the serum level of urea and creatinine, and also the complement activity and the phagocytic capacity of peripheral neutrophils (PC) [34,35].

2.5. *The Investigation of the Chitosan-Coated Liposomes Entrapping DEX on the Nociceptive Reactivity in Rats*

The somatic nociceptive reactivity was assessed using the hot plate test (Ugo-Basile apparatus), in order to measure the latency time response to thermal paw noxious stimulation [36]. The noted baseline latency (before the substances administration) was 4.3 ± 0.2 s (mean ± standard deviation of mean, S.D.). The suggested cut-off time of 12 s was considered to avoid damage to the paws. The paws' withdrawal latency (seconds) was counted before the test, and 15, 30, 60, 90, 120 min, and 4, 6, 8, 10, 12 h after administration of the tested substances. Variances between the measured and baseline latencies were estimated as an index of analgesia. The prolongation of the latency time reactivity was suggestive for the analgesic effect, while the diminution in the paws' reaction was indicative of hyperalgesia [37,38].

Moreover, the withdrawal threshold data from paws response measurements were transformed to the percentage maximum possible antinociceptive effect (%MPE) using the following formula [39]:

$$\%MPE = [(\text{measured latency-baseline latency})/(\text{cut-off time} - \text{baseline latency})] \times 100 \quad (1)$$

The protocol of the study was approved (Protocol No. 19157/19.10.2009) by the Ethic Committee on Research of 'Grigore T. Popa' University of Medicine and Pharmacy, in Iasi, Romania, according to the European Ethical Regulations and to the guidelines of IASP Committee for Research and Ethical Issues [40,41]. The data were statistically evaluated using the SPSS 17.0 software for Windows and the ANOVA one-way method. *p*-values less than 0.05 were interpreted as statistically significant compared to the control group.

3. Results

3.1. Characterization of Liposomes

New carrier systems containing DEX were designed. Microscopic observations of liposomes revealed their spherical shape, and the image of nanovesicles containing DEX in optical microscopy is presented in Figure 1.

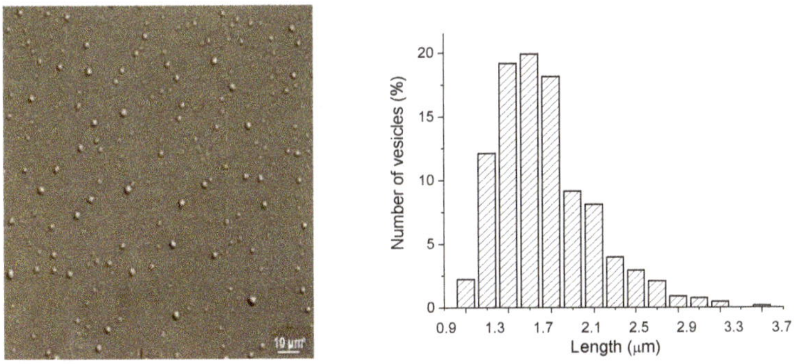

Figure 1. The optical microscopy image of nano-dexketoprofen (DEX) (sizing scale represents 10 µm).

The image obtained through optical DIC microscopy presents nano-DEX having regular shape (close to spherical), and well-dispersed due to stabilization with chitosan, prior to dyalization. The size distribution histogram was obtained by measuring the size diameter of vesicles from the microscopic image; this suggests formation of vesicles with size 1.4–1.5 µm, in concordance with hydrodynamic size measurements (Figure 1).

The analysis of liposome solutions revealed their mid-range of polydispersity (between 0.435 and 0.636). During the preparation of liposomes, a pH value of 7.12 was noted for DEX encapsulated in liposome solution, and a pH value of 3.84 for the obtained solution after adding chitosan. Because the stabilization of the vesicles with chitosan led to a major decrease in pH value, prolonged dialysis was required to achieve an optimal pH value for gastric absorption; finally, a solution with a pH value of 6.89 was obtained (Table 1).

Table 1. Characteristics of liposome solutions.

Liposomes Types	pH	Colloidal Solution Characterization			
		Polydispersity Index	Z-Average Diameter (nm)	Z Potential (mV)	Stability Criterion
Liposomes entrapping DEX	7.12	0.472	368	0.64	Threshold of agglomeration
Chitosan-coated liposomes entrapping DEX	4	0.636	1470	61.7	Very good stability
Dialyzed chitosan-coated liposomes entrapping DEX	6.89	0.435	699	3.89	Threshold of light dispersion

DEX: dexketoprofen.

From the analysis of hydrodynamic size distributions and of the Zeta potentials, a stiffening of the lipid bilayer of the vesicles was noted. Thus, the suspension stability increased, accompanied by an increase in the size of nano-DEX vesicles. Positive charging of the vesicle surface due to protonation created repulsion forces between the vesicles, preventing agglomeration. It can be said that these vesicular systems meet the criteria of colloidal solutions, because a Zeta potential of 61.7 mV for the vesicles containing DEX stabilized with chitosan falls in a very good stability domain. Dialyzation led to a diminution of the vesicle size, and also to a significant decrease in suspension stability up to the agglomeration threshold of 3.89 mV. These phenomena are a consequence of the deprotonation of chitosan amino groups at the surface of the vesicle bilayer in dialysis conditions [42,43] (Figures 2 and 3).

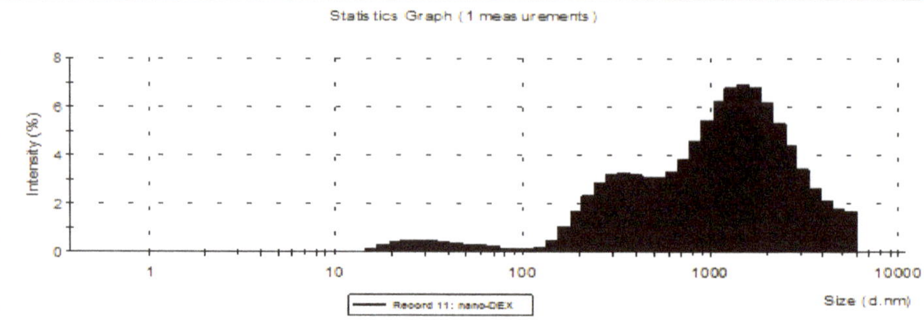

			Diam. (nm)	% Intensity	Width (nm)
Z-Average (d.nm):	680	Peak 1:	1770	75.0	1160
PdI:	0.540	Peak 2:	307	20.7	95.2
Intercept:	0.922	Peak 3:	39.9	4.3	20.8

Figure 2. Size distribution by number of nano-DEX in aqueous solution.nano-DEX: liposomes containing dexketoprofen.

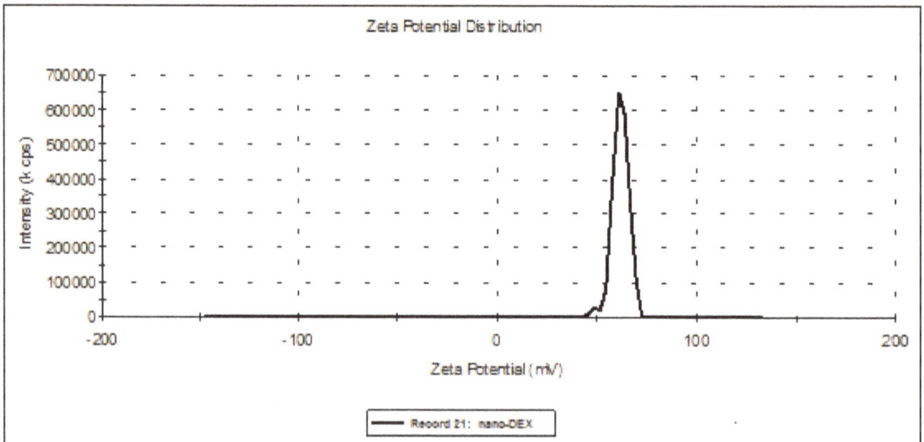

Figure 3. Distribution of the nano-DEX Zeta potential. nano-DEX: liposomes containing dexketoprofen.

3.2. In Vitro Release of DEX from Vesicles Stabilized with Chitosan

The spectrophotometric assessment of the active substance encapsulation degree was determined from the calibration curve.

For obtaining the release kinetics curve, we took into consideration that the medium volume remained constant over time (solvent evaporation during the experiment was negligible). The etalonation curve of variation of DEX concentration over time was built using the UV absorption spectra (Figure 4). The measurements were performed in the UV domain, at 259 nm, the maximum absorption spectrum for DEX, because neither the solvents, nor the other polymer ingredients used, absorbed at the given wavelength.

The analysis of the in vitro release curve revealed a slower release of DEX from chitosan-coated liposomes compared to the release from DEX solution. Detailing the aspects of the drug kinetic profile, it was found that, after one hour, 43% of DEX was released from DEX crystals, while only 20% was released from the nanovesicles. Drug release reached a percentage of 74% from DEX solution, and 41% from the nano-DEX after three hours. Finally, the DEX release from nanovesicles appeared to be effective, reaching a maximum of 92% after eight hours, while almost 80% of active substance was released from the DEX solution (Figure 5). The lower percentage of the drug released from the nano-DEX was attributed to the high dispersion of the active substance as isolated molecules well-entrapped into vesicles stabilized with chitosan.

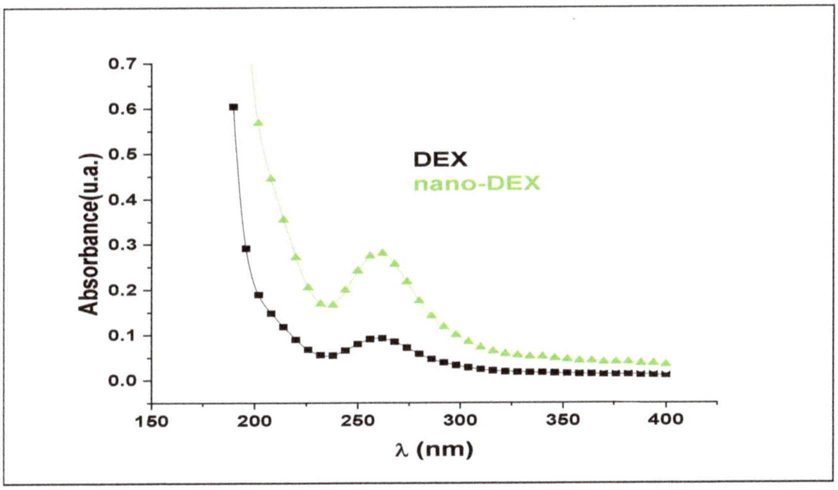

Figure 4. The absorption spectra of DEX and nano-DEX. DEX: dexketoprofen; nano-DEX: liposomes containing dexketoprofen.

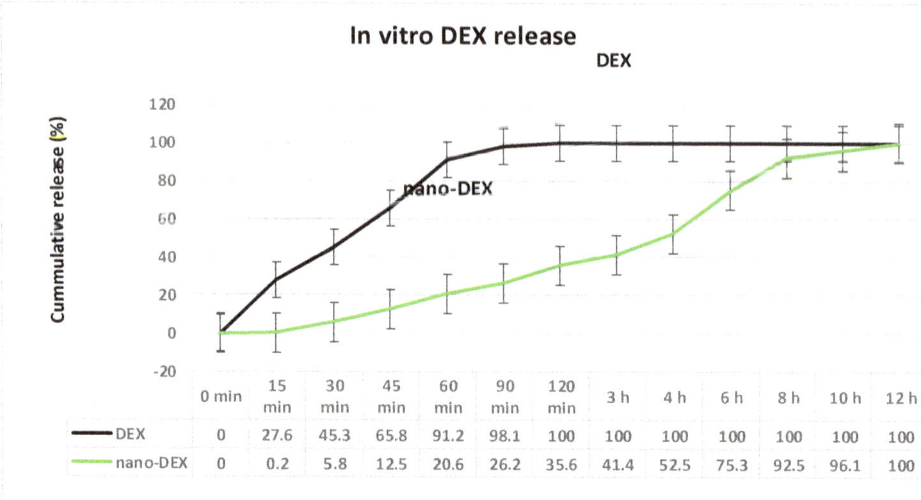

Figure 5. The release kinetics of DEX from nano-DEX *vs.* DEX solution by the permeation method. DEX: dexketoprofen; nano-DEX: liposomes containing dexketoprofen.

3.3. *The In Vivo Biocompatibility Testing*

No significant differences in the percentage of the blood count formula elements between DEX, nano-DEX and the DW group were observed during the experiment (Table 2).

Table 2. The effects of DEX and nano-DEX on the percentage of the blood count formula elements (values are expressed as mean ± S.D. for 6 rats in a group).

		Blood Count Formula (% Values)				
		PMN	**Ly**	**E**	**M**	**B**
DW	1 day	27.5 ± 3.3	68.1 ± 9.5	0.4 ± 0.1	3.8 ± 0.05	0.2 ± 0.05
	7 days	28.9 ± 3.5	66.7 ± 8.3	0.5 ± 0.1	3.7 ± 0.2	0.2 ± 0.05
DEX	1 day	28.2 ± 3.5	67.7 ± 8.7	0.4 ± 0.05	3.5 ± 0.2	0.2 ± 0.1
	7 days	29.5 ± 4.1	66.2 ± 9.1	0.5 ± 0.05	3.5 ± 0.1	0.3 ± 0.05
nano-DEX	1 day	29.3 ± 3.7	66.8 ± 8.3	0.4 ± 0.05	3.3 ± 0.05	0.2 ± 0.05
	7 days	29.7 ± 3.1	66.1 ± 8.5	0.4 ± 0.1	3.6 ± 0.1	0.2 ± 0.05

DW: distilled water; DEX: dexketoprofen; nano-DEX: liposomes containing dexketoprofen; polymorphonuclear neutrophils, PMN; lymphocytes, Ly; eosinophils, E; monocyte, M; basophils, B.

The treatment with DEX and nano-DEX did not induce major variations in the AST, ALT and LDH activity, compared to control, after one day, or seven days (Table 3).

The use of DEX and nano-DEX did not show substantial modifications in the serum level of urea and creatinine compared to the control group, at both time points of the determinations (Table 4).

No considerable variations were revealed in the serum complement activity, nor in the PC between DEX, nano-DEX groups, and control, one day or seven days after substance administration (Table 5).

Table 3. The effects of DEX and nano-DEX on the AST, ALT and LDH activity (values are expressed as mean ± S.D. for 6 rats in a group).

		AST (U/mL)	ALT (U/mL)	LDH (U/L)
DW	1 day	54.5 ± 6.7	67.3 ± 8.3	336.42 ± 70.17
	7 days	52.3 ± 5.9	68.1 ± 9.3	346.29 ± 71.45
DEX	1 day	54.2 ± 6.3	66.5 ± 7.5	341.33 ± 71.67
	7 days	56.5 ± 7.1	69.3 ± 9.1	353.37 ± 69.29
nano-DEX	1 day	53.7 ± 5.5	66.1 ± 8.3	356.72 ± 69.55
	7 days	55.3 ± 6.5	68.7 ± 8.7	360.29 ± 68.13

DW: distilled water; DEX: dexketoprofen; nano-DEX: liposomes containing dexketoprofen; aspartate transaminase, AST;alanine aminotransferase, ALT; lactate dehydrogenase, LDH.

Table 4. The effects of DEX and nano-DEX on the blood level of urea and creatinine (values are expressed as mean ± S.D. for 6 rats in a group).

		Urea (mg/dL)	Creatinine (mg/dL)
DW	1 day	35.8 ± 6.33	<0.2
	7 days	36.7 ± 6.72	<0.2
DEX	1 day	36.4 ± 6.65	<0.1
	7 days	37.6 ± 7.29	<0.2
nano-DEX	1 day	35.5 ± 6.41	<0.1
	7 days	37.9 ± 7.83	<0.1

DW: distilled water; DEX: dexketoprofen; nano-DEX: liposomes containing dexketoprofen.

Table 5. The effects of DEX and nano-DEX on the complement activity and on the PC (values are presented as mean ± S.D. for 6 rats in a group).

		Serum Complement Activity (UCH50)	PC (%)
DW	1 day	15.9 ± 3.1	54.3 ± 8.1
	7 days	16.5 ± 3.7	55.7 ± 8.5
DEX	1 day	16.3 ± 4.1	54.5 ± 8.7
	7 days	16.8 ± 4.5	56.1 ± 9.3
nano-DEX	1 day	15.6 ± 3.3	53.7 ± 7.7
	7 days	16.7 ± 3.9	55.3 ± 8.5

DW: distilled water; DEX: dexketoprofen; nano-DEX: liposomes containing dexketoprofen; PC: phagocytic capacity of peripheral neutrophils.

3.4. The Nociceptive Sensitivity Investigation

The oral administration of DEX resulted in a significant (** $p < 0.01$) lengthening of the paws' latency time response, an effect that was prolonged for four hours (* $p < 0.05$), then progressively reduced after about 10 h (* $p < 0.05$) in the hot plate test. Its maximum possible antinociceptive effect was achieved at 60 min (%MPE = 56.6 ± 8.7%) in this experimental somatic pain model in rats (Figure 6).

The treatment with DEX produced a substantial and rapid analgesic effect, even after 15 min, which was manifested for four hours. After this period, the effect progressively decreased, reaching intensity comparable with that observed in the group treated with distilled water in the hot plate. The highest intensity of the antinociceptive activity (56.6 ± 8.7%) was noted 60 min after the substance administration in this somatic pain model in rats.

The use of nano-DEX was correlated with an increase in the latency time reactivity to thermal noxious stimulation, which began after 90 min, and lasted for about 10 h (Figure 2). Its maximum possible antinociceptive effect was achieved at 6 h (%MPE = 49.4 ± 5.3%) in the experiment (Figure 7).

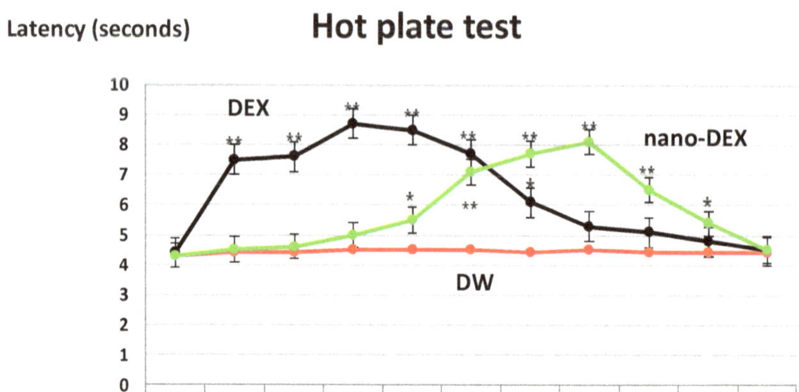

	baseline	15 min	30 min	60 min	90 min	120 min	4 h	6 h	8 h	10 h	12 h
DW	4.3	4.4	4.4	4.5	4.5	4.5	4.4	4.5	4.4	4.4	4.4
DEX	4.4	7.5	7.6	8.7	8.5	7.7	6.1	5.3	5.1	4.8	4.5
nano-DEX	4.3	4.5	4.6	5	5.5	7.1	7.7	8.1	6.5	5.4	4.5

Figure 6. The effects of nano-DEX in the hot plate test. Each point is the mean ± S.D. of the latency. * $p < 0.05$, ** $p < 0.01$ compared to control. DW: distilled water; DEX: dexketoprofen; nano-DEX: liposomes containing dexketoprofen.

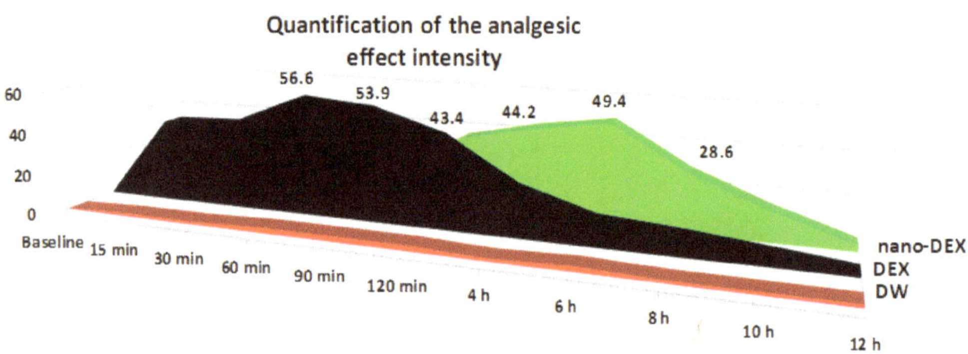

Figure 7. The percentage maximum possible antinociceptive effect (%MPE) of nano-DEX in the hot plate test. Each point is the mean ± S.D. for six rats. DW: distilled water; DEX: dexketoprofen; nano-DEX: liposomes containing dexketoprofen.

The treatment with chitosan-coated liposomes entrapping DEX was accompanied by a strong analgesic effect, which began after 120 min and persisted for up to 8 h, being the most intense (with a maximum analgesic effect of 51.7 ± 10.5%) at 6 h in this experimental somatic pain model. After this moment of determination, the effects of nano-DEX were comparable with those noted for free DEX, as well as for distilled water in the hot plate test in rats.

4. Discussion

Data in the literature show that each of the non-steroidal anti-inflammatory drug-loaded nano-formulations was demonstrated to be beneficial in correct pain management, leading to a safer and more precise medication. Furthermore, the prolonged release of the antinociceptive agents from nano-systems delivers an effective concentration of active drug for a long period of time [44–48].

The delayed onset of the analgesic activity, and the prolongation of the antinociceptive action of the nano-DEX, can be attributed to the particularities in the release of the drug entrapped into liposomes stabilized with chitosan.

Given its biological half-life and physico-chemical characteristics, dexketoprofen is a good drug candidate for prolonged release, in order to improve patient compliance by reducing dose frequency. It has been formulated in prolonged released mini-matrix tablets [49], minitablets coated with Eudragit S and ethylcellulose for colon targeting [50], and gastroretentive systems by the wet granulation method with various grades of hydroxylpropyl methyl cellulose (HPMC) [51]. All these dosage forms have proved a modified release profile.

Various methods have been implemented for preparing more evolved carriers, such as nanosystems containing dexketoprofen, all demonstrating to provide prolonged release of the drug and to maintain the analgesic effect for several hours [52–55].

We obtained original formulations of DEX loaded nanoparticle systems with chitosan and phosphatidylcholine. These new systems were physico-chemically and structurally characterized, tested for biocompatibility, and the in vitro release of drugs was studied. Additionally, we tested how DEX-loaded nanoparticles influenced nociceptive sensitivity compared to the free drug on a standard experimental somatic pain model in rats.

The advantages of these nanoparticulate systems rely on the extended release of DEX, which provides a prolongation of antinociceptive effects in the hot plate test in rats.

Ozturk et al. prepared DEX-loaded nanoparticles by spray-drying with Eudragit 100 RL, which displayed controlled release kinetics, with prospects of prolonging DEX analgesic activity [26]. Using other encapsulating agents, such as Compritol ATO 888 and Dynasan 114, the same research team prepared solid lipid nanoparticles containing DEX, which demonstrated prolonged analgesic activity for up to nine hours in the writhing test in mice [56].

DEX was also used as a model drug form obtaining modern, new systems—nanocochleates—in combination with other drugs [52]. We prepared nanosystems as unilamellar liposomes entrapping dexketoprofen, using chitosan, with good stability and uniform size range. In our experimental conditions, the use of lipid vesicles entrapping DEX did not induce substantial blood element discrepancies, biochemical changes, and immune parameters modifications; all these findings support the idea that these new systems have a good in vivo biocompatibility in rats.

The in vivo release of DEX from nanoparticles was not tested, which represents one limitation of the study; another limitation consists of the small number of animals used in the experiment.

5. Conclusions

The administration of lipid vesicles containing DEX offered the advantage of a prolonged release of the drug and was confirmed to have an extended antinociceptive effect in the hot plate test in rats.

The results obtained in this study provide important perspectives for fundamental medicine—through the design and characterization of new nanoparticles incorporating DEX —as well as for medical practice—through the possibility of using, after multiple subsequent studies, such nanoparticulate formulations—with prolonged release of the non-steroidal anti-inflammatory drug in pain therapy. The reduced frequency of administration shall increase patient compliance, and consequently, the therapy outcome.

Author Contributions: Conceptualization and methodology, L.M.-T., D.A.P. and E.G.P.; data analysis, M.B. and D.A.P.; investigation, L.M.-T., D.A.P., L.H., A.-M.P., B.R.B. and E.G.P.; writing—original draft preparation, M.B., D.A.P., L.H., A.-M.P., B.R.B., L.A.D. and E.G.P.; writing—review and editing, L.M.-T., M.B. and L.A.D.; supervision and project administration, L.M.-T. and E.G.P. All authors have read and agreed to the published version of the manuscript.

Funding: This research was funded by "Grigore T. Popa" University of Medicine and Pharmacy, Iasi, Romania, Internal Research Grant number 16407—Project director Liliana Mititelu-Tartau.

Institutional Review Board Statement: The protocol of the study was approved (Protocol No. 19157/19.10.2009) by the Ethic Committee on Research of 'Grigore T. Popa' University of Medicine and Pharmacy, in Iasi, Romania, according to the European Ethical Regulations and to the guidelines of IASP Committee for Research and Ethical Issues.

Informed Consent Statement: Not applicable.

Data Availability Statement: Not applicable.

Conflicts of Interest: The authors declare no conflict of interest.

References

1. Boisseau, P.; Loubaton, B. Nanomedicine, nanotechnology in medicine. *Comptes Rendus Phys.* **2011**, *12*, 620–626. [CrossRef]
2. Hughes, G.A. Nanostructure-mediated drug delivery. *Nanomedicine* **2005**, *1*, 22–30. [CrossRef] [PubMed]
3. Singh, R.; Lillard, J.W., Jr. Nanoparticle-based targeted drug delivery. *Exp. Mol. Pathol.* **2009**, *86*, 215–223. [CrossRef] [PubMed]
4. Stoicea, N.; Fiorda-Diaz, J.; Joseph, N.; Shabsigh, M.; Arias-Morales, C.; Gonzalez-Zacarias, A.; Mavarez-Martinez, A.; Marjoribanks, S.; Bergese, S. Advanced Analgesic Drug Delivery and Nanobiotechnology. *Drugs* **2017**, *77*, 1069–1076. [CrossRef]
5. Kenyon, N.J.; Bratt, J.M.; Lee, J.; Luo, J.; Franzi, L.M.; Zeki, A.A.; Lam, K.S. Self-assembling nanoparticles containing dexamethasone as a novel therapy in allergic airways inflammation. *PLoS ONE* **2013**, *8*, e77730. [CrossRef]
6. Lu, X.; Howard, M.D.; Mazik, M.; Eldridge, J.; Rinehart, J.J.; Jay, M.; Leggas, M. Nanoparticles containing anti-inflammatory agents as chemotherapy adjuvants: Optimization and in vitro characterization. *AAPS J.* **2008**, *10*, 133–140. [CrossRef] [PubMed]
7. Machelska, H.; Celik, M.Ö. Advances in Achieving Opioid Analgesia Without Side Effects. *Front. Pharmacol.* **2018**, *9*, 1388. [CrossRef]
8. Andreu, V.; Arruebo, M. Current progress and challenges of nanoparticle-based therapeutics in pain management. *J. Control. Release* **2018**, *269*, 189–213. [CrossRef]
9. Al-Lawati, H.; Binkhathlan, Z.; Lavasanifar, A. Nanomedicine for the effective and safe delivery of non-steroidal anti-inflammatory drugs: A review of preclinical research. *Eur. J. Pharm. Biopharm.* **2019**, *142*, 179–194. [CrossRef] [PubMed]
10. Rodríguez, M.J.; Arbós, R.M.; Amaro, S.R. Dexketoprofen trometamol: Clinical evidence supporting its role as a painkiller. *Expert Rev. Neurother.* **2008**, *8*, 1625–1640.
11. Balani, M.; Gawade, P.; Maheshgauri, S.; Ghole, S.; Shinde, V.; Sathe, V.J. Results of two multicentric, comparative, randomized, parallel group clinical trials to evaluate the efficacy and safety of dexketoprofen trometamol in the treatment of dental pain and dysmenorrhoea in Indian patients. *Clin. Diag. Res.* **2008**, *2*, 1086–1091.
12. Sweetman, B.J. Development and use of the quick acting chiral NSAID dexketoprofen trometamol (keral). *Acute Pain* **2003**, *4*, 109–115. [CrossRef]
13. Moore, R.A.; Barden, J. Systematic review of dexketoprofen in acute and chronic pain. *BMC Clin. Pharmacol.* **2008**, *8*, 11. [CrossRef] [PubMed]
14. Carne, X.; Rios, J.; Torres, F. Postmarketing cohort study to assess the safety profile of oral dexketoprofen trometamol for mild to moderate acute pain treatment in primary care. *Methods Find. Exp. Clin. Pharmacol.* **2009**, *31*, 533–540. [PubMed]
15. Walczak, J.S. Analgesic properties of dexketoprofen trometamol. *Pain Manag.* **2011**, *1*, 409–416. [CrossRef]
16. Barbanoj, M.J.; Antonijoan, R.M.; Gich, I. Clinical pharmacokinetics of dexketoprofen. *Clin. Pharmacokinet.* **2001**, *40*, 245–262. [CrossRef]
17. Barbanoj, M.J. Clinical pharmacokinetics of dexketoprofen trometamol: Recent studies. *Methods Find. Exp. Clin. Pharmacol.* **2006**, *28*, 3. [PubMed]
18. Matsui, H.; Shimokawa, O.; Kaneko, T.; Nagano, Y.; Rai, K.; Hyodo, I. The pathophysiology of non-steroidal anti-inflammatory drug (NSAID)-induced mucosal injuries in stomach and small intestine. *J. Clin. Biochem. Nutr.* **2011**, *48*, 107–111. [CrossRef] [PubMed]
19. Ahmad, A.; Mubharak, N.M.; Naseem, K.; Tabassum, H.; Rizwan, M.; Najda, A.; Kashif, M.; Bin-Jumah, M.; Hussain, A.; Shaheen, A.; et al. Recent Advancement and development of chitin and chitosan -based nanocomposite for Drug Delivery: Critical approach to clinical research. *Arab. J. Chem.* **2020**, *13*, 8935–8964. [CrossRef]
20. Ali, E.A.; Ismail, M.N.; Elsabee, M. Chitosan based polyelectrolyte complexes development for anionic and cationic dyes adsorption. *Egypt J. Chem.* **2020**, *63*, 537–554. [CrossRef]
21. Quemeneur, F.; Rinaudo, M.; Pépin-Donat, B. Influence of Molecular Weight and pH on Adsorption of Chitosan at the Surface of Large and Giant Vesicles. *Biomacromolecules* **2008**, *9*, 396–402. [CrossRef] [PubMed]

22. Freitas, E.D.; Moura, C.F., Jr.; Kerwald, J.; Beppu, M.M. An Overview of Current Knowledge on the Properties, Synthesis and Applications of Quaternary Chitosan Derivatives. *Polymers* **2020**, *12*, 2878. [CrossRef]
23. Yaneva, Z.; Ivanova, D.; Nikolova, N.; Tzanova, M. The 21st century revival of chitosan in service to bio-organic chemistry. *Biotechnol. Biotechnol. Equip.* **2020**, *34*, 221–237. [CrossRef]
24. Degim, I.T.; Kadioglu, D. Cheap, suitable, predictable and manageable nanoparticles for drug delivery: Quantum dots. *Curr. Drug Deliv.* **2013**, *10*, 32–38. [CrossRef] [PubMed]
25. Sevim, B.B.; Cerci, B.H.; Kucuk, Z.; Bahadori, F.; Kazdal, F.; Eskandari, Z.; Tiris, G.; Demiray, M. Optimization of a new controlled release oral dexketoprofen formulation. *J. Nanomater. Mol. Nanotechnol.* **2018**, *7*, 109.
26. Öztürk, A.A.; Yenilmez, E.; Yazan, Y. Dexketoprofen trometamol-loaded Eudragit®RL 100 nanoparticle formulation, characterization and release kinetics. *Acta Pharm. Sci.* **2019**, *57*, 69–84.
27. Öztürk, A.A.; Kiyan, H.T. Treatment of oxidative stress-induced pain and inflammation with dexketoprofen trometamol loaded different molecular weight chitosan nanoparticles: Formulation, characterization and anti-inflammatory activity by using in vivo HET-CAM assay. *Microvasc. Res.* **2020**, *128*, 103961. [CrossRef] [PubMed]
28. Gârlea, A.; Popa, M.I.; Pohoaţă, V.; Melnig, V. Ibuprofen/ketoprofen entrapment in chitosan based vesicle carrier. *Rom. J. Biophys.* **2007**, *17*, 157–168.
29. Tartau, L.; Cazacu, A.; Melnig, V. Ketoprofen-liposomes formulation for clinical therapy. *J. Mater. Sci. Mater. Med.* **2012**, *23*, 2499–2507. [CrossRef] [PubMed]
30. Vasile, C.; Stoleru, E.; Darie-Nita, R.N.; Dumitriu, R.P.; Pamfil, P.; Tartau, L. Biocompatible materials based on plasticized poly(lactic acid), chitosan and Rosemary ethanolic extract I. Effect of chitosan on the properties of plasticized poly(lactic acid) materials. *Polymers* **2019**, *11*, 941. [CrossRef] [PubMed]
31. Stefan, M.; Melnig, V.; Pricop, D.; Neagu, A.; Mihasan, M.; Tartau, L.; Hritcu, L. Attenuated effects of chitosan-capped gold nanoparticles on LPS-induced toxicity in laboratory rats. *Mater. Sci. Eng. C Mater.* **2013**, *33*, 550–556. [CrossRef] [PubMed]
32. Peniche, C.; Argüelles-Monal, W.; Peniche, H.; Acosta, N. Chitosan: An attractive biocompatible polymer for microencapsulation. *Macromol. Biosci.* **2003**, *3*, 511–520. [CrossRef]
33. Tiyaboonchai, W. Chitosan nanoparticles: A promising system for drug delivery. *Naresuan Univ. J.* **2003**, *11*, 51–66.
34. Li, X.; Wang, L.; Fan, Y.; Feng, Q.; Cui, F.Z. Biocompatibility and toxicity of nanoparticles and nanotubes. *J. Nanomater.* **2012**, *2012*, 548389.
35. Wolf, M.F.; Andwraon, J.M. Practical approach to blood compatibility assessments: General considerations and standards. In *Biocompatibility and Performance of Medical Devices*; Boutrand, J.-P., Ed.; Woodhead Publishing: Cambridge, UK, 2012; pp. 159–201, 201e–206e.
36. Bannon, A.W.; Malmberg, A.B. Malmberg, A.B. Models of nociception: Hot-plate, tail-flick, and formalin tests in rodents. *Curr. Protoc. Neurosci.* **2007**, *41*, 9.1–9.15. [CrossRef] [PubMed]
37. Ma, C. Animal models of pain. *Int. Anesthesiol. Clin.* **2007**, *45*, 121–131. [CrossRef] [PubMed]
38. Mogil, J.S. Animal models of pain: Progress and challenges. *Nat. Rev. Neurosci.* **2009**, *10*, 283–294. [CrossRef] [PubMed]
39. Walker, E.A.; Butelman, E.R.; Decosta, B.R. Opioid thermal antinociception in rhesus monkeys: Receptor mechanisms and temperature dependency. *J. Pharmacol. Exp. Ther.* **1993**, *267*, 280–286.
40. European Union. *DIRECTIVE 2010/63/EU of the European Parliament and of the Council of 22 September 2010 on the Protection of Animals Used for Scientific Purposes*; European Union: Brussels, Belgium, 2010.
41. Zimmerman, M. Ethical guidelines for investigations of experimental pain in conscious animals. *Pain* **1983**, *16*, 109–110. [CrossRef]
42. Ailincai, D.; Tartau-Mititelu, L.; Marin, L. Drug delivery systems based on biocompatible imino-chitosan hydrogels for local anticancer therapy. *Drug Deliv.* **2018**, *25*, 1080–1090. [CrossRef] [PubMed]
43. Kumar, A.; Dixit, C.K. Methods for characterization of nanoparticles. In *Advances in Nanomedicine for the Delivery of Therapeutic Nucleic Acids*; Nimesh, S., Chandra, R., Gupta, N., Eds.; Woodhead Publishing: Cambridge, UK, 2017; Chapter 3; pp. 43–58.
44. Sashmal, S.; Mukherjee, S.; Ray, S.; Thakur, R.S.; Ghosh, L.K.; Gupta, B.K. Design and optimization of NSAID loaded nanoparticles. *Pak. J. Pharm. Sci.* **2007**, *20*, 157–162.
45. Tomić, M.; Micov, A.; Pecikoza, U.; Stepanović-Petrović, R. Clinical Uses of Nonsteroidal Anti-Inflammatory Drugs (NSAIDs) and Potential Benefits of NSAIDs Modified-Release Preparations. In *Microsized and Nanosized Carriers for Nonsteroidal Anti-Inflammatory Drugs Formulation Challenges and Potential Benefits*; Čalija, B., Ed.; Elsevier Inc.: Amsterdam, The Netherlands; Academic Press: Cambridge, MA, USA, 2017; pp. 1–29.
46. Krajišnik, D.; Čalija, B.; Cekić, N. Polymeric Microparticles and Inorganic Micro/Nanoparticulate Drug Carriers: An Overview and Pharmaceutical Application. In *Microsized and Nanosized Carriers for Nonsteroidal Anti-Inflammatory Drugs Formulation Challenges and Potential Benefits*; Čalija, B., Ed.; Elsevier Inc.: Amsterdam, The Netherlands; Academic Press: Cambridge, MA, USA, 2017; pp. 31–67.
47. Moradkhani, M.R.; Karimi, A.; Negahdari, B. Nanotechnology application for pain therapy. *Artif. Cells Nanomed. Biotechnol.* **2018**, *46*, 368–373. [CrossRef]
48. Nita, L.E.; Chiriac, A.P.; Nistor, M.T.; Tartau, L. Indomethacin uptake into poly(2-hydroxyethyl methacrylate-co-3,9-divinyl-2,4,8,10-tetraoxaspiro [5.5]-undecane) network: In vitro and in vivo controlled release study. *Int. J. Pharm.* **2012**, *426*, 90–99. [CrossRef]

49. Sweed, N.M.; Basalious, E.B.; Nour, S.A. Combined site-specific release retardant mini-matrix tablets (C-SSRRMT) for extended oral delivery of dexketoprofen trometamol: In vitro evaluation and single versus multiple doses pharmacokinetic study in human volunteers. *Drug Dev. Ind. Pharm.* **2019**, *45*, 1777–1787. [CrossRef]
50. Ugurlu, T.; Ilhan, E. Development and in vitro evaluation of a novel pulsatile drug delivery system containing dexketoprofen trometamol. *J. Pharm. Innov.* **2020**, 1–13. [CrossRef]
51. Hossain, M.D.R.; Ahsan, A.; Hossain, M.R.; Dey, R. Formulation & In Vitro Evaluation of Gastroretentive Floating Drug Delivery System of Dexketoprofen. Available online: https://www.researchgate.net/profile/Md_Hossain367/publication/332670659_USTC_RoAR_FDDS_POSTER/links/5cc297c092851c8d220508cf/USTC-RoAR-FDDS-POSTER.pdf (accessed on 8 February 2021).
52. Çoban, Ö.; Değim, Z.; Yılmaz, Ş.; Altıntaş, L.; Arsoy, T.; Sözmen, M. Efficacy of targeted liposomes and nanocochleates containing imatinib plus dexketoprofen against fibrosarcoma. *Drug Dev. Res.* **2019**, *80*, 556–565. [CrossRef] [PubMed]
53. Çulcu, Ö.; Tunçel, E.; İlbasmış Tamer, S.; Tırnaksız, F. Characterization of Thermosensitive Gels for the Sustained Delivery of Dexketoprofen Trometamol for Dermal Applications. *GUHES* **2020**, *2*, 1–12.
54. Öztürk, A.A.; Yenilmez, E.; Arslan, R.; Şenel, B.; Yazan, Y. Dexketoprofen trometamol-loaded Kollidon®SR and Eudragit®RS 100 polymeric nanoparticles: Formulation and in vitro-in vivo evaluation. *Lat. Am. J. Pharm.* **2017**, *36*, 2153–2165.
55. Öztürk, A.A.; Banderas, L.M.; Otero, M.D.C.; Yenilmez, E.; Şenel, B.; Yazan, Y. Dexketoprofen trometamol-loaded poly-lactic-co-glycolic acid (PLGA) nanoparticles: Preparation, in vitro characterization and cyctotoxity. *Trop. J. Pharm. Res.* **2019**, *18*, 1–11. [CrossRef]
56. Öztürk, A.A.; Yenilmez, E.; Arslan, R.; Şenel, B.; Yazan, Y. Dexketoprofen trometamol loaded solid lipid nanoparticles (SLNs): Formulation, in vitro and in vivo evaluation. *J. Res. Pharm.* **2020**, *24*, 82–89. [CrossRef]

polymers

MDPI

Article

The Effect of Poly (Ethylene glycol) Emulation on the Degradation of PLA/Starch Composites

Sarieh Momeni [1], Erfan Rezvani Ghomi [2,*], Mohamadreza Shakiba [1,*], Saied Shafiei-Navid [3], Majid Abdouss [1,*], Ashkan Bigham [4], Fatemeh Khosravi [2], Zahed Ahmadi [1], Mehdi Faraji [5], Hamidreza Abdouss [6] and Seeram Ramakrishna [2,*]

1 Department of Chemistry, Amirkabir University of Technology, Tehran 15875-4413, Iran; sara.momeni181@yahoo.com (S.M.); zahmadi@aut.ac.ir (Z.A.)
2 Center for Nanotechnology and Sustainability, Department of Mechanical Engineering, National University of Singapore, Singapore 117581, Singapore; fatemeh_khosravi22@yahoo.com
3 Department of Organic Chemistry, Faculty of Chemistry, University of Mazandaran, Babolsar 47416-95447, Iran; s.shafiei@stu.umz.ac.ir
4 Institute of Polymers, Composites and Biomaterials—National Research Council (IPCB-CNR), Viale J.F. Kennedy 54—Mostra d'Oltremare pad. 20, 80125 Naples, Italy; ashkan.bigham@ipcb.cnr.it
5 School of Chemistry, College of Science, University of Tehran, Tehran 14155-6455, Iran; m4.faraji@gmail.com
6 Department of Polymer, Amirkabir University of Technology, Tehran 15875-4413, Iran; hamidrezaabdouss@gmail.com
* Correspondence: erfanrezvani@u.nus.edu (E.R.G.); rezashakiba011@gmail.com (M.S.); phdabdouss44@aut.ac.ir (M.A.); seeram@nus.edu.sg (S.R.)

Citation: Momeni, S.; Rezvani Ghomi, E.; Shakiba, M.; Shafiei-Navid, S.; Abdouss, M.; Bigham, A.; Khosravi, F.; Ahmadi, Z.; Faraji, M.; Abdouss, H.; et al. The Effect of Poly (Ethylene glycol) Emulation on the Degradation of PLA/Starch Composites. *Polymers* **2021**, *13*, 1019. https://doi.org/10.3390/polym13071019

Academic Editor: José Miguel Ferri

Received: 27 February 2021
Accepted: 23 March 2021
Published: 25 March 2021

Abstract: As a hydrophilic renewable polymer, starch has been widely used in biocompatible plastics as a filler for more than two decades. The present study aimed at investigating the effects of polyethylene glycol (PEG), as a plasticizer, on the physicochemical properties of a hybrid composite—polylactic acid (PLA) and thermoplastic starch (TPS). A solvent evaporation process was adopted to gelatinize the starch and disparate PEG contents ranging from 3 to 15 wt.% (with respect to the sample weight) were examined. It was revealed that the increase in the PEG content was accompanied by an increment in the starch gelatinization degree. Referring to the microstructural analyses, the TPS/PLA mixture yielded a ductile hybrid composite with a fine morphology and a uniform phase. Nevertheless, two different solvents, including acetone and ethanol, were used to assess if they had any effect on the hybrid's morphology, tensile strength and thermal properties. It was found that ethanol culminated in a porous hybrid composite with a finer morphology and better starch distribution in the PLA structure than acetone. As the result of PEG addition to the composite, the crystallinity and tensile strength were decreased, whereas the elongation increased. The hydrolytic degradation of samples was assessed under different pH and thermal conditions. Moreover, the microbial degradation of the PLA/TPS hybrid composite containing different PEG molar fractions was investigated in the soil for 45 days. The rate of degradation in both hydrolytic and biodegradation increased in the samples with a higher amount of PEG with ethanol solvent.

Keywords: polyethylene glycol; polylactic acid; starch mixture; hydrolytic degradation; biodegradation

1. Introduction

The properties of polymers can be manipulated through an alteration in their chemical structure and preparation process [1]. The polymers are being applied in different areas such as packaging [2], automobile manufacturing [3], computer [4], clothing, etc., and also they are being used as additives in dyes, glues, coatings [5] and concrete [6]. Despite the fact that they are now an inseparable part of our lives, their low degradation rate makes those polymers a threat to the environment [7–9]. Degradation refers to any irreversible process—light, heat, moisture, chemical conditions and biological activity—leading to a change in the polymers molecular structure followed by influencing the physical and chemical properties [10–12].

Biodegradation is a phenomenon by which organic substances are cleaved by living organisms, followed by being degraded with oxygen (aerobic) or without oxygen (anaerobic) [13,14]. The biodegradation process is vital because it causes the removal of polymer products from the environment, or in other words, to return to the life cycle [15]. Microorganisms such as bacteria and fungi are involved in the degradation of natural and synthetic plastics [16]. The key parameters affecting the degradation kinetic are the polymer, organism type and environmental conditions. To penetrate a microorganism's cell, most polymers are too large. Therefore, they need to depolymerize into smaller monomers to be uptaken and degraded [13,17]. The chemical structure is a decisive factor showing whether a polymer is biodegradable. There are some practical ways to promote biodegradation, including changing the polymers' molecular structure and developing some microorganisms capable of consuming a particular carbon source [18].

The degradation process is caused in polymers when a change in their physical and/or chemical properties occurs [19]. Physical changes include reducing the molecular weight, tensile strength, elongation at break and teardrop in the level of transparency [20,21]. Chemical changes are referred to the alterations in the chemical structure of a material [22]. Taking sunlight as an example into account, the physical and chemical changes in polymers as the result of that show themselves as discolored cracks, loss of mechanical properties and loss of gloss [23]. Generally, when the degradation is caused by an optical source like sunlight, UV, etc., the polymer will be converted into a low molecular weight polymer followed by turning into carbon dioxide [24–26].

Biodegradable plastics can be referred to the thermoplastic materials such as polylactic acid (PLA), polycaprolactone (PCL), polyhydroxybutyrate (PHB), starch and polyvinyl alcohol (PVA) [27,28]. PLA is a linear aliphatic polyester obtained from renewable sources such as corn and sugar cane, making it an appropriate biological polymer [29]. The amounts of lactic acid entered into the human body through packaging are less than the lactic acid usually taken through eating food. Therefore, lactic acid-based polymers as environment-compatible materials are suitable alternatives for food packaging [30]. Considering the good processability of PLA, it has better potential than other biodegradable polymers such as PCL and it can be processed through injection molding, blow molding and thermoforming [31]. The diffusion of water, oxygen and carbon dioxide is weak through PLA [32]. The addition of an emulsifier to PLA increases the transient molecule's solubility by increasing the free volumes and as the result, the toughness will be improved. To reach a biodegradable PLA with a desirable toughness, alloying with other biodegradation polymers is required, but it is costly [33].

To increase the biodegradability and decrease costs, PLA is often mixed with starch [34]. The PLA and starch mixtures are used in thermal products such as drinking cups, food trays, dishes and boxes, food packaging, children's clothing and medical applications—dental implants, suture and bone screws [14,35]. Starch is a carbohydrate belonging to the polysaccharides family and it is composed of many glucose units. Starch has two molecular architectures: linear amylose molecules and branched amylopectin molecules [36]. Natural starch is not thermoplastic and has limited processability due to the large size of the particles (5–100 μm). In the presence of a plasticizer, thermoplasticity is provided at a high temperature and sheared/cut by overcoming the strong inter-molecular and intra-molecular hydrogen bonds of starch [37]. During this process called gelatinization, the melting temperature and glass transition temperature of starch decrease and become injectable, just like the traditional synthetic plastics [38,39]. It is worth mentioning that the branches with short chains decrease the crystallinity degree of aliphatic polyesters and the branches with long chains decrease the melting viscosity yielding a rigid starch [40,41]. Hence, the brittleness of PLA and starch mixtures is the main problem in most of the applications. A few plasticizers with low molecular weights, such as glycerol and sorbitol, are used in the mixtures [42]. Polyethylene glycol (PEG) is a common plasticizer and emulsifier used to reduce crystallinity and improve the mechanical properties of PLA/starch [43]. PEG is among the most applicable stabilizers and intermediates used in different industrial processes [44]. In addition, to decreasing the crystallinity degree, this plasticizer not only decreases Tm but also provides more thermal stability by producing sufficient interactions with the polymer's structure and it is more efficient than other plasticizers—sorbitol, glycerol, etc. [45].

The present study aimed to investigate the physical, chemical and biological properties of PLA/TPS when different contents of PEG had been added. Under laboratory conditions, the degradation of PEG-added PLA/TPS was tested and tracked-in soil. The weight loss, molecular weight change and surface erosion were assessed. Moreover, the effects of time and surface morphology on the degradation rate were studied.

2. Materials and Methods

2.1. Materials

Granules of PLA (3 mm nominal granule size) and corn starch were purchased from Sigma-Aldrich (St. Louis, MO, USA). Corn starch was dried in an oven to remove excess moisture for 24 h at 70 °C before being used. Chloroform ($CHCl_3$), acetone (C_3H_6O), PEG, ethanol, polyethylene (PE), phosphate buffer (PBS, pH = 7.4), sodium hydroxide (NaOH), hydrochloric acid 37 wt.% (HCl) were purchased from Merck, Darmstadt, Germany, and used without any further purification.

2.2. Film Preparation

The solution-casting technique was adopted to fabricate the films. Pure PLA (2.18 g) was dissolved in 30 mL of chloroform in a 150 mL flask. Then, the reaction mixture was magnetically stirred at a rate of 150 rpm. PLA was dissolved entirely in 2 h and starch (0.93 g) was added to the reaction mixture at ambient temperature. After dissolving the starch, a colorless solution was formed. The solution was degassed using ultrasonic irradiation for 2 h. Next, the solution was poured onto a glass plate and allowed to evaporate slowly at ambient temperature to form a film. The film was dried under vacuum to a constant weight and kept at room temperature for over a week to reach an equilibrium crystallinity [39].

Other samples were prepared with PEG as an emulsifier. These samples contained from 3 to 15 wt.% PEG (with respect to the sample weight) [46,47]. In brief and referring to the mentioned experimental procedure, the desired weight of PLA was dissolved in chloroform followed by pouring the dissolved starch into it. After obtaining a homogenous mixture, the reaction temperature was raised to 70 °C and then PEG in a 1:1 ratio mixture of acetone: chloroform (10 mL) was added to the reaction flask. Finally, the films were prepared the same as the procedure mentioned in the former paragraph. The thickness of films was 20–50 μm. The all-prepared composite samples are listed in Table 1.

The PSE_{22}, PSE_{32}, PSE_{42} and PSE_{52} films were prepared similarly to the PSE_2, PSE_3, PSE_4, and PSE_5 and their compositions were identical, but ethanol just replaced acetone. The resulting samples were compared with the reference sample of PE and starch with the combination of 30% starch and 70% PE (PES) [48]. All the stages were performed on the reference samples as well.

2.3. Photodegradation of the Samples

The films were cut (5 cm × 1 cm) and placed in a UV cabinet up to 500 h. Then, the samples were taken out of the cabinet and their characteristics were investigated.

2.4. The Microstructures of the Samples

The scanning electron microscopy (SEM, AIS 2100, Uiwang-si, Korea) was applied to assess the microstructure of samples and also the surface of samples during the degradation period was observed to check if any changes occurred. It is important to notice that the micrographs were taken from the samples' surfaces before and after landfilling. In the case of landfilled samples, after removing the soil, they were rinsed carefully and dried in a vacuum oven at 60 °C. The experiment was performed on the samples after the sixth week.

Table 1. Composition of PLA/Starch/PEG samples.

Sample	PLA (wt.%)	PE (wt.%)	Starch (wt.%)	PEG (wt.%)	PEG Solvent	UV Exposure (h)
PES	–	70	30	–	–	–
PES_{500}	–	70	30	–	–	500
PLA	100	–	–	–	–	–
PLA_{500}	100	–	–	–	–	500
PSE_S	70	–	30	–	–	–
$PSE_{S, 500}$	70	–	30	–	–	500
PSE_1	70	–	27	3	acetone	–
PSE_2	70	–	24	6	acetone	–
PSE_3	70	–	21	9	acetone	–
$PSE_{3, 500}$	70	–	21	9	acetone	500
PSE_4	70	–	18	12	acetone	–
PSE_5	70	–	15	15	acetone	–
PSE_{22}	70	–	24	6	ethanol	–
PSE_{32}	70	–	21	9	ethanol	–
$PSE_{32,500}$	70	–	21	9	ethanol	500
PSE_{42}	70	–	18	12	ethanol	–
PES_{52}	70	–	15	15	ethanol	–

2.5. Mechanical Performance

The tensile strength of prepared films was tested through an Instron 5566 tensile machine (Norwood, MA, US) with a stretch rate of 10 mm/min at room temperature. Three identical samples were tested and the average value of 5 samples was reported with a standard deviation.

The results belonging to the change in the breakpoint length besides the tensile strength are known by most researchers as a key factor in determining the degradation rate of polymeric compounds [49]. The below Equations (1) and (2) are applied to reach such information as follows:

$$\text{Retained elongation at break } (\varepsilon_r) = \frac{\varepsilon_t}{\varepsilon_0} \tag{1}$$

$$\text{Retained tensile strength } (\sigma_r) = \frac{\sigma_t}{\sigma_0} \tag{2}$$

In this respect, ε_t and ε_0 are the change in the breakpoint length of samples with and without being exposed to UV at different time intervals, respectively. Similarly, σ_t and σ_0 are the tensile strength of samples with and without being exposed to light at different time intervals.

To track the effect of degradation on the mechanical properties, the tensile strength of samples before and after the UV radiation was tested at different time periods at ambient temperature. For this purpose, all prepared films were cut into six pieces (5 cm × 1 cm) and each piece was placed under UV radiation up to 500 h and at each specific time interval, the tensile strength test was taken. The mechanical tests were accomplished on the samples without being exposed to UV radiation to make a better comparison.

2.6. Differential Scanning Calorimetry (DSC) Study

Thermodynamic behaviors and non-isothermal crystallinity of the samples were studied through DSC in the presence of nitrogen gas. The samples (15 mg) were placed into the container and the applied heat was set at a rate of 10 °C/min. After reaching to the final temperature (200 °C), the samples were kept for 2 min and then again cooled at the same rate down to the room temperature.

2.7. Fourier Transform Infrared Spectroscopy (FTIR) Study

Changes in the chemical structure of samples after being exposed to UV were measured by a Fourier transform infrared spectrophotometer (Thermo, Nocolet 8700, Waltham, MA, USA). The test was conducted according to the KBr pellets method and in the wavelength range of 4000–400 cm^{-1}.

2.8. Hydrolytic Degradation of the Samples

The samples with a dimension of 3 cm × 3 cm were cut, dried and placed in the glass vials of PBS and distilled water (pH = 7.4 ± 0.2). They were then placed in an incubator at 25 °C (room temperature) and 40 °C for three weeks. After the selected immersion periods, the selected samples were collected and rinsed with distilled water several times. Afterward, the rinsed samples were put into the oven to dry. This hydrolytic degradation was assessed in the neutral pH; the hydrolytic degradation in acidic and alkaline mediums were also investigated. To this end, the net weight of samples before and after hydrolytic degradation was measured.

The samples with the same dimensions were placed into the glass vials of NaOH solution (pH = 13). They were taken out at specific time intervals, rinsed with distilled water just like the previous step and placed in an oven at 70 °C for 3 h. Thus, hydrolytic degradation was investigated at the water bath at temperatures of 23 °C and 40 °C for three weeks. The sample weights were then measured precisely and an acid environment was created with hydrochloric acid. Afterward, hydrolytic degradation was analyzed according to alkaline and acid environments.

The residual weight after hydrolytic degradation (Q) was calculated based on Equation (3) as follows:

$$Q = \frac{W_t}{W_0} \times 100 \quad (3)$$

W_0 and W_t represent the samples' initial weight (g) and the samples' weight (g) after being degraded at each specific time, respectively.

To make a better understanding of the samples' hydrolytic degradation, the hydrolytic degradation (R, %h) was defined according to the following Equation (4):

$$R = \frac{Q_0 - Q_t}{t} \quad (4)$$

Q_0 and Q_t represent the weight of the samples before and after degradation at specific time periods.

2.9. Biodegradation of the Samples

In this test, the samples were landfilled beneath 10 cm gardening soil in glass vases. Some samples were placed into the soil before exposure to UV and the others were exposed to UV up to 500 h first, weighted and then landfilled horizontally. It should be noted that the samples were placed into an incubator at 40 °C. During the procedure, the water content was preserved at about 20%. The weight loss test evaluated the biodegradation rate of the samples. Every two weeks, the samples were removed from the soil, rinsed with distilled water and dried for 10 h at 70 °C; after weighing the samples, they were returned to the soil.

The rate of biodegradation (Q′) was calculated based on Equation (5) as follows:

$$Q' = \frac{W_1 - W_2}{W_1} \times 100 \quad (5)$$

In this respect, W_1 and W_2 (Equation (5)) are the weights of samples before and after landfilling at different time intervals, respectively.

3. Results and Discussion

3.1. Mechanical Properties Evaluation

Tensile strength is one of the most important criteria for degradable polymers, specifically in the case of optical–oxidative degradation [50]. The rapid decrease in elongation of polymers is a meaningful sign giving valuable information about polymers' structure. If a polymer's elongation decreases, the rupture and fragmentation rate will increase under the environmental stresses [51]. The effects of PEG addition on the mechanical properties—tensile strength and elongation at break—of samples are tabulated through Table 2. PSE samples showed higher tensile strength than the PES sample, indicating that polyester has higher tensile strength than PE [52].

Table 2. The elongation at break and tensile strength of samples without subjecting to UV.

Sample	Elongation at Break (%)	Tensile Strength (MPa)
PLA	9.44 ± 5.23	47.04 ± 11.32
PES	8.14 ± 1.02	25.13 ± 6.08
PSE_S	5.44 ± 0.70	39.91 ± 7.89
PSE_2	8.66 ± 0.65	26.32 ± 6.32
PSE_3	12.00 ± 0.25	23.99 ± 6.92
PSE_4	8.22 ± 0.23	24.05 ± 4.87
PSE_5	9.09 ± 0.81	23.01 ± 5.65
PSE_{22}	9.50 ± 1.22	13.04 ± 1.12
PSE_{32}	14.94 ± 1.85	15.50 ± 1.15
PSE_{42}	21.11 ± 4.44	9.02 ± 0.58
PSE_{52}	23.94 ± 6.02	8.23 ± 0.51

The decrease in the tensile strength of PLA when starch was added could be attributed to the incomplete miscibility of both hydrophobic (PLA) and hydrophilic (starch) phases [53]. By incorporation of PEG as an emulsifier to PLA/starch composite, the elongation at break could be increased through the improvement of PLA and starch miscibility and stability [54]. Comparing the effect of both solvents, acetone and ethanol, it was observed that using ethanol instead of acetone as PEG solvent caused more homogeneous emulsifier distribution in the polymer's structure, which was why here the increase in elongation was observed. This phenomenon could be attributed to better emulsifier's effect on the films through the gelatinization of starch [55]. Among the PSE_2, PSE_3, PSE_4 and PSE_5 samples, PSE_3 was the one with the highest elongation percentage and the strength percentage of PSE_2 almost equaled PSE_4. The PSE_{52}, PSE_{42}, PSE_{32} and PSE_{22} samples had the least tensile strength attributed to the solvent changing. The highest percentage of elongation belonged to the PSE_{52}, which had the least tensile strength percentage.

As the results showed better mechanical performance for acetone-based samples, those were selected to be evaluated after UV exposure. Figure 1 shows the effect of filler on the film's tensile strength after exposure to UV light. According to Figure 1, the samples' tensile strength dwindled from the beginning. The PSEs film experienced a sharper decrease than the other samples ascribed to the particle size and their distribution in the polymer's matrix. The PSEs film sample showed a significant decrease in tensile properties after exposure to light and this reduction in tensile properties for PSE_2, PSE_3, PSE_4 and PSE_5 samples was less than PSE_S. The reason could be attributed to the pores in the PSEs sample. These pores need the energy to get out of their normal size and stretched out. They must consume energy because of the particle's smallness, but if the particle size is large, quite the opposite is the case and the tensile strength drops. There is a direct relationship between being exposed to UV irradiation and the decrease in samples' mechanical properties [56]. The mechanical properties of the samples were declined sharply up to 500 h. The observed reduction in the mentioned properties was due to the degradation caused by UV irradiation. In the case of starch-containing samples, the reduction of tensile properties compared to the pure PLA films was due to the incompatibility of hydrophilic starch and hydrophobic PLA.

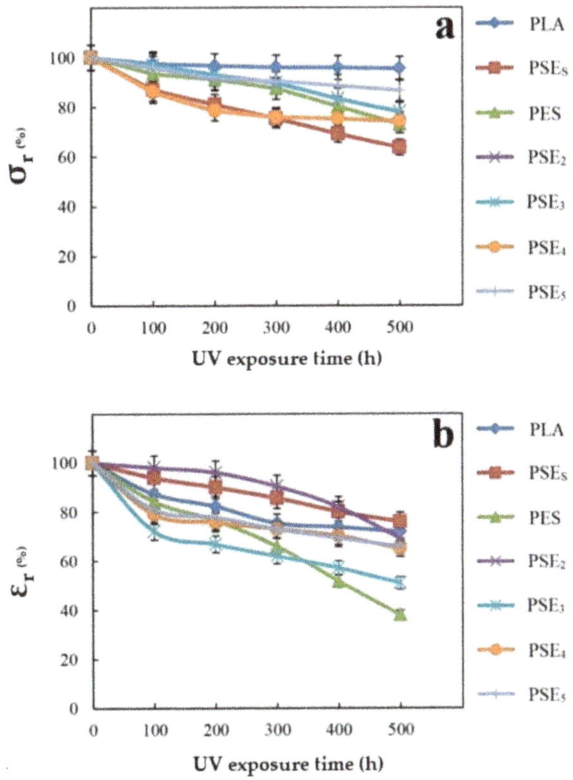

Figure 1. Changes of retained (**a**) tensile strength and (**b**) elongation to rupture point of polymer films after exposure to UV irradiation.

No significant change was observed in the tensile strength of PSE_3 comparing to PSE_4 and PSE_5, but PSE_3 showed the highest elongation at break among acetone-based samples. On the other hand, PSE_{32} exhibited the highest tensile strength among ethanol-based samples. Therefore, PSE_3 and PSE_{32} are more acceptable and optimal composites for further investigations.

3.2. Morphology and Structural Properties

The SEM was applied to study the samples' surface morphology before and after landfills for up to two weeks, as shown in Figure 2. The SEM micrographs of the pure PLA sample show a flat and smooth surface. However, some scratches occurred on the PLA surface when UV irradiation up to 500 h was applied. It is clear that the polymer's structure was degraded under UV light radiation and so the degradation process in the soil environment would become much more intense. According to the SEM results of PES, starch seeds were relatively well distributed in the structure. After being exposed to UV light for 500 h, the PES sample's surface was affected negatively and the changes in the morphology are observable. As observed in the micrographs of PSE_S, starch particles had acceptable distribution in the PLA phase, considering no emulsifier was used. PSE_s became sensitive after UV exposure, with some bubbles formed on the surface leading to a destruction in the surface after the landfill period. The PSE_3 film micrograph before being treated with UV displayed a very smooth and uniform morphology with no pore or bubble on the surface. However, after exposure to UV light, the PSE_3 film was mostly degraded and displayed some pores in the range of 1.5 to 2 μm on the surface.

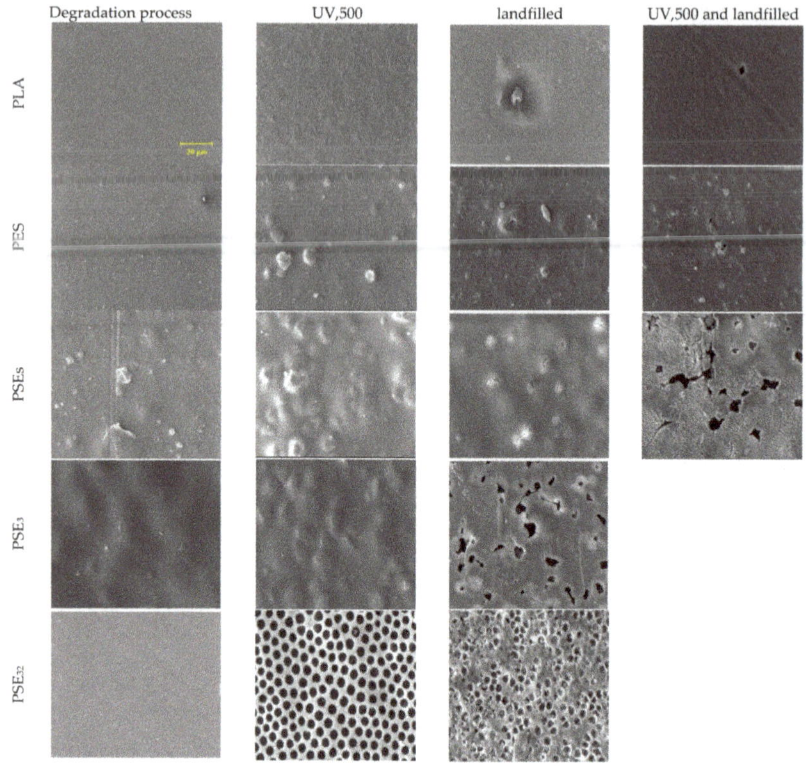

Figure 2. SEM micrographs of PLA film composites before and after degradation at 1.5 k× magnification.

After six weeks of landfilling, the SEM micrographs of samples showed progressive biodegradation for the PLA and starch samples (PSE_S, PSE_3 and PSE_{32}), whereas PES experienced less degradation than them. The reason might be found in the type of bonds in the PES sample, which is single covalent bonds [57]. Therefore, microorganisms showed less tendency towards the PES sample. The addition of starch to the PLA ester groups caused the compound more prone to degradation [58]. In contrast, it is known that the addition of polar groups to PE makes it more resistant to biological degradation [59]. Comparing the SEM images of PSE_3 and PSE_{32} indicates the effect of PEG solvent on forming the pores in the film structure. As can be seen, the use of ethanol as a solvent resulted in better PEG dispersion and smoother surface in PSE_{32} [60]. It is important to notice that the UV irradiation caused a dramatic change in the films' surface like forming bubbles and irregularities, leading to a faster degradation rate when they had been landfilled. It can be seen that the surface of samples after landfill became porous and also cracks were formed rooting in the microorganisms' activity and their effect on the morphology. The degradation rate depends on different factors, among which the nature of microorganisms, soil humidity, pH, temperature and the physicochemical properties of the substrate can be enumerated [61]. The pure PLA sample did not experience a high degradation rate compared to the PLA/starch films. Therefore, it cannot be considered as a suitable biodegradable polymer. However, the addition of a biopolymer such as starch increased its biodegradability properties dramatically.

3.3. Thermal Behavior Assessment

DSC studies were performed to investigate the effect of PEG addition on the PLA/starch composite. In addition, changes in the thermal properties of composites after using ethanol as the PEG solvent instead of acetone were observed. Figure 3 presents DSC analysis of PSE_S, PSE_2, PSE_3, PSE_{22} and PSE_{32}. In the case of PSE_2, PSE_3, PSE_{22} and PSE_{32} samples, there are three peaks visible—50–70, 100–120 and 140–160 °C. The first one (50–70 °C) pertained to the glass transition temperature (T_g) followed by changes in the PEG phase. The second one (100–120 °C) is related to the crystallization temperature (T_c) of PLA and the third one (140–160 °C) was attributed to the melting temperature (T_m). As there is no significant change in the T_g and T_m in all samples, it can be concluded that the addition of PEG did not considerably affect the mobility of the polymeric chains [62]. A former study had shown that starch content in the PLA/starch blend had little effect on the T_m [63]. However, after the incorporation of PEG, T_g decreased slightly, which can be attributed to PEG's emulsifier properties as well as its role as binding molecules between the hydrophobic (PLA) and the hydrophilic (starch) phases [64,65]. In the DSC curve of the PSE_s sample, two peaks are observable, referring to the T_g and T_m and the reason why no endothermic peak was seen pertained to using no emulsifier in this sample. The obtained results showed that PLA and starch were combined well through the solvent evaporation technique. The area under the T_g and T_m peaks gives information about the crystallinity percentage of samples. Therefore, it can be concluded that the addition of the emulsifier caused a decrease in the sample's crystallinity, making the modified samples susceptible to a faster degradation [66]. With the addition of PEG to PLA/starch, the T_c peak appeared in the DSC diagram of PSE_2. Then, the T_c of PSE_3 increased with an increment in the content of PEG compared to PSE_2, indicating an increase in the T_c. These results are in accordance with SEM findings which showed that PEG-containing samples have faster degradation rates. After using ethanol instead of acetone as PEG solvent (PSE_{22} and PSE_{32}), the peak at T_m was separated and reduced. This reduction in T_m intensity indicates a decrease in the crystallinity of the samples using ethanol solvent. It was also proved before that the amorphous structures show faster degradation rates [67]. This is again in accordance with SEM and mechanical properties results that ethanol-based samples showed higher degradation rates. On the other hand, the T_c was increased and the corresponding peak was broadened. These observations could be due to the uniformity of the composites and the better distribution of PEG after the application of ethanol as the solvent, which has increased the emulsifying effect of PEG on PLA/starch.

3.4. FTIR

The FTIR was deployed to identify the chemical groups of hybrid composites including PLA (a), PSE_3 (b), PSE_{32} (c), PSE_{3500} (d) and $PSE_{32,500}$ (e) (Figure 4); through the taken analysis, the effects of both solvent and UV were assessed. The formation of oxygen-containing groups is a way by which the effect of UV on the chemical groups can be tracked [68]. The sample code $PSE_{3,\,500}$ signifies that the PSE_3 sample was placed under UV radiation for 500 h. Changes in the carbonyl groups and the absorption changes were investigated through the band at 1716 cm^{-1}. To reduce the error caused by the thickness of samples, the band at 1895 cm^{-1} was chosen corresponding to the CH groups as the reference; the absorption rate of each group was determined by dividing the absorption rate of that group to 1895 cm^{-1}. The resulting films before and after the light seeing were taken at specified intervals of the FTIR spectra [47,69].

Optical-oxidative degradation of polymer samples is usually accompanied by the appearance of unsaturated bonds and various carbonyl groups by which the progression of degradation increases [70]. Two major observations—the formation of C=O (2000–1500 cm^{-1}) and O-H groups in (4000–3000 cm^{-1}) can also be found in the FTIR spectrum of polyolefin and other polymers [71].

When organic polymers are oxidized, carbonyl and hydroxyl groups are the predominant degradation products. These groups are easily characterized by an increase in bands' intensity at 1710 cm^{-1} and 3400 cm^{-1}. In the PLA spectrum, the band in the range of 3600–3500 cm^{-1} belonged to the C=O and the CH_3 stretching vibrations were observed at 3000–2940 cm^{-1}. The C=O stretching vibrations (1780–1761 cm^{-1}) and C-O stretching vibrations (1190–1033 cm^{-1}) were attributed to ester bonds. The broadband in the range of 3760–3010 cm^{-1} was attributed to the hydrogen bonds of starch hydroxyl groups. The stretching vibrations of CH_2 and hydro glucose starch ring appeared in the range of 1180–960 cm^{-1} and 861–575 cm^{-1}.

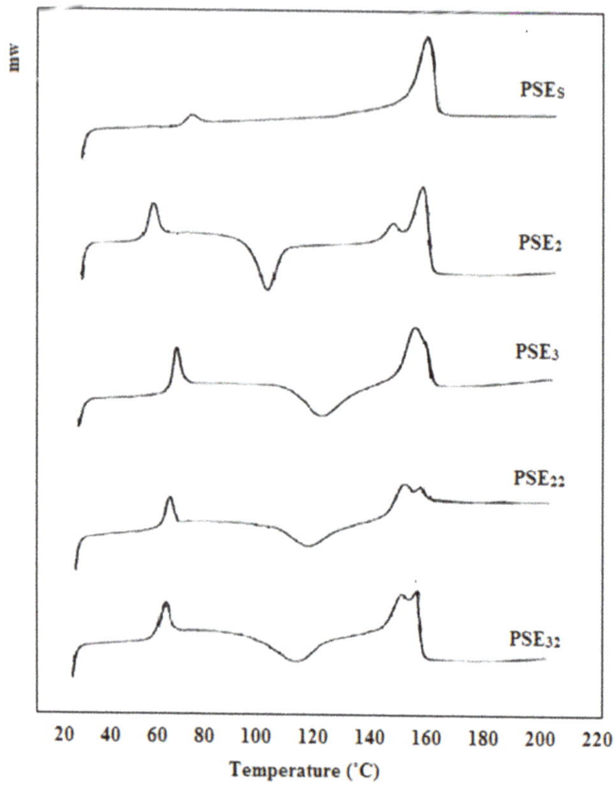

Figure 3. DSC curves of PSE_S, PSE_2, PSE_3, PSE_{22} and PSE_{32} samples.

As can be seen in the PSE_3 films FTIR spectrum, the absorption band related to the carbonyl group (1750–1640 cm^{-1}) was more intense than PLA; this phenomenon could be attributed to the presence of optimal amounts of filler. On the other hand, with an increase in the amount of filler in the polymer matrix, the OH groups with reductive properties at the end of the chains were increased and, hence, it would partially prevent the polymer's oxidation. By comparing the spectra of PSE_3 before and after UV radiation, it is visible that prolonging the irradiation time was synchronized with an increase in the carbonyl absorption peaks; this phenomenon is an indicator of increasing in the oxidation rate of polymers [72]. Nonetheless, the effect of solvents—acetone and ethanol—on the chemical groups of samples was investigated. It is noteworthy that using ethanol instead of acetone did not cause the appearance of new peaks in the FTIR spectra, but a shift towards lower wavenumbers (1736 to 1718 cm^{-1}). The possible phenomenon can be resulted from ethanol to increase the hydrogen bonds between PLA (carbonyl groups) and starch. This increment led to an increase in the bond length, followed by shifting the carbonyl peak towards lower wavenumbers [73]. Moreover, the FTIR peaks (oxygenated functional groups) of $PSE_{32,500}$ experienced an increase in their intensities due to the better distribution of starch particles all over the structure.

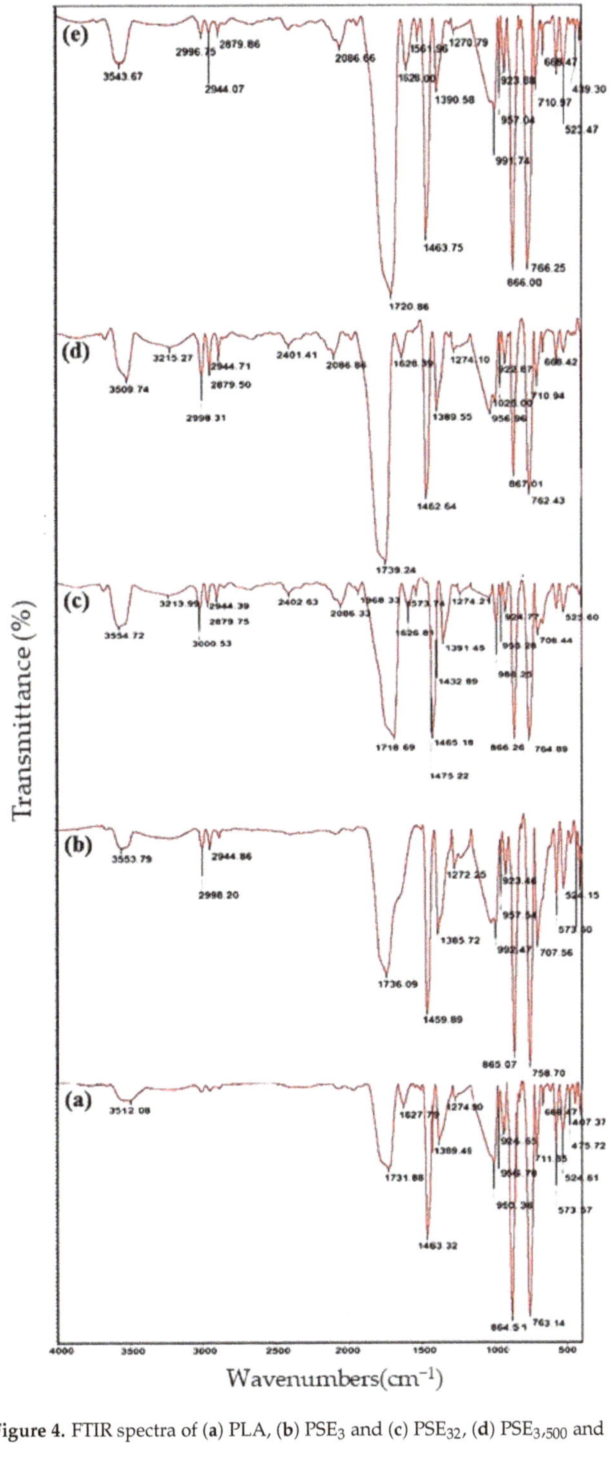

Figure 4. FTIR spectra of (**a**) PLA, (**b**) PSE_3 and (**c**) PSE_{32}, (**d**) $PSE_{3,500}$ and (**e**) $PSE_{32,500}$.

3.5. Hydrolytic Degradation

Figure 5 shows the hydrolytic degradation percentage of samples at pH 4, 7 and 13 and two temperatures of 23 °C (room temperature) and 40 °C in order to evaluate the effect of pH and temperature increase on degradation rate. Considering Figure 5a, degradation kinetic is faster when the temperature is increased. The highest degradation at pH = 4 occurred for the $PSE_{32,500}$ according to the mechanism shown in Figure 6b. Therefore, the time duration of being exposed to the light is effective in hydrolytic degradation. Another important parameter affecting the degradation results is the solvent, which showed higher degradation in the case of ethanol. Both the effect of UV exposure and the solvent was is in accordance with mechanical and SEM results. Thus, by increasing UV exposure duration, hydrolytic degradation of the samples is increased and hydrolytic degradation is similarly increased at higher temperatures. The lowest degradation belonged to the PES, indicating that PE has a low potential for hydrolytic degradation. The difference between PES and PSE_3 is in the main polymeric phase in the samples due to the potential for hydrolysis [74,75].

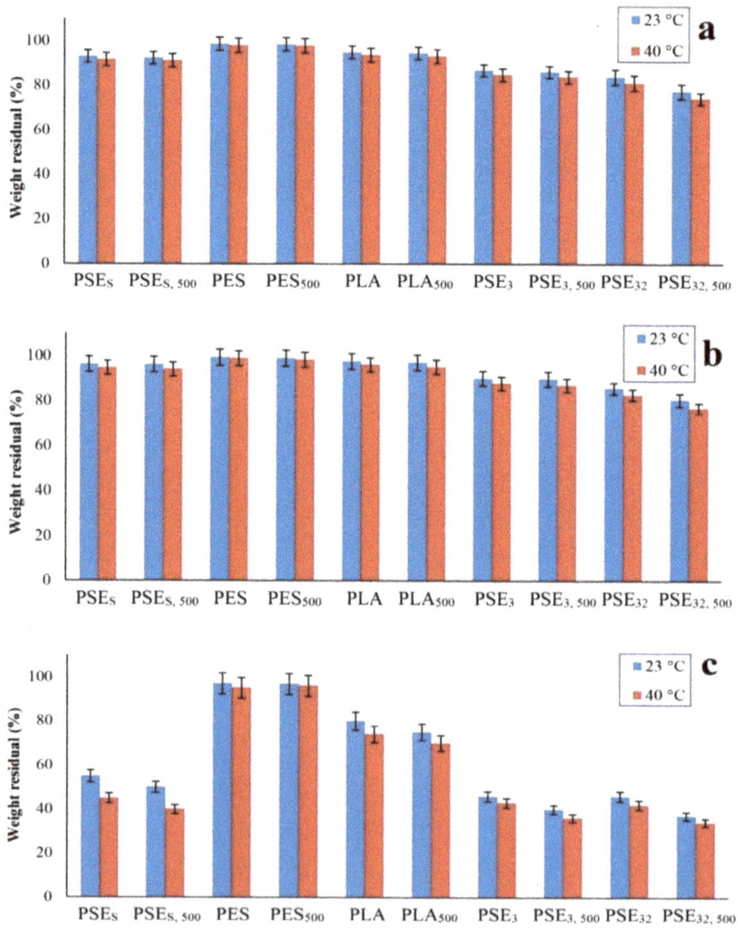

Figure 5. The weight residual after hydrolytic degradation of samples at (**a**) pH = 4, (**b**) pH = 7 and (**c**) pH = 13.

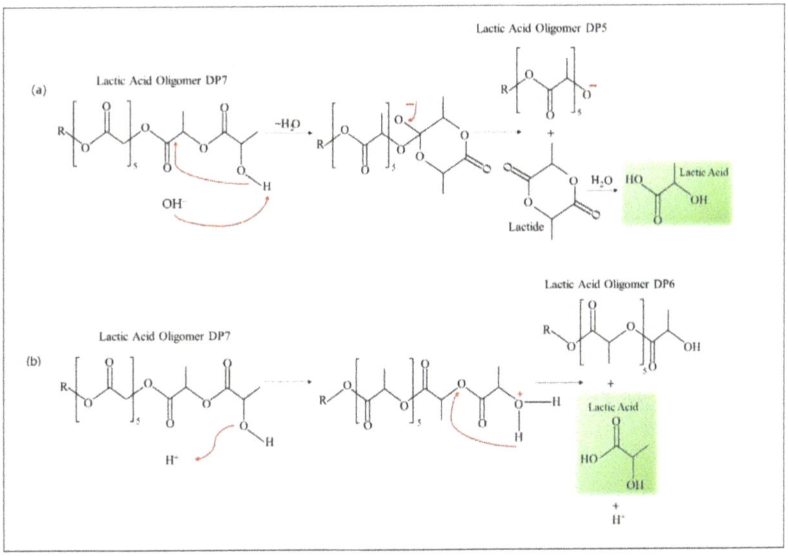

Figure 6. Hydrolytic degradation mechanism in (**a**) alkaline environment and (**b**) acidic environment.

The percentage of hydrolytic degradation at pH = 7 was less than pH = 4. It could be attributed to the concentration of H^+ ions and their role in the hydrolysis of films [76]. As shown in Figure 5b, there was no significant difference between hydrolytic degradation percentages at both environments and the highest degradation was related to the $PSE_{32,500}$, which experienced a higher hydrolytic degradation in 40 °C. The lowest hydrolytic degradation was related to PES having less degradation relative to the main sample of PSEs.

According to Figure 5c, the hydrolytic degradation percentage in alkaline environments (pH = 13) was more than the degradation percentage in neutral and acidic environments (pH = 4 and pH = 7). It can be due to the high hydrolysis power of OH^- ions which can hydrolyze polymer films at a high rate. Among the samples, $PSE_{32,500}$ had the highest degradation percentage and the effect of UV radiation on film's degradation was evident. Additionally, the lowest degradation percentage belonged to the PES sample. Based on the obtained results, the hydrolytic degradation of samples at pH = 13 occurred much faster than at pH = 4 and pH = 7. Making a comparison between Figure 5 revealed that the hydrolytic degradation of samples at 40 °C was more than that of 23 °C.

Figure 6 shows the mechanism of oligolactate degradation. The degradation in alkaline environments was accompanied by an intra-molecular ester exchange. The nucleophilic attack of the hydroxyl group to the second carbonyl causes the formation of a sustainable six-member interface loop and this reaction is the base of catalysis (mechanism a). According to this mechanism, oligomer lactic acid DP5 and lactic acid were produced during hydrolytic degradation. During the experiment, significant hydrolytic degradation was observed for polylactic/starch/PEG with acetone and PSEs samples which were almost fully degraded. The rupture of ester bonds in the hydroxyl groups of oligomers happens in low pH through protonation of the OH groups and through an intramolecular hydrogen bond [13]. A five-member ring may be one of the most stable intermediate structures. The formation of hydrogen bonds increases the electrophilic nature of carbonyl groups and hydrolysis through water molecules would occur through the site. According to this experiment, the degradation of various ester groups in lactic acid oligomers was independent of pH. Considering the low concentration of OH groups, the primary phase of these systems is stochastically degraded in polymer chains such that adding starch in polymer degradation increase hydrolytic degradation. Increasing starch in the matrix leads to an increase in hydrophilicity which can be attributed to the constant increase of dielectric. A higher dielectric constant leads to faster degradation and protecting the hydroxyl groups leads to a significant decrease in the hydrolytic degradation [77–80].

3.6. Biodegradation

The biodegradation of samples was assessed during optical oxidative degradation. Due to the complexity of the biodegradation process and many factors affecting this process, various methods have been reported by researchers and some standards have been developed in this area [81–83]. However, applying them into practice often requires the use of special equipment. Biodegradation through the landfill is an easy way by which the degradation rate can be estimated [84]. The soil contains different microorganisms leading to the digestion of polymers followed by converting them into water and CO_2 and subsequently the sample's weight loss. The results are presented in Figure 7.

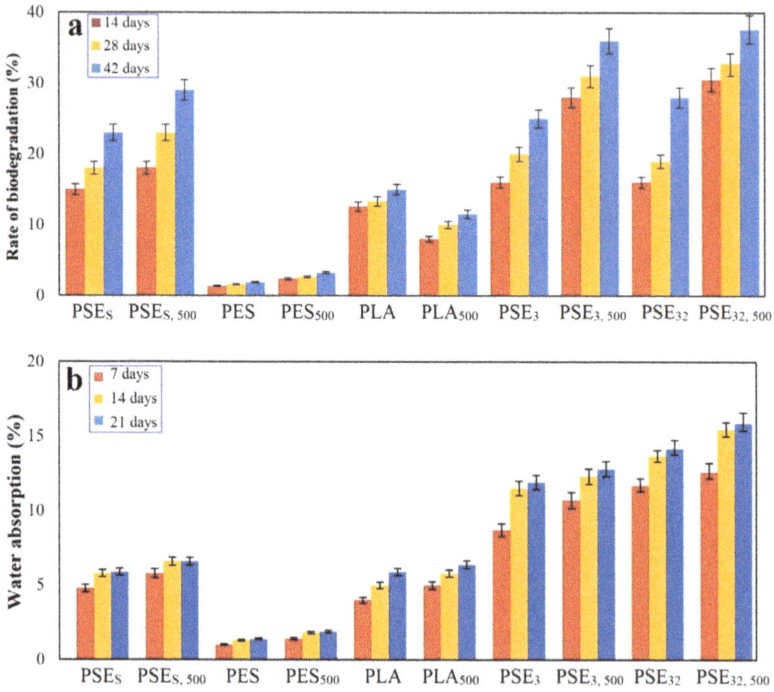

Figure 7. Polymer samples (**a**) rate of biodegradation and (**b**) water absorption after landfill at different times.

As can be observed in the case of control samples (PES), their weight loss was less than PLA samples. In addition, those samples, which were not in the exposure of UV, experienced less change in their weight during the period of the landfill than the samples treated with UV. The weight loss of samples was impressive, particularly after two weeks of landfill. The main mechanism governing this degradation process starts with the formation of pores in the polymers' structure after being exposed to UV and so it leaves behind more surface area leading to an increase in the oxidation and digestion of substrates. Moreover, the breakage of polymer chains due to mechanical stresses is another parameter that may lead to the creation of active radicals and the oxidation of polymer in the environment.

The hydrophobic property of polyolefin is a preventive factor that impedes the penetration of enzymes into polymers' structure and, thus, makes them useless. Moreover, many microorganisms and fungi require a suitable substrate besides a food source in order to proliferate and develop their colonies. Hence, polyolefin films are not suitable for the growth of microorganisms [39]. To make polyolefin biodegradable, their molecular mass must be dropped massively. Although environmental factors such as light, heat and oxygen are effective for this purpose, they cannot cause massive molecular mass drop within a relatively short period of time [85].

The addition of starch to the PLA structure would improve its wettability culminating in a faster water molecules penetration. The water absorption of samples is shown in Figure 7b, where the PSEs sample showed a higher absorption capability than the PES sample. Considering that no compatibilizer was used to prepare these samples, the PLA and starch compound was found to be mixed well with a proper distribution of starch through the whole structure. It is noteworthy that the PESs polymer film should have shown at least 20% water absorption, while only about 5% was obtained. The results are in accordance with SEM studies of composites (Figure 2). As can be seen from the SEM images, the highest degradation rate after the landfill is related to PSE_{32} due to the better PEG dispersion in the PLA/starch blend.

In the case of using ethanol as the solvent, higher water absorption in the prepared samples was observed. The reason for this phenomenon is rooted in the lack of compatibilizer, which can lead to bonding PLA and starch. Since these two are naturally hydrophobic (PLA) and hydrophilic (starch), their mixture would yield a physical bond even without using a compatibilizer leading to an increase in the compound's water absorption rate. The maximum water absorption occurred in the first week, followed by a decrease in the rate up to the second week; this can be due to the saturated hydrophilic groups and saturated pores found in the compounds.

4. Conclusions

In summary, the emulsifier's effect in the distribution and compatibility of thermoplastic starch and PLA phases containing the different percentages of PLA, Starch and emulsifier was investigated. According to the mechanical test and SEM results, the use of ethanol led to a better distribution of the PEG in the matrix and a higher degradation rate. In addition, FTIR results showed the effect of UV radiation on increasing the hydrophilicity of the samples and the degradation rate. The hydrolytic degradation of these samples was investigated under different pH and thermal conditions before and after UV irradiation. The results of the hydrolytic degradation of samples showed that the pH of the basic degradation rate was much higher than the acidic and neutral environments and by increasing the temperature, the degradation became more intense. Moreover, the PLA and starch films with varying PEG percentages were landfilled for 45 days and then studied by SEM. It showed some clear evidence of degradation, such as higher surface roughness, grooves, pores and decay. Various holes and grooves and a coordinating degradation of consortium microbial colonies, including fungi, bacteria and actionists, were observed on the film surfaces. The primary mechanism for biodegradation of polymers with a high molecular weight was oxidation or hydrolysis by enzymes which led to the creation of functional groups and improved the hydrophobic properties. The polymer main chains were then degraded and the samples showed lower molecular weight and mechanical properties. Therefore, it will be more prone to microbial uptake.

Author Contributions: Conceptualization, S.M., E.R.G., M.S., S.S.-N., M.A., A.B., F.K., Z.A., M.F., H.A.; methodology, S.M., M.S., S.S.-N., M.A., Z.A., M.F., H.A.; validation, S.M., E.R.G., M.S., S.S.-N., M.A., A.B., F.K., Z.A., M.F., H.A.; investigation, S.M., E.R.G., M.S., S.S.-N., M.A., A.B., F.K., Z.A., M.F., H.A.; writing—original draft preparation, S.M., E.R.G., M.S., S.S.-N., M.A., A.B., F.K., Z.A., M.F., H.A.; writing—review and editing, E.R.G., M.S., S.S.-N., M.A., A.B., F.K., Z.A., M.F., H.A., S.R.; supervision, M.A. and S.R. All authors have read and agreed to the published version of the manuscript.

Funding: This research received no external funding.

Institutional Review Board Statement: Not applicable.

Informed Consent Statement: Not applicable.

Data Availability Statement: The data presented in this study are available on request from the corresponding author.

Conflicts of Interest: The authors declare no conflict of interest.

References

1. Shakiba, M.; Nabavi, S.R.; Emadi, H.; Faraji, M. Development of a superhydrophilic nanofiber membrane for oil/water emulsion separation via modification of polyacrylonitrile/polyaniline composite. *Polym. Adv. Technol.* **2021**, *32*, 1301–1316. [CrossRef]
2. Cinelli, P.; Coltelli, M.B.; Signori, F.; Morganti, P.; Lazzeri, A. Cosmetic packaging to save the environment: Future perspectives. *Cosmetics* **2019**, *6*, 26. [CrossRef]
3. Sarvankar, S.G.; Yewale, S.N. Additive Manufacturing in Automobile Industry. *Int. J. Res. Aeronaut. Mech. Eng.* **2019**, *7*, 1–10.
4. AlSalhi, M.S.; Alam, J.; Dass, L.A.; Raja, M. Recent advances in conjugated polymers for light emitting devices. *Int. J. Mol. Sci.* **2011**, *12*, 2036–2054. [CrossRef] [PubMed]
5. Rezvani Ghomi, E.; Esmaeely Neisiany, R.; Nouri Khorasani, S.; Dinari, M.; Ataei, S.; Koochaki, M.S.; Ramakrishna, S. Development of an epoxy self-healing coating through the incorporation of acrylic acid-co-acrylamide copolymeric gel. *Prog. Org. Coat.* **2020**, *149*, 105948. [CrossRef]
6. Wang, F.; Yang, J.; Cheng, H.; Wu, J.; Liang, X. Study on mechanism of desorption behavior of saturated superabsorbent polymers in concrete. *Aci Mater. J.* **2015**, *112*, 463. [CrossRef]
7. Krzan, A.; Hemjinda, S.; Miertus, S.; Corti, A.; Chiellini, E. Standardization and certification in the area of environmentally degradable plastics. *Polym. Degrad. Stab.* **2006**, *91*, 2819–2833. [CrossRef]
8. Rezvani Ghomi, E.; Khosravi, F.; Tahavori, M.A.; Ramakrishna, S. Circular Economy: A Comparison Between the Case of Singapore and France. *Mater. Circ. Econ.* **2021**, *3*, 2. [CrossRef]
9. Foroughi, F.; Rezvani Ghomi, E.; Morshedi Dehaghi, F.; Borayek, R.; Ramakrishna, S. A Review on the Life Cycle Assessment of Cellulose: From Properties to the Potential of Making It a Low Carbon Material. *Materials* **2021**, *14*, 714. [CrossRef]
10. Göpferich, A. Mechanisms of polymer degradation and erosion. *Biomaterials* **1996**, *17*, 103–114. [CrossRef]
11. Gagliardi, M.; Lenarda, P.; Paggi, M. A reaction-diffusion formulation to simulate EVA polymer degradation in environmental and accelerated ageing conditions. *Sol. Energy Mater. Sol. Cells* **2017**, *164*, 93–106. [CrossRef]
12. Rezvani Ghomi, E.; Khosravi, F.; Mossayebi, Z.; Saedi Ardahaei, A.; Morshedi Dehaghi, F.; Khorasani, M.; Neisiany, R.E.; Das, O.; Marani, A.; Mensah, R.A.; et al. The Flame Retardancy of Polyethylene Composites: From Fundamental Concepts to Nanocomposites. *Molecules* **2020**, *25*, 5157. [CrossRef]
13. Shah, A.A.; Hasan, F.; Hameed, A.; Ahmed, S. Biological degradation of plastics: A comprehensive review. *Biotechnol. Adv.* **2008**, *26*, 246–265. [CrossRef]
14. Hakkarainen, M. Aliphatic polyesters: Abiotic and biotic degradation and degradation products. In *Degradable Aliphatic Polyesters*; Springer: Berlin, Germany, 2002; pp. 113–138.
15. Eubeler, J.P.; Zok, S.; Bernhard, M.; Knepper, T.P. Environmental biodegradation of synthetic polymers I. Test methodologies and procedures. *Trac Trends Anal. Chem.* **2009**, *28*, 1057–1072. [CrossRef]
16. Kawai, F. The biochemistry and molecular biology of xenobiotic polymer degradation by microorganisms. *Biosci. Biotechnol. Biochem.* **2010**, *74*, 1743–1759. [CrossRef]
17. Li, K.; Al-Rudainy, B.; Sun, M.; Wallberg, O.; Hulteberg, C.; Tunå, P. Membrane Separation of the Base-Catalyzed Depolymerization of Black Liquor Retentate for Low-Molecular-Mass Compound Production. *Membranes* **2019**, *9*, 102. [CrossRef]
18. Kinane, J.A.; Benakanakere, M.R.; Zhao, J.; Hosur, K.B.; Kinane, D.F. Porphyromonas gingivalis influences actin degradation within epithelial cells during invasion and apoptosis. *Cell. Microbiol.* **2012**, *14*, 1085–1096. [CrossRef]
19. Shakiba, M.; Kakoei, A.; Jafari, I.; Rezvani Ghomi, E.; Kalaee, M.; Zarei, D.; Abdouss, M.; Shafiei-Navid, S.; Khosravi, F.; Ramakrishna, S. Kinetic Modeling and Degradation Study of Liquid Polysulfide Resin-Clay Nanocomposite. *Molecules* **2021**, *26*, 635. [CrossRef]
20. Houchin, M.; Topp, E. Physical properties of PLGA films during polymer degradation. *J. Appl. Polym. Sci.* **2009**, *114*, 2848–2854. [CrossRef]
21. Wei, B.; Qi, H.; Zou, J.; Li, H.; Wang, J.; Xu, B.; Ma, H. Degradation mechanism of amylopectin under ultrasonic irradiation. *Food Hydrocoll.* **2021**, *111*, 106371. [CrossRef]
22. Beltrán-Sanahuja, A.; Casado-Coy, N.; Simó-Cabrera, L.; Sanz-Lázaro, C. Monitoring polymer degradation under different conditions in the marine environment. *Environ. Pollut.* **2020**, *259*, 113836. [CrossRef] [PubMed]
23. Moetazedian, A.; Gleadall, A.; Han, X.; Ekinci, A.; Mele, E.; Silberschmidt, V.V. Mechanical performance of 3D printed polylactide during degradation. *Addit. Manuf.* **2020**, *38*, 101764.
24. Yazdan Mehr, M.; Bahrami, A.; van Driel, W.D.; Fan, X.; Davis, J.L.; Zhang, G. Degradation of optical materials in solid-state lighting systems. *Int. Mater. Rev.* **2020**, *65*, 102–128. [CrossRef]

25. Yuan, J.; Ma, J.; Sun, Y.; Zhou, T.; Zhao, Y.; Yu, F. Microbial degradation and other environmental aspects of microplastics/plastics. *Sci. Total Environ.* **2020**, *715*, 136968. [CrossRef]
26. Lu, T.; Solis-Ramos, E.; Yi, Y.; Kumosa, M. UV degradation model for polymers and polymer matrix composites. *Polym. Degrad. Stab.* **2018**, *154*, 203–210. [CrossRef]
27. Saalah, S.; Saallah, S.; Rajin, M.; Yaser, A.Z. Management of Biodegradable Plastic Waste: A Review. In *Advances in Waste Processing Technology*; Springer: Berlin, Germany, 2020; pp. 127–143.
28. Khosravi, F.; Nouri Khorasani, S.; Khalili, S.; Esmaeely Neisiany, R.; Rezvani Ghomi, E.; Ejeian, F.; Das, O.; Nasr-Esfahani, M.H. Development of a Highly Proliferated Bilayer Coating on 316L Stainless Steel Implants. *Polymers* **2020**, *12*, 1022. [CrossRef] [PubMed]
29. Avinc, O.; Khoddami, A. Overview of poly (lactic acid)(PLA) fibre. *Fibre Chem.* **2009**, *41*, 391–401. [CrossRef]
30. Winursito, I. Biodegradabilitas Polikarboksilat Dari Asam Alginat dan Tapioka. *J. Litbang Ind.* **2013**, *3*, 39–47. [CrossRef]
31. Rasal, R.M.; Janorkar, A.V.; Hirt, D.E. Poly (lactic acid) modifications. *Prog. Polym. Sci.* **2010**, *35*, 338–356. [CrossRef]
32. Drieskens, M.; Peeters, R.; Mullens, J.; Franco, D.; Lemstra, P.J.; Hristova-Bogaerds, D.G. Structure versus properties relationship of poly (lactic acid). I. Effect of crystallinity on barrier properties. *J. Polym. Sci. Part B: Polym. Phys.* **2009**, *47*, 2247–2258. [CrossRef]
33. Jeong, J.; Ayyoob, M.; Kim, J.-H.; Nam, S.W.; Kim, Y.J. In situ formation of PLA-grafted alkoxysilanes for toughening a biodegradable PLA stereocomplex thin film. *RSC Adv.* **2019**, *9*, 21748–21759. [CrossRef]
34. Sun, Y.; Lee, D.; Wang, Y.; Li, S.; Ying, J.; Liu, X.; Xu, G.; Gwon, J.; Wu, Q. Thermal decomposition behavior of 3D printing filaments made of wood-filled polylactic acid/starch blend. *J. Appl. Polym. Sci.* **2021**, *138*, 49944. [CrossRef]
35. Ghomi, E.R.; Khosravi, F.; Neisiany, R.E.; Singh, S.; Ramakrishna, S. Future of additive manufacturing in healthcare. *Curr. Opin. Biomed. Eng.* **2021**, *17*, 100255. [CrossRef]
36. Zou, W.; Yu, L.; Liu, X.; Chen, L.; Zhang, X.; Qiao, D.; Zhang, R. Effects of amylose/amylopectin ratio on starch-based superabsorbent polymers. *Carbohydr. Polym.* **2012**, *87*, 1583–1588. [CrossRef]
37. Nafchi, A.M.; Moradpour, M.; Saeidi, M.; Alias, A.K. Thermoplastic starches: Properties, challenges, and prospects. *Starch Stärke* **2013**, *65*, 61–72. [CrossRef]
38. Huneault, M.A.; Li, H. Preparation and properties of extruded thermoplastic starch/polymer blends. *J. Appl. Polym. Sci.* **2012**, *126*, E96–E108. [CrossRef]
39. Rhim, J.-W.; Ng, P.K.W. Natural Biopolymer-Based Nanocomposite Films for Packaging Applications. *Crit. Rev. Food Sci. Nutr.* **2007**, *47*, 411–433. [CrossRef] [PubMed]
40. Wang, L.; Jing, X.; Cheng, H.; Hu, X.; Yang, L.; Huang, Y. Rheology and Crystallization of Long-Chain Branched Poly(l-lactide)s with Controlled Branch Length. *Ind. Eng. Chem. Res.* **2012**, *51*, 10731–10741. [CrossRef]
41. Nozue, Y.; Kawashima, Y.; Seno, S.; Nagamatsu, T.; Hosoda, S.; Berda, E.B.; Rojas, G.; Baughman, T.W.; Wagener, K.B. Unusual Crystallization Behavior of Polyethylene Having Precisely Spaced Branches. *Macromology* **2011**, *44*, 4030–4034. [CrossRef]
42. Esmaeili, M.; Pircheraghi, G.; Bagheri, R.; Altstädt, V. Poly(lactic acid)/coplasticized thermoplastic starch blend: Effect of plasticizer migration on rheological and mechanical properties. *Polym. Adv. Technol.* **2019**, *30*, 839–851. [CrossRef]
43. Yu, Y.; Cheng, Y.; Ren, J.; Cao, E.; Fu, X.; Guo, W. Plasticizing effect of poly(ethylene glycol)s with different molecular weights in poly(lactic acid)/starch blends. *J. Appl. Polym. Sci.* **2015**, *132*. [CrossRef]
44. Yue, H.; Zhao, Y.; Ma, X.; Gong, J. Ethylene glycol: Properties, synthesis, and applications. *Chem. Soc. Rev.* **2012**, *41*, 4218–4244. [CrossRef]
45. Jiang, X.; Jiang, T.; Zhang, X.; Dai, H.; Zhang, X. Melt processing of poly(vinyl alcohol) through adding magnesium chloride hexahydrate and ethylene glycol as a complex plasticizer. *Polym. Eng. Sci.* **2012**, *52*, 2245–2252. [CrossRef]
46. Feng, J.; Yang, G.; Zhang, S.; Liu, Q.; Jafari, S.M.; McClements, D.J. Fabrication and characterization of β-cypermethrin-loaded PLA microcapsules prepared by emulsion-solvent evaporation: Loading and release properties. *Environ. Sci. Pollut. Res.* **2018**, *25*, 13525–13535. [CrossRef]
47. Huneault, M.A.; Li, H. Morphology and properties of compatibilized polylactide/thermoplastic starch blends. *Polymers* **2007**, *48*, 270–280. [CrossRef]
48. Walker, A.M.; Tao, Y.; Torkelson, J.M. Polyethylene/starch blends with enhanced oxygen barrier and mechanical properties: Effect of granule morphology damage by solid-state shear pulverization. *Polymers* **2007**, *48*, 1066–1074. [CrossRef]
49. Lambert, S.; Sinclair, C.J.; Bradley, E.L.; Boxall, A.B. Effects of environmental conditions on latex degradation in aquatic systems. *Sci. Total. Environ.* **2013**, *447*, 225–234. [CrossRef] [PubMed]
50. Sheela, T.; Bhajantri, R.; Ravindrachary, V.; Rathod, S.G.; Pujari, P.; Poojary, B.; Somashekar, R. Effect of UV irradiation on optical, mechanical and microstructural properties of PVA/NaAlg blends. *Radiat. Phys. Chem.* **2014**, *103*, 45–52. [CrossRef]
51. Sionkowska, A.; Płanecka, A.; Lewandowska, K.; Michalska, M. The influence of UV-irradiation on thermal and me-chanical properties of chitosan and silk fibroin mixtures. *J. Photochem. Photobiol. B Biol.* **2014**, *140*, 301–305. [CrossRef] [PubMed]
52. Wüst, D.M.; Meyer, D.C.; Favre, P.; Gerber, C. Mechanical and Handling Properties of Braided Polyblend Polyethylene Sutures in Comparison to Braided Polyester and Monofilament Polydioxanone Sutures. *Arthrosc. J. Arthrosc. Relat. Surg.* **2006**, *22*, 1146–1153. [CrossRef]
53. Pukánszky, B.; Tüdős, F. Miscibility and mechanical properties of polymer blends. *Makromol. Chemie. Macromol. Symp.* **1990**, *38*, 221–231. [CrossRef]

54. Sharma, S.; Singh, A.A.; Majumdar, A.; Butola, B.S. Tailoring the mechanical and thermal properties of polylactic ac-id-based bionanocomposite films using halloysite nanotubes and polyethylene glycol by solvent casting process. *J. Mater. Sci.* **2019**, *54*, 8971–8983. [CrossRef]
55. Rogovina, S.Z.; Aleksanyan, K.V.; Loginova, A.A.; Ivanushkina, N.E.; Vladimirov, L.V.; Prut, E.V.; Berlin, A.A. Influence of PEG on Mechanical Properties and Biodegradability of Composites Based on PLA and Starch. *Starch Stärke* **2018**, *70*, 1700268. [CrossRef]
56. Signor, A.W.; VanLandingham, M.R.; Chin, J.W. Effects of ultraviolet radiation exposure on vinyl ester resins: Charac terization of chemical, physical and mechanical damage. *Polym. Degrad. Stab.* **2003**, *79*, 359–368. [CrossRef]
57. Whiteley, K.S.; Heggs, T.G.; Koch, H.; Mawer, R.L.; Immel, W. Polyolefins. *Ullmann's Encycl. Ind. Chem.* **2000**. [CrossRef]
58. Acioli-Moura, R.; Sun, X.S. Thermal degradation and physical aging of poly(lactic acid) and its blends with starch. *Polym. Eng. Sci.* **2008**, *48*, 829–836. [CrossRef]
59. Zhao, C.; Qin, H.; Gong, F.; Feng, M.; Zhang, S.; Yang, M. Mechanical, thermal and flammability properties of polyeth-ylene/clay nanocomposites. *Polym. Degrad. Stab.* **2005**, *87*, 183–189. [CrossRef]
60. Chen, J.; Spear, S.K.; Huddleston, J.G.; Rogers, R.D. Polyethylene glycol and solutions of polyethylene glycol as green reaction media. *Green Chem.* **2005**, *7*, 64–82.
61. Walse, C.; Berg, B.; Sverdrup, H. Review and synthesis of experimental data on organic matter decomposition with respect to the effect of temperature, moisture, and acidity. *Environ. Rev.* **1998**, *6*, 25–40. [CrossRef]
62. Buzarovska, A.; Grozdanov, A. Biodegradable poly(L-lactic acid)/TiO_2 nanocomposites: Thermal properties and degra-dation. *J. Appl. Polym. Sci.* **2012**, *123*, 2187–2193. [CrossRef]
63. Jang, W.Y.; Shin, B.Y.; Lee, T.J.; Narayan, R. Thermal properties and morphology of biodegradable PLA/starch com-patibilized blends. *J. Ind. Eng. Chem.* **2007**, *13*, 457–464.
64. Liu, J.; Jiang, H.; Chen, L. Grafting of Glycidyl Methacrylate onto Poly(lactide) and Properties of PLA/Starch Blends Compatibilized by the Grafted Copolymer. *J. Polym. Environ.* **2012**, *20*, 810–816. [CrossRef]
65. Xiong, Z.; Li, C.; Ma, S.; Feng, J.; Yang, Y.; Zhang, R.; Zhu, J. The properties of poly(lactic acid)/starch blends with a functionalized plant oil: Tung oil anhydride. *Carbohydr. Polym.* **2013**, *95*, 77–84. [CrossRef]
66. Pegoretti, A.; Fambri, L.; Migliaresi, C. In vitro degradation of poly (L-lactic acid) fibers produced by melt spinning. *J. Appl. Polym. Sci.* **1997**, *64*, 213–223. [CrossRef]
67. Zhou, Q.; Xanthos, M. Nanoclay and crystallinity effects on the hydrolytic degradation of polylactides. *Polym. Degrad. Stab.* **2008**, *93*, 1450–1459. [CrossRef]
68. Naderi, M.; Aghabararpour, M.; Najafi, M.; Motahari, S. An investigation into resorcinol formaldehyde carbon aero-gel/epoxy coatings: Exploring mechanical properties, ultraviolet stability, and corrosion resistance. *Polym. Compos.* **2020**, *41*, 121–133. [CrossRef]
69. Wang, X.-L.; Yang, K.-K.; Wang, Y.-Z. Properties of Starch Blends with Biodegradable Polymers. *J. Macromol. Sci. Part C* **2003**, *43*, 385–409. [CrossRef]
70. Therias, S.; Rapp, G.; Masson, C.; Gardette, J.-L. Limits of UV-light acceleration on the photooxidation of low-density polyethylene. *Polym. Degrad. Stab.* **2021**, *183*, 109443. [CrossRef]
71. Gharehdashli, A.; Mortazavi, S.; Rashidi, H. Photodegradation of low-density polyethylene with prooxidant and pho-tocatalyst. *J. Appl. Polym. Sci.* **2020**, *137*, 48979. [CrossRef]
72. Rosu, D.; Rosu, L.; Cascaval, C.N. IR-change and yellowing of polyurethane as a result of UV irradiation. *Polym. Degrad. Stab.* **2009**, *94*, 591–596. [CrossRef]
73. Nie, B.; Stutzman, J.; Xie, A. A Vibrational Spectral Maker for Probing the Hydrogen-Bonding Status of Protonated Asp and Glu Residues. *Biophys. J.* **2005**, *88*, 2833–2847. [CrossRef]
74. ElSawy, M.A.; Kim, K.-H.; Park, J.-W.; Deep, A. Hydrolytic degradation of polylactic acid (PLA) and its composites. *Renew. Sustain. Energy Rev.* **2017**, *79*, 1346–1352. [CrossRef]
75. Bos, H.L.; Meesters, K.P.H.; Conijn, S.G.; Corré, W.J.; Patel, M.K. Accounting for the constrained availability of land: A comparison of bio-based ethanol, polyethylene, and PLA with regard to non-renewable energy use and land use. *Biofuels, Bioprod. Biorefin.* **2012**, *6*, 146–158. [CrossRef]
76. Xu, L.; Crawford, K.; Gorman, C.B. Effects of Temperature and pH on the Degradation of Poly(lactic acid) Brushes. *Macromol.* **2011**, *44*, 4777–4782. [CrossRef]
77. Albertsson, A.-C.; Varma, I.K. Aliphatic polyesters: Synthesis, properties and applications. In *Degradable Aliphatic Polyesters*; Springer: Berlin, Germany, 2002; pp. 1–40.
78. Liu, H.; Zhang, J. Research progress in toughening modification of poly(lactic acid). *J. Polym. Sci. Part B Polym. Phys.* **2011**, *49*, 1051–1083. [CrossRef]
79. Muroga, S.; Hikima, Y.; Ohshima, M. Visualization of hydrolysis in polylactide using near-infrared hyperspectral imaging and chemometrics. *J. Appl. Polym. Sci.* **2018**, *135*, 45898. [CrossRef]
80. Avérous, L. Biodegradable Multiphase Systems Based on Plasticized Starch: A Review. *J. Macromol. Sci. Part C* **2004**, *44*, 231–274. [CrossRef]
81. Li, G.; Sarazin, P.; Orts, W.J.; Imam, S.H.; Favis, B.D. Biodegradation of Thermoplastic Starch and its Blends with Poly(lactic acid) and Polyethylene: Influence of Morphology. *Macromol. Chem. Phys.* **2011**, *212*, 1147–1154. [CrossRef]

82. Torres, A.V.; Zamudio-Flores, P.B.; Salgado-Delgado, R.; Bello-Pérez, L.A. Biodegradation of low-density polyethylene-banana starch films. *J. Appl. Polym. Sci.* **2008**, *110*, 3464–3472. [CrossRef]
83. Panahi, L.; Gholizadeh, M.; Hajimohammadi, R. Investigating the degradability of polyethylene using starch, oxo-material, and polylactic acid under the different environmental conditions. *Asia-Pacific J. Chem. Eng.* **2019**, *15*, 2402. [CrossRef]
84. Da Silva, A.P.; Pereira, M.D.P.; Passador, F.R.; Montagna, L.S. PLA/Coffee Grounds Composites: A Study of Photodegradation and Biodegradation in Soil. *Macromol. Symp.* **2020**, *394*, 2000091. [CrossRef]
85. Chamas, A.; Moon, H.; Zheng, J.; Qiu, Y.; Tabassum, T.; Jang, J.H.; Abu-Omar, M.M.; Scott, S.L.; Suh, S. Degradation Rates of Plastics in the Environment. *ACS Sustain. Chem. Eng.* **2020**, *8*, 3494–3511. [CrossRef]

polymers

MDPI

Article

Preparation and Antifouling Property of Polyurethane Film Modified by PHMG and HA Using Layer-by-Layer Assembly

Huihui Yuan [1], Chenli Xue [1], Jiaqian Zhu [1], Zhaogang Yang [2] and Minbo Lan [1,*]

1 Shanghai Key Laboratory of Functional Materials Chemistry, School of Chemistry & Molecular Engineering, East China University of Science and Technology, Shanghai 200237, China; yuanhuihui@ecust.edu.cn (H.Y.); Y30180191@mail.ecust.edu.cn (C.X.); Y30190263@mail.ecust.edu.cn (J.Z.)

2 Department of Radiation Oncology, The University of Texas Southwestern Medical Center, Dallas, TX 75390, USA; zhaogang.yang@utsouthwestern.edu

* Correspondence: minbolan@ecust.edu.com; Tel.: +86-21-6425-3574

Abstract: To reduce the possibility of bacterial infection and implant-related complications, surface modification on polyurethane (PU) film is an ideal solution to endow hydrophobic PU with antibacterial and antifouling properties. In this work, a variety of polyhexamethylene guanidine/hyaluronic acid (PHMG/HA) multilayer films were self-assembled layer-by-layer on PU films using polyanions, carboxyl-activated HA, and polycations PHMG by controlling the concentration of these polyelectrolytes as well as the number of layers self-assembled. Attenuated total reflection Fourier transform infrared spectroscopy (ATR-FTIR) spectra, water contact angle (WCA), and A Atomic force microscope (AFM) of PU and modified PU films were studied. Protein adsorption and bacterial adhesion as well as the cytotoxicity against L929 of the film on selected PU-$(PHMG/HA)_5$/5-5 were estimated. The results showed that PU-$(PHMG/HA)_5$/5-5 had the best hydrophilicity among all the prepared films, possessing the lowest level of protein adsorption. Meanwhile, this film showed efficient broad-spectrum antibacterial performance as well as significant resistance of bacterial adhesion of more than a 99.9% drop for the selected bacteria. Moreover, almost no influence on cell viability of L929 enhanced the biocompatibility of film. Therefore, the modified PU films with admirable protein absorption resistance, antimicrobial performance, and biocompatibility would have promising applications in biomedical aspect.

Keywords: modified PU film; protein adsorption; antibacterial; hyaluronic acid (HA); polyhexamethylene guanidine (PHMG)

Citation: Yuan, H.; Xue, C.; Zhu, J.; Yang, Z.; Lan, M. Preparation and Antifouling Property of Polyurethane Film Modified by PHMG and HA Using Layer-by-Layer Assembly. *Polymers* **2021**, *13*, 934. https://doi.org/10.3390/polym13060934

Academic Editor: José Miguel Ferri

Received: 26 February 2021
Accepted: 14 March 2021
Published: 18 March 2021

Publisher's Note: MDPI stays neutral with regard to jurisdictional claims in published maps and institutional affiliations.

1. Introduction

Polyurethane (PU) ureteral stents are wildly used in urological clinics for their good flexibility and elasticity, biocompatibility, and low cost compared to what or within which range of materials [1,2]. However, the hydrophobic surface of PU reduces the antifouling and antimicrobial properties, which results in the increasing amount of protein adsorption, bacteria adhesion, and salt deposition in a urine environment [3,4]. Thus, encrustation, infection, and implant-related post-complications such as ureteral stricture, perforation, and mucosal injury [5,6] are observed during the implantation in vivo. Thus, the antimicrobial property of the stent's surface is vital for its service life in clinic. It is believed that an ideal antibacterial surface possesses properties of repelling protein adsorption in order to prevent initial bacteria attachment [7,8], repelling direct bacteria adhesion and killing the attached bacteria during the period of implantation in human body. With the aim of achieving these targets, various materials are applied to modify the PU surface to improve its hydrophilicity or to confer its antibacterial properties. Yuan et al. [9] modified chondroitin sulfate onto the PU surface to improve its hydrophilicity and reduce the protein adsorption. Manohar et al. [4] covalently crosslinked papain onto PU to prevent bacterial adhesion. Fischer et al. [10] attached a hydrogel coating loaded with Ag nanoparticles to a PU conduit to improve its antibacterial activity. However most material modifications can only improve one aspect of performance of the surface. Therefore, modification with an antibacterial agent and antifouling material on the surface is a good strategy to endow the PU surface with both antifouling properties and bactericidal properties [11,12], which would be a desirable antibacterial surface for clinical usage.

A surface with good hydrophilicity has been proven to effectively prevent non-specific protein adsorption. Hyaluronic acid (HA) is one of the most hydrophilic molecules in nature with non-toxic, non-immunogenic, non-inflammatory, and biodegradable properties [13–16]. HA is also a polyanion glycosaminoglycan that can repel most negatively-charged proteins and bacteria with negatively-charged cell membranes by electrostatic repulsive force. Conversely, it would be able to electrostatically attract cationic antimicrobial such as chitosan [17,18], quaternary ammonium salts [19,20], and cationic antimicrobial peptides [21] to integrate antibacterial function. Hence, HA is suitable for the surface modification of materials to reach the ultimate purpose of reducing bacterial adhesion [11,22]. Polyhexamethylene guanidine (PHMG) is a highly water-soluble, colorless, and odorless positively-charged antimicrobial [23]. Due to its broad spectrum activity against bacteria and fungi [24,25] and its low toxicity to mammals [26], PHMG has been successfully applied in several products such as topical wound solutions, contact lens cleaning products, and cosmetics [26–28]. Wei et al. [29] demonstrated that aqueous solutions of PHMG with concentrations as low as 1.0 ppm showed more than a 90.0% antibacterial rate. Ding et al. [30] bonded PHMG to resins to generate antibacterial acrylic coatings. The inhibitory factors against both *Escherichia coli* (*E. coli*) and *Staphylococcus aureus* (*S. aureus*) were over 99.99% at a PHMG content of 1.0 wt%. Therefore, the combination of HA and PHMG provides the modified surface with the desirable multifunction of hydrophilicity and antibacterial activity.

In this study, we created PHMG/HA multilayer films on PU by using layer-by-layer self-assembly with HA and PHMG as polyanions and polycations to render the surface of PU films with both antifouling and antibacterial properties. PHMG was chemically bonded to the PU surface via the reaction between amide groups of PHMG and isocyanato groups modified on PU films. Negatively-charged HA was assembled by electrostatic adsorption with positively-charged PHMG modified on PU films. Simultaneously, HA and PHMG were also covalently combined by the reaction of the partial activated carboxyl group of HA and amide groups of PHMG. The different concentrations of HA and PHMG as well as and the number of assembled layers were studied to attain PU-$(PHMG/HA)_n$ films with different properties. The surface properties of modified and unmodified PU films were characterized by attenuated total reflection Fourier transform infrared spectroscopy (ATR-FTIR), water contact angle (WCA), and atomic force microscopy (AFM). The antifouling and antibacterial properties of the surface were detected by a bicinchoninic acid (BCA) protein detection kit and bacterial assay. Finally, the cytotoxicity of L929 cells was estimated for the improvement in the biocompatibility of the material.

2. Materials and Methods

2.1. Materials

PU (pellethane 2363-80AE) was provided by Lubrizol Corporation (Wickliffe, OH, USA). Methylenediphenyl 4,4′-Diisocyanate (MDI) was bought from Aladdin Chemical Co. Ltd. (Shanghai, China). N,N-dimethylformamide (DMF), dimethyl sulfoxide (DMSO), toluene, and triethylamine were purchased from Sinopharm Chemical Reagent Co. Ltd. (Shanghai, China), and HA was obtained from Xianding Biotechnology Co. Ltd. (Shanghai, China). 1-(3-Dimethylaminopropyl)-3-ethylcarbodiimide hydrochloride (EDC·HCl), N-Hydroxysuccinimide (NHS), Lysozyme (LYS), bovine fibrinogen (BFG), human serum albumin (HSA), and 3-(4,5-dimethyl-2-thiazolyl)-2,5-diphenyl-2-H-tetrazolium bromide (MTT) were obtained from Yuanye Bio-Technology Co. Ltd. (Shanghai, China). Sodium dodecyl sulfate (SDS) and the Micro BCA Protein Assay Kit were obtained from Sangon Biotech (Shanghai, China) Co. Ltd. Fetal bovine serum (FBS) and Dulbecco's modified

Eagle medium (DMEM) medium were obtained from Gibco (Grand Island, NY, USA). *E. coli* (DH5 alpha), *S. aureus* (ATCC 6538), and *P. aeruginosa* (ATCC 2785) and lysogenic broth (LB) medium were provided by Professor Cui, East China University of Science and Technology (ECUST, Shanghai, China). PHMG was provided by Professor Guan, ECUST.

2.2. The Fabrication of the PU-(PHMG/HA)$_n$ Films

Ten grams of commercial PU was dissolved in 100 mL DMF by magnetic stirring. Then, the PU solution was vacuum dried at 60 °C for 72 h to obtain thin PU films, and the films were cut into discs (6 mm in diameter and 1 mm in thickness). Afterward, PU films were put into MDI solution for a 3 h reaction to obtain PU-NCO films. Finally, the PU-PHMG film was prepared by the reaction of amino groups of PHMG with PU-NCO.

Carboxyl activated HA intermediate was prepared according to the literature [31]. A total of 0.24 g EDC·HCl and 0.15 g NHS were slowly added to 50 mL HA solution (1%, m/v) in turn under the condition of pH 4.75 with stirring for a 2.5 h reaction. After that, the reaction was terminated by increasing the pH value to 7.5. The reaction product was further dialyzed to remove EDC and NHS at room temperature for two days. Finally, carboxyl-activated HA was obtained and freeze-dried by a vacuum freeze dryer.

PU-(PHMG/HA)$_n$ films were prepared by sequential assembling HA and PHMG onto the PU-PHMG film layer by layer. Figure 1 illustrates the fabrication process of the PU-(PHMG/HA) films. The assembling time per polymer layer was 20 min [32]. In between the PHMG and HA assembling steps, the assembled films were put into deionized (DI) water to remove the unassembled molecules and dried at 60 °C. Different samples were obtained by adjusting the polyelectrolyte concentration and the number of bilayers self-assembled, and Table 1 lists the formulations for various samples.

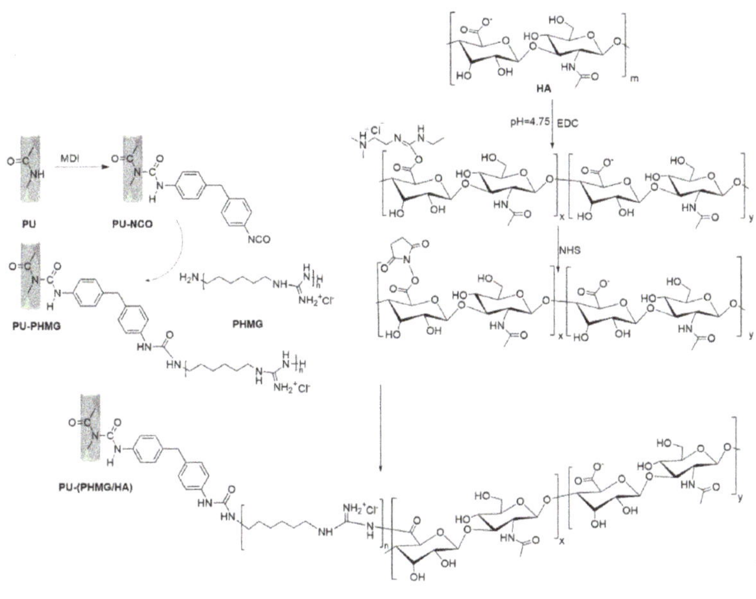

Figure 1. Synthesis procedure for the PU-(PHMG/HA) films.

Table 1. The formulations of the PU-(PHMG/HA)$_n$ films synthesized.

Sample Code	HA (mg/mL)	PHMG (mg/mL)	Number of Bilayers
PU-(PHMG/HA)$_5$/2-2	2	2	5
PU-(PHMG/HA)$_5$/5-2	5	2	5
PU-(PHMG/HA)$_5$/5-5	5	5	5
PU-(PHMG/HA)$_5$/5-10	5	10	5
PU-(PHMG/HA)$_{10}$/5-5	5	5	10
PU-(PHMG/HA)$_1$/5-5	5	5	1

2.3. *Characterizations of the Films*

The surface chemical structure of the film was identified by a FTIR Spectrometer (Thermo Nicolet 6700, Madison, WI, USA) with an ATR device. The spectra were collected at the following instrument parameters: scan range 400–4000 cm^{-1}, resolution 4 cm^{-1}, and scan times 16 [33]. The WCA of the surface was measured by a contact angle meter (Powereach JC 2000D, Shanghai, China). The surface morphology of the films was analyzed by AFM (Veeco DI3100, Plainview, NY, USA). The surface roughness of the films was the average roughness of three areas.

2.4. *Protein Adsorption*

BFG, HSA, and LYS were selected for protein adsorption experiments to investigate the anti-protein adsorption properties of the films. The films were soaked in 1 mL protein solution (1 mg/mL) and incubated at 37 °C for 1 h, and then washed with PBS buffer solution (pH 7.4) and ultrapure water in order to remove the unabsorbed proteins on the surface. A total of 0.5 mL SDS (1% *w*/*v*) solution was used to elute the adsorbed proteins and incubated with films at 37 °C for 2 h. The amount of protein was calculated by measuring the absorbance of the eluate at 562 nm with a UV–Vis spectrophotometer (Thermo Fisher Evolution 220, Waltham, MA, USA) using the BCA assay.

2.5. *Bacteria Adhesion*

Gram-positive bacteria (*S. aureus*) and Gram-negative bacteria (*E. coli* and *P. aeruginosa*) were selected to test the antibacterial activity of the films. A single pure colony was cultured in LB medium at 37 °C for 18 h. The supernatant was removed by centrifugation and the bacteria were diluted to 10^8 CFU/mL with PBS. The films were immersed in 1 mL bacterial solution and incubated at 37 °C with shaking for 24 h. Nonadherent bacteria on the film surface were removed by washing with PBS three times, and then the adherent bacteria were eluted into 1 mL PBS by sonication for 10 min. The amount of bacteria was measured by the flat colony counting method. Additionally, the adherent bacteria on film were studied by confocal laser scanning microscopy (CLSM, NIKON A1R, Tokyo, Japan) after fluorescein diacetate (FDA)/propidium iodide (PI) staining. PI stain selectively binds to the dead bacteria and stains them fluorescent red, whereas the FDA stain selectively binds to the live bacteria and stains them fluorescent green during CLSM imaging.

2.6. *Cytotoxicity Test*

Cytotoxicity of films was determined by the MTT assay [34,35]. The conditioned cell culture medium was obtained by immersing films in 1 mL cell culture medium, which was applied to assay the effect of the films on cultured cells. L929 cells were seeded into 96-well plates (7000 cells per well) and cultured in an incubator with 5% CO_2 at 37 °C for 24 h before they were incubated with 200 µL conditioned cell culture medium for 24 h. A total of 200 µL MTT reagent was added to each well for a further 4 h incubation at 37 °C. Then, the formazan precipitate was extracted by 150 µL DMSO, and the absorbance (492 nm) was recorded using a Multiskan MK3 ELISA reader (Thermo Fisher, Waltham, MA, USA).

2.7. Statistical Analysis

Statistical analyses were performed by SPSS for Windows software, version 18 (SPSS, Chicago, IL, USA). Data are presented as mean ± standard deviations (SD) of at least three replicates. p value < 0.05 was considered statistically significant.

3. Results and Discussion

3.1. Characterization of the Films

The ATR-FTIR spectra of films at different preparation stages are shown in Figure 2. PU presented peaks originating from C=O and C–N at 1700 cm^{-1} and 1530 cm^{-1}, respectively (Figure 2a) [36]. A dominant absorption peak was observed at 2285 cm^{-1}, which suggested that the –NCO group was successfully grafted on the PU surface (Figure 2b) [37,38]. However, Figure 2c shows that the peak of –NCO disappeared, and two symmetric and asymmetric –CH_2 stretching vibrations attributed to PHMG were noted at 2854 cm^{-1} and 2924 cm^{-1} [39], respectively, which confirmed that –NCO totally reacted with the –NH_2 of PHMG. Nevertheless, the spectra of PU-(PHMG/HA) (Figure 2d) showed no obvious change compared with that of PU-PHMG, suggesting that the first layer of HA might have little influence on the improvement of the surface properties. The wide peaks at 3324 cm^{-1} assigned to the –OH group in HA increased (Figure 2e,f), indicating that the PHMG/HA bilayers were successfully assembled on the PU film [40]. Furthermore, the relatively broad peak at 1150 cm^{-1} belonging to the ester group [41] in COOH– activated HA was found in the spectra of PU-$(PHMG/HA)_5$/5-5 and PU-$(PHMG/HA)_{10}$/5-5, while it did not appear in that of PU-PHMG and PU-$(PHMG/HA)_1$/5-5. This phenomenon indicated that HA partially covalently bonded on the surface as expected. The reason might be due to the following: The activated –COOH provided by HA was not sufficient and completely reacted with the –NH_2 of PU-PHMG to –CO–NH during the preparation of the first bilayer PHMG/HA. Additionally, the –CO–NH was not able to be distinguished due to its original existence in any of the PU and modified PU films. With the increase in bilayer number, more HA provided more reactive ester groups, which could meet the demand in crosslinking of HA-PHMG. Nevertheless, the peak of the ester group was reduced with the increased bilayer, according to the spectra comparison of PU-$(PHMG/HA)_5$/5-5 and PU-$(PHMG/HA)_{10}$/5-5. This might be attributed to molecular rearrangement during the proceeding of assembly, which created more chances for the ester group to react with –NH_2. In addition, the peak of –CH_2 weakened with the increased number of bilayers, but still existed on the surface of all assembled films in Figure 2d–f. It was supposed that the molecules of HA and PHMG were assembled in an entangled manner, which resulted in incomplete coverage of the HA chains on the surface [42].

The variation of the WCA was likewise related to the introduction of functional groups/molecules onto the surface. The WCA of the original PU was 90.1° due to its hydrophobicity [43,44]. The successful grafting of hydrophobic isocyanate on PU resulted in the WCA of PU-NCO increasing to 96.8° [37,45]. However, the succeeding PHMG onto the surface led to a low WCA (82.3°) of PU-PHMG because of the introduction of the hydrophilic –NH_2 group. After HA was covalently bonded and electrostatic self-assembled onto PU-PHMG films, the surface became more hydrophilic. Subsequently, PHMG and HA alternately assembled onto the surface, which contributed to the WCA of corresponding films with a zig-zag effect (Figure 3). The HA (odd) layer achieved smaller WCA than that of the PHMG layer (even), suggesting higher hydrophilicity of HA than PHMG and proving that films with alternating deposition of polyelectrolyte were successfully obtained. In addition, the concentration of polyelectrolyte had an obvious effect on the WCA of the films. The increase or decrease in PHMG concentration both caused the WCA of the surface with relatively high value based on the comparison of preparation groups PHMG (10 mg/mL), PHMG (5 mg/mL), and PHMG (2 mg/mL) when HA was fixed at 5 mg/mL (Figure 3). One explanation might be less PHMG, leading to less HA loaded. The other might be the excessive PHMG providing more –CH_2 exposed on the surface when they entangled with HA. Therefore, it was found that the combination of HA (5 mg/mL)-PHMG

(5 mg/mL) achieved the greatest reduction in the WCA of the modified films among those prepared by other concentration combination of HA-PHMG. At this HA-PHMG concentration pair, HA and PHMG were well matched and reached a dynamic balance during the assembly process. Nevertheless, the WCA of PU-(PHMG/HA)$_n$/5-5 had almost no apparent reduction, indicating that excessive assembled layers might have little impact on the function promotion of the surface.

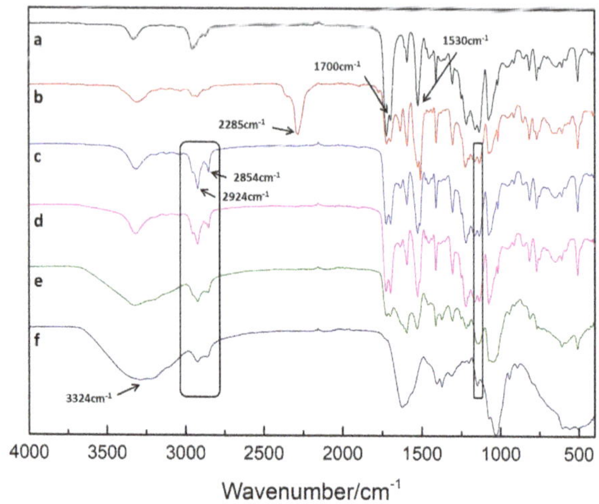

Figure 2. ATR–FTIR spectra of (**a**) PU; (**b**)PU-NCO; (**c**)PU-PHMG; (**d**) PU-(PHMG/HA)$_1$/5-5; (**e**) PU-(PHMG/HA)$_5$/5-5; (**f**) PU-(PHMG/HA)$_{10}$/5-5.

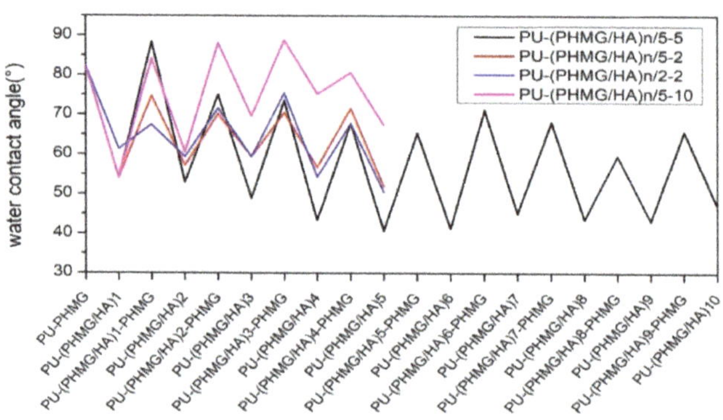

Figure 3. Water contact angle (WCA) of modified PU films. n in PU-(PHMG/HA)$_n$/5-5, PU-(PHMG/HA)$_n$/5-2, PU-(PHMG/HA)$_n$/2-5 and PU-(PHMG/HA)$_n$/5-10 was the number of bilayer.

The surface topography of the film was determined by AFM. The surface of the original PU was fairly flat and smooth with a root-mean-square (RMS) roughness of 36.4 ± 2.5 nm (Figure 4). However, the RMS of PU-PHMG surfaces increased significantly to 177.7 ± 2.3 nm ($p < 0.001$) compared to the PU films. One layer of HA assembled on PU-PHMG made little contribution to lower roughness of surface (173.9 ± 3.3 nm), which was consistent to the result of the ATR-FTIR spectra. However, after alternating PHMG and HA modification on PU films a few times, the surface roughness of PU-(PHMG/HA)$_n$ (e.g., PU-(PHMG/HA)$_5$/5-5) decreased in comparison with that of PU-PHMG, but was still rougher than that of PU. Table 2 lists the surface roughness value for various samples. With the increase of HA concentration, the roughness of the films showed no obvious change based on the comparison of PU-(PHMG/HA)$_5$/2-2 and PU-(PHMG/HA)$_5$/5-2. With the increase in PHMG concentration, the roughness of the films decreased in comparison with PU-(PHMG/HA)$_5$/5-2, PU-(PHMG/HA)$_5$/5-5, and PU-(PHMG/HA)$_5$/5-10. This influence might be related to the molecular weight of HA (>10 kDa) and PHMG (~600 Da). PHMG with far lower molecular weight than HA had relative flexibility and more PHMG was able to fill the void, which resulted in the lower roughness of the surface. Additionally, the number of assembled layers positively influenced the roughness of the modified films at the fixed preparation concentration based on the comparison of PU-(PHMG/HA)$_1$/5-5, PU-(PHMG/HA)$_5$/5-5, and PU-(PHMG/HA)$_{10}$/5-5, whereas the increase in the bilayer number had a minor contribution to lower the roughness when the number of bilayers was more than five. PU-(PHMG/HA)$_{10}$/5-5 possessed the smoothest surface with a RMS roughness value of 130.8 ± 2.6 nm, followed by PU-(PHMG/HA)$_5$/5-10 and PU-(PHMG/HA)$_5$/5-5. The roughness of the above three films had no remarkable differences. Therefore, PU-(PHMG/HA)$_5$/5-5 was the optimum film when taking into account the preparation costs.

Figure 4. Atomic force microscopy (AFM) images of (**a**) PU; (**b**) PU-PHMG; (**c**) PU-(PHMG/HA)$_5$/5-5.

Table 2. The surface roughness value of the PU-(PHMG/HA)$_n$ films.

Sample Code	PU-(PHMG/HA)$_1$/5-5	PU-(PHMG/HA)$_5$/2-2	PU-(PHMG/HA)$_5$/5-2	PU-(PHMG/HA)$_5$/5-5	PU-(PHMG/HA)$_5$/5-10	PU-(PHMG/HA)$_{10}$/5-5
RMS Roughness Value (nm)	173.9 ± 3.3	157.0 ± 1.9	156.5 ± 1.7	136.5 ± 3.4	131.9 ± 4.2	130.8 ± 2.6

3.2. Protein Adsorption

Adsorption of protein on the surface works as the initial step of biofouling when implanted in vivo and further compromises the surface properties, promotes cell attachment, and initiates the final foreign body response [46,47]. Thus, the resistance to protein adsorption of PU films is first taken into consideration.

Due to the inherent hydrophobic and electrostatic interactions between the surface and proteins, generally, the adsorption of proteins on the hydrophobic surface of PU films has no obvious inhibitory effect on the adhesion of any kind of contaminant [48,49]. Therefore, improving hydrophilicity is a valid approach to enhance the antifouling property of film to some extent.

The adsorption amount of BFG, HSA, and LYS on different films is listed in Figure 5. The PU-(PHMG/HA)$_5$/5-5 surface exhibited the best resistance to protein adsorption, on which the adsorption levels of BFG, HSA and LYS were 2.43, 0.49, 0.16 μg/cm^2, respectively. The adsorption amount followed the order of the molecular weight of proteins, that is BFG > HSA > LYS, as the high molecular weight resulted in the high amount of protein adsorbed on to patch at the same adsorption sites. The corresponding protein adsorption level was reduced 67.85%, 85.33%, and 80.31% compared with that on PU film. This is attributed to the higher hydrophilicity and lower surface roughness of PU-(PHMG/HA)$_5$/5-5.

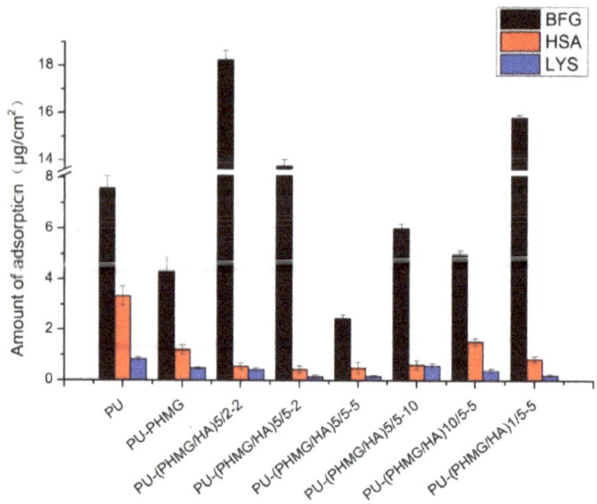

Figure 5. Protein adsorption on films.

Besides hydrophilicity and surface roughness, the nanotopography structure also influenced the amount of protein adsorbed on these modified PU films. In Figure 5, BFG adsorption level on PU-PHMG films was markedly lower than that on PU-(PHMG/HA)$_5$/2-2, PU-(PHMG/HA)$_5$/5-2, and PU-(PHMG/HA)$_1$/5-5, though the hydrophilicity and roughness of PU-PHMG were higher than that of the three films. The possible reason might be the brush structure of PHMG on PU films repulsing parts of the proteins. The surface entropy of PU-PHMG increased when PHMG brushes were compressed by BFG, which was disadvantageous to thermodynamic stability and led to repelling BFG adsorption on the surface [50,51]. Similarly, the amount of HSA and LYS adsorbed on PU-PHMG were comparatively low in comparison with that on PU-(PHMG/HA)$_{10}$/5-5 and PU-(PHMG/HA)$_5$/5-10, respectively. Thus, the brush structure dominated the protein adsorption on PU-PHMG. Likewise, the anomalous observation was the relatively low level of BFG adsorption on PU in contrast with that on films with high roughness and medium hydrophilicity. This result might embody the importance of the protein sharp in adsorption. Fibrinogen is known as a cylinder (diameter = 6 nm, length = 45 nm) [52], where the side-on orientations were difficult to adsorb stably on PU with a roughness of 36.4 nm because of stereohindrance. In addition, more assembled layers negatively affected the protein repelling property (typical case illustrated in PU-(PHMG/HA)$_{10}$/5-5), which provided more internal space to capture smaller sized proteins such as HSA into the swelling inner films through the microstructure [52]. Among the three model proteins, LYS possessed the smallest size, and was a positively-charged (isoelectric point at 11.0) and 'hard' one. The electrostatic repulsion between LYS and PHMG was noticeable, which reflected in lower LYS adsorption quantity on the PU-(PHMG/HA)$_1$/5-5 with high roughness and medium hydrophilicity compared with other PU-(PHMG/HA)$_n$ films. After the formation of the first bilayer of PHMG and HA, the two molecules might be coiled and the positive charge of PHMG was not well covered by HA, as explained in th ATR-FTIR spectra, which resulted in a surface with good LYS repelling performance. As seen in the results above, the hydrophilicity, roughness, charges, and nanotopography structure of the surface as well as the size, shape, and charges of the proteins were the important factors for the protein adsorption property. Among these facts, surface hydrophilicity was a dominating one for protein adsorption.

3.3. Bacteria Adhesion

Bacterial infection is the main complication after stent implantation, and the adhesion and colonization of bacteria on the stent play an essential part in scaling. Therefore, antibacterial functionalization becomes the key target for surface modification. Three typical uropathogens were selected to evaluate the broad-spectrum antimicrobial properties of PU, PU-PHMG, and PU-$(PHMG/HA)_5$/5-5 (best resistance to protein adsorption). The results are shown in Table 3. The amount of bacteria adhered to naked PU was 29.2×10^5 CFU/cm^2 for *E. coli*, 14.0×10^5 CFU/cm^2 for *P. aeruginosa*, and 24.3×10^5 CFU/cm^2 for *S. aureus*. Compared with PU, PU-PHMG showed excellent antibacterial effect against the three strains, and the adherence levels of the corresponding bacteria were as low as 0.0345×10^5, 0.0153×10^5, and 0.160×10^5 CFU/cm^2 with inhibitory rates of 99.88%, 99.89%, and 99.34%, respectively. The effective antibacterial activity of PU-PHMG was attributed to the bactericidal capacity of PHMG. The interaction between PHMG and the anionic components of bacterial cell wall compromises membrane integrity, further causing cell membrane rupture and leads to microbial death [53,54]. The inhibition rates of PU-$(PHMG/HA)_5$/5-5 on *E. coli*, *P. aeruginosa*, and *S. aureus* were 99.99%, 99.96%, and 99.99%, respectively, indicating that the film had outstanding antibacterial activity. The inhibition rate was slightly higher than that of PU-PHMG, indicating that the improvement of surface hydrophilicity and roughness also affect the antibacterial effect.

Table 3. Antibacterial test of films against *E. coli*, *P. aeruginosa*, and *S. aureus*.

Samples	*E. coli*		*P. aeruginosa*		*S. aureus*	
	Colonies ($\times 10^5$, CFU/cm^2)	Inhibition (%)	Colonies ($\times 10^5$, CFU/cm^2)	Inhibition (%)	Colonies ($\times 10^5$, CFU/cm^2)	Inhibition (%)
PU	29.2 ± 8.77	/	14.0 ± 0.283	/	24.3 ± 0.778	/
PU-PHMG	0.0345 ± 0.00127	99.88	0.0153 ± 0.00148	99.89	0.160 ± 0.0247	99.34
PU-$(PHMG/HA)_5$/5-5	0	99.99	0.00514 ± 0.000247	99.96	0	99.99

Aside from its antimicrobial activities, biofouling resistance is another crucial element affecting the long-term property of the films. Generally, bacteria will both adhere to the film to form colonies and participate in the formation of subsequent biofilms, covering up the function of antibacterial substances, and subsequently causing inevitable biological contamination. After incubation with bacteria for one day, the antifouling ability of the films was assessed by imaging bacterial adhesion on the surface. Figure 6 illustrates the bacterial adhesion on PU, PU-PHMG, and PU-$(PHMG/HA)_5$/5-5, respectively. As observed, most of the live bacteria and few dead bacteria accumulated on the PU surface (Figure 6a) because of its hydrophobic property. PU-PHMG, in contrast, adhered to most of the dead bacteria (Figure 6b), showing that it had efficient antibacterial property but nearly no antifouling performance due to electrostatic adsorption and hydrophobic interaction [55]. Therefore, PU-PHMG merely maintained the antibacterial properties at the initial stage, but was gradually covered by dead bacteria and lost its function during long-term incubation with bacteria. To our delight, bacteria were barely observed on the surface of PU-$(PHMG/HA)_5$/5-5 (Figure 6c), indicating no biofilm had formed. The high bactericidal efficiency was attributed to two aspects. On one hand, PHMG can kill bacteria temporarily adhered to the surface. On the other hand, the size of almost all of the bacteria was larger than 500 nm, which made the bacteria unable to be entrapped in the rough area. Therefore, the killed bacteria, gently adsorbed on the surface, can be easily stripped by simple hydraulic turbulence [46] due to the hydrophilicity of HA. The result of the antifouling property of three test films indicated that the antifouling property of the surface was important for an antibacterial effect. PU-$(PHMG/HA)_5$/5-5 with good hydrophilicity containing PHMG and HA exhibited excellent antibacterial and antifouling properties, suggesting that it had an ideal antibacterial surface for future biomedical usage.

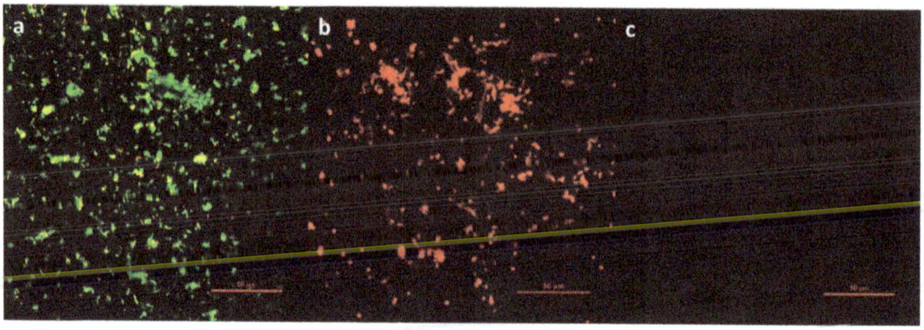

Figure 6. Confocal laser scanning microscopy (CLSM) images of *E. coli* adherent onto (**a**) PU, (**b**) PU-PHMG, and (**c**) PU-$(PHMG/HA)_5$/5-5 film. The live bacteria appear as green fluorescence and dead bacteria appear as red fluorescence. Scale bars are 50 μm.

3.4. Cytotoxicity Test

Biocompatibility is an essential requirement in bio-materials for their potential biomedical application [56]. Cytotoxicity testing can generally be performed in two ways—contact (direct) and extraction (indirect) [35]. The extraction method was applied due to the anti-adhesion property of the film surface, which was difficult for cells to adhere on. The conditioned cell culture medium mimicked the effect of the film on the physiological environment. The results of the cytotoxicity of L929 cultured in leaching solution of films are shown in Figure 7. PU and PU-PHMG films had high cell viability (over 88%) and the PU-$(PHMG/HA)_5$/5-5 film had no cytotoxicity against L929 cell compared to the control, which indicated that the final surface modification was favorable to cell viability.

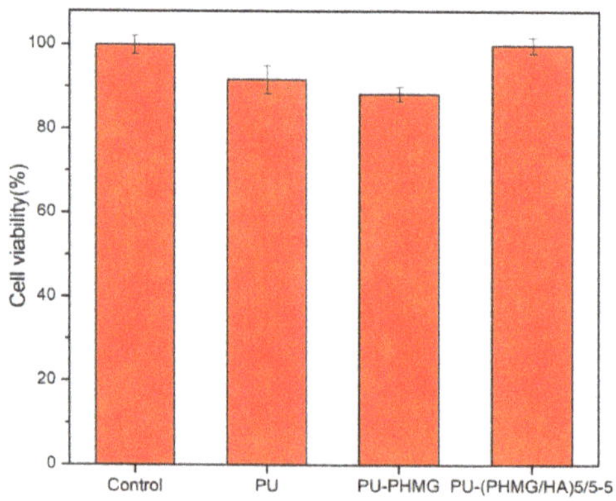

Figure 7. Cell viability of L929 after being cultured in PU, PU-PHMG, and PU-$(PHMG/HA)_5$/5-5 extracts.

4. Conclusions

In this study, we focused on surface modification with hydrophilic material and an antibacterial agent to simultaneously improve the antifouling and antibacterial properties of the PU film. We successfully created PHMG/HA multilayer films on PU by using layer-by-layer self-assembly with COOH-activated HA and PHMG as polyanions and polycations. An optimal film named as PU-$(PHMG/HA)_5$/5-5 with the lowest WCA and medium roughness was obtained, which possessed excellent protein repelling performance. The adsorption levels of BFG, HSA, and LYS reduced 67.85%, 85.33% and 80.31%, respectively, compared with that on the PU film. In addition, the high bacteriostatic rate of over 99.9% against the three tested bacteria and excellent antibacterial adhesion property showed that PU-$(PHMG/HA)_5$/5-5 possessed high antimicrobial and anti-biofouling performance.

Furthermore, the film had nearly no cytotoxicity against L929 cells, which made it possible for biomedical applications in the future.

Author Contributions: Conceptualization, H.Y. and M.L.; Funding acquisition M.L.; Methodology, H.Y., C.X., and J.Z.; Resources C.X. and Z.Y.; Software, C.X.; Validation, C.X. and J.Z.; Investigation, C.X.; Resources, C.X.; Writing—original draft preparation, H.Y., C.X., Z.Y., and M.L.; Writing—review and editing, H.Y., C.X. and M.L. All authors have read and agreed to the published version of the manuscript.

Funding: This research was funded by the Science and Technology Commission of Shanghai Municipality (STCSM, 20520712500) and the Fundamental Research Funds for the Central Universities (50321102117022).

Acknowledgments: This work was supported by the Science and Technology Commission of Shanghai Municipality (STCSM, 20520712500) and the Fundamental Research Funds for the Central Universities (50321102117022). Additional support was provided by the Feringa Nobel Prize Scientist Joint Research Center.

Conflicts of Interest: The authors declare no conflict of interest.

References

1. Gultekinoglu, M.; Kurum, B.; Karahan, S.; Kart, D.; Sagiroglu, M.; Ertas, N.; Ozen, A.H.; Ulubayram, K. Polyethyleneimine brushes effectively inhibit encrustation on polyurethane ureteral stents both in dynamic bioreactor and in vivo. *Mater. Sci. Eng. C Mater. Biol. Appl.* **2017**, *71*, 1166–1174. [CrossRef]
2. Gorbunova, M.A.; Anokhin, D.V.; Badamshina, E.R. Recent Advances in the Synthesis and Application of Thermoplastic Semicrystalline Shape Memory Polyurethanes. *Polym. Sci. Ser. B* **2020**, *62*, 427–450. [CrossRef]
3. Yuan, H.H.; Qian, B.; Chen, H.Y.; Lan, M.B. The influence of conditioning film on antifouling properties of the polyurethane film modified by chondroitin sulfate in urine. *Appl. Surf. Sci.* **2017**, *426*, 587–596. [CrossRef]
4. Manohar, C.M.; Doble, M. Papain immobilized polyurethane as an ureteral stent material. *J. Biomed. Mater. Res. Part B* **2016**, *104*, 723–731. [CrossRef]
5. Traxer, O.; Thomas, A. Prospective Evaluation and Classification of Ureteral Wall Injuries Resulting from Insertion of a Ureteral Access Sheath during Retrograde Intrarenal Surgery. *J. Urol.* **2013**, *189*, 580–584. [CrossRef] [PubMed]
6. Cooper, J.L.; Francois, N.; Sourial, M.W.; Miyagi, H.; Rose, J.R.; Shields, J.; Sharma, N.; Domino, P.; Otto, B.; Box, G.N.; et al. The Impact of Ureteral Access Sheath Use on the Development of Abnormal Postoperative Upper Tract Imaging after Ureteroscopy. *J. Urol.* **2020**, *204*, 976–981. [CrossRef]
7. Felgueiras, H.P.; Wang, L.M.; Ren, K.F.; Querido, M.M.; Jin, Q.; Barbosa, M.A.; Ji, J.; Martins, M.C.L. Octadecyl Chains Immobilized onto Hyaluronic Acid Coatings by Thiol-ene "Click Chemistry" Increase the Surface Antimicrobial Properties and Prevent Platelet Adhesion and Activation to Polyurethane. *ACS Appl. Mater. Interfaces* **2017**, *9*, 7979–7989. [CrossRef]
8. Mohan, T.; Cas, A.; Bracic, M.; Plohl, O.; Vesel, A.; Rupnik, M.; Zemljic, L.F.; Rebol, J. Highly Protein Repellent and Antiadhesive Polysaccharide Biomaterial Coating for Urinary Catheter Applications. *ACS Biomater. Sci. Eng.* **2019**, *5*, 5825–5832. [CrossRef]
9. Yuan, H.; Xue, J.; Qian, B.; Chen, H.; Zhu, Y.; Lan, M. Preparation and antifouling property of polyurethane film modified by chondroitin sulfate. *Appl. Surf. Sci.* **2017**, *394*, 403–413. [CrossRef]
10. Fischer, M.; Vahdatzadeh, M.; Konradi, R.; Friedrichs, J.; Maitz, M.F.; Freudenberg, U.; Werner, C. Multilayer hydrogel coatings to combine hemocompatibility and antimicrobial activity. *Biomaterials* **2015**, *56*, 198–205. [CrossRef]
11. Caykara, T.; Sande, M.G.; Azoia, N.; Rodrigues, L.R.; Silva, C.J. Exploring the potential of polyethylene terephthalate in the design of antibacterial surfaces. *Med. Microbiol. Immunol.* **2020**, *209*, 363–372. [CrossRef]
12. Moreno-Couranjou, M.; Mauchauffe, R.; Bonot, S.; Detrembleur, C.; Choquet, P. Anti-biofouling and antibacterial surfaces via a multicomponent coating deposited from an up-scalable atmospheric-pressure plasma-assisted CVD process. *J. Mater. Chem. B* **2018**, *6*, 614–623. [CrossRef]
13. Mutlu, E.C.; Yildirim, A.B.; Yildirm, M.; Ficai, A.; Ficai, D.; Oktar, F.N.; Ţîţu, M.; Çetinkaya, A.; Demir, A. Improvement of antibacterial and biocompatibility properties of electrospray biopolymer films by ZnO and MCM-41. *Polym. Bull.* **2020**, *77*, 3657–3675. [CrossRef]
14. Salguero, Y.; Valenti, L.; Rojas, R.; Garcia, M.C. Ciprofloxacin-intercalated layered double hydroxide-in-hybrid films as composite dressings for controlled antimicrobial topical delivery. *Mater. Sci. Eng. C Mater. Biol. Appl.* **2020**, *111*, 110859. [CrossRef]
15. Wu, Z.; Ma, X.; Ma, Y.; Yang, Z.; Yuan, Y.; Liu, C. Core/Shell PEGS/HA Hybrid Nanoparticle Via Micelle-Coordinated Mineralization for Tumor-Specific Therapy. *ACS Appl. Mater. Interfaces* **2020**, *12*, 12109–12119. [CrossRef]
16. Wang, L.; Dong, S.; Liu, Y.; Ma, Y.; Zhang, J.; Yang, Z.; Jiang, W.; Yuan, Y. Fabrication of Injectable, Porous Hyaluronic Acid Hydrogel Based on an In-Situ Bubble-Forming Hydrogel Entrapment Process. *Polymers* **2020**, *12*, 1138. [CrossRef]

17. Hartmann, H.; Hossfeld, S.; Schlosshauer, B.; Mittnacht, U.; Pego, A.P.; Dauner, M.; Doser, M.; Stoll, D.; Krastev, R. Hyaluronic acid/chitosan multilayer coatings on neuronal implants for localized delivery of siRNA nanoplexes. *J. Control. Release* **2013**, *168*, 289–297. [CrossRef] [PubMed]
18. del Hoyo-Gallego, S.; Perez-Alvarez, L.; Gomez-Galvan, F.; Lizundia, E.; Kuritka, I.; Sedlarik, V.; Laza, J.M.; Vila-Vilela, J.L. Construction of antibacterial poly(ethylene terephthalate) films via layer by layer assembly of chitosan and hyaluronic acid. *Carbohydr. Polym.* **2016**, *143*, 35–43. [CrossRef] [PubMed]
19. Ayaz, P.; Xu, B.; Zhang, X.; Wang, J.; Yu, D.; Wu, J. A pH and hyaluronidase dual-responsive multilayer-based drug delivery system for resisting bacterial infection. *Appl. Surf. Sci.* **2020**, *527*, 146806. [CrossRef]
20. Yan, X.J.; Chen, Q.X.; An, J.X.; Liu, D.E.; Huang, Y.K.; Yang, R.; Li, W.; Chen, L.; Gao, H. Hyaluronic acid/PEGylated amphiphilic nanoparticles for pursuit of selective intracellular doxorubicin release. *J. Mater. Chem. B* **2019**, *7*, 95–102. [CrossRef] [PubMed]
21. Zhang, Y.M.; Wang, F.M.; Huang, Q.L.; Patil, A.B.; Hu, J.J.; Fan, L.L.; Yang, Y.; Duan, H.P.; Dong, X.; Lin, C.J. Layer-by-layer immobilizing of polydopamine-assisted epsilon-polylysine and gum Arabic on titanium: Tailoring of antibacterial and osteogenic properties. *Mater. Sci. Eng. C Mater. Biol. Appl.* **2020**, *110*, 13. [CrossRef]
22. Agarwal, S.; Riffault, M.; Hoey, D.; Duffy, B.; Curtin, J.; Jaiswal, S. Biomimetic Hyaluronic Acid-Lysozyme Composite Coating on AZ31 Mg Alloy with Combined Antibacterial and Osteoinductive Activities. *ACS Biomater. Sci. Eng.* **2017**, *3*, 3244–3253. [CrossRef] [PubMed]
23. Choi, H.; Kim, K.-J.; Lee, D.G. Antifungal activity of the cationic antimicrobial polymer-polyhexamethylene guanidine hydrochloride and its mode of action. *Fungal Biol.* **2017**, *121*, 53–60. [CrossRef]
24. Allen, M.J.; Morby, A.P.; White, G.F. Cooperativity in the binding of the cationic biocide polyhexamethylene biguanide to nucleic acids. *Biochem. Biophys. Res. Commun.* **2004**, *318*, 397–404. [CrossRef] [PubMed]
25. Chindera, K.; Mahato, M.; Sharma, A.K.; Horsley, H.; Kloc-Muniak, K.; Kamaruzzaman, N.F.; Kumar, S.; McFarlane, A.; Stach, J.; Bentin, T.; et al. The antimicrobial polymer PHMB enters cells and selectively condenses bacterial chromosomes. *Sci. Rep.* **2016**, *6*, 23121. [CrossRef] [PubMed]
26. Sowlati-Hashjin, S.; Carbone, P.; Karttunen, M. Insights into the Polyhexamethylene Biguanide (PHMB) Mechanism of Action on Bacterial Membrane and DNA: A Molecular Dynamics Study. *J. Phys. Chem. B* **2020**, *124*, 4487–4497. [CrossRef]
27. Sousa, I.; Maia, F.; Silva, A.; Cunha, A.; Almeida, A.; Evtyugin, D.V.; Tedim, J.; Ferreira, M.G. A novel approach for immobilization of polyhexamethylene biguanide within silica capsules. *RSC Adv.* **2015**, *5*, 92656–92663. [CrossRef]
28. Kamaruzzaman, N.F.; Pina, M.d.F.; Chivu, A.; Good, L. Polyhexamethylene Biguanide and Nadifloxacin Self-Assembled Nanoparticles: Antimicrobial Effects against Intracellular Methicillin-Resistant Staphylococcus aureus. *Polymers* **2018**, *10*, 521. [CrossRef] [PubMed]
29. Wei, D.F.; Ma, Q.X.; Guan, Y.; Hu, F.Z.; Zheng, A.N.; Zhang, X.; Teng, Z.; Jiang, H. Structural characterization and antibacterial activity of oligoguanidine (polyhexamethylene guanidine hydrochloride). *Mater. Sci. Eng. C Mater. Biol. Appl.* **2009**, *29*, 1776–1780. [CrossRef]
30. Ding, X.X.; Chen, F.X.; Guan, Y.; Zheng, A.N.; Wei, D.F.; Xu, X. Preparation and properties of an antimicrobial acrylic coating modified with guanidinium oligomer. *J. Coat. Technol. Res.* **2020**, *17*, 1505–1513. [CrossRef]
31. Gao, Y.; Kong, W.; Li, B.; Ni, Y.; Yuan, T.; Guo, L.; Lin, H.; Fan, H.; Fan, Y.; Zhang, X. Fabrication and characterization of collagen-based injectable and self-crosslinkable hydrogels for cell encapsulation. *Colloids Surf. B-Biointerfaces* **2018**, *167*, 448–456. [CrossRef]
32. Pieper, J.S.; Hafmans, T.; Veerkamp, J.H.; van Kuppevelt, T.H. Development of tailor-made collagen-glycosaminoglycan matrices: EDC/NHS crosslinking, and ultrastructural aspects. *Biomaterials* **2000**, *21*, 581–593. [CrossRef]
33. Song, J.; Xie, J.; Li, C.; Lu, J.-H.; Meng, Q.-F.; Yang, Z.; Lee, R.J.; Wang, D.; Teng, L.-S. Near infrared spectroscopic (NIRS) analysis of drug-loading rate and particle size of risperidone microspheres by improved chemometric model. *Int. J. Pharm.* **2014**, *472*, 296–303. [CrossRef] [PubMed]
34. Martina, S.; Rongo, R.; Bucci, R.; Razionale, A.V.; Valletta, R.; D'Anto, V. In vitro cytotoxicity of different thermoplastic materials for clear aligners. *Angle Orthod.* **2019**, *89*, 942–945. [CrossRef] [PubMed]
35. Chen, Z.; Patwari, M.; Liu, D. Cytotoxicity of orthodontic temporary anchorage devices on human periodontal ligament fibroblasts in vitro. *Clin. Exp. Dent. Res.* **2019**, *5*, 648–654. [CrossRef] [PubMed]
36. Gradinaru, L.M.; Vlad, S.; Spiridon, I.; Petrescu, M. Durability of polyurethane membranes in artificial weathering environment. *Polym. Test.* **2019**, *80*, 10. [CrossRef]
37. Liu, Y.; Liu, Z.; Gao, Y.; Gao, W.; Hou, Z.; Zhu, Y. Facile Method for Surface-Grafted Chitooligosaccharide on Medical Segmented Poly(ester-urethane) Film to Improve Surface Biocompatibility. *Membranes* **2021**, *11*, 37. [CrossRef]
38. Sohail, M.; Ashfaq, B.; Azeem, I.; Faisal, A.; Dogan, S.Y.; Wang, J.J.; Duran, H.; Yameen, B. A facile and versatile route to functional poly(propylene) surfaces via UV-curable coatings. *React. Funct. Polym.* **2019**, *144*, 14. [CrossRef]
39. Chen, F.X.; Ding, X.X.; Jiang, Y.C.; Guan, Y.; Wei, D.F.; Zheng, A.N.; Xu, X. Permanent Antimicrobial Poly(vinylidene fluoride) Prepared by Chemical Bonding with Poly(hexamethylene guanidine). *ACS Omega* **2020**, *5*, 10481–10488. [CrossRef]
40. Bachimam, K.; Emul, E.; Saglam, N.; Korkusuz, F. Baicalein Nanofiber Scaffold Containing Hyaluronic Acid and Polyvinyl Alcohol: Preparation and Evaluation. *Turk. J. Med. Sci.* **2020**, *50*, 1139–1146. [CrossRef]

41. Revelou, P.-K.; Pappa, C.; Kakouri, E.; Kanakis, C.D.; Papadopoulos, G.K.; Pappas, C.S.; Tarantilis, P.A. Discrimination of botanical origin of olive oil from selected Greek cultivars by SPME-GC-MS and ATR-FTIR spectroscopy combined with chemometrics. *J. Sci. Food Agric.* **2020**. [CrossRef]
42. Xia, Y.; Adibnia, V.; Shan, C.; Huang, R.; Qi, W.; He, Z.; Xie, G.; Olszewski, M.; De Crescenzo, G.; Matyjaszewski, K.; et al. Synergy between Zwitterionic Polymers and Hyaluronic Acid Enhances Antifouling Performance. *Langmuir* **2019**, *35*, 15535–15542. [CrossRef] [PubMed]
43. Kopec, K.; Wojasinski, M.; Ciach, T. Superhydrophilic Polyurethane/Polydopamine Nanofibrous Materials Enhancing Cell Adhesion for Application in Tissue Engineering. *Int. J. Mol. Sci.* **2020**, *21*, 6798. [CrossRef] [PubMed]
44. Davoudi, P.; Assadpour, S.; Derakhshan, M.A.; Ai, J.; Solouk, A.; Ghanbari, H. Biomimetic modification of polyurethane-based nanofibrous vascular grafts: A promising approach towards stable endothelial lining. *Mater. Sci. Eng. C Mater. Biol. Appl.* **2017**, *80*, 213–221. [CrossRef]
45. Zheng, F.L.; Jiang, P.P.; Hu, L.; Bao, Y.M.; Xia, J.L. Functionalization of graphene oxide with different diisocyanates and their use as a reinforcement in waterborne polyurethane composites. *J. Macromol. Sci. Part A Pure Appl. Chem.* **2019**, *56*, 1071–1081. [CrossRef]
46. Zhu, M.-M.; Fang, Y.; Chen, Y.-C.; Lei, Y.-Q.; Fang, L.-F.; Zhu, B.-K.; Matsuyama, H. Antifouling and antibacterial behavior of membranes containing quaternary ammonium and zwitterionic polymers. *J. Colloid Interface Sci.* **2021**, *584*, 225–235. [CrossRef]
47. Pan, F.; Lau, K.H.A.; Messersmith, P.B.; Lu, J.R.; Zhao, X. Interfacial Assembly Inspired by Marine Mussels and Antifouling Effects of Polypeptoids: A Neutron Reflection Study. *Langmuir* **2020**, *36*, 12309–12318. [CrossRef]
48. Wang, F.; Cong, H.; Xing, J.; Wang, S.; Shen, Y.; Yu, B. Novel antifouling polymer with self-cleaning efficiency as surface coating for protein analysis by electrophoresis. *Talanta* **2021**, *221*, 121493. [CrossRef] [PubMed]
49. Xu, Z.; Hwang, D.-G.; Bartlett, M.D.; Jiang, S.; Bratlie, K.M. Alter macrophage adhesion and modulate their response on hydrophobically modified hydrogels. *Biochem. Eng. J.* **2021**, *165*, 107821. [CrossRef]
50. Yuan, H.; Qian, B.; Zhang, W.; Lan, M. Protein adsorption resistance of PVP-modified polyurethane film prepared by surface-initiated atom transfer radical polymerization. *Appl. Surf. Sci.* **2016**, *363*, 483–489. [CrossRef]
51. Wu, C.Q.; Zhou, Y.D.; Wang, H.T.; Hu, J.H.; Wang, X.L. Formation of antifouling functional coating from deposition of a zwitterionic-co-nonionic polymer via "grafting to" approach. *J. Saudi Chem. Soc.* **2019**, *23*, 1080–1089. [CrossRef]
52. Wang, L.C.; Chen, X.G.; Xu, Q.C.; Liu, C.S.; Yu, L.J.; Zhou, Y.M. Plasma protein adsorption pattern and tissue-implant reaction of poly(vinyl alcohol)/carboxymethyl-chitosan blend films. *J. Biomater. Sci. Polym. Ed.* **2008**, *19*, 113–129. [CrossRef]
53. Bouattour, Y.; Chennell, P.; Wasiak, M.; Jouannet, M.; Sautou, V. Stability of an ophthalmic formulation of polyhexamethylene biguanide in gamma-sterilized and ethylene oxide sterilized low density polyethylene multidose eyedroppers. *PeerJ* **2018**, *6*, e4549. [CrossRef] [PubMed]
54. Bueno, C.Z.; Moraes, A.M. Influence of the incorporation of the antimicrobial agent polyhexamethylene biguanide on the properties of dense and porous chitosan-alginate membranes. *Mater. Sci. Eng. C Mater. Biol. Appl.* **2018**, *93*, 671–678. [CrossRef] [PubMed]
55. Valverde, A.; Perez-Alvarez, L.; Ruiz-Rubio, L.; Olivenza, M.A.P.; Blanco, M.B.G.; Diaz-Fuentes, M.; Vilas-Vilela, J.L. Antibacterial hyaluronic acid/chitosan multilayers onto smooth and micropatterned titanium surfaces. *Carbohydr. Polym.* **2019**, *207*, 824–833. [CrossRef] [PubMed]
56. Thukkaram, M.; Vaidulych, M.; Kylian, O.; Rigole, P.; Aliakbarshirazi, S.; Asadian, M.; Nikiforov, A.; Biederman, H.; Coenye, T.; Du Laing, G.; et al. Biological activity and antimicrobial property of Cu/a-C:H nanocomposites and nanolayered coatings on titanium substrates. *Mater. Sci. Eng. C Mater. Biol. Appl.* **2021**, *119*, 111513. [CrossRef]

polymers

MDPI

Article

On the Use of OPEFB-Derived Microcrystalline Cellulose and Nano-Bentonite for Development of Thermoplastic Starch Hybrid Bio-Composites with Improved Performance

Di Sheng Lai [1,2], Azlin Fazlina Osman [1,2,*], Sinar Arzuria Adnan [1,2], Ismail Ibrahim [1,2], Awad A. Alrashdi [3,*], Midhat Nabil Ahmad Salimi [1,2] and Anwar Ul-Hamid [4]

1 Faculty of Chemical Engineering Technology, University Malaysia Perlis (UniMAP), Arau 02600, Malaysia; ericson_1206@hotmail.com (D.S.L.); sinar@unimap.edu.my (S.A.A.); ismailibrahim@unimap.edu.my (I.I.); nabil@unimap.edu.my (M.N.A.S.)
2 Biomedical and Nanotechnology Research Group, Center of Excellence Geopolymer and Green Technology (CEGeoGTech), Universiti Malaysia Perlis (UniMAP), Arau 02600, Malaysia
3 Chemistry Department, Umm Al-Qura University, Al-Qunfudah University College, Al-qunfudah Center for Scientific Research (QCSR), Al Qunfudah 21962, Saudi Arabia
4 Core Research Facilities, King Fahd University of Petroleum & Minerals, Dhahran 31261, Saudi Arabia; anwar@kfupm.edu.sa
* Correspondence: azlin@unimap.edu.my (A.F.O.); aarashdi@uqu.edu.sa (A.A.A.)

Citation: Lai, D.S.; Osman, A.F.; Adnan, S.A.; Ibrahim, I.; Alrashdi, A.A.; Ahmad Salimi, M.N.; Ul-Hamid, A. On the Use of OPEFB-Derived Microcrystalline Cellulose and Nano-Bentonite for Development of Thermoplastic Starch Hybrid Bio-Composites with Improved Performance. *Polymers* **2021**, *13*, 897. https://doi.org/10.3390/polym13060897

Academic Editor: José Miguel Ferri

Received: 4 February 2021
Accepted: 12 March 2021
Published: 15 March 2021

Abstract: Thermoplastic starch (TPS) hybrid bio-composite films containing microcrystalline cellulose (C) and nano-bentonite (B) as hybrid fillers were studied to replace the conventional non-degradable plastic in packaging applications. Raw oil palm empty fruit bunch (OPEFB) was subjected to chemical treatment and acid hydrolysis to obtain C filler. B filler was ultra-sonicated for better dispersion in the TPS films to improve the filler–matrix interactions. The morphology and structure of fillers were characterized by scanning electron microscope (SEM), Fourier transform infrared spectroscopy (FTIR) and X-ray diffraction (XRD). TPS hybrid bio-composite films were produced by the casting method with different ratios of B and C fillers. The best ratio of B/C was determined through the data of the tensile test. FTIR analysis proved the molecular interactions between the TPS and the hybrid fillers due to the presence of polar groups in their structure. XRD analysis confirmed the intercalation of the TPS chains between the B inter-platelets as a result of well-developed interactions between the TPS and hybrid fillers. SEM images suggested that more plastic deformation occurred in the fractured surface of the TPS hybrid bio-composite film due to the higher degree of stretching after being subjected to tensile loading. Overall, the results indicate that incorporating the hybrid B/C fillers could tremendously improve the mechanical properties of the films. The best ratio of B/C in the TPS was found to be 4:1, in which the tensile strength (8.52MPa), Young's modulus (42.0 MPa), elongation at break (116.4%) and tensile toughness of the film were increased by 92%, 146%, 156% and 338%, respectively. The significantly improved strength, modulus, flexibility and toughness of the film indicate the benefits of using the hybrid fillers, since these features are useful for the development of sustainable flexible packaging film.

Keywords: oil palm empty fruit bunch (OPEFB); microcrystalline cellulose; nano-bentonite; bio-composites; tensile properties

1. Introduction

Conventional plastics bring so much convenience to our life. We consume plastics in our day-to-day life and in many applications, including food packaging and drinking straws [1]. Conventional plastic or fossil-fuel plastic has many unique characteristics that make it one of the most essential materials in manufacturing a product, such as having a wide range of processing temperatures, high chemical resistance, a high strength to weight ratio, ease of processing and low costs [2]. These properties indicate the versatility of plastic materials, and therefore they are being used to substitute other materials such as metal and ceramic in certain applications. Unfortunately, plastic's great properties are sometimes overshadowed by its non-degradable properties which cause disposal problems and pollution on earth [3]. Nowadays, the plastic disposal problem is getting more severe. Thus, researchers are focusing on producing bio-plastics from natural sources such as starch which can be obtained from rice, corn, potato, and pea to solve the problem highlighted above.

Among all the biopolymers, starch has drawn significant attention from the researchers owing to its unique characteristics. Starch can be processed using the same processing techniques as conventional plastic, such as injection, extrusion and thermoforming. Starch is biodegradable and can be plasticized into a more useful form of plastic, which is called thermoplastic starch (TPS) [4]. However, to compete with conventional plastic, TPS-based plastics should possess comparable mechanical properties and barrier properties to conventional plastic. Yet, this is very difficult to achieve since thermoplastic starch inherited hydrophilic properties; thus, it has high sensitivity toward moisture absorption from the surroundings. The mechanical properties of thermoplastic starch decrease exponentially with the increase in humidity [5]. In order to improve the mechanical and barrier properties of thermoplastic starch, inorganic fillers or organic fillers are added to thermoplastic starch to increase its mechanical strength and barrier properties [6].

Palm oil is considered as one of the most important edible oils in the world, next only to soybean. However, palm oil extraction will generate an enormous amount of palm oil biomass wastes that consume up a large landfill area and endangering the environment [6]. Oil palm empty fruit bunch (OPEFB), as one of the major palm oil biomass wastes, contains a high percentage of cellulose of about 40–60%, which is the highest compared to kenaf, corn, and bagasse fiber [7]. Therefore, extraction of microcrystalline cellulose from the OPEFB is one of the effective strategies to add value the waste material.

In the search of new fillers for TPS that capable of enhancing its properties for various applications, the idea of using hybrid fillers (more than one type of filler) arrived due to the demands of consumers for biodegradable plastic packaging with combination of strength and toughness properties. The use of hybrid fillers may allow more comprehensive improvement in the properties of the TPS film by providing synergistic effect through the combination of both filler's properties [8].

In this study, we have investigated the efficiency of using hybrid nano-bentonite (in-organic nano-clay) and OPEFB derived microcrystalline cellulose (organic fiber) fillers in improving the mechanical properties of the TPS bio-composite plastic film, particularly its tensile strength and flexibility (toughness). The microcrystalline cellulose and ultra-sonicated nano-bentonite were expected to have good interaction with the TPS due to their hydrophilic properties, thus enhancement in the TPS film's mechanical properties could be gained. However, the TPS bio-composite film's optimal strength can only be achieved when the nano-bentonite and microcrystalline cellulose are well dispersed in the matrix of the biopolymer. Nonetheless, nano-bentonite particles always tend to agglomerate when dispersed in the polymer matrix. To encounter this, ultra-sonication treatment is applied to improve dispersion of the nano-clay filler in a polymer matrix [6].

Tensile toughness property of a material relates to its capability to possess both high tensile strength and high flexibility when stretched. So far, there have been no published papers reporting an increase in both the strength and toughness of the bio-polymeric film with the addition of fillers; in fact, most articles report a decrease in the flexibility of the film along with the increase in its strength [9–14]. This is because high-modulus filler can restrict the molecular motions of the biopolymer matrices, lowering its elongation at break value and thus making it more brittle. Unfortunately, this characteristic is not suitable for flexible film packaging applications because the ability to stretch without breaking (tough) is a key requirement of such a film. Therefore, there is the need to discover new hybrid fillers that are capable of improving the toughness property of the bio-composite film by elevating both tensile strength and elongation at break values of the material. In this work, we have proved that enhancement in both tensile strength and elongation at break can be gained with the use of B/C hybrid fillers. As far as the bio-composite with a dual-type filler system (hybrid) is concerned, this discovery is the first to be reported. Furthermore, the dual hybrid filler system which comprised the OPEFB-derived microcrystalline cellulose and nano-bentonite was first to be introduced in the TPS bio-composite system. Both fillers were characterized by scanning electron microscope (SEM), X-ray diffraction (XRD) and Fourier transform infrared spectroscopy (FTIR). Bio-composite films with several ratios of B/C filler were produced using the casting technique. A tensile test was employed to determine the TPS bio-composite films' mechanical properties containing a different ratio of microcrystalline cellulose and nano-bentonite. The optimal ratio of the OPEFB microcrystalline cellulose/nano-bentonite of the hybrid fillers that was capable of providing the best reinforcing and toughening effects to the TPS bio-composite film was determined.

2. Materials and Methods

2.1. Materials

The starch granules and nano-bentonite were supplied by Sigma-Aldrich (St. Louise, MO, USA). It was supplied by Euroscience Sdn. Bhd (Kuala Lumpur, Malaysia). The glycerine, sodium chlorite ($NaClO_2$) and sodium bicarbonate were supplied by HmbG Chemicals (Hamburg, Germany) purchased through A.R Alatan Sains Sdn Bhd (Alor setar, Malaysia). Sulfuric acid was supplied by Fisher Chemical™ (Concord, NH, USA) purchased through A.R Alatan Sains Sdn Bhd (Alor setar, Malaysia). OPEFB fiber was purchased from United Oil Palm Industries Sdn Bhd (Nibong Tebal, Malaysia). Deionized water was used as a solvent for all the experimental works.

2.2. Preparation of Treated OPEFB Cellulose from Raw OPEFB

The chemical isolation processes of the OPEFB were in accordance with the one reported by Zailuddin et al. [15]. The OPEFB was treated with 4% of NaOH solution at a temperature of 80 °C with constant stirring for one hour. The treated filler was washed with distilled water and filtered several times to ensure the removal of alkaline. This process was repeated four times to totally remove the hemicellulose. After that, the treated OPEFB fiber underwent bleaching treatment using dilute sodium chlorite ($NaClO_2$) solution at a temperature of 80 °C under stirring conditions for 1 h. The treated fiber was washed with distilled water and filtered for several times to ensure the removal of $NaClO_2$. These bleaching and washing procedures were repeated three times to ensure the removal of lignin and hemicellulose. The resultant cellulose is called treated OPEFB cellulose.

2.3. Preparation of Microcrystalline Cellulose (C) from the Treated OPEFB Cellulose

The treated OPEFB cellulose was subjected to concentrated acid hydrolysis with 64% H_2SO_4 at 45 °C for 1 h under stirring conditions. The suspension was added with 5-fold of cold water to stop the reaction. The cellulose suspension was centrifuged for 15 min at the speed of 7500 rpm and washed with distilled water to remove the acid content. This procedure was repeated at least ten times or until the pH of the solutions reached the value of 7 measured by using a pH meter. The produced microcrystalline cellulose was homogenized at 5000 rpm for 30 s and dried in the oven at 45 °C for 24 h.

2.4. Plasticization of Starch to Thermoplastic Starch (TPS)

Five grams of corn starch powder was added with 100 mL water into 250 mL beaker. After that, 2 g of glycerine was added into the starch solution to plasticize the starch. Then, the mixture was continuously stirred using a magnetic stirrer (500 rpm) for 45 min in a water bath, in which the temperature was maintained to be around 75–85 °C. A gel-like suspension of starch was obtained, showing that the starch has been transformed into the TPS.

2.5. Ultra-Sonication of Nano-Bentonite (B)

Nano-bentonite (B) was dispersed in deionized water in the ratio 1:20 and kept stirring under constant speed with a magnetic stirrer at room temperature. The (B) suspension was ultra-sonicated with an ultrasonic probe for 120 min with 20 s pulse on and 10s pulse off under 20% amplitude by using Branson Digital Ultrasonic Disruptor, Model 450D supplied by ProSciTech (Queensland, Australia). The suspension was filtered and dried in the oven. The dried powder was ground and sieved through a 53-micron sieve for de-agglomeration.

2.6. Preparation of TPS/B and TPS/C Bio-Composite and TPS/B/C Hybrid Bio-Composite Films

The hybrid fillers of microcrystalline cellulose (C) and nano-bentonite (B) were added into the TPS solution and continuously stirred at 75–85 °C. After that, 0.5 g of sodium bicarbonate was added into 20 mL water in a 50mL beaker and then added into the TPS mixture. The mixture was continuously stirred for about 5 min to obtain viscous and transparent suspension. Then, the mixture was poured into Teflon-coated mold and placed in the oven at 45 °C for 24 h for drying. The unfilled TPS, TPS/B and TPS/C bio-composite films were also produced as control samples. The formulation of all samples and acronyms is summarized in Table 1.

Table 1. The formulation of thermoplastic (TPS) bio-composites and TPS hybrid bio-composites containing hybrid filler (C/B).

TPS (wt%)	Microcrystalline Cellulose (C) (wt%)	Nano-Bentonite (B) (wt%)	Acronym
100	-	-	TPS
95	5	-	TPS/5C *
95	1	4	TPS/4B1C⁺
95	2	3	TPS/3B2C ⁺
95	3	2	TPS/2B3C ⁺
95	4	1	TPS/1B4C ⁺
95	5	-	TPS/5B *

* TPS bio-composite ⁺ TPS hybrid bio-composite.

2.7. Fourier Transform Infrared Spectroscopy (FTIR)

The FTIR analysis of the raw OPEFB, OPEFB-treated cellulose, OPEFB microcrystalline cellulose and nano-bentonite was performed by using the KBr pellet method. The functional groups of all the samples were analyzed using a Perkin Elmer spectrum 65 FTIR spectrometer (Waltham, MA, USA) with the wavenumber range of 650–4500 cm^{-1}, 16 scans and resolution of 4 cm^{-1}. FTIR transmitted infrared radiation through the sample and resulted in the infrared spectrum of the sample.

2.8. Scanning Electron Microscope (SEM)

The surface structure and morphology of fillers were analyzed using scanning electron microscope (SEM) (JEOL JSM-6460LA) (JEOL. Ltd., Tokyo, Japan). The morphology of raw OPEFB fiber, treated OPEFB cellulose and microcrystalline cellulose was studied under SEM to allow for comparison. Before images were captured, the fillers were coated with platinum using JFC-1600 Auto Fine Coater (JEOL Ltd., Tokyo, Japan) for getting the optimal SEM image of fillers. The surface and morphology of the raw OPEFB fiber and the treated OPEFB cellulose samples were taken at a magnification of ×500 and ×4000. Meanwhile, the morphologies of the microcrystalline cellulose and the ultra-sonicated nano-bentonite were taken at a magnification of ×30k and ×10k, respectively.

2.9. X-ray Diffraction (XRD)

Raw OPEFB, OPEFB cellulose and OPEFB microcrystalline cellulose were analyzed by using a Bruker D2 Phaser X-Ray diffractometer (Billerica, MA, USA) using Cu Kα X-rays.

The samples were tested by using a scan rate of 0.1 s per step from 2θ = 10–40°. The crystallinity index (CrI) of all samples was calculated according to the following equation:

$$\text{Crystallinity Index (CrI)} = \frac{I - I'}{I} \times 100 \quad (1)$$

where: I = the height of the intensity for the crystalline peak measured at 2θ = 22–23°.; I′= the height of the intensity for the amorphous peak measured at 2θ = 18–19°.

The basal spacing and crystallinity of the nano-bentonite (before and after ultrasonication) were analyzed using a Bruker D2 Phaser X-ray diffractometer with Cu Kα X-rays. The samples were tested using a scan rate of 0.1 s per step from 2θ = 5–40°. The XRD data were analyzed by using high score plus software. The basal spacing of the nano-bentonite was calculated by using Bragg's Law (nλ = 2d sin θ).

2.10. Tensile Test

The tensile tests of the TPS, TPS bio-composite and TPS hybrid bio-composite films were performed by Instron machine model-5582 (Norwood, MA, USA),according to ASTM D 638 Type V. The samples were tested with a crosshead speed of 10 mm/min to obtain the stress–strain curves. From the data, the tensile properties, such as tensile strength, Young's modulus, elongation at break and tensile toughness, were determined.

3. Results and Discussion

3.1. Morphological Characterization of Raw OPEFB Fiber, Treated cellulose, Microcrystalline Cellulose and Nano-Bentonite

SEM analysis was performed to have a better understanding of the microstructural changes of the OPEFB caused by the pre-treatment and acid hydrolysis process. Figure 1a shows the surface morphology of the raw OPEFB fiber, while Figure 1b presents the structure of the OPEFB cellulose (after it underwent alkaline pre-treatment and the bleaching process). Apparently, there are changes in the OPEFB morphology detected through the comparison of both micrographs indicating the success of the chemical pre-treatment procedure in removing non-cellulosic materials such as hemicellulose, lignin and waxy substances from the raw OPEFB fiber.

The raw OPEFB fiber in Figure 1a possesses an irregular shape and the surface of the microfibrils cellulose (pointed by the blue arrow) was covered up by non-cellulosic materials which act as the protective surface (as marked by red circles). In contrast, Figure 1b shows the smooth, clear and individualized rod-like microfibrils surface of cellulose. The different morphology of the surface structure in the raw OPEFB (Figure 1a) and the treated OPEFB cellulose (Figure 1b) indicates that chemical pretreatment has successfully removed the non-cellulosic materials such as hemicellulose, lignin and waxy substances from the raw OPEFB fiber, which is in line with the FTIR results. Due to these removals, the microfibrils surface of the cellulose has been exposed and the crystallinity region of the cellulose was increased, as confirmed through the XRD analysis. Figure 1c represents the morphology of the microcrystalline cellulose after being dried up and observed under SEM with ×30k magnification. It can be observed that the structure of the OPEFB cellulose's microfibril was completely destroyed and the size of the cellulose was significantly being reduced. However, the SEM diagram suggests that the microcrystalline cellulose tends to aggregate together into large particles of cellulose. Agglomeration of cellulose usually occurs after being dried from the suspension form [15].

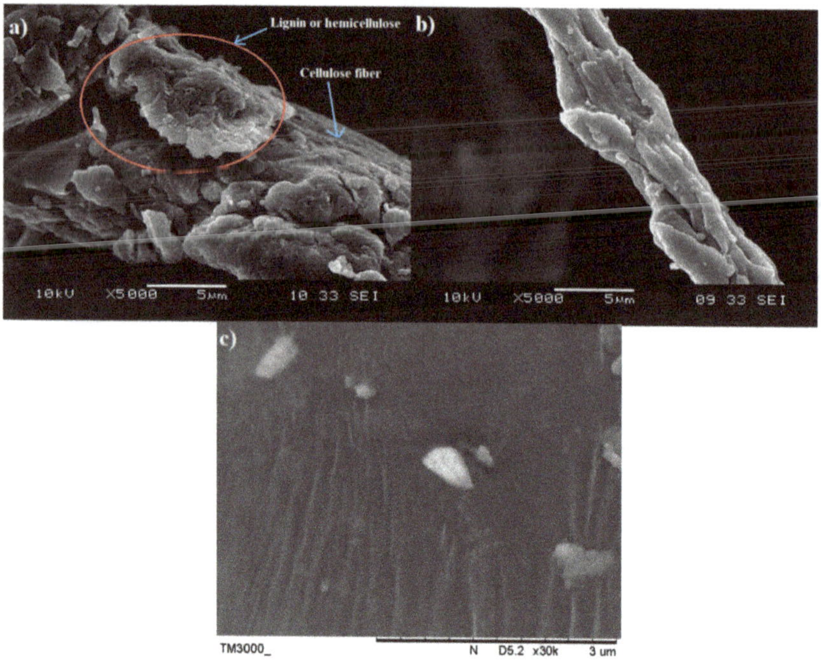

Figure 1. Scanning electron microscope (SEM) micrographs of: raw oil palm empty fruit bunch (OPEFB) fiber taken at ×5000 (**a**); treated OPEFB cellulose taken at ×5000 (**b**); microcrystalline cellulose taken at ×30k (**c**).

Figure 2a,b display the SEM images of the pristine nano-bentonite and ultra-sonicated bentonite, respectively. By comparing both figures, we can prove that the particle size of the nano-bentonite had been reduced significantly upon the ultra-sonication process. Before ultra-sonication, the nano-bentonite appears in large particles with an irregular shape and size. This was due to the high stacking of platelets forming large tactoids of the phyllo-silicate structure. After being ultra-sonicated, the particle size of the nano-bentonite was seen to greatly reduce, with rougher surface morphology. This observation was due to the tactoid size reduction of the nano-clay as a result of the delamination of the nano-platelets. However, in agglomerated form, it is difficult to get a clear image of the individual platelet of the nano-bentonite and determine its particle size. However, the results can confirm that the ultra-sonication process has successfully reduced the tactoid size of the nano-bentonite.

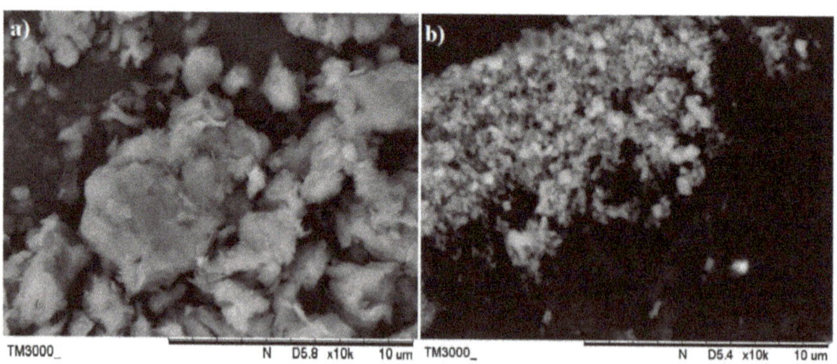

Figure 2. SEM picture of: nano-bentonite (before ultrasonication) (**a**); nano-bentonite (after ultrasonication) (**b**), taken at ×10k.

3.2. Chemical Structure of Raw OPEFB Fiber, Treated Cellulose, Microcrystalline Cellulose and Nano-Bentonite

FTIR spectroscopy is a non-destructive method to study the chemical composition of material by analyzing its functional groups. The FTIR spectra of the raw OPEFB fiber, treated OPEFB cellulose and OPEFB microcrystalline cellulose are illustrated in Figure 3. Generally, the FTIR spectra of all samples were almost the same, indicating that they possess the same chemical composition and structure.

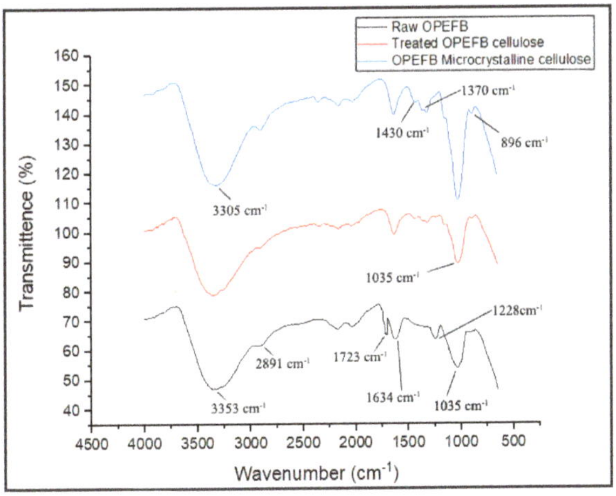

Figure 3. Fourier transform infrared spectroscopy (FTIR) spectra of raw OPEFB fiber, treated OPEFB cellulose fiber and OPEFB microcrystalline cellulose.

The FTIR spectra of the raw OPEFB and treated OPEFB cellulose represent a broad peak at the region of 3353 and 2891 cm^{-1} due to the –OH stretching of the aliphatic structure and aliphatic saturated C-H stretching of CH_2 and CH_3. The 3353 cm^{-1} can also be attributed to moisture content where the hydroxyl group is found in cellulose, hemicellulose and lignin. This shows that the cellulose is hydrophilic in nature [16]. The peak at 1035 cm^{-1} appears in both samples and represents the characteristics of anhydroglucose chains with the C-O-C stretch [17]. The treated cellulose has a sharper peak than the raw OPEFB due to its higher surface area caused by the removal of non-cellulosic components in the cellulose. Meanwhile, the peaks at 1634 and 2891 cm^{-1} represent the –OH bending vibration of absorbed water from the strong interaction between cellulose and water [18]. A peak at 896 cm^{-1} is related to β-glycosidic linkage between the glucose rings of cellulose. [19] This indicates that the cellulose is made up of a linkage bond of saccharides that formed polysaccharides. Meanwhile, the peak that appears at 1330–1369 cm^{-1} represents the bending vibration of the C-H and C-O groups of aromatic rings in polysaccharides [20].

The disappearance of a peak at 1723 cm^{-1} in the spectrum of the treated cellulose and microcrystalline cellulose samples verifies the removal of hemicellulose and lignin. This is because the presence of lignin and hemicellulose can be detected through the appearance of those peaks due to the functional group of carbonyl ester and the C=O acetyl group of the uronic ester, respectively [4,8,21]. In addition, the peak at 1228 cm^{-1}, which represents the syringyl ring unit and C-O stretching at lignin and xylan, disappeared in the spectra of the treated cellulose, suggesting the removal of lignin and hemicellulose after various treatments [15,22]. These clearly show that most non-cellulosic materials have been removed upon the alkaline treatment, bleaching and acid hydrolysis processes.

Figure 4 shows the FTIR spectra of the nano-bentonite (before and after ultra-sonication). By comparing the results of both samples, no significant change in the pattern of the FTIR spectra can be observed. This indicates that the ultra-sonication process does not alter the chemical structure of the nano-bentonite. Both samples exhibit peaks at 3614 and 3435 cm^{-1}, which indicate the possibility of the hydroxyl linkage from alumino-silicate layers [23]. The peak at 3614 cm^{-1} represents the interlayer hydrogen bonding, corresponding to the stretching vibration of the OH coordinated octahedral layer of Al+Mg [18]. The bands at 1636 and 3435 cm^{-1} indicate the –OH stretching vibration due to the absorption of water in the bentonite di-octahedral surface [23]. The sharp and intense peak near 1000 cm^{-1} represents the Si-O-Si stretching in the nano-bentonite [24]. The peaks at 790 and 997 cm^{-1} are the characteristic signals of Al-O-Si and Si-O stretching vibrations, respectively [24]. Meanwhile, the peaks at 3614 cm^{-1}, 3435 cm^{-1}, 1636 cm^{-1}, 997 cm^{-1} and 790 cm^{-1} imply the presence of montmorillonite (MMT) in the bentonite [23]. The ultra-sonicated nano-bentonite has a more intense peak related to Si-O stretching around 790 cm^{-1}, indicating that the ultra-sonication treatment has increased the surface area of the nano-filler by delamination or the stacking platelets, which is also proved through the SEM and XRD analyses.

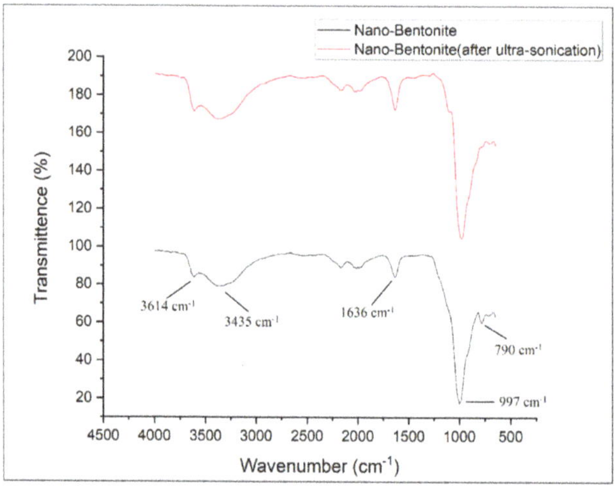

Figure 4. FTIR spectra of nano-bentonite (before and after ultra-sonication).

3.3. *Structure of Raw OPEFB Fiber, Treated Cellulose, Microcrystalline Cellulose and Nano-Bentonite as Observed by XRD*

The XRD patterns of raw OPEFB fiber, treated OPEFB cellulose and OPEFB microcrystalline cellulose are shown in Figure 5. The XRD diffractograms of all the samples revealed a broad amorphous hump and crystalline peaks, indicating that OPEFB is a semi-crystalline material in nature. Generally, all samples show peaks at around 2θ = 15–16°, 2θ = 21–22° and 2θ = 34–35°, which indicates that the OPEFB contains a crystalline structure of cellulose I with (110), (200) and (004) diffraction planes, respectively, which is typical for natural plant cellulose [25]. The lowest intensity peak at 2θ = 18° is related to amorphous region arrangement, while 2θ = 22° is related to the crystalline structure of cellulose. The CrI was calculated based on the Segal formula [26]. The raw OPEFB possesses 37.80% CrI value, which is the lowest among all samples. This can be associated with its high amorphous region. Conversely, the microcrystalline cellulose possesses the highest CrI value, which is 65.9% due to the removal of the amorphous phase from its structure.

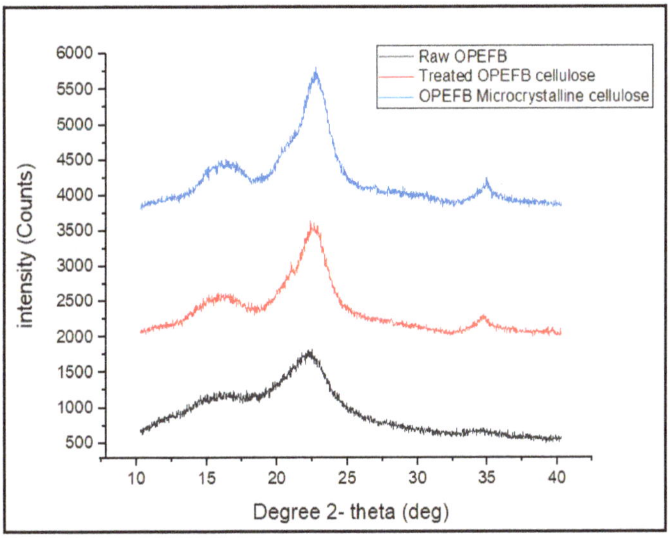

Figure 5. X-ray diffraction (XRD) pattern of raw OPEFB cellulose, treated cellulose and microcrystalline cellulose.

The low Crl value of the raw OPEFB indicates that it contains a high content of hemicellulose and lignin. The same results were obtained through previous studies [27]. As expected, crystallinity of the OPEFB increased after the removal of lignin and hemicellulose. After the alkaline and bleaching pre-treatment, the peak of the treated OPEFB cellulose became sharper and more intense than the raw OPEFB, suggesting that the crystallinity of the treated OPEFB cellulose becomes higher after the treatment. However, the CrI value has been further increased when the treated OPEFB cellulose was acid hydrolyzed to form microcrystalline cellulose. This clearly indicates that more amorphous linkages within the cellulose structure have been broken down, releasing the high crystalline cellulose [28].

Figure 6a displays the XRD pattern of the nano-bentonite (before and after ultra-sonication) in the region of $2\theta = 5°$ to $50°$. However, the signal from 5° to 10° was focused to clearly observe the changes in d_{001} basal spacing of the nano-bentonite upon ultra-sonication (Figure 6b). The nano-bentonite's basal spacing represents the stacking distance and order between its platelets and provides information about the platelets dispersion. Figure 6b exhibits that the d_{001} basal spacing of raw nano-bentonite is 1.44 nm. However, the d_{001} of the nano-bentonite has been increased to 1.47 nm after the ultra-sonication process. This indicates that the ultra-sonication process does slightly increase the d_{001} of the nano-bentonite, by altering the bentonite layer's stacking through de-agglomeration of the bentonite's tactoids. The small increase in the basal spacing of bentonite can facilitate the intercalation of the TPS chain during the mixing process and improve the film's strength. This outcome was the same with the result of Abdul Hamid et al., where they showed that the ultra-sonication process could produce more loosely packed MMT and encourage the copolymer chains' intercalation inter-platelets [29].

Apparently, there are clear changes in the XRD pattern of the nano-bentonite after it is subjected to the ultra-sonication process (Figure 6a). The diffraction peaks at $2\theta = 21.8°$, 25.4° and 26.6° disappeared, suggesting that the ultra-sonication process has successfully and effectively disrupted and at a certain level destroyed the organization of the nano-bentonite platelets [30]. Moreover, the diffraction peaks at $2\theta = 19.76°$ and 35° are shifted to lower angles to 18.9° and 34.0°, respectively, due to the expanding of basal spacing of the nano-bentonite platelets.

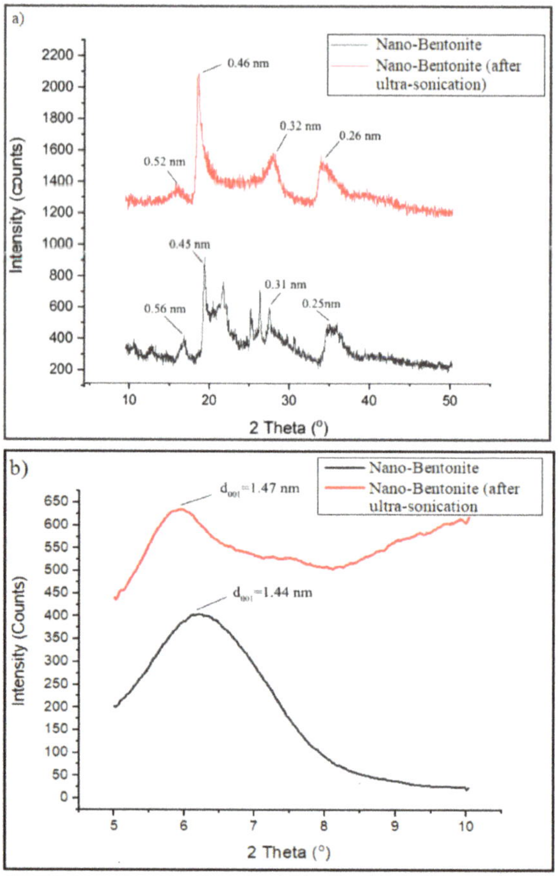

Figure 6. XRD pattern of nano-bentonite (before and after ultra-sonication) at 2θ = 10–50° (**a**); XRD pattern of nano-bentonite (before and after ultra-sonication) at 2θ = 5–10° (**b**).

3.4. Mechanical Analysis of TPS, TPS/B and TPS/C Bio-Composites, TPS/B/C Hybrid Bio-Composite Films

Figure 7a shows the tensile strength of the TPS and TPS bio-composite film with different nano-bentonite (B) and microcrystalline cellulose (C) ratios. Regardless of the types of filler used (single or hybrid), the tensile strength value of the film increased with the incorporation of 5 wt% fillers into the TPS matrix. These results were in accordance with many reviews where phyllosilicate clay and cellulose can improve the TPS composite films' tensile strength due to high compatibility and positive interactions such as hydrogen bonding between the TPS matrix and nanofiller, leading to better stress transfer [31,32]. The unfilled TPS exhibits the lowest tensile strength, which is 4.44 MPa, while the TPS/4B1C hybrid bio-composite film possesses the highest tensile strength (8.52 MPa), showing 92% increment when benchmarked with the unfilled TPS. This signifies that the inclusion of the hybrid B/C fillers at the optimum ratio can bring a positive synergy effect within the TPS matrix and enhance the tensile strength of the film with a more efficient load transfer mechanism. This explanation was in line with the results of Zakuwan et al., in which tensile strength improvement was obtained when cellulose nanocrystal and modified MMT was incorporated into the biopolymer. They claimed that the network forming of nanocellulose and MMT enforcement in a different dimension in the biopolymer matrix can improve bio-composite films' reinforcement capability [33].

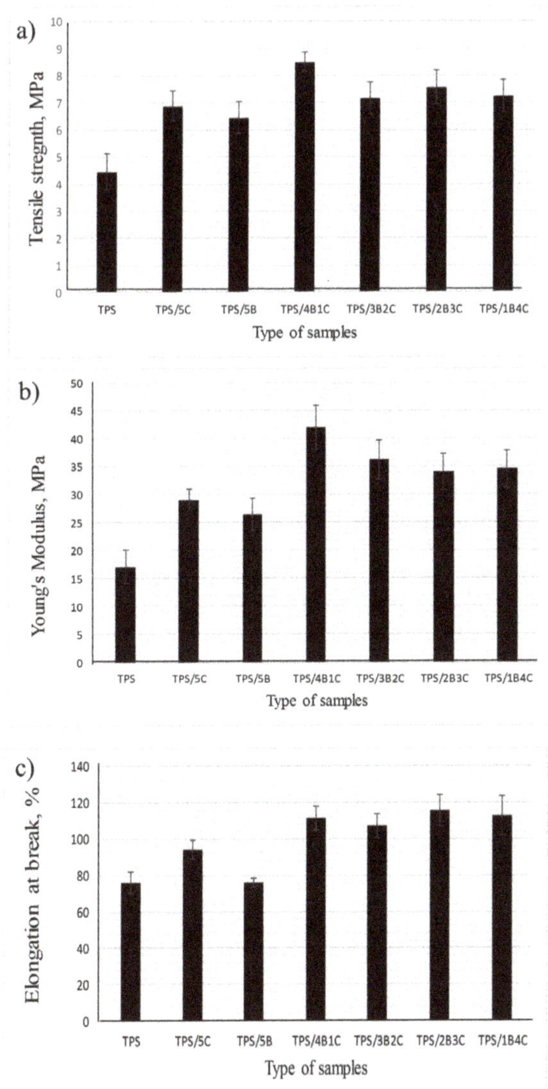

Figure 7. Tensile properties of TPS, TPS bio-composites and TPS hybrid bio-composite films: (**a**) tensile strength; (**b**) Young's Modulus; (**c**) elongation at break.

The results also suggest that the use of single filler (either C alone or B alone) could not allow the film to achieve higher tensile strength than the hybrid fillers 4B1C. However, the use of C resulted in a higher increment in tensile strength as compared to B. As compared to C, B tends to experience a higher degree of agglomeration in the biopolymer matrix due to its platy particles that are easily stacked together to form tactoids. Poor dispersion of tactoids reduces the efficiency of the B filler in reinforcing the matrix. C can better reinforce the TPS due to its fibrous-like particles that can transfer the load more efficiently [34]. However, when used as hybrid filler with the B, C cannot be added in more than 1 wt%. We have postulated that the high content of C in the hybrid filler system can reduce the reinforcing effect because overcrowding of the C particles may occur and inhibit good dispersion and distribution of the B's nanoplatelets in the matrix. Poorly dispersed hybrid fillers may form a stress concentration point in the TPS matrix and reduce the tensile properties of the TPS [35,36].

The Young's modulus values of the TPS, TPS bio-composites and TPS hybrid bio-composites with different B and C ratios are compared in Figure 7b. It can be seen that the Young's modulus values of all the TPS bio-composites are higher than the unfilled TPS, showing that incorporating single or hybrid fillers into the TPS may improve the stiffness of the film. The unfilled TPS exhibits the lowest Young's modulus, which is 17.1 MPa, while TPS/4B1C hybrid bio-composite demonstrates the highest Young's modulus, which is 42.0 MPa. However, the Young's modulus of the film has slightly decreased to 36.22 MPa when the C loading increased to 2 wt% (3B2C). Moreover, the Young's modulus values of the TPS/2B3C and TPS/1B4C films show no significant difference with the TPS/3B2C, indicating that increases in the C loading in the matrix does not further improve the stiffness of the film. Agglomeration of C particles may occur when added in high content, reducing the matrix–filler interactions [37]. In agreement with the tensile strength data, the Young's modulus of the TPS/5B is lower than the TPS/5C bio-composite film. As mentioned earlier, the platy particles of B can stack together in great numbers, poorly dispersed and distributed in the matrix. This reduces matrix–filler interactions that play the main role in stiffening the matrix.

Based on Figure 7c, the TPS/1B4C hybrid bio-composite film demonstrates the elongation at a break value of 117%, which is 156% higher than the unfilled TPS. This is somewhat impressive as the addition of natural filler or inorganic filler always reduces the elongation at break of the matrix. This is because filler with a stiffer characteristic reduces the flexibility of the biopolymer [9–14].

Tensile toughness is calculated based on the area under the stress–strain curve. The greater the area under the curve, the greater the toughness value will be, showing that the polymer performs greater tensile strength and elongation at break due to a more efficient energy absorption mechanism [38]. Figure 8 illustrates the tensile toughness of the TPS, TPS bio-composite and TPS hybrid bio-composite films. Overall, the results suggest that the film's tensile toughness has been tremendously improved upon incorporation of the hybrid fillers. TPS/4B1C displays the highest value of tensile toughness with an increment of ~338%, when compared with the unfilled TPS. The difference in physical characteristics (size, shape, etc.) and chemistry of B and C fillers might contribute to a greater toughening effect on the film. We are still investigating these factors and perhaps will come out with theory and a mechanism in our next publications after finishing all the analyses. For now, we are postulating that the fibrous particles of C, when dispersed well in the matrix, not only can contribute to good stress transfer mechanism, but will also produce more free volume in the matrix when the film is stretched. Figure 9 illustrates the proposed mechanism. During tensile deformation, the small size of B nanoplatelets and the TPS molecular chains will arrange and stretch according to strain direction. This will produce greater interface bonding between the B and the TPS molecular chains, thus producing a more effective stress transferring mechanism. The bigger and longer C particles are not as mobile as the B nanoplatelets and, therefore, will not exactly follow the strain direction, but rather arrange themselves vertically between the stretched TPS molecular chains. This resulted in the formation of more free volumes in the hybrid bio-composite structure, allowing for greater starch chain mobility (entropy) and conformational freedom. The conformation of the TPS chains will stimulate chain relaxation in the stress concentrated area and allow the matrix to be more flexible and more efficient in absorbing energy through molecular motions. These interactions might responsible for contributing to the toughening effect on the film.

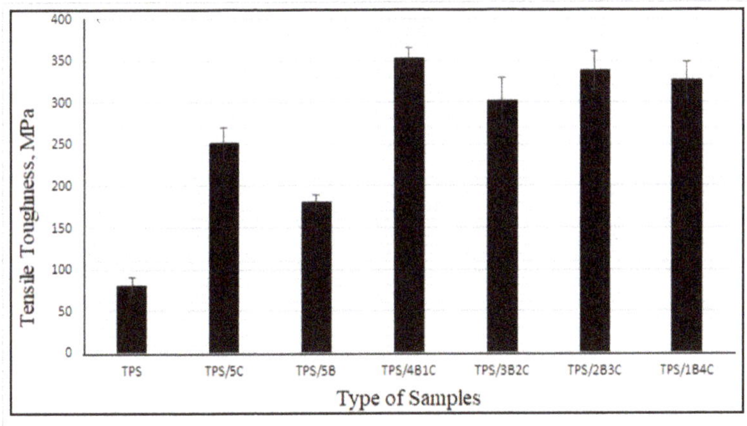

Figure 8. Tensile toughness of TPS, TPS bio-composites and TPS hybrid bio-composite films.

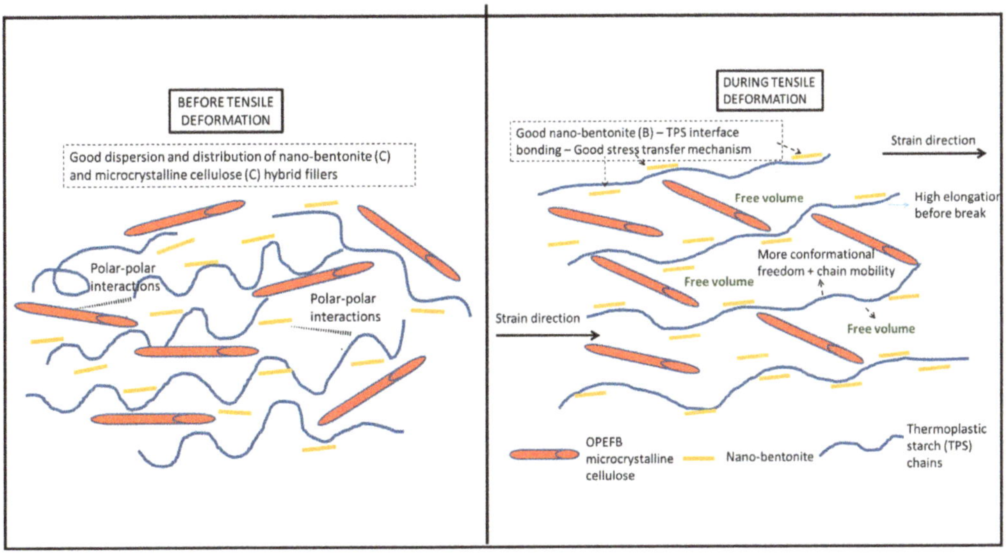

Figure 9. Mechanism of interactions between the TPS and the hybrid fillers during tensile deformation.

Comparative study was done with other TPS bio-composite systems investigated by other researchers. Based on the data obtained from recent literatures, our TPS hybrid bio-composite film's tensile strength is on par with the values reported, while the flexibility of our film is much better than them [13,14,39]. For instance, Fazeli et al. have used cellulose fibers to reinforce the TPS derived from corn. They have applied surface modification on the cellulose fiber using the air plasma treatment to improve the matrix/fiber adhesion in the TPS bio-composite structure. The maximum tensile strength was achieved when the plasma treated cellulose fiber was added in 6 wt%. However, elongation at break value has been dropped significantly from 48.2% to 9.2% only [13]. In another work, Chen et al. have prepared TPS bio-composite films containing oxidized microcrystalline cellulose (MCC) as a reinforcing filler via a hot-compression molding technique. As compared to the original MCC, the oxidized MCC was more capable of enhancing the tensile strength of the TPS.

However, the tensile strength obtained was 6.61 MPa while the elongation at break was only 40% [14]. In more recent work, Yin et al. have prepared the TPS bio-composites filled with dialdehyde lignocellulose (DLC) in the loading of 0 to 12 wt%. Maximum tensile strength was obtained when 3 wt% DLC was employed as filler with the value of 5.26 MPa. However, the bio-composite only managed to achieve elongation at a break value of 91.60% [39]. Both tensile strength and elongation at break values reported for the TPS bio-composites in the above literatures were lower than the values of our optimum hybrid bio-composite system (TPS/1B4C). As mentioned in the above discussion, the maximum tensile strength and elongation at break of the TPS hybrid bio-composite film in this study were 8.52 MPa and 117%, respectively. Apparently, the improvement in toughness value upon the inclusion of the hybrid fillers was impressive (+338%). The work by Li et al. has shown that they have successfully obtained very high tensile strength starch-based bio-composites when using montmorillonite and cellulose nanofiber as hybrid fillers. However, the elongation at break values of the samples was very low (less than 7%) compared to our hybrid TPS bio-composite films (117%). The above findings proved that that the origin of starch, method of processing and types of filler/hybrid fillers used will determine the obtained tensile strength and elongation at break of the bio-composite film. These are in agreement with review papers wrote by Pérez-Pacheco et al. [11] and Xie et al. [31]. In our case, the unfilled TPS has shown very moderate tensile strength, but the hybrid B/C fillers have played their role in simultaneously enhancing the tensile strength and elongation at break of the TPS matrix. Thus, significant enhancement in the tensile toughness of the film could be observed. This trend is rarely obtained and reported. Nevertheless, we believe that our films that possess good flexibility and toughness will suit for certain applications such as for short term flexible packaging and wrapping plastic.

3.5. The Interactions of Single and Hybrid Fillers with the TPS Matrix as Observed through FTIR and XRD Analyses

FTIR analysis was used to interpret the chemical functionalities' presence in the bio-composite's chemical structure and analyze a potential interaction between the TPS matrix and single filler and with the hybrid filler system. FTIR is sensitive to detecting the change of TPS structure at the molecular level, such as chain conformation, crystallinity, water content, and TPS and filler interaction [40]. According to Figure 10, the FTIR spectra of all the samples indicate a broad band located at 3000 to 3700 cm^{-1}, corresponding to free, inter-and intramolecular O-H stretching. This indicates the presence of a high amount of O-H functional group of the TPS and TPS bio-composite films' structure [41]. For the unfilled TPS film, wavelength from 800 to 1200 cm^{-1} is the fingerprint region of the TPS, contributing to glucan ring vibration by C-OH stretching and bending vibration and the C-O-C glycoside bond vibration [42]. A single peak observed at 1640 cm^{-1} represents the water—bound tightly in TPS film due to its hygroscopic nature [42].

The characteristic peak at 2929 cm^{-1} was attributed to asymmetrical C-H stretching and vibration. Meanwhile, the characteristic peak at 1375 cm^{-1} represents the –CH_2 bending. Furthermore, there is a small peak at 1153 and 1080 cm^{-1} representing the C-O stretch in the C-O-H group in the TPS film, whereas 1240 cm^{-1} shows the C-O stretch of C-O-C bond in the structure of the film [41,43]. The peaks at 927 and 862 cm^{-1} represent the starch glycosidic linkage of glucose in starch [41]. Overall, the FTIR spectra for all the TPS bio-composite films are almost similar to the unfilled TPS films but there was a slight change in the intensity of some peaks. Furthermore, the shifting of certain bands was also noticed.

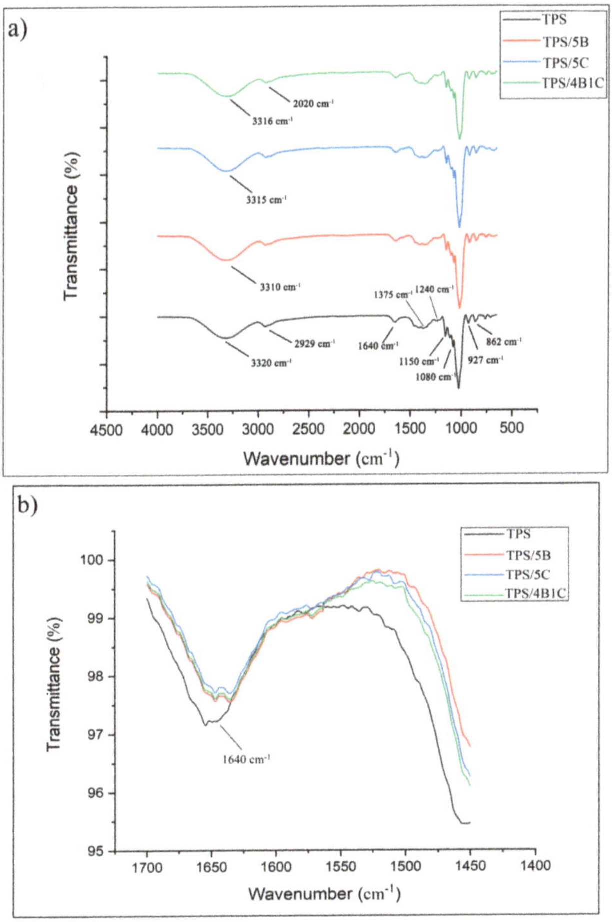

Figure 10. FTIR of unfilled TPS, TPS bio-composites and TPS hybrid bio-composite, in the region of (**a**) 650–4000 cm^{-1} and (**b**) 1450–1700 cm^{-1}.

In the region of 3000 to 3700 cm^{-1}, the band appears due to O-H stretching that provides information related to hydrogen bonding between TPS and fillers. For TPS/5B bio-composite, the disappearance of the nano-bentonite peak at 3435 cm^{-1} is associated with the hydroxyl linkage formation within the alumino-silicate layered structure of the clay, indicating that the filler is forming new hydrogen bonding with the TPS. XRD analysis has further proved this by showing a broadening of d_{001} peak for the nano-bentonite. For the TPS/5B sample, there is a shift of peak at 3320 cm^{-1} to a lower wavenumber, which is 3310 cm^{-1}, indicating new and stable hydrogen bonds formed in the TPS bio-composite films [44]. Hydrogen bonding has been developed due to compatibility between the TPS and B filler.

Previous research showed that there is good interfacial bonding between the cellulose and TPS due to their chemical similarity and good compatibility that allow for the formation of hydrogen bonding between them [45]. Based on the FTIR spectra of TPS/5C, it can be observed that incorporation of C into the TPS matrix has slightly sharpened the peak and also shifted the peak to a lower wavenumber (3315 cm^{-1}). This can be associated with the O-H vibration of the high crystalline structure of the C filler. This outcome was in accordance with the study of Zhang et al. They have concluded that the O-H stretching vibration shifts to a lower wavenumber in the FTIR spectrum due to new hydrogen bonding between TPS and nanocellulose [46]. For the analysis of the TPS/4B1C hybrid bio-composite film, interactions between the TPS and hybrid fillers can also be realized through the FTIR data. Since both of the single fillers showed interaction with TPS films by forming hydrogen bonding, hybrid fillers are expected to interact with TPS films in the same ways. Peaks at 2929 (C-H) and 3320 cm^{-1} of the TPS were shifted to 2020 and 3316 cm^{-1} due to new hydrogen forming between the TPS and hybrid fillers (B and C). This indicates that hybrid fillers have good compatibility with the TPS matrix by forming strong polar hydroxyl interactions.

Figure 11a shows the XRD pattern of the unfilled TPS, TPS bio-composite and TPS hybrid bio-composite films. TPS has experienced a reduction in peak intensity in the region of 2θ = 10–40° when added with filler (either single or hybrid fillers) due to the reduction in retrogradation rate that occurs in the matrix. Interface interactions between the matrix and fillers can slow down the retrogradation process of the starch. During the TPS matrix cooling, the amylose and amylopectin chains are starting to arrange back into an ordered structure different from native starch granules. The retrogradation process involved a few steps: extrusion of water, an increase in viscosity, gel formation and forming a crystalline structure. The typical retrogradation peaks of TPS can be seen at 17° and 22.6°.

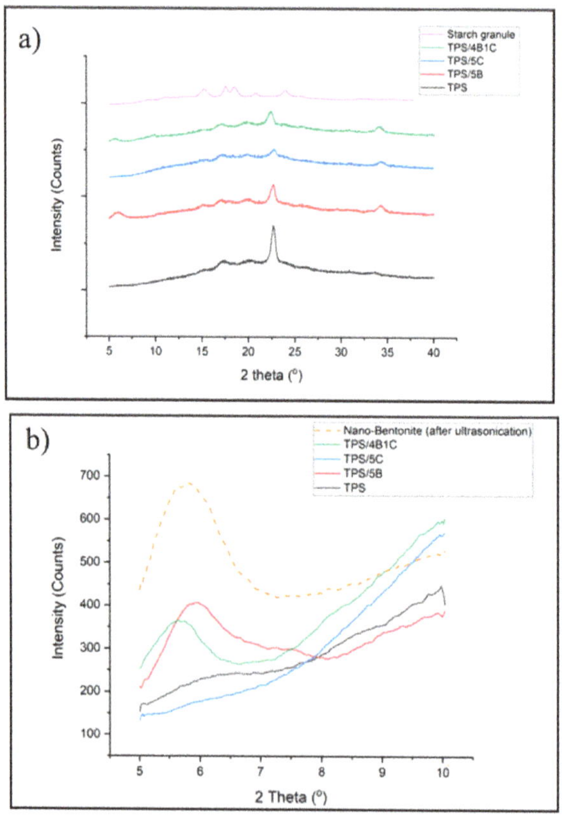

Figure 11. XRD diagram of TPS, TPS bio-composite and TPS hybrid bio-composite films, at the region of (**a**) 2θ = 5–40° and (**b**) 2θ = 5–10°.

The XRD pattern of the TPS/5C film presents the typical peak of retrograded starch structure (type B and type Vh peak) and the signal of the C filler. However, due to the similar chemical structure of TPS and microcrystalline cellulose in the TPS/5C composite, the XRD pattern shows superimposition of both parent components balanced by the composition. This is in line with the study of Dufresne et al. [37]. Incorporation of C into the TPS has brought a reduction in the intensity of Vh peaks in the XRD spectrum. This was possibly due to the transcrystallization of amylose and amylopectin on the microcrystalline cellulose surface, leading to a reduction of the starch chain's recrystallization. The same observation was reported in the study of Fourati et al. They found that the Vh-type structure of the TPS reduced with the increase in the cellulose nanofiber content [47].

It can also be observed that the intensity of the peak related to the Vh-type structure of the TPS is lower in the TPS/4B1C hybrid bio-composite as compared to the unfilled TPS. The Vh-type structure formed in TPS was due to amylose's recrystallization with plasticizer in the helix channel. Reduction in the peak intensity of the Vh type crystalline structure indicates that the retrogradation of the TPS was hindered with the incorporation of the single B filler or hybrid B/C fillers. The anti-retrogradation of the TPS films with the nano-bentonite was due to the hydrophilicity of bentonite clay's surface that increased the interaction between the starch chain and bentonite platelets, affecting the dynamic rearrangement of the starch chain. Consequently, the ability of the TPS chains to recrystallize was reduced. This trend was also reported by Lara et al., where they found that the retrogradation of starch plasticized by water and glycerol was reduced by incorporating MMT [48]. Interestingly, the TPS/4B1C hybrid bio-composite film exhibits a smaller and broader peak of the Vh-type structure as compared to the TPS/5B bio-composite. This proved that the use of hybrid filler with a low content of C (1 wt%) may result in greater efficiency in preventing retrogradation than the use of single C or single B filler.

Next, the XRD signal from 5° to 10° was focused to have a clear comparison on the d_{001} basal spacing of the nano-bentonite clay before and after being incorporated into the TPS matrix. As expected, there is no peak that can be observed in the XRD signal of the unfilled TPS and TPS/5C bio-composite in the low-angle region (2θ = 5–10°) since TPS is amorphous. It is known that the peak that appears at a low angle value between 5–10° represents the d_{001} basal spacing of bentonite clay [8]. When single B filler or hybrid B/C fillers were incorporated into the TPS, the d_{001} peak of B had shifted to a lower angle, indicating an increase in the interlayer basal spacing of the clay due to the intercalation of TPS chains and microcrystalline cellulose into the layered silicate structure of bentonite without complete exfoliation.

The TPS/4B1C hybrid bio-composite film shows the highest increment in basal spacing in which the d_{001} increased from 1.47 to 1.55 nm. Moreover, broadening of the d_{001} peak can clearly be seen. It seems that the inter-platelets of the nano-bentonite have been well intercalated by the TPS chains with the presence of a small amount of the microcrystalline cellulose. This could be due to the capability of bentonite to interact with the C filler, other than the host biopolymer. As mentioned earlier, there are strong polar interactions between the TPS, C and B fillers due to the hydroxyl group composition within their structure. The interface interactions between B and C help to pull the B platelets away from its tactoid structure, making a wider space for the TPS chains intercalation. This factor explains why the TPS/4B1C performed the greatest mechanical properties compared to other TPS bio-composite films.

3.6. Fractured Surface Morphology of the TPS Bio-Composite and TPS Hybrid Bio-Composite Films as Observed through Scanning Electron Microscopy (SEM) Analyses

The tensile fractured surface morphologies of the TPS, TPS bio-composite and TPS hybrid bio-composite were studied by Scanning Electron Microscopy (SEM). Figure 12a shows that the fractured surface of the TPS films was uniform and no residual for the starch granule structure can be observed. The smooth surface indicates that the origin starch granule structure was completely disrupted and broken when starch was plasticized with glycerol and water at a high temperature. The TPS film possesses a smooth surface showing that the unfilled TPS underwent a very low degree of matrix deformation and easily fracture upon the application of a small tensile load/force. Both TPS/5C (Figure 12e) and TPS/5B (Figure 12b) also show a clean and smooth fracture surface, indicating that both types of the TPS bio-composites have experienced some degree of matrix deformation upon tensile loading, but not much. For the TPS/5B bio-composite film, no image of B filler could be captured due to the very small size of the nano-bentonite. However, for the TPS/5C bio-composite film, the fractured surface shows some aggregation of C (circle in red color) inside the matrix of the TPS due to the strong intramolecular hydrogen bonding between the particles of C. This filler–filler interaction caused aggregation of C in the TPS matrix and prohibited the bio-composite to achieve a higher tensile strength. This was in line with the study of Zhang et al. where the nanocellulose agglomeration phenomenon in the host TPS had caused uneven loading distribution, void formation in the matrix and subsequently led to premature failure [33]. Small voids were observed due to the agglomerated C fiber pullout from the matrix phase. The C fiber was pulled out in the same direction of the tensile force applied. By adding 1% of C as co-filler with B, the fractured surface morphology of the TPS film has changed drastically. The surface roughness increased due to the occurrence of significant matrix deformation. This tallies with the tensile data which indicate that the elongation at break of this TPS/4B1C film was higher than the unfilled TPS and TPS bio-composite with single filler. However, there is no agglomeration and aggregation of C spotted in the SEM images of the TPS/4B1C, indicating that the C particles have been well dispersed inside the TPS matrix due to the good compatibility and interactions with the matrix and B filler. Furthermore, there is no fiber pullout that can be observed in the surface of the TPS/4B1C bio-composite film. This could be attributed to better interface interaction between C and the matrix with the presence of B. The C fiber was well wetted and embedded in the TPS matrix and led to high contact surface area between C and the TPS. In addition, the appearance of a tiny fibril structure bridging the small voids (blue color circle in Figure 12g,h) was noticed, suggesting that the C filler has been vertically oriented to the force applied during the tensile test. A similar structure of nanocellulose bridging in TPS was also observed by Li et al. [40].

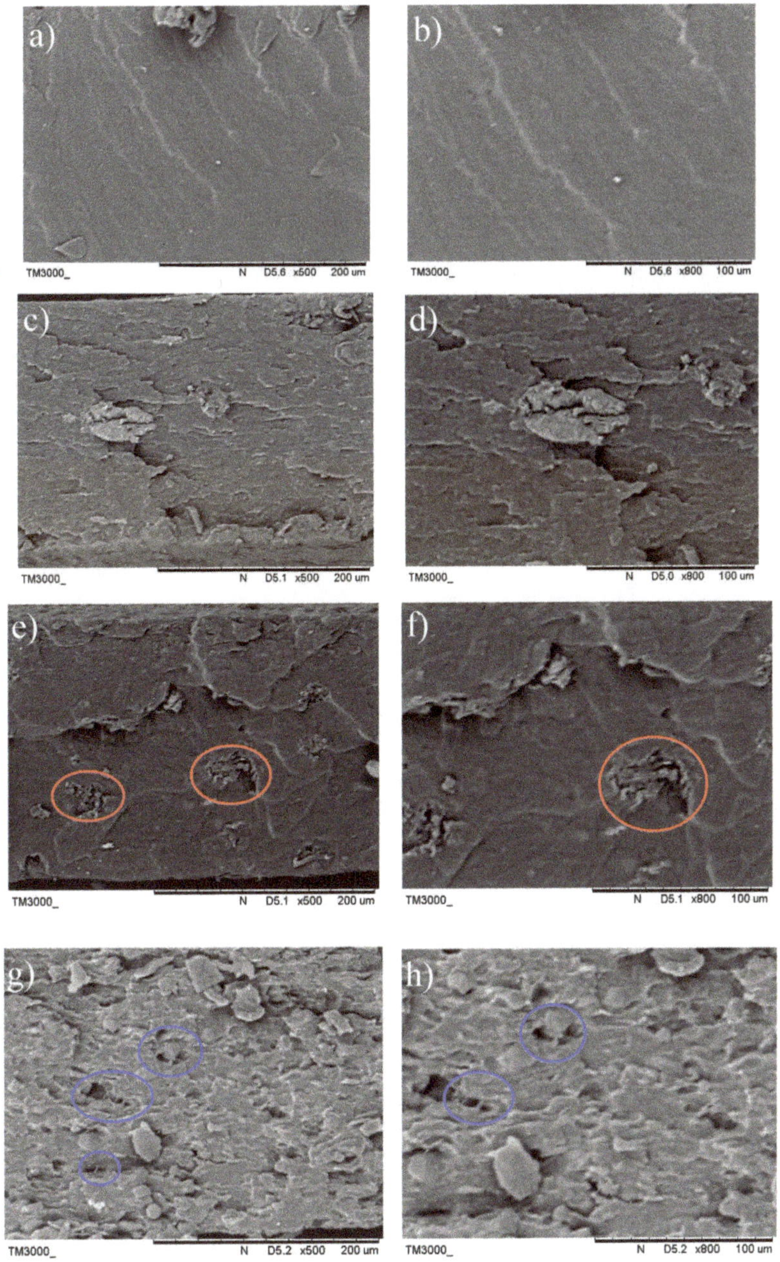

Figure 12. SEM images of the tensile fractured surface of the (**a**,**b**) TPS films, (**c**,**d**) TPS/5B, (**e**,**f**) TPS/5C, (**g**,**h**) and TPS/4B1C at 500× magnification (**left**) and 800× magnification (**right**).

4. Conclusions

Microcrystalline cellulose (C) was extracted from the raw OPEFB fiber and nano-bentonite (B) was ultra-sonicated to reduce the large tactoid structure. The changes in structure and morphology of the fillers from their original form after being subjected to chemical treatment and the ultra-sonication process have been witnessed and confirmed through FTIR, XRD and SEM analyses. Both fillers were used to form hybrid fillers in the production of the TPS hybrid bio-composite film. The TPS bio-composite films and TPS hybrid bio-composite films were produced by the casting method and the effects of the OPEFB microcrystalline cellulose (C)/nano-bentonite (B) ratio on the mechanical properties of the resultant TPS bio-composite films were studied. The highest tensile strength, Young's modulus and tensile toughness was achieved when the TPS was incorporated with 4 wt% B and 1% of C filler (TPS/4B1C). The use of this hybrid filler system resulted in a more significant enhancement in the tensile properties of the film as compared to the single B or C filler. The results signify that B/C hybrid fillers have brought a positive synergistic effect to the TPS matrix. The good compatibility between C, B and TPS was observed through the FTIR, XRD and SEM analyses. The findings demonstrate that the combination of the OPEFB microcrystalline cellulose and nano-bentonite can produce an efficient hybrid filler system that allows for greater enhancement in the mechanical properties of the TPS-based film for packaging applications. Our work signifies that this hybridization approach is a powerful method to toughen the TPS film. This encourages further exploration on the potential of this hybrid filler system for toughening other bio-polymers/polymers. More importantly, all the ingredients used to form the hybrid TPS bio-composite are environmentally friendly and renewable, thus contributing to the development of sustainable packaging materials.

Author Contributions: Conceptualization, A.F.O. and S.A.A.; data curation, D.S.L. and A.A.A.; formal analysis, D.S.L. and A.A.A.; investigation, A.F.O. and D.S.L.; methodology, A.F.O., D.S.L. and I.I.; project administration, A.F.O., S.A.A. and I.I.; software, M.N.A.S. and A.U.-H.; writing—original draft, D.S.L.; writing—review and editing, A.F.O., M.N.A.S. and A.U.-H. All authors have read and agreed to the published version of the manuscript.

Funding: The author would like to acknowledge the support from the Fundamental Research Grant Scheme (FRGS) under a grant no: FRGS/1/2019/TK10/UNIMAP/03/2 from the Ministry of Education Malaysia.

Institutional Review Board Statement: Not applicable.

Informed Consent Statement: Not applicable for studies not involving humans.

Data Availability Statement: The data presented in this study are available on request from the corresponding author.

Conflicts of Interest: The authors declare no conflict of interest.

References

1. Rambonnet, L.; Vink, S.C.; Land-Zandstra, A.M.; Bosker, T. Making citizen science count: Best practices and challenges of citizen science projects on plastics in aquatic environments. *Mar. Pollut. Bull.* **2019**, *145*, 271–277. [CrossRef] [PubMed]
2. Vieira, M.G.A.; Silva, M.A.D.; Santos, L.O.D.; Beppu, M.M. Natural-based plasticizers and biopolymer films: A review. *Eur. Polym. J.* **2011**, *47*, 254–263. [CrossRef]
3. Galgani, L.; Beiras, R.; Galgani, F.; Panti, C.; Borja, A. Editorial: Impacts of Marine Litter. *Front. Mar. Sci.* **2019**, *6*, 208. [CrossRef]
4. Glenn, G.M.; Orts, W.; Imam, S.; Chiou, B.S. *Starch Polymers*; Elsevier: San Diego, CA, USA, 2014; pp. 421–452.
5. Ribba, L.; Garcia, N.L.; Accorso, N.D.; Goyanes, S. *Starch-Based Materials in Food Packaging*; Elsevier: London, UK, 2018; pp. 37–76.
6. Osman, A.F.; Siah, L.; Alrashdi, A.A.; Ul-Hamid, A.; Ibrahim, I. Improving the Tensile and Tear Properties of Thermoplastic Starch/Dolomite Biocomposite Film through Sonication Process. *Polymers* **2021**, *13*, 274. [CrossRef]
7. Husseinsyah, S.; Zailuddin, N.L.I.; Osman, A.F.; Lili, C.; Al-rashdi, A.A.; Alakrach, A. Methyl methacrylate (MMA) treatment of empty fruit bunch (EFB) to improve the properties of regenerated cellulose biocomposite films. *Polymers* **2020**, *12*, 2618. [CrossRef]
8. Osman, A.F.; Ashafee, A.M.T.; Adnan, S.A.; Alakrach, A. Influence of Hybrid Cellulose/Bentonite Fillers on Structure, Ambient and Low Temperature Tensile Properties of Thermoplastic Starch Composites. *Polym. Eng. Sci.* **2020**, *60*, 810–822. [CrossRef]
9. Amanda, G.; Tadeusz, S.; Magdalena, Z. Thermoplastic starch/wood biocomposites processed with deep eutectic solvents. *Compos. Part A Appl. Sci. Manuf.* **2019**, *121*, 517–524. [CrossRef]
10. El, H.S.; Colussi, R.; Deon, V.G.; Pinto, V.Z.; Villanova, F.A.; Carreño, N.L.V.; Dias, A.R.G.; Zavareze, E.D.R. Films based on oxidized starch and cellulose from barley. *Carbohydr. Polym.* **2015**, *133*, 644–653. [CrossRef]
11. Emilio, P.P.; Jorge, C.C.P.; Víctor, M.M.H.; Iván, A.E.M.; Raciel, J.E.L.; Luis, C.G. Thermoplastic Starch (TPS)-Cellulosic Fibers Composites: Mechanical Properties and Water Vapor Barrier: A Review, Composites from Renewable and Sustainable Materials. *IntechOpen* **2016**. [CrossRef]

12. Hermansyah, H.; Carissa, R.; Faiz, M.B.; Deni, P. Food Grade Bioplastic Based on Corn Starch with Banana Pseudostem Fibre/Bacterial Cellulose Hybrid Filler. *Adv. Mater. Res.* **2014**, *997*, 158–168. [CrossRef]
13. Fazeli, M.; Florez, J.P.; Simão, R.A. Improvement in adhesion of cellulose fibers to the thermoplastic starch matrix by plasma treatment modification. *Compos. Part B Eng.* **2019**, *163*, 207–216. [CrossRef]
14. Chen, J.; Chen, F.; Meng, Y.; Wang, S.; Long, Z. Oxidized microcrystalline cellulose improve thermoplastic starch-based composite films: Thermal, mechanical and water-solubility properties. *Polymer* **2019**, *168*, 228–235. [CrossRef]
15. Zailuddin, N.L.I.; Osman, A.F.; Rahman, R. Morphology, mechanical properties, and biodegradability of all-cellulose composite films from oil palm empty fruit bunch. *SPE Polym.* **2020**, *1*, 4–14. [CrossRef]
16. Lamaming, J.; Hashim, R.; Sulaiman, O.; Leh, C.P.; Sugimoto, T.; Nordin, N.A. Cellulose nanocrystals isolated from oil palm trunk. *Carbohydr. Polym.* **2015**, *127*, 202. [CrossRef]
17. Chieng, B.W.; Lee, S.H.; Ibrahim, N.A.; Then, Y.Y.; Loo, Y.Y. Isolation and Characterization of Cellulose Nanocrystals from Oil Palm Mesocarp Fiber. *Polymers* **2017**, *9*, 355. [CrossRef]
18. Bano, S.; Negi, Y.S. Studies on cellulose nanocrystals isolated from groundnut shells. *Carbohydr. Polym.* **2017**, *157*, 1041–1049. [CrossRef] [PubMed]
19. Rashid, S.; Dutta, H. Characterization of nanocellulose extracted from short, medium and long grain rice husks. *Ind. Crops Prod.* **2020**, *154*, 112627. [CrossRef]
20. Haafiz, M.K.M.; Hassan, A.; Khalil, H.P.S.A.; Owolabi, A.F.; Marliana, M.M.; Arjmandi, R.; Inuwa, I.M.; Fazita, M.M.N.; Hussin, M.W. *Cellulose-Reinforced Nanofibre Composites*; Woodhead Publishing: Duxford, UK, 2017; pp. 241–259.
21. Ahmad Fauzi, A.A.; Osman, A.F.; Abdullah, M.A.A.; Mandal, S.; Ananthakrishan, R. Ethylene vinyl acetate nanocomposites with hybrid silicate nanofillers of destabilized natural and commercial bentonites and organomontmorillonites. *Vinyl Addit. Technol.* **2019**, *25*, 396–411. [CrossRef]
22. Miz, M.E.; Akichoh, H.; Berraaouan, D.; Salhi, S.; Tahani, A. Chemical and physical characterization of moroccan bentonite taken from nador (north of Morocco). *Am. J. Chem.* **2017**, *7*, 105–112. [CrossRef]
23. Berraaouan, D.; Elmiz, M.; Essifi, K.; Salhi, S.; Tahani, A. Adsorption of carvacrol on modified bentonite in aqueous solutions. *Mater. Today Proc.* **2020**, *31*, 28–32. [CrossRef]
24. Rasheed, U.; Ain, Q.U.; Yaseen, M.; Fan, X.; Yao, Z.; Tong, Z.; Liu, B. Modification of bentonite with orange peels extract and its application as mycotoxins' binder in buffered solutions and simulated gastrointestinal fluids. *J. Clean. Prod.* **2020**, *267*, 122105. [CrossRef]
25. Okahisa, Y.; Furukawa, Y.; Ishimoto, K.; Narita, C.; Intharapichai, K.; Ohara, H. Comparison of cellulose nanofiber properties produced from different parts of the oil palm tree. *Carbohydr. Polym.* **2018**, *198*, 313–319. [CrossRef] [PubMed]
26. Segal, L.; Creely, J.; Martin, A., Jr.; Conrad, C.M. An empirical method for estimating the degree of crystallinity of native cellulose using the X-ray diffractometer. *Sage* **1959**, *29*, 786–794. [CrossRef]
27. Nordin, N.A.; Sulaiman, O.; Hashim, R.; Mohamad Kassim, M.H. Oil Palm Frond Waste for the Production of Cellulose Nanocrystals. *Phys. Sci.* **2017**, *28*, 115–126. [CrossRef]
28. Kargarzadeh, H.; Ioelovich, M.; Ahmad, I.; Thomas, S.; Dufresne, A. *Handbook of Nanocellulose Cellulose Nanocomposites*; Wiley, V.C.H: New Jersey, NJ, USA, 2017; pp. 1–51.
29. Abdul Hamid, A.R.; Osman, A.F.; Mustafa, Z.; Mandal, S.; Ananthakrishnan, R. Tensile, fatigue and thermomechanical properties of poly (ethylene-co-vinyl acetate) nanocomposites incorporating low and high loadings of pre-swelled organically modified montmorillonite. *Polym. Test.* **2020**, *85*, 106426. [CrossRef]
30. Hamid, A.R.; Osman, A.F. Effects of Ultrasonication Time on Thermal Stability and Swelling Behaviour of the Commercial Organo-Montmorillonite (O-MMT). *Int. Eng. Technol.* **2018**, *7*, 288–291. [CrossRef]
31. Xie, F.; Pollet, E.; Halley, P.J.; Avérous, L. Starch-based nano-biocomposites. *Prog. Polym. Sci.* **2013**, *38*, 1590–1628. [CrossRef]
32. Xie, F.; Avérous, L.; Halley, P.J.; Liu, P. *Biocomposites*; Woodhead Publishing: Cambridge, UK, 2015; pp. 53–92.
33. Zhang, B.; Huang, C.; Zhao, H.; Wang, J.; Yin, C.; Zhang, L.; Zhao, Y. Effects of Cellulose Nanocrystals and Cellulose Nanofibers on the Structure and Properties of Polyhydroxybutyrate Nanocomposites. *Polymers* **2019**, *11*, 2063. [CrossRef] [PubMed]
34. Shen, R.; Xue, S.; Xu, Y.; Liu, Q.; Feng, Z.; Ren, H.; Zhai, H.; Kong, F. Research Progress and Development Demand of Nanocellulose Reinforced Polymer Composites. *Polymers* **2020**, *12*, 2113. [CrossRef] [PubMed]
35. Madhumitha, G.; Fowsiya, J.; Mohana, R.S.; Thakur, V.K. Recent advances in starch–clay nanocomposites. *Int. J. Polym. Anal. Charact.* **2018**, *23*, 331–345. [CrossRef]
36. Bhattacharya, M. Polymer Nanocomposites-A Comparison between Carbon Nanotubes, Graphene, and Clay as Nanofillers. *Materials* **2016**, *9*, 262. [CrossRef]
37. Zakuwan, S.Z.; Ahmad, I. Synergistic Effect of Hybridized Cellulose Nanocrystals and Organically Modified Montmorillonite on Kappa-Carrageenan Bionanocomposites. *Nanomaterials* **2018**, *8*, 874. [CrossRef]
38. Miles, K.B.; Ball, R.L.; Matthew, H.W.T. Chitosan films with improved tensile strength and toughness from N-acetyl-cysteine mediated disulfide bonds. *Carbohydr. Polym.* **2016**, *139*, 1–9. [CrossRef]
39. Yin, P.; Zhou, W.; Zhang, X.; Guo, B.; Li, P. Bio-Based Thermoplastic Starch Composites Reinforced by Dialdehyde Lignocellulose. *Molecules* **2020**, *25*, 3236. [CrossRef] [PubMed]
40. Li, J.; Zhou, M.; Cheng, G.; Cheng, F.; Lin, Y.; Zhu, P.X. Fabrication and characterization of starch-based nanocomposites reinforced with montmorillonite and cellulose nanofibers. *Carbohydr. Polym.* **2019**, *210*, 429–436. [CrossRef]

41. Castillo, L.A.; Lopez, O.V.; Garcia, M.A.; Barbosa, S.E.; Villar, M.A. Crystalline morphology of thermoplastic starch/talc nanocomposites induced by thermal processing. *Heliyon* **2019**, *5*, e01877. [CrossRef] [PubMed]
42. Pozo, C.; Llamazares, R.S.; Barral, L.; Catano, J.; Muller, N.; Restrepo, I. Study of the structural order of native starch granules using combined FTIR and XRD analysis. *Polym. Res.* **2018**, *25*, 266. [CrossRef]
43. Travalini, A.P.; Lamsal, B.; Magalhaes, W.L.E.; Demiate, I.M. Cassava starch films reinforced with lignocellulose nanofibers from cassava bagasse. *Int. J. Biol. Macromol.* **2019**, *139*, 1151–1161. [CrossRef] [PubMed]
44. Ren, J.; Dang, K.M.; Pollet, E.; Averous, L. Preparation and Characterization of Thermoplastic Potato Starch/Halloysite Nano-Biocomposites: Effect of Plasticizer Nature and Nanoclay. *Content. Polym.* **2018**, *10*, 808. [CrossRef] [PubMed]
45. Dufresne, A.; Castaño, J. Polysaccharide nanomaterial reinforced starch nanocomposites: A review. *Starch-Stärke* **2017**, *69*, 1500307. [CrossRef]
46. Zhang, C.W.; Nair, S.S.; Chen, H.; Yan, N.; Farnood, R.; Li, F.Y. Thermally stable, enhanced water barrier, high strength starch bio-composite reinforced with lignin containing cellulose nanofibrils. *Carbohydr. Polym.* **2020**, *230*, 115626. [CrossRef] [PubMed]
47. Fourati, Y.; Magnin, A.; Putaux, J.L.; Boufi, S. One-step processing of plasticized starch/cellulose nanofibrils nanocomposites via twin-screw extrusion of starch and cellulose fibers. *Carbohydr. Polym.* **2020**, *229*, 115554. [CrossRef] [PubMed]
48. Lara, S.C.; Salcedo, F. Gelatinization and retrogradation phenomena in starch/montmorillonite nanocomposites plasticized with different glycerol/water ratios. *Carbohydr. Polym.* **2016**, *151*, 206–212. [CrossRef] [PubMed]

polymers

MDPI

Article

Nano/Micro Hybrid Bamboo Fibrous Preforms for Robust Biodegradable Fiber Reinforced Plastics

Junsik Bang [1], Hyunju Lee [1], Yemi Yang [1], Jung-Kwon Oh [1,2] and Hyo Won Kwak [1,2,*]

[1] Department of Agriculture, Forestry and Bioresources, College of Agriculture & Life Sciences, Seoul National Universtiy, 1 Gwanak-ro, Gwanak-gu, Seoul 08826, Korea; bangsik.e@snu.ac.kr (J.B.); leehj3030@naver.com (H.L.); yemi98@snu.ac.kr (Y.Y.); jungoh@snu.ac.kr (J.-K.O.)

[2] Research Institute of Agriculture and Life Sciences, Seoul National University, 1 Gwanak-ro, Gwanak-gu, Seoul 08826, Korea

* Correspondence: bk0502@snu.ac.kr

Abstract: The focus on high-strength and functional natural fiber-based composite materials is growing as interest in developing eco-friendly plastics and sustainable materials increases. An eco-friendly fibrous composite with excellent mechanical properties was prepared by applying the bamboo-derived nano and microfiber multiscale hybridization phenomenon. As a result, the cellulose nanofibers simultaneously coated the micro-bamboo fiber surface and adhered between them. The multiscale hybrid phenomenon implemented between bamboo nano and microfibers improved the tensile strength, elongation, Young's modulus, and toughness of the fibrous composite. The enhancement of the fibrous preform mechanical properties also affected the reinforcement of biodegradable fiber-reinforced plastic (FRP). This eco-friendly nano/micro fibrous preform can be extensively utilized in reinforced preforms for FRPs and other green plastic industry applications.

Keywords: bamboo fiber; cellulose nanofiber; adhesion; multiscale hybridization; fibrous composite

Citation: Bang, J.; Lee, H.; Yang, Y.; Oh, J.-K.; Kwak, H.W. Nano/Micro Hybrid Bamboo Fibrous Preforms for Robust Biodegradable Fiber Reinforced Plastics. *Polymers* **2021**, *13*, 636. https://doi.org/10.3390/polym13040636

Academic Editor: José Miguel Ferri

Received: 4 February 2021
Accepted: 18 February 2021
Published: 20 February 2021

1. Introduction

Problems related to petroleum resource depletion and the potential dangers of microplastics are continuously raised, and the need for eco-friendly materials continues to rise. The consumption of reusable or biodegradable natural polymers is essential, especially when taking raw materials into account. Since ancient times, natural fibers, such as wool, silk, and cotton, have been used in clothing, and cellulose, which is a wood-based polymer, has been actively employed in the paper industry. In addition, since the inception of FRPs, studies using natural fibers as reinforcing agents have been repeatedly reported.

Natural fibers are not only abundant and inexpensive, but are also biodegradable due to their nature-derived properties. Therefore, researchers have examined their application in FRP reinforcement [1]. However, the hydrophilic properties of natural fiber reduce compatibility with synthetic polymer matrices. Chemical modification of natural fiber surfaces has been attempted to improve the compatibility between natural fibers andmatrix polymers [2]. Although, this surface modification process mostly inhibits the economical and environmentally-friendly properties of natural fibers.

Nanocellulose is a nanomaterial obtained from wood and agricultural byproducts [3–6], as well as from bacteria [7,8], and has a diameter numbering tens of nanometers [9]. This nanomaterial has low density, high strength, and excellent transparency [10]. As a result, research is being performed to utilize nanocellulose in various applications ranging from food packaging materials, composite-material nanofillers, electronic devices, and biomaterials [11]. Due to the nanomorphology of nanocellulose and the resulting strong mechanical strength, composite materials with various general-purpose polymers, such as polyolefin and polyester [9], are also being developed [12]. In addition, as biodegradable polyesters such as polylactic acid (PLA), polycaprolactone (PCL), and polybutylene succinate (PBS) were developed, studies to apply nanocellulose to compensate for the insufficient mechanical strength were also actively conducted [7]. However, nanocellulose also has hydrophilic properties similar to natural fibers and wood-based biomass, which are their raw materials. Thus, insufficient compatibility with a polymer matrix is considered a limitation [13]. Biomimetic methods have been tried to overcome this drawback. When bacteria biosynthesized nanocellulose, a water-soluble polymer, such as hydroxyethyl cellulose or polyvinyl alcohol, was introduced to prepare a nanocellulose/polymer composite [7,14]. These biomimetic nanocellulose/polymer composites showed higher tensile properties than composites prepared by simple mixing [12]. However, this method has the disadvantage of interfering with the cellulose biosynthesis of bacteria, and mass production is impossible [15,16].

Recently, attempts have been made to use nanocellulose as a "green" binder in the manufacture of preforms for hierarchical composites. Lee et al. used bacterial cellulose as a binder to prepare sisal fiber preforms for the reinforcement of polyacrylated epoxidized soybean oil (AESO) [16]. As a result, it is evident that the mechanical properties of the preform improved significantly. The mechanical properties of poly AESO/sisal preforms manufactured by thermal curing following vacuum-assisted resin injection are also extensively improved. It is known that the excellent binding and adhesion ability of these nanocelluloses can be clearly expressed in natural fibers, including sisal and silk fibers [17,18]. However, most of the research results indicate that bacterial cellulose was used as a binder [19], and there has been no effort to manufacture a hierarchical fibrous-composite preform using microfibers and nanocellulose produced from the same tree species.

Bamboo is a plant resource that is cultivated worldwide [20], especially in Asia and South America. Since ancient times, bamboo has been steadily used in furniture and construction materials due to its rigid mechanical properties [20]. In addition, as bamboo fiber extraction technology advances, attempts to use it as a sustainable raw material for high-performance composite materials and advanced carbon materials are steadily progressing. [21–23]. Furthermore, various types of nanocellulose manufacturing technologies utilize bamboo fibers as raw materials [24–26]. Therefore, in the development of eco-friendly composite materials, it is very attractive to develop a facile preforming process using the hierarchical nano/micro morphology of bamboo fiber [27,28].

This study's primary purpose is to fabricate a fibrous composite preform with improved mechanical properties through a multiscale, hybridizing process using bamboo-derived microfibers and cellulose nanofibers. First, the optimal preparation conditions (temperature and time) were established to minimize the energy consumption required to manufacture the fibrous composite preform. The physicochemical properties of the prepared micro/nano bamboo-fibrous composite preform were analyzed according to the nanocellulose content, and the changes in mechanical properties were confirmed accordingly. Finally, FRP was prepared using PBS, which is an actual biodegradable polyester, and changes in mechanical properties of FRP due to the reinforcing effect of nano/microfiber preforms were observed.

2. Materials and Methods

2.1. Materials

The bamboo fiber (BF) used in the experiment was purchased from Studio Buntu (Seoul, Korea) and cut into 2 cm lengths; a bamboo cellulose nanofiber (CNF) dispersion (1 wt.%) was purchased from Chuetsu Pulp & Paper Co., Ltd. (Toyama, Japan); Direct Red 80 (analytical grade, dye content, 25%) was purchased from Sigma Aldrich (Yongin, Korea); and polybutylene succinate (PBS) pellets (BG5000M, 100% PBS) were kindly provided by Anko Bioplastics (Wonju, Korea).

2.2. Preparation of BF/CNFs

The bamboo fiber and CNF composite were readied using a thermocompression method. First, 2 g of bamboo fiber was immersed in 50 mL of a CNF dispersion solution of different concentrations (0–2 $w \cdot v^{-1}$ %), stirred sufficiently, and then adequately dehydrated by hand in a 3 × 10 cm frame. Afterward, the BF/CNF mixture was thermocompressed with a mini hot press (QM900S, Qmasys, Uiwang, Korea). BF/CNF preforms were prepared by varying the compression temperatures at 25 °C, 60 °C, 80 °C, 120 °C, 160 °C, and 180 °C to find the optimum thermocompression temperature. Moreover, BF/CNF preforms were produced with different compression times for 1, 2, 3, 5, 10, and 15 min to optimally establish the thermocompression time. The thermocompression pressure in all experiments was fixed at 20 MPa. BF/CNFs with different CNF contents were denoted as BF/CNFn, where n denotes the wt.% of CNFs based on BF weight. Ten samples of BF/CNFs were prepared through single preforming process and used for tensile testing and characterization.

2.3. Preparation of BF/CNFs Preform Reinforced PBS Composite

In preparing the FRP sample, a bamboo fiber preform and a PBS pellet were directly impregnated using a hot press. At this time, the process temperature was set at 120 °C, and the pressure was fixed at 5 MPa. The PBS pellet and bamboo fiber preform were put in a ratio of 5:5 in reference to the total weight. The FRP composites were named PBS-BF/CNFn. Subsequently, distilled water was used to wash off unabsorbed dye molecules present in the BF/CNF, and an image of the BF/CNF preform was obtained.

2.4. Characterization of BF/CNFs

Optical images of BF and CNF were obtained using a Galaxy Note 10 (Samsung, Seoul, Korea). The morphology of the BFs was observed through a Field emission scanning electron microscope (FE-SEM, SUPRA 55VP, Carl Zeiss, Oberkochen, Germany), and the morphology of the CNFs was detected using a Field emission transmission electron microscope (FE-TEM, JEM2100F, JEOL, Tokyo, Japan). The thickness of the BF/CNF preform was measured using a thickness measuring device (CD-20CP, Mitutoyo, Kawasaki, Japan). The preform's moisture content was measured using a moisture meter (MB120, OHAUS, New Jersey, United States). The BF/CNF preform was immersed in 60 mL of Direct Red 80 dye solution (0.0025 wt.%) and incubated for 30 min to visualize the distribution of CNFs within the BF nonwoven structure. Next, distilled water was applied to remove the unreacted dye molecules from the sample until no red color was observed on the paper towel, which was pressed onto the BF/CNF fibrous composite sample.

The BF/CNF preform's physical structure was confirmed by measuring the porosity, and the absolute density was determined by He pycnometry (AccuPyc 1330, Micromeritics, Georgia, United States). Samples were weighed before being loaded into the measuring chamber. The nonwoven preform's apparent density (ρe) was calculated from the mass and sheath volume of the nonwoven preform. The porosity of the preform was computed with the following Equation (1):

$$Porosity\ (\%) = \left\{ \frac{1 - Enveloped\ density}{Absolute\ Density} \right\} \times 100 \quad (1)$$

The morphological properties of the BF/CNF preform were observed through FE-SEM. A pull-to-break experiment was performed using a universal testing machine (34SC-1, Instron, Massachusetts, United States) to evaluate the BF/CNF preform's mechanical properties. A preform sample with a width of 10 mm, a length of 20 mm, and a thickness of 0.8 mm was utilized. A 1 kN load cell was applied, and the experiment was carried out at a tensile speed of 0.5 $mm \cdot min^{-1}$. The experiment was repeated five times to obtain the average value and standard deviation of tensile strength, modulus of elasticity, elongation, and toughness. The PBS-BF/CNF FRP mechanical properties were measured through a tensile test, and the experiment was performed under the same conditions as the BF/CNF preform.

3. Results

3.1. Morphology of Bamboo Fibers and Bamboo-Derived Nanocellulose

Figure 1 displays an optical image and electron microscope image of BF and its derived cellulose nanofibrils (CNFs). The single microfiber consists of many elementary fibrils in bamboo fibers, and rough and flat oval-shaped fibers can be observed. The rough surface of these bamboo fibers is known to cause the mechanical reinforcement effect of FRP by increasing the interfacial active bonding sites between the fiber surface and the polymer matrix. However, the effect is insignificant due to the roughness on the micrometer scale and the extreme hydrophilicity difference between the matrix (hydrophobic polymer) and filler (hydrophilic fiber) [29]. On the other hand, bamboo nanofibers are dispersed in water and have viscosity like a sticky glue substance. In addition, it can be seen from the transmission electron microscope image that it is composed of nanofibers with a diameter of about 10–20 nm. The viscous flow of CNFs means that the aqueous CNFs can act as a general adhesive binder and penetrate the fiber surface [17,30]. Additionally, the CNF's long aspect ratio with nanofibrillar morphology is expected to enable the CNFs to impart a nano brush structure to the BF fiber surface in the BF/CNF system and enable bonding between the BF fibers.

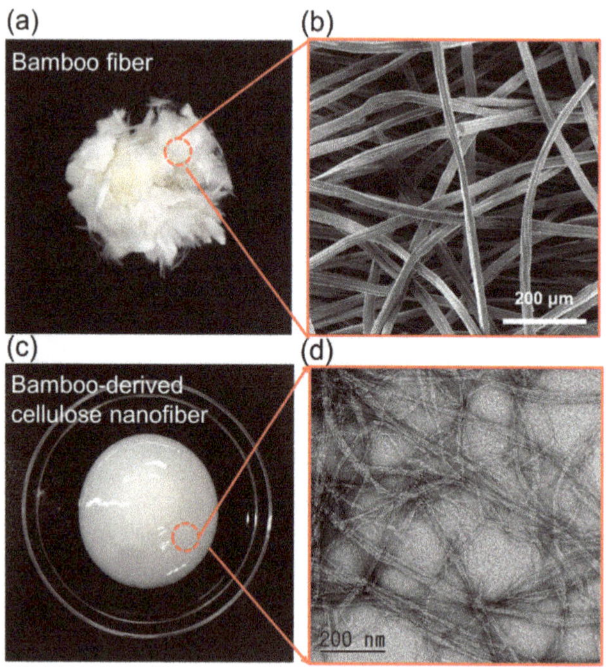

Figure 1. Morphology of bamboo fibers (BFs) and cellulose nanofibrils (CNFs). (**a**) Optical and (**b**) FE-SEM images of BFs. (**c**) Optical and (**d**) FE-TEM images of CNFs.

3.2. Optimization of the Preform Bamboo Fibrous Composite Process

Manufacture of preform-type fiber reinforcement not only minimizes separation due to the difference in compatibility between the fiber and the matrix polymer, but also makes it possible to easily manufacture fiber-reinforced plastic materials through methods such as laminating and impregnation of polymer melt [31,32]. When preparing a fibrous composite preform using nanocellulose, the manufacturing process proceeds to wet the CNF aqueous solution to microfibers, dehydration, and thermocompression processes. Among these processes, thermocompression greatly influences the thickness and moisture content of the fiber preform. In addition, to design energy-efficient polymer processing, it is necessary to optimize the heat treatment temperature and compression time in the thermocompression process. To find the optimum preforming condition for the rigid fibrous preform along with an effective dehydration process, the physical properties of the BF/CNFs determined according to the thermocompression temperature were confirmed through the thickness and moisture content (Figure 2). As seen from the figure, the color change of BF/CNF did not occur significantly until heat treatment at 120 °C, whereas yellowing occurred when thermal compression was performed at high temperatures of 160 °C and 200 °C. This yellowing phenomenon of the BF/CNF preform due to the high heat treatment temperature was caused by thermal degradation of cellulose [33]. The more severe the yellowing, the more degradation proceeded [34]. Since sufficient dehydration does not occur in the low temperature condition of 20–80 °C, the thickness of the preform is thick and not uniform, which means that it does not provide sufficient structural stability when used as a reinforcing preform for fiber reinforced plastics. As the heat treatment temperature increases, the thickness and water content of the BF/CNF decreases, meaning that heat treatment at a sufficient temperature is required for dehydration of the absorbed water molecules in the wetting process. Through a sufficient dehydration process, the gap between BF and CNFs becomes close, which means that various types of interactions, including hydrogen bonds between nano and microfibers, can easily occur. In summarizing the above results, it was found that 120 °C could be an optimal temperature condition necessary to minimize the preform's moisture content while preventing thermal decomposition of both BFs and CNFs.

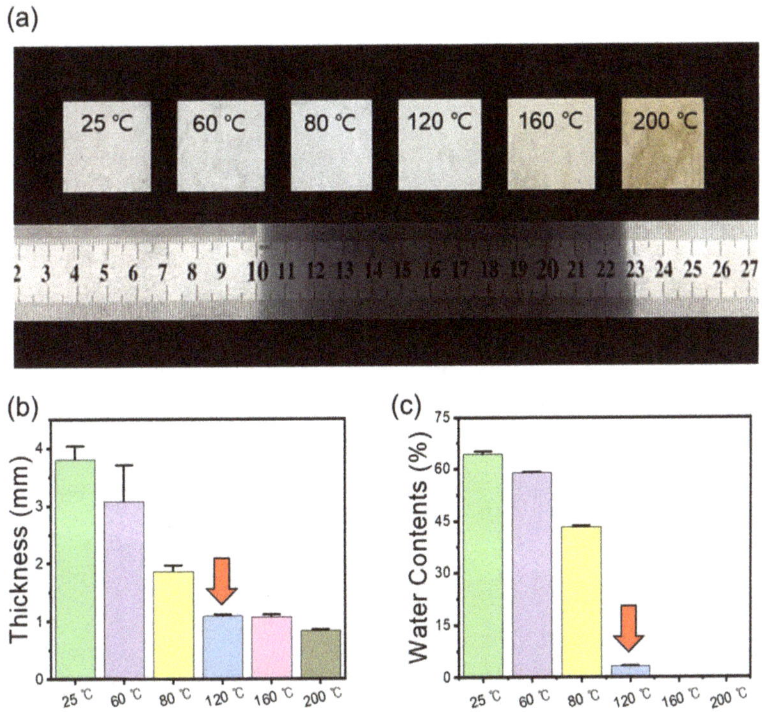

Figure 2. Effect of thermocompressing temperature on the physical properties of BF/CNF fibrous preforms. (**a**) Optical images, (**b**) thickness, and (**c**) water contents of BF/CNF fibrous preforms.

In fibrous preform manufacturing, the pressing time is a crucial factor along with the pressing temperature. The preform was manufactured by varying the thermocompression time from 0 to 15 min at 120 °C to establish an energy-friendly manufacturing process by minimizing the time while ensuring the BF/CNF composite's structural stability. Figure 3 depicts the thickness and moisture content results of the BF/CNF preform according to the thermocompression time. The optimum temperature is confirmed as 120 °C because the color change does not appear during the 15-min thermocompression process. As the thermal compression time lengthens, the moisture content and thickness of the preform fall because the water molecules that participate in the wetting process of BF/CNF evaporate. This drop in moisture content and thickness decreases continuously until a thermal compression time of 10 min when equilibrium is reached. Compared with the initial BF/CNF preforms, which only proceeded with 1-min thermal compression (1.47 mm thickness and 21% moisture content), the preforms that underwent 10 min of thermal compression reduced the moisture content to 1.8%. The thickness was reduced to 1.18 mm because the BF/CNFs could penetrate the evaporated water space. Taken together, the optimal condition for minimizing energy consumption, and ensuring sufficient dehydration and a rigid preform structure, is thermal compression bonding at 120 °C for 10 min.

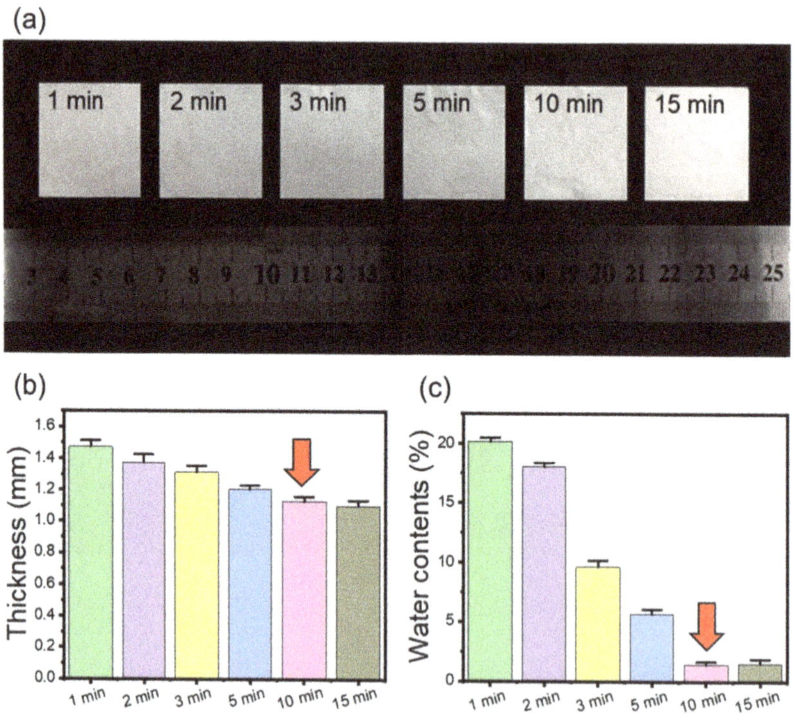

Figure 3. Effect of thermocompression time on the physical properties of BF/CNF fibrous preforms. (**a**) Optical images, (**b**) thickness, and (**c**) water contents of BF/CNF fibrous preforms.

3.3. Effect of CNF Addition on the Physical Properties of BF/CNF Fibrous Preforms

It is essential to deposit a nanomaterial on a microfibrous structure when preparing a nano/micro hybrid fibrous preform. The nanomaterial deposition process can be primarily carried out through impregnation and spraying methods [35]. Separate equipment is required for the spraying method, and penetration into the fiber structure may be challenging. BF/CNFs hybrid preforms can be constructed through a dehydration technique after thoroughly mixing and stirring the CNF aqueous solution and BF in the immersion method. This immersion mixing method also has the advantage of being able to easily control the added CNFs in the BF/CNF framework by varying the concentration of aqueous CNFs. However, a sufficient number of CNFs may not penetrate the BF nonwoven structure when a larger volume of aqueous CNFs is added over the microfiber's absorption properties or when high-viscosity CNFs are input according to a high concentration. Figure 4 illustrates the dyeing images and weight results of BF/CNF fibrous preforms prepared by varying the content of CNFs. Even with the same bamboo cellulose, there is a difference in dyeing kinetics depending on the fiber morphology. For micrometer-scale BFs, an adequate dyeing process does not undergo because the dye molecules cannot be adsorbed on the BF microfiber's surface and diffuse into the interior in a short dyeing time of 10 min [36]. On the other hand, for nanoscale CNFs with a high surface area of the nanofibrillar morphology, the dye molecules easily and rapidly adsorb onto the CNF surface, and the color remains satisfactory even after the washing process.

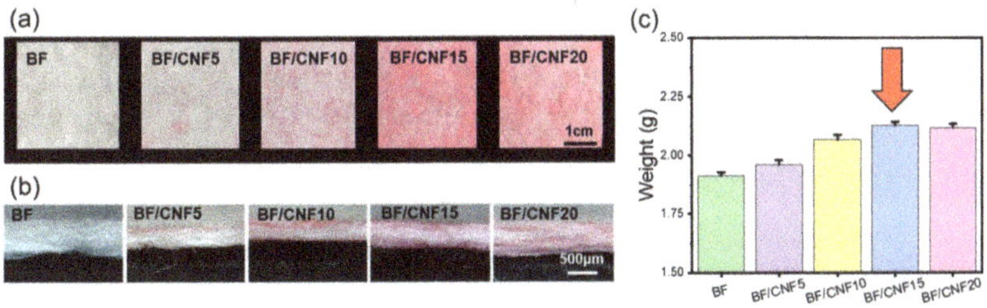

Figure 4. Effect of CNF addition on the visual inspection of BF/CNF fibrous preforms. (**a**,**b**) Optical images of Direct Red 80 dyed BF/CNF fibrous preforms, and (**c**) weight-gain ratio results.

This difference in dyeing properties between BFs and CNFs made it possible to confirm the distribution of CNFs present in the BF/CNF preform. As shown in Figure 4, as the content of CNFs increased to 15 wt.%, the red color of the BF/CNFs gradually became darker, meaning that more CNFs were present in the BF/CNFs. On the other hand, when 20 wt.% CNFs were included, there was no significant difference in color from BF/CNF15, implying that more than 15 wt.% of CNFs can no longer participate in BF fiber coating and fiber-to-fiber bonding; these CNFs are separated in the dehydration process. In addition, the difference in dyeability of the BF/CNFs preform according to the amount of CNFs added was consistent with the actual weight increase result.This pattern was similar to the results of manufacturing silk fiber/CNF fibrous preforms [37]. Altogether, the amount of CNFs in the BF/CNF preform can be easily controlled by varying the concentration of aqueous CNF solutions.

When CNFs are added to the BF fiber nonwoven fabric manufacturing process, the nanofibrils coat the microfibers, and the interfacial bonding between the microfibers is more strongly induced [38]. Since CNFs have a higher surface area (14.5 $m^2 \cdot g^{-1}$) than wood pulp (1.8 $m^2 \cdot g^{-1}$) [17], they can be expected to act as a sufficient binder even with a small amount of CNFs. Furthermore, among various nanocellulose morphologies, nanofibrillated CNFs with a high aspect ratio could be more effective in bonding BF microfibers than road-like cellulose nanocrystals (CNCs). The absolute and apparent densities were measured to examine the effect of CNF binder addition on the BF/CNF fibrous preform physical properties (Figure 5). Even when CNFs were added, the absolute density did not change significantly because both BFs and CNFs were composed of the same cellulose. However, the apparent density is calculated by including the pore volume of materials, so the volume increases as the CNF rises. The quantity of CNFs grows because CNFs are located in the void structure of the BF nonwoven structure. This rise in the apparent density and the resulting fall in porosity have a high correlation. The BF nonwoven fabric has the highest porosity of 43.4%, but when the CNFs were included as a binding material, the coating and interfibrillar bonding of CNFs onto the BF nonwoven structure were able to occur; therefore, BF/CNF15 has the lowest porosity of 35.4%.

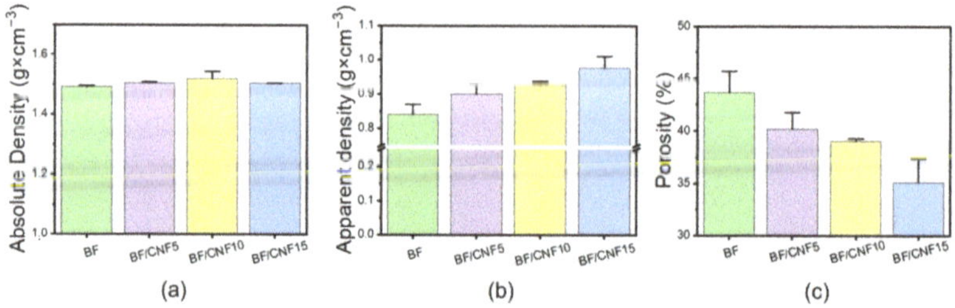

Figure 5. Effect of CNF addition level on the physical properties of BF/CNF fibrous preforms. (**a**) Absolute density, (**b**) apparent density, and (**c**) calculated porosity.

To investigate the effect of CNF addition on the morphology of the BF/CNF fibrous preform, FE-SEM observations were conducted, and the corresponding images are presented in Figure 6. In the case of nonwoven BF, no change in fiber diameter and morphology was observed despite its thermocompression. It has a porous, nonwoven structure, and the fibers are connected smoothly. This means that the previously selected preforming conditions (120 °C for 10 min) do not cause physical damage to the preform. For BF/CNF5, the CNF bundle is coated on the surface of the BF fiber. More CNF-coated BFs were observed for BF/CNF10, and CNF bundles began to distribute in the interfibrillar space of the preform structure. CNF bundles cover the entire surface of the BF/CNF preform in BF/CNF15, and the pore structure is significantly lost. Based on the morphology findings, the addition of CNFs increases BF fiber coating and the interfacial bonding between microfibers.

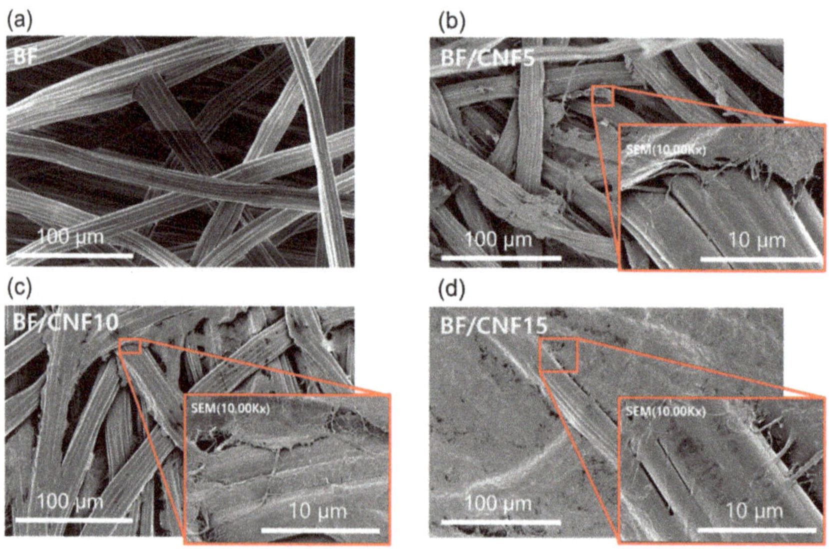

Figure 6. FE-SEM images of BF/CNF fibrous preforms with various amounts of CNFs. The insets represent high-magnitude images. (**a**) BF, (**b**) BF/CNF5, (**c**) BF/CNF10, and (**d**) BF/CNF15.

3.4. Effect of CNF Addition on the Mechanical Properties of BF/CNF Fibrous Preforms

The fibrous preform's mechanical properties are closely related to the final FRP product manufactured as a reinforcing agent. Figure 7 summarizes the tensile properties of the fabricated BF/CNF nonwoven preforms. The tensile test results could not be obtained for BF nonwoven samples without CNFs because the fibers loosened as soon as the tension load was applied, and the nonwoven structure was damaged. Alternatively, a tensile test was possible for BF/CNF5. As a result, it was determined that the tensile strength was 8.1 MPa, and Young's modulus was 570 MPa. CNFs begin to act as coatings and binders as they penetrate the BF nonwoven structure. The mechanical reinforcing effect of CNFs increased as the CNF quantity rose. BF/CNF15 demonstrated a tensile strength of 13.8 MPa, and Young's modulus of 770 MPa. In addition, as the number of CNFs increased, the tensile elongation of the BF/CNF preform also grew. As CNFs are input, the breakage of BF/CNFs is converted from BF-oriented brittle fracture behavior to elastic and progressive failure due to the energy dissipation effect of the CNF bundle. Vilchez prepared chicken hair fiber/CNF preforms and confirmed that as the quantity of CNFs enlarged, progressive failure appeared, increasing tensile elongation [18]. This progressive failure behavior of the BF/CNFs by CNFs consequently contributed to a substantial surge in the BF/CNF preform toughness.

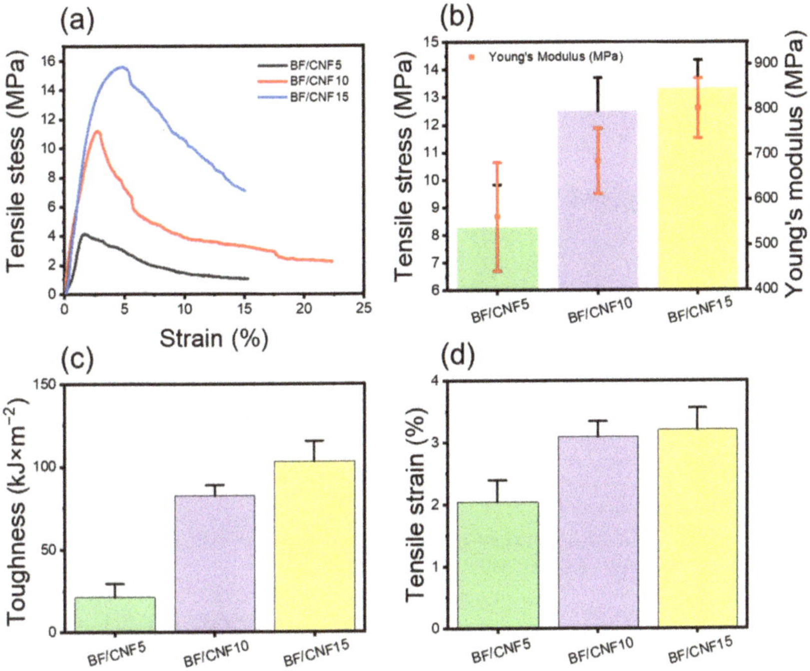

Figure 7. Mechanical properties of BF/CNF fibrous preforms. (**a**) Representative tensile stress–strain curve, (**b**) tensile stress and Young's modulus, (**c**) toughness, and (**d**) tensile strain.

3.5. Mechanical Properties of BF/CNF Preform Reinforced PBS Composites

Polybutylene succinate (PBS) was used as a biodegradable polymer matrix to examine the effect of the physical properties of BF/CNF preform on the mechanical properties of the final biodegradable FRP. PBS is known to be biodegradable in a composting environment and using natural succinic acid as a monomer maximizes its valorization [39–41]. Since PBS has a lower melting temperature (~120 °C) than PLA and Polypropylene (PP), it has the advantage of preventing the thermal decomposition of natural fiber preforms when manufacturing FRP composites [2,39,42]. Figure 8a displays a direct impregnation process for PBS-BF/CNF FRP composite preparation considering cost-efficiency [31]. As shown in Figure 8b, the PBS-BF/CNF FRP composite exhibited a color similar to the BF/CNF preform, implying that the preform did not undergo thermal decomposition and subsequent yellowing during the direct PBS impregnation process. Figure 8c shows the representative stress–strain curves of the PBS-BF/CNF FRP composite tested under uniaxial tension. PBS mainly exhibited elastic deformation behavior, low tensile strength, and Young's modulus of 13.4 MPa and 270 MPa, respectively. When the BF/CNF preform was used as a reinforcing agent [43,44], the FRP composite modulus began to increase rapidly, and this phenomenon was evident in BF/CNF15 with its high CNF content. As a result, PBS-BF/CNF15 improves the tensile strength by 240% (26.4 MPa), and Young's modulus by 700% (1.6 GPa), compared to PBS. The improved mechanical properties mean that the BF fiber's properties do not merely exhibit the BF/CNF preform's reinforcing effect, but the improved mechanical properties of nano/micro fibrous preforms directly participate in the reinforcing effect of FRP [35]. In addition, the nanobrush surface morphology formed on the CNFs coated on BF could maximize the interaction with the PBS matrix [19,45,46].

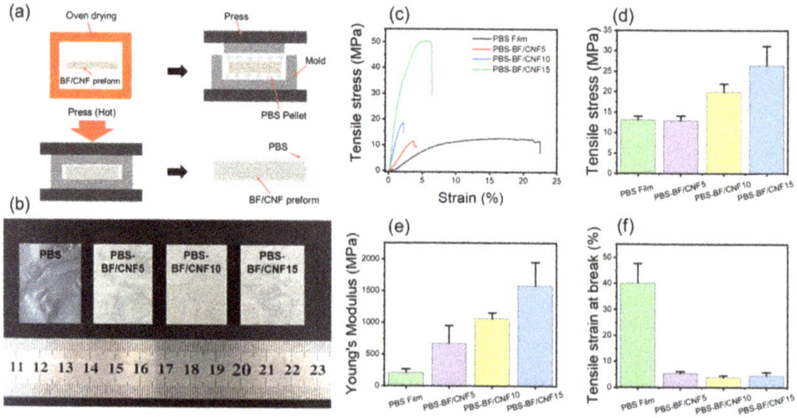

Figure 8. Mechanical properties of BF/CNF reinforced PBS composites. (**a**) Schematic of the PBS-BF/CNF composite fabrication process, (**b**) optical images of PBS-BF/CNF composites, and (**c–f**) tensile properties of PBS-BF/CNF composites.

4. Conclusions

In this study, using a hybridization process of bamboo-derived nano/microfibers, a green and mechanically robust fibrous preform for fiber-reinforced plastics was successfully fabricated. The optimal thermocompression environment for minimizing fiber thermal degradation and obtaining a dense fibrous preform was 120 °C and 10-min compression conditions. As CNF was introduced into the BF/CNF preforming process, the structural stability of the fibrous preform was significantly enhanced. The hybrid phenomenon of BF/CNFs provided a nano-brush morphology that maximized the interaction with the matrix polymer on the surface of the fibrous preform and, at the same time, increased the tight bonding between the BF fibers, making the internal structure of the preform more compact. Ultimately, even when the BF/CNFs preform was used as an eco-friendly reinforcing material in biodegradable fiber reinforced plastic (PBS-BF/CNFs composites), the improved mechanical properties of the fibrous preforms could be effectively expressed. As a result, the PBS-BF/CBF15 composite material improved the tensile strength by 240% (26.4 MPa) and Young's modulus by 700% (1.6 GP) compared to the neat PBS plastics. The synthetic binder or crosslinker-free and energy-efficient micro/nano hybridized green fibrous preforms manufacturing process has excellent potential to be widely utilized in the field of eco-friendly biodegradable fiber-reinforced composite materials in the future.

Author Contributions: Conceptualization, H.W.K.; methodology, J.B. and H.L.; validation, H.W.K.; data curation, J.B. and Y.Y.; writing—original draft preparation, H.W.K.; writing—review and editing, J.-K.O. and H.W.K.; funding acquisition, H.W.K. All authors have read and agreed to the published version of the manuscript.

Funding: This study was carried out with the support of 'R&D Program for Forest Science Technology (2020224D10-2122-AC02)' provided by Korea Forest Service (Korea Forestry Promotion Institute). This work was supported by Creative-Pioneering Researchers Program through Seoul National University (SNU).

Institutional Review Board Statement: Not applicable.

Informed Consent Statement: Not applicable.

Data Availability Statement: The data presented in this study are available on request from the corresponding author.

Conflicts of Interest: The authors declare that there are no conflict of interest. The funders had no role in the design of the study; in the collection, analyses, or interpretation of data; in the writing of the manuscript, or in the decision to publish the results.

References

1. Chin, S.C.; Tee, K.F.; Tong, F.S.; Ong, H.R.; Gimbun, J. Thermal and mechanical properties of bamboo fiber reinforced composites. *Mater. Today Commun.* **2020**, *23*. [CrossRef]
2. Han, Q.; Zhao, L.; Lin, P.; Zhu, Z.; Nie, K.; Yang, F.; Wang, L. Poly(butylene succinate) biocomposite modified by amino functionalized ramie fiber fabric towards exceptional mechanical performance and biodegradability. *React. Funct. Polym.* **2020**, *146*. [CrossRef]
3. García, A.; Gandini, A.; Labidi, J.; Belgacem, N.; Bras, J. Industrial and crop wastes: A new source for nanocellulose biorefinery. *Ind. Crops Prod.* **2016**, *93*, 26–38. [CrossRef]
4. Jonoobi, M.; Mathew, A.P.; Oksman, K. Producing low-cost cellulose nanofiber from sludge as new source of raw materials. *Ind. Crops Prod.* **2012**, *40*, 232–238. [CrossRef]
5. Moslemi, A.; Zolfagharlou koohi, M.; Behzad, T.; Pizzi, A. Addition of cellulose nanofibers extracted from rice straw to urea formaldehyde resin; effect on the adhesive characteristics and medium density fiberboard properties. *Int. J. Adhes. Adhes.* **2020**, *99*. [CrossRef]
6. Tibolla, H.; Pelissari, F.M.; Rodrigues, M.I.; Menegalli, F.C. Cellulose nanofibers produced from banana peel by enzymatic treatment: Study of process conditions. *Ind. Crops Prod.* **2017**, *95*, 664–674. [CrossRef]
7. Fernandes, M.; Souto, A.P.; Gama, M.; Dourado, F. Bacterial Cellulose and Emulsified AESO Biocomposites as an Ecological Alternative to Leather. *Nanomaterials* **2019**, *9*, 1710. [CrossRef] [PubMed]
8. Lee, K.Y.; Buldum, G.; Mantalaris, A.; Bismarck, A. More than meets the eye in bacterial cellulose: Biosynthesis, bioprocessing, and applications in advanced fiber composites. *Macromol. Biosci.* **2014**, *14*, 10–32. [CrossRef] [PubMed]
9. Hao, W.; Wang, M.; Zhou, F.; Luo, H.; Xie, X.; Luo, F.; Cha, R. A review on nanocellulose as a lightweight filler of polyolefin composites. *Carbohydr. Polym.* **2020**, *243*, 116466. [CrossRef]
10. Siró, I.; Plackett, D. Microfibrillated cellulose and new nanocomposite materials: A review. *Cellulose* **2010**, *17*, 459–494. [CrossRef]
11. Chen, W.; Yu, H.; Lee, S.Y.; Wei, T.; Li, J.; Fan, Z. Nanocellulose: A promising nanomaterial for advanced electrochemical energy storage. *Chem. Soc. Rev.* **2018**, *47*, 2837–2872. [CrossRef] [PubMed]
12. Wang, L.; Roach, A.W.; Gardner, D.J.; Han, Y. Mechanisms contributing to mechanical property changes in composites of polypropylene reinforced with spray-dried cellulose nanofibrils. *Cellulose* **2017**, *25*, 439–448. [CrossRef]
13. Chu, Y.; Sun, Y.; Wu, W.; Xiao, H. Dispersion Properties of Nanocellulose: A Review. *Carbohydr. Polym.* **2020**, *250*, 116892. [CrossRef]
14. Neves, R.M.; Ornaghi, H.L., Jr.; Zattera, A.J.; Amico, S.C. Recent studies on modified cellulose/nanocellulose epoxy composites: A systematic review. *Carbohydr. Polym.* **2021**, *255*, 117366. [CrossRef]
15. Wang, J.; Tavakoli, J.; Tang, Y. Bacterial cellulose production, properties and applications with different culture methods—A review. *Carbohydr. Polym.* **2019**, *219*, 63–76. [CrossRef]
16. Moniri, M.; Boroumand Moghaddam, A.; Azizi, S.; Abdul Rahim, R.; Bin Ariff, A.; Zuhainis Saad, W.; Navaderi, M.; Mohamad, R. Production and Status of Bacterial Cellulose in Biomedical Engineering. *Nanomaterials* **2017**, *7*, 257. [CrossRef] [PubMed]
17. Fortea-Verdejo, M.; Lee, K.-Y.; Zimmermann, T.; Bismarck, A. Upgrading flax nonwovens: Nanocellulose as binder to produce rigid and robust flax fibre preforms. *Compos. Part A* **2016**, *83*, 63–71. [CrossRef]

18. Vilchez, V.; Dieckmann, E.; Tammelin, T.; Cheeseman, C.; Lee, K.-Y. Upcycling Poultry Feathers with (Nano)cellulose: Sustainable Composites Derived from Nonwoven Whole Feather Preforms. *ACS Sustain. Chem Eng.* **2020**, *8*, 14263–14267. [CrossRef]
19. Lee, K.-Y.; Ho, K.K.C.; Schlufter, K.; Bismarck, A. Hierarchical composites reinforced with robust short sisal fibre preforms utilising bacterial cellulose as binder. *Compos. Sci. Technol.* **2012**, *72*, 1479–1486. [CrossRef]
20. Mohammed, L.; Ansari, M.N.M.; Pua, G.; Jawaid, M.; Islam, M.S. A Review on Natural Fiber Reinforced Polymer Composite and Its Applications. *Int. J. Polym. Sci.* **2015**, *2015*, 1–15. [CrossRef]
21. Ju, Z.; Zhan, T.; Zhang, H.; He, Q.; Hong, L.; Yuan, M.; Cui, J.; Cheng, L.; Lu, X. Strong, Durable, and Aging-Resistant Bamboo Composites Fabricated by Silver In Situ Impregnation. *ACS Sustain. Chem. Eng.* **2020**, *8*, 16647–16658. [CrossRef]
22. Cui, R.; Pan, L.; Deng, C. Synthesis of carbon nanocoils on substrates made of plant fibers. *Carbon* **2015**, *89*, 47–52. [CrossRef]
23. Le Phuong, H.A.; Izzati Ayob, N.A.; Blanford, C.F.; Mohammad Rawi, N.F.; Szekely, G. Nonwoven Membrane Supports from Renewable Resources: Bamboo Fiber Reinforced Poly(Lactic Acid) Composites. *ACS Sustain. Chem. Eng.* **2019**, *7*, 11885–11893. [CrossRef]
24. Abe, K.; Yano, H. Comparison of the characteristics of cellulose microfibril aggregates isolated from fiber and parenchyma cells of Moso bamboo (Phyllostachys pubescens). *Cellulose* **2009**, *17*, 271–277. [CrossRef]
25. Hu, Z.; Zhai, R.; Li, J.; Zhang, Y.; Lin, J. Preparation and Characterization of Nanofibrillated Cellulose from Bamboo Fiber via Ultrasonication Assisted by Repulsive Effect. *Int. J. Polym. Sci.* **2017**, *2017*, 1–9. [CrossRef]
26. Wang, H.; Zhang, X.; Jiang, Z.; Li, W.; Yu, Y. A comparison study on the preparation of nanocellulose fibrils from fibers and parenchymal cells in bamboo (Phyllostachys pubescens). *Ind. Crops Prod.* **2015**, *71*, 80–88. [CrossRef]
27. Borri, A.; Castori, G.; Corradi, M.; Speranzini, E. Durability Analysis for FRP and SRG Composites in Civil Applications. *Key Eng. Mater.* **2014**, *624*, 421–428. [CrossRef]
28. Chiu, H.-H.; Young, W.-B. Characteristic study of bamboo fibers in preforming. *J Compos. Mater.* **2020**, *54*, 3871–3882. [CrossRef]
29. Miao, M.; Shan, M. Highly aligned flax/polypropylene nonwoven preforms for thermoplastic composites. *Compos. Sci. Technol.* **2011**, *71*, 1713–1718. [CrossRef]
30. Mautner, A.; Kwaw, Y.; Weiland, K.; Mvubu, M.; Botha, A.; John, M.J.; Mtibe, A.; Siqueira, G.; Bismarck, A. Natural fibre-nanocellulose composite filters for the removal of heavy metal ions from water. *Ind. Crops Prod.* **2019**, *133*, 325–332. [CrossRef]
31. Kim, S.H.; Park, C.H. Direct impregnation of thermoplastic melt into flax textile reinforcement for semi-structural composite parts. *Ind. Crops Prod.* **2017**, *95*, 651–663. [CrossRef]
32. Codispoti, R.; Oliveira, D.V.; Olivito, R.S.; Lourenço, P.B.; Fangueiro, R. Mechanical performance of natural fiber-reinforced composites for the strengthening of masonry. *Compos. Part B* **2015**, *77*, 74–83. [CrossRef]
33. Manasoglu, G.; Kanik, M.; Yildirim, K. Effect of fixation conditions on yellowing behavior of cellulose powder-coated fabrics. *J. Eng. Fibers Fabr.* **2019**, *14*. [CrossRef]
34. Matsuo, M.; Umemura, K.; Kawai, S. Kinetic analysis of color changes in cellulose during heat treatment. *J. Wood Sci.* **2012**, *58*, 113–119. [CrossRef]
35. Uribe, B.E.B.; Soares-Pozzi, A.C.; Tarpani, J.R. Nanocellulose-coated carbon fibers towards developing hierarchical polymer matrix composites. *Mater. Today Proc.* **2019**, *8*, 820–831. [CrossRef]
36. Sribenja, S.; Saikrasun, S. Adsorption behavior and kinetics of lac dyeing on poly (ethyleneimine)-treated bamboo fibers. *Fibers Polym.* **2015**, *16*, 2391–2400. [CrossRef]
37. Eom, J.; Park, S.; Jin, H.-J.; Kwak, H.W. Multiscale hybridization of natural silk-nanocellulose fibrous composites with exceptional mechanical properties. *Front. Mater.* **2020**, *7*, 98. [CrossRef]
38. Lu, Z.; Hu, W.; Xie, F.; Hao, Y. Highly improved mechanical strength of aramid paper composite via a bridge of cellulose nanofiber. *Cellulose* **2017**, *24*, 2827–2835. [CrossRef]
39. Xu, J.; Guo, B.H. Poly (butylene succinate) and its copolymers: Research, development and industrialization. *Biotechnol. J.* **2010**, *5*, 1149–1163. [CrossRef]
40. Nghiem, N.; Kleff, S.; Schwegmann, S. Succinic Acid: Technology Development and Commercialization. *Fermentation* **2017**, *3*, 26. [CrossRef]
41. Song, H.; Lee, S.Y. Production of succinic acid by bacterial fermentation. *Enzyme Microb. Technol.* **2006**, *39*, 352–361. [CrossRef]
42. Petchwattana, N.; Sanetuntikul, J.; Sriromreun, P.; Narupai, B. Wood plastic composites prepared from biodegradable poly (butylene succinate) and Burma Padauk sawdust (Pterocarpus macrocarpus): Water absorption kinetics and sunlight exposure investigations. *J. Bionic Eng.* **2017**, *14*, 781–790. [CrossRef]
43. Li, Y.; Jiang, L.; Xiong, C.; Peng, W. Effect of Different Surface Treatment for Bamboo Fiber on the Crystallization Behavior and Mechanical Property of Bamboo Fiber/Nanohydroxyapatite/Poly(lactic-co-glycolic) Composite. *Ind. Eng. Chem. Res.* **2015**, *54*, 12017–12024. [CrossRef]
44. Manalo, A.C.; Wani, E.; Zukarnain, N.A.; Karunasena, W.; Lau, K.-t. Effects of alkali treatment and elevated temperature on the mechanical properties of bamboo fibre–polyester composites. *Compos. Part B* **2015**, *80*, 73–83. [CrossRef]
45. Heng, J.Y.Y.; Pearse, D.F.; Thielmann, F.; Lampke, T.; Bismarck, A. Methods to determine surface energies of natural fibres: A review. *Compos. Interfaces* **2012**, *14*, 581–604. [CrossRef]
46. Juntaro, J.; Pommet, M.; Kalinka, G.; Mantalaris, A.; Shaffer, M.S.P.; Bismarck, A. Creating Hierarchical Structures in Renewable Composites by Attaching Bacterial Cellulose onto Sisal Fibers. *Adv. Mater.* **2008**, *20*, 3122–3126. [CrossRef]

polymers MDPI

Article

Steroid Eluting Esophageal-Targeted Drug Delivery Devices for Treatment of Eosinophilic Esophagitis

Alka Prasher [1,2], Roopali Shrivastava [1,2], Denali Dahl [1,2], Preetika Sharma-Huynh [3], Panita Maturavongsadit [1,2], Tiffany Pridgen [4], Allison Schorzman [5,6,7,8], William Zamboni [5,6,7,8], Jisun Ban [5,6,7,8], Anthony Blikslager [4], Evan S. Dellon [9] and Soumya Rahima Benhabbour [1,3,*]

1 Department of Biomedical Engineering, UNC Chapel Hill & North Carolina State University, Chapel Hill, NC 27599-3290, USA; alkap@email.unc.edu (A.P.); roopalis@email.unc.edu (R.S.); ddahl@email.unc.edu (D.D.); panita@med.unc.edu (P.M.)
2 Department of Chemistry, University of North Carolina at Chapel Hill, Chapel Hill, NC 27599-3290, USA
3 Division of Pharmacoengineering and Molecular Pharmaceutics, UNC Eshelman School of Pharmacy, University of North Carolina at Chapel Hill, Chapel Hill, NC 27599-3290, USA; preetika.sharma@unc.edu
4 Department of Clinical Sciences, College of Veterinary Medicine, North Carolina State University, Raleigh, NC 27606, USA; tladams3@ncsu.edu (T.P.); anthony_blikslager@ncsu.edu (A.B.)
5 Division of Pharmacotherapy and Experimental Therapeutics, UNC Eshelman School of Pharmacy, University of North Carolina, Chapel Hill, NC 27599-3290, USA; aschorz@email.unc.edu (A.S.); zamboni@email.unc.edu (W.Z.); jsharie@email.unc.edu (J.B.)
6 UNC Lineberger Comprehensive Cancer Center, Chapel Hill, NC 27599-3290, USA
7 Carolina Institute for Nanomedicine, Chapel Hill, NC 27599-3290, USA
8 UNC Advanced Translational Pharmacology and Analytical Chemistry Lab, Chapel Hill, NC 27599-3290, USA
9 Division of Gastroenterology and Hepatology, UNC School of Medicine, University of North Carolina, Chapel Hill, NC 27599-3290, USA; evan_dellon@med.unc.edu
* Correspondence: benhabs@email.unc.edu; Tel.: +1-(919)-843-6142

Citation: Prasher, A.; Shrivastava, R.; Dahl, D.; Sharma-Huynh, P.; Maturavongsadit, P.; Pridgen, T.; Schorzman, A.; Zamboni, W.; Ban, J.; Blikslager, A.; et al. Steroid Eluting Esophageal-Targeted Drug Delivery Devices for Treatment of Eosinophilic Esophagitis. *Polymers* **2021**, *13*, 557. https://doi.org/10.3390/polym13040557

Academic Editor: José Miguel Ferri

Received: 25 January 2021
Accepted: 8 February 2021
Published: 13 February 2021

Abstract: Eosinophilic esophagitis (EoE) is a chronic atopic disease that has become increasingly prevalent over the past 20 years. A first-line pharmacologic option is topical/swallowed corticosteroids, but these are adapted from asthma preparations such as fluticasone from an inhaler and yield suboptimal response rates. There are no FDA-approved medications for the treatment of EoE, and esophageal-specific drug formulations are lacking. We report the development of two novel esophageal-specific drug delivery platforms. The first is a fluticasone-eluting string that could be swallowed similar to the string test "entero-test" and used for overnight treatment, allowing for a rapid release along the entire length of esophagus. In vitro drug release studies showed a target release of 1 mg/day of fluticasone. In vivo pharmacokinetic studies were carried out after deploying the string in a porcine model, and our results showed a high local level of fluticasone in esophageal tissue persisting over 1 and 3 days, and a minimal systemic absorption in plasma. The second device is a fluticasone-eluting 3D printed ring for local and sustained release of fluticasone in the esophagus. We designed and fabricated biocompatible fluticasone-loaded rings using a top-down, Digital Light Processing (DLP) Gizmo 3D printer. We explored various strategies of drug loading into 3D printed rings, involving incorporation of drug during the print process (pre-loading) or after printing (post-loading). In vitro drug release studies of fluticasone-loaded rings (pre and post-loaded) showed that fluticasone elutes at a constant rate over a period of one month. Ex vivo pharmacokinetic studies in the porcine model also showed high tissue levels of fluticasone and both rings and strings were successfully deployed into the porcine esophagus in vivo. Given these preliminary proof-of-concept data, these devices now merit study in animal models of disease and ultimately subsequent translation to testing in humans.

Keywords: eosinophilic esophagitis (EoE), esophageal drug delivery systems; drug-eluting string; 3D printing; photopolymerizable resins; Drug loading strategies; steroids; fluticasone

1. Introduction

Eosinophilic esophagitis (EoE) is an emerging chronic allergic disease characterized by eosinophilic infiltration of the esophageal mucosa [1,2]. The most common symptoms of EoE include dysphagia and food impaction in adolescents and adults, and abdominal pain, vomiting, heartburn, feeding intolerance, and failure to thrive in children. EoE represents a major healthcare burden as the estimated prevalence is at least 1 in 2000 Americans, and EoE-associated healthcare costs approach up to $1.4 billion per year [3,4]. Because there are no FDA-approved medications for EoE, asthma preparations, such as a fluticasone (FTS) from an inhaler, have traditionally been swallowed rather than inhaled to coat the esophagus [5]. The European Medical Agency (EMA) has approved a dissolvable budesonide tablet (Jorveza) for both acute and a long-term treatment. While these treatment approaches can be effective [6–8], because the medication delivery does not target the esophagus, it leads to suboptimal outcomes and adherence [9]. Our prior data, however, suggest that increased esophageal dwell time of pharmacologic agents is closely associated with subsequent treatment response [10]. To address these limitations, we aimed to produce strategies to more rationally provide drug exposure to the esophagus. Herein, we report the development of two esophageal-specific drug delivery platforms, a drug-eluting string and a 3D-printed ring for local rapid or sustained release of FTS in the esophagus.

These two delivery devices shared a common rationale—to deliver medication to the esophagus. A drug-eluting string could be swallowed, allowing a sustained release of drug along the entire length of the esophagus, perhaps during an overnight dwell, rather than the usual one-time bolus from a swallowed medication. While the minimally invasive esophageal string test has been used to absorb inflammatory factors related to EoE and characterize disease activity [11–13], there are no reports of delivering medication to the esophagus in this manner. A recent report outlined how a membrane deployed from a capsule might also target drug delivery to the esophagus. [14]. Our research strategy involved developing a FTS-eluting string and showing a target release of 1 mg/day of fluticasone in vitro for 24 h period of time. Fluticasone was coated on the surface of the string via drug adsorption using a dip coating process [15–17]. Studies were conducted to determine drug loading, in vitro drug release kinetics, ex vivo PK, and in vivo safety and PK in a porcine model [18].

In our second approach, we developed rings that can be inserted in the proximal esophagus to provide a long-acting sustained delivery of fluticasone. Rings were fabricated using 3D printing with additive manufacturing (AM), which involves a process of transforming digital models into real objects by placing materials layer by layer [19–23]. Recently, this emerging technology has been widely applied in the fields of drug delivery [24], biomaterials [20,25,26], medical [27] and food industry [28], and customized consumer products [29]. There are a number of 3D printing methodologies available which differ in how the 2D layers of materials are deposited. These include stereolithography (SLA) [30], two-photon polymerization [31], selective laser sintering (SLS) [32], fused deposition modeling (FDM) [33], and laminated object manufacturing (LOM) [34]. Among the existing AM technologies, photopolymerization offers the advantage of using a variety of synthetic polymers that can be tailored to have a desired structural property [35–37]. A photoprintable system mainly consists of an initiating system, photopolymerizable monomer or oligomer, and a light absorber [38]. Photoinitiators provide the required energy to initiate the photopolymerization, and a suitable light absorber is usually added to have an effective control over cure-depth [39]. The final properties of the printed part can be tailored by changing the constituents of photocurable resin formulation [40]. Recent advances in photopolymerization-based 3D printing involves spatially controlled curing of a photopolymer by interacting with a digitally modulated UV or visible light beam, defined by layer profiles of an Standard Tessellation Language (STL) file [41]. The process involves projection of an array of light spots on the surface of the resin, and propagation of light within each illuminated voxel in direction of the beam. The thickness of the photopolymerized layer is dependent on the interaction of photopolymer and the incident beam. In our system, we used digital light processing (DLP) projection as the method of illumination which has the advantages of high printing precision and fast printing speed [42]. 3D printing, which includes the ability to construct structures with complex geometries via computer-aided design (CAD) models, has opened the door for rapid and facile production of controlled drug release devices and polymer biomaterials [43–47].

In recent years, drug delivery devices have been developed to release medication at a controlled rate, and upon implantation, allow treatment without requiring oral or intravenous dosing [48–50]. We demonstrated in vitro that our novel 3D printed ring eluted fluticasone at a constant rate. All rings were fabricated using a variety of biocompatible photopolymerizable resin formulations, with various strategies of incorporating fluticasone in the ring device. Fluticasone-loaded rings were tested ex vivo and in vivo in the porcine model and showed high levels of fluticasone in the esophageal tissue in the ex vivo studies. To the best of our knowledge, these two approaches—the fluticasone-eluting string and ring—are among the first such esophageal-specific devices developed for the treatment of EoE. These promising results will need to be extended into disease models, with the ultimate goal of translating this technology into human disease.

2. Materials and Methods

2.1. Materials

Anhydrous dichloromethane (DCM), triethylamine (TEA), poly (ethylene glycol) Mn = 530 g mol^{-1}, sodium chloride (NaCl), hydroxyethyl methacrylate (HEMA), diphenyl (2,4,6-trimethylbenzoyl)phosphine oxide (TPO), 2-tert-butyl-6(5-chloro-2H-benzo-triazol-2-yl)-4-methylphenol (BLS-1326), methacryloyl chloride, Solutol-HS 15, phosphate-buffered saline (0.01 M PBS, pH 7.4), HPLC-grade acetonitrile, water, and chloroform were purchased from Sigma Aldrich (St. Louis, MO, USA). Poly(ε-caprolactone) (PCL) (3 mm OD) string was purchased from Lactel absorbable polymers (Birmingham, Al, USA), and was made from poly (ε-caprolactone) ester terminated (Mn = 68 kDa). Cotton strings (1 mm OD) were kindly provided by Dr. Dellon. The therapeutic drug fluticasone propionate (FTS) was purchased from Letco medical (Decatur, Al, USA). All materials were stored as directed by the supplier and used as received.

HeLa (ATCC® CCL-2™) cells were kindly provided by Dr. Chris Luft, University of North Carolina, Chapel Hill. HeLa cells were cultured in Dulbecco's Modified Eagle's Medium from Gibco Gaithersberg MD, USA) with 10% fetal calf serum from Gibco (Gaithersberg, USA) and 100 units/mL each of penicillin and streptomycin from Life Technologies (Grand Island, NY, USA). Cell lines were maintained at 37 °C in a humidified atmosphere containing 5% CO_2, and ATCC guidelines were used for sub-culturing cells. For quantification of fluticasone in pig serum and esophageal homogenate in in vivo and ex vivo studies, pig plasma (K_2EDTA anti-coagulant was purchased from BioIVT (Hicksville, NY, USA), fluticasone propionate, dimethylsulfoxide, and zinc sulfate solution were purchased from Sigma (St. Louis, MO, USA); d5-fluticasone propionate was purchased from Toronto Research Chemicals (North York, ON, CA); ammonium hydroxide, methanol, and acetonitrile were purchased from VWR (Radnor, PA, USA); and HLB μElution plates were purchased from Waters (Milford, MA, USA)

2.2. Fluticasone-Loaded Strings

2.2.1. Optimization of Drug Loading

Poly(ε-caprolactone) (PCL) and fabric strings were cut into a length of 4 cm and marked up to a height of 2 cm (Supplementary Figure S1A). The dipping solution (2 mg/mL fluticasone in acetone/PLGA 1:5 *w/w*; 10 mL) was introduced in a 20 mL scintillation vial and the strings were immersed in the solution and incubated at room temperature for different time durations ranging from 1 min to 24 h (Supplementary Figure S1D). The strings were subsequently removed from the solution and allowed to dry at room temperature (RT) for 24 h. To investigate the effect of successive dipping of strings in the drug solution on percent drug loading, strings were subjected to a repeated dipping process. In this process, strings were incubated in the drug solution for a predetermined duration, removed, then reintroduced again in the solution and finally dried at RT. Various incubation times (1 min, 24 h), time between the repeated incubations (3, 10, 30 s), and number of incubation iterations (1 vs. 6) were investigated to optimize the amount of fluticasone loading onto the string.

2.2.2. In Vitro Drug Release

To test the in vitro drug release, FTS-coated strings were individually placed in straight sided glass jars containing 20 mL of phosphate-buffered saline (PBS) (0.01 M with 2% Solutol HS, pH 7.4) under sink conditions at 37 °C. Sink conditions were defined as fluticasone (FTS) concentration at or below 1/5 of its maximum solubility (i.e., $\leq$0.11 mg/mL FTS) in PBS + 2% Solutol at 37 °C. The saturation solubility of FTS in PBS + 2% Solutol was determined by HPLC analysis. Sample aliquots (1 mL) were collected at 24 h time point, and FTS concentration was quantified by HPLC analysis. The amount of FTS released was quantified by HPLC analysis and was normalized to the total concentration of FTS coated on the string. All experiments were performed in triplicates.

2.2.3. Ex Vivo Pharmacokinetic (PK) Studies

Drug elution from the strings was investigated ex vivo in a porcine model. In these studies, fresh normal esophageal tissue explants of pigs were used (Supplementary Figure S2A). These studies were performed on FTS coated strings with dimensions of 20 mm height and 2 mm cross-sectional diameter. Strings were inserted into approximately 20 mm of esophageal tissue section and incubated in 10 mL of PBS (to simulate the in vitro environment under sink conditions) at 4 °C for various durations (1 day and 3 days). A temperature of 4 °C was chosen for the study in order to preserve the fresh esophageal tissue. At each respective timepoint, the strings were removed, and the entire esophageal tissue was collected for PK analysis by LC-MS/MS analysis. In addition, residual FTS in the strings was extracted using acetonitrile and quantified by HPLC analysis.

2.3. Fluticasone Loaded Rings

2.3.1. Synthesis of Poly(caprolactone dimethacrylate)

Poly(caprolactone-dimethacrylate) (PCL_{700}-DMA) Mn = 700 $gmol^{-1}$ was synthesized as described in the literature (Supplementary Scheme S1) [51]. In brief, poly-ε-caprolactone (PCL) diol (50 g, 96.22 mmol) was added to an oven-dried round-bottom flask. The reaction flask was sealed with a rubber septum and backfilled with nitrogen (N_2). Anhydrous dichloromethane (DCM, 200 mL) and triethylamine (TEA, 222.5 mmol) were added and the flask was placed in an ice bath. A 10% molar excess of methacryloyl chloride (210 mmol) in 100 mL DCM was added dropwise at 0 °C over the course of 1 h and the reaction was stirred for 24 h at room temperature (RT). The TEA.HCl salt formed was removed by filtration and the filtrate was diluted 3 times with DCM, washed three times with a NaCl solution and three times with deionized water. The organic layer was collected, dried over magnesium sulfate, and filtered through a fritted funnel. DCM was removed under reduced pressure using a rotatory evaporator.

2.3.2. Resin formulations for 3D Printing

For all experiments, a base resin containing no drug was formulated at the given ratios. Resin solutions with known amounts of photoinitiator, UV absorber, and reactive diluents dissolved directly into the liquid oligomeric monomers at different ratios. All resins were formulated by combining the components in an amber glass bottle and mixing with magnetic stirring overnight at room temperature.

2.3.3. Device Design and Fabrication

3D-printed rings were fabricated using a top down, Digital Light Processing (DLP) Gizmo® 3D printer (Gizimate® 130, Queensland, Australia) equipped with a UV light source. Computer-aided design (CAD) files were downloaded from Thingiverse and modified using Meshmixer to design a ring suitable for the esophagus. The final dimensions of the ring CAD files were 24 mm outer diameter (OD), 20 mm height (H), and 2.5 mm cross section (CS). Standard tessellation language (STL) files were sliced at 50 μm thickness using the Gizmetor software, and printed at a continuous speed of 4 mm min^{-1} and a light intensity of 23 mW cm^{-1}.

2.3.4. Saturation Solubility of FTS in Resin Formulations

The saturation solubility of FTS was determined in all resin formulations used to print rings. For each resin formulation, 100 mg of drug was weighed into individual vials, and 200 mg of resin was added and allowed to stir at RT overnight to dissolve the drug. The mixture was then centrifuged for 15 min at 14,000 rpm (Eppendorf Centrifuge 5417C, Marshall Scientific, Hampton, NH, USA) to remove undissolved drug. Sample aliquots (1–2 mg, n = 4) were collected from the saturated supernatant and diluted with acetonitrile (ACN). Drug concentration in the saturated aliquots was determined by HPLC analysis.

2.3.5. High-Performance Liquid Chromatography (HPLC)

A reversed-phase HPLC method was developed and validated to quantify the concentration of FTS released from strings and 3D printed rings. The HPLC analyses were carried out with a Thermo Finnigan Surveyor HPLA, (Thermo Finnigan, San Jośe, CA, USA) on an Intersil, ODS-3 column (4 μm, 4.6 Å ~150 nm (GL Sciences, Torrance, CA), maintained at 40 °C, with a flow rate of 1.0 mL/min, a 25 μL sample injection, and an acetonitrile/water mobile phase, each modified with 0.1% trifluoroacetic acid. A gradient method was utilized to achieve separation (0–20 min: 5–100% acetonitrile; 20–22 min: 100% acetonitrile; 23–25 min: 5% acetonitrile). Fluticasone was eluted at a retention time of 17.3 min. Data were collected at 265 nm and computed using Chromequest software. Concentrations were derived from a calibration curve generated using fluticasone standards prepared in 100% acetonitrile (250 μg/mL to 61 ng/mL).

2.3.6. Gel Fraction and Percent Swelling

Swelling and gel fraction of 3D-printed rings was carried out in chloroform to determine the monomer incorporation and crosslink density. To determine gel fraction, rings (n = 3) were first weighed to record their initial mass at time zero (Mo) and then placed in 50 mL jars containing chloroform for one week. After one week, rings were removed and carefully cleaned with Kimwipe and the swollen mass (Ms) was recorded. The rings were subsequently air dried for three days and the dry weight (Md) was recorded. The gel fraction and degree of swelling was calculated using the following equations:

$$\text{Gel Fraction} = \frac{\text{Md}}{\text{Mo}} \tag{1}$$

$$\%\ \text{Solvent uptake} = \frac{\text{Ms} - \text{Md}}{\text{Md}} \times 100 \tag{2}$$

2.3.7. Rheology Analysis

The dynamic viscosity of various resin formulations (placebo and drug loaded) was measured using a Brookfield Cone and Plate Digital Rheometer (Model: LVDV-III + CP, Middleboro, MA, USA) and the reading was recorded at 25 °C with a spindle speed of 10 rpm.

2.3.8. Mechanical Testing

The mechanical properties of 3D-printed rings were tested using Micro-strain analyzer TA instrument RSA III. The mechanical testing involved compression of the ring from the top in a radial fashion. The effect of post-fabrication UV cure on the mechanical properties of 3D printed rings was investigated. Rings were fabricated with biodegradable resins (PCL_{700}-DMA, PCL_{700}-DMA/HEMA) and tested in triplicates to determine the compressive force. The average force required to achieve 10%, 20%, and 50% compression of the ring diameter was measured at the proximal, center, and distal ends of the rings. Additionally, the rings were rotated clockwise at three different angles to get the average compression force value of the entire ring.

2.3.9. Drug Loading Studies

FTS-loaded rings were fabricated using a pre-loading or post-loading process. In the pre-loading method, the drug was added into the resin precursor at a given weight % (wt %) and stirred at room temperature overnight to fully dissolve the drug (Figure 1). In the post-loading process, drug was incorporated into the ring post 3D printing ring fabrication (Figure 3). Placebo rings were fabricated with a base resin formulation using the 3D printing process described above. Pre-weighed placebo rings were incubated in acetone (50 mL) at RT for 24 h to remove unreacted resin components (leachables/extractables) from the rings. The rings were subsequently air dried overnight, and their final weight recorded. The dried rings were then incubated in a saturated solution of FTS in acetone at RT for 24 h. The drug was absorbed into the polymer network via swelling of the ring matrix in the concentrated drug solution. The post-loaded rings were subsequently air-dried overnight at RT to remove all solvent and the final mass of rings was recorded.

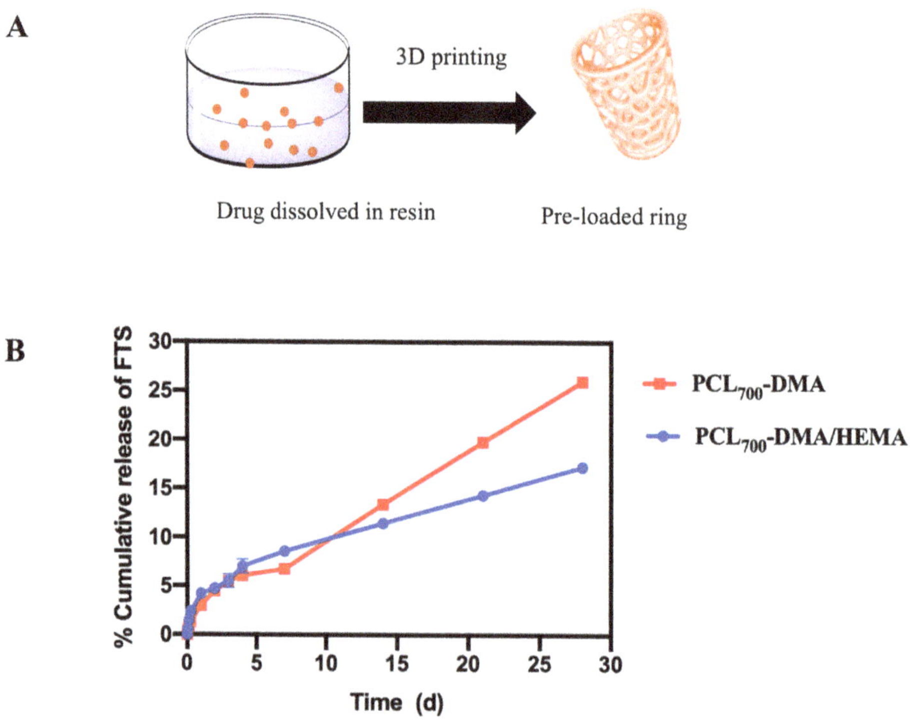

Figure 1. Effect of crosslink density of resin formulations on drug release kinetics (**A**) A pictorial representation of pre-loading FTS into 3D printed rings. (**B**) In vitro release kinetics of FTS-loaded rings incubated in PBS at 37 °C for 30 days. All error bars represent standard deviation for $n = 3$.

2.3.10. In Vitro Cumulative Drug Release

To test the in vitro drug release, FTS-loaded rings (pre- and post-loaded) were individually placed in straight sided glass jars containing 200 mL of PBS (0.01 M with 2% (*w/w*) Solutol, pH 7.4) and incubated at 37 °C under sink conditions for 30 days. Sample aliquots (1 mL) were collected at various time points and replaced with fresh release medium. The release medium was completely removed and replaced with fresh medium every week to maintain sink conditions. FTS concentration in release samples was quantified by HPLC analysis. Cumulative drug release was normalized to the total concentration of drug in the ring determined by HPLC analysis. All experiments were performed in triplicates.

2.3.11. In Vitro Cell Viability Studies

The safety and tolerability of the resin materials used to fabricate the rings was tested in vitro in HeLa cells. The HeLa cell line is the most robust and widely used cell line to assess in vitro cell viability [52]. HeLa cells were plated at a density of 17,500 cells/well in 96-well opaque, flat-bottomed, white plates from Corning Inc. (Corning, NY, USA) 48 h prior to treatment. The cell lines were treated with 3D-printed small discs (Supplementary Figure S4A). Cell viability was assessed after a 48 h treatment exposure using CellTiter-Glo® Luminescent Cell Viability Assay from Promega (Madison, WI, USA) and plates were read using the Synergy 2 Multi-mode Plate Reader from BioTek Instruments (Winooski, VT, USA). The CellTiter-Glo Luminescent Cell Viability Assay is a homogeneous method to determine the number of viable cells in culture based on quantitation of the ATP present (*Renilla* luciferase units (RLU)), which signals the presence of metabolically active cells. Cells grown in growth media alone were used as controls. Cell viability was calculated as a percentage of RLU Experimental/RLU Control.

2.3.12. Ex Vivo Pharmacokinetic (PK) Studies

3D-printed rings were tested ex vivo in fresh porcine esophagus to determine the concentration of FTS released into the esophageal tissue overtime. These studies were performed with both pre- and post-loaded FTS-loaded rings (10 mm H, 10 mm OD, and 2 mm CS) (Supplementary Figure S5A). Drug-loaded rings were inserted into approximately 10 mm of esophageal tissue explants. The esophageal tissue explants were subsequently placed in 10 mL of PBS at 4 °C and incubated for 1, 3, 7, and 14 days (n = 3 per timepoint). At each timepoint, the rings (n = 3) were removed from the media and the esophageal tissue was collected and stored at −80 °C for pharmacokinetics analysis by LC-MS/MS. To determine the total amount of fluticasone delivered to the esophageal tissue ex vivo, fluticasone released in PBS and residual drug in the ring were quantified by HPLC analysis.

2.4. In Vivo PK Analysis by LC-MS/MS

2.4.1. Sample Preparation—Solid-Phase Extraction of Serum and Plasma Samples

Blood was harvested and centrifuged to isolate the serum, which was stored at −80 °C until time of analysis. Fluticasone stock solutions (1 mg/mL) were prepared and stored in DMSO at −80 °C. The standard curve and quality controls (QCs) were prepared by spiking fluticasone intermediate solutions prepared in methanol into K2-EDTA pig plasma to final concentrations of 10, 30, 50, 100, 300, 500, 1000, and 3000 pg/mL and 25, 250, and 2500 pg/mL, respectively. It was determined that plasma served as a surrogate matrix for serum for measurement of fluticasone. The quantitative method described here was adapted from a previously published method [53]. In short, 300 µL of 40 mM $ZnSO_4$/10% ammonium hydroxide was added to 300 µL of plasma standard, QC, blank, or serum sample. The mixture was vortexed briefly and centrifuged at 10,000× *g* for 15 min. An HLB µElution plate was conditioned with 100 µL methanol and equilibrated with 2 × 200 µL ddH_2O before 550 µL of plasma or serum supernatant was loaded to the appropriate well of the plate. Each well was washed with 2 × 200 µL of 25% methanol and the fluticasone eluted into a polypropylene 96-well plate with 2 × 50 µL of acetonitrile: methanol (1:1). The eluent was diluted with 150 µL ddH_2O before analysis by LC-MS/MS.

2.4.2. Protein Precipitation of Esophagus Tissue Homogenate

Proximal, mid, and distal sections of esophagus tissue were frozen after harvest, as above, and stored at −80 °C prior to homogenization. Representative pieces of tissue were excised from each section, weighed, and homogenized in PBS (1:3) using three 30 s cycles at 5500 rpm with six 2.8 mm ceramic beads in a bead mill homogenizer (VWR). Homogenates were stored at −80 °C until time of analysis. The standard curve and QCs were prepared by spiking fluticasone propionate intermediate solutions prepared in methanol into esophagus homogenate from untreated animals at 1, 3, 5, 10, 30, 50, 100, 300, 500, 1000, 3000, and 5000 ng/mL and 4, 40, 400, and 4000 ng/mL, respectively. Each 50 μL standard, QC, blank, or sample was extracted with 200 μL of acetonitrile containing 5 ng/mL of d_5-fluticasone propionate internal standard. Samples were vortexed for 5 min, centrifuged at 10,000× g for 10 min at 4 °C. The supernatant containing fluticasone propionate was transferred to a clean tube and evaporated under nitrogen at 40 °C. Samples were reconstituted in 500 μL of 25% methanol and analyzed by LC-MS/MS.

2.4.3. LC-MS/MS Analysis

Liquid chromatography for fluticasone propionate was accomplished using a Shimadzu LC-20AD liquid chromatograph with an XBridge BEH C18 (2.5 μm 2.1 × 50 mm) analytical column (Waters) protected with a guard column. The mobile phase consisted of 0.1% ammonium hydroxide (mobile A) and methanol (mobile phase B), and used a gradient of 50% to 95% mobile phase B over 1.5 min. The flow rate was 0.3 mL/min and the total run time was 5 min. Compounds were detected using a Thermo TSQ Ultra triple quadrupole mass spectrometer equipped with a heated electrospray ionization source in the positive ion mode. The spray voltage was held at 4.0 kV and the vaporizer temperature at 300 °C. Fluticasone propionate and d_5 fluticasone propionate (internal standard) were detected by multiple-reaction monitoring (MRM) using the transitions 501 ≥ 293 *m/z* and 506 ≥ 293 *m/z*, respectively. Calibration curves were fit using linear regression with $1/X^2$ weighting in Xcalibur® v. 2.0 (Thermo Fisher Scientific, Waltham, MA, USA).

2.5. In Vivo Pharmacokinetic (PK) Studies

All in vivo animal studies were approved by the North Carolina (NC) State IACUC. Yorkshire-cross pigs (~25–35 kg bodyweight to allow an optimal esophagus size approximating a human esophagus size) were obtained from campus-maintained breeding herds at NC State University's Swine Educational Unit, and then transferred to the care of Laboratory Animal Resources (LAR) at the College of Veterinary Medicine (CVM). CVM facilities are AAALAC- and USDA-approved, and are managed by the University Attending Veterinarian and a team of board-certified Laboratory Animal Veterinarians (LAVs), Registered Veterinary Technicians (RVTs), and animal care workers. Endoscopy was performed with the pigs under general anesthesia with assistance from RVTs for anesthesia, using standard of care protocols. Pigs were sedated with a combination of xylazine (0.25 mg/kg) and ketamine (11 mg/kg) administered IM, followed by mask induction of anesthesia with isoflurane vaporized in 100% O_2. Pigs were then orotracheally intubated and maintained on isoflurane throughout the procedure. All post-endoscopic care of the animals was performed by LAVs, RVTs, and animal caretakers within LAR, in accordance with protocols established in the IACUC application. Any evidence of postoperative pain was managed with administration of buprenorphine (0.05 mg/kg, IM, q8h as needed).

For endoscopic ring implantation, three sets of 5 pigs were used. In the first set, 3D-printed rings were used without fluticasone (placebo rings) to confirm that the rings could be placed. In order to deploy the rings, the rings were back-loaded onto the inner section of an esophageal overtube (Gardus, Steris Healthcare, Mentor, OH, USA), through which the endoscope was inserted and the esophagus was intubated with the tube and ring in place. Then, the outer section of the overtube was gently advanced over the inner tube to push the ring into position in the proximal esophagus. As this technique was successful, fluticasone-eluting rings were deployed in the two additional sets of pigs. Rings were allowed to indwell for seven days during which time the pigs were provided with an ad lib diet, and then repeat endoscopy was performed to determine if the ring was still seated at day seven, serum, and plasma specimens were obtained for PK analysis and the animals were euthanized per the standard CVM/LAR protocol consistent with 2009 American Veterinary Medical Association (AVMA) Guidelines on Euthanasia. Pigs were initially sedated as before with xylazine/ ketamine, after which they were administered an overdose of pentobarbital (60 mg/kg, IV). The esophagus was immediately explanted once euthanasia had been confirmed, divided into proximal, mid, and distal segments, and flash-frozen for subsequent PK analysis.

For the endoscopic string implantation, an additional three sets of 5 pigs were used. For the first set, the thicker fluticasone-coated string was advanced through a previously placed esophageal overtube, and then held in place as the overtube was removed. The string was then brought through the posterior aspect of the pig's cheek (caudal to the last molar) after making a small incision and retrieving the string with forceps. Strings were externally fixed in place with 2-0 nylon suture and covered with a section of iodine-impregnated self-adhering drape material. The string was allowed to dwell for 24 h while solid feed was withheld, and buprenorphine analgesia was provided. At 24 h, the animals were euthanized, and serum, plasma, and esophageal samples were obtained as above. The procedure for the second set of pigs was identical to the first, with the exception of a 72 h string dwell time, and after the first 24 h the pigs were allowed to eat a liquid diet. For the last set of pigs, the FTS-impregnated fabric string was endoscopically placed. For this, biopsy forceps were used to grasp a fabric tape tag at the end of the string, and gently advanced this into position using direct visualization. After external fixation as above, this string had a 24 h dwell time, with the same sample procurement protocol as for the first set of pigs.

3. Results

3.1. *Fluticasone-Eluting Esophageal String*

3.1.1. Dip Coating Parameters Optimization

In the dipping process, the amount of drug coated onto a device can be optimized by balancing the viscosity of the drug solution, the incubation and drying times, and the number of dipping/drying iterations [54,55]. A comprehensive study was carried out to investigate the aforementioned effects on the amount of fluticasone loaded onto the string. Fluticasone (FTS)-loaded strings were designed to cover the entire human esophagus length (~25 cm) and release a clinically relevant dose of FTS (1–2 mg) in 24 h [54]. Initial optimization studies were accomplished with a commercially available biocompatible PCL string. The coating solution was comprised of FTS in a biocompatible polymer and organic solvent. The solvent used to load FTS onto the PCL string using a dip-coating process was selected based on high solubility of FTS in the solvent, its volatility, and ease of removal post drug loading. A number of solvents were screened, and acetone was selected as a dipping solvent based on its low boiling point and high saturation solubility of FTS (~20 mg/mL). To optimize the dip-coating process, poly(lactide-*co*-glycolide) (PLGA, 50:50) of various molecular weights (MW 11, 27, and 53 kDa) was used as a model to investigate the effect of dipping solution viscosity on percent PLGA loaded onto the string (Supplementary Figure S1D). Results from these experiments were used to extrapolate the amount of FTS that can be loaded on the strings using the drug-loaded dipping solution and minimize the use of an expensive drug for preliminary optimization studies.

A series of control experiments were performed to evaluate the effect of polymer molecular weight (MW) on the mass of polymer loaded onto strings under the same dipping conditions. The prototype PCL strings (3 mm OD, 20 mm L, n = 3), were pre-weighed (M_i) before the dip coating process. The dipping conditions involved incubating PCL strings in the dipping solution for 1 min followed by drying at RT for 24 h. After drying, the strings were weighed again (M_f) and mass of polymer loaded was calculated as the difference in mass ($M_f - M_i$) (Supplementary Figure S1D). Results showed that the amounts of PLGA loaded onto the string increased with increasing the dipping solution viscosity. The viscosity of the dipping solution significantly increased with increasing PLGA MW from 1.4 cP (PLGA MW 11 kDa) to 15.7 cP (PLGA MW 53 kDa). This increase in viscosity resulted in a 2.67-fold increase in PLGA loading onto the string (4.67 ± 0.09 vs. 1.75 ± 0.17 mg PLGA) with the highest viscosity solution (15.7 cP) compared to the lowest viscosity solution (1.4 cP). To investigate the effect of incubation time on PLGA loading, strings were incubated in PLGA solutions at RT for 1 min or 24 h (Supplementary Figure S1D). Results showed that there was no significant difference in PLGA loading with longer incubation time. The number of dipping iterations on PLGA loading was investigated over a 1 min incubation period. Strings were dipped once and incubated for the entire 1 min in PLGA solutions or dipped for 6 consecutive times (10 s each) over 1 min. Results showed that increasing the number of dipping iterations resulted in a significant increase in PLGA loading onto the strings for all three PLGA MWs. PLGA loading increased from 1.75 mg to 2.61 mg, 3.07 to 4.6 mg, and 4.67 to 7.27 mg for PLGA MW 11, 27, and 53 kDa, respectively (Supplementary Figure S1D). Finally, the effect of drying time between iterative dipping on PLGA loading was investigated. Results showed that increasing the drying time between iterative dips from 3 s to 30 s resulted in a significant increase in PLGA (MW 53 kDa) loading from 4.67 mg to 10.73 mg (Supplementary Figure S1E). Based on the collective data from the aforementioned optimization steps, the optimized drug loading conditions using the dip-coating process were set to a total of 1 min incubation time with 6 consecutive dipping steps (10 s each) and 30 s drying time between consecutive dipping steps. The loading solution used in subsequent studies with fluticasone (FTS) consisted of 1:5 *w/w* PLGA/acetone. The saturation solubility of FTS in the dipping solutions was ~10 mg/mL for solutions made with PLGA MW 11 and 27 kDa and ~6 mg/mL in the solution containing the highest MW PLGA (53 kDa) (Supplementary Figure S1E).

As the string would ultimately be for human use with an average esophageal length of 20–25 cm (the same length as in the porcine model), and for an overnight dwell time, the drug release was tested for a period of 24 h. Optimization studies were carried out with a 2 cm PCL string and extrapolated to a 25 cm string based on the homogenous and consistent drug loading data obtained from the optimization studies (Table 1). In vitro release studies showed that FTS had a slow burst release within 24 h (≤1%) corresponding to ~15, ~23, and ~44 μg for solutions containing PLGA MW 10, 27, and 53 kDa, respectively. These FTS concentrations were much lower than a targeted release of 1–2 mg FTS in 24 h (Table 1). The low FTS burst release within 24 h was attributed to a high affinity of FTS to PLGA given its hydrophobic nature (LogP = 2.78) resulting in a slow diffusion from the polymer layer to the release medium. To enhance FTS release from the strings, PLGA was removed from the loading solution and a saturated solution of FTS in acetone (20 mg/mL) was used. The drug loading process using the saturated FTS solution in acetone was further optimized to include a 4 min incubation time, with 24 consecutive dipping steps and 30 s drying steps between each consecutive dipping steps. These new loading parameters resulted in significantly higher FTS release within 24 h meeting the targeted 1 mg/day release. This optimized process was used to test FTS release both ex vivo and in vivo in a porcine model.

To design a clinically translatable fluticasone eluting string for human use and easy swallowing, a fabric (cotton) string was loaded with FTS using the optimized dip-coating process (4 min, 24 dips, and 30 s drying) and investigated for FTS release in vitro, ex vivo, and in vivo. In vitro release studies showed that FTS was released from the fabric string at the target rate with ~1.4 mg/day for a 25 cm string (Table 1).

Table 1. In vitro release kinetics of FTS from PCL strings and fabric strings (A) In vitro drug release profile of FTS loaded PCL string (2 cm L) and predicted profile for a 25 cm L string. (B) In vitro release profile of FTS loaded PCL string at varying dipping conditions. (C) In vitro release profile of FTS loaded fabric string. * Dipping Time = 2 min, Number of dips = 12, Time between dips = 30 s. ** Dipping Time = 4 min, Number of dips = 24, Time between dips = 30 s.

(A)

Dipping solution PLGA 50:50 Mw (kDA): Acetone (1:5)	Saturation solubility (FTS) (mg/mL)	Amount of FTS loaded on a 2 cm string * (μg)	% FTS released in 24 h	FTS released in 24 h (μg)	Predicted FTS release from 25 cm PCL string (μg)
10.6	10.14 ± 0.15	149.01 ± 5.40	0.81 ± 0.05	1.20 ± 0.009	15
27.2	10.13 ± 1.13	250.31 ± 82.24	0.74 ± 0.15	1.85 ± 0.30	23.12
53.4	6.17 ± 1.09	4.67 ± 0.09	0.61 ± 0.01	3.50 ± 1.13	43.75

(B)

Dipping solution (Saturated solution of FTS in Acetone)	Amount of FTS loaded on a 2 cm PCL string * (μg)	% FTS released in 24 h	FTS released in 24 h (μg)	Predicted FTS release from 25 cm PCL string (μg)
Dipping condition I *	935.08 ± 114.8	4.73 ± 0.3	43.89 ± 2.9	548.62
Dipping condition II **	1610.16 ± 40.32	5.08 ± 0.15	81.11 ± 1.46	1021.93

(C)

Dipping solution (Saturated solution of FTS in Acetone)	Amount of FTS loaded on a 2 cm fabric string * (μg)	% FTS released in 24 h	FTS released in 24 h (μg)	Predicted FTS release from 25 cm fabric string (μg)
Dipping condition II **	1756.6 ± 20.29	5.54 ± 1.1	111.81 ± 10.17	1397.8

3.1.2. In Vivo Pharmacokinetic (PK) Studies

In vivo studies were carried out in a porcine model to assess the local and systemic pharmacokinetics of FTS from drug-eluting strings (Supplementary Figure S6). FTS-loaded strings (25 cm L) were prepared using the dip-coating method described above. The strings were placed along the entire length of the pig's esophagus as described above. FTS level was measured in both esophagus tissue and plasma using LC-MS/MS analysis. The serum samples collected at 1 and 3 days post string administration showed minimum FTS concentrations (FTS < 30 pg/mL). There was only one exception of FTS concentration (57.6 pg/mL) detected in serum sample of one pig at day 3 post string administration. In contrast, FTS was detected in all in vivo esophageal tissue explants with higher FTS concentrations detected in day 1 (105.89 ±10.87 μg) compared to day 3 (14.42 ± 9.9 μg) (Supplementary Figure S6B) for FTS-loaded PCL strings. Ex vivo studies with fluticasone coated PCL strings (2 cm L) on a fresh porcine esophageal section incubated for 1 or 3 days in PBS showed an opposite trend to in vivo studies with an increase in FTS tissue accumulation after 3 days incubation in PBS (102.2 ± 27.6 μg) compared to a 24 h incubation (31.78 ± 13.19 μg) (Supplementary Figure S2B). These results demonstrate that FTS was released from the strings at a continuous rate and was highly accumulated in the esophageal tissue both ex vivo and in vivo with minimal or no accumulation in the blood in vivo.

To evaluate a more clinically relevant string material, additional in vivo PK studies were carried out with FTS-loaded fabric strings in the porcine model (Supplementary Figure S6A). An average of 22.08 ± 0.31 (n = 5) mg fluticasone was loaded on 25 cm long (L) strings. Strings were then implanted in the esophagus as described above for 1 day of dwell time. Blood (plasma) samples collected at 1 day showed no absorption of FTS, with an exception of 106 pg/mL detected in one pig. FTS was detected in 4 out of 5 pigs' blood (serum) samples with an average of 83 ± 21.6 pg/mL (n = 4). On the other hand, FTS was detected in all in vivo esophageal tissue explants at significantly higher concentrations of 23 ± 12.9 μg, which is orders of magnitude higher compared to serum levels. These results showed that FTS accumulated in the porcine esophageal tissue explants at high levels (~2×10^5 serum levels) detected at each collection time point, with minimum systemic absorption of FTS detected in blood (serum/plasma) samples. To corroborate the in vitro release studies and in vivo PK studies, ex vivo studies were performed with FTS-loaded fabric strings (2 cm L, $n = 3$) to determine FTS accumulation in porcine esophagus tissue explants over 24 h incubation in PBS at 4 °C (Supplementary Figure S2). An average of 103.9 ± 35.4 µg FTS was accumulated in the esophageal tissue explants, which corroborates the in vivo results demonstrating the absorption of FTS in the esophagus.

3.2. Fluticasone-Eluting Esophageal 3D Printed Rings

Fluticasone-loaded esophageal rings were fabricated using biocompatible resins. PCL is commonly used for biomedical applications owing to its biocompatibility and robust mechanical properties [56–59]. To fabricate esophageal rings PCL was chemically functionalized to introduce methacrylate groups and enable photopolymerization cross-linking during the 3D printing process via radical-mediated cross-linking reactions. PCL-diol was successfully functionalized with methacrylate groups to produce PCL dimethacrylate (PCL-DMA) using a previously reported method (Supplementary Scheme S1) [51]. The structure of PCL-DMA was confirmed with nuclear magnetic resonance (NMR) analysis by the presence of peaks at δ5.8 ppm (1H) and δ6.1 ppm (1H), corresponding to the vinyl protons and a peak δ1.5 ppm corresponding to methyl groups (3H) of the methacrylate moiety (Supplementary Figure S3).

Fluticasone (FTS) was formulated in the rings via a pre-loading or post-loading process. In the pre-loading process, FTS was dissolved in a liquid resin prior to 3D printing fabrication. The resin formulation composed of a photoinitiator (TPO), UV absorber (BLS), and PCL_{700}-DMA added at predetermined ratios (Table 2) [51]. The mole percent (%) of the various resin components was calculated relative to the methacrylate groups. The resin components were mixed and stirred overnight in an amber glass bottle at room temperature (RT). Resin viscosity in the range of 200 to 600 cP was required in order to allow fabrication accuracy and part shape fidelity using the 3D printing process [60,61]. For photopolymerization-based 3D printing, a light reactive diluent is usually added to reduce the viscosity of the resin [59,62]. Resin formulations used to fabricate the rings had viscosities in the aforementioned printable range. The viscosities of resin formulations were 490 cP and 350 cP for PCL_{700}-DMA resin without diluent and PCL_{700}-DMA with a diluent (HEMA) respectively.

Table 2. Composition of resin formulations (PCL_{700}-DMA) and (PCL_{700}-DMA/HEMA) used to 3D print rings.

Formulation	PCL_{700}DMA (mole%)	HEMA (mole%)	TPO (mole%)	BLS (mole%)
PCL_{700}-DMA	100	-	0.04	0.007
PCL_{700}-DMA/HEMA	23	73	0.05	0.001

The saturation solubility of FTS in the PCL_{700}-DMA and PCL_{700}-DMA/HEMA resin formulation was determined by HPLC analysis and quantified at 14.62 ± 0.82 mg/g and 12.91 ± 0.09 mg/g, respectively. To ensure FTS was homogenously dissolved within the resin, sample aliquots ($n = 4$) were collected from different areas in the solution and analyzed by HPLC. The FTS-resin formulation was deemed homogenous if the average concentration of FTS in the sample aliquots ($n = 4$) had a standard deviation ≤5%. The final resin formulations contained 1.4% *w/w* FTS in the resin. FTS-loaded rings were fabricated using 3D printing as described above and the amount of FTS loaded in 3D-printed rings was determined by incubating rings in acetonitrile (ACN, 50 mL) at 37 °C for 24 h to extract FTS and quantifying FTS concentration in ACN by HPLC analysis.

The stability of FTS in the resin formulations (PCL_{700}-DMA and PCL_{700}-DMA/HEMA) was determined under three storage conditions of varying temperature and relative humidity (RH) (25 °C, 25 °C/60% RH, 40 °C/75%RH) over 6 months by quantifying FTS concentration in the resin by HPLC analysis. Results showed that FTS was stable under all three storage conditions for up to 6 months (Supplementary Figure S7).

3.2.1. In Vitro Release Studies

In order to achieve the target drug release rate of 1 mg/day over 30 days, rings were loaded with 30 mg of FTS (Figure 1A). The average amount of drug loaded in the rings was 32.11 ± 1.21 mg as determined by drug extraction and HPLC analysis. Results showed FTS had a minimum burst release of ~3% (958 µg) in the first 24 h followed by zero-order kinetics at 282 µg/day over 30 days (Figure 1B, Table 3). The daily drug release from these rings was lower than the target drug release (1 mg/day). An important advantage of utilizing 3D printing is the ability to use a variety of photopolymerizable resins and ring designs. The rate of diffusion and drug release can be controlled by changing the crosslink density of the resin as well as the chemical structure and properties of the polymer. Hydroxyethylmethacrylate (HEMA) was added as a hydrophilic chain extending diluent to study the effect of crosslink density of resin formulation on the release kinetics of FTS (Table 2).

Table 3. Total amount of FTS in rings normalized to the weights of rings (n = 3), release rate of FTS at zero order kinetics (µg/day).

Formulation	Amount of FTS in Rings (mg)	FTS Burst in 24 h (%)	FTS Burst in 24 h (µg)	FTS Zero Order Release Rate (µg/day)
PCL_{700}-DMA	32.11 ± 1.12	2.98 ± 0.24	958.17 ± 74.45	282.0 (R^2 = 0.99)
PCL_{700}-DMA/HEMA	31.09 ± 0.62	4.21 ± 0.14	1310.0 ± 10.18	207.0 (R^2 = 0.98)

Results showed that FTS exhibited a minimum burst release in the first 24 h (~4.5%, 1.31 mg) and slower zero-order kinetics with 207.0 µg/day over 30 days compared to 282 µg/day obtained with FTS rings fabricated without HEMA (Table 3). The slower release kinetics when HEMA was included as a diluent in the resin was attributed to the presence of hydrogen bonding between the hydroxyl groups of HEMA and the carboxyl groups in FTS (Supplementary Figure S8). The hydrogen bonding resulted in a tighter association between the drug and the polymeric network of the resin formulation and thus slower release rate in vitro. Collectively, these results showed that FTS release kinetics could be fine-tuned by changing the resin composition.

3.2.2. Mechanical Properties of 3D Printed Rings

To enable successful implantation in the esophagus, the mechanical properties of the FTS-eluting esophageal rings were tested and optimized. In general, a post-fabrication UV or thermal curing process is implemented in 3D printing in order to improve the mechanical properties of 3D printed parts [63–68]. These post-fabrication curing treatments lead to improved compression strength and elastic modulus of the 3D-printed parts [69–71]. The effect of post-fabrication UV curing on the mechanical properties of placebo rings fabricated without (PCL_{700}-DMA) or with diluent (PCL_{700}-DMA/HEMA) was investigated. UV-treated and non-UV-treated rings were compressed to a distance of 10, 20, and 50% of their outer diameter. The load at these percent compressions in Newtons (N) was reported (Table 4). All measurements were done in triplicates. Results showed that using both resins (PCL_{700}-DMA, PCL_{700}-DMA/HEMA), rings exposed to UV treatment post fabrication had significantly higher compression forces at all compression distances compared to rings that did not undergo post-fabrication UV treatment (Table 4).

Table 4. The effect of UV-curing on 3D printed rings: Radial compression force at 10%, 20%, and 50% outer diameter compression of non-UV- and UV-cured FTS-loaded EoE rings printed with two different resin formulations (PCL_{700}-DMA, PCL_{700}-DMA/HEMA).

Ring Diameter Displacement (%)	Compression Force PCL_{700}-DMA (non-UV) (N)	Compression Force PCL_{700}-DMA (UV) (N)	Compression Force PCL_{700}DMA/HEMA (non-UV) (N)	Compression force PCL_{700}DMA/HEMA(UV) (N)
10	0.04 ± 0.01	0.72 ± 0.09	0.10 ± 0.03	0.93 ± 0.1
20	0.10 ± 0.01	1.22 ± 0.06	0.26 ± 0.01	1.6 ± 0.03
50	0.26 ± 0.08	1.99 ± 0.14	0.96 ± 0.02	2.9 ± 0.01

The Young's modulus measures the resistance of a ring to elastic deformation under a specific load and its ability to recover its original shape once the stress force is removed. Results showed that rings exposed to UV curing post fabrication were stiffer and had a higher Young's modulus (1736.33 ± 34.8 Pa for PCL_{700}-DMA and 1830.9 ± 60.3 Pa for PCL_{700}-DMA/HEMA) compared to 8.61 ± 3.5 Pa (PCL_{700}-DMA) and 10.93 ± 1.2 Pa (PCL_{700}-DMA/HEMA) for rings that did not undergo post-fabrication UV cure (Supplementary Table S1). Collectively, these results demonstrate that UV curing post fabrication significantly improved the mechanical properties of 3D printed rings. The increase in compression force and Young's modulus was attributed to an increase in crosslink density of the polymer matrix upon exposure to UV and formation of a tighter network. This effect was confirmed by determining the gel fraction and percent (%) swelling of the rings in organic and aqueous solvents. Rings that were exposed to a UV curing treatment post fabrication had higher gel fraction and lower percent swelling compared to rings that did not undergo UV curing (Supplementary Table S1) due to higher crosslink density of the polymer matrix as a result of UV exposure.

To assess the stability of FTS when exposed to post-fabrication UV curing, in vitro release kinetics of fluticasone from UV cured rings was investigated (Figure 2). Results showed that rings exposed to a post-fabrication UV cure had slightly higher burst of FTS in the first 24 h (4%, 1179.9 µg) and higher zero-order release rate (370 µg/day) compared to non-UV cured rings (Table 5). These results showed that UV curing improved the mechanical properties of rings and resulted in higher release rates of FTS in vitro.

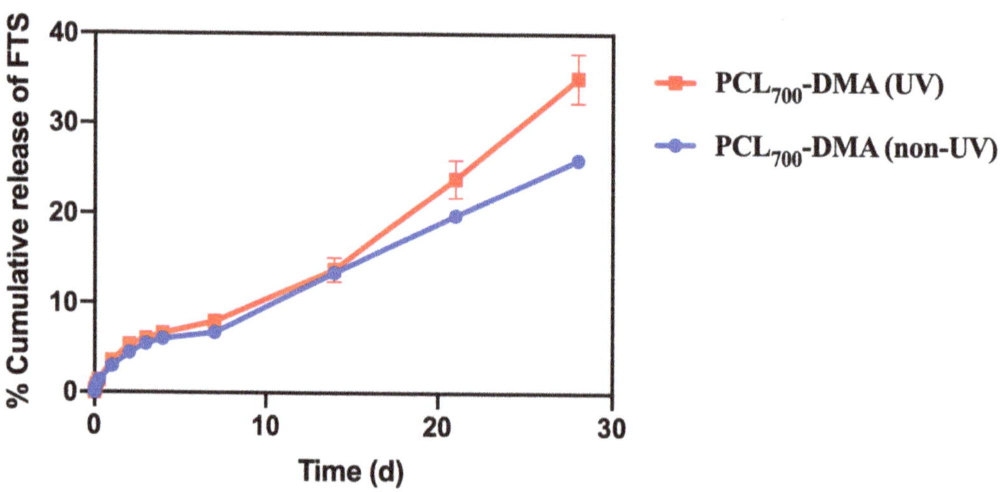

Figure 2. In vitro release kinetics of FTS from non-UV- and UV-cured rings printed with PCL_{700}-DMA resin formulation. Rings were incubated in PBS at 37 °C for 30 days and sample aliquots (1 mL) were collected and analyzed by HPLC analysis. All error bars represent standard deviation for $n = 3$.

Table 5. Total amount of FTS in rings normalized to the weights of rings (n = 3), release rate of FTS at zero order kinetics (µg/day).

Rings	Amount of FTS in Rings (mg)	FTS Burst in 24 h (%)	FTS Burst in 24 h (µg)	FTS Zero Order Release Rate (µg/day)
PCL_{700}-DMA (non-UV)	32.11 ± 1.12	2.98 ± 0.24	958.17 ± 74.45	282.0 (R^2 = 0.99)
PCL_{700}-DMA (UV)	32.77 ± 1.11	3.59 ± 0.41	1179.9 ± 180.8	370.0 (R^2 = 0.98)

3.2.3. Effect of Drug Incorporation Process

Fabrication of drug-loaded rings can be done by either incorporating drug(s) in the initial resin formulation (termed pre-loading, Figure 1A) or by adding drug(s) to a 3D printed ring post fabrication (termed post-loading, Figure 3A). We sought out to investigate the effect of FTS incorporation step on in vitro release kinetics from rings. A number of solvents were screened to determine a suitable solvent for post-loading FTS, and acetone was selected based on high solubility of FTS and ease of removal. Placebo rings were post-loaded with FTS by incubating them in a near saturated solution of FTS in acetone (20 mg/mL; 50 mL) at RT for 24 h.

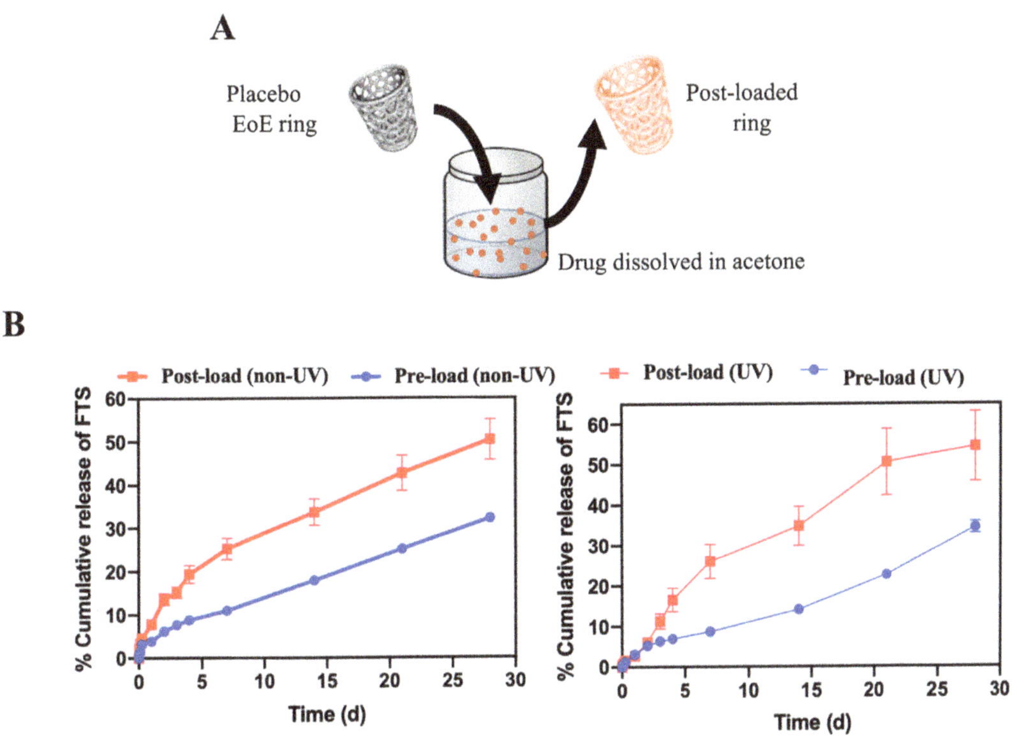

Figure 3. Effect of drug incorporation process onto drug release kinetics of rings printed with the PCL_{700}-DMA resin formulation. (**A**) A pictorial representation of post-loading of FTS into 3D printed placebo ring. (**B**) In vitro release kinetics of FTS from pre-loaded (non-UV- and UV-cured rings) and post-loaded (non-UV- and UV-cured rings) printed using the PCL_{700}-DMA resin formulation. Rings were incubated in PBS at 37 °C for 30 days and sample aliquots (1 mL) were collected and analyzed by HPLC analysis. All error bars represent standard deviation for n = 3.

Results demonstrated that rings post-loaded with FTS exhibited ~2-fold increase in burst release within the first 24 h (~8%, 550 μg) compared to rings pre-loaded with FTS (~4%, 280 μg). Rings post-loaded with FTS also exhibited greater zero-order release kinetics over 30 days, with 90 μg/day compared to 50 μg/day for pre-loaded rings (Figure 3B, Table 6). These results demonstrate the ability to fine-tune drug release kinetics using different drug loading processes.

Table 6. Total amount of FTS in rings normalized to the weights of rings (n = 3), release rate of FTS at zero order kinetics (μg/day).

Rings	FTS Loading Per Ring (mg/g)	FTS Burst in 24 h (%)	FTS burst in 24 h (μg)	FTS Zero Order Release Rate (μg/day)
Pre-loaded (non-UV)	7.13	3.99 ± 0.11	280.0 ± 70.0	50.0 (R^2 = 0.99)
Post-loaded (non-UV)	6.84	7.94 ± 0.50	550.0 ± 1.0	90.0 (R^2 = 0.99)
Pre-loaded (UV)	11.70	3.12 ± 0.13	450.0 ± 7 0.1	237 (R^2 = 0.978)
Post-loaded (UV)	12.56	2.78 ± 0.21	420.0 ± 10.6	270 (R^2 = 0.98)

Similar to the pre-loaded rings, the effect of post-fabrication UV curing on fluticasone release rate in post-loaded rings was investigated. 3D-printed placebo rings were UV cured using the method described above and subsequently incubated in a near saturated solution of fluticasone in acetone at RT for 24 h. Results showed that post-loaded UV-cured rings exhibited a lower burst release of FTS in the first 24 h (~3%, 420 μg) but faster zero order release rates ~270 μg/day over 30 days compared to non-UV cured post-loaded rings (Table 6). For both drug loading strategies, fluticasone exhibited a sustained zero-order release profile over 30 days, demonstrating the potential use of 3D printed rings as a long-acting esophageal drug delivery device.

3.2.4. In Vitro Cytotoxicity Studies

Biocompatibility of 3D-printed rings was assessed in vitro in HeLa cells. Small discs (3 mm OD) were fabricated with PCL_{700}-DMA resin and tested with HeLa cells in 96-well plates. Placebo, FTS-loaded, and UV-cured FTS-loaded discs were tested for cytotoxicity using the CellTiter-Glo® luminescent viability assay. Results showed that all discs were biocompatible and well tolerated by the HeLa cells (Supplementary Figure S2).

3.2.5. Ex Vivo Pharmacokinetic Studies

To assess the pharmacokinetics of FTS in local tissue, ex vivo studies were performed using fresh porcine esophageal tissue with pre-loaded and post-loaded fluticasone rings (n = 3 per timepoint). The FTS concentration in esophageal tissue was determined using LC-MS/MS. Results demonstrated that sustained levels of FTS were detected in all samples over 14 days (Supplementary Figure S3). These results demonstrate that FTS was effectively diffused from the rings and highly accumulated in the esophageal tissue.

3.2.6. In Vivo Pharmacokinetic Studies

Additionally, in vivo studies were carried out in a porcine model to assess the local and systemic pharmacokinetics of FTS-loaded rings. To ensure that rings were properly placed in the proximal part of the esophagus, initial studies were carried out using placebo rings (without FTS). We have shown successful endoscopic ring implantation into the proximal esophagus of one set of five pigs. After successful implantation of placebo rings, FTS-loaded rings were fabricated using pre-loaded PCl_{700}-DMA resin formulation. Additionally, a post-fabrication UV curing was implemented to fabricate rings with mechanical properties appropriate for esophageal implantation. These FTS-loaded rings were deployed in the proximal esophagus of two sets of five pigs each for a period of 7 days. A daily endoscopy was performed to monitor the placement of rings. During endoscopic screening, it was observed that the rings were not present in the proximal esophagus of pigs after day 1. This observation was corroborated by the absence of FTS in esophageal tissue and plasma collected at euthanasia at day 7 post ring administration. As the pigs were provided with an ad lib diet during the study, we hypothesize that the rings slipped into the pig's stomach along with the food. The PK analysis showed no accumulation of FTS in the esophageal tissue and no detectable plasma levels at day 7. These studies demonstrated that rings can be successfully inserted in the proximal part of the esophagus; however, further optimization in order to keep these rings in place for a longer time is needed.

4. Discussion

Development of new and innovative esophageal medication delivery systems are urgently needed for the treatment of EoE. In this study, we provide a comprehensive overview of the development of two types of drug delivery systems: (1) an FTS-eluting string for a rapid but sustained overnight release, and (2) a 3D-printed FTS-eluting ring to allow a long-acting constant drug release. In vitro release studies of FTS from drug-coated strings achieved the target drug release for a 24 h time period. FTS-eluting 3D-printed rings are capable of providing drug release for a period of 1 month or longer in vitro. FTS can be incorporated into the rings via two drug loading processes. Pre-loading involves addition of FTS into the resin formulation prior to ring fabrication. FTS loaded in these resin formulations was stable under accelerated stability storage conditions over six months, and FTS retained its physical and chemical properties. Our results have suggested that this method of drug loading exhibited controlled and sustained release of FTS over a period of 1 month (Figure 1). Alternatively, FTS can be incorporated into rings via post-loading, i.e., after the 3D printing fabrication process. Results demonstrated post-loading of FTS after ring fabrication led to increase in burst release compared to rings loaded with FTS using the pre-loading process. Additionally, post-loaded rings showed faster zero-order release kinetics over 30 days in comparison to pre-loaded rings (Figure 3). Drug release kinetics can also be controlled by changing the composition and crosslink density of resin formulations. Our results showed that addition of a hydrophilic diluent (HEMA) to hydrophobic resin formulation PCL_{700}-DMA lead to slower zero-order kinetics of FTS at 207 μg/day over 30 days compared to 282 μg/day obtained with FTS rings fabricated without HEMA (Figure 1). Mechanical properties of 3D printed rings can be optimized without impacting drug release kinetics of FTS. A post-fabrication UV-curing led to significant improvement in the mechanical properties of FTS loaded rings and higher FTS release rate in vitro compared to non-UV-cured rings. Collectively, these results demonstrate the flexibility of two fluticasone releasing platform technologies to achieve a controlled release kinetics and drug concentrations that can be translated to human use for treatment of EoE.

5. Conclusions

We have successfully demonstrated two platforms for delivery of fluticasone (FTS) for the treatment of eosinophilic esophagitis (EoE). An FTS-eluting string is an innovative approach to drug delivery that is potentially very appealing as it can be swallowed and release drug along the entire length of the esophagus. Additionally, FTS-loaded rings were successfully fabricated using 3D-DLP printing to provide controlled and prolonged drug release, though refinements will be need for more durable implantation. In vitro release studies demonstrated the ability to fine-tune drug release kinetics by changing the crosslink density of resin formulations. Post-printing UV curing improved the mechanical properties of these rings. Additionally, results showed that the process of drug loading into the rings had an effect on drug release kinetics. While there is still room to improve the drug release rate from these rings to achieve the target daily dose of 1 mg/day, and to improve the design for a more durable esophageal implantation, the approaches for esophageal drug delivery overall are very promising. To our knowledge, this is the first report of fabrication and development of string- and ring-based drug delivery devices for the treatment of EoE. Moreover, these drug delivery technologies hold the potential to also serve as a platform for treatment of other esophageal conditions, ranging from inflammatory to fibrotic to neoplastic diseases.

Supplementary Materials: The following are available online at https://www.mdpi.com/2073-4360/13/4/557/s1.

Author Contributions: Conceived and designed the experiments: S.R.B., A.P., A.S., A.B., and E.S.D.; Performed the experiments: A.P., R.S., D.D., P.S.-H., P.M., T.P., A.B., and E.S.D.; Contributed reagents/materials/analysis tools: J.B., S.R.B., A.B., W.Z., and E.S.D.; Wrote the paper: S.R.B., A.P., A.S., and E.S.D. All authors have read and agreed to the published version of the manuscript.

Funding: This research received no external funding.

Data Availability Statement: All other data supporting the findings of this manuscript are available from the corresponding author (S.R.B.) upon reasonable request.

Acknowledgments: This work was supported by the NC TraCS/NC State Collaborative Translational Research Pilot Grant Award (UNCSUR11602) and the UNC CTSA grant (UL1TR001111), a UNC Center for Health Innovation Pilot Award, and used resources from the Large Animal and Histology Cores of the UNC Center for GI Biology and Disease (P30 DK034987).

Conflicts of Interest: The authors declare no conflict of interest.

References

1. Furuta, G.T.; Liacouras, C.A.; Collins, M.H.; Gupta, S.K.; Justinich, C.; Putnam, P.E.; Bonis, P.; Hassall, E.; Straumann, A.; Rothenberg, M.E. Eosinophilic Esophagitis in Children and Adults: A Systematic Review and Consensus Recommendations for Diagnosis and Treatment: Sponsored by the American Gastroenterological Association (AGA) Institute and North American Society of Pediatric Gastroenterology, Hepatology, and Nutrition. *Gastroenterology* **2007**, *133*, 1342–1363.
2. Dellon, E.S.; Liacouras, C.A.; Molina-Infante, J.; Furuta, G.T.; Spergel, J.M.; Zevit, N.; Spechler, S.J.; Attwood, S.E.; Straumann, A.; Aceves, S.S.; et al. Updated International Consensus Diagnostic Criteria for Eosinophilic Esophagitis: Proceedings of the AGREE Conference. *Gastroenterology* **2018**, *155*, 1022–1033.e10. [CrossRef] [PubMed]
3. Dellon, E.S.; Jensen, E.T.; Martin, C.F.; Shaheen, N.J.; Kappelman, M.D. Prevalence of eosinophilic esophagitis in the United States. *Clin. Gastroenterol Hepatol* **2014**, *12*, 589–596.e1. [CrossRef] [PubMed]
4. Jensen, E.T.; Kappelman, M.D.; Martin, C.F.; Dellon, E.S. Health-care utilization, costs, and the burden of disease related to eosinophilic esophagitis in the United States. *Am. J. Gastroenterol.* **2015**, *110*, 626–632. [CrossRef]
5. Dellon, E.S.; Liacouras, C.A. Advances in clinical management of eosinophilic esophagitis. *Gastroenterology* **2014**, *147*, 1238–1254. [CrossRef]
6. Cotton, C.C.; Eluri, S.; Wolf, W.A.; Dellon, E.S. Six-Food Elimination Diet and Topical Steroids are Effective for Eosinophilic Esophagitis: A Meta-Regression. *Dig. Dis. Sci.* **2017**, *62*, 2408–2420. [CrossRef]
7. Butz, B.K.; Wen, T.; Gleich, G.J.; Furuta, G.T.; Spergel, J.; King, E.; Kramer, R.E.; Collins, M.H.; Stucke, E.; Mangeot, C.; et al. Efficacy, dose reduction, and resistance to high-dose fluticasone in patients with eosinophilic esophagitis. *Gastroenterology* **2014**, *147*, 324–333.e5. [CrossRef] [PubMed]
8. Dellon, E.S.; Woosley, J.T.; Arrington, A.; McGee, S.J.; Covington, J.; Moist, S.E.; Gebhart, J.H.; Tylicki, A.E.; Shoyoye, S.O.; Martin, C.F.; et al. Efficacy of Budesonide vs Fluticasone for Initial Treatment of Eosinophilic Esophagitis in a Randomized Controlled Trial. *Gastroenterology* **2019**, *157*, 65–73.e5. [CrossRef] [PubMed]
9. Dellon, E.S. Management of refractory eosinophilic oesophagitis. *Nat. Rev. Gastroenterol. Hepatol.* **2017**, *14*, 479–490. [CrossRef]
10. Dellon, E.S.; Sheikh, A.; Speck, O.; Woodward, K.; Whitlow, A.B.; Hores, J.M.; Ivanovic, M.; Chau, A.; Woosley, J.T.; Madanick, R.D.; et al. Viscous Topical Is More Effective Than Nebulized Steroid Therapy for Patients With Eosinophilic Esophagitis. *Gastroenterology* **2012**, *143*, 321–324.e1. [CrossRef] [PubMed]
11. Fillon, S.A.; Harris, J.K.; Wagner, B.D.; Kelly, C.J.; Stevens, M.J.; Moore, W.; Fang, R.; Schroeder, S.; Masterson, J.C.; Robertson, C.E.; et al. Novel device to sample the esophageal microbiome—The esophageal string test. *PLoS ONE* **2012**, *7*, e42938. [CrossRef] [PubMed]
12. Furuta, G.T.; Kagalwalla, A.F.; Lee, J.J.; Alumkal, P.; Maybruck, B.T.; Fillon, S.; Masterson, J.C.; Ochkur, S.; Protheroe, C.; Moore, W.; et al. The oesophageal string test: A novel, minimally invasive method measures mucosal inflammation in eosinophilic oesophagitis. *Gut* **2013**, *62*, 1395–1405. [CrossRef] [PubMed]
13. Ackerman, S.J.; Kagalwalla, A.F.; Hirano, I.; Gonsalves, N.; Katcher, P.M.; Gupta, S.; Wechsler, J.B.; Grozdanovic, M.; Pan, Z.; Masterson, J.C.; et al. One-Hour Esophageal String Test: A Nonendoscopic Minimally Invasive Test That Accurately Detects Disease Activity in Eosinophilic Esophagitis. *Am. J. Gastroenterol.* **2019**, *114*, 1614–1625. [CrossRef]
14. Krause, J.; Rosenbaum, C.; Grimm, M.; Rump, A.; Kessler, R.; Hosten, N.; Weitschies, W. The EsoCap-system—An innovative platform to drug targeting in the esophagus. *J. Control. Release* **2020**, *327*, 1–7. [CrossRef] [PubMed]

15. Mayer, H.C.; Krechetnikov, R. Landau-Levich flow visualization: Revealing the flow topology responsible for the film thickening phenomena. *Phys. Fluids* **2012**, *24*, 5. [CrossRef]
16. Gibson, M.; Frejlich, J.; Machorro, R. Dip-coating method for fabricating thin photoresist films. *Thin Solid Films* **1985**, *128*, 161–170. [CrossRef]
17. Brinker, C.J.; Frye, G.C.; Hurd, A.J.; Ashley, C.S. Fundamentals of sol-gel dip coating. *Thin Solid Films* **1991**, *201*, 97–108. [CrossRef]
18. Plundrich, N.J.; Lila, M.A.; Plundrich, N.J.; Smith, A.R.; Borst, L.B.; Snider, D.B.; Kaser, T.; Blikslager, A.T.; Odle, J.; Lila, M.A.; et al. Oesophageal eosinophilia accompanies food allergy to hen egg white protein in young pigs. *Clin. Exp. Allergy* **2020**, *50*, 95–104. [CrossRef]
19. Barner-Kowollik, C.; Bastmeyer, M.; Blasco, E.; Delaittre, G.; Muller, P.; Richter, B.; Wegener, M. 3D Laser Micro- and Nanoprinting: Challenges for Chemistry. *Angew. Chem. Int. Ed. Eng.* **2017**, *56*, 15828–15845. [CrossRef]
20. Jungst, T.; Smolan, W.; Schacht, K.; Scheibel, T.; Groll, J. Strategies and Molecular Design Criteria for 3D Printable Hydrogels. *Chem. Rev.* **2016**, *116*, 1496–1539. [CrossRef]
21. Murphy, S.V.; Atala, A. 3D bioprinting of tissues and organs. *Nat. Biotechnol.* **2014**, *32*, 773–785. [CrossRef]
22. Matos, F.; Godina, R.; Jacinto, C.; Carvalho, H.; Ribeiro, I.; Peças, P. Additive Manufacturing: Exploring the Social Changes and Impacts. *Sustainability* **2019**, *11*, 3757. [CrossRef]
23. Gao, W.; Zhang, Y.; Ramanujan, D.; Ramani, K.; Chen, Y.; Williams, C.B.; Wang, C.C.L.; Shin, Y.C.; Zhang, S.; Zavattieri, P.D. The status, challenges, and future of additive manufacturing in engineering. *Comput. Aided Des.* **2015**, *69*, 65–89. [CrossRef]
24. Prasad, L.K.; Smyth, H. 3D Printing technologies for drug delivery: A review. *Drug Dev. Ind. Pharm.* **2016**, *42*, 1019–1031. [CrossRef]
25. O'Brien, C.M.; Holmes, B.; Faucett, S.; Zhang, L.G. Three-dimensional printing of nanomaterial scaffolds for complex tissue regeneration. *Tissue Eng. Part B Rev.* **2015**, *21*, 103–114. [CrossRef]
26. Piard, C.M.; Chen, Y.; Fisher, J.P. Cell-Laden 3D Printed Scaffolds for Bone Tissue Engineering. *Clin. Rev. Bone Miner. Metab.* **2015**, *13*, 245–255. [CrossRef]
27. Dawood, A.; Marti Marti, B.; Sauret-Jackson, V.; Darwood, A. 3D printing in dentistry. *Br. Den. J.* **2015**, *219*, 521–529. [CrossRef] [PubMed]
28. Sun, J.; Zhou, W.; Huang, D.; Fuh, J.Y.H.; Hong, G.S. An Overview of 3D Printing Technologies for Food Fabrication. *Food Bioprocess Technol.* **2015**, *8*, 1605–1615. [CrossRef]
29. Ambrosi, A.; Pumera, M. 3D-printing technologies for electrochemical applications. *Chem. Soc. Rev.* **2016**, *45*, 2740–2755. [CrossRef] [PubMed]
30. Melchels, F.P.; Feijen, J.; Grijpma, D.W. A review on stereolithography and its applications in biomedical engineering. *Biomaterials* **2010**, *31*, 6121–6130. [CrossRef] [PubMed]
31. Xing, J.F.; Zheng, M.L.; Duan, X.M. Two-photon polymerization microfabrication of hydrogels: An advanced 3D printing technology for tissue engineering and drug delivery. *Chem. Soc. Rev.* **2015**, *44*, 5031–5039. [CrossRef]
32. Shirazi, S.F.; Gharehkhani, S.; Mehrali, M.; Yarmand, H.; Metselaar, H.S.; Adib Kadri, N.; Osman, N.A. A review on powder-based additive manufacturing for tissue engineering: Selective laser sintering and inkjet 3D printing. *Sci. Technol. Adv. Mater.* **2015**, *16*, 033502. [CrossRef] [PubMed]
33. Zein, I.; Hutmacher, D.W.; Tan, K.C.; Teoh, S.H. Fused deposition modeling of novel scaffold architectures for tissue engineering applications. *Biomaterials* **2002**, *23*, 1169–1185. [CrossRef]
34. Naito, T.; Nakamura, M.; Kaji, N.; Kubo, T.; Baba, Y.; Otsuka, K. Three-Dimensional Fabrication for Microfluidics by Conventional Techniques and Equipment Used in Mass Production. *Micromachines (Basel)* **2016**, *7*, 82. [CrossRef]
35. Chatani, S.; Kloxin, C.J.; Bowman, C.N. The power of light in polymer science: Photochemical processes to manipulate polymer formation, structure, and properties. *Polym. Chem.* **2014**, *5*, 2187–2201. [CrossRef]
36. Credi, C.; Fiorese, A.; Tironi, M.; Bernasconi, R.; Magagnin, L.; Levi, M.; Turri, S. 3D Printing of Cantilever-Type Microstructures by Stereolithography of Ferromagnetic Photopolymers. *ACS Appl. Mater. Interfaces* **2016**, *8*, 26332–26342. [CrossRef]
37. Tumbleston, J.R.; Shirvanyants, D.; Ermoshkin, N.; Janusziewicz, R.; Johnson, A.R.; Kelly, D.; Chen, K.; Pinschmidt, R.; Rolland, J.P.; Ermoshkin, A.; et al. Continuous liquid interface production of 3D objects. *Science* **2015**, *347*, 1349. [CrossRef] [PubMed]
38. Bagheri, A.; Jin, J. Photopolymerization in 3D Printing. *ACS Appl. Polym. Mater.* **2019**, *1*, 593–611. [CrossRef]
39. O'Neill, P.F.; Kent, N.; Brabazon, D. Mitigation and control of the overcuring effect in mask projection micro-stereolithography. *AIP Conf. Proc.* **2017**, *1896*, 200012.
40. Ge, Q.; Sakhaei, A.H.; Lee, H.; Dunn, C.K.; Fang, N.X.; Dunn, M.L. Multimaterial 4D Printing with Tailorable Shape Memory Polymers. *Sci. Rep.* **2016**, *6*, 31110. [CrossRef] [PubMed]
41. Sun, C.; Fang, N.; Wu, D.M.; Zhang, X. Projection micro-stereolithography using digital micro-mirror dynamic mask. *Sens. Actuators A Phys.* **2005**, *121*, 113–120. [CrossRef]
42. Hwang, H.H.; Zhu, W.; Victorine, G.; Lawrence, N.; Chen, S. 3d Printing: 3D-Printing of Functional Biomedical Microdevices via Light- and Extrusion-Based Approaches (Small Methods 2/2018). *Small Methods* **2018**, *2*, 1870021. [CrossRef]
43. Acosta-Vélez, G.F.; Zhu, T.Z.; Linsley, C.S.; Wu, B.M. Photocurable poly(ethylene glycol) as a bioink for the inkjet 3D pharming of hydrophobic drugs. *Int. J. Pharm.* **2018**, *546*, 145–153. [CrossRef]
44. Holländer, J.; Hakala, R.; Suominen, J.; Moritz, N.; Yliruusi, J.; Sandler, N. 3D printed UV light cured polydimethylsiloxane devices for drug delivery. *Int. J. Pharm.* **2018**, *544*, 433–442. [CrossRef] [PubMed]

45. Khaled, S.A.; Alexander, M.R.; Wildman, R.D.; Wallace, M.J.; Sharpe, S.; Yoo, J.; Roberts, C.J. 3D extrusion printing of high drug loading immediate release paracetamol tablets. *Int. J. Pharm.* **2018**, *538*, 223–230. [CrossRef] [PubMed]
46. Zhang, J.; Feng, X.; Patil, H.; Tiwari, R.V.; Repka, M.A. Coupling 3D printing with hot-melt extrusion to produce controlled-release tablets. *Int. J. Pharm.* **2017**, *519*, 186–197. [CrossRef]
47. Chia, H.N.; Wu, B.M. Recent advances in 3D printing of biomaterials. *J. Biol. Eng.* **2015**, *9*, 1–33. [CrossRef] [PubMed]
48. Baum, M.M.; Butkyavichene, I.; Churchman, S.A.; Lopez, G.; Miller, C.S.; Smith, T.J.; Moss, J.A. An intravaginal ring for the sustained delivery of tenofovir disoproxil fumarate. *Int. J. Pharm.* **2015**, *495*, 579–587. [CrossRef]
49. Kimball, A.B.; Javorsky, E.; Ron, E.S.; Crowley, W., Jr.; Langer, R. A novel approach to administration of peptides in women: Systemic absorption of a GnRH agonist via transvaginal ring delivery system. *J. Control Release* **2016**, *233*, 19–28. [CrossRef]
50. Nel, A.; Bekker, L.G.; Bukusi, E.; Hellstrm, E.; Kotze, P.; Louw, C.; Martinson, F.; Masenga, G.; Montgomery, E.; Ndaba, N.; et al. Safety, Acceptability and Adherence of Dapivirine Vaginal Ring in a Microbicide Clinical Trial Conducted in Multiple Countries in Sub-Saharan Africa. *PLoS ONE* **2016**, *11*, e0147743. [CrossRef]
51. Bloomquist, C.J.; Mecham, M.B.; Paradzinsky, M.D.; Janusziewicz, R.; Warner, S.B.; Luft, J.C.; Mecham, S.J.; Wang, A.Z.; DeSimone, J.M. Controlling release from 3D printed medical devices using CLIP and drug-loaded liquid resins. *J. Control Release* **2018**, *278*, 9–23. [CrossRef]
52. Lyapun, I.N.; Andryukov, B.G.; Bynina, M.P. HeLa Cell Culture: Immortal Heritage of Henrietta Lacks. *Mol. Genet. Microbiol. Virol.* **2019**, *34*, 195–200. [CrossRef]
53. Mather, J.; Rainville, P.D.; Graham, K.S.; Plumb, R.S. *A High Sensitivity UPLC/MS/MS Method for the Analysis of Fluticasone Propionate in Plasma*; Application Note; Waters Corporation: Milford, MA, USA, 2018; pp. 1–5.
54. Quéré, D. Fluid coating on a fiber. *Annu. Rev. Fluid Mech.* **1999**, *31*, 347–384. [CrossRef]
55. Shim, E.; Park, J.O.; Srinivasarao, M. Forced coating of polypropylene fibers with non-wetting fluids: The scaling of the film thickness. *Mod. Phys. Lett. B* **2008**, *22*, 2043–2053. [CrossRef]
56. Hollister, S.J. Erratum: Porous scaffold design for tissue engineering. *Nat. Mater.* **2006**, *5*, 590. [CrossRef]
57. Kweon, H.; Yoo, M.K.; Park, I.K.; Kim, T.H.; Lee, H.C.; Lee, H.-S.; Oh, J.-S.; Akaike, T.; Cho, C.-S. A novel degradable polycaprolactone networks for tissue engineering. *Biomaterials* **2003**, *24*, 801–808. [CrossRef]
58. Williams, J.M.; Adewunmi, A.; Schek, R.M.; Flanagan, C.L.; Krebsbach, P.H.; Feinberg, S.E.; Hollister, S.J.; Das, S. Bone tissue engineering using polycaprolactone scaffolds fabricated via selective laser sintering. *Biomaterials* **2005**, *26*, 4817–4827. [CrossRef] [PubMed]
59. Gunatillake, P.A.; Adhikari, R. Biodegradable synthetic polymers for tissue engineering. *Eur. Cells Mater.* **2003**, *5*, 1–16. [CrossRef]
60. Jansen, J.; Melchels, F.P.W.; Grijpma, D.W.; Feijen, J. Fumaric Acid Monoethyl Ester-Functionalized Poly(d,l-lactide)/N-vinyl-2-pyrrolidone Resins for the Preparation of Tissue Engineering Scaffolds by Stereolithography. *Biomacromolecules* **2009**, *10*, 214–220. [CrossRef]
61. Mu, Q.; Wang, L.; Dunn, C.K.; Kuang, X.; Duan, F.; Zhang, Z.; Qi, H.J.; Wang, T. Digital light processing 3D printing of conductive complex structures. *Addit. Manuf.* **2017**, *18*, 74–83. [CrossRef]
62. Choi, J.-W.; Wicker, R.; Lee, S.-H.; Choi, K.-H.; Ha, C.-S.; Chung, I. Fabrication of 3D biocompatible/biodegradable micro-scaffolds using dynamic mask projection microstereolithography. *J. Mater. Process. Technol.* **2009**, *209*, 5494–5503. [CrossRef]
63. Salmoria, G.V.; Leite, J.L.; Ahrens, C.H.; Lago, A.; Pires, A.T.N. Rapid manufacturing of PA/HDPE blend specimens by selective laser sintering: Microstructural characterization. *Polym. Test.* **2007**, *26*, 361–368. [CrossRef]
64. Zhang, W.; Sita, L.R. Highly efficient, living coordinative chain-transfer polymerization of propene with ZnEt2: Practical production of ultrahigh to very low molecular weight amorphous atactic polypropenes of extremely narrow polydispersity. *J. Am. Chem. Soc.* **2008**, *130*, 442–443. [CrossRef]
65. Karalekas, D.; Rapti, D. Investigation of the processing dependence of SL solidification residual stresses. *Rapid Prototyp. J.* **2002**, *8*, 243–247. [CrossRef]
66. Puebla, K. Effects of environmental conditions, aging, and build orientations on the mechanical properties of ASTM type I specimens manufactured via stereolithography. *Rapid Prototyp. J.* **2012**, *18*, 374–388. [CrossRef]
67. Stansbury, J.W.; Idacavage, M.J. 3D printing with polymers: Challenges among expanding options and opportunities. *Dent Mater.* **2016**, *32*, 54–64. [CrossRef] [PubMed]
68. Gross, B.C.; Erkal, J.L.; Lockwood, S.Y.; Chen, C.; Spence, D.M. Evaluation of 3D printing and its potential impact on biotechnology and the chemical sciences. *Anal. Chem.* **2014**, *86*, 3240–3253. [CrossRef]
69. Lee, J.-Y.; An, J.; Chua, C.K. Fundamentals and applications of 3D printing for novel materials. *Appl. Mater. Today* **2017**, *7*, 120–133. [CrossRef]
70. Salmoria, G.V.; Ahrens, C.H.; Fredel, M.; Soldi, V.; Pires, A.T.N. Stereolithography somos 7110 resin: Mechanical behavior and fractography of parts post-cured by different methods. *Polym. Test.* **2005**, *24*, 157–162. [CrossRef]
71. Steyrer, B.; Neubauer, P.; Liska, R.; Stampfl, J. Visible Light Photoinitiator for 3D-Printing of Tough Methacrylate Resins. *Materials* **2017**, *10*, 1445. [CrossRef] [PubMed]

polymers MDPI

Article

Characterization and Biodegradability of Rice Husk-Filled Polymer Composites

Saw Yin Yap [1], Srimala Sreekantan [1,*], Mohd Hassan [1], Kumar Sudesh [2] and Ming Thong Ong [3]

1 School of Materials and Mineral Resources Engineering, Universiti Sains Malaysia, Nibong Tebal 14300, Penang, Malaysia; yapsawyin@hotmail.com (S.Y.Y.); mohdhassan43@gmail.com (M.H.)
2 School of Biological Sciences, Universiti Sains Malaysia, Gelugor 11800, Penang, Malaysia; ksudesh@usm.my
3 Institute for Research in Molecular Medicine (INFORMM), Universiti Sains Malaysia, Gelugor 11800, Penang, Malaysia; omt@usm.my
* Correspondence: srimala@usm.my; Tel.: +60-04-5995255

Abstract: The fabrication of affordable biodegradable plastics remains a challenging issue for both the scientific community and industries as mechanical properties and biodegradability improve at the expense of the high cost of the material. Hence, the present work deals with fabrication and characterization of biodegradable polymer with 40% rice husk waste filler and 60% polymer-containing mixture of polybutylene succinate (PBS) and poly butylenes adipate-Co-terephthalate (PBAT) to achieve good mechanical properties, 92% biodegradation in six months, and competitive pricing. The challenge in incorporating high amounts of hydrophilic nature filler material into hydrophobic PBS/PBAT was addressed by adding plasticizers such as glycerol and calcium stearate. The compatibilizers such as maleic anhydride (MA) and dicumyl peroxide (DCP) was used to improve the miscibility between hydrophobic PBS/PBAT and hydrophilic filler material. The component with the formulation of 24:36:40 (PBS/PBAT/TPRH) possessed the tensile strength of 14.27 MPa, modulus of 200.43 MPa, and elongation at break of 12.99%, which was suitable for the production of molded products such as a tray, lunch box, and straw. The obtained composite polymer achieved 92% mass loss after six months of soil burial test confirming its biodegradability.

Keywords: rice husk waste; polybutylenes adipate-Co-terephthalate (PBAT); polybutylene succinate (PBS)

Citation: Yap, S.Y.; Sreekantan, S.; Hassan, M.; Sudesh, K.; Ong, M.T. Characterization and Biodegradability of Rice Husk-Filled Polymer Composites. *Polymers* **2021**, *13*, 104. https://doi.org/10.3390/polym13010104

Received: 2 December 2020
Accepted: 24 December 2020
Published: 29 December 2020

Publisher's Note: MDPI stays neutral with regard to jurisdictional clai-ms in published maps and institutio-nal affiliations.

1. Introduction

Petroleum-based plastics have been extensively used in numerous fields such as packaging bags, consumer goods, medical equipment, the automotive sector, and construction sites. The global production of plastics was valued at around 52.9 million tons in 2017. Asia commanded up to 31.4% of the global market in 2018, with a value of 16.61 million tons [1]. The World Wide Fund for Nature (WWF) reported that Malaysia was one of the top plastic consumers in Asia, with 16.8 kg of plastic consumption per person reported annually. In 2020, the consumption of plastic in Malaysia was 543.5 kilo tons [2]. Polyolefin plastics dominate 35 to 45 percent of the synthetic polymer produced in total [3]. The heavy usage of plastics produces a hefty amount of non-degradable wastes, which induces harmful effects on the ecosystem. The environmental pollution incurred due to the use of these traditional polymers has introduced the development of biodegradable polymers.

Biodegradable polymers such as polylactic acid (PLA) [4], polycaprolactone (PCL) [5], poly (3-hydroxybutyrate-co-3-hydroxyvalerate) (PHBV) [6], polybutylene succinate (PBS) [7], and polybutylene adipate-co-terephthalate (PBAT) [8] were used by researchers to obtain cost-effective biocomposites with superior properties. PBS is considered as one of the promising alternatives because of its virtues in strength, toughness, excellent biodegradability, and good processing parameters [9]. Nevertheless, PBS shows insufficient impact strength and gas barrier issues for certain applications. This problem can be addressed by the physical blending of PBS with highly flexible PBAT [9]. PBAT is a 100% biodegradable polymer exhibiting good thermal and mechanical properties [10]. The tensile strength is comparable with low-density polyethylene [11]. The studies conducted by Muthuraj et al. [12] showed good compatibility was achievable in the PBS/PBAT (40/60 wt%) blends, which was further proved by even dispersion of the PBS phase in the PBAT phase. However, poor cost performance existed when a comparison was done with a polymer like polyethylene (PE) and polypropylene (PP), thus restricting its wide-scale applications for practical usage. Hence in this work, high loading of waste material as fillers was used to reduce the cost of the production of biodegradable plastic. Blending with biodegradable materials was considered as an effective strategy to overcome the costs incurred during the material processing.

Natural fibers possess economic advantages in comparison with synthetic fibers. Besides this, they are lightweight, renewable, and biodegradable. One of the most commonly used biodegradable materials is starch. It is regarded as the optimal additive due to its cheap cost and availability. It also derives many advantages, such as renewability and complete biodegradability from nature [13,14]. However, starch is a hydrophilic polymer, while PBS and PBAT are hydrophobic polymers. Hydrophilic starch and hydrophobic polyesters are thermodynamically incompatible, having improper adhesion characteristics [15]. These properties can be enhanced by adding a plasticizer in starch to generate thermoplastic starch (TPS) [16]. The plasticizers such as water, glycerol, and polyvinyl alcohol are used to generate thermoplastic starch [15]. Considering this fact, the high tensile properties and thermal characteristics of the blend can be attained if TPS is well dispersed in PBS/PBAT matrix, hence possessing good phase interaction between TPS and PBS/PBAT [17].

Previous researchers have reported the potential of using waste material as filler in the composite fabrication [18–20]. Hence, rice husk waste was utilized in this study for the fabrication of a biodegradable polymer. It was reported that total rice production worldwide in 2018/2019 was valued around 495.9 metric tons [21]. The rice production approximately generates 123.87 metric tons of rice husk; out of that, some proportion is used for cattle feeding, while the remaining is dumped as waste in landfill and later burned openly. Burning of rice husk contributes to high CO_2 emission and environmental pollution, which further causes health issues [22]. Hence, utilization of this renewable agriculture waste material to form biodegradable polymers would resolve environmental issues, and hence contribute towards Sustainable Development Goal 12, which ensures sustainable consumption and production patterns. It could also be a way to resolve cost and waste disposal issues.

Several reported works on the utilization of rice husks in polymers are summarized in Table 1. As seen, the rice husk has been dominantly utilized in polyolefin rather than in bioplastics polymers. For bioplastic polymers, filler loading up to 5 and 30% have been utilized with PLA and PBAT, respectively. The PBAT/RH with a 70:30 weight ratio exhibited tensile strength, Young's modulus, and elongation at break as 14.5 MPa, 54 MPa, and 820%, respectively [19]. It possessed lower tensile strength and Young's modulus than PP/RH 70:30 wt%. Thus, more rice husk loading is required to achieve better tensile strength and Young's modulus. Hydrophilic rice husk and hydrophobic PBAT and PBS are thermodynamically incompatible, having poor miscibility. Therefore, in this work, rice husk waste (up to 40%) was used after modifying it with glycerol to form thermoplastic rice husk (TPRH). The outcome of 40% loading of TPRH with PBS/PBAT is investigated in this study.

Table 1. The summary of biodegradable polymer/rice husk composites with their respective composition, plasticizer, and mechanical properties.

Material	Ratio	Compatibilizer	Tensile Strength (MPa)	Young's Modulus (MPa)	Elongation at Break (%)	Reference
PP/rice husk	60/40	-	21.5	-	4.7	[18]
PBAT/rice husk silica	70/30	-	14.5	54	820	[19]
PLA/AT rice husk	75/25	-	7.35	-	0.99	[20]
PP/rice husk	90/10	-	32.50	1944	-	[23]
	80/20	-	31.78	2209	-	[23]
	70/30	-	31.42	2540	-	[23]
PP/Rice husk	70/30	-	24	2100	-	[24]
PP/rice husk	70/30	-	29.1	2000	-	[25]
PP/rice husk	50/50	-	17.76	2134	-	[26]
PP/rice husk	60/40	struktol	20	5000	-	[27]
PLA/rice husk	100/0	-	0.22	0.2	45	[28]
	95/5		0.24	2	7	[28]

AT: Alkaline treated, PP: polypropylene, PBAT: poly butylenes adipate-Co-terephthalate, PLA: polylactic acid.

The bare PBAT was soft, and therefore, blending with PBS to mold a product with a stiffness of 78.13 MPa [29] was an essential step. However, blending filler and polymers at nearly equal proportions resulted in an immiscibility issue. Hence, compatibilizer is required to reduce the interfacial tension and to form a co-continuous structure [30]. Therefore, in this work, compatibilizer such as MA and DCP was used to avoid phase separation and to promote an excellent interfacial adhesion for improved mechanical properties of PBS/PBAT with TPRH. The results of TPRH samples were compared with PBS/PBAT filled with TPS.

2. Materials and Methods

2.1. Materials

The corn starch (particle size: 14.89 µm) was provided by Thye Huat Chan Sdn Bhd (Penang, Malaysia). Rice husk waste (particle size: 16.59 µm) was obtained from Fragstar Corporation Sdn. Bhd. (Kedah, Malaysia). The polymer poly (butylene adipate co-terephthalate) (PBAT) with T_m = 115–125 °C and MRF (190 °C/2.16 kg): 3.0–5.0 g/10 min) was purchased from ZhuHai WanGo Chemical Co., Ltd. (Zhuhai, China) under the commercial name of A400. Polybutylene succinate (PBS; T_m: 120 °C; MRF: 10–15 g/10 min) (injection molding grade) was obtained from Hefei TNJ Chemical Industry Co., Ltd. (Hefei, China). Glycerol was obtained from Aldrich Chemical Company (St. Louis, MO, USA) with a purity of 85%. Calcium stearate was purchased from Strem Chemical Company (Massachusetts, United States). The reactive modification of PBAT/PBS was performed using maleic anhydride (98% pure), which was purchased from Fluka (Jawa Timur, Indonesia). Dicumyl peroxide with a purity of 99% obtained from Sigma-Aldrich (St. Louis, MO, USA), was used as an initiator.

2.2. Polymer Composite Preparation

Starch, rice husk, PBS, and PBAT were dried at 70 °C overnight before use. TPS was prepared using a Haake internal mixer at 100 °C with a rotor speed of 50 rpm by mixing 75 wt% starch with 23 wt% glycerol. TPRH was also prepared by mixing 90.0 wt% rice husk with 9.1 wt% glycerol using a Haake internal mixer at 100 °C with a rotor speed of 50 rpm. The PBS, PBAT, maleic anhydride (MA), dicumyl peroxide (DCP), calcium stearate, and TPRH/TPS (with the amount shown in Table 2) were physically mixed in the Haake internal mixer at 160 °C with a rotor speed of 50 rpm for 15 min. Subsequently, the samples were fabricated by a compression molding machine at 160 °C for 6 min preheating time, 3 min heating time, and 2 min cooling time into 140 mm × 180 mm × 0.5 mm sheets. The compositions of the melt-blended specimens are summarized in Table 2.

Table 2. The compositions of the PBAT/PBS/TPS and PBAT/PBS/TPRH for the fabrication of biodegradable polymers.

Designation	Composition	Parts
PBS	PBS	100
PBAT	PBAT	100
TPRH48/12	PBAT/PBS/TPRH/CS/MA/DCP	48/12/40/0.5/2/0.4
TPS48/12	PBAT/PBS/TPS/CS/MA/DCP	48/12/40/0.5/2/0.4
TPRH 36/24	PBAT/PBS/TPRH/CS/MA/DCP	36/24/40/0.5/2/0.4
TPS36/24	PBAT/PBS/TPS/CS/MA/DCP	36/24/40/0.5/2/0.4
Commercial PBAT	-	-

CS: calcium stearate, MA: maleic anhydride, DCP: Dicumyl peroxide, PBS: polybutylene succinate, TPS: thermoplastic starch, TPRH: thermoplastic rice husk.

2.3. *Material Characterization*

2.3.1. Fourier Transforms Infrared Spectroscopy (FT-IR)

Vector-33 FTIR spectrometer (Bruker, Germany) was used to carry out FT-IR measurement to analyze the reaction between PBS, PBAT, and maleic anhydride. The scanning range varied from 4000 to 500 cm^{-1} with a resolution of 16 cm^{-1}.

2.3.2. Mechanical Properties

Tensile testing was conducted with a speed of 5 mm$^{-1,}$ according to ASTM D638. The samples were injection molded with standard Type V samples with a thickness of 0.5 mm. The samples were sealed packed in the plastic bags after molding and conditioned at room temperature for 24 h before testing. Five samples were tested for each composition, and the average value with the standard deviation was recorded.

2.3.3. Thermal Behavior

The crystallization kinetics were investigated using a Perkin-Elmer DSC 7 system. Samples were heated from 30 to 200 °C at 10 °C/min. The samples were kept in the molten state for 5 min to remove thermal history and then cooled down to 30 °C at the rate of 10 °C/min. Then, the samples were heated back again to 200 °C at a rate of 10 °C/min to analyze the crystallization characteristic after heating. The percent crystallinity of the PBS and PBAT was calculated using Equation (1).

$$Xc = \left(\frac{\Delta H_m}{\Delta H_{m100}(1 - w_P)} \right) \times 100\,\% \tag{1}$$

where ΔH_{m100} is the theoretical enthalpy of melting for 100% crystalline PBS (110.3 J/g) and PBAT (114 J/g), w_p is the weight fraction of the PBS, TPS, and TPRH in the PBS/PBAT blend.

2.3.4. Morphology Analysis

The morphology of the fractured specimen was observed using scanning electron microscopy (SEM) operated at an acceleration voltage of 5 kV. The samples were vacuum coated with a thin layer of gold before testing.

2.3.5. Water Absorption Test

The molded samples (size 10 × 10 × 0.5 mm) were immersed in distilled water for a different interval of time at room temperature. For each interval, the samples were gently wiped with soft tissue paper to remove the excess water on the surface. The water absorption (%) was calculated using Equation (2):

$$Waterabsorption\ (\%) = \frac{W_f - W_i}{W_f} \quad (2)$$

where w_f is the weight of the sample after immersion and w_i is the sample weight before immersion.

2.3.6. Soil Burial Test

Compost soil was collected from agriculture field at the USM engineering campus, Malaysia for soil burial test. Molded samples (size 10 × 10 × 0.1 mm) were buried at a depth of 15 cm in the ground in USM engineering campus. Five specimens of each sample were taken out in 2, 4, and 6 months for testing. Then, the samples were rinsed with distilled water and blotted with tissue paper to remove dirt. The samples were dried until a constant weight was achieved. The percentage of weight loss was calculated using Equation (3):

$$\%ofweightloss = \frac{W_X - W_O}{W_O} \times 100\ \% \quad (3)$$

where W_x and W_o indicate the weights of the collected samples and initial weight of the samples.

3. Results

3.1. FT-IR Analysis

The performance of the grafting process within PBS/PBAT/TPS and PBS/PBAT/TPRH was evaluated through the FT-IR technique, and the results are shown in Figure 1. For comparison, FT-IR results of bare PBS and bare PBAT are also displayed in Figure 1. Figure 1a displays a peak at 1160 cm^{-1}, which belonged to the aliphatic ester groups of PBS samples, confirming its aliphatic structure [31]. The bands at the 810, 964, 1116, and 1694 cm^{-1} confirmed the C-H bending of alkane, -C-OH blending of carboxylic acids group, -C-O- stretching vibrations, and C = O stretching vibration in the ester linkages of PBS [32,33]. The band at 1338 cm^{-1} and 2906 cm^{-1} were attributed to the symmetric and asymmetric deformational vibrations of -CH_2- groups of the PBS structure [34]. Figure 1b exhibits the peaks of bare PBAT at 1260 and 1162 cm^{-1}, which were ascribed to O = C-O-C stretching of aromatic and aliphatic ester groups, respectively, validating its aliphatic-aromatic structure. The peaks at 732, 936, and 1112 cm^{-1} are attributed to = C-H bending of benzene ring, -C-OH blending of carboxylic acids group, and -C-O- stretching vibrations of PBAT [32,35]. The band at 1694 and 2952 cm^{-1} are attributed to the C = O stretching vibration and -CH_2- groups of PBAT, respectively.

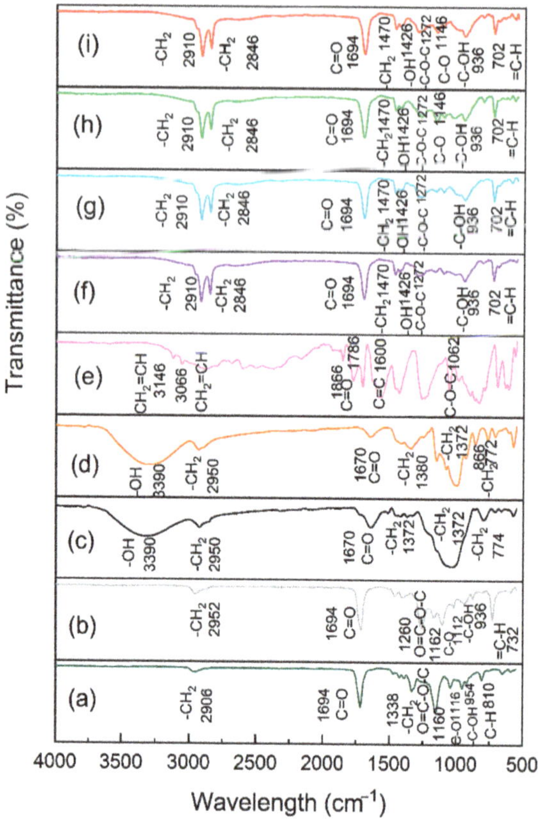

Figure 1. FT-IR spectra of the PBS/PBAT/TPS composites and PBS/PBAT/TPRH composites (**a**) PBS, (**b**) PBAT, (**c**) rice husk, (**d**) starch, (**e**) maleic anhydride, (**f**) TPRH48/12, (**g**) TPS48/12, (**h**) TPRH36/24, and (**i**) TPS36/24.

Figure 1c,d represent the FT-IR spectra of rice husk and starch. The broad absorption band at 3390 cm^{-1} was ascribed to the stretching occurring in the -OH group [36,37]. The band indicated rice husk and starch had a considerable amount of surface absorbed moisture [38]. The peak at 2950 cm^{-1} was assigned to C-H stretching vibration. The presence of a band at 1670 cm^{-1} affirmed the stretching vibration of the C = O group in rice husk and starch. The absorption band at 1372 cm^{-1} was attributed to -CH_2 scissoring vibrations [39,40]. The peak around 774 cm^{-1} showed the presence of -CH_2 blending [41]. Figure 1e displays the FT-IR spectra of maleic anhydride. The absorption bands at 3146, 3066, 1600, and 1062 cm^{-1} were assigned to asymmetrical C-H stretching vibration (CH_2 = CH_2), symmetrical C-H stretching vibration (CH_2 = CH_2), C = C stretching band, and C-O-C symmetrical stretching band, respectively [42]. The peaks at 1866 cm^{-1} and 1786 cm^{-1} were assigned to the C = O stretching vibration of maleic anhydride [43].

Figure 1f–i show the FT-IR spectra of TPRH48/12, TPS48/12, TPRH36/24, and TPS36/24 blends, respectively. The spectra show a similar peak for bare PBS and PBAT. However, an additional band at 2846 cm^{-1} suggests the -CH_2 group from the TPRH or TPS and CH_2 = CH vibration in the cyclic MA. Since the MA was only applied in 2 p/hr, which is considered a small amount. Thus, this bond was corresponded to the -CH_2 group from the TPRH or TPS. This confirms the reaction between PBS/PBAT and TPRH or TPS and addresses the incompatibility between polymer matrix and filler material [44]. In summary, Figure 1 shows an insignificant difference between PBS/PBAT/TPRH and PBS/PBAT/TPS blends as rice husk and starch are organic-based fillers having similar functional groups. This indicated the potential of using waste material such as rice husk to replace starch in biodegradable plastics. The presence of strong absorption bands at 1694 and 1470 cm^{-1} is associated with the stretching vibration of the C = O group and -CH_2 scissoring vibrations of TPRH and TPS, respectively [37]. Besides, the band at 1426 cm^{-1} confirmed the -OH group of glycerol, which was used to modify the surface of rice husk and starch to form TPRH and TPS [33].

3.2. Mechanism between MA, DCP, PBS, PBAT, and TPRH/TPS

Figure 2 presents the mechanism between MA, DCP, PBS, PBAT, and TPRH/TPS. The reaction between PBS and PBAT was formed via the hydrolysis reaction and forming the C-O bond. DCP decomposed at the initial step to form primary radicals. These primary radicals attracted the hydrogen atom from the PBS/PBAT backbone and yielded PBS/PBAT radicals at the initiation step. The propagation step shows the maleic anhydride (MA) molecules grafted onto the PBS/PBAT radicals to form PBS/PBAT-MA radicals and followed by termination reaction. The PBS/PBAT-MA radicals might undergo hydrogen transfer from TPS and TPRH and form the final product. The termination step showed the reaction between PBS/PBAT-g-MA with TPS and TPRH.

3.3. Mechanical Properties

Typical stress–strain curves for bare PBS, bare PBAT, PBS/PBAT/TPRH, and PBS/PBAT/TPS composites are shown in Figure 3. Bare PBS, bare PBAT, and commercial PBAT are highly elastic polymers with high elongation at break. With the introduction of 40% TPRH or TPS, a significant change for both tensile strength and elongation at break was observed compared to bare PBS and bare PBAT. Table 3 represents the tensile data obtained from TPRH composites and TPS composites at a different ratio. For the testing, both TPRH and TPS (as filler materials) were fixed at 40% by weight. The tensile strength, Young's modulus, and elongation at break of bare PBAT were found to be 38.99 MPa, 16.1 MPa, and 1421.9%, respectively. The bare PBS possessed lower tensile strength (30.63 MPa), lower elongation at break (547.45%), and higher Young's modulus (166.23 MPa) when compared with PBAT. The higher incorporation of PBS caused an increment in tensile strength and Young's modulus, but decrement in elongation at break. The increment of Young's modulus and decrement in elongation at break is because of PBS, which has a lower elongation at break (547.45%) and higher Young's modulus 166.23 MPa. The TPRH36/24 and TPS36/24 were prepared using a PBAT:PBS ratio of 36:24. Both the composites with more incorporation of PBS exhibited better tensile strength and Young's modulus, but lower elongation at break when compared with TPRH48/12 and TPS48/12, which is consistent with the work of Boonprasertpoh et al. [29]. This shows that when both polymers (PBS and PBAT) were at a comparable amount, the co-continuous phase occured. This was confirmed with SEM morphology, which is elaborated in the later section. Furthermore, the DSC result shows that TPRH36/24 and TPS36/24 have higher relative crystallinity than TPRH48/12 and TPS48/12, supporting PBS contribution towards a crystallization process. This subsequently affects the tensile strength of the polymer matrix.

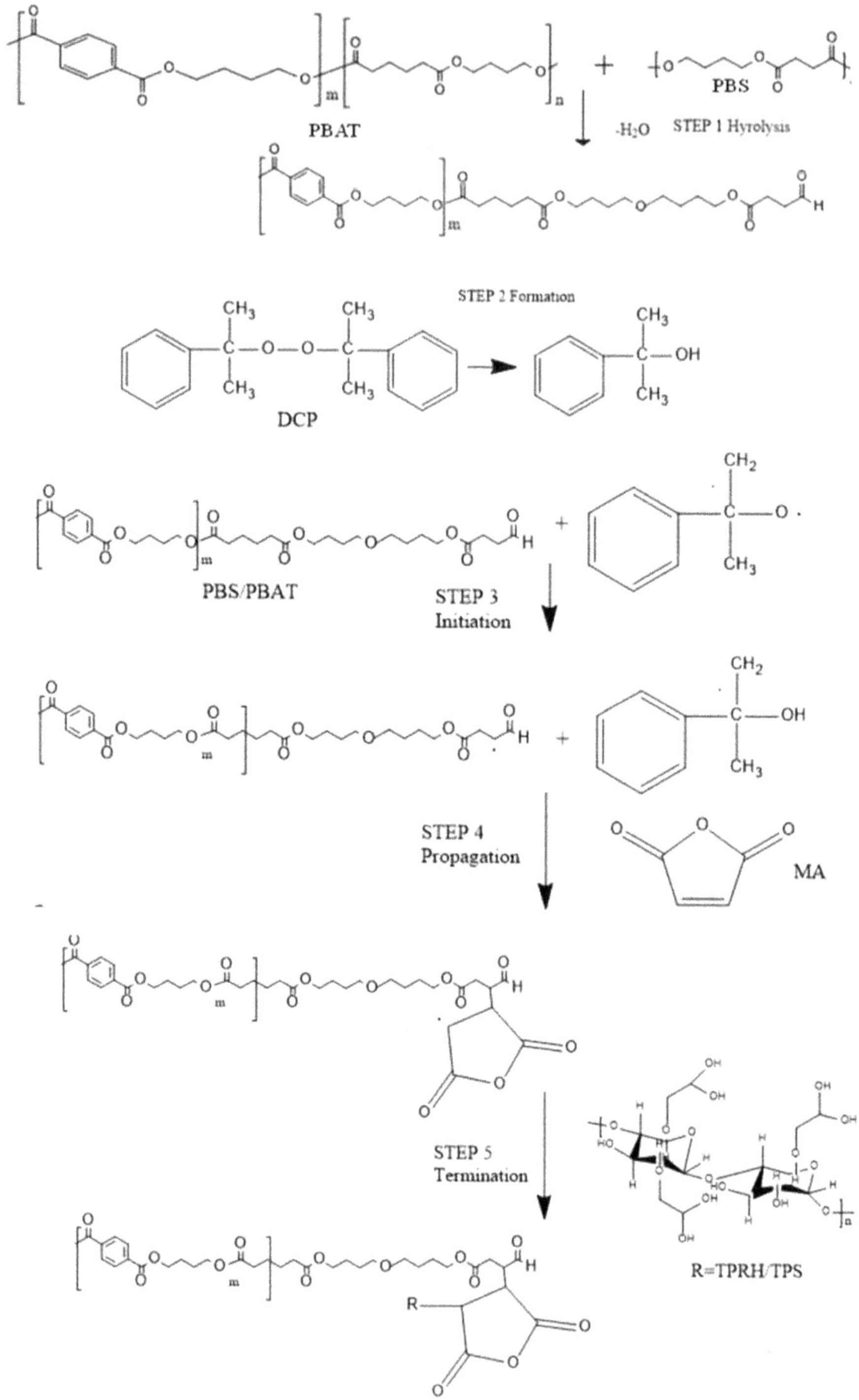

Figure 2. The mechanism between maleic anhydride, DCP, PBS, PBAT, and TPRH/TPS.

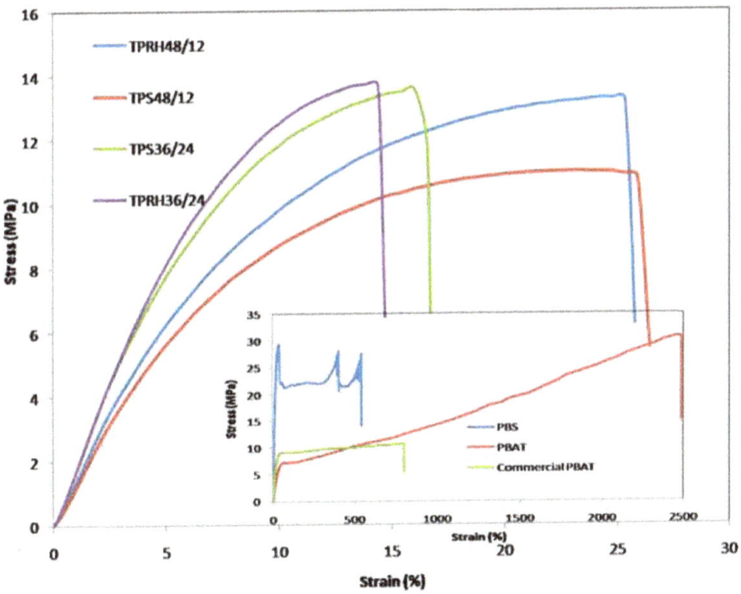

Figure 3. Stress–strain curves of PBS/PBAT/TPRH and PBS/PBAT/TPS composites.

Table 3. Tensile strength, Young's modulus, and elongation at break for PBS/PBAT/TPS and PBS/PBAT/TPRH composites.

Sample Code	Tensile Strength (MPa)	Young's Modulus (MPa)	Elongation at Break (%)
PBS	30.63 ± 2.4	166.23 ± 2.71	547.45 ± 21.61
PBAT	38.99 ± 7.25	16.1 ± 2.63	1421.93 ± 123
TPRH48/12	12.43 ± 0.48	150.34 ± 4.43	24.70 ± 2.53
TPS48/12	10.42 ± 0.52	134.38 ± 3.63	28.62 ± 3.16
TPRH36/24	14.27 ± 1.13	200.43 ± 14.73	12.99 ± 2.34
TPS36/24	14.21 ± 0.81	199.49 ± 9.03	15.39 ± 0.98
Commercial PBAT	10.07 ± 0.99	55.84 ± 2.77	716.95 ± 125.44

Table 3 reveals that both TPRH and TPS composites showed lower tensile strength, elongation at break, but higher Young's modulus when compared with bare PBS and PBAT. This result was in agreement with the study of Hardinnawirda and Aisha [45], who claimed that when the rice husk loading exceeds 15 wt%, the tensile strength shows a remarkable decrement. Incorporation of 40% TPS or TPRH caused linear decrement in the tensile strength and elongation at break of PBS/PBAT matrix, which was same as the results of Garalde et al. [46]. The decrement of tensile strength was due to the stiffness of TPS or TPRH, causing the steric hindrance effect ascribed to cross-linked aromatic structures of PBS and PBAT. The presence of filler material caused reinforcing effects on the properties of the composites, thus reducing the mobility of polymer chains. The improved Young's modulus compared to bare PBAT was due to the enhanced interaction between the carbonyl group of PBS/PBAT matrix and OH groups of TPRH or TPS. The enhanced interaction allows efficient stress transfer from semi-crystalline TPRH to PBS/PBAT [46].

PBAT/PBS/TPRH composite blends with 40% filler prepared in this work, attained remarkable mechanical properties when compared to reported work by Sabetzadeh and his colleagues with just 15% loading of starch filler [47]. The tensile strength and elongation at break of the 15% filler in LDPE were reported to be between 9–12 MPa and 260–360%, respectively [47]. The PBAT/PBS/TPRH composites prepared in this work exhibited better tensile strength and Young's modulus but lower elongation at break than commercial PBAT. For the injection molding process, tensile strength, Young's modulus, and elongation at break required are 11.70 MPa [29], 78.13 MPa [29] and 9% [48]. Thus, all the samples prepared in this work are applicable for the injection molding process as they possess the required mechanical properties.

3.4. Thermal Behavior

Table 4 summarizes the melting point, enthalpy of melting at 100% crystallinity, and crystallinity of the compounded blends. Figure 4 shows that all samples possessed endothermic peaks. The bare PBAT displayed a broad peak at 121.68 °C, while bare PBS exhibited a sharp peak at 114.04 °C. The direct proof of polymer miscibility was obtained by observing the change in the melting point (T_m) of both polymers in the blends. One melting point endotherm was observed for TPRH48/12, TPS48/12, TPRH36/24, and TPS36/24. This scenario indicated that PBS and PBAT were miscible. The T_m of rice husk and starch was absent because the melting point was beyond 200 °C [49,50]. The melting point of corn starch and rice husk is 256 °C–258 °C [50] and 1440 °C, respectively [49].

The bare PBS with a high degree of crystallinity (64.03%) was less susceptible to water absorption because of smaller gaps present between the polymer chains. It was evident from the test results that the addition of filler material slightly reduced the melting point of PBS/PBAT blends. The composites TPRH48/12, TPS48/12, TPRH36/24, and TPS36/24 showed a lower degree of crystallization than PBS. This was due to fillers in polymer matrix reduces the mobility of polymer chains, thus causing the steric hindrance effect ascribed to the cross-linked aromatic structure, leading to a reduction in the extent of crystallinity [51].

Relatively, the composites TPRH36/24 and TPS36/24 exhibited a higher degree of crystallization than TPRH48/12 and TPS48/12. This was due to the high amount of PBS, which possesses higher crystallinity. This phenomenon was consistent with the results of the tensile test, whereby TPRH36/24 and TPS36/24 yield higher tensile strength, as shown in Table 3. However, there was no significant difference found in crystallinity between PBS/PBAT/TPRH blends and PBS/PBAT/TPS blends, indicating the potential of using rice husk waste to substitute starch in the polymer matrix.

Table 4. Thermal properties of bare PBAT, bare PBS, PBAT/PBS/RH blends, and PBAT/PBS/TPS blends.

Code	T_m (°C)	Enthalpy of Melting of 100% Crystalline, ΔH_{m100} (J/g)	X_c (%)
PBS	114.04	70.62	64.03
PBAT	121.68	9.76	8.56
TPRH48/12	110.29	8.49	15.52
TPS48/12	111.27	7.92	14.47
TPRH36/24	111.38	14.78	36.01
TPS36/24	111.72	17.69	43.10
Commercial PBAT	119.77	1.4	1.75

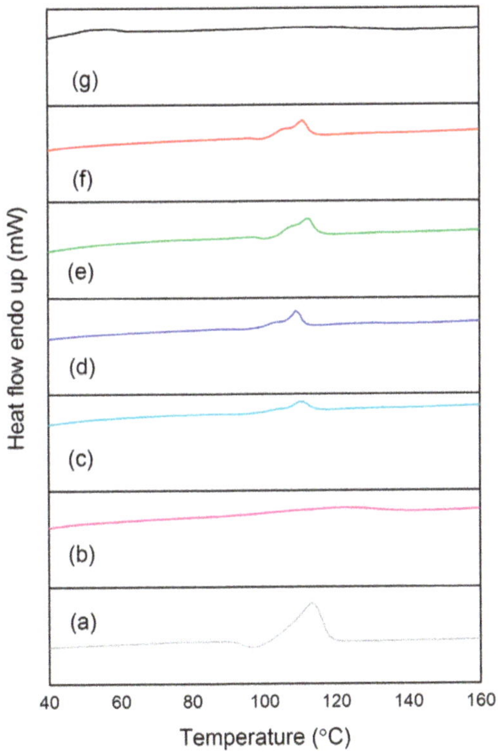

Figure 4. Second-heating Differential Scanning Calorimetry curves of (**a**) PBS, (**b**) PBAT, (**c**) TPRH48/12, (**d**) TPS48/12, (**e**) TPRH36/24, (**f**) TPS36/24, and (**g**) commercial PBAT.

3.5. Morphology Analysis

Figure 5 shows the surface morphology of TPRH and TPS with irregular (Figure 5a) and spherical (Figure 5b) shape, respectively. The TPS granules with spherical shape caused a reduction in the contact area of the polymer matrix when compared with the irregular shape of TPRH [52]. For bare polymers of PBS (Figure 6a) and PBAT (Figure 6b), smooth surface morphology was observed while with filler loading, a rough surface was observed as shown in Figure 6c–f. When 40% of TPRH was added into PBAT/PBS matrix, the TPRH particles dispersed homogeneously in the PBAT/PBS matrix, as shown in Figure 6c,e. Some voids in the interfacial boundary were observed in TPRH48/12 due to the pull out of the rice husk particle from the PBAT/PBS matrix. This is caused by the difference in polarity between the hydrophilic rice husk and the hydrophobic PBAT. Fewer filler pullouts with good fibers-matrix adhesion were observed in TPRH36/24 (Figure 6e) as compared to TPRH48/12 (Figure 6c). For TPS blends, a similar trend was observed in TPS48/12 (Figure 6d), which showed a higher filler pullout than the TPS36/24 (Figure 6f). The SEM morphology of TPRH36/24 and TPS36/24 exhibited the co-continuous phase as PBS and PBAT are at a comparable amount. Hence, this explains the less pull out circumstances and better tensile strength in TPRH36/24 and TPS36/24 in comparison to TPRH48/12 and TPS48/12.

Figure 5. Surface morphology of (**a**) irregular TPRH granules and (**b**) spherical TPS granules.

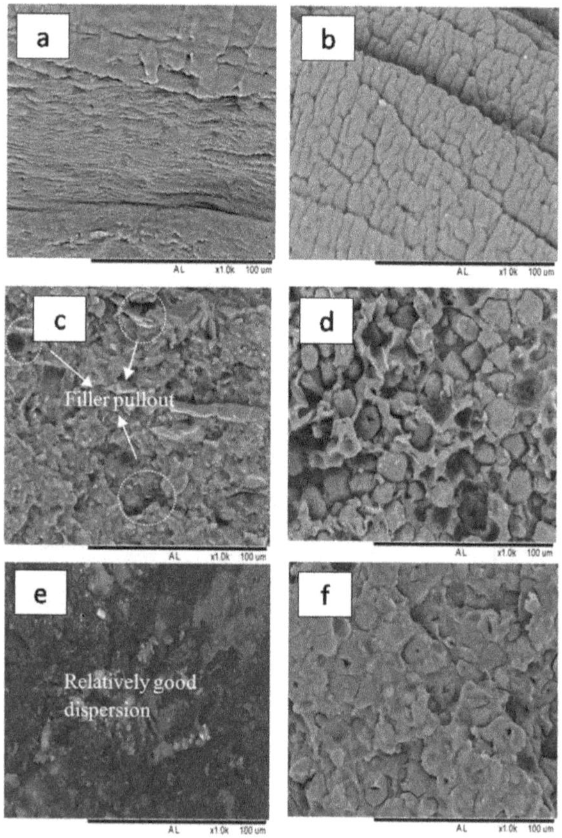

Figure 6. Fractured morphology of (**a**) PBS, (**b**) PBAT, (**c**) TPRH48/12, (**d**) TPS48/12, (**e**) TPRH36/24, and (**f**) TPS36/24.

3.6. Water Absorption

The water absorption of the bare PBS, PBAT, and PBS/PBAT blends are shown in Figure 7. All polymers undergo three stages during the water absorption which are absorption, saturation, and swelling. The water absorption of PBAT and PBS film increased slightly and achieved saturation of around 0.35% after 10 days of immersion in water.

The low water absorption capacity was due to PBAT and PBS being hydrophobic polymers with the presence of acyl groups [31]. However, at the initial stage (day 3), it was obvious that PBS, having higher crystallinity, possessed a water absorption capacity of 0 wt%, while PBAT had a capacity of 0.39%, which indicated that the degree of crystallinity is an important factor for water absorption of the polymer. Besides this, PBS with a high degree of crystallinity was less prone to water absorption as compared to PBAT due to the smaller amorphous region accessible for water intake.

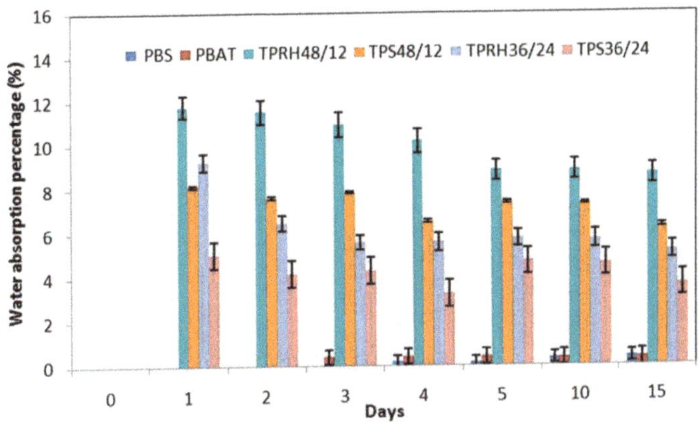

Figure 7. Water absorption of bare PBS, bare PBAT, and PBS/PBAT blends with immersion times.

The PBS/PBAT/TPRH composites showed a larger water absorption capacity than bare PBAT and PBS. The water absorption rose tremendously in the first 24 h, and reached the saturation limit after 24 h of immersion. The TPRH48/12 and TPRH36/24 showed water absorption capacity at 11.78 and 9.24%, respectively after 24 h. The composite TPRH36/24 possessed a lower water absorption capacity than TPRH48/12 due to the intrinsic nature of PBS with high crystallinity. Moreover, the water absorption capacity of PBS/PBAT/TPRH was also found to be more than PBS/PBAT/TPS, which is an attribute to the lumen and cell wall of rice husk that provide more room for the water absorption [18]. Besides, the hydrophilic nature of rice husk that favors the intermolecular hydrogen bonding enhanced the water absorption of the film.

The swelling in composites occured due to the presence of internal stresses that prevents polymer matrix from absorbing water [53]. The swelling effect was observed for both TPRH and TPS blends after 3 days of immersion, which in agreement with the work was reported by Muthuraj et al. [54]. Nevertheless, TPRH36/24 and TPRH48/12 showed lower water absorption capacity (5.17 and 8.68%) than the reported LDPE by Sabezadeh et al. [47] with 11% absorption capacity at day 15. Water absorption characteristics influence the water vapor barrier properties of a material. Thus, the results indicated that the fabricated polymer in this work has good water vapor barrier property, which would extend the shelf life of the food.

3.7. Soil Burial Test

The entire composites showed a smoother surface before the degradation process. The mass change in the PBS/PBAT blended composites as a function of degradation time is shown in Table 5. The macroscopic appearance of biodegradation in PBS/PBS blended composites at different burying time is shown in Figure 8. Matting of the sample surface and color change was noticeable after the degradation process. The mass loss percentage increased with increasing burying time for the entire samples, which affirmed the biodegradation properties of PBS, PBAT, and PBS/PBAT blends. Progressive fragmentation was noticed in all the samples as they gradually degraded within 6 months. At the end of 6 months, only a small amount of residual debris remained. The color of the film gradually became brownish with increasing burying time. For TPS36/24, the mass loss percentage reached 97.06% after 6 months, which was in agreement with the results of Dammak et al. [55], indicating a fully biodegradable characteristic of this material. As for TPRH36/24, a similar result was obtained with a maximum mass loss of 92%. TPS48/12 and TPRH48/12 showed a much lower mass loss, indicating higher PBS content expedite the degradation process. The bare PBAT, PBS, and commercial PBAT showed mass loss at 8.9, 9.2, and 32.51%, respectively, after 6 months.

Table 5. Mass loss percentage of PBS/PBAT blends after soil burial test for six months.

Sample Code	Mass Loss Percentage (%)			
	0 month	2 months	4 months	6 months
PBS	0	4.19 ± 0.29	6.54 ± 0.10	8.90 ± 0.06
PBAT	0	6.23 ± 0.31	6.82 ± 0.11	9.20 ± 0.05
TPRH48/12	0	20.94 ± 0.14	64.56 ± 0.21	79.78 ± 0.10
TPS48/12	0	48.74 ± 0.21	83.30 ± 0.27	86.88 ± 0.19
TPRH36/24	0	31.52 ± 0.30	88.91 ± 0.08	92.00 ± 0.08
TPS36/24	0	53.36 ± 0.17	92.07 ± 0.08	97.06 ± 0.03
Commercial PBAT	0	14.27 ± 0.15	18.16 ± 0.19	32.51 ± 0.17

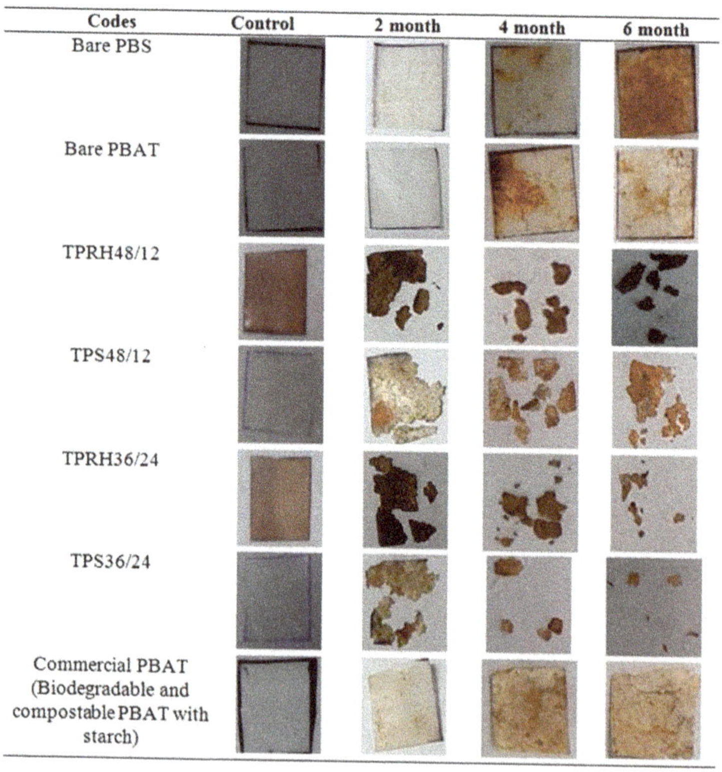

Figure 8. Macroscopic appearance of biodegradation in the soil at different burying times.

As it was evident from the appearance, PBS/PBAT/TPRH and PBS/PBAT/TPS samples showed faster degradation rates than bare PBAT and PBS. TPS and TPRH are the nutrient source for microorganisms, thus it provides more degradation sites to be attacked by microorganisms [56]. When microorganisms consume the TPS and TPRH, they leave the polymer matrix more porous, which accelerates the biodegradation rate of PBS/PBAT blends [11]. This causes the polymer chains to split into lower molecular weight oligomers, monomers, dimmers, and finally mineralized to CO_2 and H_2O [57]. The result also indicated that the utilization of rice husk with a high amount of PBS has the potential to degrade faster and is comparable with TPS (refer to the mass loss for TPRH36/24 and TPS48/12). Although TPS36/24 and TPRH36/24 showed higher crystallinity and lower moisture absorption, they exhibited a higher mass loss percentage than TPS48/12 and TPRH48/12. This is because of the aromatic structure of PBAT, which decreases the mobility of polymer chains, reducing the degradation rate of polymer matrix [31].

The sample prone for degradation (TPRH36/24) was selected for FT-IR analysis to observe the changes in the intensity of certain transmittance peak, the formation of new peaks, or migration of the peak position before and after degradation. Figure 9a shows the FT-IR spectra of TPRH36/24 before biodegradation, while Figure 9b depicts FT-IR spectra of TPRH36/24 after 6 months of degradation. It was found that the highly intense -CH_2 stretching vibration position of the intrinsic polymer diminishes and migrated to 2974 cm^{-1}, indicating a significant degradation process of the sample. Moreover, the less intense carbonyl region of the C = O group migrated from 1694 cm^{-1} to a highly intense and broader peak at 1728 cm^{-1}, affirming the process of degradation has occurred. The emergence of the new peak was noticed at 3708 cm^{-1} after degradation, which corresponds to the O-H group of absorbed water in the polymer matrix [38]. The disappearance of the -CH_2 peak at 2846 cm^{-1} indicated the degradation of TPRH and TPS. The changes observed in the FT-IR spectra are in agreement with the polymer oxidation degradation process reported by Celina et al. [58], which showed the dominant carbonyl formation with the diminishing C-H bands. The FTIR spectra of TPS36/24 before and after biodegradation are shown in Figure 10a,b, respectively. The results show that there was no significant difference between TPRH36/24 and TPS36/24 before and after biodegradation. This suggests the potential of using rice husk waste to swap starch in biodegradable polymer composites.

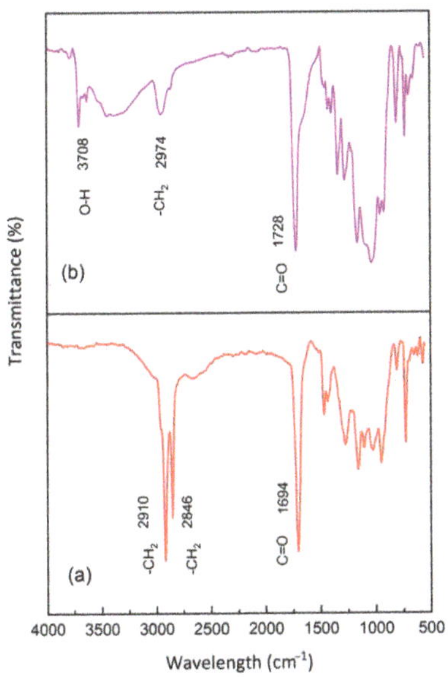

Figure 9. FT-IR spectra of the PBS/PBAT/TPRH composites (**a**) TPRH36/24 before degradation, (**b**) TPRH36/24 after degradation with changes in functional group.

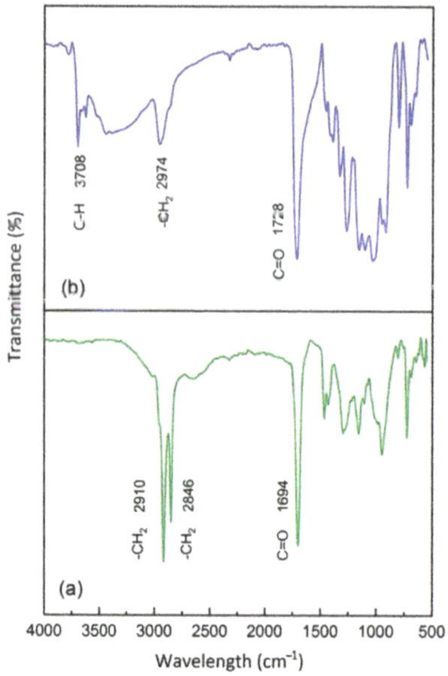

Figure 10. FT-IR spectra of the PBS/PBAT/TPS composites (**a**) TPS36/24 before degradation, (**b**) TPS36/24 after degradation with changes in the functional group.

4. Conclusions

In this study, it was found that the incorporation of 40% rice husk was able to substitute starch-based biodegradable polymer. Optimization of the ratio PBAT:PBS to 36:24 expedited the biodegradation rate of the samples. PBAT:PBS blends with a 36:24 ratio showed 97.06% mass loss for TPS and 92% for TPRH. A comparable amount of PBAT and PBS allowed the formation of co-continuously phases to improve the mechanical properties. The bio-composite TPRH36/24 possessed good mechanical properties such as tensile strength (14.27 MPa), Young's modulus (200.43 MPa), and elongation at break (12.99%), which is adequate for the manufacturing of molded products such as a tray, lunch box, and straw. Finally, it achieved a 92% mass loss after six months, evidencing itself as a biodegradable material. The test results from this study indicated an accomplishment in the fabrication of cost-efficient biodegradable polymer using waste fillers, which has tremendous potential for practical use in various industrial applications.

Author Contributions: Conceptualization, S.Y.Y. and S.S.; formal analysis, S.Y.Y., K.S., and M.T.O.; investigation, S.Y.Y.; methodology, S.Y.Y.; project administration, S.S.; supervision, S.S.; writing—original draft, S.Y.Y.; writing-review and editing, S.S. and M.H. All authors have read and agreed to the published version of the manuscript.

Funding: This research was funded by the Ministry of Education for funding under the Public-Private Research Network (PPRN) 2.0 grant number 304/PBAHAN/6314053.

Institutional Review Board Statement: Not applicable.

Informed Consent Statement: Not applicable.

Data Availability Statement: Not applicable.

Acknowledgments: The authors are thankful to Ministry of Education for providing Public-Private Research Network (PPRN) 2.0 grant no.304/PBAHAN/6314053. The authors are grateful to Universiti Sains Malaysia (USM) for providing necessary facilities for this research work. We also would like to acknowledge Fragstar Corporation Sdn Bhd, an industrial partner in this project who partially had contributes financial and facilities support.

Conflicts of Interest: The authors declare no conflict of interest.

References

1. Strategies. 2017. Available online: https://www.statista.com/statistics/255977/total-global-riceconsumption/ (accessed on 4 August 2019).
2. Taylor, M. The Star. 2020. Available online: https://www.thestar.com.my/news/nation/2020/02/17/malaysia-top-plastic-ocean-polluter-in-asia-says-wwf (accessed on 4 August 2019).
3. Beckman, E. The World's Plastic Problem in Numbers. 2018. Available online: https://www.weforum.org/ (accessed on 4 August 2019).
4. Díez-Rodríguez, T.M.; Blázquez-Blázquez, E.; Pérez, E.; Cerrada, M.L. Composites Based on Poly(Lactic Acid) (PLA) and SBA-15: Effect of Mesoporous Silica on Thermal Stability and on Isothermal Crystallization from Either Glass or Molten State. *Polymers* **2020**, *12*, 2743. [CrossRef] [PubMed]
5. Rezk, A.I.; Kim, K.-S.; Kim, C.S. Poly(ε-Caprolactone)/Poly(Glycerol Sebacate) Composite Nanofibers Incorporating Hydroxyapatite Nanoparticles and Simvastatin for Bone Tissue Regeneration and Drug Delivery Applications. *Polymers* **2020**, *12*, 2667. [CrossRef] [PubMed]
6. Gigante, V.; Cinelli, P.; Righetti, M.C.; Sandroni, M.; Polacco, G.; Seggiani, M.; Lazzeri, A. On the Use of Biobased Waxes to Tune Thermal and Mechanical Properties of Polyhydroxyalkanoates–Bran Biocomposites. *Polymers* **2020**, *12*, 2615. [CrossRef] [PubMed]
7. Costa, A.R.M.; Crocitti, A.; De Carvalho, L.H.; Carroccio, S.C.; Cerruti, P.; Santagata, G. Properties of Biodegradable Films Based on Poly(butylene Succinate) (PBS) and Poly(butylene Adipate-co-Terephthalate) (PBAT) Blends. *Polymers* **2020**, *12*, 2317. [CrossRef]
8. Kirsh, I.; Frolova, Y.; Beznaeva, O.; Bannikova, O.; Gubanova, M.; Tveritnikova, I.; Romanova, V.; Filinskaya, Y. Influence of the Ultrasonic Treatment on the Properties of Polybutylene Adipate Terephthalate, Modified by Antimicrobial Additive. *Polymers* **2020**, *12*, 2412. [CrossRef]
9. Muthuraj, R.; Misra, M.; Mohanty, A.K. Hydrolytic degradation of biodegradable polyesters under simulated environmental conditions. *J. Appl. Polym. Sci.* **2015**, *132*, 42189. [CrossRef]
10. Vroman, I.; Tighzert, L. Biodegradable Polymers. *Materials* **2009**, *2*, 307–344. [CrossRef]
11. Ferreira, F.V.; Cividanes, L.S.; Gouveia, R.F.; Lona, L.M.F. An overview on properties and applications of poly (butylene adipate-co-terephthalate)–PBAT based composites. *Polym. Eng. Sci.* **2017**, *59*, 7–15. [CrossRef]
12. Muthuraj, R.; Misra, M.; Mohanty, A.K. Binary blends of poly(butylene adipate-co-terephthalate) and poly(butylene succinate):A new matrix for biocomposites applications. *AIP Conf. Proc.* **2015**, *1664*, 15009.
13. Jantanasakulwong, K.; Leksawasdi, N.; Seesuriyachan, P.; Wongsuriyasak, S.; Techapun, C.; Ougizawa, T. Reactive blending of thermoplastic starch and polyethylene-graft-maleic anhydride with chitisan as compatibilizer. *Carbohydr. Polym.* **2016**, *153*, 89–95. [CrossRef]
14. Ghavimi, S.A.A.; Ebrahimzadeh, M.H.; Abu Osman, N.A.; Solati-Hashjin, M. Polycaprolactone/starch composite: Fabrication, structure, properties, and applications. *J. Biomed. Mater. Res. Part A* **2015**, *103*, 2482–2498. [CrossRef] [PubMed]
15. Meng, L.; Yu, L.; Khalid, S.; Liu, H.; Zhang, S.; Duan, Q.; Chen, L. Preparation, microstructure and performance of poly (lactic acid)-Poly (butylene succinate-co-butyleneadipate)-starch hybrid composites. *Compos. Part B Eng.* **2019**, *177*, 107384. [CrossRef]
16. Zhang, S.; He, Y.; Yin, Y.; Jiang, G. Fabrication of innovative thermoplastic starch bio-elastomer to achieve high toughness poly(butylene succinate) composites. *Carbohydr. Polym.* **2019**, *206*, 827–836. [CrossRef] [PubMed]
17. Panda, B.P.; Mohanty, S.; Nayak, S.K. Mechanism of Toughening in Rubber Toughened Polyolefin—A Review. *Polym. Technol. Eng.* **2015**, *54*, 462–473. [CrossRef]
18. Prachayawarakorn, J.; Niracha, Y. Effect of recycling on properties of rice husk-filled-polypropylene. *J. Sci. Technol.* **2005**, *27*, 2.
19. Phosee, J.; Wittayakun, J.; Suppakarn, N. Mechanical Properties and Morphologies of Rice Husk Silica (RHS)/ Poly (butylene adipate-co-terephthalate) (PBAT) Composites: Effect of Silane Coupling Agent. *Adv. Mater. Res.* **2010**, *125*, 141–144. [CrossRef]
20. Dina, A.F.; Zaleha, S.S.; Najmi, B.N.; Azowa, I.N. The Influence of Alkaline Treatment on Mechanical Properties and Morphology of Rice Husk Fibre Reinforced Polylactic Acid. *Adv. Mater. Res.* **2014**, *911*, 13–17. [CrossRef]
21. Statista. Total Global Rice Consumption 2008–2020. 2020. Available online: https://www.statista.com/statistics/255977/total-global-rice-consumption/ (accessed on 2 September 2020).
22. Kumar, A.; Sengupta, B.; Dasgupta, D.; Mandal, T.; Datta, S. Recovery of value added products from rice husk ash to explore an economic way for recycle and reuse of agricultural waste. *Rev. Environ. Sci. Technol.* **2015**, *15*, 47–65. [CrossRef]
23. Burgstaller, C. A comparison of processing and performance for lignocellulosic reinforced polypropylene for injection moulding applications. *Compos. Part B Eng* **2014**, *69*, 192–198. [CrossRef]
24. Arjmandi, R.; Hassan, A.; Majeed, K.; Zakaria, Z. Rice Husk Filled Polymer Composites. *Int. J. Polym. Sci.* **2015**, *2015*, 1–32. [CrossRef]
25. Salazar, M.A.H.; Salinas, E. Mechanical, thermal, viscoelastic performance and product application of PP—Rice husk Colombian biocomposites. *Compos. Part B Eng.* **2019**, *176*, 107135. [CrossRef]
26. Yeh, S.-K.; Hsieh, C.-C.; Chang, H.-C.; Yen, C.C.; Chang, Y.-C. Synergistic effect of coupling agents and fiber treatments on mechanical properties and moisture absorption of polypropylene–rice husk composites and their foam. *Compos. Part A Appl. Sci. Manuf.* **2015**, *68*, 313–322. [CrossRef]
27. Aridi, N.A.M.; Sapuan, S.M.; Zainudin, E.S.; Oqla, F.M.A. Mechanical and morphological properties of injection-molded rice husk polypropylene composites. *Int. J. Polym. Anal. Charact.* **2016**, *21*, 305–313. [CrossRef]
28. Adeosun, S.O.; Aworinde, A.K.; Diwe, I.V.; Olaleye, S. Mechanical and Microstructural Characteristic of Rice Husk Reinforced Polylactide Nano Composite. *West Indian J. Eng.* **2016**, *39*, 63–71.
29. Boonprasertpoh, A.; Pentrakoon, D.; Junkasem, J. Investigating rheological, morphological and mechanical properties of PBS/PBAT blends. *J. Met. Mater. Miner.* **2017**, *27*, 1–11. [CrossRef]

30. Savas, L.A.; Tayfun, U.; Doğan, M. The use of polyethylene copolymers as compatibilizers in carbon fiber reinforced high density polyethylene composites. *Compos. Part B Eng.* **2016**, *99*, 188–195. [CrossRef]
31. Arslan, A.; Çakmak, S.; Cengiz, A.; Gümüşderelioğlu, M. Poly(butylene adipate-co-terephthalate) scaffolds: Processing, structural characteristics and cellular responses. *J. Biomater. Sci. Polym. Ed.* **2016**, *27*, 1841–1859. [CrossRef]
32. Brian, C.S. The C=O Bond, Part III: Carboxylic Acids. *Spectroscopy* **2018**, *33*, 14–20.
33. Merck. 2020. Available online: https://www.sigmaaldrich.com/technical-documents/articles/biology/ir-spectrum-table.html (accessed on 28 November 2020).
34. Phua, Y.J. Reactive processing of maleic anhydride-grafted poly(butylene succinate) and the compatibilizing effect on poly(butylene succinate) nanocomposites. *Express Polym. Lett.* **2013**, *7*, 340–354. [CrossRef]
35. Cai, Y.; LV, J.; Liu, Y.; Wang, Z.; Zhao, M.; Shi, R. Discrimination of Poly (butylenes adipate-co-terephathalate) andPoly(ethylene terepthalate) with Fourier Transform Infrared Microscope and Raman Spectroscope. *Spectrosc. Lett.* **2012**, *45*, 280–284. [CrossRef]
36. Mohkami, M.; Taleipour, M. Investigation of the chemical structure of carboxylated and carboxymethylated fibers from waste paper via XRD and FTIR analysis. *Bioresources* **2014**, *6*, 1988–2003.
37. Gupta, H.; Kumar, H.; Kumar, M.; Gehlaut, A.K.; Gaur, A.; Sachan, S.; Park, J.-W. Synthesis of biodegradable films obtained from rice husk and sugarcane bagasse to be used as food packaging material. *Environ. Eng. Res.* **2019**, *25*, 506–514. [CrossRef]
38. Amigo, N.; Palza, H.; Canales, D.; Sepúlveda, F.; Vasco, D.A.; Sepúlveda, F.; Zapata, P.A. Effect of starch nanoparticles on the crystallization kinetics and photodegradation of high density polyethylene. *Compos. Part B Eng.* **2019**, *174*, 106979. [CrossRef]
39. Rachtanapun, P.; Rattanaponone, N. Synthesis and Characterization of Carboxymethyl Cellulose powder and Films from Mimosa pigra. *J. Appl. Polym. Sci.* **2011**, *122*, 3218–3226. [CrossRef]
40. Yeasmin, M.S.; Mondal, M.I.H. Synthesis of highly substituted carboxymethyl cellulose depending on cellulose particle size. *Int. J. Biol. Macromol.* **2015**, *80*, 725–731. [CrossRef]
41. Sarma, P.; Kumar, R.; Parshirajan, K. Batch and Continuous Removal of Copper and Lead from Aqueous Solution using Cheaply Available Agriculture Waste Materials. *Int. J. Environ. Res.* **2015**, *9*, 635–648.
42. Derkus, B.; Emregui, K.; Mazi, H.; Emreg, L.; Yumak, T.; Sinag, A. Protein A immunosensor for the detection of immunoglobulin G by impedance spectroscopy. *Bioprocess Biosyst. Eng.* **2014**, *37*, 965–976. [CrossRef]
43. Rahman, M.R.; Hamdan, S.; Hasan, M.; Baini, R.; Salleh, A.A. Physical, Mechanical, and Thermal Properties of Wood Flour Reinforced Maleic Anhydride Grafted Unsaturated Polyester (UP) Biocomposites. *Bioresources* **2015**, *10*, 10. [CrossRef]
44. Mendes, J.; Paschoalin, R.; Carmona, V.B.; Neto, A.R.D.S.; Marques, A.; Marconcini, J.M.; Mattoso, L.H.C.; Medeiros, E.S.; Oliveira, J.E. Biodegradable polymer blends based on corn starch and thermoplastic chitosan processed by extrusion. *Carbohydr. Polym.* **2015**, *137*, 452–458. [CrossRef]
45. Hardinnawirda, K.; Aisha, I.S. Effect of Rice Husks as Filler in Polymer Matrix Composites. *J. Mech. Eng. Sci.* **2014**, *2*, 181–186. [CrossRef]
46. Garalde, R.A.; Thipmanee, R.; Jariyasakoolroj, P.; Sane, A. The effects of blend ratio and storage time on thermoplastic starch/poly(butylene adipate-co-terephthalate) films. *Heliyon* **2019**, *5*, 01251. [CrossRef] [PubMed]
47. Sabetzadeh, M.; Bagheri, R.; Masoomi, M. Effect of nanoclay on the properties of low density polyethylene/linear low density polyethylene/thermoplastic starch blend films. *Carbohydr. Polym.* **2016**, *141*, 75–81. [CrossRef] [PubMed]
48. Farotti, E.; Natalini, M. Injection molding. Influence of process parameters on mechanical properties of polypropylene polymer. A first study. *Procedia Struct. Integr.* **2018**, *8*, 256–264. [CrossRef]
49. Onojah, A.D.; Agbendeh, N.A.; Mbakaan, C. Rice Husk Ash Refractory: The Temperature Dependent Crystalline Phase Aspects. *Int. J. Recent Res. Appl. Stud.* **2013**, *15*, 2.
50. Bath, H. 2020. Available online: https://alisbathbomblab.weebly.com/index.html (accessed on 2 September 2020).
51. Mohammadi-Rovshandeh, J.; Davachi, S.M.; Kaffashi, B.; Hassani, A.; Bahmeyi, A.; Pouresmaeel-Selakjani, P. Effect of lignin removal on mechanical, thermal, and morphological properties of polylactide/starch/rice husk blend used in food packaging. *J. Appl. Polym. Sci.* **2014**, *131*, 22. [CrossRef]
52. Meincke, D.K.; de Oliveira Ogliari, A.; Ogliari, F.A. Influence of difference fillers on the properties of an experimental vinyl polysiloxane. *Braz. Oral Res.* **2016**, *30*, 1. [CrossRef]
53. Derrien, K.; Gilormini, P. The effect of moisture-induced swelling on the absorption capacity of transversely isotropic elastic polymer–matrix composites. *Int. J. Solids Struct.* **2009**, *46*, 1547–1553. [CrossRef]
54. Muthuraj, R.; Misra, M.; Mohanty, A.K. Biocomposite consisting of miscanthus fiber and biodegradable binary blend matrix: Compatibilization and performance evaluation. *RSC Adv.* **2017**, *7*, 27538–27548. [CrossRef]
55. Dammak, M.; Fourati, Y.; Tarrés, Q.; Delgado-Aguilar, M.; Mutjé, P.; Boufi, S. Blends of PBAT with plasticized starch for packaging applications: Mechanical properties, rheological behaviour and biodegradability. *Ind. Crop. Prod.* **2020**, *144*, 112061. [CrossRef]
56. Sun, E.; Liao, G.; Zhang, Q.; Qu, P.; Wu, G.; Huang, H. Biodegradable copolymer-based composites made from straw fiber for biocomposite flowerpots application. *Compos. Part B Eng.* **2019**, *165*, 193–198. [CrossRef]
57. Tokiwa, Y.; Calabia, B.P.; Ugwu, C.U.; Aiba, S. Biodegradability of plastics. *Int. J. Mol. Sci.* **2009**, *10*, 3722–3742. [CrossRef] [PubMed]
58. Celina, M.; Ottesen, D.; Gillen, K.; Clough, R. FTIR emission spectroscopy applied to polymer degradation. *Polym. Degrad. Stab.* **1997**, *58*, 15–31. [CrossRef]

MDPI
St. Alban-Anlage 66
4052 Basel
Switzerland
Tel. +41 61 683 77 34
Fax +41 61 302 89 18
www.mdpi.com

Polymers Editorial Office
E-mail: polymers@mdpi.com
www.mdpi.com/journal/polymers

www.ingramcontent.com/pod-product-compliance
Lightning Source LLC
Chambersburg PA
CBHW041935260926
51850CB00018B/178
* 9 7 8 3 0 3 6 5 7 4 0 6 6 *